W9-CFV-295

Encyclopedia of Biopharmaceutical Statistics

Encyclopedia of Biopharmaceutical Statistics

edited by Shein-Chung Chow

StatPlus, Inc.
Yardley, and
Temple University
Philadelphia, Pennsylvania

MARCEL DEKKER, INC. NEW YORK • BASEL

ISBN: 0-8247-6001-8

This book is printed on acid-free paper.

Headquarters
Marcel Dekker, Inc.
270 Madison Avenue, New York, NY 10016
tel: 212-696-9000; fax: 212-685-4540

Eastern Hemisphere Distribution
Marcel Dekker AG
Hutgasse 4, Postfach 812, CH-4001 Basel, Switzerland
tel: 41-61-261-8482; fax: 41-61-261-8896

World Wide Web
http://www.dekker.com

The publisher offers discounts on this book when ordered in bulk quantities. For more information, write to Special Sales/Professional Marketing at the headquarters address above.

Current printing (last digit):
10 9 8 7 6 5 4 3 2 1

PRINTED IN THE UNITED STATES OF AMERICA

Preface

The process of pharmaceutical development entails drug discovery, laboratory development, animal studies, clinical development, regulatory registration, and postmarketing surveillance. To ensure the efficacy, safety, and quality of the pharmacetucal product, regulatory agencies have developed guidances and guidelines for good pharmaceutical practices to assist the sponsors and researchers in drug research and development. Even after a pharmaceutical product has been approved, it must be tested for its identity, strength, quality, and purity before it can be released for use. Good pharmaceutical practices include guidances and guidelines for good laboratory practice (GLP), good clinical practice (GCP), and current good manufacturing practice (cGMP). For the implementation of good pharmaceutical practices, statistical methods are usually employed as a useful tool in design and analysis at various stages of drug research and development. Good statistics practice (GSP) is the foundation of good pharmaceutical practice. The purpose of this encyclopedia is to provide a comprehensive and unified presentation of designs and analyses utilized at different stages of pharmaceutical development and to give a well-balanced summary of current regulatory requirements and recently developed statistical methods in this area.

It was my goal to provide a useful reference book for the chemical scientist, pharmaceutical scientist, development pharmacist, clinical scientist, and biostatistician, in the pharmaceutical industry, regulatory agencies, and academia, and other scientists in the associated areas of pharmaceutical development and health-related fields. The primary focus of this encyclopedia is biopharmaceutical statistical applications. All statistical methods and their interpretations are illustrated through examples of real instances that have occurred during various stages of pharmaceutical development. The book was developed to help scientists understand the statistical language in pharmaceutical development. It provides clear, illustrated explanations of how statistical design and methodology can be used for the demonstration of quality, safety, and efficacy in pharmaceutical development. Detailed information is given on many popular topics in biopharmaceutical applications. In addition, a clear systematic account of the state-of-the-art statistical knowledge in pharmaceutical research and development has been provided.

This encyclopedia attempts to make entries self-contained and self-explanatory to a reader who is not an expert in the subject matter of each topic. Each entry provides a brief history or background, regulatory requirements (if any), statistical design and analysis, recent development, and related references.

From Marcel Dekker, Inc., I would like to thank M. Allegra for providing the opportunity to work on this encyclopedia, and K. Baldonado and E. Cacciarelli for their outstanding efforts in preparing this volume for publication. I am deeply indebted to the United States Food and Drug Administration (FDA) and many major pharmaceutical companies including Merck, Pfizer, Lilly, Bristol-Myers Squibb, Wyeth-Ayerst, and Procter & Gamble for their support. I would like to express my gratitude to C. Wagner, J. Wagner, and M. Filippi of Covance for their adminsitrative assistance at the early stages.

Finally, the views expressed in this volume are those of the contributors and are not necessarily those of their respective companies or organizations. Any comments or suggestions that you may have are very much appreciated for the preparation of future volumes of this encyclopedia.

Shein-Chung Chow

Contents

Contributors

Marilyn A. Agin, Ph.D. Pfizer Central Research, Groton, Connecticut

Shan Bai, M.S. Medfocus, Des Plaines, Illinois

William H. Barr, Pharm.D., Ph.D. Center for Drug Studies, School of Pharmacy, Virginia Commonwealth University, Richmond, Virginia

James Bergum, Ph.D. Department of Non-Clinical Biostatistics, Bristol-Myers Squibb Company, New Brunswick, New Jersey

Joseph C. Cappelleri, Ph.D., M.P.H. Pfizer Central Research, Groton, Connecticut

Patricia B. Cerrito, Ph.D. Department of Mathematics, University of Louisville, Louisville, Kentucky

Cheng-Tao Chang, Ph.D. Department of Biometrics, Baxter Pharmaceutical Products Inc., Liberty Corner, New Jersey

James J. Chen, Ph.D. Division of Biometry and Risk Assessment, National Center for Toxicological Research, U.S. Food and Drug Administration, Jefferson, Arkansas

Mon-Gy Chen, M.S. Department of Statistics, Covance, Princeton, New Jersey

Shu-Lin Cheng, Ph.D. Pfizer Central Research, Groton, Connecticut

Robert D. Chew, Ph.D. Pfizer Central Research, Groton, Connecticut

Vernon M. Chinchilli, Ph.D. Department of Health Evaluation Sciences, The Pennsylvania State University, Hershey, Pennsylvania

Shein-Chung Chow, Ph.D. Department of Technical Operations, StatPlus, Inc., Yardley, and Temple University, Philadelphia, Pennsylvania

Mario Comelli, Dr. Dipartimento di Scienze Sanitarie Applicate e Psicocomportamentali, Università di Pavia, Pavia, Italy

John R. Cook, Ph.D. Health Economic Statistics, Merck Research Laboratories, Blue Bell, Pennsylvania

Stacy R. David, Ph.D. Department of Statistical and Mathematical Sciences, Eli Lilly and Company, Indianapolis, Indiana

Michael F. Drummond, Ph.D. Centre for Health Economics, University of York, York, England

Susan S. Ellenberg, Ph.D. Center for Biologics Evaluation and Research, U.S. Food and Drug Administration, Rockville, Maryland

Douglas E. Faries, Ph.D. Department of Statistical and Mathematical Sciences, Eli Lilly and Company, Indianapolis, Indiana

Michael L. Garriott, Ph.D. Department of Investigative Toxicology, Eli Lilly and Company, Greenfield, Indiana

Patrick Genyn, M.B.A. Department of Clinical Data Processing, Janssen Research Foundation, Beerse, Belgium

A. Lawrence Gould, Ph.D. Department of Biometrics Research, Merck Research Laboratories, Rahway, New Jersey

Dieter Hauschke, Ph.D. Department of Biometry, ByK Gulden Pharmaceuticals, Konstanz, Germany

Joseph F. Heyse, Ph.D. Vaccine Biostatistics and Research Data Systems, Merck Research Laboratories, Blue Bell, Pennsylvania

Wherly P. Hoffman, Ph.D. Department of Statistical and Mathematical Sciences, Eli Lilly and Company, Greenfield, Indiana

A. Dale Horne, Ph.D. Center for Biologics Evaluation and Research, U.S. Food and Drug Administration, Rockville, Maryland

Alice T. M. Hsuan, Ph.D. Department of Biostatistics and Data Operations, Janssen Research Foundation, Titusville, New Jersey

William J. Huster, Ph.D. Department of Biostatistics, Eli Lilly and Company, Indianapolis, Indiana

John P. A. Ioannidis, M.D. Department of Hygiene and Epidemiology, University of Ioannina, Ioannina, Greece

Ronald K. Knickerbocker, Ph.D. Department of Biostatistics, Eli Lilly and Company, Indianapolis, Indiana

Paul F. Kramer, Pharm.D. Department of Medical Operations, Bristol-Myers Squibb Company, Princeton, New Jersey

Anant M. Kshirsagar, D.Sc. Department of Biostatistics, University of Michigan, Ann Arbor, Michigan

Peter A. Lachenbruch, Ph.D. Center for Biologics Evaluation and Research, U.S. Food and Drug Administration, Rockville, Maryland

Mani Y. Lakshminarayanan, Ph.D. Pfizer Central Research, Groton, Connecticut

Joseph Lau, M.D. Division of Clinical Care Research, New England Medical Center, Boston, Massachusetts

C. Gordon Law, Ph.D. Pfizer Central Research, Groton, Connecticut

T. Y. Lee, Sc.D. Pharmaceutical Consultant, Glen Ridge, New Jersey

David Li, Ph.D. Department of Biometrics, Clinical Research, Solvay Pharmaceuticals, Inc., Marietta, Georgia

Junfang Li, Ph.D. Department of Biometrics and Statistical Sciences, Procter and Gamble Pharmaceuticals, Cincinnati, Ohio

Karl K. Lin, Ph.D. Division of Biometrics 2, Office of Biostatistics, Center for Drug Evaluation and Research, U.S. Food and Drug Administration, Rockville, Maryland

Melody H. Lin, Ph.D. Office for Protection from Research Risks, National Institutes of Health, Bethesda, Maryland

Jen-pei Liu, Ph.D. Division of Biostatistics, Department of Research Resources, National Health Research Institutes, Taipei, and Department of Statistics, National Cheng-Kung University, Tainan, Taiwan

Herman G. Luus, Ph.D. Department of Biometry, Quintiles ClinData, Bloemfontein, South Africa

Cornelius J. Lynch, Ph.D. Center for Biologics Evaluation and Research, U.S. Food and Drug Administration, Rockville, Maryland

Iain J. McGilveray, Ph.D., F.C.I.C. McGilveray Pharmacon Inc., Ottawa, Ontario, Canada

Helen McGough, M.A. University of Washington, Seattle, Washington

Laura J. Meyerson, Ph.D. Department of Biostatistics, Quintiles, Inc., Kansas City, Missouri

Jorge G. Morel, Ph.D. Department of Biometrics and Statistical Sciences, The Procter and Gamble Company, Mason, Ohio

John R. Murphy, Ph.D. Department of Statistical and Mathematical Sciences, Eli Lilly and Company, Indianapolis, Indiana

Nagaraj K. Neerchal, Ph.D. Department of Mathematics and Statistics, University of Maryland Baltimore County, Baltimore, Maryland

Clet Niyikiza, Ph.D. Department of Statistical and Mathematical Sciences, Eli Lilly and Company, Indianapolis, Indiana

Earl Nordbrock, Ph.D. Anesta Corporation, Salt Lake City, Utah

Walter W. Offen, Ph.D. Statistical and Mathematical Sciences, Eli Lilly and Company, Indianapolis, Indiana

Sofia Paul, Ph.D. Evista Biostatistics, Eli Lilly and Company, Indianapolis, Indiana

Thomas Permutt, Ph.D. U.S. Food and Drug Administration, Rockville, Maryland

Edward F. C. Pun, Ph.D. Pfizer Central Research, Groton, Connecticut

M. Mushfiqur Rashid, Ph.D. Division of Biometrics 2, Office of Biostatistics, Center for Drug Evaluation and Research, U.S. Food and Drug Administration, Rockville, Maryland

Cindy Rodenberg, Ph.D. Pfizer Central Research, Groton, Connecticut

Robert Schall, Ph.D. Department of Biometry, Quintiles ClinData, Bloemfontein, South Africa

Kenneth B. Schechtman, Ph.D. Department of Biostatistics, Washington University School of Medicine, St. Louis, Missouri

Timothy L. Schofield, M.S. Merck Research Laboratories, West Point, Pennsylvania

Stephen Senn, Ph.D. Departments of Statistical Science and of Epidemiology and Public Health, University College London, London, England

Weichung Joseph Shih, Ph.D. Department of Biostatistics and Research Data Systems, Merck Research Laboratories, Rahway, and University of Medicine and Dentistry of New Jersey, Piscataway, New Jersey

Brian P. Smith, Ph.D. Department of Statistical and Mathematical Sciences, Eli Lilly and Company, Indianapolis, Indiana

Catharine B. Stack, Ph.D. Pfizer Central Research, Groton, Connecticut

Thomas R. Stiger, Ph.D. Pfizer Central Research, Groton, Connecticut

Naitee Ting, Ph.D. Pfizer Central Research, Groton, Connecticut

Jawahar Tiwari, Ph.D. Center for Biologics Evaluation and Research, U.S. Food and Drug Administration, Rockville, Maryland

William M. K. Trochim, Ph.D. Department of Policy Analysis and Management, Cornell University, Ithaca, New York

Merlin L. Utter, Ph.D. Q. A. Statistical Services, Wyeth-Ayerst Pharmaceuticals, Pearl River, New York

Julia A. Varshavsky, Ph.D. Department of Statistical and Mathematical Sciences, Eli Lilly and Company, Indianapolis, Indiana

Greg C. G. Wei, Ph.D. Pfizer Central Research, Groton, Connecticut

Michael G. Wilson, M.S. Department of Statistical and Mathematical Sciences, Lilly Research Laboratories, Indianapolis, Indiana

Robert Wong, M.D. Division of Rheumatology, Department of Medicine, University of Medicine and Dentistry of New Jersey–Robert Wood Johnson Medical School, New Brunswick, New Jersey

Aileen L. Yam, M.A. Department of Biostatistics and Data Management, PharmaNet, Inc., Princeton, New Jersey

Lilly Q. Yue, Ph.D. Center for Devices and Radiological Health, U.S. Food and Drug Administration, Rockville, Maryland

Donghui Zhang, Ph.D. Biometrics Research, Merck Research Laboratories, Rahway, New Jersey

Shu Zhang, M.D., Sc.D. Pfizer Central Research, Groton, Connecticut

Adjustment for Covariates*

See also *Confounding and Interaction*

I. INTRODUCTION

The techniques of analysis of covariance are employed in three mathematically similar but conceptually very different kinds of problem. Examples of all three kinds arise in connection with the development of pharmaceutical products.

In the first case, a regression model is expected to fit the data well enough to serve as the basis for prediction. In testing the stability of a drug product, for example, the potency may be modeled as a linear function of time, and the possibility of different lines for different batches of the product needs to be allowed for. The purpose of the statistical analysis is to ensure, with a stated degree of confidence, that the potency at a given time will be within given limits.

The second and perhaps widest application of analysis of covariance is in observational studies, such as arise in the postmarketing phase of drug development. It may be desired, for example, to study the association of some outcome with exposure to a drug. It is necessary to adjust for covariates that may be systematically associated both with the outcome and with the exposure and so induce a spurious relationship between the outcome and the exposure. In such studies the unexplained variation is typically high, so the model is not expected to fit the individual observations well. It must, however, include all the important potential confounders and must have at least approximately the right functional form, if a causal relationship, or the absence of one, between the outcome and the exposure is to be inferred.

The third kind of application of analysis of covariance, although the first historically (1), is to randomized, controlled experiments such as clinical trials of the efficacy of new drugs. In such experiments adjustment for covariates is optional in a sense, because the validity of unadjusted comparisons is ensured by randomization. Adjustments properly planned and executed, however, can reduce the probabilities of inferential errors and so help to control the size, cost, and time of clinical trials.

The modeling problem is straightforward, well covered in textbooks, and, strictly speaking, not a matter of "adjustment." The observational problem, in contrast, is essentially intractable from the standpoint of formal statistical inference; but heuristic methods have had wide application and discussion. We focus here on the adjustment for covariates in the experimental setting. This problem has had relatively little attention in the literature, partly because early writings (1) are largely complete, correct, and still sufficient. Unfortunately, the more recent literature on modeling and on observational studies has been misapplied to the experimental problem. Either a well-fitting model is though to be required, as in the first problem, or the analysis is supposed to be heuristic, as in the second. In fact, a rigorous theory of analysis of covariance in controlled experiments can be developed, even in the absence of a good model for the covariate effects.

*The opinions expressed are those of the author and not of the Food and Drug Administration.

II. ADJUSTING FOR BASELINE VALUES

Consider the case of a randomized trial of two treatments, with a continuous measure of outcome (Y) that is also measured at baseline (X). If the populations are normal or the samples are large, the treatments might be compared by a two-sample t-test on the difference in mean outcome $\bar{Y}$. Alternatively, the change from baseline, $Y - X$, might be analyzed in the same way. The difference between groups in $\bar{Y}$ and the difference between groups in $\overline{Y - X}$ have the same expectation, because the expected difference between groups in $\bar{X}$ is zero. We therefore have two unbiased estimators of the same parameter. They have different variances, according to how well the baseline predicts the outcome. If the variances (within treatment groups) of baseline and outcome are the same and the correlation is ρ, then the standard errors are in the ratio $(2 - 2\rho)^{1/2}$. The adjusted estimator is better if $\rho > 0.5$.

Of course, there is no need to choose. The average of the two estimators has standard error proportional to $(1.25 - \rho)^{1/2}$, which is less than either of the two whenever $0.25 < \rho < 0.75$. This average can be written as the difference between treatment groups in $\bar{Y} - 0.5\bar{X}$. So $\bar{Y} - 0.5\bar{X}$ is a less variable measure of outcome than either the mean raw score $\bar{Y}$ or the mean difference from baseline $\overline{Y - X}$, whenever the correlation is between 0.25 and 0.75. This can, but need not, be viewed as fitting parallel straight lines with slope 0.5 to the two groups and measuring the vertical distance between them.

Naturally, there is no need to choose 0.5, either. The difference in group means of any statistic of the form $Y - \beta X$ can be used to estimate the treatment effect. The smallest variance, and so the most sensitive test, is achieved when β happens to coincide with the least-squares common slope, but the variance does not increase steeply as β moves away from this optimal value. Thus, even a very rough a priori guess for β is likely to perform better than either of the special cases $\beta = 0$ (no adjustment) and $\beta = 1$ (subtract the baseline).

Finally, there is no need to guess. The least-squares slope, calculated from the data, can be used for β, without any consequences beyond the loss of a degree of freedom for error. Asymptotic theory for the resulting adjusted estimator of the treatment effect was given by Robinson (2), and an exact, small-sample theory by Tukey (3).

In general, then, the best way to adjust for a baseline value is neither to ignore it nor to subtract it, but to subtract a fraction of it. The fraction will be estimated from the data, simultaneously with the treatment effect, by analysis of covariance. There is no need to check the "assumption" that the outcome is linearly related to the baseline value, because this assumption plays no role in the analysis. If it did, not only the analysis of covariance would be tainted: after all, the unadjusted analysis also "assumes" a linear relationship, with slope 0, and the change-from-baseline analysis "assumes" a slope of 1.

III. OTHER COVARIATES

Any single, prespecified covariate can be adjusted for in much the same way as a baseline measurement of the outcome variable. That is, the mean of a linear function $Y - \beta X$ may be compared across treatment groups, the coefficient β being estimated, simultaneously with the treatment effect, by least squares. Again, the much-tested assumption of a linear relationship between Y and X is superfluous. Two other, critical assumptions are sometimes neglected, however.

First, the covariate must be unaffected by treatment. While it is possible to give an interpretation of analysis of covariance adjusting for intermediate causes, this interpretation is not often useful in clinical trials. Any covariate measured before randomization is acceptable. With care, some covariates measured later may be assumed to be unaffected by treatment: the weather, for example, in a study of seasonal allergies. It may be noted that, while analysis of covariance is not usually appropriate for variables in the causal pathway, some of the advantages of analysis of covariance are shared by instrumental-variables techniques (4) that are appropriate.

Second, the covariate is assumed to be prespecified. Model-searching procedures are unavoidable in observational studies, for there are typically many potential confounding variables whose effects must be considered and eliminated if necessary. Alarmingly little is known about the statistical properties of such procedures, however, and what is known is not generally encouraging. It is usual, though unjustifiable, to ignore the searching process in reporting the results, presenting simply the chosen model, its estimates, and its optimistic estimates of variability.

Randomized trials are radically different from observational studies in this respect. There is no confounding, because a covariate cannot be systematically associated with treatment if it is not affected by treatment and if treatment is assigned at random. The purpose of analysis of covariance in randomized studies is to reduce the random variability of the estimated treatment effects by eliminating some of what would otherwise be unexplained variance in the observations. This difference has implications for the choice of covariates, which will be discussed in the next section.

IV. CHOICE OF COVARIATES

Whereas a confounder in an observational study is a variable correlated both with the outcome and with the treatment, a useful covariate in a randomized trial is a variable correlated just with the outcome. The greater the absolute correlation, the more the reduction in residual variance and so also in the standard error of the estimated treatment effect. This benefit is realized whether the treatment groups happen to be balanced with respect to the covariate or not. It is neither necessary nor useful, therefore, to choose covariates retrospectively, on the basis of imbalance (5).

It is accordingly safe to prespecify, in the protocol for a randomized trial, a covariate, or a few covariates, unaffected by treatment but likely to be correlated with the outcome. Analysis of covariance, adjusting for these covariates, may then be carried out and relied on, without any justification after the fact. The probability of Type I error will be controlled by significance testing, and the probability of Type II error will be less than if covariates were not used.

The improvement, however, depends on the correlations (and partial correlations) between the covariates and the outcome, and these may not be perfectly known ahead of time. It might therefore seem advantageous to determine the correlations for some candidate covariates with the data in view, and select a subset that explains a high proportion of the variance of the outcome. With care, it is possible to specify an unambiguous algorithm for selecting a model and to control the probability of Type I error (3). It is not known, however, whether such procedures have any advantage with respect to Type II error over simply prespecifying the model. In practice, in critical efficacy trials the relevant covariates will often be apparent in advance; and when they are not, it may not be any easier or better to specify a set of candidates and an algorithm for choosing among the than to specify a single model.

The properties of models with large numbers of covariates are not well understood. Various rules of thumb relating the number of variables to the sample sizes have been given, but none has any compelling theoretical justification. Furthermore, searches in large sets of potential models probably share some of the defects of models with many covariates, even if the chosen model has only a few covariates.

V. NONLINEAR MODELS

The word *linear* in the context of the analysis of covariance may be understood in two senses. In many applications, the model is linear in the covariates. However, a model with polynomial or other nonlinear covariate effects is still linear in the *coefficients*, and the least-squares estimators are consequently linear functions of the outcome measurements, so the theory of the general linear model applies. In contrast, logistic, proportional-hazards, and Poisson regression models all involve covariates in a more fundamentally nonlinear way.

Nonlinear covariate effects can be added to an analysis of covariance without difficulty. The most common examples are the 1/0 variables used to represent categorical covariates, but polynomial, logarithmic, exponential, and other functions may sometimes be useful. It is important to bear in mind, however, that in randomized trials the purpose of the covariate model is to reduce unexplained variance. Thus, nonlinear terms should be introduced when they are expected to explain substantial variance in the outcome, and not simply because it is feared that the "assumption" of a linear relationship between the outcome and the covariate may be violated.

Conversely, trials with outcomes that are successes or failures, survival times, or small numbers of events are analyzed by methods that are nonlinear in the second sense. Recent theoretical developments (the generalized linear model) and computer programs have tended to emphasize the analogies between these methods and the linear model. Some of the same principles undoubtedly apply when such methods are used to analyze randomized trials. For example, if a model selection procedure is used, it is vital to understand the statistical properties of the procedure as a whole, rather than simply to report the nominal standard errors and p-values of the model that happens to be chosen. On the other hand, the similarity in form may conceal important differences in mathematical structure between linear and nonlinear models, and the linear results must not be casually assumed to have nonlinear analogs. It is not clear, for example, that the robustness of linear models against misspecification in randomized trials carries over to all the nonlinear cases.

VI. INTERACTION

If the difference in mean outcome between treatments changes as a covariate changes, there is said to be a treatment-by-covariate interaction. In a drug trial, such a finding would have important implications. In the extreme case, the treatment effect might change direction as the covariate changed. That is, a drug that was beneficial in one subset of patients, identified by the covariate, would be harmful in a different subset. Clearly such a drug would be "effective." Equally clearly, for such a drug to be useful, the populations in which it was beneficial and harmful would need to be characterized. In less extreme cases, where the magnitude but not the direction of the treatment effect changes, considerations of risk and benefit might also make it very desirable to estimate the effect in different subgroups.

The question of interaction often arises in connection with analysis of covariance, but it really has little to do with adjustment for covariates. Everything in the preceding paragraph is equally true whether the covariate in question is adjusted for, ignored, or even unmeasured. Furthermore, if the treatment main effect is to be estimated, it is still better to estimate it by analysis of covariance, even without an interaction term, than by the unadjusted difference in means. As with the "assumption" of linearity, the analysis of covariance is not invalidated by violation of the "assumption" of parallelism, for this assumption plays no role in the analysis. Also, as with linearity, if this assumption were crucial, its failure would taint as well the unadjusted analysis, which also "assumes" parallel regressions of the outcome on the covariate, but forces them to have slope 0.

The possibility of interaction should be taken into account whenever it appears at all probable that different groups may respond differently. The reason for this is practical and concerns the interpretation and application of the results of a successful trial. However, the presence of interaction or, what is more common, the inability to rule interaction in or out with confidence, should not be seen as invalidating analysis of covariance nor, especially, as a reason to prefer unadjusted analysis.

REFERENCES

1. RA Fisher. Statistical Methods for Research Workers. 14th ed. Edinburgh: Oliver and Boyd, 1970, pp 272–286.
2. J Robinson. J Roy Statist Soc (Series B) 35:368–376, 1973.
3. JW Tukey. Controlled Clin Trials 14:266–285, 1993.
4. JD Angrist, GW Imbens, DB Rubin. J Am Statist Assoc 91:444–455, 1996.
5. T Permutt. Statist in Med 9:1455–1462, 1990.

Thomas Permutt

Ames Test*

See also *Carcinogenicity Studies*

I. INTRODUCTION

Genetic toxicology tests are among the early studies designed to establish a safety profile for a compound. Of a battery of short-term genetic assays, the *Salmonella typhimurium*/microsome test developed by Ames and associates (1,2) is the most commonly used genotoxicity test. The design, statistical analyses, and interpretation of the results will be discussed in this entry. Section II includes background for the relevance of the Ames test to the toxicology safety profile. Section III describes the basic design of the test. Section IV provides a review of some analysis methods employed for the evaluation of the mutagenicity of a compound. In Sec. V, we discuss the methods reviewed and some important issues in the evaluation of mutagenicity based on Ames test results, and in Sec. VI we draw concluding remarks to summarize this entry.

II. BACKGROUND

In this section, a brief summary is presented of why the Ames test was developed and how this assay is conducted. The procedure for conducting the Ames test is described without too much detail. This section is not intended to provide enough information to conduct

*Adapted from WP Hoffman. In: SC Chow and JP Liu, eds. Design and Analysis of Animal Studies in Pharmaceutical Development. New York: Marcel Dekker, 1998, pp 357–372.

the assay; it is intended to provide readers with a general understanding of how the responses are obtained and thus leads naturally into the following sections on the design and statistical methods employed for analysis of the test data.

A. Why the Ames Test

When developing a potentially beneficial compound, one has to establish its safety profile. Animal in vivo and in vitro tests are designed to serve this purpose. In general, the testing includes acute toxicity tests in rodents, genetic toxicity tests, developmental toxicity tests, general animal toxicity tests from 30 days up to one year in duration, and lifetime carcinogenicity tests in rodents. Of these general tests that are needed to demonstrate the safety of a compound and to ensure its registration worldwide, the most costly and time-consuming test is the carcinogenesis bioassay in rodents. Assays for genotoxicity were developed initially in an attempt to predict in a shorter time frame the eventual outcome of the bioassay. The basic premise revolved around the somatic theory of carcinogenesis, which suggested that mutational events were causative factors in the development of cancers. Additionally, there was evidence that mutagenic events were associated with embryonic mortality, birth defects, and genetic disease. Therefore the use of genotoxicity tests seemed to have broad application in the overall safety assessment. Although bacterial mutation tests had been used since the 1950s, they were not efficient in the detection of compounds acting by various mutagenic mechanisms. Efforts by Dr. Bruce Ames and his colleagues at the University of California, Berkeley, to develop a more efficient bacterial screening system culminated in 1971 with the test that now bears his name. Despite its utility, it was realized that bacteria are not mammalian cells and that the organization of the genetic material was different such that other short-term mutagenicity tests would be required to ensure the detection of the majority of rodent carcinogens (3). While over 200 mutagenicity tests have been developed, few have been well validated, and current regulatory requirements for product registration include only a battery of three to five standard tests. Of these required tests, the Ames test is the most commonly used and for this reason alone merits discussion.

B. The Ames Test

The Ames bacterial mutation test is an in vitro assay in that it does not use live animals; instead, it uses bacterial cells. Typically, the bacterial tester strains used in the test are *Salmonella typhimurium* strains TA1535, TA1537, TA98, and TA100 and *Escherichia coli* strain WP2uvrA. Although WP2uvrA was added to the bacterial test system after the Ames test was developed, it is handled the same as the four *S. typhimurium* strains in all aspects of the conduct and analysis of the experiment. Therefore, in this chapter, Ames test will refer to the assay of the four *S. typhimurium* strains and the *E. coli* strain. For each strain of the bacteria, a mixture of proper amounts of three ingredients is first prepared in a tube. See Fig. 1. The three ingredients are: (1) from 10 to 100 million bacterial cells from a particular strain, (2) a compound, which can be a test article, negative control, or positive control, and (3) minimal agar, which includes trace amounts of histidine, tryptophan, and biotin. The mixture is then poured into a plate containing a layer of solidified minimal agar for the growth of revertant colonies. Some compounds are not directly mutagenic in vivo but are metabolized to a compound that is. Therefore, all in vitro genotoxicity tests, including the Ames test, are conducted so that cells are exposed to the test article both in the presence and in the absence of an exogenous metabolic activation system. This metabolic activation system consists of a postmitochondrial supernatant (S9) of homogenized liver tissue from rats. This cell-free extract contains a variety of drug-metabolizing enzymes. The S9 is supplemented with cofactors, salts, and a buffer system collectively known as S9-mix.

Histidine and tryptophan are amino acids essential for the growth of the tester strains of *S. typhimurium* and *E. coli*, respectively. The Ames tester strains contain mutations in the genes that allow normal bacteria (prototrophs) to manufacture their own histidine or tryptophan. The tester strains are dependent upon the inclusion of the amino acids into the minimal agar to support their growth. These mutant bacteria are called auxotrophs. The Ames test is often referred to as a *reverse* mutation assay, because what is measured is the reverse mutation rate from histidine/tryptophan dependency (auxotrophy) to histidine/tryptophan independence (prototrophy). Bacterial cells that have undergone the reverse mutational event are called *revertants*. Trace amounts of histidine and tryptophan are provided at the start of the experiment to sustain the bacterial cells at the beginning and thus to allow the possible expression of reverse mutation events later. After a 48-hour incubation period, the revertant cells will form visible colonies, which are counted. Because reverse mutation may occur without treatment with a mutagen,

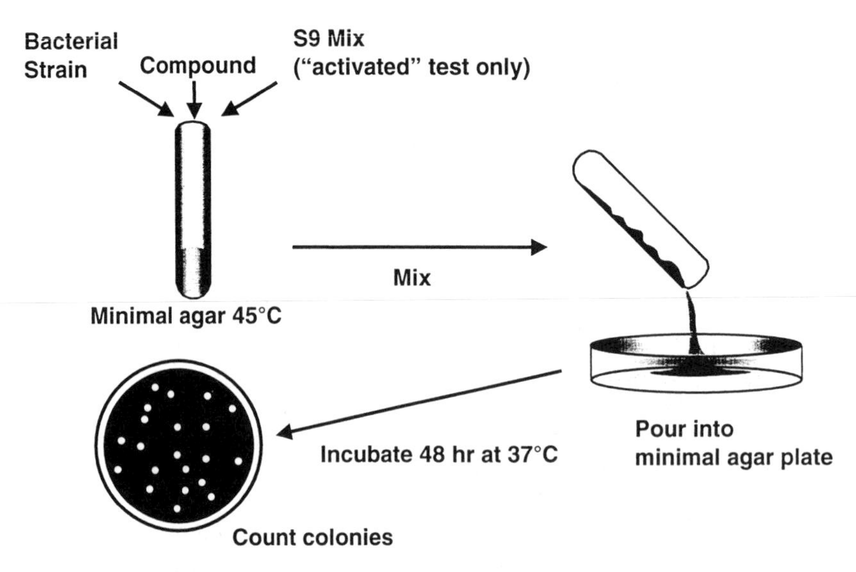

Fig. 1 Description of the conduct of the Ames test.

a negative control is always included in the assay to account for the spontaneous reverse mutation rate. If a compound is not a mutagen, then one would not expect a large increase in the number of revertant colonies compared to the number of spontaneous revertant colonies in the negative control. Therefore, the number of revertant colonies is the response evaluated to determine the mutagenic effect of a test article. A significant increase in the number of revertant colonies on the compound-treated plates is an indication of a positive response for bacterial mutation.

III. DESIGN

The Ames test procedure is rather strictly defined for regulatory purposes (4,5). The assay is typically conducted both with and without metabolic activation for each of the five bacterial strains. For each bacterial strain, five or more concentration levels of a compound and proper concentration levels for the negative and positive controls are determined by scientists. A positive control is required for the validity of the assay but is not included in the evaluation of the mutagenicity of a compound. A negative control is needed to account for spontaneous reverse mutation in the evaluation of the mutagenicity of the test article. Historical control data are useful for further evaluation of positive effects associated with a compound. Each dose level is tested in triplicate, and each of the triplicate plates is prepared independently. Therefore, if five concentrations are selected for each bacterial strain, then a total of 50 sets of triplicate data from the compound (five concentrations, two types of activation, and five strains of bacteria) and 10 sets of triplicate data from each of the negative controls and positive controls are the results of the assay.

The purpose of the Ames test is to evaluate the mutagenic effects of a compound. A suitable selection of the concentration levels of a compound for testing is essential. Logarithmically spaced concentration levels are determined by scientists in an attempt to capture an increasing trend in the response. However, if the concentration of the compound is too high, then one would expect a downturn in the number of revertant colonies as a result of toxicity. For determination of the concentration levels for the Ames test, a compound is assayed over a wide range of concentration levels up to 5000 μg/plate for each bacterial strain. Based on the preliminary results, five to seven concentration levels are then selected using a top concentration of 5000 μg/plate, adjusted for toxicity. The highest concentration may be very close to the toxic level.

IV. ANALYSIS METHODS FOR AMES TEST DATA

Proper analysis methods should account for various sources of variability in the response. Therefore, one needs to identify different sources of variability in the assay data before considering the analysis. Based on the variability and some assumptions, we discuss various statistical and nonstatistical methods for the analysis of revertant colonies obtained from the Ames test. Elder (6) and Mahon and associates (7) gave a good overview of statistical methods for the Ames *Salmonella* test. For a more recent review of statistical methods, see Lin (8). Here we will begin the discussion with some basic information on data in Sec. IV.A. Then selected methods will be discussed to bring about different approaches for the analysis of Ames test data. The methods discussed are modified two- and threefold rules in Sec. IV.B, nonparametric methods in Sec. IV.C, and parametric methods in Sec. IV.D.

A. Understanding Data

What is the distribution of the number of revertants from the Ames test? Before suggesting an answer, let us understand the data first and try to identify various sources of variability in it. For each replicate plate, a separate minimal agar mixture of the selected strain of bacterial cells, the test article, and a solution of histidine, tryptophan and biotin is made in a tube. Therefore, the numbers of revertants from each of the triplicate plates are independent. Because auxotrophs need to grow in a histidine/tryptophan-sufficient environment, trace amounts of these amino acids are supplied at the beginning of the test. For each strain of bacteria, variability in the number of revertants can be a result of the number of bacterial cells that are plated at the beginning of the test, the amount of histidine/tryptophan provided initially, the concentration of the test article, the contents of the minimal agar, the amount of S9 mix, the duration of the incubation time, or other factors. In a laboratory, attempts are made to control all factors so that changes in the revertant counts would reflect only the effect of the test article through the concentration levels.

To demonstrate that a compound is not mutagenic, one has to test it at sufficiently high concentrations to rule out doubts that the bacterial cells may not have been challenged enough. Pilot studies are usually conducted first to select proper concentration levels for each strain of bacteria. The final selection often includes concentration levels that are very close to the toxicity threshold of the bacterial cells. Consequently, one may observe a downturn in the number of revertants when some of the concentration levels of a compound are higher than the toxicity threshold of the bacterial cells.

Because the number of bacterial cells plated initially is large (about 10^7) and the probability of a cell undergoing a reverse mutation from auxotrophy to prototrophy is small, the resulting revertant counts of such rare events may have a Poisson distribution. Does the number of revertants X follow a Poisson distribution $p(X = x, \lambda)$ where

$$p(X = x, \lambda) = \frac{\lambda^x e^{-\lambda}}{x!} \quad \text{for } x = 0, 1, 2, \ldots \tag{1}$$

and λ is the mean rate of reverse mutation? One can examine this either graphically or by performing a statistical test. To get a good assessment of the distribution assumption, both approaches require more data than are available from one assay in general. When sufficient amount of data are available, one can check this Poisson assumption by plotting the sample variance against the sample mean on a logarithmic scale. A linear relation with slope near 1 is an indication of Poisson distribution. Or one can perform Fisher's dispersion test (9) for a sample of n counts, $x_1, x_2, \ldots, x_n$, which compares the test statistic

$$\frac{\sum_{i=1}^{n} (x_i - \bar{x})^2}{\bar{x}} \tag{2}$$

to a chi square distribution with $n - 1$ degrees of freedom, where $\bar{x}$ is the average of the n revertant counts. Stead and associates (10) reported no overdispersion in their experience, whereas Vollmar (11) and Margolin and associates (12) observed overdispersion. Because each laboratory may have its unique features in conducting the assay and controlling the assay environment, an assessment of overdispersion should be performed based on its own historical data, and reevaluation should be performed periodically.

If the rate of reverse mutation λ from replicate to replicate is not a constant, then the excess of variability in replicates over the mean would suggest that the revertant counts are a random sample from a mixture of Poisson distributions. For ease of further references in later sections, we will follow the parameterization of Margolin and associates (12). If λ is a random variable with a gamma distribution, then the revertant counts have a negative binomial distribution

$$p(X = x|\mu, c) = \binom{x + c^{-1} - 1}{x} \left(\frac{\mu}{\mu + c^{-1}}\right)^x \cdot \left(\frac{c^{-1}}{\mu + c^{-1}}\right)^{c^{-1}} \quad \text{for } x = 0, 1, 2, \ldots \tag{3}$$

where $c \geq 0$ and $\mu > 0$. The mean of this distribution is μ, and the variance is $\mu(1 + c\mu)$. The extra variability in the Poisson counts is reflected by the dispersion parameter c. Large values of c indicate inadequacy of modeling without accounting for the extra variability in the sampling.

B. Modified Two- and Threefold Rules

The two- and threefold rules declare a given test article to have induced a positive response if a concentration-related increase in the number of revertant colonies is at least twice the control count for strains TA98, TA100, and WP2uvrA or three times the control count for strains TA1535 and TA1537 (13). The comparison is based on the average number of revertants from the replicates for the control and each concentration level. The treated averages are compared to the control average. In the event that there is a downturn in the number of revertants owing to toxicity, the evaluation criteria are only applied to the results at the nontoxic level.

C. Nonparametric Methods

Vollmar (11) examined the results from the European Collaborative Ames-test Study 1977–78, and the plots of log-range versus log-mean indicated that the number of revertants did not follow a Poisson distribution. Overdispersion was evident between laboratories as well as within each laboratory. It was recommended that nonparametric methods be applied to Ames test results. Two rank-based tests were considered: the Jonckheere test (14) for monotonic concentration effects and the Kruskal–Wallis test (15) for other effects, including a downturn at the toxic levels. Given 713 Ames test results, the Jonckheere test was superior to the Kruskal-Wallis test and was recommended as the routine statistical method for qualitative evaluations. Vollmar pointed out that, in general, three to five replicate samples were necessary for achieving a statistical significance level of 0.05.

Wahrendorf and associates (16) proposed a nonparametric approach to the analysis of Ames test data. Assume that there are K concentration levels with n_k replicates at the kth level and the total number of replicates is N. In their approach, the concentration levels were arranged in K increasing levels, with $K = 1$ for the control. For a given j, $1 \leq j \leq K$, the revertant counts from all K concentration levels are assigned to one of two categories: 1 is the "below" category for levels $1 - j$; 2 is the "above" category for levels $j + 1$ to K. The revertant counts are ranked from low to high. If there are no concentration-related effects, then the sum of the ranks, L_j, from the "below" category should be the product of the midrank, $(N + 1)/2$, and the number of observations in the "below" category. Define L as the sum of the "below" categories corresponding to j for $1 \leq j \leq K - 1$. If a compound does induce a positive trend in the number of revertants, then the ranks in the "above" category would be relatively higher than those in the "below" category. Therefore, positive trends are associated with smaller values of L, and a test statistic for positive trends is L. Wahrendorf and associates provided critical values for L based on Monte Carlo simulations for standard designs in their paper. Incorporating the expectation, $E_0(L)$, and variance, $\text{Var}_0(L)$, of L under the null hypothesis of no concentration-related effects, the resulting standardized test T for the alternative hypothesis of an increasing trend is a one-sided test:

$$T = \frac{E_0(L) - L}{\sqrt{\text{Var}_0(L)}} \tag{4}$$

where $E_0(L)$ and $\text{Var}_0(L)$ are defined (16) by the total sample N and s_j, the number of observations in the "below" category for a given j as follows:

$$E_0(L) = \frac{N + 1}{2} \sum_{j=1}^{K-1} s_j \tag{5}$$

$$\text{Var}_0(L) = \frac{N + 1}{12} \left[\sum_{j=1}^{K-1} s_j(N - s_j) + 2 \sum_{j=1}^{K-2} \sum_{k=j+1}^{K-1} s_j(N - s_k) \right] \tag{6}$$

The T statistic is approximated by a normal distribution. In addition to the test proposed in the foregoing, Wahrendorf and associates also provided a calculation for the probability $\hat{q}_j$ that a revertant count from the "below" category is smaller than that from the "above" category for a given j. Under the null hypothesis of no concentration-related effects, $\hat{q}_j$ is expected to have a value of 0.5. This probability gives insight to the change of trends in the revertant counts and is helpful for further understanding of the type and strength of the trends.

D. Parametric Modeling

Parametric methods include: (1) regression on transformed data, assumed to be normally distributed (17);

(2) nonlinear regression on data assumed to be Poisson distributed, either with or without extra variation, using a full likelihood or quasi-likelihood approach; and (3) biologically based models incorporating extra-Poisson variation (11,18,19).

1. Linear Regression on Logarithmically Transformed Revertant Counts

Chu and associates (17) at the In Vitro Program of the National Cancer Institute (NCI)/National Toxicology Program (NTP) evaluated the statistical methods for Ames test data based on results from 2362 tests performed in four laboratories on 17 test articles. In defining the screening criteria for adequate data, they defined *toxic dose level* as: "any dose level which was greater than the dose eliciting the highest average response and which had every response less than the lowest single response in the highest average response dose level." Evidence of toxicity is a decrease in the number of revertants after reaching the toxic level. Therefore, when a downturn is present in the revertant counts and the peak mean revertant count occurs at concentration level c_i, then the toxic levels are those concentration levels, c_j, for $j > i$, that have all revertant counts smaller than the lowest revertant count at concentration level c_i.

Once the toxic concentration levels are identified, evaluation of the mutagenic effects will be based on data obtained in the nontoxic concentration range. One approach to the statistical evaluation of the mutagenic effects is regressing the logarithmically transformed revertant count on the logarithmically transformed concentration level (17). Because the control concentration level is 0 and the log transformation of 0 does not exist, the recommendation was to increase all concentration levels by 1. The validity of this regression approach lies in the appropriateness of the normality assumption resulting from the log transformation on the revertant counts. Mutagenicity is established by a statistically significant positive slope of the regression line. Chu and associates required a test to have at least five concentration levels, including a negative control, and each with a minimum of two replicates. When concentration levels are approximately equally spaced, instead of adding 1 to all concentration levels to allow for the log transformation on the 0 concentration level, an alternative was recommended by Margolin and associates (20). The concentration level of the negative control is selected so that all concentration levels are equally spaced on the logarithmic scale.

2. Nonlinear Regression on Revertant Counts

Although rare events are often modeled by Poisson models, and it seems reasonable to consider the small number of revertants in about 10 million bacterial cells a Poisson random variable, there are statistical concerns in the assumption of simple Poisson distribution. Evidence of extra-Poisson variation was observed by some, but not all. Because replicates are required at each concentration level, accumulating historical control data for examination of the adequacy of the Poisson assumption is essential. Under the Poisson assumption, the revertant counts can be modeled by a generalized linear model (21) with Poisson distribution and logarithmic link. Positive trends in the treated groups can be tested in a sequential fashion by a trend test (22) if identifying the concentration level associated with no observable effect is of interest. For laboratories that exhibit extra-Poisson variation in the historical control data, the generalized linear model can be adjusted using a quasi-likelihood method for the extra-Poisson variation. The extra-Poisson variation is accounted for by including an extra factor in the variance of the model. The extra factor can be estimated by the square root of Pearson's chi square divided by the degrees of freedom. This analysis can be carried out using PROC GENMOD in SAS 6.11 (23).

If the mean rate for reverse mutation, λ, in Eq. (1) is a random variable with gamma distribution, then the posterior distribution of the revertant counts is a negative binomial distribution, as in Eq. (3). Thus, the extra-Poisson variation can be incorporated in the modeling and evaluation of the mutagenicity of compound using a full likelihood approach.

3. Biologically Based Models

Nonmonotonicity observed as a downturn in the revertant counts is not uncommon in practice. Bacterial cells subject to toxicity are inhibited from full expression of mutagenicity. Once a definition of a toxic concentration level is established, then one can eliminate the toxic concentration levels and model the mutagenic effects of a compound using the remaining data. An alternative to this two-step modeling approach is to model both mutagenicity and toxicity of a compound simultaneously. Margolin and associates (12) developed biologically based models for modeling the mutagenicity and toxicity while accounting for the extra variation in the Poisson distribution in the revertant counts from the Ames test. The choice of their models depends on the number of generations the bacterial cells have for their histidine/tryptophan supply, the number of generations the compound is mutagenic, and the number of generations the compound is toxic. Two basic assumptions of the biologically based models are that mutagenic and toxic effects are independent, and durations of bacterial cells, being mutagenic and toxic, are not affected by the concentration of the compound. The mean, μ, of the negative binomial distribution, is related to the number of bacterial cells exposed to the test article, N_0, and the probability of resulting a revertant colony when exposed to the compound at concentration level d, $P(d)$, as

$$\mu = N_0 P(d) \tag{7}$$

We will discuss the two simplest cases, when the bacterial cells have enough histidine/tryptophan supply for just one generation and the compound is mutagenic for one generation. The first case is when the compound is toxic for one generation. The second case is when it is toxic for many generations. If a compound is only toxic for one generation, then $P(d)$ is

$$P(d) = (1 - e^{-(\alpha+\beta d)})e^{-\gamma d} \tag{8}$$

where α and β are positive and γ is nonnegative. However, if it has long-lasting toxicity, then $P(d)$ is

$$P(d) = (1 - e^{-(\alpha+\beta d)})[2 - e^{-\gamma d}]_+ \tag{9}$$

where $[y]_+ = \max(y,0)$. The spontaneous reverse mutation rate, mutagenicity, and toxicity are modeled through parameters α, β, and γ. The evaluation of the Ames data is based on Models (8) and (9) in combination with Eq. (7) and the negative binomial distribution in Eq. (3) through parameters α, β, γ, and c. If there is no toxicity, then $\gamma = 0$ and Models (8) and (9) are identical. Mutagenicity of a compound is evaluated by testing the null hypothesis of $H_0{:}\beta = 0$ against the alternative of $H_a{:}\beta > 0$. Margolin and associates (12) approximated the distribution of the ratio of the maximum likelihood estimate of β, $\hat{\beta}$, and its standard error, s.e.($\hat{\beta}$), by a standard normal distribution. The s.e.($\hat{\beta}$) was obtained based on the Fisher information matrix evaluated at the maximum likelihood estimates of α, β, γ, and c. However, further research by Margolin and associates (18) identified problems with the two models. Inflated Type I error rates were reported for both models. Their proposal to contain the Type I error rates was performing a pretest on $\gamma = 0$ to identify the presence of toxicity. The pretest of $H_0{:}\gamma = 0$ against $H_\alpha{:}\gamma > 0$ is a likelihood ratio rest:

$$\lambda_\gamma = 2(L(\alpha, \beta, \gamma; c) - L(\alpha, \beta, 0; c)) \tag{10}$$

where $L(\alpha, \beta, \gamma; c)$ is the maximum of the log-likelihood of the data. If the toxicity is not supported by the data, then the mutagenicity parameter β will be tested by the likelihood ratio test in Eq. (10) constrained on $\gamma = 0$:

$$\lambda_\gamma = 2(L(\alpha, \beta, 0; c) - L(\alpha, 0, 0; c)) \tag{11}$$

Otherwise, the mutagenicity parameter β will be tested by

$$\lambda_\gamma = 2(L(\alpha, \beta, \gamma; c) - L(\alpha, 0, \gamma; c)) \tag{12}$$

Histidine/tryptophan is necessary for the growth of auxotrophic cells. In the biologically based models described in the foregoing, amino acid diffusion was not considered and each cell will have a local supply of the essential amino acid. However, if the amino acids do diffuse through the plate agar, then the revertant counts should be a function of the amount of histidine/tryptophan as well. Krewski and associates (19) proposed to generalize the biologically based models by Margolin and associates (12) to allow for diffusion of the amino acids in an agar plate.

V. DISCUSSION

The original and modified two- or threefold rules are simple to use and have been compared to other methods in (17,24). Chu and associates reported favorable performance of the modified twofold rule based on results from 2362 tests performed in four laboratories on 17 test articles. Define the false positive and false negative as follows:

False positive = negative tests determined by consensus of microbiologists that is concluded positive using the decision rule.
False negative = positive tests determined by consensus of microbiologists that is concluded negative using the decision rule.

Among the methods that gave higher false-positive conclusions than false-negative conclusions is the modified twofold rule, which gave 4.1% false-positive and 1.8% false-negative results and the positive linear trend test described in Section IV.D.1, which gave 20.0% false-positive and 0.4% false-negative results. If the modified twofold rule is supplemented with a modified threefold rule, one would expect the proportion of false positives to decrease further.

For situations where there is no convincing evidence for the assumptions of Poisson distribution, overdispersion in the Poisson distribution, and the normality distribution of the transformed data, the nonparametric methods discussed in Section IV.C—namely, the Jonkeere test and the method proposed by Wahrendorf and associates—seem plausible. The latter also provides a descriptive quantity for assessment of the types of increasing trends and the possibility of downturn due to toxicity. An example of revertant counts of strain TA100 in triplicate of (66, 82, 64), (66, 73, 87), (89, 84, 86), (76, 83, 87), (87, 103, 91), and (89, 98, 82) corresponding to the concentration levels of 0, 10, 30, 50, 100, and 300 μg per plate, respectively, was taken from Stead and associates (10). The nonparametric method by Wahrendorf concluded a highly significant mutagenic effect with p-value = 0.0028 based on normal approximation and $p = 0.0016$ based on critical values established by Monte Carlo simulations. If the modified twofold rule were applied to the data, no mutagenic effects would be concluded.

Parametric methods with the assumptions of Poisson distribution with or without extra variation and negative binomial distribution of the revertant counts utilize statistical tools in modeling the counts for mutagenicity, toxicity, and dispersion, whereas the modified twofold rule only calls for simple averages of the revertant counts. While the modified twofold rule demonstrated good agreement with the consensus of the microbiologists in the NCI/NTP study (17), it was considered too conservative by statisticians in general. It is clear that if the statistical methods are adopted, then there will be a lot more statistically significant findings than using the modified twofold and threefold rules. To address the differences in the mutagenicity of compounds using statistical evaluation and biological evaluation, several issues merit further discussion here:

1. The use of historical control data
2. The toxicity threshold of a bacterial cell to a test article
3. Extra-Poisson variation in the revertant counts
4. The amount of histidine/tryptophan provided
5. The relation between statistical significance and biological significance

Clearly, these issues are all interrelated and, therefore, the discussion may not be in the order listed here. Because, in practice, most laboratories conduct Ames tests in triplicate, it is difficult to identify if extra-Poisson variation is present in the revertant counts within each experiment. A historical database for negative control data should be a must for examining the assumption of excessive variation in the revertant counts. In addition, a historical database can provide a base range for the evaluation of compounds that may yield statistical significance that is not biologically important. Because there may be variations among various laboratories, it is best for each laboratory to establish its own database for the historical negative control.

The toxicity threshold of a bacterial cell to a test article may be determined using the rule described by Chu and associates or by 50–90% lethality. In other words, the number of revertants in the treated plates has to be reasonably low to indicate a toxic effect. Statistical methods can then be applied without modeling the toxic effects. However, if the rules are not accepted by the regulatory agencies, statistical modeling may be the best way to determine the presence of the toxic effects.

Biologically based models for different combinations of mutagenicity and toxicity lead to different models. How one determines the number of generations that the amino acid supply will last, the number of generations the compound will be mutagenic, and the number of generations the compound will be toxic is unclear.

VI. CONCLUSIONS

At this point, no consensus has been reached on statistical methods for analyzing and interpreting the Ames data. This is reflected in the ICH guideline for genotoxicity (25,26), which does not mention any statistical methods for the evaluation of the test results. Considering the small set of numbers that the Ames test provides for each bacterial strain, statistical methods designed to evaluate the mutagenic effects should reflect the biological significance to be of any help to microbiologists. Statistical methods that find biologically unimportant increases statistically significant with small p-values are not desirable and may slow down the development of a compound unnecessarily. Any statistical methods recommended for the routine data analysis for the Ames test should be validated by a collection of compounds with known mutagenicity. A historical database is a must in the evaluation of the mutagenicity of compounds and should be established in each laboratory. When a historical database is not available, nonparametric approaches may be a better choice than parametric methods. Statistical methods should support microbiologists' findings when the effects, mutagenic or nonmutagenic, are clear, and should guide scientists when the effects are not clear. Since the modified twofold and threefold rules seem to agree well with the consensus of microbiologists (17), further work in this area is needed to compare various proposed statistical methods with the modified twofold and threefold rules based on known compounds.

REFERENCES

1. Ames BN, Lee FD, Durston WE. An improved bacterial test system for the detection and classification of mutagens and carcinogens. Proc Natl Acad Sci USA 70: 782–786, 1973.
2. Ames BN, McCann J, Yamasaki E. Methods for detecting carcinogens and mutagens with the *Salmonella*/mammalian microsome mutagenicity test. Mutat Res 31:347–364, 1975.
3. Casciano DA. Introduction: Historical Perspectives of Genetic Toxicology. In: Li AP, Heflich RH, eds. Genetic Toxicology. Boca Raton: CRC Press, 1991, pp 1–12.
4. Claxton LD, Allen J, Auletta A, Mortelmans K, Nestmann E, Zeiger E. Guide for the *Salmonella typhimurium*/mammalian microsome tests for bacterial mutagenicity. Mutat Res 189:83–91, 1987.
5. Gatehouse D, Haworth S, Cebula T, Gocke E, Kier L, Matsushima T, Melcion C, Nohmi T, Ohta T, Venitt S, Zeiger E. Recommendations for the performance of bacterial mutation assays. Mutat Res 312:217–233, 1994.
6. Elder L. Statistical methods for short-term tests in genetic toxicology: The first fifteen years. Mutat Res 277: 11–33, 1992.
7. Mahon GAT, Middleton B, Robinson WD, Green MHL, Mitchell I, Tweats DJ. Analysis of data from microbial count assays. In: Kirkland DJ, ed. Statistical Evaluation of Mutagenicity Test Data. Cambridge: Cambridge University Press, 1989, pp 26–65.
8. Lin KK. Statistical review and evaluation of in vitro mutagenicity study data. Drug Information J 31:335–344, 1997.
9. Fisher RA. The significance of deviations from expectation in a Poisson series. Biometrics 6:17–24, 1950.
10. Stead AG, Hasselblad V, Greason JP, Claxton L. Modelling the Ames test. Mutat Res 85:13–27, 1981.
11. Vollmar J. Statistical problems in the Ames test. In: Kappas A, ed. Progress in Mutation Research. Vol. 2. Amsterdam: Elsevier/North-Holland Biomedical Press, 1981, pp 179–186.
12. Margolin BH, Kaplan N, Zeiger E. Statistical analysis of the Ames *Salmonella*/microsome test. Proc Natl Acad Sci USA 78:3779–3783, 1981.
13. Kier L, Brusick DJ, Auletta AE, Von Halle ES, Brown MM, Simmon VF, Dunkel V, McCann J, Mortelmans K, Prival M, Rao TK, Ray V. The *Salmonella typhimurium*/mammalian microsomal assay. A report of the U.S. Environmental Protection Agency Gene-Tox Program. Mutat Res 168:69–240, 1986.
14. Jonckheere AR. A distribution-free *k*-sample test against ordered alternative. Biometrika 41:133–145, 1954.
15. Kruskal WH, Wallis WA. Use of ranks in one-criterion variance analysis. J Am Stat Assoc 47:583–621, 1952.
16. Wahrendorf J, Mahon GAT, Schumacher M. A nonparametric approach to the statistical analysis of mutagenicity data. Mutat Res 147:5–13, 1985.
17. Chu KC, Patel KM, Lin AH, Tarone RE, Linhart MS, Dunkel VC. Evaluating statistical analyses and reproducibility of microbial mutagenicity assays. Mutat Res 85:119–132, 1981.
18. Margolin BH, Kim BS, Risko KJ. The Ames *Salmonella*/microsome mutagenicity assay: issues of inference and validation. J Am Stat Assoc 84:651–661, 1989.
19. Krewski D, Leroux BG, Bleuer SR, Broekhoven LH. Modeling the Ames *Salmonella*/microsome assay. Biometrics 49:499–510, 1993.
20. Margolin BH, Resnick MA, Rimpo JY, Archer P, Galloway SM, Bloom AD, Zeiger E. Statistical analysis for in vitro cytogenetic assays using Chinese hamster ovary cells. Environ Mutat 8:183–204, 1986.
21. McCullagh P, Nelder JA. Generalized linear models. 2nd ed. London: Chapman & Hall, 1989, pp 193–214.
22. Tukey JW, Ciminera JL, Heyse JF. Testing the statistical certainty of a response to increasing doses of a drug. Biometrics 41:295–301, 1985.
23. SAS/STAT software: changes and enhancements, through release 6.11. Cary, NC: SAS Institute, 1996, pp 231–314.
24. Margolin BH. Statistical studies in genetic toxicology: a perspective from the U.S. National Toxicology Program. Environ Health Perspective 63:187–194, 1985.
25. International Conference on Harmonisation; Guidance on specific aspects of regulatory genotoxicity tests for pharmaceuticals. Federal Register 61:18198–18202, 1996.
26. International Conference on Harmonisation; Guidance on genotoxicity: A standard battery for genotoxicity testing of pharmaceuticals. Federal Register 62:62472–62475, 1997.

Wherly P. Hoffman
Michael L. Garriott

Assay Development

See also *Assay Validation; Drug Development*

I. INTRODUCTION

Assays are utilized in the pharmaceutical industry to characterize drugs and biological products. Assay development is coordinated with pharmaceutical development to furnish the tools necessary to guide the research process. Measures of content, potency, and purity are combined to warrant the safety and effectiveness of marketed products. Thus special care is taken to develop assays that are sensitive to changes in the important properties of pharmaceutical products.

Statistics plays a primary role in the processing of assay results and in the development of experimental strategies to optimize an analytical method. Most of these strategies employ basic statistical moieties, and they are thus accessible to the analytical researcher and biostatistician alike. Their simplicity does not absolve the user, however, from careful attention to their implementation and interpretation.

This entry will discuss the varieties of technologies and statistical methodologies used in the development of an assay. A review of standard calibration methodologies will illustrate the synergy between the scientific and statistical properties of an assay. Sound statistical experimental design can be utilized to explore the effect of operational factors on assay response and to establish the levels of the factors that yield reliable results.

II. TERMINOLOGY

Several features common to all assays performed in pharmaceutical research, development, and quality control are associated with specific terminology. These terms will be defined here and illustrated with specific cases.

The *analyte* is the substance being measured. This may be the active drug in the product or that compound in clinical samples obtained during clinical development. The analyte can also be a known impurity or a degradate of the active chemical species. Biotechnological and biological products are subject to the introduction of adventitious agents, such as retroviruses associated with the biological system used to produce the product. In vaccine studies, products of the biological response of the vaccinee, such as antibodies or responder cells, are measured to establish successful immunization with the immunogen.

The *sample matrix* is the environment in which the analyte exists. For a drug product, this will be the inactive excipients that comprise the formulation, while for a biological it will be these as well as byproducts of the process, such as cellular debris, surfactants, and other process components. The clinical sample is the matrix for specimens tested from pharmacokinetic studies as well as from immunogencity studies.

Various terms are used to describe the units of measure of an assay. Most assays of a drug substance measure the *content* or *gravimetric mass* of the compound, which is reported in units of mass (e.g., micrograms) or mass per volume of sample (e.g., micrograms per milliliter). The *potency* of a drug represents the biological activity of the compound, and it is usually reported as the amount of the analyte that induces a specific response (e.g., the amount of a vaccine that is effective in inducing antibody response in half of immunized animals is called the 50% effective dose, or ED_{50}). Many potency assays involve a titration of a drug or biological (i.e., a series of doses of the analyte), yielding a *titer* for the sample. Finally, many assays measure the physical properties of a drug or biological, such as pH, polydispersity, and solubility.

The term *assay* is used synonymously to represent the particular analytical method, an individual run of the procedure, or the composite result obtained for a sample in multiple runs of the procedure. Where not otherwise specified in this and related entries, the term *assay* will refer to the analytical method being discussed.

In this entry, a calibration *standard* represents the material in an assay, which is utilized to construct the "ruler" against which test samples are measured, while an assay *control* is a material that is measured alongside test samples and that is used to monitor the performance of the procedure.

III. ASSAY TECHNOLOGIES

A variety of technologies form the foundation of analytical method development. A basic understanding of the mechanism and operation of these technologies is useful to their successful implementation.

Separation techniques, such as high-performance liquid chromatography (HPLC), are among the most widely employed assay technologies in the pharmaceutical industry. The basis of HPLC is the separation of molecular species, which are forced by a mobile phase through a resin-packed column (solid phase) under fixed pressure. The separation can be used on different properties of the analyte, including size, ionization, hydrophobicity, and affinity to a ligand. The output from an HPLC assay is a chromatographic trace, displaying spiked peaks with areas that are directly proportional to the amount of each molecular species in the sample. High-performance liquid chromatography is used either to detect or to quantify the amount of an analyte in a product or clinical sample. Quantitation of the analyte is usually accomplished using standard calibration techniques, in correspondence to a reference preparation.

Immunoassays employ the principle of antibody–antigen binding as a means to quantify an analyte. There are a variety of technologies associated with immunoassay, but common to all of these is a *detector*. Among the most common detectors are radioactive isotopes (associated with radioimmunoassay, or RIA), which discharge radioactive emissions in counts per minute (CPM), and enzyme substrates (associated with enzyme immunoassay, or EIA), which yield a change in the optical density (OD) of the solution at a specific wavelength of light. In either case, the level of detector signal is directly proportional to the amount of analyte in the sample. As in HPLC, quantitation of the analyte is accomplished using standard calibration techniques in correspondence to a reference preparation.

Many pharmacologically active compounds yield a response in vivo, in a biological milieu. Activity can be tested in a specified tissue, in cell culture, or in animals, utilizing biological assay techniques. A typical *bioassay* in animals is comprised of groups of animals, with each group receiving a graded dose of the test compound. The number of animals eliciting a particular response in each group is noted, and an endpoint is calculated from the dose series using one of a variety of quantal calculation methods.

A number of emerging technologies have been introduced into the analytical portfolio of many drugs and vaccines. In most cases these technologies fit into the traditional analytical paradigm. Thus fluorescence and refractory index can now easily be measured in a sample, which serve as the detector in a traditional calibration assay design. Powerful methods for exploring the structure of a drug molecule or biological derivative, such as nuclear magnetic resonance (NMR) and polymerase chain reaction (PCR), bring with them new challenges for assay development and validation.

IV. CALIBRATION

The most common assay design used to quantify the amount or potency of analyte in a drug or clinical sample is a standard curve, or calibration, design. A typical run of a calibration assay is comprised of a series of known amounts of a calibration, or reference, standard, run alongside one or several levels of a test preparation. The responses from the standard series are statistically fit using one of a variety of mathematical equations; this fit is then used to predict the amount of the analyte in the test sample. A typical calibration assay is illustrated in Fig. 1.

A. Linear Calibration

One of several strategies can be employed in selecting an appropriate calibration equation for an assay. During development, data from several runs of the standard preparation can be used to identify the "linear" range of the standard curve, in order to utilize a linear equation to fit the standard curve. This range is usually determined graphically, and can be verified statistically using common goodness-of-fit techniques. Some of the factors related to the implementation of the linear standard curve design include: (a) the range of concentrations used in the standard curve; (b) the number of standard curve concentrations; and (c) the concentration scaling—i.e., arithmetic (e.g., 10, 20, 30, 40, . . .) or geometric (e.g., 10, 20, 40, 80, . . .).

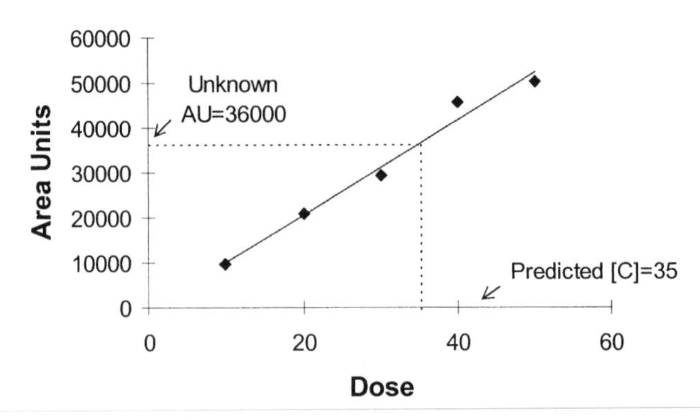

Fig. 1 Calibration using a linear fit to a standard curve; inverse regression is employed to predict the concentration corresponding to the response in an unknown.

These factors will, for the most part, be selected on the basis of the intended use and specific characteristics of the assay. The range of the standard curve should embrace the expected range of concentrations in test samples, with the exception that high-concentration samples can usually be manipulated (either by dilution or by loading a different amount of the test specimen) to achieve a level that falls within the standard curve range. The number of standard curve concentrations can be selected on the basis of the impact on the margin of error of measurement of the analyte concentration in the test sample. The concentration scaling can be selected on the basis of the behavior in the variability of the instrument readout (e.g., CPM, OD); if the variability of the readout is well behaved (i.e., there is no association between variability and response level), then an arithmetic scaling of standard curve concentrations can be used. If the variability of the instrument readout increases with increased response, a geometric scaling of response may be preferred. In this case, the standard curve equation should be fit using a log-log transformation of the data, thereby achieving both equal spacing among the standard curve concentrations and homogeneity of variability across the (log) responses. A linear-linear fit and a log-log fit to a standard curve with geometric scaling are illustrated in Fig. 2.

Once a linear equation has been fit to the standard curve data, and prior to predicting the concentrations in unknowns, the statistical "goodness of fit" of the linear equation can be assessed using common regression diagnostics. Customarily, a restriction is placed upon the coefficient of determination (R^2, or R-square) associated with the regression fit, and an assay run that fails to meet this restriction is judged invalid and is usually repeated. This criterion is frequently and inappropriately used to select the linear range of the standard curve during development of an assay as well as during its routine use. The dangers of relying exclusively on R^2 as a measure of goodness of fit are illustrated in an article by Anscombe (1). Several conditions yielding satisfactory R^2 values, such as nonlinear standard curve kinetics, a regression outlier, or heterogeneity of variability across the range of the standard curve, might still require remedial action. These conditions might more aptly be addressed through graphical diagnostic techniques, such as residual plots. Figure 3 illustrates residual patterns associated with a properly behaved regression fit, along with several commonly occurring alternatives. Residual plots can, in turn, be supplemented with more sophisticated diagnostic criteria, such as statistical tests of patterned residuals, of curvature, or of regression outliers (2).

Another graphical device used to diagnose linearity of the standard curve is a "response factor" plot (3). The "response factor" is calculated as the ratio of the instrument readout to the nominal concentration of each standard curve point, and it is plotted against concentration. When the regression intercept is zero, the "response factor" corresponds to the slope of the standard curve, and it should be constant across levels. If a trend is observed, this indicates either a nonlinear calibration curve or a nonzero intercept. For more complex calibration functions, or when a nonzero intercept can be tolerated, the "response factor" plot of the standard can be compared with that of a concentration series of a test sample; similarly shaped "response factor" profiles indicate an assay that is "linear," in the sense that measurements at each level of the test sample will be directly proportional to the test sample concentrations (see Linearity in the entry *Assay Validation*).

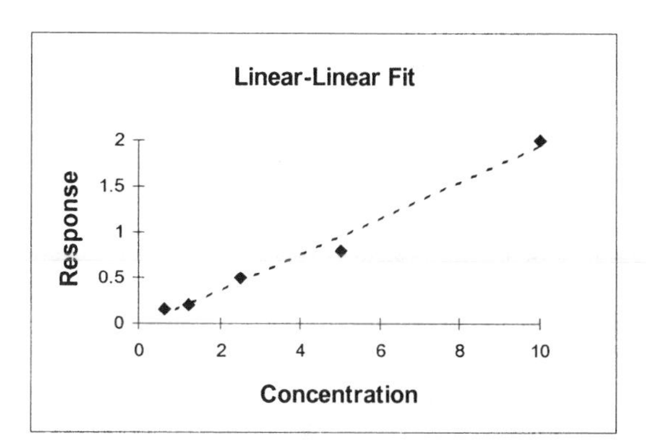

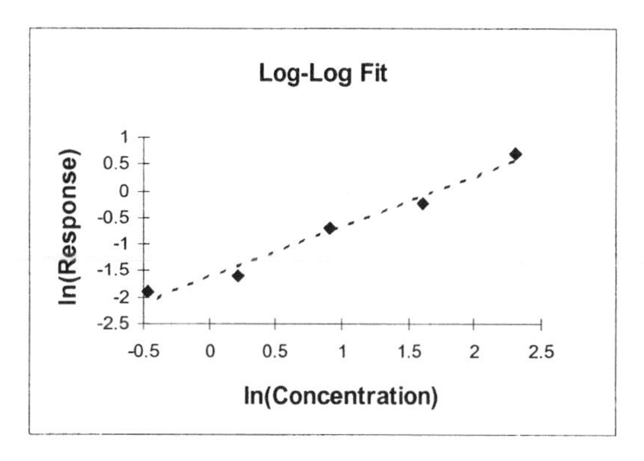

Fig. 2 A linear-linear fit to a geometrically scaled standard curve alongside a log-log fit, illustrating equal spacing between log concentrations.

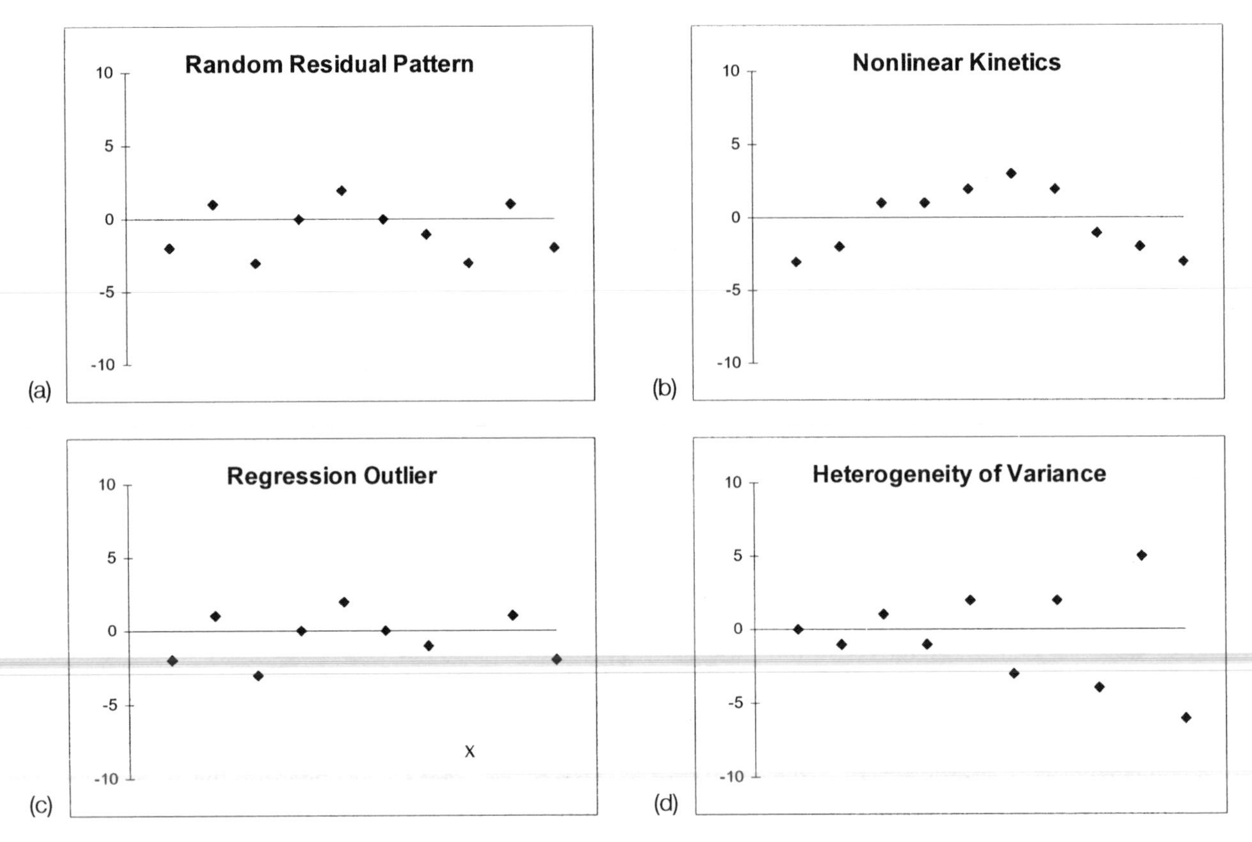

Fig. 3 Residual plots for standard curve fits exhibiting various behaviors: (a) normal behavior with random variability, (b) nonlinear kinetics, (c) a regression outlier, and (d) heterogeneity of variance.

Having established goodness of fit of the standard curve, parameter estimates for the intercept (a) and the slope (b) can be utilized to predict the concentration of analyte in an unknown sample:

$$\hat{X} = \frac{Y_{\text{unknown}} - a}{b}$$

where Y_{unknown} represents the response in the unknown (note that this pertains to the case of a "linear-linear" fit to the standard curve; prediction from a "log-log" fit of the standard curve can be developed similarly). The reliability of the estimate can be ascertained through the prediction interval on the result (4):

$$\frac{\hat{X} \pm \frac{t_{n-2} \cdot \hat{\sigma}}{b} \sqrt{(1 + 1/n) \cdot (1 - g) + \hat{X}^2/SXX}}{1 - g}$$

where

$$g = \frac{t_{n-2}^2}{t_b^2} = \frac{t_{n-2}^2 \cdot \hat{\sigma}^2}{b^2 \cdot SXX},$$

t_{n-2} is the appropriate percentile from the t-distribution with $n - 2$ degrees of freedom, $\hat{\sigma}$ is the standard error of the regression, t_b is the statistic associated with a test of significance of the regression slope, and SXX is the sum of squared deviations in the concentrations. In most cases, "g" will be small, and can be treated as if it were equal to "0" in the calculation of the standard error of estimate of the predicted concentration. In this case, the standard error can be utilized in conjunction with the predicted concentration to calculate the relative standard deviation (RSD) of the result:

$$\text{RSD} = 100 \cdot \frac{\frac{\hat{\sigma}}{b} \cdot \sqrt{1 + 1/n + \hat{X}^2/SXX}}{\hat{X}} \%$$

In this way, the variability of each prediction can be determined, and results that meet a prespecified restriction (e.g., $\leq$20% RSD) can be reported.

B. Nonlinear Calibration

Many analytical methodologies, primarily immunoassays, generate nonlinear standard curves. As indicated earlier, the linear range of the standard curve can be identified, and linear calibration can be used to fit the standard curve and interpolate sample results. This portion of the standard curve is only approximately linear, however, and it might be more suitable to fit the entire curve using one of a variety of concentration kinetics equations. In fact, basic immunochemistry principles predict that the standard curve can be modeled using the four-parameter logistic regression equation (Fig. 4). Letting Y be the instrument readout associated with an individual standard curve concentration X, the four-parameter logistic equation is as follows:

$$Y = D + \frac{A - D}{1 + \left(\frac{X}{C}\right)^B}$$

In this equation, the parameter "D" represents the minimum response (usually the background), while "A" represents the level in response associated with saturation of the "binder" (5). The parameters "B" and "C" are related to the slope and intercept from the linear regression of logistic transformed response on log concentration. In fact, "C" corresponds specifically to EC_{50}, the concentration of the standard predicted to yield 50% activity in the assay. This parameter can be used to monitor the stability of the standard.

The nonlinear equation can be fit using conventional nonlinear curve-fitting methodology. Similar regression diagnostics as employed with linear calibration can be utilized in conjunction with nonlinear regression (e.g., residuals plots, regression outliers), while the interpolated concentration for an unknown sample can be supplemented by its standard error, yielding an assessment of the quantifiable range for an individual assay run:

$$\hat{X} = C \cdot \left[\frac{A - Y}{Y - D}\right]^{1/B}$$

The standard error of a predicted value can be established via estimates of the variances and covariances of the parameter estimates, and measurement variability, using a first-order Taylor series approximation (6):

$$\text{var}(\hat{X}) = \sum \left(\frac{\partial \hat{X}}{\partial P}\right)^2 \cdot \text{var}(P) + \sum \sum \left(\frac{\partial \hat{X}}{\partial P}\right) \cdot \left(\frac{\partial \hat{X}}{\partial Q}\right) \cdot \text{cov}(P, Q) + \left(\frac{\partial \hat{X}}{\partial Y}\right) \cdot \hat{\sigma}^2$$

where $\hat{X}$ is the predicted concentration, P and Q are parameters of the nonlinear model, and var(P) and cov(P, Q) represent the approximate variances and covariances among the parameters.

The design of an assay utilizing a nonlinear standard curve will differ from that of a linear calibration system in several ways. Typically more standard concentrations will be necessary to support the estimation of additional parameters in the nonlinear model. Immunoassay design usually requires that standard curve points be extended to measure the asymptotes of the curve ("A" and "D" in the four-parameter logistic model). A common concern is that this implies that test samples can be read from these "flat" portions of the standard

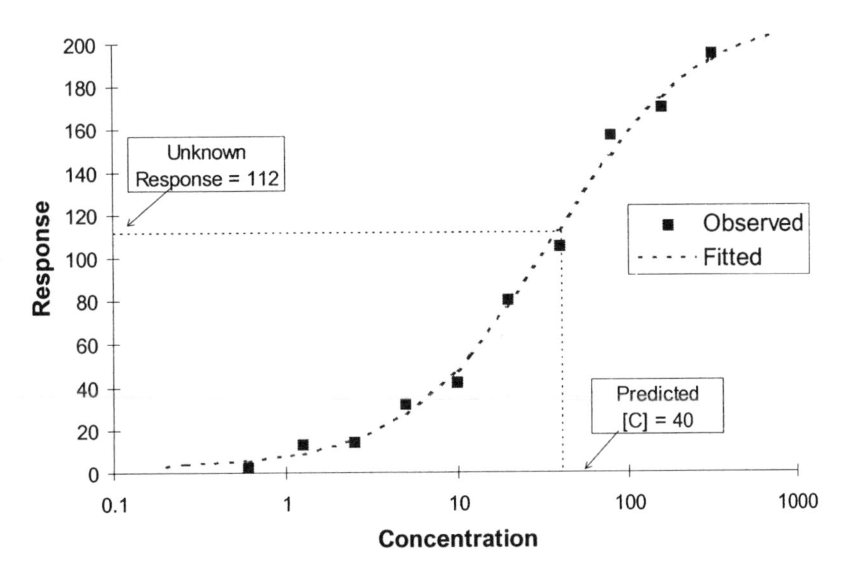

Fig. 4 Illustration of a standard curve fit using four-parameter logistic regression; inverse prediction is utilized to estimate the concentration corresponding to the response in an unknown.

curve; however, the reliable portion of the standard curve can be calculated for each run using methods discussed previously for calculating the standard error of a predicted concentration from the nonlinear curve fit.

C. Other Calibration Conventions

One means to ameliorate the effects of increasing variability with increasing response is to transform the data. Commonly used transformations are the log and square root. Another widely used solution is to implement a weighted regression, where the weight assigned each standard curve concentration is a function of the predicted response. This method is called *iteratively reweighted least squares* (7); it can easily be implemented with computation software such as EXCEL®.

The weighting functions of the predicted response (y) that are commonly used are: (a) weight = $1/y$, when the response is believed to be generated by a Poisson process (e.g., when the response is a count, such as in RIAs; note that this is analogous to performing a square root transformation of the instrument readout); and (b) weight = $1/y^2$, when it can be established that the % RSD is a constant (note that this is analogous to performing a log transformation of the instrument readout).

An adaptation of the standard calibration method is a parallel-line design. In the parallel-line design, a test sample is diluted in conjunction with the standard preparation, forming simultaneous dose–response curves. Statistical conformance to the assumptions of the parallel-line model is tested prior to the measurement of the *relative potency* (RP) of the test preparation to the standard. The assumptions tested are specific to the mathematical model utilized to fit the simultaneous dose–response profiles. When a simple linear model is used to fit the data, tests of linearity and parallelism of the dose–response curves usually precede potency estimation (8). The measured relative potency corresponds to the horizontal distance between parallel response profiles, at any level of response; thus, when the response profiles are parallel, this horizontal distance is the same at all levels of response. When the response profiles are not parallel, the estimated relative potency (horizontal distance) is not unique and can vary depending upon the level of response used to predict the relative potency (see Fig. 5).

The four-parameter logistic regression equation is frequently used to model response in a parallel-line immunoassay. The goodness of fit of various parameterizations of the simultaneous four-parameter logistic equation can be tested, preliminary to estimating the relative potency (5).

In either case, when the assumptions are met, the relative potency of the test preparation to the standard can be estimated, along with an appropriate confidence interval:

$$M = \log(\mathrm{RP}) = \bar{x}_S - \bar{x}_T - \frac{\bar{y}_S - \bar{y}_T}{b}$$

and

$$(\bar{x}_S - \bar{x}_T) + \frac{\left[M - \bar{x}_S + \bar{x}_T \pm \dfrac{t_{\mathrm{df}}}{b} \{\hat{\sigma}^2((1-g)(1/n_S + 1/n_T) + (M - \bar{x}_S + \bar{x}_T)^2 \Big/ \sum SXX)\}^{1/2} \right]}{1-g}$$

where

$$g = \frac{t_{\mathrm{df}}^2}{t_b^2} = \frac{t_{\mathrm{df}}^2 \cdot \hat{\sigma}^2}{b^2 \sum SXX}$$

"x" represents the dose, "y" is the response, "b" is

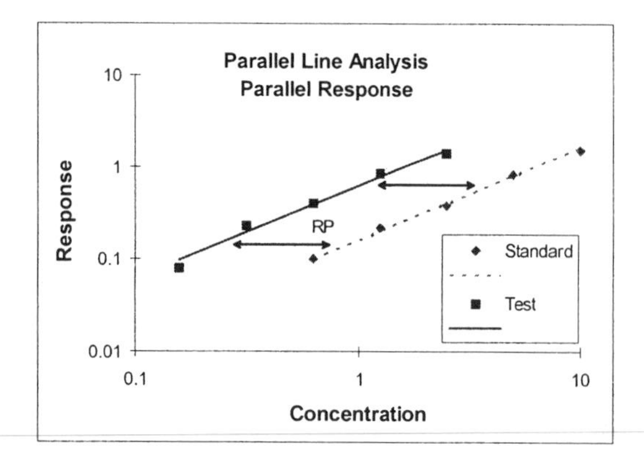

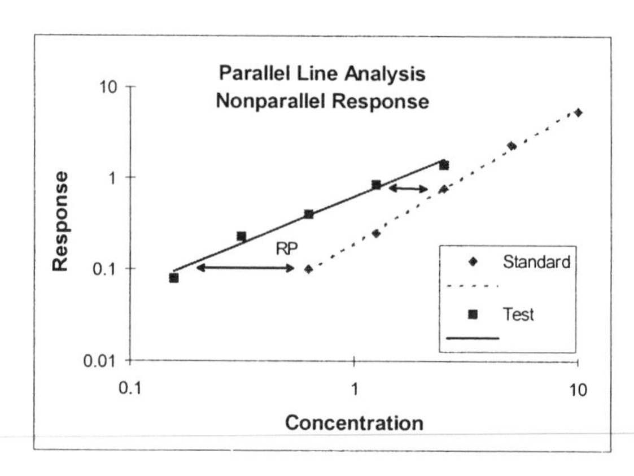

Fig. 5 The relative potency represents the horizontal distance between dose–response profiles. The relative potency is unique when the profiles are parallel, but it is different for different levels of response when the profiles are not parallel.

the pooled slope from the standard and test preparations, "t_{df}" is the appropriate percentile from the t-distribution with $n_S + n_T - 2$ degrees of freedom, $\hat{\sigma}$ is the standard error of the regression, t_b is the statistic associated with a test of significance of the pooled regression slope, and SXX is the sum of squared deviations in the concentrations, which is summed between standard and test preparations.

The relative potency is unitless and represents the "shift" in concentration of the dose–response curve for the test sample relative to the standard sample. A relative potency of test to standard equal to 1.00 represents "equipotence" of the two samples. The reported potency of the test preparation can be converted to an absolute scale by multiplying the known potency of the standard by the relative potency.

Design components of a parallel-line assay are the same as those for a calibration assay, with the additional requirements implicit in the reliable estimation of a relative potency. Thus in addition to the choice of the dose range and the number of doses of the test and standard preparations, these should be selected so as to achieve approximately equal average responses across doses for the two preparations.

V. ASSAY OPTIMIZATION

Assay development commences with scientifically reasonable levels of factors that might affect the assay (factors such as pressure, pH, temperature of incubation, time of incubation, and levels of certain key reagents) and then proceeds with empirical studies that seek to optimize the levels of these factors. The goal of the optimization is to achieve satisfactory performance in one or several key attributes of the assay. Examples of performance attributes include: background reactivity in an immunoassay, linearity in an HPLC, and repeatability and precision of the method (see the entry *Assay Validation*).

Key to the successful optimization of a method is a strategic plan for studying the affects of the assay factor(s). This can be accomplished using tools such as factorial and fractional-factorial experimental designs (9). Consider an example in which an experiment is conducted to minimize the background in an immunoassay. The factors that are thought to influence background reactivity (Y in optical density, OD) are incubation temperature (X1, ranging from 30°C to 35°C), incubation time (X2, ranging from 6 to 8 hours), and dilution of conjugate (X3, ranging from 1:1000 to 1:4000). A 2^3 factorial design is used to study simultaneously the effects of these three factors in Table 1. By inspection it appears that conjugate dilution (X3) has an effect on background reactivity, with the 1:1000 dilution yielding higher background than the 1:4000 dilution. The effect of dilution of conjugate can be estimated from experimental results:

$$X3 = \frac{0.29 + 0.34 + 0.30 + 0.35}{4} - \frac{0.16 + 0.22 + 0.19 + 0.20}{4} = 0.32 - 0.19 = 0.13$$

The interaction between factors can also be estimated; for example, the interaction of temperature with dilution of conjugate (X1 * X3) is:

$$X1 * X3 = \frac{0.29 + 0.34 + 0.19 + 0.20}{4} - \frac{0.16 + 0.22 + 0.30 + 0.35}{4} = 0.255 - 0.2575 = -0.0025$$

The effects of these factors and their interactions can be plotted in order of largest to smallest absolute effect in a Pareto plot (9), as shown in Fig. 6. This plot shows that dilution of conjugate (X3) asserts the greatest effect on background OD, while incubation time (X2) has the next largest effect. Temperature of incubation (X1) and the two- and three-factor interactions appear to have little affect.

Another aspect of assay optimization is the replication strategy used to test a sample. This is discussed in the entry on *Assay Validation*.

VI. BIOLOGICAL ASSAYS AND ASSAYS OF ANALYTES IN COMPLEX MATRICES

Many biologicals and some drugs are tested using biological assays (bioassays), which measure the biological activity of the analyte. This is typically true of vaccines, which induce an antibody response in animals as well as in man. In a bioassay for a vaccine, a fixed number of mice (or some other animal species that responds to the vaccine; vaccines are also tested ex vivo, in tissue cells that demonstrate a prescribed response, usually cytophathicity due to infection with a vaccine virus) are injected, each with one of a graded series of doses of the vaccine. After a prescribed period of time, the mice are bled and the presence or absence of antibody is determined in each animal. The data become a series of response rates across doses. Assays of this sort are often called "quantal response" assays. A

Table 1 A 2^3 Factorial Experiment Studying the Effects of Temperature, Time, and Dilution of Conjugate on Background Level in an EIA

Run	Temperature (X1)	Time (X2)	Dilution (X3)	OD
1	30°C	6 hours	1:1000	0.29
2	30°C	6 hours	1:4000	0.16
3	30°C	8 hours	1:1000	0.34
4	30°C	8 hours	1:4000	0.22
5	35°C	6 hours	1:1000	0.30
6	35°C	6 hours	1:4000	0.19
7	35°C	8 hours	1:1000	0.35
8	35°C	8 hours	1:4000	0.20

Runs are usually performed in randomized order.

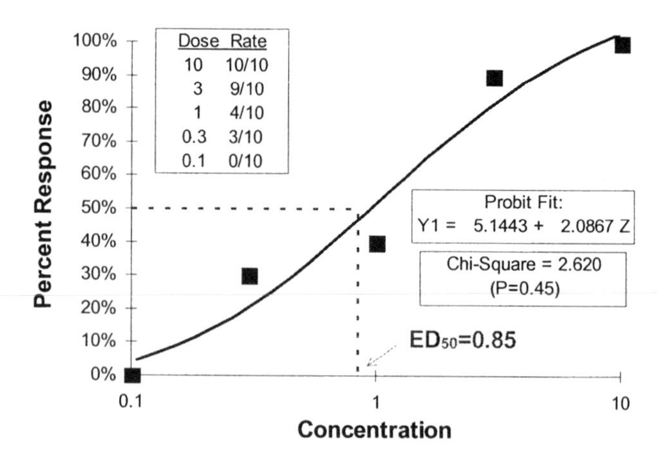

Fig. 7 Quantal response curve fit using probit analysis; the estimated ED_{50} represents the level of the test preparation that is predicted to yield 50% response in animals.

typical quantal response curve is illustrated in Fig. 7. The reported potency is usually either the dose predicted to yield responses in 50% of immunized animals (usually called either the ED_{50} in an in vivo animal model or $TCID_{50}$ when tissue culture infection is the analytical endpoint) or the relative potency of the test preparation to a standard (this will be mathematically equal to the ratio of ED_{50}'s).

There are several ways to calculate these endpoints, among them methods due to Reed and Muensch (10) and to Spearman and Kärber, as well as probit and logit analyses. The Reed and Muensch method is easy to table and was popular before computers were enlisted to perform the complex calculations associated with other methods of analysis. It does not provide, however, any confirmation of the reliability of the measured potency. The Spearman and Kärber method furnishes an estimate of the standard error of the estimate; the probit method offers this along with a test of goodness of fit of the probit model.

Several design factors affect the statistical properties of an in vivo assay. The number of doses, the dose range, and the number of animals or tissue culture wells per dose affect the variability of the measured potency. Care should be taken in defining the cutoff when an immunoassay is used to score positive responders. A cutoff that yields a high "false-negative" rate in the immunoassay will result in a potency that is negatively biased. A parallel-quantal-response design can be used to ameliorate the affects of the operating characteristics of the immunoassay. As with parallel-line design in immunoassay, this requires that the quantal response curves be "parallel," and it results in the measurement of a relative potency of the test to the standard.

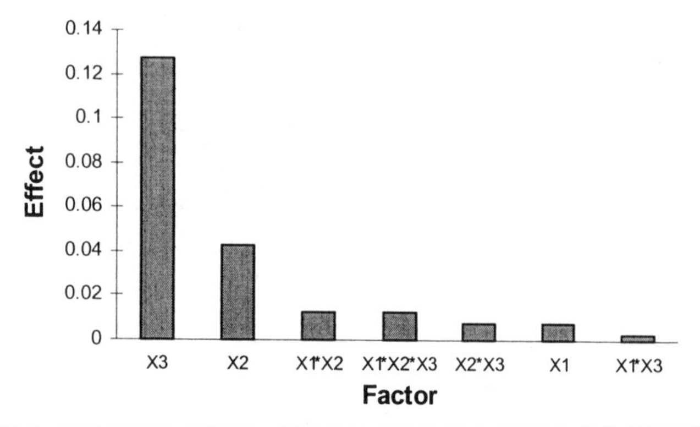

Fig. 6 Pareto plot showing absolute effects of temperature (X1), time (X2), and dilution of conjugate (X3) on background reactivity in an immunoassay.

REFERENCES

1. F Anscombe. Title/ Am Statistician 27:17–21, 1973.
2. V Barnett, T Lewis. Outliers in Statistical Data. New York: Wiley, 1978, pp 252–256.
3. CM Riley, TW Rosanske, eds. Development and Validation of Analytical Methods. Tarrytown, NY: Publisher, 1996, pp 263–270.
4. G Snedecor, W Cochran. Statistical Methods. 6th Edition. Ames, IA: Iowa State University Press, 1967, pp 159–160.
5. A De Lean, P. Munson, D Rodbard. Title. Am J Physiol 235:97–102, 1978.
6. B Morgan. Analysis of Quantal Response Data. New York: Chapman & Hall, 1992; pp 370–371.

7. R Myers. Classical and Modern Regression with Applications. Boston: Duxbury Press, 1986, pp 204–211.
8. D Finney. Statistical Method in Biological Assay. 3rd ed. New York: Macmillan, 1978, pp 69–99.
9. P Haaland. Experimental Design in Biotechnology. New York: Marcel Dekker, 1989, pp 5–7, 45.
10. L Reed, H Muench. Am J Hygiene 27:493–497, 1938.
11. B Morgan. Analysis of Quantal Response Data. New York: Chapman & Hall, 1992, pp 306–310.

Timothy L. Schofield

Assay Validation

See also *Assay Development; Process Validation*

I. INTRODUCTION

Upon completion of the development of an assay, and prior to implementation, a well conceived assay validation is carried out to demonstrate that the procedure is *fit for use*. Regulatory requirements specify the assay characteristics that are subject to validation and suggest experimental strategies to study these properties. These requirements, when united with sound statistical experimental design, furnish an effective and efficient validation plan. That plan should also address the form of statistical processing of the experimental results and thereby ensure reliable and meaningful measures of the properties of the method.

This entry will introduce the validation parameters used to describe the characteristics of an analytical method. Performance metrics associated with those parameters will be defined, with illustrations of their utility in application. Realistic and relevant acceptance criteria are proposed, as a basis for evaluating a procedure. An example of a validation protocol will illustrate the design of a comprehensive exploration, in conjunction with the statistical methods used to extract the relevant information.

II. TERMINOLOGY

Most of the terminology related to assay validation can be found in the entry on *Assay Development*. Some terms that are unique to this topic are *relative standard deviation, spike recovery,* and *dilution effect*.

The *relative standard deviation* (RSD) is a measure of the proportional variability of an assay, and is usually expressed as a percentage of the measured value:

$$\text{RSD} = 100 \cdot \frac{\hat{\sigma}\ \text{assay}}{\hat{x}}\ \%$$

Note that the RSD is the same as the coefficient of variation (CV) in statistics. Other conventions for calculating the RSD, which are appropriate for measurements that are log-normally distributed, utilize $\sigma_{\log_2}$ the log standard deviation or the fold variability:

$$\text{RSD} = 100\hat{\sigma}_{\log \hat{x}}\% \qquad \text{or}$$

$$\text{RSD} = 100(\text{FV}-1)\% = 100(e^{\hat{\sigma}_{\log \hat{x}}}-1)\%$$

Note that the log standard deviation provides an approximation to the RSD; it should be restricted to levels below 20% RSD. An important feature of the RSD is that it is calculated from the predicted measurements for a sample, such as concentrations, and not from an intermediate, such as the instrument readout. These are the same only in the case of linear calibration.

Spike recovery is the process of measuring analyte that has been distributed as a known amount into the sample matrix, with results usually reported as percent recovery or percent bias. *Dilution effect* is a measure of the proportional bias measured across a dilution series of a validation sample; it is usually reported as percent bias per dilution increment (usually percent bias per twofold dilution).

III. REGULATORY REQUIREMENTS

The U.S. Pharmacopeia (USP) specifies that the "Validation of an analytical method is the process by which it is established, in laboratory studies, that the performance characteristics of the method meet the requirements for the intended analytical applications" (11). Key to this definition are the treatment of a validation study as an *experiment*, with the goal of that experiment being to demonstrate that the assay is *fit for use*. As an experiment, a validation study should be suitably designed to meet its goal, while the acceptance criteria should yield opportunities to judge the assay as unsuitable as well as suitable, possibly requiring further development. This does not suggest that it is the role of the validation study to optimize the assay; this should be completed during assay development. In fact, the properties of the assay are probably well understood by the time the validation experiment is undertaken. The validation experiment should serve as a "snapshot" of the long-term routine performance of the assay, thereby documenting its characteristics so that we can make reliable decisions during drug development and control.

The International Conference on Harmonization (ICH) has devoted two topics of its guidelines to analytical method validation. The first topic, entitled "Guideline on Validation of Analytical Procedures: Definitions and Terminology" (2), offers a list of validation parameters that should be considered when implementing a validation experiment and guidance toward the selection of those parameters. The second topic, entitled "Validation of Analytical Procedures: Methodology" (3), provides direction on the design and analysis of results from a validation experiment.

IV. ASSAY VALIDATION PARAMETERS AND VALIDATION DESIGN

It's convenient for purposes of planning a validation experiment to dichotomize assay validation parameters into two categories: parameters related to the accuracy of the analytical method, and parameters related to variability. In this way, the analyst will choose the test sample set that will be carried forward into the validation, in accord with accuracy parameters, and devise a replication plan for the validation to satisfy the estimation of the variability parameters.

The validation parameters related to the accuracy of an analytical method are accuracy, linearity, and specificity. The *accuracy* of an analytical method "expresses the closeness of agreement between the value which is accepted either as a conventional value or an accepted reference value, and the value found." Accuracy is usually established by spiking known quantities of analyte (see the entry *Assay Development* for definitions of terms) into the sample matrix and demonstrating that these can be completely recovered. Conventional practice is to spike using five levels of the analyte: 50%, 75%, 100%, 125%, and 150% of the declared content of analyte in the drug (note that this is more stringent than the ICH, which requires "a minimum of 9 determinations over a minimum of 3 concentration levels"). The experimenter is not restricted to these levels but should plan to spike through a region that embraces the range of expected measurements from samples that will be tested during development and manufacture. Accuracy can also be determined relative to a "referee" method; in the case of complex mixtures, such as combination vaccines, accuracy can be judged relative to a monovalent control.

The *linearity* of an analytical procedure is "its ability (within a given range) to obtain test results that are directly proportional to the concentration (amount) of analyte in the sample." This parameter is often confused with the "graphical linearity" of the standard curve; in fact, the guidelines encourage the experimenter to evaluate linearity "by visual inspection of a plot of signals as a function of concentration or content." The principle of analytical linearity, however, should not be confused with an abstract property, such as the "straightness of the standard curve," but rather should be based upon the attribute that graded concentrations of the analyte yield measurements that are in proper proportion. Thus, for example, if a sample is tested in twofold dilution, the measured concentrations in those samples should yield a twofold series. In this way, linearity can be established from the spiking experiment outlined earlier, while a dilution series is usually employed when the assessment of linearity cannot be achieved through spiking (i.e., when the purified analyte does not exist, such as in vaccines).

Specificity is "the ability to assess unequivocally the analyte in the presence of components which may be present." Thus specificity speaks to the variety of sample matrices containing the analyte, including process intermediates and stability samples. The ICH guidelines provide a comprehensive description of the means to establish specificity in identity tests and in analytical methods for impurities and content. When the samples are available, specificity can be established through testing of the analyte-free matrix.

The assay validation parameters related to variability are precision, repeatability, ruggedness, limit of detection, and limit of quantitation. The *precision* of an assay "expresses the closeness of agreement (degree of scatter) between a series of measurements obtained from multiple sampling of the same homogeneous sample under the prescribed conditions." Precision is frequently called "interrun" variability, where a run of an assay represents the independent preparation of assay reagents, tests samples, and a standard curve. The *repeatability* of an assay "expresses the precision under the same operating conditions over a short interval of time," and is frequently called "intrarun" variability. Repeatability and precision can be studied simultaneously, using the levels of a sample required to establish accuracy and linearity, in a consolidated validation study design such as that depicted in Table 1. Here, precision is associated with the multiple experimental runs, while repeatability is associated with the run-by-level interaction. In many cases the experimenter may wish to test true within-run replicates rather than employ this subtle statistical artifact. The runs in this design can be strategically allocated to *ruggedness* parameters, such as laboratories, operators, and reagent lots, using a combination of nested and factorial experimental design strategies. Thus *operators* might be nested within *laboratory*, while *reagent lot* can be crossed with *operator*.

The *robustness* of an analytical method "is a measure of its capacity to remain unaffected by small, but deliberate, variations in method parameters," and might be more suitably established during the development of the assay (see the entry on *Assay Development*). Note that while ruggedness parameters represent uncontrollable factors affecting the analytical method, robustness parameters can be varied and should be controlled when it has been observed that they have an effect on assay measurement.

The *limit of detection* (LOD) and *limit of quantitation* (LOQ) of an assay can be obtained from replication of the standard curve during implementation of the consolidated validation experiment. The LOD of an analytical procedure "is the lowest amount of analyte in a sample that can be detected but not necessarily quantitated as an exact value," while the LOQ "is the lowest amount of analyte in a sample that can be quantitatively determined with suitable precision and accuracy." The limit of quantitation need not be restricted to a lower bound on the assay, but might be extended to an upper bound of a standard curve, where the fit becomes flat (for example, when using a four-parameter logistic regression equation to fit the standard curve).

Table 1 Strategic Validation Design Employing Five Levels of the Analyte Tested in k Independent Runs of the Assay

Run	Level 1	Level 2	Level 3	Level 4	Level 5
1	y_{11}	y_{12}	y_{13}	y_{14}	y_{15}
⋮	⋮	⋮	⋮	⋮	⋮
n	y_{n1}	y_{n2}	y_{n3}	y_{n4}	y_{n5}

Finally, the composite of the ranges identified with acceptable accuracy and precision is called the *range* of the assay.

V. CHOICE OF VALIDATION PARAMETERS

The choice of parameters that will be explored during an assay validation is determined by the intended use as well as by the practical nature of the analytical method. For example, a biochemical assay using a standard curve to establish drug content might require the exploration of all of these parameters if the assay is to be used to determine low as well as high levels of the analyte (such as an assay used to determine drug level in clinical samples or an assay that will be used to measure the content of an unstable analyte). If the assay is used, however, to determine content in a stable preparation, the LOD and LOQ need not be established. On the other hand, an assay for an impurity requires adequate sensitivity to detect and/or quantify the analyte; thus the LOD and LOQ become the primary focus of the validation of an impurity assay.

Many of these assay validation parameters have limited meaning in the context of biological assay and various other potency assays. There is no means to explore the accuracy and linearity of assays in animals or tissue culture, where the scale of the assay is defined by the assay; thus validation experiments for this sort of analytical method are usually restricted to a study of the specificity and ruggedness of the procedure. As discussed previously, the accuracy of a potency assay may be limited to establishing linearity with dilution when the purified analyte is unavailable to conduct a true spike-recovery experiment.

VI. VALIDATION METRICS AND ACCEPTANCE CRITERIA

As previously defined, the goal of the validation experiment is to establish that the analytical method is fit for use. Thus, for an impurity assay, for example, the goal of the validation experiment should be to show that the sensitivity of the analytical procedure (as measured by the LOD for a limit assay and by the LOQ for a quantitative assay) is adequate to detect a meaningful level of the impurity. Assays for content and potency are usually used to establish that a drug conforms to specifications that have been determined either through clinical trials with the drug or from process/product performance. It's important to point out that process capability limits should include the manufacturing distribution of the product characteristic under study, the change in that characteristic due to instability under recommended storage conditions, as well as affects on the measurement of the characteristic due to the performance attributes of the assay. In the end, the analytical method must be capable of reliably discriminating satisfactory from unsatisfactory product against these limits. The portion of the process range that is due to measurement, including measurement bias as well as measurement variability, serves as the foundation for setting acceptance criteria for assay validation parameters.

Validation parameters related to accuracy are rated on the basis of recovery or bias (bias = 100 − recovery, usually expressed as a percentage), while validation parameters related to random variability are appraised on the basis of variability (usually expressed as % RSD). These can be combined with process variability to establish the "process capability" of the measured characteristic (4):

$$CP = \frac{\text{Specification range}}{6 \cdot \text{Product variation}} = \frac{\text{Specification range}}{6\sqrt{\sigma^2_{\text{product}} + \sigma^2_{\text{assay}} + \text{Bias}^2}}$$

Process capability, in turn, is related to the percent of measurements that are likely to fall outside of specifications. Thus acceptance criteria on the amount of bias and analytical variability can be established, based upon knowledge of the product variability and the desired process capability.

When limits have not been established for a particular product characteristic, such as during the early development of a drug or biological, acceptance criteria for an assay attribute might be specified on the basis of a typical expectation for the analytical method as well as on the nature of the measurement in the particular sample. Thus, for example, high-performance liquid chromatography (HPLC) for the active compound in the final formulation of a drug might be typically capable of yielding measurements equal to or less than 10% RSD, while an immunoassay used to measure antigen content in a vaccine might only be capable of achieving up to 20% RSD. Measurement of a residual or an impurity by either method, on the other hand, may only achieve up to 50% RSD, owing to the variable nature of measurement at low analyte concentrations.

VII. VALIDATION DATA ANALYSIS

A. Analysis of Parameters Related to Accuracy

Analytical accuracy and linearity can be parameterized as a simple linear model, assuming either absolute or relative bias. In one model, if we let μ represent the known analyte content and x is the measured amount, then accuracy can be expressed as $x = \mu$ at a single concentration, or as $x = \alpha + \beta\mu$, where $\alpha = 0$ and $\beta = 1$, across a series of concentrations. This linear equation can be rewritten as

$$x = \mu + [\alpha + (\beta - 1)\mu]$$

In this parameterization, α is related to the accuracy of the analytical method and represents the constant bias in an assay measurement (usually reported in the units of measurement of the assay), while $(\beta - 1)$ is related to the linearity of the analytical method and represents the proportional bias in an assay measurement (usually reported in the units of measurement in the assay, per unit increase in that measurement; e.g., 0.02 μg per μg increase in content). Data from a spiking experiment can be utilized to estimate α and $\beta - 1$, and these (with statistical confidence limits if the validation experiment has been designed to establish statistical conformance) can be compared to the acceptance criteria established for these parameters.

To illustrate, consider the example in Table 2. A validation study has been performed in which samples have been prepared with five levels of an analyte ([C] in mg), and their content is determined in six runs of an assay. These data are depicted graphically, along with results from their analysis, in Fig. 1. A linear fit

Table 2 Example of Results from a Validation Experiment in Which Five Levels of an Analyte Were Tested in Six Runs of the Assay

[C]	1	2	3	4	5	6	Avg.
3	3.3	3.4	3.2	3.1	3.4	3.3	3.3
4	4.2	4.2	4.2	4.4	4.0	4.3	4.2
5	5.0	4.9	4.9	5.0	4.8	4.9	4.9
6	5.8	5.8	5.6	5.9	5.9	5.7	5.8
7	6.4	6.7	6.7	6.8	6.6	6.4	6.6

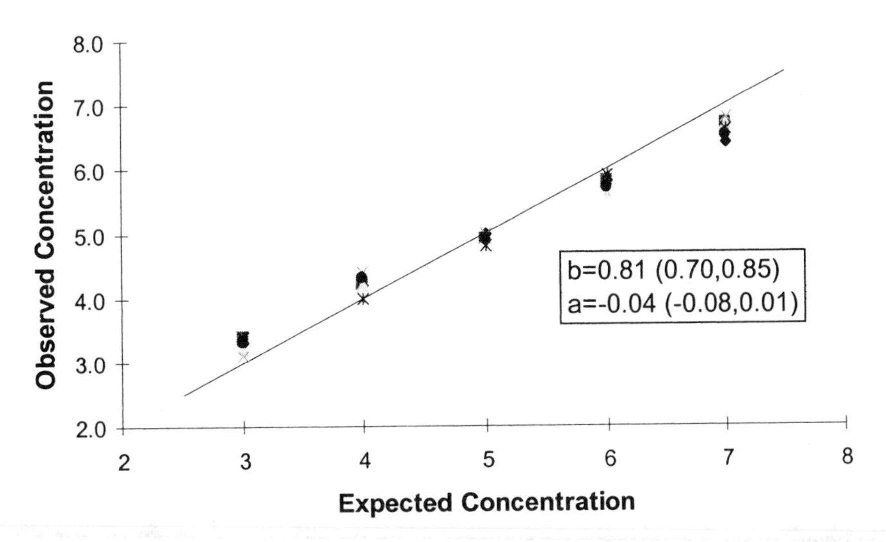

Fig. 1 Validation results from an assay demonstrating proportional bias, with low-concentration samples yielding high measurements and high-concentration samples yielding low measurements.

to the data yields the following estimates of intercept (note that the data have been centered in order to obtain a "centered" intercept) and slope:

$$\hat{\alpha}' = -0.04\ (-0.08,0.01),\quad \hat{\beta} = 0.81\ (0.70,0.85).$$

The slope can be utilized to estimate the bias per unit increase in drug concentration:

$$\hat{\beta} - 1 = -0.19\ (-0.30,-0.15).$$

This proportional bias is significant, owing to the fact that the confidence interval excludes 0. It is more desirable, however, to react to the magnitude of the bias and to judge that there is unacceptable nonlinearity when the bias is in excess of a prespecified acceptance limit. Thus there is a 0.2-mg decrease per unit increase in concentration (as much as −0.3-mg-per-unit decrease in concentration based on the confidence interval). If the range in concentration of the drug is typically 4–6 mg (i.e., a 2-mg range), this would predict that the bias due to "nonlinearity" is 0.4 mg (0.6 mg in the confidence interval). This can be judged to be of consequence or not in testing product; for example, if the specification limits on the analyte are 5 ± 1 μg/dose, much of this range will be consumed by the proportional bias, potentially resulting in an undesirable failure rate in the measurement of that analyte.

When purified analyte is unavailable for spiking, and levels of the analyte in the validation have been attained through dilution, then the data generated from that series can be alternatively evaluated using the model $x = \alpha\mu^{\beta}$ (note that μ can represent either the expected concentration upon dilution or the actual dilution). Taking logs, the data from the dilution series is used to fit the linear model,

$$\log(x) = \log(\alpha) + \beta \log(\mu).$$

The intercept of this model has no practical meaning, while the proportional bias (sometimes called "dilution effect") can be estimated as $2^{\beta-1} - 1$ if μ is in units of concentration and as $2^{\beta+1} - 1$ if μ is in units of dilution (note that the units of "dilution effect" are percent bias per dilution increment; the dilution increment used here is twofold).

To illustrate, consider the example in Table 3. Dilution series of a sample are performed in each of five runs of an assay. These data are depicted graphically, along with results from their analysis, in Fig. 2. A linear fit of log-titer to log-dilution yields an estimated slope equal to: $\hat{\beta} = -1.01\ (-1.04,-0.98)$. The corresponding dilution effect is equal to $2^{-1.01+1} - 1 = -0.007$ (i.e., a 0.7% decrease per twofold dilution).

Table 3 Example of Results from a Validation Experiment in Which Five Twofold Dilutions of an Analyte Were Tested in Five Runs of the Assay

Dilution	1	2	3	4	5	Titer
1	75	90	93	79	72	81
2	43	45	46	35	40	83
4	20	23	22	18	21	83
8	11	10	12	10	10	85
16	4	5	6	4	5	78

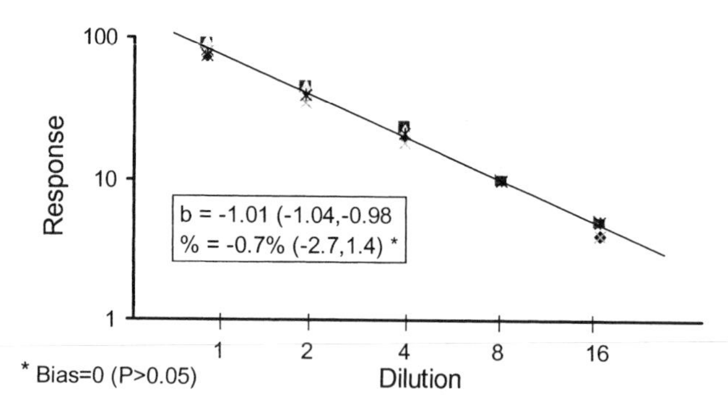

Fig. 2 Results from a validation series demonstrating satisfactory linearity, i.e., approximately a twofold decrease in response with twofold dilution.

The dilution effect can also be calculated as a function of the estimated slopes for the standard (slope as a function of dilution rather than concentration) and the test sample:

$$\text{Dilution effect} = 2^{1-\hat{\beta}_T/\hat{\beta}_T} - 1$$

with the corresponding standard error a function of the standard error of a ratio:

$$\text{SE}\left(\frac{\hat{\beta}_T}{\hat{\beta}_S}\right) = \sqrt{\frac{1}{\hat{\beta}_S^2}\,[\text{var}(\hat{\beta}_T) + (\hat{\beta}_T/\hat{\beta}_S)^2 \cdot \text{var}(\hat{\beta}_S)]}$$

This has the advantage over the earlier calculation of being a more reliable estimate of the underlying variability in the measurements, coming from the contribution to the estimate of variability from the data for the standard.

A special case of accuracy comes from a comparison of an experimental analytical method with a validated "referee" method. Paired measurements in the two assays can be made on a panel of samples, and these can be utilized to establish the "linearity" and accuracy of the experimental method relative to the "referee" method. The paired measurements are a sample from a multivariate population, and the "linearity" of the experimental method to the "referee" method can be estimated using principal-component analysis (5). A "concordance slope" can be estimated from the elements of the first characteristic vector of the sample covariance matrix (a_{21}, a_{11}):

$$\hat{\beta}_C = \frac{a_{21}}{a_{11}}$$

The standard error of the "concordance slope" is obtained from these along with other elements of the characteristic value evaluation (a_{i1} is the ith element of the first characteristic vector and l_i is the corresponding characteristic roots):

$$\text{SE}(\hat{\beta}_C) = \sqrt{\left[\frac{\text{V}(a_{21})}{a_{21}} + \frac{\text{V}(a_{11})}{a_{11}} - 2\cdot\frac{\text{cov}(a_{21}, a_{11})}{a_{21}a_{11}}\right]\cdot\hat{\beta}_C^2}$$

where

$$\text{V}(a_{i1}) = \left[\frac{l_1 l_2}{n}\cdot(l_2 - l_1)^2\right] a_{i2}^2$$

and

$$\text{cov}(a_{21}, a_{11}) = \left[\frac{l_1 l_2}{n}\cdot(l_2 - l_1)^2\right] a_{12}a_{22}$$

The "concordance slope" can be used, along with the standard error of that estimate, to measure the discordance of the experimental analytical method relative to the "referee" method and its corresponding confidence interval:

$$\text{Discordance} = 2^{\beta_C - 1}$$

This is similar to the "dilution effect" discussed earlier and represents the percentage difference in reported potency per twofold increase in result as measured in the "referee" method.

When the two assays are scaled alike, the accuracy of experimental method relative to the "referee" method can be assessed through a simple paired variate analysis. When the percent difference is expected to be constant across the two methods, the data are analyzed on a log scale. An estimate of the simple paired difference in the logs can be used, in conjunction with the standard error of the paired difference, to estimate the percentage difference in measurements obtained from the experimental procedure relative to the "referee" method (6):

$$\bar{d} \pm t_{n-1}\cdot s_d/\sqrt{n},$$

$$\text{Percent difference} = 100(e^{\bar{d}} - 1)\%$$

Note that as usual, the significance of the percentage difference should be judged primarily on the basis of practical considerations rather than statistical significance.

The specificity of an analytical procedure can be assessed through an evaluation of the results obtained from a "placebo" sample (i.e., a sample containing all constituents except the analyte of interest) relative to background. A statistical evaluation might be composed of either a comparison of instrument measurements obtained for this "placebo," with background measurements (i.e., through background-corrected "placebo" measurements), or from measurements made on the "placebo" alone. In either case, an increase over the detection level of the assay would indicate that some constituent of the sample, in addition to analyte, is contributing to the measurement in the sample. This should be assessed by noting that the upper bound on a confidence interval falls below some prespecified level in the assay.

B. Analysis of Parameters Related to Variability

Estimates of repeatability (intrarun, or within-run, variability) and precision (interrun, or between-run, variability) can be established by performing a variance component analysis on the results obtained from the data generated to assess accuracy and linearity (7). Other sources of variability (e.g., laboratories, operators, reagent lots) can be incorporated into the pattern of runs performed during the validation, to estimate the affects of these sources of variability. The expected mean squares are shown in Table 4. From this, $\hat{\sigma}^2_{\text{within-run}} = \text{MSE}$, and $\hat{\sigma}^2_{\text{between-run}} = (\text{MSR} - \text{MSE})/5$ (note that the MSE is a composite of pure error and the interaction of run and level; pure error could be separated from the interaction by performing replicates at each level; however, the within-run variability will still be reported as the composite of pure error and the level-by-run interaction). The total variability associated with testing a sample in the assay can be calculated from these variance component estimates as:

$$\hat{\sigma}^2_{\text{Total}} = \frac{\hat{\sigma}^2}{r} + \frac{\hat{\sigma}^2_W}{nr}$$

Note that this is the variance of $\bar{y}$, where "r" represents the number of independent runs performed on the sample, and "n" represents the number of replicates within a run.

Table 4 Analysis of Variance Table Showing Expected Mean Squares as Functions of Intrarun and Interrun Components of Variability

Effect	df	MS	Estimated mean square	F
Level	4	MSL	$\hat{\sigma}^2_W + Q(L)$	
Run	5	MSR	$\hat{\sigma}^2_W + 5\hat{\sigma}^2$	MSR/MSE
Error	20	MSE	$\hat{\sigma}^2_W$	

A confidence interval can be established on the total variability as follows (8):

$$\begin{aligned}\hat{\sigma}^2_{\text{Total}} &= \frac{\hat{\sigma}^2}{r} + \frac{\hat{\sigma}^2_W}{nr} \\ &= \left(\frac{1}{Jr}\right)\cdot \text{MSR} + \left(\frac{n-J}{Jnr}\right) + \text{MSE} \\ &= c_1 \cdot \text{MSR} + c_2 \cdot \text{MSE} \\ &= \sum_q c_q S_q\end{aligned}$$

where "J" represents the number of replicates of a sample tested in each run of the assay validation experiment (here $J = 5$ for the number of levels of the sample). The $1 - 2\alpha$ confidence interval on $\hat{\sigma}^2_{\text{Total}}$ becomes:

$$\left[\hat{\sigma}^2_{\text{Total}} - \sqrt{\sum_q G_q^2 c_q^2 S_q^4},\ \hat{\sigma}^2_{\text{Total}} + \sqrt{\sum_q H_q^2 c_q^2 S_q^4}\right]$$

where

$$G_q = 1 - \frac{1}{F_{\alpha,\text{df}_q,\infty}}, \qquad H_q = \frac{1}{F_{1-\alpha,\text{df}_q,\infty}} - 1,$$

$$F_{\alpha,\text{df}_q,\infty}, \qquad F_{1-\alpha,\text{df}_q,\infty}$$

represent percentiles of the F-distribution.

The limit of detection (LOD) and the limit of quantitation (LOQ) are calculated using the standard deviation of the responses ($\hat{\sigma}$) and the slope of the calibration curve ($\hat{\beta}$) as $\text{LOD} = 3.3\hat{\sigma}/\hat{\beta}$ (note, the 3.3 is derived from the 95th percentile of the standard normal distribution, 1.64; here, $2 \cdot 1.64 = 3.3$, thereby limiting both the false-positive and false-negative error rates to be equal to or less than 5%), and $\text{LOQ} = 10\hat{\sigma}/\hat{\beta}$ (note, the 10 corresponds to the restriction of no more than 10% RSD in the measurement variability; $\text{RSD}/100 = \hat{\sigma}/\bar{x} < 0.10$ gives $\bar{x} < 10\hat{\sigma}$; thus if the restriction were changed to "not more than 20% RSD," then the LOQ could be calculated as $\text{LOQ} = 5\hat{\sigma}/\hat{\beta}$).

These estimates do not account for the calibration curve, and they can be improved upon by acknowledging the variability in prediction from the curve (9). For a linear fit, the LOD is depicted graphically in Fig. 3. The LOQ can be calculated as prescribed in the ICH guideline, solving for x_Q in the following equation:

$$\text{LOQ} = 10\left\{\hat{\sigma}\sqrt{1 + \frac{1}{n} + \frac{(x_Q - \bar{x})^2}{SXX}}\right\}$$

The limit of quantitation is depicted graphically in Fig. 4.

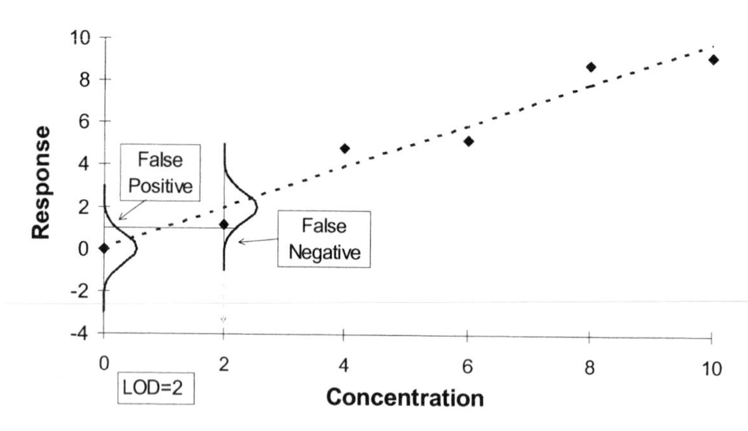

Fig. 3 Determination of the LOD using the linear fit of the standard curve plus restrictions on the proportions of false positives and false negatives.

The LOQ is more complicated to calculate from a nonlinear calibration function, such as the four-parameter logistic regression equation, and involves the first-order Taylor series expansion of the solution equation for x_Q (10) (see the entry on *Assay Development*). The calculation of the LOQ for this equation in illustrated in Fig. 5.

VIII. FOLLOW-UP

The information from the analytical method validation can be utilized to establish a testing format for samples in the assay. In particular, variance component estimates of the within-run and between-run variabilities can be used to predict the total variability associated with a variety of testing formats. For example, consider the case where component estimates have been determined to be equal to $\hat{\sigma}_B = 0.05$ (5%) and $\hat{\sigma}_W = 0.03$ (3%). Letting "r" represent the number of independent runs that will be performed on a sample, and "n" be the number of replicates within a run, the variabilities are listed in Table 5. Suppose, due to process capability, the analytical variability must be restricted to RSD $\leq$ 3%. As can be observed from the table, this cannot be accomplished using a single run of the assay; it must therefore be achieved through a combination of multiple runs with multiple replicates within each run.

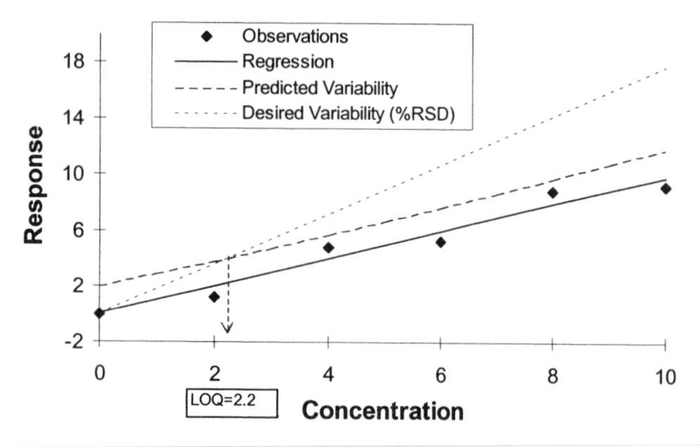

Fig. 4 Determination of the LOQ using the linear fit of the standard curve and a restriction on the variability (% RSD) in the assay.

IX. ANALYTICAL METHOD QUALITY CONTROL

Assay validation does not end with the formal validation experiment but is continuously assessed through assay quality control. Control measures of variability and bias are utilized to bridge to the assay validation and to warrant the continued reliability of the procedure.

Extra variability in replicates can be detected utilizing standard statistical process control metrics (11); detection and prespecified action upon detection can help maintain the variability characteristics of an assay. For example, when the measurement of analyte is obtained from multiple runs of an assay, those results can be examined by noting the range (either "log(max) − log(min)" or "max − min," depending upon whether the measurements are scaled proportionally or absolutely). A bound on the range can be derived:

$$\text{Range} \leq \hat{\sigma}_{\text{dimension}}(d_2 + kd_3)$$

where $\hat{\sigma}_{\text{dimension}}$ is the variability in the dimension of

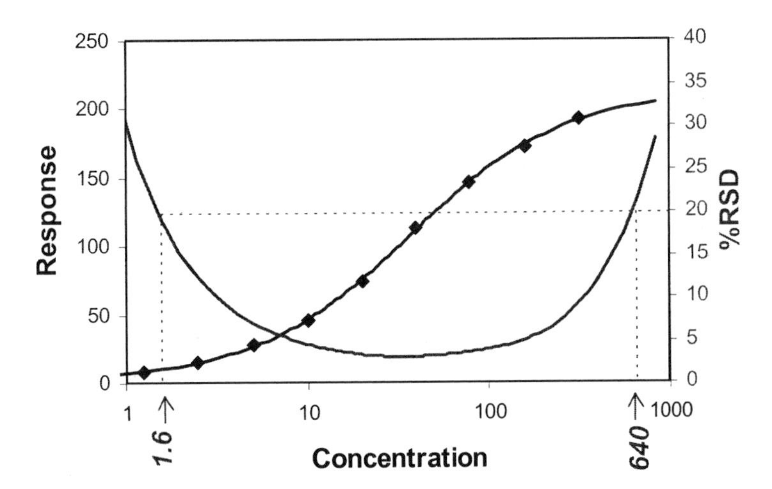

Fig. 5 Quantifiable range determined as limits within which assay variability (% RSD) is less than or equal to 20%.

replication (here, run to run), d_2 and d_3 are factors used to scale the standard deviation to a range, and "k" is a factor representing the desired degree of control (e.g., $k = 2$ represents approximately 95% confidence). Having detected "extra variability," additional measurements can be obtained, and a result for the sample can be determined from the remainder of the measurements made after the maximum and minimum results have been eliminated from the calculation. When "extra variability" has been identified in instrument replicates of a standard concentration or a dilution of a test sample, that concentration of the standard or dilution of the test sample might be eliminated from subsequent calculations.

Measurements on control sample(s) help identify shifts in scale and, therefore, bias in a run of an assay. Information in multiple control samples can typically be reduced to one or two meaningful scores, such as an average and/or a slope. These can help isolate problems and provide information regarding the cause of a deviation in a run of an assay. Suppose, for example, that three controls of various levels are used to monitor shifts in sensitivity in the assay. The average can be used to detect absolute shifts, due to some underlying source of constant variability, such as inappropriate preparation of standards or the introduction of significant background. The slope can be used to detect proportional shifts due to some underlying source of proportional variability.

Table 5 Predicted Variability (% RSD) for a Given Number of Runs (r) and a Given Number of Replicates Within Each Run (n)

r/n	1	2	3	∞
1	6%	5%	5%	5.0%
2	4%	4%	4%	3.5%
3	3%	3%	3%	2.9%

Suitability criteria on the performance of column peaks can detect degeneration of a column, while characteristics of the standard curve, such as the slope or EC_{50} (EC_{50} corresponds to "C" in a four-parameter logistic regression) help predict an assay with atypical or substandard performance.

Care must be taken, however, to avoid either meaningless or redundant controls. The increased false-positive rate inherent in the multiplicity of a control strategy must be considered, and a carefully contrived quality control scheme should acknowledge this.

REFERENCES

1. USP XXII, General Chapter 1225, Validation of Compendial Methods: pp. 1982–1984.
2. Federal Register, International Conference on Harmonization; Guideline on Validation of Analytical Procedures: Definitions and Terminology, March 1, 1995, pp. 11259–11262.
3. International Conference on Harmonization; Guideline on Validation of Analytical Procedures: Methodology, November 6, 1996.
4. D Montgomery. Introduction to Statistical Quality Control. 2nd ed. New York: Wiley, 1991, pp 208–210.
5. D Morrison. Multivariate Statistical Methods. New York: McGraw-Hill, 1967, pp 221–258.
6. G Snedecor, W Cochran. Statistical Methods. 6th ed. Ames, IA: Iowa State University Press, 1967, pp 91–100.

7. B Winer. Statistical Principles in Experimental Design. 2nd ed. New York: McGraw-Hill, 1971, pp 244–251.
8. R Burdick, F Graybill. Confidence Intervals on Variance Components. New York: Marcel Dekker, 1992, pp 28–39.
9. L Oppenheimer, T Capizzi, R Weppelman, H Mehta. Determining the lowest limit of reliable assay measurement. Analyt Chem 55:638–643, 1983.
10. B Morgan. Analysis of Quantal Response Data. New York: Chapman & Hall, 1992, pp 370–371.
11. D Montgomery. Introduction to Statistical Quality Control, 2nd ed. New York: John Wiley & Sons, 1991, pp. 203–206.
12. DL Massart, A Dijkstra, L Kaufman. Evaluation and Optimization of Laboratory Methods and Analytical Procedures. New York: Elsevier, 1978.
13. LA Currie. Limits for qualitative detection and quantitative determination: Application to radiochemistry. Analyt Chem 40:586–593, 1968.
14. MJ Cardone. Detection and determination of error in analytical methodology. Part I. the method verification program. J Assoc Official Analyt Chemists 66:1257–1281, 1983.
15. MJ Cardone. Detection and determination of error in analytical methodology. Part II. correction for corrigible systematic error in the course of real sample analysis. J Assoc Official Analyt Chemists 66:1283–1294, 1983.
16. JC Miller, JN Miller. Statistics for Analytical Chemistry. 2nd ed. New York: Wiley, 1988.
17. R Caulcutt, R Boddy. Statistics for Analytical Chemists. New York: Chapman & Hall, 1983.

Timothy L. Schofield

B

Bayesian Statistics

I. INTRODUCTION

The Bayesian approach to statistical inference has received increased attention in the scientific community over the past several decades due to the general framework it provides for incorporating new experimental evidence into scientific conclusions. Due to the core principles it rests upon, Bayesian analysis can be particularly useful in the field of biomedical research and development. There is an abundant literature on Bayesian analysis in general, as well as on Bayesian applications to the biopharmaceutical industry. Breslow's review, "Biostatistics and Bayes" (1), can serve as an introduction to the subject and links Bayesian analysis to numerous issues arising in clinical practice. Berry and Stangl (2) provide an extensive collection of papers pertaining to the use of Bayesian methodology in health-related research. Among excellent general references are Berger (3), Bernardo and Smith (4), Press (5), and Gelman et al. (6). Specific applications with suggested references include design of biomedical experiments (7), clinical trial monitoring (8–12), evaluation of drug pharmacokinetics and pharmacodynamics (13–15), planning trials with multiple endpoints (16), meta-analysis (17,18), and addressing ethical concerns associated with clinical trials (19).

In this entry we will present an introduction to the Bayesian school of thought, providing definitions of its basic notions and core concepts. Key components of the methodology will be demonstrated using examples from the biomedical field. We also present main ideas on how to use Bayesian analysis for clinical trial monitoring and sample size determination. The literature cited contains many references to the other areas of application mentioned.

We will not attempt to discuss differences between Bayesian and classical approaches in this entry. In the authors' opinion, the two approaches address completely different sets of scientific questions and should be considered complimentary. For a thorough comparison of both philosophies, see Berger (3), Carlin and Louis (20), or Gelman et al. (6).

It should be noted that the implementation of Bayesian ideas has become increasingly prevalent in many areas of application due to recent advances in computational tools. Problems that were once just too difficult or too time consuming to tackle from a Bayesian point of view are now becoming tractable. The last section of the entry contains a brief description of Bayesian computational methods.

II. BASIC NOTIONS AND BAYES THEOREM

Suppose a researcher conducts an experiment to test certain scientific beliefs, Q. Bayesian analysis provides a tool by which all prior knowledge about Q, $P(Q)$, can be combined with information about Q contained in data from an experiment, resulting in a posterior probability, or distribution of Q conditional on the data, $P(Q|\text{data})$. The core of this informational update is formed by Bayes' theorem. The theorem was first discovered by Rev. Thomas Bayes in the first half of the 17th century and published posthumously in 1763 (21). In its simplest form it states that

$$P(Q|\text{data}) = \frac{P(Q) \times P(\text{data}|Q)}{P(\text{data})} \tag{1}$$

This result is a direct consequence of the laws of probability. In general, Q can be a statistical hypothesis, a distribution of a quantity of interest, or a parameter of this distribution. In the case of a parameter vector θ, Eq. (1) can be rewritten as:

$$\pi(\theta|y) = \frac{\pi(\theta) \times f(y|\theta)}{m(y)} \quad (2)$$

where y represents a vector of data, $f(y|\theta)$ is the likelihood, or conditional density, of the data given θ, $\pi(\theta)$ is the prior distribution of θ, and

$$m(y) = \int f(y|\theta)\pi(\theta)\, d\theta \quad (3)$$

is called the marginal distribution of Y. The following two examples demonstrate the use of Bayes' theorem for estimation purposes in both a discrete and a continuous case.

Example 1. Suppose a 40-year-old woman arrives for a routine OB/GYN visit and her physician performs a mammogram to test for the presence of breast cancer. Given a positive result of the test (the data), what is the chance that the patient indeed has breast cancer? Assuming that the woman has no family history of breast cancer and given the patient's age, the chance (prior probability) that she has the breast cancer is small—assume it is 0.1%. Let $\theta = 1$ if the cancer is present and $\theta = 0$ otherwise. Using Bayes' theorem,

$$P(\theta = 1|\text{test is positive}) = \frac{P(\text{test is positive}|\theta = 1) \times P(\theta = 1)}{P(\text{test is positive})}$$

On the other hand,

$$P(\text{test is positive}) = P(\text{test is positive}|\theta = 1) \times P(\theta = 1) + P(\text{test is positive}|\theta = 0) \times P(\theta = 0)$$

Assuming further that the test procedure produces 5% false positives and 1% false negatives,

$$P(\text{test is positive}) = 0.99 \times 0.001 + 0.05 \times (1 - 0.001) = 0.051$$

and, finally,

$$P(\theta = 1|\text{test is positive}) = 0.99 \times 0.001/0.051 \cong 0.0194, \text{ or } 1.94\%$$

In addition to illustrating the use of Bayes' theorem, this example highlights another important point. While having highly precise medical procedures (with high true-positive and low false-negative rates) is important, it is the "inverse" question, namely, the chance of having the disease given a positive (or negative) result of the test, that is usually of the most interest. The Bayesian paradigm allows us to address this question directly. This seeming discrepancy between the precision of the diagnostic test and the posterior probability becomes especially pronounced for rare diseases (e.g., cancer, AIDS, and tuberculosis).

Example 2. Zyprexa® (olanzapine), a novel drug for the treatment of schizophrenia, was approved by the FDA and EMEA (European Medicinal Evaluation Agency) in September 1996. After approval, the sponsor, Eli Lilly & Company, was interested in collecting additional safety information. Given that serum prolactin elevations associated with older conventional neuroleptics were well documented in the literature, the sponsor wanted to estimate the impact of prolonged exposure to olanzapine on serum prolactin levels in a schizophrenic population. The mean prolactin level in a schizophrenic population taking olanzapine, θ, was chosen as the safety summary measure. The investigator's prior knowledge of θ was well represented by $N(\mu,\tau^2)$, with μ = 18.6 ng/mL and τ = 20.7 ng/mL, based on the data previously summarized in the new drug application to the FDA. Let $y = (y_1, \ldots, y_n)$ represent collected prolactin levels (ng/mL) from a prospective trial on n patients assessed after six months of drug exposure. For simplicity, we shall assume that $Y \sim N(\theta,\sigma^2)$, with σ = 19.1 ng/mL. Bayes' theorem yields a posterior distribution of θ that is also normal, with the mean $\mu(y)$ and variance υ^2 given by:

$$\mu(\bar{y}) = \mu \left(\frac{\sigma^2}{\sigma^2 + n\tau^2}\right) + \bar{y}\left(\frac{n\tau^2}{n\tau^2 + \sigma^2}\right)$$
$$\text{and} \quad \upsilon^2 = \frac{\tau^2\sigma^2}{n\tau^2 + \sigma^2} \quad (4)$$

where $\bar{y} = n^{-1} \sum_{i=1}^{n} y_i$. Results from a multicenter double-blind clinical trial (22) yielded $\bar{y}$ = 20.97 ng/mL for n = 144 patients with a 6-month postbaseline serum prolactin measurement. The resulting posterior distribution of θ given the prior knowledge and the data is $N(20.96,1.59^2)$. Hence, the investigator could report with 95% probability that, given currently available knowledge, the mean prolactin level in this patient population is contained between 17.84 ng/mL and 24.08 ng/mL. Note the straightforward interpretation of this interval, born of conditioning on the data, in contrast to intervals constructed under the repeated sampling paradigm of classical statistics.

The marginal density $m(y)$ that appears in the denominator of Eq. (2) is sometimes called the *predictive distribution of* the data Y, which is also an important notion in Bayesian statistics. This distribution represents information about the value of an observation y given the prior knowledge and the likelihood function. Using the same idea as in Eq. (3) we can arrive at the posterior predictive distribution of new, yet-to-be-observed data, y_{new}, given observed data y. This type of distribution could be valuable, for example, in evaluating safety concerns of drugs under investigation.

Example 3. Clozapine is an atypical antipsychotic that has been associated with an increased risk of agranulocytosis, a potentially fatal blood disorder defined as an absolute neutrophil count (ANC) < 500 μL and the presence of symptoms of an infectious process. During clinical trials before the approval of olanzapine, there was interest in assessing whether olanzapine was also associated with an increased risk of agranulocytosis. Given the severity of the adverse event and the large amount of data being collected, it is reasonable to assume that physicians would want the ability to predict, for example, the probability of ANC < 500 for future patients conditional on all data collected up to a visit. To answer this question, one could average $f(y_{\text{new}}|\theta)$ with respect to the posterior distribution $\pi(\theta|y)$ that now plays the role of an updated prior, resulting in

$$p(y_{\text{new}}|y) = \int f(y_{\text{new}}|\theta)\pi(\theta|y)\, d\theta$$

Given this distribution, one could then calculate any probability of interest, for example,

$$p(y_{\text{new}} < 500|y) = \int_0^{500} p(y_{\text{new}}|y)\, dy_{\text{new}}$$

or report estimates such as 95% probability intervals, to aid in monitoring the safety of patients. One could even imagine setting objective bounds such as, say, $p(y_{\text{new}} < 500|y) > .0001$ to assist decisions of a monitoring board. Had this approach actually been used prior to approval, the physician monitoring the predictive distribution would have observed a constant decrease over time in the predictive probability of ANC < 500 as further evidence accumulated.

III. HYPOTHESIS TESTING

The problems of estimation and prediction are closely related to that of testing statistical hypotheses.

Example 4. Suppose researchers are now interested in testing the performance of olanzapine versus a conventional neuroleptic competitor, haloperidol, in terms of ability to reduce negative symptomatology. Such comparisons are often considered in pricing and reimbursement negotiations in European countries as well as for U.S. and Canadian formulary boards and marketing strategies worldwide. Consider a clinical trial where patients were assigned to olanzapine and haloperidol in a double-blinded fashion. The response variable of interest was assessed by the negative score from the Brief Psychiatric Rating Scale (BPRS). Let $Z = (z_1, \ldots, z_n)$ and $H = (h_1, \ldots, h_n)$ be observations representing the difference between endpoint and baseline of scores collected from patients on olanzapine and haloperidol, respectively (here we assumed for simplicity that an equal number of patients was assigned to each treatment). Defining $d_i = z_i - h_i$, $i = 1, \ldots, n$, suppose that $D = (d_1, \ldots, d_n) \sim f(d|\theta)$, where θ represents the difference between mean change BPRS negative scores (endpoint minus baseline) of olanzapine and haloperidol, respectively. The test of the hypothesis $H_0{:}\theta \geq 0$ versus $H_1{:}\theta < 0$ is of primary interest.

In the Bayesian paradigm, the choice between H_0 and H_1 is based upon the so-called *posterior odds ratio*:

$$\frac{P(H_1|\text{data})}{P(H_0|\text{data})} = \frac{P(\text{data}|H_1)}{P(\text{data}|H_0)} \times \frac{P(H_1)}{P(H_0)} \tag{5}$$

which represents the evidence in favor of H_1 versus H_0 after observing the data. The ratio of $P(H_1)$ to $P(H_0)$ is called the *prior odds*. The ratio of posterior to prior odds,

$$B_{10} = \frac{\{P(H_1|\text{data})/P(H_0|\text{data})\}}{\{P(H_1)/P(H_0)\}} \tag{6}$$

is called the *Bayes factor*. It comprises a measure of how much evidence in favor of H_1 has changed after the data have been observed. Note that it represents the *change* in evidence, not an absolute measure of support for a hypothesis; interpreting Bayes factor as the latter can be very misleading (23–24).

Example 4 (continued). Consistent with the literature, we will assume that the average of $n = 65$ collected observations, $\bar{D}$, is distributed as $N(\theta,\sigma_D^2/n)$, with $\sigma_D^2 = 11.0$, and that $\theta \sim N(\mu,\tau^2)$, with $\mu = -0.77$ and $\tau^2 = 17.53$. Note that, a priori, $P(H_0) = P(\theta > 0) = 0.428$ and $P(H_1) = P(\theta < 0) = 0.572$, and hence the prior odds of the investigator's beliefs about olanzapine's reducing negative symptoms, as compared to haloperidol, are $0.572/0.428 = 1.34$. Suppose that observed data from the prospective clinical trial yields $\bar{d} = -0.18$. The posterior distribution of θ [see Eq. (4)] is normal, with the mean equal to -0.186 and the variance equal to 0.168. Hence the posterior probabilities of the hypotheses are $P(H_0|\text{data}) = P(\theta > 0|\text{data}) = 0.326$ and $P(H_1|\text{data}) = P(\theta < 0|\text{data}) = 0.674$, yielding the posterior odds ratio $P(H_1|\text{data})/P(H_0|\text{data}) = 2.07$. We conclude that, based on the prior information and the evidence provided by the data, olanzapine is 2.07 times more likely than haloperidol to reduce negative symptoms as measured by BPRS. Note that the Bayes factor here is 1.55, thus implying that, after conducting the trial, the investigators' beliefs about the odds of olanzapine's reducing negative symptoms, as compared to haloperidol, have increased by 1.55.

The Bayesian approach to model selection and hypothesis testing was pioneered by Jeffreys (25) who used it to choose among competing scientific theories. Kass and Raftery (26) provide an excellent review of the uses of Bayes factors for model selection in a variety of scientific applications.

IV. CHOICE OF PRIOR DISTRIBUTIONS

The process of quantifying knowledge prior to decision making or future experimentation such as in the prolactin experiment is a struggle faced by *all* statisticians. Prior distributions are probabilistic expressions of knowledge possessed *prior* to conducting an experiment. It is perhaps easiest to think about prior distributions as falling into two broad categories, classified according to how the prior information is going to be incorporated into a distributional form or the purpose of the prior. The first category describes the scenario in which it is desirable to measure and incorporate an expert's uncertainty into posterior inference. The second category is characterized by situations in which one does not want to (or cannot) represent an expert's opinion, often referred to as "letting the data speak for themselves."

Scenarios in which one desires to incorporate prior information may arise, for instance, in a situation in which historical data exist or/and an expert opinion is available. In this case the prior distribution can be formed by direct approximation from the available data or expert opinion. Alternatively, one can first choose a reasonable parametric functional form for the shape of the prior and then estimate or elicit the parameters for this distribution. For example, suppose that a normal distribution was deemed adequate to summarize the prior knowledge. One could then elicit or estimate information such as the center and the upper or lower quartile (quartiles are usually easier to assess than the standard deviation). The elicitation process boils down to a measurement problem in which the unit of measurement is a probability. Since experts are rarely trained in probability theory (because they can be anyone possessing the knowledge prior to an experiment), the elicitation process can be a challenging task, and the elicited information should be checked often for consistency. In the case of the normal distribution, for instance, it may be useful to gather information about *both* quartiles and perhaps some other percentiles of the distribution. For an excellent review and recommendations for eliciting prior distributions for biomedical applications, see Kadane and Wolfson (27), O'Hagan (28), and Chaloner (29). For some standard ways to approximate prior distributions in general, see, for example, Berger (3, pp 74–113).

A common criticism of these types of priors is their very nature: they are based on subjective expert opinion or subjectively chosen data and thus can change, should another pool of experts or a set of data be chosen. For instance, one certainly wants to avoid the scenario in which an overzealous investigator overestimates the performance of a treatment and thus has an undue effect on the posterior conclusions drawn. To address this problem, Spiegelhalter et al. (30,31) suggested considering a series of "experts" representing a skeptic, a clinician, the interest of a regulatory agency, and an uninformed or neutral individual. Another way of addressing this criticism is to extend the single prior to a reasonable class of priors and to investigate the sensitivity of the final conclusion of the experiment to the class of priors chosen. See, for example, Greenhouse and Wasserman (11), Lavine et al. (32), and Berger (33,34) for a discussion of issues related to prior robustness. Another argument to keep in mind is that, as the data from the experiment accumulate, the effect of the prior becomes less and less pronounced on the posterior. Hence, for large sample sizes, usage of a prior becomes merely a way of invoking Bayesian machinery, and its particular choice become inessential.

The elicitation process may also result in a prior that does not prefer any single value of a parameter over any other. This scenario meets the objective of the first category of priors (thus incorporating expert opinion), although it is perhaps more common under the second category (where the impact of the prior information on the posterior is minimized). These types of priors are often called *noninformative* or objective. For example, in a case of a binomial experiment with parameters n and p, a possible noninformative prior on p is a uniform distribution on $[0,1]$. In the case of a normal distribution with a known variance, a reasonable noninformative prior for the mean parameter is a constant over the whole real line. Note that this prior is improper, in the sense that it does not integrate to a finite value, but this fact poses no problem for estimation purposes. Different statistical arguments can give rise to different noninformative priors. If the variance of the normal is unknown, for instance, reasonable choices of noninformative priors are $\pi_1(\theta,\sigma) = 1/\sigma^2$ and $\pi_2(\theta,\sigma) = 1/\sigma$. The former is an example of the so-called Jeffreys prior, based on the Fisher information matrix (25). The latter arises from the arguments of Bernardo (35) and represents a so-called reference prior for the problem. A summary of various types of noninformative priors and corresponding posterior distributions is provided in Box and Tiao (36). For a general discussion of noninformative priors, see, for example, Berger (3), Bernardo and Smith (4), Carlin and Louis (20), and Gelman et al. (6). Kass and Wasserman (37) provide a summary of the vast literature in which theoretical and practical considerations of noninformative prior distributions are explored and debated.

When employing noninformative priors it is important to keep in mind that no prior can literally be noninformative or objective; it must be judged as such on a relative basis. The "noninformativeness" it represents is always specific to the information criteria being used and to the scientific and statistical structure of a particular experimental problem. For an insightful discussion of objectivity in the role of science, see Howson and Urbach (38). Noninformative priors have long been viewed as attractive and will continue to be important as Bayesian assessments and interpretations become more prevalent in regulatory settings, due to the minimal role they play in the posterior distribution relative to the data.

In trying to accomplish either of the goals associated with the two categories of priors, sometimes priors of a specific functional form are chosen for ease of computation. In Example 2, the use of normal data $Y \sim N(\theta,\sigma^2)$, with fixed σ and $N(\mu,\tau^2)$ prior distribution for θ, resulted in a normal posterior distribution for θ, $N(\mu(y), v^2)$. This example illustrates the property of *conjugacy*, in which the posterior distribution follows the same parametric form as the prior distribution. As another example consider binomial data Y resulting from n Bernoulli trials with success probability p. A *conjugate* family of priors for p corresponding to this scenario is a Beta family, $B(a,b)$, where $a,b > 0$ are parameters of the distribution. Specifically, if $Y \sim$ Binomial(n,p), and a priori $p \sim B(a,b)$, then the posterior distribution for p is $B(a + y, b + n - y)$. More generally, we know that if we sample from the exponential family and our prior distribution is also a member of the exponential family, then the posterior, in turn, will also be from the exponential family. This knowledge removes the necessity to calculate explicitly the marginal density while still allowing the prior to be updated with information contained in the data. Clearly, mathematical convenience should not dictate the form of the prior distribution. However, *mixtures* of conjugates are often mathematically convenient and allow for approximation of prior beliefs in a more flexible manner. For a discussion of conjugate priors, as well as mixtures of conjugate priors and families, see, for example, Bernardo and Smith (4, pp 265–285).

V. CLINICAL TRIAL MONITORING

In the view of regulatory, medical, and budget constraints, most controlled clinical trials have a fixed sample size chosen prior to start-up based upon the expected magnitude of drug effectiveness and adverse event profile. Unfortunately, the precision of knowledge available regarding these factors is often minimal prior to conducting a trial, and thus the choice of the sample size may depend on possibly false assumptions. To address ethical aspects of this dilemma and to assess the possibility of stopping a trial early, partially collected data from a clinical trial are often analyzed during interim analyses. Repeated testing of the null hypothesis in a classical setting leads to an accumulation of the Type I error rate (39). Methods due to Pocock (40), O'Brien and Fleming (41), Lan and DeMets (42), and Jennison and Turnbull (43) suggest various criteria for adjusting significance levels at interim and at final analyses to address the problem. The main shortcoming of all these approaches is the fact that the final inference depends on whether or not one has entertained the possibility of rejecting the null hypothesis early at an interim analysis, i.e., the stopping rule selected for the trial. In other words, the same set of data may lead to a different conclusion about the significance of the observed results depending on whether one has looked at part of the data at an interim analysis. This concept is usually found to be counterintuitive by many practitioners. Bayesian inference, on the other hand, depends only on the observed data and not on the stopping rule used. An excellent discussion of this issue is presented in Berger and Wolpert (44). From a Bayesian perspective one could look at the data as often as necessary without the risk of affecting the final conclusion. The essence of how this monitoring can be carried out may be found in many references (see, for example, Ref. 45

or Ref. 10) and is demonstrated here through an example.

Recall the setting of Example 4. Under the assumptions of the example, the posterior distribution of d, the true treatment difference between olanzapine and haloperidol with respect to negative symptoms, given n observed pairs of observations, is normal $N(\mu(\bar{d}), \rho^{-1})$, with

$$\mu(\bar{d}) = \mu \left[\frac{\sigma^2/n}{(\tau^2 + \sigma^2/n)}\right] + \bar{d}\left[\frac{\tau^2}{(\tau^2 + \sigma^2/n)}\right]$$

$$\text{and} \quad \rho = \frac{n\tau^2 + \sigma^2}{\tau^2\sigma^2}$$

Note that, in this setting, smaller values of d—say, $\{d < d_1\}$—would speak to the advantage of olanzapine (i.e., inferiority of haloperidol), while larger values of d—say, $\{d > d_2\}$—would speak to the advantage of haloperidol (i.e., inferiority of olanzapine). Then the set $\{d_1 < d < d_2\}$ indicates a region of clinical equivalence of the two drugs (in some instances, one may be interested in setting $d_1 = d_2$). When comparing to a proven active control, it is reasonable to suggest that the trial should be stopped early if either treatment is found no worse than the competitor drug with high probability. This requirement means that we would stop after m observed pairs if either posterior probability

$$P(d < d_2|D_1, \ldots, D_m) > 1 - \varepsilon_1$$

$$\text{or} \quad P(d > d_1|D_1, \ldots, D_m) > 1 - \varepsilon_2$$

where ε_1 and ε_2 are prespecified, small, positive constants. These requirements translate into the following regions for $\mu(\bar{d})$, $\bar{d} = m^{-1} \sum_{i=1}^{m} d_i$:

$$\mu(\bar{d}) < d_2 + \rho^{(-1/2)}\Phi^{-1}(\varepsilon_1)$$

$$\text{or} \quad \mu(\bar{d}) < d_1 - \rho^{(-1/2)}\Phi^{-1}(\varepsilon_2)$$

That is, one is to stop after m observations if either of the foregoing conditions is satisfied. Here, $\Phi^{-1}(\varepsilon)$ is the εth percentile of the standard normal distribution.

VI. SAMPLE SIZE DETERMINATION

Since Bayesian analysis allows frequent monitoring of the data being collected, the choice of the sample size prior to conducting an experiment is not necessary. Under the Bayesian paradigm, one can evaluate the decision to stop an experiment with each look at the data. However, since all scientists have budgetary constraints and must plan for future expenditures associated with experiments, the choice of the sample size is often required.

A formal approach to sample size determination in Bayesian analysis requires the use of decision-theoretic framework (3). The use of this framework requires the introduction of a statistical loss function that specifies a penalty for errors in estimation and a cost of obtaining sampling information. Minimization of the loss function results in the optimal sample size. Here, however, we will present some practical ideas on how to choose the sample size without explicitly specifying loss functions, since the general concept of the latter may seem unnatural to the practitioner. In particular, these ideas rely upon the notion of a so-called highest-posterior-density (HPD) set. Their application to determining sample sizes for binomial proportions, multinomial and normal sampling, can, for example, be found in Adcock (46), Adcock (47), Pham-Gia and Turkkan (48), and Joseph et al. (49). Formally speaking, an HPD set with coverage probability $(1 - \alpha)$ for a posterior density $\pi(\theta|x)$ is defined as a set $A = \{\theta \in \Theta\}$ such that $\int_A \pi(\theta|x)\, d\theta = 1 - \alpha$, and such that the length of A, $l(A)$, or in higher dimensions, volume, is minimal. For instance, in Example 2, the interval [17.48,24.08] for mean level of serum prolactin (ng/mL) in a schizophrenic population was actually a 95% HPD set.

Typically an HPD set of length l and coverage probability $(1 - \alpha)$ is sought. Since such a set depends on the data X, which are not known before the experiment, one of the ideas is to average across all possible data. *Average coverage criterion* (*ACC*) proposes to choose the minimum number of observations, n, satisfying

$$\int_{\aleph} \left\{\int_{A(x,n)} \pi(\theta|x)\, d\theta\right\} f(x)\, dx \geq 1 - \alpha$$

$$\text{and} \quad l(A(x)) = l$$

where $f(x)$ is the data density, $\aleph$ is the observational space, and $A(x,n)$ represents the HPD set. Alternatively, one could use an *average length criterion* (*ALC*). In this case, for each data point x in $\aleph$ one would first find an HPD set of length $l^*(x,n)$ with coverage probability $(1 - \alpha)$ and then find the minimum n satisfying $\int_{\aleph} l^*(x,n)f(x)\, dx \leq l$. Another alternative, the *worst outcome criterion* (*WOC*), ensures that the coverage probability is at least $(1 - \alpha)$ and $l(A)$ is at most l, regardless of the observed data, by requiring minimum n to be chosen so that

$$\inf_{x \in \aleph} \left\{\int_{A(x,n)} \pi(\theta|x)\, d\theta\right\} \geq 1 - \alpha$$

$$\text{and} \quad l(A) = l$$

In the case of Example 2, the posterior distribution

is normal, and thus $l(x,n)$ depends only on n and not on x. In addition, $\int_{A(x,n)} \pi(\theta|x)\, d\theta$ also does not depend on x. Hence all three criteria offer the same answer:

$$n \geq \left(\frac{4}{l^2}\right) \sigma^2 \Phi^{-1}\left(1 - \frac{\alpha}{2}\right) - \frac{\sigma^2}{\tau^2}$$

where $\Phi^{-1}(1 - \alpha/2)$ represents the $(1 - \alpha/2)$th percentile of the standard normal distribution. A more computationally challenging scenario arises in the case of the binomial data with Beta (conjugate) priors. Joseph et al. (49) considered an example of a randomized two-arm parallel clinical trial designed to compare two drugs, warfarin and low-molecular-weight heparin, for the treatment of deep vein thrombosis (DVT). Equal numbers of patients, n, were to be randomized to each treatment. Suppose p_1 and p_2 are DVT rates under warfarin and heparin, respectively. Joseph et al. (49) assumed that the number of patients who would develop DVT in the trial were independent binomial random variables with respective parameter sets (p_1,n_1) and (p_2,n_2). Based on the previous data, the authors chose Beta(3,11) and Beta(11,52) priors for p_1 and p_2, respectively. Using $\alpha = 0.05$ and $l = 0.05$ yielded $n_{ACC} = 1800$, $n_{ALC} = 1770$ and $n_{WOC} = 3075$. Computation of these sample sizes were based on the exact likelihoods and were performed by drawing Monte Carlo samples from the posterior distribution (50–52). The software implementing these computations in *S*-plus can be found on the World Wide Web (WWW) at the following address:

http://www.epi.mcgill.ca/
~web2/Joseph/software.html

VII. BAYESIAN COMPUTING

Implementation of Bayesian methodology requires computation of posterior and marginal distributions that involve integral expressions. The use of conjugate priors is very appealing from that perspective, since it results in closed-form expressions for the posterior. In all other cases one can use methods based on either analytic approximations of Taylor's theory (see, for example, Schwarz (53), Tierney and Kadane (54)) or numerical quadratures (Naylor and Smith (55), Dellaportes and Wright (56,57)), importance sampling (Hammersley and Handscomb (58), Berger (3)), and Markov chain Monte Carlo techniques (Geman and Geman (59), Tanner and Wong (60), Gelfand and Smith (61), Gamerman (62), Cowles and Carlin (63)). Development of these computational methods has received a lot of attention in recent years and is an active field of current research (64).

There is abundant software available implementing some of the preceding ideas. Interested readers are referred to Goel (65) and to Carlin and Louis (20), who provide an extensive overview of the currently available Bayesian software. Here we would like to mention a few popular programs that, in the authors' view, are likely to be used by scientists in the biomedical field. One such package, BUGS ("Bayesian inference Using Gibbs Sampling"), developed by D. Spiegelhalter, A. Thomas, N. Best, and W. Gilks, is built upon the use of Markov chain Monte Carlo techniques. The software and a set of manuals, along with many references on theory and application of the tool, is available through the WWW site of MRC Biostatistics Unit (Cambridge, UK) at

http://www.mrc-bsu.cam.ac.uk/bugs/Welcome.html

for Unix, DOS, and Windows platforms. "Sbayes," developed by L. Tierney at the University of Minnesota, utilizes Laplace approximations for posterior evaluations and consists of a set of functions written in the S programming language. This program is available at

http://lib.stat.cmu.edu/S/sbayes

Both "BUGS" and "Sbayes" are free software. Commercially available packages, SAS and BMDP, also allow some Bayesian computation within their procedures PROC MIXED and BMDP-5V, respectively. These procedures are targeted for the analysis of mixed-effects models (66), often used for the analysis of clinical trials, and other scenarios that give rise to repeated-measures data.

ACKNOWLEDGMENTS

The authors wish to thank their colleagues J. W. Seaman, V. Arora, and J. D. Helterbrand for many insightful comments and discussions that led to the improvement of both the structure and the focus of the manuscript.

REFERENCES

1. N Breslow. Statist Sci 5(3):269–298, 1990.
2. DA Berry, DK Stangl, eds. Bayesian Biostatistics. New York: Marcel Dekker, 1996.
3. JO Berger. Statistical Decision Theory and Bayesian Analysis. 2nd ed. New York: Springer Verlag, 1985.

4. JM Bernardo, AFM Smith. Bayesian Theory. New York: Wiley, 1994.
5. SJ Press. Bayesian Statistics: Principles, Models, and Applications. New York: Wiley, 1989.
6. A Gelman, JB Carlin, HS Stern, DB Rubin. Bayesian Data Analysis. London: Chapman and Hall, 1996.
7. I Verdinelli. In: JM Bernardo, JO Berger, AP Dawid, AFM Smith, eds. Bayesian Statistics 4. London: Oxford University Press, 1992, pp 467–481.
8. DA Berry. Statistics in Medicine 4:521–526, 1985.
9. DA Berry. Drug Information Journal 25:345–368, 1991.
10. LS Freedman, DJ Spiegelhalter. Contr Clin Trials 10: 357–367, 1989.
11. JB Greenhouse, L Wasserman. Stat Med 14:1379–1391, 1995.
12. PF Thall, R Simon. Contr Clin Trials 15:463–481.
13. C Gatsonis, JB Greenhouse. Stat Med 11:1377–1389, 1992.
14. A Racine-Poon, Smith AFM. In DA Berry, ed. Statistical Methodology in Pharmaceutical Sciences. New York: Marcel Dekker, 1990, pp 139–162.
15. J Wakefield, A Racine-Poon. Stat Med 14:971–986, 1995.
16. DA Berry. In: JM Bernardo, MH DeGroot, DV Lindley, AFM Smith, eds. Bayesian Statistics 3. Oxford: Oxford University Press, 1989, pp 239–270.
17. DA Berry. A Bayesian approach to multicenter trials and metaanalysis. Proceedings of the Biopharmaceutical Section of the ASA, Anaheim, CA, 1990, pp 1–10.
18. W. DuMouchel. In: DA Berry, ed. Statistical methodology in the Pharmaceutical Sciences. New York: Marcel Dekker, 1989, pp 509–529.
19. JB Kadane, ed. Bayesian Methods and Ethics in a Clinical Trial Design. New York: Wiley, 1996.
20. BP Carlin, TA Louis. Bayes and Empirical Bayes Methods for Data Analysis. New York: Chapman and Hall, 1996.
21. T Bayes. Philos Trans Roy Soc London:53, 370–418, 1763. Reprinted with an introduction by G Barnard. Biometrika 45:293–315, 1958.
22. SR David, AM Crawford, A Breier. Schizophrenia Research 29(1–2):153, 1998.
23. DV Lindley, AFM Smith. J Roy Statist Soc Ser B 16: 1–42, 1972.
24. DV Lindley. Math Scientist 18:60–63, 1993.
25. H Jeffreys. Theory of Probability. 3rd ed. London: Oxford University Press, 1961.
26. RE Kass, AE Raftery. J Am Statist Assoc 90:773–796, 1995.
27. JB Kadane, LF Wolfson. Statistician 47(1):3–21, 1998.
28. A O'Hagan. Statistician 47(1):21–37, 1998.
29. K Chaloner. In: DA Berry, DK Stangl, eds. Bayesian Biostatistics. New York: Marcel Dekker, 1996, pp 141–156.
30. DJ Spiegelhalter, AP Dawid, SL Lauritzen, RG Cowell. Statist Sci 8:219–283, 1993.
31. DJ Spiegelhalter, LS Freedman, MKB Parmer. J Roy Statist Soc Ser A 157:357–416, 1994.
32. M Lavine, L Wasserman, RL Wolpert. J Am Statist Assoc 86(416):964–971, 1991.
33. JO Berger. In: Kadane, ed. Robustness of Bayesian Analysis. Amsterdam: Elsevier Science, 1984, pp 63–144.
34. JO Berger. J Statist Plann Inference 25:303–328, 1990.
35. JM Bernardo. J Roy Statist Soc 41:113–147, 1979.
36. GEP Box, G Tiao. Bayesian Inference in Statistical Analysis. New York: Wiley Classics, 1973.
37. RE Kass, LA Wasserman. Formal Rules for Selecting Prior Distributions: A Review and Annotated Bibliography. Technical Report 583, Carnegie-Mellon University, Pittsburgh, PA, 1995.
38. C Howson, P Urbach. Scientific Reasoning: The Bayesian Approach. Carnegie-Mellon University, Ill: Open Court, 1989.
39. P Armitage, CK McPherson, BC Rowe. J Roy Statist Soc Ser A 132:235–244, 1969.
40. SJ Pocock. Biometrika 64:191–199, 1977.
41. PC O'Brien, TR Fleming. Biometrics 35:549–556, 1979.
42. KKG Lan, DL DeMets. Biometrika 70:659–663, 1983.
43. C Jennison, BW Turnbull. J Roy Statist Soc Ser B 51: 305–361, 1989.
44. JO Berger, RL Wolpert. The Likelihood Principle. 2nd ed. Hayward, CA: IMS, 1988.
45. DV Lindley. Introduction to Probability and Statistics from a Bayesian Viewpoint. Cambridge: Cambridge Univ. Press, 1965.
46. CJ Adcock. Statistician 36:155–159, 1987.
47. CJ Adcock. Statistician 37:433–439, 1988.
48. T Pham-Gia, N Turkkan. Statistician 41:389–397, 1992.
49. L Joseph, D Wolfson, R Du Berger. Statistician 44(2): 143–154, 1995.
50. JH Albert. American Statistician 47:182–191, 1993.
51. M Tanner. Tools for Statistical Inference. New York: Springer-Verlag, 1991.
52. T Pham-Gia, N Turkkan. Commun Statist Theory Methods 22:1755–1771, 1993.
53. G Schwarz. Ann Statist 6:461–464, 1978.
54. L Tierney, B Kadane. J Am Statist Assoc 81:82–86, 1986.
55. JC Naylor, AFM Smith. Applied Statistics 31:214–225, 1982.
56. P Dellaportas, DE Wright. Statistics Computing 1:1–12, 1991.
57. P Dellaportas, DE Wright. In: JM Bernardo, JO Berger, AP Dawid, AFM Smith, eds. Bayesian Statistics 4. London: Oxford University Press, 1992, pp 601–606.
58. JM Hammersley, DC Handscomb. Monte Carlo Methods. New York: Wiley, 1964.
59. S Geman, D Geman. Stochastic relaxation, Gibbs distributions, and the Bayesian restoration of images.

IEEE Transactions on Pattern Analysis and Machine Intelligence 6:721–741, 1984.
60. M Tanner, W Wong. J Am Statist Assoc 82:528–550, 1987.
61. AE Gelfand, AFM Smith. J Am Statist Assoc 85:398–409, 1990.
62. D Gamerman. Markov Chain Monte Carlo: Stochastic Simulation for Bayesian Inference. London: Chapman and Hall, 1997.
63. MK Cowles, BP Carlin. J Am Statist Assoc 91:883–904, 1996.
64. AE Gelfand, AFM Smith. Bayesian Computation. New York: Wiley, in preparation.
65. PK Goel. In: JM Bernardo, MH DeGroot, DV Lindley, AFM Smith, eds. Bayesian Statistics 3. Oxford: Oxford University Press, 1988, 173–188.
66. NM Laird, JH Ware. Biometrics 38:963–974, 1982.

Julia A. Varshavsky
Stacy R. David

Bioassay

See also *Assay Development; Assay Validation*

I. INTRODUCTION

Finney (1) defines a bioassay as an experiment for estimating the potency of a drug, material, preparation, or process by means of the reaction that follows its application to living matters. The emphasis in bioassays is on comparing the potencies of treatments rather than estimating the difference between the effects of treatments. For example, Finney points out that an investigation into the effects of different samples of corticotrophin on the ascorbic acid in rat adrenals is not necessarily an assay; it becomes one if the interest lies in using the changes in ascorbic acid for estimating the potencies of the samples in standard units of corticotrophin.

The typical bioassay involves a stimulus (for example, a vitamin or a drug) applied to a subject (animal, tissue, or some such experimental unit). The level of the stimulus can be varied and the effect of the stimulus on the subject can be measured in terms of a characteristic we call "response." The relationship between stimulus and response is a statistical relation subject to a random error. The relationship can be used to study the potency of a dose from the response it produces.

The estimate of potency is always relative to a "standard" preparation of the stimulus. A new test preparation of the stimulus is then assayed to find the mean response to a selected drug. Next we find the dose of the standard preparation that produces the same mean response. The ratio ρ of the dose Z_s of the standard preparation equally effective as this dose Z_M of the test preparation is the relative potency of the test preparation. Thus if $\rho = 4$, 1 unit of the test preparation is as effective as 4 units of the standard preparation.

A statistician must choose a sound design for assaying one or more test preparations. He has to choose the number of levels of doses of each preparation, the number of subjects to be tested at each dose, and the allocation of the treatments to the units. Sound statistical principles such as randomization, replication, elimination of extraneous sources of variation blocking, and covariates must be employed in designing the experiment for the assay. The data collected are then analyzed and the relative potencies of the test preparation estimated. The statistical techniques that play an important role are analysis of variance and covariance, regression analysis, and linear statistical inference in general. It is necessary to consider carefully the implicit and explicit assumptions about the data and their effect on the conclusions.

There are three types of biological assays: direct assays, indirect assays, and assays based on quantal ("all or nothing") responses. Direct assays are of fundamental importance. Here the response is measured for a number of doses to establish a statistical dose–response relation, and from this the potency of a test preparation is determined.

II. PARALLEL-LINE BIOASSAYS

An important type of bioassay is the parallel-line bioassay, in which the dose–response relation is linear for the standard as well as the test preparation and the lines are parallel or that they have a common slope. This linearity can often be achieved by a logarithmic or Box–Cox (2) type of transformation of the dose and/or the response. The treatments are then applied to the experimental units by using a suitable experimental design, such as randomized blocks, Latin squares, cross-over, split-plot, or even an incomplete block design (such as a balanced incomplete block design, partially balanced incomplete block design, or a lattice design). Whether linearity of the dose–response relation is achieved or not, and, if so, whether the slopes of the lines for the standard and test preparations are the same or not, is tested first by what are known as tests of validity, which include the test of parallelism also.

Thus if t_S, t_M denote, respectively, the effects of the standard (S) and the test (M) preparation, the parallel lines representing the dose–response relation can be taken as

$$t_S = \alpha_S + \beta x_S$$

$$t_M = \alpha_M + \beta x_M$$

where x_S, x_M are, respectively, $\log Z_S$ and $\log Z_M$, and Z_S, Z_M are the doses of s and M in original units. If ρ is the relative potency of M with respect to s, then ρZ_M and Z_S produce the same effect, and this leads to

$$\log \rho = \frac{\alpha_M - \alpha_S}{\beta}$$

This can be expressed in terms of the treatment contrasts, using

$$\alpha_M - \alpha_S = \bar{t}_M - \bar{t}_S - \beta(\bar{x}_M - \bar{x}_S),$$

$$\beta = \frac{\sum t_M(x_M - \bar{x}_M) + \sum t_S(x_S - \bar{x}_S)}{\sum (x_M - \bar{x}_M)^2 + \sum (x_S - \bar{x}_S)^2}$$

where $\bar{t}_M$, $\bar{t}_S$ etc. are the averages and the summation is over the doses of S and M. The contrast $\alpha_m - \alpha_S$ is called the *preparation contrast*; the contrast β is called the *regression contrast*. The contrasts can be estimated from the observations in the experimental design. In complete block designs, these estimates are obtained simply by replacing the treatment effects by their average yields; in incomplete block designs, however, one needs to use adjusted treatment yields for this response, after confirming that the constrasts are estimable. In fact, designs should be chosen so that these two contrasts are unconfounded and are estimated with maximum precision. Mukerjee and Gupta (3) have studied efficient designs for this purpose and have attempted to unify the results in this area. Finney (1) and Das and Kulkarni (4) give several examples of complex and incomplete block designs to be used in parallel-line bioassays. The validity tests are based on orthogonal polynomial contrasts of the treatment effects, such as

$$\sum P_{Sr} t_S \quad \text{and} \quad \sum P_{Mr} t_S$$

where P_{Sr}, P_{Mr} are orthogonal polynomials of the rth degree in the values of the doses of S and M. In a well-planned experiment, the doses are chosen to be equal in number and also equidistant on a log scale so that tables of orthogonal polynomial can be used to construct these contrasts. If reduction in block size is inevitable, one has to resort to confounding of validity contrasts of higher degrees. Kshirsagar and Yuan (5) have developed a unified theory of parallel-line bioassays in incomplete block designs. Wang and Kshirsagar (6) give an illustration of the use of lattice squares—a two-way design with incomplete rows and columns, to estimate the relative potency of several preparations. Recovery of interblock or interrow and intercolumn information can also be utilized in increasing the precision of the estimates of the preparation and regression contrasts if the blocks (i.e., rows and columns) are chosen at random. A confidence interval for ρ can then be obtained by using Fieller's method (7) for the ratio of the two contrasts in defining $\log \rho$. The assumption of normal distribution and homogeneity of variances of errors is necessary for carrying out this analysis.

If the linearizing transformation of dose is not logarithmic but of the type $X = Z^\lambda$, the relative potency turns out to be $\rho = (\beta_M/\beta_S)^{1/\lambda}$, provided the dose–response lines of S and M intersect and have slope β_M, β_S for M and S, respectively, and provided the intercepts of α_M, α_S of the two lines are equal. The validity of these assumptions in such slope-ratio assays will have to be tested by testing the departure from linearity of the dose–response regression lines and a test of significance of $\alpha_M - \alpha_S$. It is expected that when $x_S = x_M = 0$, the response should be the same for S and M, and this is tested by introducing "blanks" (neither S nor M) in the experiment.

Vølund (8) has discussed bioassays in the case of multivariate responses. In the case of multiple responses and multiple doses (in which the test or standard preparation is made up of several component drugs or chemicals), composite functions such as discriminant functions or canonical variables could be used, as discussed by Finney (1) and by Yuan and Kshirsagar (9,10).

III. QUANTAL RESPONSE AND THE TOLERANCE DISTRIBUTION

Sometimes, it is not possible to measure the response and all that can be done is to record whether or not the subject manifests a certain reaction. The response in such cases is quantal, and can be death or any other easily recognizable change in the subjects. Such a change could be irreversible, and a subject can, therefore, be used only once. Such a situation arises in the assaying of insecticides, fungicides, or pesticides or assays of insulin or estrogenic hormones or assays of vitamins. The tolerance of a subject is then defined as the minimum dose that would produce the desired effect. In a direct assay, the tolerance is measured for each subject and the potency is estimated by comparing mean tolerance. In a quantal assay, a selected dose is used, and one can only record whether or not the change occurred and infer from that whether the tolerance was greater or less than the dose used. The potency in quantal assays is estimated by use of the relation between the dose and the proportion of subjects responding. This proportion does not necessarily have a linear regression on dose or its transform. It is customary to assume a suitable distribution for tolerance or log tolerance for approximate linearization. The potency is sometimes characterized by the median lethal dose or, more generally, the median effective dose, symbolized, respectively, by LD50 and ED50. This is a dose that, on an average, produces the desired reponse in 50% of the subjects. For estimating ED50, the parameters of the tolerance distribution are estimated first from the data, by using the maximum likelihood estimation method or the minimum χ^2 method or some such suitable method. The assumption of a normal distribution for tolerance leads to what is known as *probit analysis* and the assumption of a logistic distribution leads to *logit analysis*. If $f(z)$ is the probability density function of the tolerance distribution, the probability p that the tolerance of a subject is less than x is

$$p = \int_{-\infty}^{x} f(z)\, dz = \int_{-\infty}^{y} \phi(z)\, dz$$

where $y = (x - \mu)/\sigma$, μ, σ^2 being, respectively, the mean and variance of the normal distribution of Z and $\phi(z)$ is the p.d.f. of a standard normal variable. If the distribution function of a standard normal distribution is denoted by Φ and its inverse by Φ^{-1}, we then have

$$\Phi^{-1}(p) = \alpha + \beta x$$

where α, β are transforms of μ and σ. One can then estimate ED50 or LD50 by the probit method. The method of scoring (11) is generally used to solve the maximum-likelihood equations. Several tolerance distributions, such as normal, logistic, binomial, Wilson–Worcester, and angle sigmoid, have been discussed and employed in the literature, and one may refer to Finney (1), Govindrajulu (12), or Hubert (13) for further details. The iterative calculations needed to solve the maximum-likelihood equations are laborious, and alternative, simpler but approximate, methods were developed earlier. These include the Spearman–Kärbes, Dragstedt–Behrens, and Reed–Muench (1). However, they have lost their merit due to the availability of ready-made computer programs.

Finney (1) has discussed the principles of planning an assay. He stresses the importance of validity or consistency of an estimate of the relative potency, the economics of the assay design, the necessity of pilot investigations before choosing an optimum design, the simplicity of a symmetric design, and the cost of statistical analysis. He has given specific detailed recommendations for parallel-line and slope ratio assays as well as quantal assays.

For some assays, the measured response is time: usually the time that elapses between the application of the stimulus and the occurrence of some reaction. In such time-response assays the main difficulty is that the assay may end before a response has been measured for every subject. Finney (1) suggests converting the data to a quantal focus or assigning an arbitrary value as the response for subjects that have not reacted when the assay ends or using a mathematical model for reaction time in such cases.

The relative potency of a test preparation may also change with time due to storage or environmental conditions and the dependence on time of the relative potency can be studied using a growth-curve type of model for the preparation and slope contrast, as in Bandekar and Kshirsagar (14), for example. Multivariate bioassays have been discussed by Carter and Huber (15), Williams (16), Hui and Rosenberg (17), Vølund (8), Meisner et al. (18), and Srivastava (19). For the Bayesian analysis of parallel-line and slope-ratio bioassays, reference may be made to Darley (20) and Meudoza (21). Distribution-free methods for the estimation of relative potency have been discussed by Sen (22) and Bennett (23).

REFERENCES

1. Finney, D.J. Statistical Methods in Biological Assay. 3rd ed. London: Charles Griffin, 1978.
2. Box, G.E.P., Cox, D.R. An analysis of transformations. J. Roy. Statist. Soc. B 26:211–252, 1964.
3. Mukerjee, R., Gupta, S. A efficient design for bioassays. J. Statist. Planning Inference 48:247–259, 1995.
4. Das, M. N., Kulkarni, G.A. Incomplete block designs for bioassays. Biometrics 22:706–729, 1966.

5. Kshirsagar, A.M., Yuan W. A united theory for parallel line bioassays in incomplete block designs. Int. J. Math. Statist. Se. 1:151–171, 1992.
6. Wang, W., Kshirsagar, A.M. Use of lattice square designs in bioassays. J. Biopharmaceutical Statist. 6:185–199, 1996.
7. Fieller, E.C. A fundamental formula in the statistics of biological assay and same applications. Q. J. Pharm. Pharmacol. 17:117–123, 1994.
8. Vølund, A. Multivariate bioassay. Biometrics 36:225–236, 1980.
9. Yuan, W., Kshirsagar, A.M. Analysis of multivariate parallel-line bioassay with composite responses and composite doses, using canonical correlations. J. Biopharmaceut. Statist. 3:57–71, 1993.
10. Kshirsagar, A.M., Yuan W. Estimation of relative potency in multivariate parallel line bioassays. Commun. Statist. Theory Methods 22:3355–3361, 1993.
11. Rao, C.R. Linear Statistical Inference and Its Applications. 2nd ed. New York: Wiley, 1973.
12. Govindrajulu, Z. Statistical Techniques in Bioassay. New York: Kargesm, 1988.
13. Hubert J.J. Bioassay. Dubuque, IA: Kendall-Hunt, 1980.
14. Bandekar, R., Kshirsagar, A.M. Time-dependent relative potency. Submitted for publication.
15. Carter, E.M., Hubert, J.J. Analysis of parallel-line assays with multivariate responses. Biometrics 41:703–710, 1985.
16. Williams, D.A. An exact confidence region for relative potency. Biometrics 34:659–661, 1978.
17. Hui, S.L., Rosenberg, S.H. Multivariate slope ratio assay with repeated measurements. Biometrics 41:11–37, 1985.
18. Meisner, M., Kushner, H.B., Laska, E.M. Combining multivariate bioassays. Biometrics 42:421–427, 1986.
19. Srivastava, M.S. Multivariate bioassays, combination of bioassays, and Fieller's theorem. Biometrics 42:131–141, 1986.
20. Darley, S.C. A Bayesian approach to parallel line bioassay. Biometrika 67:607–612, 1980.
21. Meudoza, M. A Bayesian analysis of slope ratio bioassay. Biometrics 46:1059–1069, 1990.
22. Sen, P.K. On the estimation of potency in dilution (direct) assays by distribution free methods. Biometrics 19:532–552, 1963.
23. Bennett, B.M. Distribution free methods in bioassay results. J. Hyg. 76:147–162, 1970.

Anant M. Kshirsagar

Bioavailability and Bioequivalence

See also *Crossover Design; Individual Bioequivalence; Dose Proportionality*

I. INTRODUCTION AND BACKGROUND

Bioavailability can be described as the rate and extent that a drug in some dosage form becomes available at the site of the pharmacological effect. As is pointed out by Benet (1), defining the criteria in terms of site of pharmacological effect "creates an impossible dilemma.'' The site of action is unknown for many drugs, and it is often impossible to measure concentrations at the site of action. Usually one solves this dilemma by assuming that the concentration of drug in the blood (or serum or plasma) is related to availability at the site of action. So a working definition for bioavailability often is the rate and extent that a drug in some dosage form reaches the systemic circulation.

With some substances, a metabolite is as responsible for the pharmacological and/or toxicological effect as the parent compound. Thus, the bioavailability of the metabolite becomes as important as the bioavailability of the parent.

A great deal of effort in clinical pharmacology research goes into estimating central tendencies and variability for variables that are thought to describe the bioavailability of a drug. What is more, characterizing the factors that significantly influence the compound's bioavailability is of great concern. If a factor significantly influences bioavailability, then it may require either an adjustment of the dose or the restriction of the use of the compound while that factor is present. For instance, if taking the drug with food is shown to increase its bioavailability, then it may be wise to take the compound while eating in order to increase the bioavailability. (This of course would not be done if it raised the drug level into a toxic range.) On the other hand, if on average a man's bioavailability is half that of a woman's, then it may be wise for men to take twice the dose that women would in order to achieve therapeutic concentrations.

Another important area of study is when one wishes to show the bioavailability of one formulation of a drug is equivalent to another formulation of the same drug. If two formulations are shown to have equivalent bioavailability, then the formulations are said to be bioequivalent. Two bioequivalent formulations are assumed to have therapeutic equivalence. There are three situations when bioequivalence needs to be established:

1. When the dosage form used in the pivotal clinical trials differs from the proposed marketed form
2. When substantial changes are made to the manufacturing process of the marketed drug
3. When one is developing a new generic form of the compound

II. SCIENTIFIC ISSUES

A. Measuring Bioavailability

The area under the concentration-time curve (AUC), calculated by the trapezoidal rule, is generally accepted as a measure of the extent of absorption. Often a correction factor that tries to account for the area under the curve from the last time point observed to infinity is added to this quantity. Of course, the area under the curve for an individual could be estimated by fitting an appropriate nonlinear curve. The "noncompartmental" method of using a trapezoidal rule, however, is preferred by regulatory agencies.

There is no consensus for how best to measure the rate of absorption for single-dose studies. Steinijans et al. (2) mention the following as potential measures:

1. Deconvoluted absorption profiles, which they conclude are difficult to compare for bioequivalence
2. Mean absorption time, which would require an additional formulation

Also, the rate of absorption from a single-dose study could be approximated directly with the Wagner–Nelson method (3). This method is unattractive, since it applies well only to drugs with one-compartment models. "Clearly, the Wagner–Nelson method should not be applied if drug concentration-time data after oral administration indicate multicompartment characteristics" (4). This leaves measures that less directly measure rate of absorption but are easily obtained: the concentration maximum (C_{max}) and the time at which the maximum is reached (T_{max}). Endrenyi et al. (5) show that C_{max}, however, is fairly highly correlated with AUC. This helps show that C_{max} reflects not only the rate of absorption but also the extent of absorption. They propose C_{max}/AUC as an appropriate measure of rate of absorption. Yet a plethora of other measures has been suggested. Among them are area under the curve up to T_{max}(AUC_i), AUC_i/AUC, area under the curve up to the minimum of the T_{max}'s for the two formulations (AUC_e), AUC_e/AUC, area under the curve up to the reference formulation T_{max}(AUC_r), AUC_r/AUC, C_{max}/T_{max}, and C_{max}/AUC_i. All of these measures are examined in Elze et al. (6).

If a drug is given repeatedly (multiple dosing) it will eventually achieve approximately steady state. Blume et al. (7) show that measures of AUC and C_{max} have less variability when steady state is achieved. Thus, for highly variable drugs, it is advantageous to test bioequivalence in a multiple-dose setting. Yet again there is no consensus on how to measure rate of absorption. One popular measure is the degree of fluctuation, which is defined at $100\% \times (C_{max} - C_{min})/C_{av}$, where C_{av} is the average concentration over a dosing interval and C_{min} is the minimum concentration over the dosing interval. Another popular measure is the time that the concentration is greater than $0.75(C_{max} - C_{min}) + C_{min}$ (T75). C_{max} and C_{min} are also used. Stenijans et al. (8) study all of these measures.

B. Switchability and Prescribability

Prescribability and *switchability* are terms defined by Anderson and Hauck (9) that give two notions of what it means for two formulations to be bioequivalent. If the bioavailability characteristics of the reference formulation distribution are sufficiently similar to the bioavailability characteristics of the test formulation distribution, then the two formulations have demonstrated prescribability. That is, a patient who is naive to a drug would expect to have similar bioavailability with either formulation. If it can be established that most individuals who receive the reference formulation can expect a similar bioavailability if they switch to the test formulation, then the two formulations have demonstrated switchability. These notions are quite different hypotheses that require far different designs to answer the bioequivalence question. The current controversy is deciding under what circumstances prescribability is the proper hypothesis and under what circumstances switchability is the proper hypothesis. Entries in this encyclopedia under individual and population bioequivalence go into this issue in more depth.

III. REGULATORY REQUIREMENTS FOR MEASURING BIOAVAILABILITY FOR BIOEQUIVALENCE STUDIES

Certainly regulatory agencies have a significant interest in bioequivalence issues. Where there is no consensus on how to handle an issue, they are placed in the difficult position of recommending how that issue should be handled. So it goes for the continuous issue of measures of rate of absorption. In the 1992 FDA Guidance on Statistical Procedures for Bioequivalence Studies Using a Standard Two-Treatment Crossover Design (10) it states that the following need to be measured in a single-dose bioequivalence study:

1. Area under the plasma/blood concentration time curve from 0 to the last measured time point (AUC_{0-t}) using the trapezoidal rule
2. Area under the plasma/blood concentration time curve from 0 to time infinity ($AUC_{0-\text{inf}}$), where $AUC_{0-\text{inf}} = AUC_{0-t} + C_t/\lambda_z$, where C_t is the last measurable drug concentration and λ_z is the terminal elimination rate constant
3. C_{max}
4. T_{max}

For multiple-dose studies they require:

1. Area under the plasma/blood concentration time curve from 0 to time τ at steady state ($AUC_{0-\tau}$), where τ is the dosing interval
2. C_{max} and T_{max} after the last dose
3. Drug concentration at the end of each dosing interval (C_{min})
4. Average drug concentration at steady state (C_{av}), where $C_{av} = AUC_{0-\tau}/\tau$
5. Degree of fluctuation = $100\% \times (C_{max} - C_{min})/C_{av}$

IV. STATISTICAL DESIGN/METHOD/ANALYSIS

A. Design Issues

There is typically a tremendous amount of intersubject variability in bioavailability. For that reason, it is advantageous to design a bioequivalence study with a randomized crossover design. When there are two formulations, the standard 2 × 2 randomized crossover design is often appropriate. Bioequivalence studies may involve more than two formulations. Latin square design becomes attractive in this setting. Period and sequence effects become balanced. Schuirmann (11) further recommends that each formulation precede each other formulation an equal number of times. This design is said to be "balanced for the presence of residual effects."

Chow and Liu (12) point out that if a carryover effect is present when one has used a 2 × 2 randomized crossover design, then one no longer has an unbiased estimate of the formulation effect. They point out that one may wish to use a higher-order crossover design in order to obtain unbiased estimates of the formulation effect. Some designs that they discuss follow.

Balaam's Design

Sequence	Period I	Period II
1	T	T
2	R	R
3	R	T
4	T	R

The Two-sequence Dual Design

Sequence	Period I	Period II	Period III
1	T	R	R
2	R	T	T

The Four-Period Design with Two Sequences

Sequence	Period I	Period II	Period III	Period IV
1	T	R	R	T
2	R	T	T	R

The Four-Period Design with Four Sequences

Sequence	Period I	Period II	Period III	Period IV
1	T	T	R	R
2	R	R	T	T
3	T	R	R	T
4	R	T	T	R

Consult Chow and Liu (12) for a discussion of these designs. With a sufficient washout period, however, it is safe to assume no carryover effect will be present in bioequivalence studies. Thus, a 2 × 2 randomized crossover design is typically sufficient. These alternative designs would be most useful if the drug under consideration had an extremely long half-life, making it impractical to have a sufficiently long washout period.

Parametric analysis of C_{max} and AUC: It has become the accepted practice to take a log transformation of these two variables. One rationale for this is that these variables are strictly positive, which is consistent with the log-normal distribution. Second, the variability is most likely multiplicative in nature. It is natural to talk about variability in terms of coefficient of variation, not standard deviation. Third, clinical interest rests in the ratio of the formulations, rather than the difference, which is more naturally assessed with log-normality assumptions. Fourth, these data are typically skewed to the right.

Parametric analysis of T_{max}: The way T_{max} is measured makes it a categorical variable instead of a continuous variable. The default is typically to analyze T_{max} as if it had a normal or log-normal distribution. More than likely this practice performs poorly. This is a clear example of a variable that needs either a nonparametric approach or some innovative statistical procedure. Basson et al. (13) analyze this variable assuming a Poisson distribution.

Nonparametric analysis of C_{max} and AUC: Those who are unwilling to make parametric assumptions about C_{max} and AUC will be heartened to know that nonparametric alternatives are available. See Ref. 14 for a treatment that uses the rejection of two one-sided Mann–Whitney–Wilcoxon tests and a corresponding distribution-free confidence interval.

B. The Equivalence Problem

Originally, bioequivalence was determined by standard hypothesis testing. If the difference in bioavailability of two formulations was not shown to be "statistically significant," they were declared equivalent. Obviously, if the research hypothesis is "no difference," such hypothesis testing encourages undersizing studies. In the first criterion set by the FDA (15), bioequivalence could be claimed if the null hypothesis for AUC and C_{max} could not be rejected with $\alpha = 0.05$, if the sample size was chosen so that one had 80% power of detecting a 20% difference. The 20% difference is arbitrary, but it appears to have been chosen to satisfy clinical considerations. See Ref. 1 for details.

Schuirmann (16) is reponsible for the two one-sided tests procedure, which is equivalent to the current acceptable approach. Let μ_t be the mean of the test formulation and μ_r be the mean of the reference formulation. Let $\theta_1 < 0 < \theta_2$. If the null hypothesis $\mu_t - \mu_r \leq \theta_1$ and the null hypothesis $\mu_t - \mu_r \geq \theta_2$ are both rejected at some level α, then bioequivalence is concluded. This method is equivalent to assuming bioequivalence if the $(1 - 2\alpha) \times 100\%$ confidence interval is completely contained in the interval (θ_1,θ_2). Obviously, it is natural to let $\theta_2 = -\theta_1$. It is normally accepted that the data are log transformed, $\theta_2 = \ln 1.25, \theta_1 = \ln 0.8$, and $\alpha = 0.05$. One can think of this method as a field goal approach, with the experimentally determined confidence interval being the football and the interval (θ_1,θ_2) the goalposts. Bioequivalence can be concluded only when the football successfully passes between the goalposts.

Equivalence problems often seem natural for Bayesian methods. By assuming a sufficiently noninformative prior distribution, one could find the posterior probability that the ratio of the AUC or C_{max} fall in some interval (θ_1,θ_2). If that probability is sufficiently high, one could declare bioequivalence. Metzler and Huang (17) discuss both Bayesian and frequentist approaches to the equivalence problem. Current practice, however, relies on the confidence interval approach.

Notice that the foregoing discussion has focused on AUC and C_{max}. The analysis of T_{max} is complicated by the categorical nature of the data. Typically, a hypothesis test, T_{max}, is usually examined as a normally distributed variable. Basson et al. (13) may provide the best attempt at formally examining equivalence for T_{max}.

Such functions as C_{max}/AUC are also not discussed. Certainly, a confidence interval approach with log-transformed data could be used with many of these variables. Whether the interval of (ln 0.8, ln 1.25) would be acceptable is unknown.

V. REGULATORY REQUIREMENTS FOR STATISTICAL METHODS IN BIOEQUIVALENCE STUDIES

The following can be found in the 1992 FDA Guidance on Statistical Procedures for Bioequivalence Studies Using a Standard Two-Treatment Crossover Design (10). Two-period randomized crossover designs are acceptable if:

1. It is a single dose study.
2. It includes only healthy, normal subjects.
3. The drug is not an endogenous entity.
4. More than an adequate washout period has been allowed between phases.

5. The predose blood samples do not exhibit detectable drug level in any subjects.

Other recommendations include the following: C_{max} and AUC should be log transformed before the analysis, and no tests of normality should be conducted. Sequence, subject-nested-in-sequence, period, and treatment effects should be included in the model. (Although it is not explicitly stated, it is implied that the subject-nested-in-sequence effect is a random effect.) A significant sequence effect is not a cause for concern unless one of the listed conditions is violated. Bioequivalence is declared if the 90% confidence interval (CI) of the ratio of C_{max} geometric means and the 90% CI of the ratio of AUC geometric means both lie completely in the interval (0.8,1.25). Lastly, outliers cannot be eliminated solely because of a statistical test. Scientific evidence and explanation to justify exclusion of data are necessary.

VI. RECENT DEVELOPMENTS

Two issues could have a huge impact on how bioequivalence studies are conducted and when they are necessary. Individual and population bioequivalence notions try to account not only for mean differences in formulation but also for differences in variability, as well as trying to account for subject-by-formulation interactions. Anderson and Hauck (9), Sheiner (18), Hauck and Anderson (19), and Schall and Luus (20) are just a few references on these topics. Also, see encyclopedia entries on both of these topics.

The analysis of in vitro/in vivo correlations could lead to situations where additional bioequivalence studies are unnecessary. Under certain conditions a sufficiently high in vitro/in vivo correlation will allow a company to forgo a bioequivalence study when there is a change in a manufacturing process. See the encyclopedia entry on this topic for further information.

Other articles that should be consulted include: Wagner (21), which gives guidance as to why AUC measures extent and why C_{max} can measure rate; Sauter et al. (22), which discusses how one should present bioequivalence studies; and Westlake (23), who was one of the early proponents of a confidence interval approach.

REFERENCES

1. LZ Benet. In: KK Midha, HH Blume, eds. Bio-International Bioavailability, Bioequivalence and Pharmacokinetics. Stuttgart: Medpharm, 1993, pp 27–35.
2. VW Steinijans, D Hauschke, JHG Jonkman. Clin Pharmacokinet 22(4):247–253, 1992.
3. JG Wagner, E Nelson. J Pharm Sci 53:1392, 1964.
4. M Gibaldi, D Perrier. Pharmacokinetics. 2nd ed. New York: Marcel Dekker, 1982, pp 149–155.
5. L Endrenyi, S Fritsch, W Yan. Int J Clin Pharmacol Ther Toxicol 29:394–399, 1991.
6. M Elze, H Potthast, HH Blume. In: HH Blume, KK Midha, eds. Bio-International 2, Bioavailability, Bioequivalence and Pharmacokinetic Studies. Stuttgart: Medpharm, 1995, pp 61–71.
7. HH Blume, B Scheidel, M Siewert. In: KK Midha, HH Blume, eds. Bio-International Bioavailability, Bioequivalence and Pharmacokinetics. Stuttgart: Medpharm, 1993, pp 37–52.
8. VW Steinijans, R Sauter, HG Jonkman, HU Schulz, H Stricker, H Blume. Int J Clin Pharmacol Ther Toxicol 27:261–266, 1989.
9. S Anderson, WW Hauck. J Pharmacokin Biopharm 18: 259–273, 1990.
10. FDA. Guidance on Statistical Procedures for Bioequivalence Studies Using a Standard Two-Treatment Crossover Design. Rockville, MD: Food and Drug Administration, 1992.
11. DJ Schuirmann. Drug Information J 24:315–323, 1990.
12. SC Chow, JP Liu. J Biopharm Statist 2:239–256, 1992.
13. RP Basson, BJ Cerimele, KA DeSante, DC Howey. Pharm Res 13:324–328, 1996.
14. D Hauschke, VW Steinijans, E Diletti. Int J Clin Pharmacol Ther Toxicol 28:72–78, 1990.
15. Federal Register 42:1648, 1977.
16. DJ Schuirmann. J Pharmacokin Biopharm 15:657–680, 1987.
17. CM Metzler, DC Huang. Clin Res Practices & Drug Reg Affairs 1:109–132, 1983.
18. LB Sheiner. Statist Med 11:1777–1788, 1992.
19. WW Hauck, S Anderson. J Pharmacokin Biopharm 22: 551–564, 1994.
20. R Schall, HG Luus. Statist Med 12:1109–1124, 1993.
21. JG Wagner. Pharmacology 8:102–117, 1972.
22. R Sauter, VW Steinijans, E Diletti, A Böhm, HU Schulz. Int J Clin Pharmacol Ther Toxicol 30:233–256, 1992.
23. WJ Westlake. Biometrics 32:741–744, 1976.

Brian P. Smith

Biologics*

See also *Food and Drug Administration*

I. INTRODUCTION

The Food and Drug Administration (FDA) has six centers that review products for many uses:

Center for Biologics Evaluation and Research (CBER)—regulates biologic products (e.g., blood products, vaccines, monoclonal antibodies, cytokines)
Center for Devices and Radiological Health (CDRH)—regulates medical devices and radiation products (e.g., X-ray machines, cell separators, bandages)
Center for Drug Evaluation and Research (CDER)—regulates drugs
Center for Food Safety and Applied Nutrition (CFSAN)—regulates foods, food supplements
Center for Veterinary Medicine (CVM)—regulates veterinary medical products
National Center for Toxicological Research (NCTR)—research facility devoted to toxicology

Biologics include blood and blood products; therapeutic products such as cytokines, monoclonal antibodies, cellular and gene therapies; and vaccines and allergenic products. Formally, the Code of Federal Regulations (CFR) defines a biologic as "any virus, therapeutic serum, toxin, antitoxin or analogous product, applicable to the prevention, treatment, and cure of diseases or injuries of man" (1). Generally, biologics are products derived from biological sources, tend to have large molecular weight, are immunogenic, and exhibit pleiotropic effects (affect many different cell types). They have a prolonged duration of action and are more susceptible to microbial contamination than drug (CDER-regulated) products. Another important difference between biologic and drug products is that the organic materials used to make biological products are more variable than the substances used in making drug products. The Center for Biologics Evaluation and Research considers the manufacturing process to be a fundamental aspect of the product and requires strict controls on this process. The manufacture of biological products presents unique regulatory concerns in relation to quality control, consistency, stability, microbial contamination, product administration, and source of the material (serum, cell bank, and tissue). These products also present some special safety concerns related to the modulations of the immune system and inflammatory response.

Blood and blood products include such items as immune globulin products, clotting factors for treatment of hemophilia, products that test for HIV antigens, and products that promote cessation of bleeding, among others. The Center for Biologics Evaluation and Research is responsible for regulating the nation's blood banks to ensure a safe blood supply.

The spectrum of biologics for therapeutic applications includes *monoclonal antibodies* (derived from a single clone, used for diagnostic and therapeutic purposes), *cytokines* (proteins that mediate the effector phases of natural and specific immunity), *hematopoietic growth factors* (stimulate the growth and differentiation of immature leukocytes in the bone marrow), cell and gene therapies, tumor vaccines, *xenotherapies* (products made from nonhuman sources), tissue-engineered products, and cell selection devices.

Vaccines are the most widely used class of biologic products CBER regulates. These products include childhood vaccines such as polio, measles–mumps–rubella (MMR), diphtheria–tetanus–pertussis (DTP), *Haemophilus influenzae* type b (Hib), and hepatitis vaccines. There are other vaccines that are less widely used, such as vaccines to prevent cholera and typhoid. The Office of Vaccines Research and Review within CBER also regulates allergenic products to prevent or reduce the effects of hay fever, allergies to grasses, and other allergic conditions.

The origin of the regulation of biologics dates back to the Virus, Serum and Antitoxin Act of 1902, which established federal regulation of biologic products (2). Many provisions of that act are still in force today: establishment licensing, product licensing, labeling requirements, inspection requirements, revocation or suspension of licenses, and penalties for violations. Initially (and until 1944), biologics were regulated by the Hygienic Laboratory of the Public Health Service. The laboratory promulgated regulations implementing the provisions of the act. Inspections were to be unannounced and conducted by commissioned medical officers. Samples of products were to be examined for purity and potency.

*The views expressed in this document are those of the authors and do not necessarily represent those of the FDA.

In 1944, there was a major revision of federal public health laws. One change made was that "a biological license could issue only upon showing that the product and establishment met standards to ensure the continued safety, purity, and potency of such products" (2). This act also provided for the transfer of biologics regulation to the National Microbiological Institute of the National Institutes of Health (NIH). In the 1950s, biologics regulation was transferred to the Division of Biologics Standards (DBS) of the NIH. In 1972, biologics regulation was transferred to the FDA, where it remains today. In addition to merging DBS (renamed the Bureau of Biologics) into the FDA, the 1972 merger integrated the regulations of the Public Health Service Act and the Federal Food, Drug, and Cosmetic Act. In 1982, the FDA merged the Bureau of Drugs and the Bureau of Biologics into the Center for Drugs and Biologics; in the late 1980s, these separated and became the Center for Biologics Evaluation and Research (CBER) and the Center for Drug Evaluation and Research (CDER).

A recurring problem in evaluating products or procedures is that of competing risks. For example, in evaluating blood substitutes, a life-threatening safety event may cause termination of therapy prior to a favorable outcome. When a favorable event (such as shortened hospitalization) and a competing event (such as death) have different implications for efficacy of the product, assessment of product benefit may not be straightforward. Meinert noted that "a differential mortality by treatment group may influence the occurrence of nonfatal events. The treatment group with the highest mortality rate may have the lowest nonfatal event rate if death occurs before patients have a chance to develop the nonfatal event of interest" (3).

II. REGULATORY PROCESS

The regulation of a biologic begins with the investigational new drug application (IND), filed with CBER after preclinical testing suggests a product may be safe and has the desired biological activity in animals. The clinical investigation of a new biologic, as well as of a new drug, generally proceeds in three phases. Phase I studies are the first investigations in humans, and they focus on the safety, metabolism, and pharmacologic actions of the product, potential dose range, and sometimes preliminary evidence of effectiveness. Phase II studies are generally closely monitored, controlled studies to obtain preliminary data on efficacy and side effects and other risks, and to establish an optimum dose and schedule. Investigators often use data from phase II to determine appropriate endpoints for phase II studies. Phase III studies are larger studies of the efficacy and safety of the product. They provide evidence of risk/benefit and form the basis for product labeling. After completion of the investigational clinical trials in the IND phase, the sponsor submits a biological license application (BLA) to market the product. In order for a license to be issued, the application must show that the product meets standards of safety, purity, potency, and consistency of manufacture in addition to demonstrating efficacy.

The concept of a generic drug has not been applied to biologic products to date, as a result of the inherent variability in biologic products. Because the manufacturing process is a fundamental characteristic of the product, biologic products that are made differently are considered to be distinct. For example, although there are several DTP vaccines, none is considered "generic"; they are all considered to be distinctly different products for the same disease indication.

Since the mid-1990s, there has been an effort to harmonize the requirements for licensure of drugs and biologics worldwide. This effort, the International Conference on Harmonization (ICH), has developed a series of guidelines for efficacy, safety, and quality. The regulatory agencies and pharmaceutical industries of the European Union, Japan, and the United States have developed these documents. They are available on the Internet. The address is given in Section V of this entry.

III. ISSUES IN BLOOD AND BLOOD PRODUCTS

The Office of Blood Research and Review of CBER conducts research related to the development, manufacture, testing, and activities of blood products, products to ensure the safety of the blood supply (e.g., test kits for HIV, hepatitis A, B, and C, syphilis), and those prepared by genetic engineering and synthetic procedures, in order to establish standards to ensure the continued safety, purity, efficacy, and availability of blood products. The office performs scientific functions related to commercial blood-derived and analogous recombinant products distributed in the United States, such as medical devices used to collect, test, process or store donated blood; retroviral diagnostic tests, including HIV-related tests; and certain human cellular and tissue-derived products.

Many of the statistical approaches for addressing these activities are common to other biomedical evaluations and other centers at the FDA. In many conditions that blood products address, however, the target population is small, so sponsors must contend with limited numbers of patients available for conducting clinical trials. Evaluating the safety of blood and plasma products is another challenge. For many plasma products, many donations are pooled from a large number of donors. Thus, a single donation of infectious but seronegative plasma or blood can affect many recipients. Assessing diagnostic test kits without the benefit of a gold standard sometimes arises when one test kit is compared to another. This situation leads to distinct statistical problems (5).

The primary endpoint or efficacy variable in any clinical trial should provide a convincing basis for comparing treatments. Biological products sometimes address needs for which there are no currently existing products, so there is no precedent for selecting an appropriate efficacy endpoint. The difficulty in defining an appropriate endpoint also arises in several types of blood product experiments. For example, fibrin sealants are products that are "painted" on a wound to halt bleeding. Should the endpoint be time to bleeding cessation? Should it be a dichotomous variable, taking the value 1 if bleeding stops in 10 minutes (or some other time) and 0 otherwise? What are appropriate control groups?

Hemophilia is the condition of deficiency of clotting Factor VIII, and there are products that are designed to replace this deficiency. The bleeding is often internal into a joint, however, causing the patient pain; the only measure of bleeding cessation is the patient's subjective report that joint pain has stopped. Such a response is problematic, because a subjective report can be biased by the patient's knowledge of the therapy. Also, it is common to assess multiple episodes of bleeding in such patients, leading to questions of how to account for the correlation among the responses.

In some cases it has been difficult to conduct controlled trials when the investigational product was thought likely to be much better than existing therapy; thus some products have been assessed and approved on the basis of uncontrolled studies. In such cases, it has been necessary to determine a standard for comparison—either a historical control or an absolute standard.

A. Low-Prevalence Conditions

Some blood products are designed to treat rare diseases. Clotting factors for hemophilia, for example, are targeted at a total population of about 40,000 hemophiliacs in the United States. Some subgroups of this population have developed inhibitors to Factor VIII replacement products; treatments for such subgroups are aimed at much smaller target populations.

An even more extreme example is a product proposed for hemophiliacs who have developed inhibitors to the usual Factor VIII products. There were about 400 patients in North America with this condition. For this product, it was feasible to recruit at most 20 patients for a clinical trial. These limited populations make it difficult to recruit adequate numbers of patients to conduct clinical trials or to provide a substantial understanding of the safety profile of the product. Another rare condition is envenomation from snakebites. Only about 8,000 people per year have sufficiently severe snakebites to be candidates for treatment. A new product was being investigated that it was hoped would provide the same therapeutic effect as the only currently available product, but with a better adverse event profile. For this rare condition, an appropriate endpoint was needed; since mortality is uncommon, a validated index of snakebite outcome was used. There was also discussion about whether this study could be uncontrolled. The safety profile of this product and the currently available antivenom needed to be assessed. There is a great deal of information about the currently available product, so similar information was collected for the new product. The proposed product had a superior safety profile and provided similar efficacy.

Conducting clinical trials involving few patients may require flexibility regarding some of the preferred characteristics of the ideal clinical trial. For example, it may be necessary to do single-arm studies and to compare results to historical controls. If a historical control group is used, it is important that control subjects meet the same entry criteria as the treated patients. The historical controls should be as contemporary as possible so that treatment characteristics other than the therapy in question are similar. Such rare diseases/problems are not unique to CBER; indeed, the FDA has an orphan drugs office that helps to regulate therapies for these rare conditions.

B. Evaluating the Safety of Blood and Blood Products

Some blood and blood products are used to treat patients who donated the blood in the first place. These are called *autologous* products. Other blood and blood products are from donors other than the recipient (*allogeneic* products) and may contain viruses or other unwanted particles that could transmit disease to the recipient. Safety is a particular concern with allogeneic products, since they are often administered directly into the recipient's bloodstream. Several procedures are used to reduce the risk of viral transmission from blood products:

Blood collection centers screen donors for current and prior infectious diseases and high-risk lifestyle activities that could pose a risk of viral transmission from their blood or blood products.

Blood collection establishments test all blood and plasma for HIV, HTLV, hepatitis B, hepatitis C, and other agents. Even with these procedures, occasionally (about 1 per 34,000 donations) a unit is virally contaminated, primarily due to the "window period" problem. The *window period* is the interval between the time an individual becomes infected and the time the virus or antibodies to the virus are at detectable levels in the blood. A blood or plasma donation made during the window period will not be identified as virally contaminated but may still transmit infection. The probability of a window period donation depends on the length of the window period and the effectiveness of predonation screening (6).

Finally, the plasma undergoes viral inactivation treatment prior to distribution. This treatment reduces the risk of viral transmission to extremely low levels.

IV. ISSUES IN BIOLOGICAL THERAPEUTICS

The development of recombinant biological products to treat human diseases has a relatively short history, although many biological products, such as vaccines and blood products, have a history of almost 100 years. During the last 10 years, scientific progress in recombinant DNA technology and molecular biology has resulted in new classes of products to treat diseases such as multiple sclerosis, hepatitis C, diabetic ulcers, and non-Hodgkin's lymphoma and other cancers. The excitement in the development of therapeutic biologics is due to the recent ability of molecular biologists to produce large quantities of previously scarce human proteins. In the emerging field of gene therapy, new techniques are making it easier to introduce genes into somatic cells of a target organ to correct or modify cell functions. Their goal is to prevent or correct human diseases. Many clinical trials of human gene therapy are in progress.

Biological therapeutics include cytokines, monoclonal antibodies, cellular and gene therapies, and growth factors. Monoclonal antibody products have been developed for a variety of conditions, including stroke, for diagnostic imaging, and for prevention and treatment of rejection of tissue grafts. Growth factor products are used for wound healing and to support the immune system in cancer patients receiving chemotherapy. Cytokine products such as the interferons and interleukins are used to treat cancer, hepatitis C, and multiple sclerosis. Other biological products have been developed for cystic fibrosis, anemia of chronic renal failure, stroke, myocardial infarction, and genetic disorders.

The fundamental principles of clinical trial and statistical methodologies applied in the area of biological therapeutics are the same as in other areas of drug development. However, some statistical issues and design questions have been encountered in this area more frequently than others.

Between the late 1980s and the early 1990s, when alpha interferons, G-CSF (granulocyte colony stimulating factors), erythropoietin, and other agents were entering phase III trials, the clinical studies were designed as superiority trials with concurrent placebo controls. However, as these new agents became available for clinical use, the trials of second generation products were often designed with active control arms. These studies are usually designed now as noninferiority trials, with the objective being to rule out a specified (true) difference, δ, with specified power. In statistical terms, the null hypothesis is specified as $H_0\colon \mu_T < \mu_C - \delta$. Sometimes the pivotal studies are designed as equivalence trials (i.e., two-sided), with the objective being to rule out both positive and negative differences. Such bioequivalence studies have a long history in generic drugs. In this case, the null hypothesis is $H_0\colon |\mu_T - \mu_C| > \delta$. It is widely recognized that the traditional hypothesis-testing approach should not be used in such trials. Generally, these trials require much larger sample sizes than do superiority or noninferiority trials (7).

The first generation of monoclonal antibodies for therapeutic and diagnostic applications as imaging agents has entered phase III trials. Several monoclonal antibodies for therapeutic and diagnostic applications have been approved by the FDA. Clinical trials for diagnostic agents typically obtain images of patients using a conventional diagnostic method and images using the new method. Thus, there are special problems with these trials, since they are easily unblinded (images produced using radiolabeled monoclonal antibodies appear quite distinct from traditional imaging modes), and their overall performance is affected by the proportion of diseased patients in the study. Usually, the patient is imaged with both the standard method and the new method and a comparison is made. These require methods that account for the pairing of images. The traditional measures of efficacy have been sensitivity and specificity, and ROC curves. Predictive values and accuracy are affected by the prevalence of the condition. Indeed, the measures themselves may be inappropriate in some instances. For example, if the objective is to detect the presence of disease in the patient (or an organ), sensitivity and specificity are appropriate. However, if the goal is the detection of lesions (affected areas), it is possible to ascertain the probability that an identified lesion is truly positive (positive predictive value). One cannot measure the negative predictive value, since areas that are not indicated as affected are not biopsied. The gold standard is usually a histological measurement from a surgical procedure or a biopsy.

The development of biological therapeutics does not always follow the traditional paradigm of chemical drug development (preclinical–phase I–phase II–phase III). In biologics, there has been a tendency to plan a phase III trial after early results from small phase I studies. A number of factors have contributed to this trend. The available animal test models generally lack receptors for biological products (e.g., monoclonal antibodies) and, thus, preclinical testing in such models carries limited information. Biologics are endogenous proteins and polypeptides, and we have more knowledge about their structures and functions. Furthermore, many new biologics have been found to be efficacious in treating life-threatening diseases (as compared to chronic diseases requiring long treatment duration), where consideration of risk/benefit is driven by the paramount necessity to save lives. The drug development plan may move gradually toward the more traditional paradigm of safety and efficacy evaluations as we identify more biologics that are safe and effective for the treatment of chronic diseases. When there is limited (or no) phase II data, an interim analysis (specified in the protocol!) with an explicit plan for sample size adjustment is essential.

V. ISSUES IN VACCINES

Vaccines may be administered for either preventive or therapeutic indications. *Preventive vaccines*, which constitute the vast majority of vaccine products, typically are designed to elicit antibodies in healthy individuals to the causative agents for diseases such as polio, diphtheria, tetanus, pertussis, measles, mumps, rubella, influenza, and others. *Therapeutic vaccines*, on the other hand, are intended to treat an already existing disease, such as cancer or HIV, in the vaccine recipient. Since therapeutic vaccine trials are designed and statistically analyzed in the same manner as trials for other therapeutic products, they will not be considered further here. Three major concerns addressed in preventive vaccine evaluation are the extent and duration of protection from disease or infections, and the safety profile.

An efficacy trial for a vaccine compares an investigational product to a control in an appropriate population. When the control is a placebo or a vaccine that has no effect on the disease of interest, the trial can estimate absolute vaccine efficacy. When the control is an already licensed vaccine for the same disease indication, the trial can estimate relative vaccine efficacy.

For regulatory purposes, the main interest is in estimating the *direct* effect of the vaccine on disease incidence. Case definitions (criteria for the diagnosis of disease) must be resolved in the planning stage. In general terms, vaccine efficacy is customarily estimated from a vaccine efficacy trial as

$$\mathrm{VE} = 1 - \frac{\mathrm{IR}_V}{\mathrm{IR}_C}$$

where IR_V is the disease risk, incidence rate, or hazard rate in the vaccinated group and IR_C is the corresponding estimate in the control group. VE might refer to absolute or relative efficacy, depending on the control group. There are a variety of methods for constructing confidence intervals around VE (8).

In addition to direct effects, vaccines often confer *indirect* effects as well, if the disease is spread by person-to-person contact in the population. In such cases, even persons who did not complete a vaccination series or are unvaccinated still may be protected from disease because of reduction in disease transmission. Thus, undervaccinated or unvaccinated individuals may be protected from disease through *herd immunity*, a phenomenon unique to vaccines. The direct estimates of VE obtained from clinical trials may underestimate the overall reduction in disease that might be expected in well-immunized populations postlicensure, since such trials do not fully capture the likely indirect effects of vaccination. Much work on estimation of indirect effects of vaccination has been done (9).

Often an investigational vaccine targets a disease for which an already licensed vaccine is available. In this case, the disease incidence may have already been reduced to such a low level that a clinical efficacy trial in that population is not feasible. For example, the recent trials for acellular pertussis vaccines could not be conducted in the United States because the long use of whole-cell vaccines reduced the incidence of pertussis to low levels. These acellular pertussis trials were conducted in Italy, Germany, and Sweden, where vaccination with the current whole-cell pertussis vaccines had been discontinued or was infrequent because of concerns about adverse reactions. Acellular pertussis vaccines were developed to reduce the adverse-reaction rate.

There are several concerns in conducting an efficacy trial when currently licensed individual vaccines are physically combined to be given as one injection and the resulting combination product is compared to the vaccines given separately. One concern is that mixing several antigens might result in reduced vaccine efficacy because of interference in immune response to the different antigens. Another concern is that, because of low disease incidence due to the licensed component vaccines, evaluation must be based on serological endpoints rather than on cases of disease. Conducting these trials is facilitated if there are prior data showing that certain serum antibody levels are correlated with protection from disease. Unfortunately, such correlates are known for some, but not all, vaccines. In these trials, the serum antibody levels are compared to show noninferiority of the combination product.

A methodological concern is the assessment of local reactions: there is a single injection site for the combination product, while there are two (or more) injection sites for the separate products. Thus, the separate injection treatment group has more opportunity to experience a local reaction (e.g., redness or swelling at the injection site) compared to the combination group (10).

When comparing an improved vaccine or a combination vaccine to a standard, licensed vaccine, the studies are usually designed as clinical equivalence or noninferiority trials, using immune responses (serological endpoints) as the primary outcome variable. In a noninferiority trial, in which proportions attaining some prespecified level of immune response are compared, the null hypothesis has the form $H_0\colon \pi_{V1} < \pi_{V2} - \delta$, where π is a proportion seroconverting or attaining a certain level of response and δ is the amount of decrease that can be tolerated. Unfortunately, the protective level of antibody (correlate of protection) is often unknown, thus complicating the establishment of an acceptable margin of equivalence (δ). When geometric mean titers/concentrations (GMTs/GMCs) are compared, the hypothesis is expressed in terms of a ratio rather than as a difference in proportions (11).

Studies to support consistency of manufacture are required for all products. Such studies are important because, as noted earlier, the manufacturing process is such an integral part of the product's license. These evaluations may be based on the physical characteristics of the vaccines or may aim to show that successive lots of vaccine will evoke similar immune responses or bioassay results. The latter type, called *clinical lot consistency studies*, are usually designed and analyzed as equivalence studies, i.e., two-sided, since a difference in either direction is of interest. It is customary for at least three consecutively produced lots to be compared to establish consistency of manufacture.

Change in a manufacturing process may occur after completion of a successful efficacy trial. In this situation, there could be concern that the change may have resulted in a product that is no longer clinically equivalent to the one for which efficacy was demonstrated. This product change could occur if the manufacturing process affects the biological variability of the product. To allay this concern, a *bridging study* may be conducted in which persons, randomized to receive vaccine lots similar or identical to a lot used in the efficacy trial, are compared to individuals randomized to receive the newly manufactured lots. Bridging studies aim to demonstrate the clinical equivalence of the two sets of vaccine lots with respect to immune responses and various safety measures. If the lots are similar regarding these endpoints, then the efficacy and safety that were observed for the lots used in the efficacy trial are inferred for the newly manufactured lots. A bridging study might also be conducted when there is a change in formulation or dosing schedule, as well as when the "change" is a population for intended use that is different from the one in which the efficacy trial was conducted.

Such bridging studies represent one of the most common situations in which serological correlates of protection are the primary endpoint for evaluation. Again, interpretation of comparative results from such studies is more difficult if a correlate of protection has not been determined. Bridging studies are usually much smaller than clinical efficacy studies, since they rely on immunogenicity measures rather than on detecting differences in small proportions acquiring disease. Differences that may affect the vaccine's immunogenicity include ethnicity, schedule of administration (e.g., 2, 4, 6 months of age vs. 3, 5, 12 months of age), and environmental factors. Generally, the goal of bridging studies is to show that the vaccines are similar under the different conditions of use (i.e., lots, populations, combinations, etc.).

Evaluation of the safety profile of an investigational vaccine is extremely important. Since preventive vaccines (AEs) are given to healthy populations, there must be assurance that serious adverse events associated with the vaccine do not occur at an unacceptably high rate. Large studies are usually required to allay this concern; for example, if 3,000 individuals were vaccinated with no serious adverse events being observed, the upper 95% confidence limit for the true rate of serious adverse events would be 0.001. One serious AE in a study of 4,750 subjects would also yield an upper limit of 0.001. Less serious adverse events can have a somewhat higher bound of acceptability. A common local reaction (e.g., redness) at the vaccination site might be bounded at 0.05 or 0.10. The sample size requirement for the less serious adverse events (e.g., a difference of 0.05 from a base of 0.05) are almost always met and tend to be hundreds of subjects rather than thousands.

There are several open statistical issues in vaccine evaluation. They are discussed next.

A. Evaluation of Immune Correlates of Protection

These immune response levels are surrogate outcomes for clinical vaccine efficacy that permit the evaluation of new vaccines more efficiently. The usual situation is to use the data from the clinical efficacy trials of the first generation of the vaccine and to attempt to correlate one or more serum antibody levels with vaccine efficacy (12).

B. Selection of Individuals and/or Events to Include in the Primary Efficacy Analysis

It has been common practice to count cases only in subjects who received a full immunization series (usually with an additional time period to allow effective immune response to the full injection series). Cases occurring in subjects dropping out of the study or cases of disease occurring before the full series is completed have often been ignored or have been counted only in secondary evaluations. There are arguments against this type of analysis. Excluding dropouts affects the randomization and thus raises concerns about the validity of the statistical analysis. Thus, many would argue for the study to be analyzed counting all cases in all randomized individuals. This approach preserves the significance level if testing a null hypothesis that VE $\leq V_0$ (where V_0 is fairly large). In this case, the usual procedure is to compute sample sizes under the null hypothesis that VE = V_0. It can also be argued that the dropouts may be a nonrandom subset of the group that has been randomized with respect to the probability of disease, and that omitting them may lead to biased estimates of efficacy.

On the other hand, the complete series analysis aims to estimate the biological efficacy of the vaccine if given as intended. In most vaccine efficacy trials, which enroll only healthy individuals, compliance with the full immunization series is quite high (often in the 90–95% range). Moreover, dropouts due to toxicity are not common, so that an intention-to-treat and "full series" analysis often yield similar estimates of efficacy. Even when the two estimates diverge numerically, it has been observed that they tend to lead to similar conclusions about how well the vaccine works (e.g., a VE of 0.85 and a VE of 0.65 both indicate substantial efficacy). This concordance is unlike the situation in therapeutic trials in which the two analyses may lead to conflicting conclusions about efficacy, in part because effect sizes are frequently small and there is a much greater rationale for dropouts being a nonrandom subset of the study population with respect to the study outcome. It is currently common practice to perform both the full-series and the intent-to-treat analyses of preventive vaccine efficacy.

C. Phase III Studies of Preventive HIV Vaccines

Because the risk of infection with HIV is highly related to specific behavior, special problems will arise when evaluating these vaccines in efficacy trials. For example, behavioral changes could impact efficacy assessment of these vaccines: individuals who believe they are protected by a vaccine may engage in more behaviors that expose them to infections. This situation may be especially prevalent if individuals "self-unblind" by being tested for anti-HIV antibody. Specifically, if more high-risk behavior occurs (because of a belief that the vaccine is fully protective), the incidence of new HIV infection may even increase. Similar concerns arise in trials of other diseases, such as genital herpes, that have a large behavioral component in their transmission.

On the other hand, intensive counseling aimed at limiting such behavior is considered a mandatory part of the studies (either as a separate arm or in both control and vaccinated groups). Such counseling may itself reduce the incidence of infection to an extremely low level, thus requiring much larger studies to demonstrate efficacy. If such counseling were not provided in clinical practice, the vaccine efficacy might be different; for example, vaccinated individuals might engage in more high-risk behavior, leading to a lowered efficacy. In this case, a biologically effective vaccine might have a low ''use-effectiveness'' in practice.

VI. WEBSITES FOR FDA INFORMATION AND GUIDANCE DOCUMENTS

There are several guidance documents relating to biologics available on the FDA websites. The addresses for the FDA website and the ICH website are:

FDA website: http://www.fda.gov
ICH website: http://www.ifpma.org/ich5.html

ACKNOWLEDGMENTS

We thank Karen Goldenthal and Paul Aebersold for incisive comments on this document. We also thank our colleagues at CBER for providing us with many insights into the regulation of biologics.

REFERENCES

1. Code of Federal Regulations: 21 CFR 600.3(h).
2. Brady RP, Kracov DA. From diphtheria antitoxin to cytokine products: a remarkable scientific journey/an aging regulatory framework. Regulatory Affairs 3:105–132, 1991.
3. Meinert CL. Clinical Trials Design, Conduct, and Analysis. New York: Oxford University Press, 1986.
4. Code of Federal Regulations: 21 CFR 314.500.
5. Lachenbruch PA, Lynch CJ. Tests for equivalence in dichotomous variables. Stat Med 17(9):2207–2218, 1998.
6. Schreiber G, Busch MP, Kleinman SH, Korelitz JJ. The risk of transfusion-transmitted viral infections. N Eng J Med 334:1685–1690, 1996.
7. Blackwelder WC. Similarity/equivalence trials for combination vaccines. In: JC Williams, KL Goldenthal, DL Burns, BP Lewis Jr, eds. Combined Vaccines and Simultaneous Administration. Ann New York Acad Sciences 754:321–328, 1995.
8. Ewell M. Comparing methods for calculating confidence intervals for vaccine efficacy. Stat Med 15:2379–2392, 1996.
9. Haber M, Longini IM, Halloran ME. Measures of the effects of vaccination in a randomly mixing population. Int J Epidemiol 20:300–310, 1991.
10. Department of Health and Human Services. Guidance for Industry: Guidance for the Evaluation of Combination Vaccines for Preventable Diseases. Federal Register, April 1997 (Internet: http://www.fda.gov/cber/guidelines.htm).
11. Falk LA, Midthun K, McVittie LD, Goldenthal KL. The testing and licensure of combination vaccines for the prevention of infectious diseases. In: R Ellis, ed. Combination Vaccines. Totowa, NJ: Humana Press, 1998.
12. Fleming TR, DeMets D. Surrogate endpoints in clinical tirals: are we being misled? Ann Internal Med 125:605–613, 1996.

Peter A. Lachenbruch
A. Dale Horne
Cornelius J. Lynch
Jawahar Tiwari
Susan S. Ellenberg

Biopharmaceutics

See also *Drug Development*

I. INTRODUCTION

The biopharmaceutics and biopharmaceutical properties of a medicinal/drug product can be considered the bridging information between physicochemical, pharmaceutical quality, and biological, pharmacological effects that result in therapeutic as well as toxic responses. The terms should not be confused with ''biopharmaceuticals,'' a term now being increasingly applied to active substances and products derived from biotechnology, such as ''recombinant'' proteins.

Although regulatory documents, such as the Food and Drug Administration (FDA) guidelines for new drug applications (1), refer to biopharmaceutics and indeed the Code of Federal Regulations (CFR) require (2) a separate biopharmaceutic review section in the new drug application (NDA), there does not appear to be an official definition of the term. The European regulations and the ICH guidelines have no direct reference to biopharmaceutics.

The term *biopharmaceutics* appears to have been first used in 1961 by Wagner (3) to mean:

encompassing the relation between the nature and intensity of the biological effects observed in animals and man and, 1. the nature of the form of the drug (ester, salt, complex, etc.), 2. the physical state, particle size and surface area, 3. the presence or absence of adjuvants with the drug, 4. the type of dosage form in which the drug is administered, and 5. the pharmaceutical processes used to make the dosage form.

Gibaldi (4) provided a brief definition of biopharmaceutics: "a major branch of the pharmaceutical sciences concerned with the relationship between the physicochemical properties of a drug in dosage form and the pharmacologic, toxicologic or clinical response after its administration."

Biopharmaceutics is multidisciplinary, involving specialists in developing the chemistry and manufacturing controls (CMC) dossiers of applications and toxicologists and pharmacologists involved in documenting the pharmacokinetic and sometimes pharmacodynamic responses. Both groups apply appropriate statistical models with the advice of statistical departments.

II. PROPERTIES

Biopharmaceutical properties are commonly divided into those intrinsic to the active substance and those of the finished pharmaceutical dosage form. The characteristics are defined during the development stages, which may be for the first entry of the substance or with a line extension using a different dosage form or even when generic products are being prepared. There are tests that are required only during development and results of other tests that become the lot-release specifications for each batch of substance and finished product prepared. In general, many of the physicochemical tests are used for quality control, whereas the pharmacological tests are rarely repeated postapproval.

Most of the literature on biopharmaceutics has been devoted to solid oral dosage forms of drugs that are absorbed from the gastrointestinal tract and are effective systemically. However, other routes, such as inhalation, dermal, and parenteral, also require biopharmaceutical evaluation.

A. Active (Drug) Substance: Physicochemical Characterization

1. Solubility

The aqueous solubility of the drug substance is established by obtaining saturated solutions at 37°C over 24 or 48 h. In addition, pH/solubility profiles, at physiologically relevant values from 1 to 8, are measured. This information indicates if there might be pH-dependent dissolution and perhaps effects of food in delaying absorption. Substances with aqueous solubility higher than 1% are considered less likely to show solubility-limited absorption problems (5,6).

2. Intrinsic Dissolution Rate

Because this is surface area dependent, it is defined as the amount of active substance per unit time that dissolves under standardized conditions of temperature, solvent composition, and liquid/solid interface. Commonly a nondisintegrating disk of drug substance provides the standardized surface area and liquid/solid interface (7). The United States Pharmacopeia (USP) has proposed an apparatus and test to determine intrinsic dissolution (8). The cumulative amount dissolved over time per unit area is obtained and data points fitted with linear regression to obtain a slope, which is the intrinsic dissolution rate. There is general agreement that new drug candidates with intrinsic dissolution rates greater than 1 $mg/min/cm^2$ will have fewer absorption problems than those exhibiting rates below this level (5).

3. pKa, Partition Coefficient, and Permeability

As well as bearing on solubility, the dissociation constant, pKa, and partition coefficient between aqueous and organic solvents are important influences on absorption of drugs (9). The lipid solubility of drugs is tested in partition experiments between buffer and such solvents as octanol, chloroform, and ethyl acetate. The majority of organic small molecules are absorbed in the gastrointestinal tract by means of passive diffusion, with the transfer directly proportional to the magnitude of the concentration gradient across the membrane and the lipid:water partition coefficient of the drug. Because most drugs are weak acids or bases, present in solution in both ionized and unionized forms, the unionized forms are usually lipid soluble and can diffuse across the cell membrane. This was first proposed in the pH-partition hypothesis (10) and, although it remains largely true that the pKa determines the pH for optimization of the unionized form, which will be more rapidly absorbed, the much larger surface area of the intestinal mucosa leads to a greater rate of absorption, even when the unionized form of the drug prevails at gastric pH (11). In addition to partition models, attempts to assess absorbability during development have involved *in situ* experiments with segments of animal intestine, such as the everted sac (12). However, more recently the biopharmaceutics drug classification system has been promoted as being useful to predict absorbability properties (13). This system involves using measured human jejunal permeability (14) and solubility properties to classify drugs into the following categories:

high solubility–high permeability
low solubility–high permeability
high solubility–low permeability, and
low solubility–low permeability

It claims to provide a basis for estimating the absorption of drugs based on these properties of physiologic importance (13). This system is being used by the FDA to make decisions on postapproval changes (15) and is also likely to be expected with development information in NDAs (16). Currently, efforts are being made to search for other tests than human jejunal permeability to describe drug permeability. Such procedures as Caco-2 monolayer systems, rat intestinal models, and human mass balance are under consideration (16). Caco-2 increasingly is being used in fast throughput screening of drug candidates.

While many drugs are absorbed via passive diffusion, an increasing number are absorbed by special transporters (17). Examples are nutrient, vitamin, and ion transporters. Several cephalosporins and angiotensin-converting enzyme inhibitors are substrates for the intestinal dipeptide transporter. In addition, the role of P-glycoprotein countertransport processes in the intestinal mucosa is currently being elucidated and appears to be a significant influence on the bioavailability of some drugs, including verapamil and cyclosporin (18).

4. *Particle Size, Surface Area, and Shape*

Experiments in the development phase should determine if particle size will be a potential problem and if a batch specification is required. For sparingly soluble drugs, particle size is an important determinant of solubility. Enhancement of bioavailability by micronisation to increase surface area was demonstrated for griseofulvin (19) and spironolactone. However, for hydrophobic powders, micronisation does not always lead to an increase in dissolution/absorption, because aggregation may occur (20). For acid-labile drugs, an increase in particle size can lead to greater degradation in the gastric milieu.

The particle size of inhalant dosage forms is an extremely important property for delivery to airways. Solution or suspension droplets from the delivery device aim to provide most of a dose with less than 5 μm particles, sometimes termed the *respirable volume* (21). For drug powder inhalers, powder rheology is complex.

5. *Crystal Properties, Polymorphs, Solvates*

Polymorphism, the different arrangement of molecules in solid state, influencing physicochemical properties, needs to be investigated in development to decide if a batch specification is needed. Often the less thermodynamically metastable form has more desirable solubility properties than the more stable polymorph. Depending on the rate of conversion of the metastable to stable form it may be used in dosage forms. Examples include cimetidine (22) and chlorpropamide (23).

Crystalline forms can also exist, with different external crystal shapes, or "habit." Often the solvent of crystallization will confer specific crystal properties that can influence solubility. Although crystalline forms of erythromycin estolate had better dissolution properties than the anhydrous amorphous form, usually the amorphous powder of a drug substance is more soluble and bioavailable than the crystalline form. Examples of better properties with amorphous powders include chloramphenicol stearate, cortisone acetate, and novobiocin (24). Also, with ampicillin, the anhydrous form was found to be more soluble and gave greater bioavailability than the trihydrate (25). However, for carbamazepine, the dihydrate has better absorption properties than the anhydrous drug (26).

6. *Stereochemical Characterization*

Stereochemical molecules should be thoroughly investigated during drug development. In general, geometric isomers and diastereoisomers are considered separate chemical entities, unless there are interconversions among them or fixed ratios are important for clinical effect or stability. For enantiomers, the FDA has a 1992 Statement (27), and the draft ICH quality guideline 6A (28) provides some advice. During development, the chemical and biological stability of the enantiomers should be confirmed and the pharmacokinetic and other pharmacological properties of the individual compounds characterized. This will define the specifications required.

7. *Stability*

Stability, a key attribute that must be defined early in the development of a new drug candidate, includes degradative chemistry and physical form changes. These can be influenced by the route of synthesis and processing; but from the biopharmaceutical viewpoint, changes in the solubility and rate of dissolution among batches are signals of potential problems. Also, if a drug is unstable in the gastrointestinal milieu, such as degradation by hydrolysis, alternatives include careful formulation (enteric coating) or a search for an ester resistant to hydrolysis. Although this attribute is defined in the Chemistry and Manufacturing submission file, such as advised in the ICH stability guidance (29), there are concerns about potential effects on biopharmaceutics.

8. *Salts, Prodrugs*

While optimizing the stability and biopharmaceutical properties of new drug candidates toward the formulation of consistent performance, the choice of free acid or base or salt form is important. Usually the salt form of the weak base or acid drug made with the strong counterion, such as chloride or sodium, respectively, is more soluble and more rapidly dissolved than is the parent compound. Examples of basic drugs for which this prevails are the hydrochloride salts of epinephrine and quinidine; whereas for acids such as barbitals the sodium salt is better dissolved. However, examples of free bases that are more soluble than the hydrochloride salt forms at gastric pH are chlortetracycline and methacycline (30), although the related tetracycline is marketed mainly as the hydrochloride salt. An interesting example is naproxen, which is formulated for chronic pain use as the acid, whereas the sodium salt (which is more rapidly dissolved) is prescribed for fast-acting analgesia.

In the preformulation assessment of inhalant dosage forms, the salt form is important (31). Many of the drugs formulated in these devices are weak bases prepared as salts, and the pH of solution with strong counterions such as chloride or sulfate could be problematic. Solutions with pH below 2 can cause bronchoconstriction, and thus weaker acid salts may have to be chosen.

Prodrugs are chemical modifications of drugs that are rapidly metabolized to active drug in the body, often at the intestinal villi. They are usually designed to be more lipid soluble and thus provide better absorption. Ampicillin esters, bacampicillin (32) and pivampicillin (33), were developed to improve the bioavailability of the active parent, ampicillin. Enalapril is the weakly active ethyl ester, converted to the diacid enalaprilat, which is a powerful angiotension-converting enzyme inhibitor (34). Valacyclovir is the L-valine ester of the active antiviral, acyclovir, which itself is poorly absorbed, and the de-esterification yields much higher serum concenrations of acyclovir by the oral route (35).

B. Excipients, Formulation Factors, and Manufacture

1. *Overview*

Drug product development has been described (36) as a highly complex craft based on a substantial amount of empirical or semiempirical information in technical areas. The dosage form often assists or enhances bioavailability, with formulation components, but these may also hinder by accident or design (such as in prolonged-release/"sustained-action" products). Formulation development has been described as having three stages: formulation screening, formulation qualification, and production (37). For *screening*, small-scale studies examine the influence of excipients, together with physicochemical properties of active substance, to target critical qualitative components and their interactions. In the next stage, *qualification*, effects of quantitative composition and potential critical variables of process are studied. Finally, in *production*, scale-up and batch-to-batch variables are examined, with validated in-process and final product quality tests, that apply specification standards developed from the knowledge obtained in the previous two stages.

2. *Formulation Components*

In most cases these are of pharmacopeial quality, although those not described in official compendia, which may derive from processed food components, should be approved by reference to lists such as the FDA "GRAS" list—generally recognized as safe. The European, Japanese, and U.S. pharmacopeias are working on a harmonization process in parallel with ICH (ICH Q4) (38) to gain some form of mutual acceptance of excipients as well as for tests and standards (reviewed later). Major excipients are among the monographs being examined in this process. It should be noted that several formulating agents, such as starch, are multifunctional in solid oral dosage forms.

a. *Excipients/Fillers*

The amount of filler used relates to the potency of the drug. For low- and submilligram-unit-dose drugs, care is required to avoid adsorbent fillers, since the small amount of drug may not be readily released after adsorption on the large filler substance. Lactose and starch are common agents, and variations in their physical properties can be problematic. For example, with salicylic acid, the type of starch used influences dissolution (39). Also, with sparingly soluble drugs, the solubility of the excipient influences the dissolution of the active substance. This was shown by comparison of the dissolution rate of griseofulvin, which was much increased when a soluble carrier, rather than a hydrophobic carrier, was used (40).

b. Disintegrating Agents

These substances are added to solid oral dosage forms to cause disintegration of the contents when ingested and exposed to aqueous gastrointestinal fluids. Common agents are processed carboxymethylcelluloses, starch (raw or processed, e.g., sodium glycollate), and Veegum.

c. Granulating Agents and Binders

These agents are added to a powdered drug to retain the integrity of the tablet or capsule contents and also to facilitate manufacturing. However, they can also affect dissolution and, thereby, absorption. Carboxymethylcellulose, gelatins, gums (such as acacia and tragacanth), polyethylene glycols, polyvinyl pyrrolidone (PVP), starch, and Veegum are commonly used agents. Banaker (41) has reviewed the choice of such substances.

d. Lubricants

These are processing aids included to improve the flow of bulk granulations and to avoid their adhesion to tablet manufacturing parts. The amount and type of mixing can affect the dissolution properties and absorption of certain drugs, because they are predominately hydrophobic compounds such as stearates and talc. Their presence, concentration, and blending with other components have been described (42) as among the most important factors influencing dissolution/absorption, because the substances can form a water-repellent coat around each granule. More recent work (43) has confirmed that the amount of magnesium stearate and the time and type of mixing (shear) can be important.

e. Surfactants (Wetting Agents)

Surface-active agents such as sodium lauryl (dodecyl) sulfate and polysorbate 80 and hydrophilic polymers such as PVP, HPMC are used to improve the dissolution of sparingly soluble drugs in tablet or capsule formulations. Both their properties and manner of addition can influence the resulting dosage form. The effect on solubilization is concentration dependent (44). Below critical micelle concentration, drug absorption is enhanced; but above this concentration, a portion of the drug associated with the micelle is unavailable for absorption. These surfactants also influence membrane permeability, and higher concentrations should be used with caution because of potential toxic effects (45).

In ophthalmic dosage forms, water-soluble polymers, such as polyvinyl alcohol and hydroxypropylcellulose, have been used at low concentrations to prolong the residence times of solutions on the eye surface and to provide low-viscosity products that do not interfere with vision (46).

f. Ointments and Skin Preparations

Preparation of topical skin preparation combines the art of formulation with physicochemical knowledge of the components. Barry (47) provides an interesting protocol for optimizing a dermatologic formulation. The key question is whether or not percutaneous penetration is desired. The diffusant, vehicle, and state of skin treated and how these factors interact "influence" the penetration. In another monograph (48), the tremendous variety of formulation types is considered, including liquid soaks, liniments, lotions, gels, creams, and ointments. These are composed of several bases, including aqueous, powder, and oily components, with thickening agents, emulsifying agents, buffers, antioxidants, and preservatives. Among the ointment bases are hydrocarbon bases (petrolatum); soft and hard paraffins; fats and fixed oils, including mono- di- and triglycerides (such as from olive, sesame, and peanut oils); silicones; water-soluble carbonates, and absorption bases such as lanolin, cholesterol, sorbitum monostearate, and wool alcohols. These can be mixed to minimize or optimize skin penetration with particular drugs, and biopharmaceutic tests have been difficult to devise.

For transdermal penetration, both skin patches and ointments have been applied. With many drugs the limiting absorption-through-the-skin barrier does not provide sufficient dosing, but it has been excellent when low, constant dosing has been required, e.g., estrogen, nicotine, and scopolamine (49). Some transdermal systems, as well as ointments, depend on the native rate-limiting skin permeability, and others include rate-limiting membranes. It is easier to devise biopharmaceutical tests for transdermal systems than for topical dermatologic agents, since with the transcutaneous absorption, the pharmacokinetics of the drug can be studied in the systemic circulation.

g. Formulation of Inhalants

As with skin preparations, the formulation of inhalation delivery systems is complex (31). These include nebulizers, metered-dose, and drug powder inhalers, each providing their own challenge to developers. The biopharmaceutical evaluation of metered-dose inhalers has been extensively researched in the last decade. Pharmacokinetic valuation is limited, since less than 20% of the delivered dose enters through the airways, the residue being swallowed.

Atkins et al. (31) examined the four separate components of metered-dose inhalers: (a) formulation (drug, excipients), (b) container, (c) metering device, and (d) actuator mouthpiece. There are two formulation types, solution and suspension, and both contain propellant. Solutions often contain cosolvents (usually ethanol).

Traditionally, chlorofluorocarbons (CFCs) were used as propellants. But, with international banning of these ozone-destroying agents, there has had to be reformulation with more ozone-friendly hydroxy- and hydroxychlorofluoroalkanes. These have different compatibilities with the surfactant dispersing agents, and their introduction to formulations has been a challenge after the toxicology screening was complete.

3. Manufacturing Process

The workshops on scale-up and postapproval changes (SUPAC) (6,50) indicate that critical manufacturing variables should be researched during formulation development. Also, the FDA project with the University of Maryland suggested a framework to map out such process variables; the approach is outlined by Shah (51). This explains that critical manufacturing variables include manufacturing, formulation, processing, and equipment variables that can significantly affect drug release from the product. Examples of formulation variables were noted in Sec. II.B.2 and include particle size and ranges of magnesium stearate. Equipment differences, such as mixing shear, can be critical, and process variables include compression forces, impeller speed, lubricant blend time, and, indeed, order and time of addition of ingredients in the "recipe." In the complex system of dosage-form manufacture, multifunctional influences affect dissolution and bioavailability, and knowledge of the key variables and validated tests and standards to control them is vital for consistent quality manufacture. Lukas (52) has provided a useful review of examples of manufacturing variables that have been shown to influence dissolution (and sometimes absorption).

a. Granulation

Granulation usually improves the dissolution performance of sparingly soluble drugs. Wet and dry granulation procedures are both applied. Although wet granulation has had disadvantages, particularly "clumping" for hydrolytically labile drugs, advances in equipment have helped to speed the process and minimize exposure times. Water content can be critical in this process, as an example with naproxen demonstrates (53). Less granulating water exhibited faster dissolution, attributed to smaller granules and less residual moisture. The order of addition of microcrystalline cellulose for granulation was shown to influence disintegration and dissolution (43).

b. Compression

In general it might be expected that harder tablets from stronger compression would tend to decrease dissolution rates. However, in fact, more complex relationships are described. In compression of powders with different particle sizes, there was an optimum size when bonding and separation of particles were in equilibrium within a tablet (54). The manner of packing of contents also affects capsules, and modern capsule machines also involve compression. Advice from Abdou (55) is that a full dissolution/compression study for each granulated formulation is essential to determine compression parameters that will ensure good dissolution characteristics and adequate bioavailability.

c. Mixing

Augsburger (43) and previous workers (42) have indicated that the amount of lubricant, magnesium stearate, is a key variable in formulation. The type of mixing was shown (56) to influence this variable in developing cefadroxil capsules. During scale-up, with a constant 1% magnesium stearate composition, a change of type of capsule-filling machine resulted in a decrease of dissolution. This was found to occur, with the different machine, by a repeated shear in the powder tamping process, leading to greater coating of the water-repellent lubricant on the drug particles.

C. Dosage Forms, Compendial Monograph Tests and Standards

Because "Dosage Forms" is the topic of a separate entry, few details or definitions are provided here. They are defined (with some variations) in the pharmacopeias. Compendial monographs (European, Japanese, and U.S. pharmacopeias) are public standards using validated tests and standards. While applications for new drug products contain proprietary information, the ICH process encourages the use of tests and standards described in these pharmacopoeias. This is particularly desired in specifications, as noted in ICH quality guideline 5 (Q6A and B) (28).

1. Dosage Forms

a. Solutions

These are used primarily in parenteral presentations. In general, aqueous solutions are considered to have few biopharmaceutic risks. Stability is a key parameter. Unless they have different components, generic parenteral solutions are waived from in vivo bioequivalence comparisons. Aqueous oral solutions are prescribed for children and the very old. Oily solutions are more subject to biopharmaceutic problems. Cyclosporin provides an example, because its formulation in oily solution resulted in erratic absorption in patients (57).

b. Suspensions

Suspensions, which may be for oral or intramuscular administration, are less straightforward in absorption properties than are solutions. Oral suspensions are common as pediatric dosage forms, and these have demonstrated bioavailability problems, as found with phenytoin (58). Micronisation of the drug resulted in a 50% increase in bioavailability. As well as particle size, viscosity and the use of suspending agents are important quality attributes of suspensions. Although dissolution tests for suspensions have been investigated, they are not yet official.

c. Capsules

Although less studied than tablets, differences in capsule formulations provided several classic examples of bioequivalence problems. The "Australian phenytoin incident" is one that resulted in disturbing clinical consequences (59). This occurred when the filler was changed from calcium sulfate to lactose, yielding higher plasma concentrations of a narrow-therapeutic-range drug that resulted in toxicity to many patients.

There are two main types of capsules: hard gelatin and soft gelatin (soft elastic). More research has examined the properties of hard capsules, which consists of a shell encapsulating powdered or granulated ingredient, which is often precompressed. The release properties are important, because the gelatin itself dissolves rapidly [although there have been crosslinking problems (60)].

Soft gel capsules have been useful as a delivery vehicle for drugs with solubility problems, such as lipophilic compounds. They can be dissolved in some cases in aqueous media with the aid of surfactants, but more usually in polyethylene glycol or oils (61). Formulation is critical, because drugs can crystallize, and this has been a very recent problem with ritonavir (62). In drug release, the rupture or burst time of the capsule is the critical point for the dissolution of contents.

d. Tablets

As the most familiar and convenient of the dosage forms used by the consumer, the often complex formulation of tablets is not evident. However, it is possibly the most studied type of formulation, and there are many examples of bioavailability problems often detected by dissolution. Compression and disintegration are the major determinants of drug release from tablets, and they depend on the eventual release of smaller particles for dissolution. In the past, sugar-coating interfered with disintegration and dissolution; but, presently, coating is rarely problematic, except with some enteric-coated products.

e. Nonoral Dosage

Reference has already been made to the formulation of parenteral, topical, and percutaneous skin and inhalation dosage forms. These range from simple solutions with little biopharmaceutic risk to highly complex drug-device systems for transdermal or metered-dose inhalation. Thorough biopharmaceutic evaluation is therefore an important part of product development.

2. Compendial Monograph Tests and Standards

As has been noted elsewhere, new drug applications involve setting standards and specifications that are acceptable to regulatory agencies but proprietary to the applicant. However, the ICH process in guideline Q6A (28) is intended to "assist, to the extent possible, in the establishment of a single set of global specifications for new drug substances and new drug products." The importance of pharmacopeial methods is promoted as follows:

> Pharmacopoeial Tests and Acceptance Criteria: References to certain methods are found in pharmacopoeias in each region. Wherever they are appropriate, pharmacopoeial methods should be utilized. Whereas differences in pharmacopoeial methods and/or acceptance criteria have existed among the regions, as harmonized specification is possible only if the methods and acceptance cri-

> teria defined are acceptable to regulatory authorities in all regions, this guideline is dependent on the successful completion of harmonisation of pharmacopoeial methods for several attributes commonly considered in the specification for new drug substances or new drug products.

Among the tests that are "essentially harmonized" are dissolution and disintegration apparatuses, and there are efforts to agree on the media and acceptance criteria for those tests. However, it has been noted that the goal of 100% harmonization "would be unrealistic and unattainable" (63).

Although, clearly, quality control tests on finished product, as well as in-process controls, have bearing on the biopharmaceutical properties, e.g., content uniformity, the primary in vitro batch tests for biopharmaceutical quality are dissolution and disintegration and, when appropriate, particle size. Dissolution and disintegration have been applied to nonoral as well as oral tablet, capsule, and suspension formulations. Thus the British Pharmacopoeia (BP) has disintegration tests for suppositories and pessaries (64), and the USP has dissolution tests for transdermal patches (65) and indomethacin suppositories (66).

a. Content Uniformity

As with most pharmacopeial tests, there is no statistically designed sampling plan for the content uniformity test. Production lot units are to be tested, and the sample size is the same for large or small production runs. Good manufacturing practice expectations and shelf life derived from stability investigations will lead to increased sampling.

The USP has a statement on the lack of sampling plan and statistical advice with their standards. It is noted that "repeats, replicates, statistical rejection of outliers or extrapolation of results to larger populations are neither specified nor proscribed by the compendia" (67). However, with its legal status in the United States and Canada, any specimen tested as directed in a monograph, at any time before its date of expiry, shall comply. The tests are somewhat arbitrary in the use of statistics. As an example, the content uniformity tests for some articles, but not others, use relative standard deviations. On the other hand, some dissolution tests have tolerances based on single-unit test results, whereas others have pooled samples.

The USP Content Uniformity tests (68) are to be applied to products containing 50 mg or less of active ingredient. For amounts above 50 mg, comprising 50% or more by weight of the dosage units, weight variation tests are substituted. For the content uniformity test, 30 units are chosen for testing, from which 10 individual units are first measured by appropriate assay. For tablets, suppositories, and suspensions, the individual unit amounts (% label claim) are obtained and the mean and standard deviation calculated. If 10 units all lie within the range of 85–115% and the relative standard deviation is 6% or less, the batch is accepted. If 1 unit is outside the 85–115% range but is within 75–125%, then the further 20 units are tested. The second stage requires that no more than 1 unit in the 30 be outside the range 85–115% but not 75–125%. The relative standard deviation for the 30 shall be not more than 7.8%. [The BP test has the same 1 in 30 tolerance but no relative standard deviation (69)].

For capsules, transdermal systems, and inhalants, the tolerance in the USP (68) is met at stage 1 if 9 of the 10 units are within 85–115% and no unit is outside 75–125%, then stage 2 requires testing of the remaining 20 units. The stage 2 tolerance is that not more than 3 of the 30 units be outside the range of 85–115% and no unit outside 75–125%. The same relative standard deviation limits prevail as for tablets.

For metered-dose inhalers (68) in the first stage, 10 units are allowed a wider tolerance of 75–125% and 1 unit up to the range 65–135%. In that case: If 2 or 3 units are outside the 75–125% range, a further 20 units are assayed and not more than 3 of the 30 are outside the 75–125% range and no unit is outside the 65–135% label claim. There are no relative standard deviation limits provided for this dosage form.

b. Particle Size

This is an active substance test applied to relatively few drugs of limited aqueous solubility when it has been determined to influence absorption from the final dosage form of the product. Griseofulvin is the classic example (USP). The decision tree II 3 in the ICH Q6A (28) document is useful for this test with the first question: Is the particle size critical to dissolution solubility or bioavailability? If the answer is yes, criteria have to be developed. Unfortunately, different tests and apparatuses for specific surface area or particle size distribution are not directly comparable. The acceptance criteria are set based on the observed range of variation of three or more lots, taking into account potential for particle growth. Acceptance criteria will include acceptable particle size distribution in terms of the percent of total particles in given size ranges. The mean, upper, and/or lower particle size limits should be well defined.

c. *Aerosol Droplet Size and Distribution*

This is a variation of the drug powder, active substance test that defines particle in its broadest sense (31) to include primary drug particles and multiparticulate aggregates. In most of the USP aerosol monographs, particle size is measured, after some priming shots, by actuating the aerosol onto a microscope slide. Particles are examined under standard magnification and field conditions to ensure that most measure less than 5 μm long and to limit those over 10 μm. The USP (70) also describes two single-stage impactor and one multistage impactor devices for measuring particle size distribution. These tests for respirable volume or dose are being introduced for better quality control of these devices. For salbutamol (albuterol in USA) pressurized inhalation, the BP (71) applies a twin-impinger device (one of the single-stage devices in the USP). The number of replicates is not stated, but not less than 35% of the average amount delivered per actuation is deposited in the lower impingement chamber (71). The European Pharmacopeia (EP) adopted the BP twin impinger as a general test for aerosols (72), and an adaptation has been proposed for nasal sprays (73).

d. *Disintegration*

The relationship between disintegration, which is a visual test, and dissolution, in which the amount of drug is measured, is complex. Since the 1970s, the USP has attempted to apply dissolution for as many as possible of its solid oral dosage form monographs and has expanded this to many nonoral formulations. In Europe, the issue of dissolution, as discussed by the BP (74), is limited to products for which bioavailability problems are more likely. This has led to ICH guideline Q6A, which states that "dissolution testing for immediate-release solid oral dosage products of very water-soluble drug substances may be replaced by disintegration testing" (28). It is apparent by comparing the USP and the BP that the former has many fewer monographs, which include a disintegration test, since generally it has been replaced by dissolution.

The disintegration apparatus for tablets and capsules is virtually identical in the United States (75), European (76), and Japanese (77) pharmacopeias. It is called the *basket rack assembly*, with six cells (tubes) held in place with two circular plastic plates and a wire mesh attached to the lower plate to retain particles above a certain size. For some tests plastic disks are inserted like pistons in the tubes. Water is the usual immersion fluid, but buffers are used in some monographs.

The set time for passing the visual endpoint test of complete disintegration varies with the monograph; in general, 30 minutes or less for uncoated tablets, sometimes longer for coated tablets, and with an "acid-resistance" time test for delayed release, enteric-coated tablets.

In stage 1 of the test (USP), six tablets are tested. If these are all disintegrated in the time stated in the monograph, the lot passes (again there is no given sampling plan). If up to two tablets fail to disintegrate, then (stage 2) a further 12 tablets are tested and the lot passes only if no more than 2 of the 18 total fail to disintegrate completely (an 11% allowance for failure).

The test for buccal tablets allows 4 hours, with the same tolerance as just stated for tablets. For sublingual tablets, each monograph gives a time, but the acceptance criteria are as just stated for tablets.

For delayed-release (enteric-coated) tablets there is a survival test for 1 hour in simulated gastric fluid. There is zero tolerance for failure among the six tablets, which should show no evidence of disintegration, cracking, or softening. The second part of the test is with simulated intestinal fluid buffer, when, according to the time in the individual monograph, the same acceptance criteria (all six in the first stage or 16 of 18 in the second stage) prevail as for tablets.

For capsules, whether hard or soft gelatin, there is a modification of the apparatus, with a removable wire mesh cloth being attached to the top (as well as the bottom) circular plates. The tolerances for all six to pass in stage 1 or 16–18 in stage 2 are the same as for tablets, according to the times in the individual monograph.

The BP describes a different apparatus used in Europe for suppositories and pessaries (78). Although the usual time is 30 minutes, for indomethacin three suppositories are tested and weighed at 90 minutes, when more than 75% must be dissolved.

e. *Dissolution*

This may be regarded as the most sensitive of the in vitro biopharmaceutic tests. It is regarded by the USP as a useful surrogate for in vivo bioequivalence (80). However, the FDA scale-up and postapproval changes (SUPAC) guidelines (15,81) apply dissolution for only minor changes of formulation, process, or equipment. Its use for bioequivalence comparisons between products is controversial. Nonetheless, it is the primary lot-to-lot quality control test for biopharmaceutic properties. Also, it is applied universally during formulation development.

The apparatuses are described in the USP (82) and the European Pharmacopoeia (EP) (83). However, because the EP does not include monographs on specific dosage forms, the BP is an example of a European compendium with specific dissolution standards.

Conventional/Immediate-Release. First introduced in the USP during the 1970s, with a modification of the disintegration apparatus just described, two major apparatuses are applied for dissolution of conventional tablets and capsules, including those with coatings (82). These are apparatus 1, called the *rotating basket* method, in which a removable cylindrical wire mesh base is attached to a spindle, and apparatus 2, the *paddle* apparatus, in which a spindle is attached to a paddle. Both rotate in water, buffer, or other medium (according to individual monographs) held at 37 ± 0.5°C. The volume is usually 900 mL; the stirring rate, usually 50 rpm (paddle) or 100 rpm (basket), is given in the individual monographs. Six cells are usually used in one assembly.

The test procedure involves a sampling plan similar to that for the disintegration tests, with sequential acceptance stages. Six tablets or capsules are tested first; if necessary, up to a total of 24 units may be tested. The pharmacopeias do not impose any other sampling plan but expect the acceptance criteria to be met with any sample from a lot (67).

For immediate- (conventional-) release units, the percentage of dose (label claim) to be dissolved (Q) and the time are stated in the individual monographs. The USP, however, for some time has had the acceptance standards shown in Table 1 (82). There is no request for relative standard deviation or any other statistical manipulation.

Table 1 USP Dissolution Test Acceptance Criteria

Stage	Number tested	Acceptance criteria
S_1	6	Each unit is not less than Q + 5%.
S_2	6	Average of 12 units ($S_1 + S_2$) is equal to or greater than Q, and no unit is less than Q − 15%.
S_3	12	Average of 24 units ($S_1 + S_2 + S_3$) is equal to or greater than Q, not more than 2 units are less than Q − 15%, and no unit is less than Q − 25%.

S_1 first stage; S_2 second stage; S_3 third stage

The USP has recently introduced the pooled-sample tolerance for some products, avoiding individual assays (84). There is again three-stage testing. The first stage is complete if the average for all six units (pooled sample) is not less than Q + 10%. If necessary, the second stage, with another 6 units added, requires the pooled average to equal or be more than Q + 5%. Finally, if necessary, a further 12 units are tested, and the pooled average of the total 24 units must equal or be more than Q.

Many manufacturers plot profiles of percent dissolved vs. time and SUPAC.MR (15) has suggested a formula for comparison of profiles f_2, the similarity factor. Polli et al. (85) have discussed other methods for comparing dissolution profiles. However these have rarely been validated as discriminating between lots of acceptable and unacceptable bioequivalence, and they are arbitrary.

For delayed-release (enteric-coated) dosage forms, an acid (gastric fluid) survival test is applied before the conventional-release test for which the acceptance criteria Table 1 are applied (82). The other type of modified-release dosage forms, extended-release, is described shortly.

Dissolution of Semisolid Dosage Forms (Such as Creams, Gels, Lotions, Ointments). This in vitro release test is suggested in the SUPAC guideline for nonsterile semisolid dosage forms (86). The apparatus includes an open-chamber diffusion cell (Franz cell) fitted with a synthetic membrane. Diffusion of drug from the topical product to and across the standard area of the membrane is obtained by sequential assay of samples taken from the receptor fluid over appropriate time intervals, usually to 6 hours. Typically, six cells are tested, and there is a random distribution of test and reference. (In this case the changed formulation vs. the original.) Usually, drugs in the aliquots are quantitated by high-pressure liquid or gas–liquid chromatography.

Because the release of drug from a standard surface is proportional to the square root of time ($\sqrt{t}$), the results are plotted as the amount of drug measured per unit area ($\mu g/cm^2$) vs. $\sqrt{t}$, yielding a straight line. The slope of this line reflects the release rate and is formulation specific. Changes of the slope can be used to monitor between-batch or between-formulation effects. Slopes should involve at least five sample times, and six replicates of each test (and reference) batch are recommended.

The results are obtained in a two-stage manner. In the first stage, six slopes each of test and reference release rates are obtained and a 90% confidence interval (90% CI) for the ratio of the median in vitro release for test and reference computed. If this 90% CI falls within the limits of 75–133.33%, no further testing is necessary. If this tolerance is not met, 12 additional slopes are obtained for each product and the 90% CI computed for the total of 18 slopes of each product (original 6 + 12 repeats). Computation from all results should yield a 90% CI of 75–133.33%. It is emphasized in the guideline (86) that the in vitro release test is useful, with the following caveat:

> However, at this time the evidence available for the in vitro:in vivo correlation of (the test) is not as convincing as that for in vitro dissolution as a surrogate for in vivo bioavailability of solid oral dosage forms. Its utility is limited to assess product "sameness" under certain limited scale-up and postapproval changes.

Dissolution of Transdermal Systems. Three apparatuses (5, 6, and 7) are described in the USP (87) for measuring the release of drug for transdermal patches. They are also described in the EP (88). The general release standards suggested include three time points, in terms of the dosing interval *D*, expressed in hours.

Apparatus 5, "paddle over disk," is a modification of the USP apparatus 2 (paddle) in which the patch is retained on the bottom of the vessel with a stainless steel disk.

Apparatus 6, "cylinder," is a modification of USP apparatus 1 in which the rotating basket is replaced with a stainless steel cylinder stirring element, with the patch being retained within the cylinder.

In apparatus 7, "reciprocating disk," a series of calibrated glass (or inert plastic) vials is set up in rows. The patch is attached to a sample holder that is then suspended from a vertically reciprocating shaker run at about 30 cycles/min. After a set time, the holder moves to the next row of calibrated vessels containing the dissolution medium. Samples are then obtained across time from the rows of calibrated vessels.

All the transdermal release media are kept at 32 ± 0.5°C. The only patch described in the USP to which these release rate tests have been applied is nicotine, and it is product specific, because all three apparatuses are applied. However, the standards differ for four products. One patch has stated limits of 35–75% in 1 h, 55–95% in 2 h, and not less than 73% in 4 h. The other extreme is a patch set to deliver 31–87% in 2 h, 62–191% in 12 h, and 85–261% in 24 h.

The tolerances are given, as with the immediate-release tablets and capsules, as a possible three-stage testing. Thus, if six units are tested and all are within the stated ranges, no further testing is needed. If some units are outside the ranges, stage 2 is run, and of the total of 12 units, no individual value can be outside by more than 10%. If not more than 2 are outside the ranges, then a further 12 units are tested; not more than 2 of the total 24 units can be outside the stated ranges by more than 10%, and none can be outside the average of the stated range by more than 20%.

Release Tests for Extended-Release Articles. The two apparatuses, 1 and 2, just described in detail are applied to testing the release from some extended-release products. However, usually the profile of the release of drug is obtained over the labeled dosage interval, usually 12 or 24 h. A minimum of three time points is chosen to test the usual 6 (first stage) up to 24 (third stage) units. In general, an early time point test for not more than 20% of label to dissolve, a second (mid dosage interval) will check for 50–60% release; a third sampling, toward the end of the dose interval, will test for 80% release. Tolerances are established in the individual monographs for accepting or rejecting the profile.

In the USP, beyond apparatuses 1 and 2, two additional apparatuses are described for extended-release articles (89). These are the *reciprocating cylinder* and *flow-through* apparatuses. The four apparatuses may be used to develop in vitro:in vivo correlations, also described in the USP. A general tolerance table is provided for extended release, with the ranges being specified in each monograph.

III. BIOAVAILABILITY AND BIOEQUIVALENCE

Bioavailability and bioequivalence of drug products emerged as important attributes and have been applied to the regulation of medicines for about 25 years. Bioequivalence was first used in the approval of generic products in Canada in the 1970s (90) and was defined there in 1973 (91). However, the publication in 1977 (2) of the U.S. regulations on bioavailability and bioequivalence was a landmark that led to abbreviated new drug applications, in which product efficacy is determined by bioequivalence to a reference listed product.

A. Definitions

Different authorities offer slight variations on the U.S. Federal Register definitions (2):

Bioavailability: The rate and extent to which the active drug ingredient or therapeutic ingredient is absorbed from a drug product and becomes available at the site of action.

Bioequivalent drug products: Pharmaceutically equivalent products that display comparable bioavailability when studied under similar experimental conditions. *Note*: The rate and extent of absorption of the test drug do not show a significant difference from the rate and extent of absorption of the reference drug when administered at the same molar dose of the therapeutic ingredient under similar experimental conditions in either a single dose or multiple doses.

Bioavailability is a pharmacokinetic attribute of a drug that depends on absorption, distribution, metabolism, and excretion after administration. *Absolute bioavailability* refers to the comparison of oral with intravenous administration of a drug. Thus, presystemic metabolism and first-pass metabolism can be distinguished from poor absorption properties. Drugs with poor absorption will show limited bioavailability. However, drugs that are completely absorbed can also demonstrate poor bioavailability, because of extensive metabolism of the parent. In the latter case, the oral dose shows only a fraction of the intravenous dose in plasma profiles.

B. Determination of Bioavailability and Bioequivalence

1. Determination of Bioavailability

Bioavailability is determined in humans after administration of a drug by measuring the concentration of the active drug and its (major) metabolites in whole blood, plasma, or serum as a function of time after dosing. It can also be estimated by measurement of the amount of drug or its metabolites excreted in urine as a function of time. As just noted, absolute bioavailability is one such key study. Also, a study of the blood and urine profiles related to dose is required to examine dose proportionality (or disproportionality) or change in pharmacokinetic parameters with change in dose (1). Other bioavailability studies submitted in NDAs apply the bioequivalence model described next. These include: comparison of pharmaceutical alternatives, such as clinical trial capsule dose form of a drug with the to-be-marketed formulation; also, dosage-strength-equivalent studies to show that equivalent doses of different strengths deliver the same amount of drug (e.g., 3 × 100-mg vs. 1 × 300-mg tablets) (1).

2. Determination of Bioequivalence

The 1992 amendments to the CFR (92) provide the following in vivo and in vitro approaches, in descending order of accuracy, sensitivity, and reproducibility, for determining the bioequivalence of a drug product.

1a. An in vivo test in humans in which the concentration of the active ingredient or active moiety and its active metabolites, in whole blood, plasma, serum, is measured as a function of time

1b. An in vitro test that has been correlated with and is predictive of human in vivo bioavailability data

1c. An in vivo test in animals that has been correlated with and is predictive of human bioavailability

2. An in vivo test in humans in which the urinary excretion of the active moiety and its active metabolites are measured as a function of time
3. An in vivo test in humans in which an appropriate acute pharmacologic effect of the active moiety and its active metabolites are measured as a function of time, if such an effect can be measured with sufficient accuracy, sensitivity, and reproducibility
4. Appropriately designed comparative clinical trials, for the purposes of demonstrating bioequivalence

In general, for drugs that are absorbed systemically, approach 1 is the expected study. With some products active against urinary tract infections, approach 2 might be acceptable. For inhalant dosage forms, such as bronchodilators, approach 3 has been applied. Clinical trials (approach 4) have been required for topical drug comparisons.

C. Design and Implementation of Bioequivalence Studies

Guidances on bioequivalence study design and implementation are available from the FDA on specific drugs. General guidelines are provided by Health Canada (93); these have also been described for the United States by Dighe and Adams (94).

The most common type of study for systemically available drugs is the *randomized, single-dose crossover in healthy volunteers*. Its major attributes include:

Crossover with a washout period

A small number of healthy adults, usually 24–36 (12 or more in Canada) (93)

Single doses of test and reference product
Appropriate sampling of blood to measure: >80% of total area under the time–plasma (blood) concentration curve (AUC)
Measures of AUC and peak plasma (blood) concentration (C_{max}) examined by statistical procedures

For cases *when urine is measured*, again in healthy volunteers, samples are collected until 95% of the drug is eliminated based on half-life. Usually urine is collected every 2 hours when a short-half-life drug is administered, but this interval can lengthen to 6 hours if sampling extends to 2 days or more. The Canadian guideline (93) indicates that >40% of unchanged drug should be eliminated in urine. The amount of drug per sampling period and the cumulative amount of drug excreted are calculated. Also the rate of excretion in urine is obtained.

For extended-release drugs and some drugs with nonproportional kinetics, *multiple-dose studies* are required:

Crossover.
Healthy volunteers (patients if necessary for ethical reasons).
Dosed to steady state, with a washout period between test and reference dosings.
Predose (C_{min}) concentrations obtained daily.
At steady-state, plasma concentrations are obtained to establish profile of drug (and sometimes metabolites) over time from last dose.
Measures of AUC, C_{max}, C_{min}, and fluctuation are obtained and examined by statistical procedures.

For *pharmacodynamic evaluations of equivalence*, studies may be carried out in patients. An example is for albuterol (95). In the case of this bronchodilator, asthmatic patients with moderate disease were admitted to the study. Details of another pharmacodynamic bioequivalence test, the vasoconstrictor assay for topical corticosteroids, is given later.

D. Statistical Analysis of Bioequivalence Studies

The most comprehensive instructions for statistical analysis of bioequivalence studies is given in the Canadian guidelines (93,96). However, Dighe and Adams (94) also provide a full description of the U.S. FDA expectations.

1. *Single-Dose Crossover Study (93)*

Analysis of variance (ANOVA) must include the randomization scheme for the design, where all subjects randomized into the study are included and identified by code, sequence, and dates of dosing periods for both test and reference.
The ANOVA should include the appropriate statistical tests of all effects in the model.
All data for all subjects, including any "dropouts" with measured results, should be presented. Supplementary analysis may be carried out with excluded subjects or data points. Exclusion must be justified.
ANOVA should be carried out on t_{max} and λ (terminal slope) data and on the logarithmically (ln) transformed AUC_T and AUC_I and C_{max} data. (AUC_T is to be the last quantifiable point; AUC_I is to infinity).
Reported results should include means and CVs (across subjects) for each product.
The ANOVA containing source, degrees of freedom, sum of squares, mean square, *F*- and p-values and the derived intra- and intersubject CVs.

Table 2 (from Ref. 93) provides the ANOVA for the crossover design for ln (AUC_T) for a 16-subject study and derived CVs.

Table 3 provides the AUC ratio estimate, at a 90% confidence interval, showing how these were calculated. ($\bar{X}$ represents the geometric mean.)

The standards for bioequivalence in the United States require that the 90% confidence interval for test product vs. reference (innovator) product for AUC and C_{max} be within the limits of 0.8 and 1.25 (or 80 and 125%). In Canada, only the AUC is required to be within this interval (for uncomplicated drugs) and the C_{max} ratio only (no confidence interval) within 80–125%.

Table 2 AUC_T (ng*h/mL) Analysis—ANOVA for ln (AUC_T)

Source	df	SS	MS	*F*	PR > *F*
Seq	1	0.0535	0.0535	0.09	0.770
Subject (Seq)	14	8.4375	0.6027	8.26	<0.001
Period	1	0.0241	0.0241	0.33	0.574
Form	1	0.1373	0.1373	1.88	0.192
Residual	14	1.0211	0.0729	—	—

Intrasubject CV = 100 × (MS Residual)$^{0.5}$ 2 = 100 × (0.0729)$^{0.5}$ = 27%
Intersubject CV = 100 × (MS Subject (Seq) − MS Residual/2)$^{0.5}$ = 100 × (0.6027 − 0.0729/2)$^{0.5}$ = 51%
Source: Ref. 93.

Table 3 AUC_T (ng*h/mL) Analysis—Calculations

Difference = Test $\bar{X}$ − Reference $\bar{X}$ = 5.39 − 5.52 = − 0.13
$SE_{Difference}$ = $(2MSResidual/n)^{0.5}$ = $(2 \times 0.0729/16)^{0.5}$ = 0.0955
AUC ratio = $100 \times e^{Difference}$ = $100 \times e^{(5.39-5.52)}$ = 88%
90% Confidence Limits:
Lower, Upper = $100 \times e^{(Difference \pm t_{0.05,14} \times SE_{Difference})}$
Lower = $100 \times e^{(-0.13-1.761\times0.955)}$ = 74%
Upper = $100 \times e^{(0.13-1.761\times0.09555)}$ = 104%

Table 4 AUC_τ (μg·h/mL) Analysis—ANOVA for ln (AUC_τ)

Source	df	SS	MS	F	PR > F
Seq	1	0.44521	0.44521	1.6849	0.21525
	14	3.69921	0.26423	12.4240	0.00001
	1	0.02422	0.02422	1.1389	0.30394
	1	0.00405	0.00405	0.1905	0.66917
	14	0.29775	0.02127	—	—

Intrasubject CV − $100 \times (MSResidual)^{0.5}$ = $100 \times (0.0213)^{0.5}$ = 15%

2. *Multiple-Dose Studies*

Again the Canadian guideline for modified release (96) provides details from multiple-dose comparisons. Similar to single-dose analysis, information on the ANOVA randomization and sequence is required. The difference is in the number of variables to be listed and tested:

Analysis of t_{max}, λ, and fluctuation = $(C_{max} - C_{min})/(AUC_\tau) \times 100\%$ should be carried out on the raw scale, while calculations for AUC_τ (over dosage interval), AUC_T, and AUC_I, C_{max} should use the logarithmic (ln) scale.

The 90% confidence interval about the mean AUC_τ, C_{min}, C_{max}, and fluctuation parameters should be provided.

Table 4 gives the analysis of variance (ANOVA) for the crossover design model for ln (AUC_τ). This analysis gives the appropriate intrasubject variance estimate, MS (Residual), for the calculation of the 90% confidence interval. Any significant effects in the model, other than Subject (Seq), should be investigated. The intrasubject and intersubject CVs should also be calculated.

The AUC_τ ratio estimate and its 90% confidence interval are derived in the calculations shown in Table 5 ($\bar{X}$ refers to geometric means). If this study had a balanced design (i.e., an equal number of subjects per sequence), the difference would simply be the difference in the arithmetic means of the ln (AUC)s. Since the study was not balanced, the least-squares mean estimate for each formulation is used to form this difference, together with the appropriate standard error.

The standards for bioequivalence (U.S.) are that AUC_τ, C_{max}, and C_{min} 90% confidence intervals for test vs. reference (innovator) product be within the limits of 0.8 and 1.25 (80–125%). In Canada, for C_{max} and C_{min} only, ratios are required to be within 80–125% and the AUC within the 90% confidence limits.

3. *Food-Effect Bioavailability and Bioequivalence Studies*

The presence of food in the gastrointestinal tract about the time of drug administration can affect absorption and other bioavailability measures for many drugs and drug products. Most regulatory authorities require food challenge information for any new extended-release product, since the dose-release mechanism may be either enhanced or impaired by food. This is particularly true of mechanisms influenced by the change of pH in moving down the gut. Karim (97) presented an excellent paper, providing several decision trees, that discussed the importance of food effect studies early in drug development. The FDA proposed guidance (98), in fact, describes the factors Karim discussed.

Both U.S. and Canadian guidelines (96) for extended-release dosage forms require single-dose studies of food effects for approval of new products. Generally a high-fat meal would be given in one leg of a study and the volunteers would be fasted in the other leg. In the case of a generic extended-release product, the study could be a four-way crossover of test and reference, both with and without food.

In the Canadian guidelines, there is no limit for bioavailability, because clinical trials will be provided for a new drug or for a first-entry modified release. However, the food challenge is informational for labeling.

Table 5 AUC_τ (μg·h/mL) Analysis—Calculations

Difference = Test $\bar{X}$ − Reference $\bar{X}$ = 8.0768 − 8.0992 = −0.0225
$SE_{Difference}$ = 0.0516
AUC_τ ratio = $100 \times e^{Difference}$ = $100 \times e^{ms0.235}$ = 98%
90% Confidence Limits:
Lower, Upper = $100 \times e^{(Difference \pm t_{0.025,14} \times SE_{Difference})}$
Lower = $100 \times e^{(-0.0225-1.761\times0.0516)}$ = 89%
Upper = $100 \times e^{(0.0225+1.761\times0.0516)}$ = 107%

Also, in the case of "dose-dumping," the product may not be accepted. In the Canadian guidelines for bioequivalence, the product is expected to meet the standards for single-dose studies with and without food, i.e., AUC_T, 90% confidence interval of test vs. reference within the range 80–125%, C_{max} mean ratio from 80% to 125%.

The U.S. draft guidelines of December 1997 examine food effects in much more detail (98). An important section in these guidelines discusses when food effect studies (in bioavailability/bioequivalence) may not be important and covers some "diagnostic factors" to determine whether additional studies of food effect are important. For a drug, such factors include:

High solubility across the intended dose range
High permeability across the intended dose range (e.g., oral absorption >90%)
Minimal or no effect of inactive ingredients or absorption of the drug substance in the fasted state (i.e., equivalent profiles in vivo of solution or suspension and test formulation)

For a drug product, the factors include:

Rapidly dissolving drug product
In vitro dissolution characteristics similar in different pH media and rotation speeds in USP apparatuses 1 and 2.

If all of these factors are positive, absence of a food effect study may be justified. If they are not all positive, then the sponsor must provide justification concerning why a food effect study is not important.

The general design for both BA and BE studies in the draft U.S. guidelines (98) is the randomized, balanced, single-dose, two-treatment, two-period, two-sequence crossover. The reference for the bioavailability study of a new drug at the Investigational New Drug (IND) stage would be solution or suspension, and this would be tested with and without food to show whether the drug substance is affected. In the preapproval phase, if the drug substance is affected greatly by food, the proposed marketing formulation should also be tested with and without food. For bioequivalence of a substantially changed formulation or a generic, the test formulation and the reference listed drug (or original formulation) should be tested in the presence of food.

The U.S. proposed guidelines (96) suggest that the decision documenting whether a food effect is absent or occurs be based on the ratio of means (population geometric means, based on log-transformed data) of fed and fasted treatments. It is absent if:

AUC, 90% CI fall within 80–125%
C_{max}, 90% CI fall within 70–143%

A food effect is concluded when either or both AUC and C_{max}, 90% CI are found to be outside the given ranges, and the sponsor is asked to indicate the clinical relevance of the magnitude of the observed data.

The Canadian modified-release (MR) guidelines (96) require only that the AUC in food effect studies be within 80–125% and that, as for fasted studies, the C_{max} ratio be within 80–125%.

It is of interest that the BioInternational meetings in 1992 (99) and 1996 (100) both had conference statements on the determination of food effects. In the latter it was noted that ethnic differences in food are less important in the determination of such effects than caloric content and protein, carbohydrate, and fat ratio (kCal distribution).

4. *Narrow-Therapeutic-Range Drugs in Canada*

For drugs designated to be of narrow therapeutic range, bioequivalence requirements are tightened in the NTR guidelines. Thus 95% confidence intervals are imposed and, for both AUC and C_{max} in fasted and fed conditions, should have test vs. reference means (ln) with 95% CI within 80–125% (101).

5. *Individual Bioequivalence*

Currently new guidelines for individual bioequivalence are under review (102,103). Studies will involve replicate design, i.e., two treatments of test and two of reference formulation (104). This will allow an estimate of intrasubject variaton (subject-by-formulation interaction). It is evident that the statistics are complicated (105) and will involve more assumptions than current regulatory models for bioequivalence. It may lead to improved standards for NTR drugs and a means to deal with highly variable drugs, especially if some scaling factor can be agreed upon.

6. *Pharmacodynamic (PD) Measures of Bioequivalence*

The U.S. CFR (2) and EU guidelines on bioavailability and bioequivalence allow use of pharmacodynamic comparisons to show equivalence of products. Other uses of pharmacodynamics involve surrogate endpoints for efficacy, such as blood pressure reduction and CD4 cell counts, and also for investigating the relationship between pharmacokinetic and pharmacodynamic variables. This section will discuss the use of PD in bioequivalence. Examples of PD use in bioequivalence include dermatologic corticosteroids (106) and albuterol metered-dose inhalers (95). The FDA guidelines on corticosteroids (106) provide great detail on setup and interpretation of studies, and the albuterol studies were discussed at an FDA Advisory Committee meeting in August 1996, where some information was made available (95).

The corticosteroid guidelines provide information on the use of the Stoughton–McKenzie vasoconstrictor bioassay (107) for assessment of topical corticosteroids. The approach relies on the effect of corticosteroids, which produce vasoconstriction, resulting in blanching of the skin. This is presumed to relate to the amount of drug entering the skin.

The method involves the application of a corticosteroid skin preparation (ointment, cream, or lotion) to skin (forearm) of healthy human subjects, where it is left for 6–18 hours. The vasoconstriction (blanching) of the skin is then monitored by a trained, blinded observer who notes the degree of blanching using a visual rating scale (either 0–3 or 0–4) at the single time point after removal of the formulation (2 hours is common). Chromameter readings of the blanching may offer a more objective assessment than the human observer (106).

The key problem of pharmacodynamic studies is that the dose response reaches an asymptote, or effect saturation, and this may be within the range of the application dose. It is therefore important to identify the linear dose response region. Also, some subjects are more sensitive in showing effects than others; i.e., responders have to be differentiated from nonresponders. The FDA recognizes this, and therefore studies to validate the method and choose responders, "pilot studies," define the pivotal study dose duration on the vasoconstriction assay:

$$E = E_0 + \frac{E_{max} \times D}{E_{max} + D}$$

where E = effect, E_0 = baseline effect, E_{max} = maximal effect, ED_{50} = half-maximal effect, and dose D is the dose at which the effect is half maximal. The pilot study seeks to establish ED_{50}, D, and D_2, which are the doses approximately one-half and two times ED_{50}, respectively. The values bracket ED_{50}, corresponding to approximately 33% and 67% of maximal response, and they are in the sensitive portion of the dose duration–response curve. Responders are those who provide visual scale readings above 1 on a scale of 0 to 3 or 4.

The data are obtained by chromameter or visual observation and corrected for baseline value at the site; and, from the individual points on the area under the effect curve, AUEC is measured, such as from 0 to 24 h. The model is fit by using all observations of individual subjects to provide ED_{50} and E_{max} values pooled from the 12 subjects.

The dose-duration information, such as ED_{50} = 2 h, D_1 = 1 h, and D_2 = 4 h, is taken from this pilot study (based on reference product values) to apply to up to 60 volunteers in the pivotal study to compare test product with reference. As well as using the ED_{50} from the pilot study, within-study day replicates are applied to volunteers, and individual dose-duration responses are obtained. The D_1/D_2 ratios of individual AUEC values obtained are accepted only if they exceed 1.25; values below this designate the volunteer as a "nondetector." Only "responders" and "detectors" are used in the comparisons of dosage forms.

The data treatment from the study includes only responders, but results from "nondetectors" must be reported.

The AUEC of, say, 0–24 h is computed for each baseline-adjusted, untreated control, site-corrected dose duration.

"Detectors" are identified from individual subjects whose AUEC values for D_1 and D_2 are both negative (responder) and meet the dose-duration response criterion of 1.25, i.e.,

$$\frac{\text{AUEC at } D_2}{\text{AUEC at } D_1} \geq 1.25$$

where AUEC at D_2 = 0.5 (AUEC at $D_{2\text{left arm}}$ + $\text{AUEC}_{\text{right arm}}$) and AUEC at D_1 = 0.5(AUEC at $D_{1\text{ left arm}}$ + $\text{AUEC}_{\text{right arm}}$)

Only subjects with complete data sets, i.e., duplicated values of D_1 and D_2 and quadriplicate values of test, reference, and untreated, should be included in the data analysis.

All data from all subjects, however, including nondetectors, should be reported (nonresponders would have been eliminated, because this is a subject exclusion criterion).

The statistical analysis can be applied only to untransformed results, because although AUEC values are usually negative, some are positive. The presence of positive and negative values prevents the use of conventional statistical transformations. Confidence intervals are calculated for Locke's exact confidence interval method.

The 90% confidence interval is obtained from the ratio of the AUEC response from test to that of reference. Both of the AUEC figures are averages of four replicates. The example from the guidelines given next shows that only 7 of 12 volunteers were found to be "detectors" (see Table 6).

The calculation of the 90% confidence interval for the data set from 7 to 12 subjects from the FDA guidelines (Chapter 1046) is given here. The data used to calculate the confidence interval are the average AUEC values of "detectors" (evaluable subjects) only. The calculation of the confidence interval is facilitated by the calculation of the following intermediate quantities:

$$\bar{X}_T = \frac{1}{n}\sum_{i=1}^{n} X_{Ti} \qquad \bar{X}_R = \frac{1}{n}\sum_{i=1}^{n} X_{Ri}$$

where n is the number of evaluable subjects, 7 in this example.

$$\hat{\sigma}_{TT} = \frac{\sum_{i=1}^{n}(X_{Ti} - \bar{X}_T)^2}{n-1} \qquad \hat{\sigma}_{RR} = \frac{\sum_{i=1}^{n}(X_{Ri} - \bar{X}_R)^2}{n-1}$$

$$\hat{\sigma}_{TR} = \frac{\sum_{i=1}^{n}(X_{Ti} - \bar{X}_T)(X_{Ri} - \bar{X}_R)}{n-1}$$

These are the sample means, sample variances, and sample covariance for the individual evaluable-subject average AUEC data. For the example, these are

$$\bar{X}_T = -23.43 \qquad \bar{X}_R = -21.56$$

$$\hat{\sigma}_{TT} = 323.13 \qquad \hat{\sigma}_{RR} = 80.10 \qquad \hat{\sigma}_{TR} = 78.83$$

We define t as the 95th percentile of the t-distribution for $n - 1$ degrees of freedom. For example, for $n = 7$, t (6 degrees of freedom) is 1.9432. Now we define

Table 6 Average AUEC Values for Subjects Meeting the Dose Duration–Response Criterion $D_2D_1 \geq 1.25$

Subject	$AUEC_{(0-24)}$ test product (average)	$AUEC_{(0-24)}$ reference product (average)
2	−48.52	−22.20
3	−38.99	−18.65
4	−7.62	−22.42
7	0.98	−10.96
9	−32.05	−37.40
11	−26.18	−26.73
12	−11.62	−12.56

$$G = \frac{t^2\hat{\sigma}_{RR}}{n\bar{X}_R^2}$$

$G < 1$ is required to have a proper confidence interval. If $G \geq 1$, the study does *not* meet the in vivo bioequivalence requirements. In the example, $G = 0.0930$.

Under the assumption that $G < 1$, we calculate

$$K = \left(\frac{\bar{X}_T}{\bar{X}_R}\right)^2 + \frac{\hat{\sigma}_{TT}}{\hat{\sigma}_{RR}}(1 - G) + \frac{\hat{\sigma}_{TR}}{\hat{\sigma}_{RR}}\left(G\frac{\hat{\sigma}_{TR}}{\hat{\sigma}_{RR}} - 2\frac{\bar{X}_T}{\bar{X}_R}\right)$$

In the example, $K = 2.791$.

The confidence interval limits may now be calculated:

$$\frac{\left(\frac{\bar{X}_T}{\bar{X}_R} - G\frac{\hat{\sigma}_{TR}}{\hat{\sigma}_{RR}}\right) \mp \frac{t}{\bar{X}_R}\sqrt{\frac{\hat{\sigma}_{RR}}{n}K}}{1 - G}$$

In the example, 90% confidence interval limits are 53.6% and 165.9%, based on the data of seven evaluable subjects.

The corticosteroid guidelines (106) state that "the Office of Generic Drugs has not determined at this time (June 1995) the equivalence interval for bioequivalence. The Office requires that an equivalence interval wider than 80–125%, as a public standard, may be necessary." In fact, this was recognized in the albuterol discussion of the Pharmaceutical Sciences Advisory Committee (95), which recommended 70–143% as an acceptable 90% CI for these particular pharmacodynamic response studies (bronchoprovocation).

There is another set of topical skin products guidelines (108) that advocates dermatopharmacokinetic measurements in place of clinical trials. The statistical calculations are identical to those for AUC, C_{max}, etc. under the preceding immediate-release bioequivalence studies, as is the 90% confidence interval width of 80–125%.

IV. DEVELOPMENT, SCALE-UP AND POSTAPPROVAL CHANGES (SUPAC), AND IN VITRO (DISSOLUTION) IN VIVO (PLASMA CONCENTRATION PROFILE) CORRELATIONS (IV/IV CORRELATIONS)

A. Introduction

The preceding sections have documented, with some examples, the in vitro properties that may influence the biopharmaceutical performance of drugs. It has been noted that solubility, permeability, and rate of dissolution are key properties of the drug substance and that dissolution behavior is the key property of the solid dosage form in leading to consistency of in vivo performance as measured by bioavailability or bioequivalence studies. During new drug or new product development (either by changes by innovator or toward introduction of a generic), there will be attempts to minimize expensive in vivo studies with intensive in vitro evaluation. Possibly, some pilot developmental bioequivalence studies will be completed before choosing the pivotal batch for bioequivalence. Such information is helpful in designing specifications (Q6A) (28) and can also assist in decisions postapproval.

With the SUPAC guidelines, the FDA has attempted to formalize decisions on postapproval changes. The Europeans give much less detail in their guidelines on variations (107), but they approach the same problems.

B. SUPAC Guidelines

1. Common Elements

The three SUPAC guidelines issued—immediate-release solid oral dosage forms (SUPAC-IR) (15), modified-release solid oral dosage forms (SUPAC-MR) (81), and nonsterile semisolid dosage forms (SUPAC-SS) (86)—contain some common features as well as advice specific to the dosage-form type. Some common features are:

- Currently relates only to scale-up and postapproval changes. (However, principles may be applied to other preapproval situations.)
- Level of changes defined—subgroups:
  - Components and composition
  - Site of manufacture
  - Batch size (scale-up or scale-down)
  - Manufacturing equipment
  - Manufacturing process

Each level of change, as well as being defined, has sections listing:

- Test documentation, comprising;
  - Chemistry
  - Dissolution
  - Bioequivalence
- Filing documentation, which includes all stability information

a. Changes in Components and Composition

Level 1. These are changes that are unlikely to have any detectable impact on formulation quality and performance.

Level 2. These are changes that could have a significant impact on formulation quality and performance.

Level 3. These are changes that are likely to have a significant impact on formulation quality and performance.

b. Changes in Site of Manufacture

Level 1. These changes consist of site changes within the same facility, but with no changes in:

- Manufacturing, components, or composition scale-up
- Equipment
- Standard operating procedures
- Environmental conditions and controls
- Personnel
- Manufacturing batch records

Level 2. These changes consist of site changes within a contiguous campus or between facilities in adjacent city blocks. There are no other changes—see Level 1.

Level 3. These changes consist of changing the manufacturing site to a different campus. Except for personnel, no other change should be made, as in Level 1.

c. Changes in Batch Size (Scale-Up/Scale-Down)

Level 1 includes changes up to and including a factor of 10 times the size of the pilot/biobatch. *However*:

- Equipment used is of the same design and operating principles.
- Manufactured in full compliance with good manufacturing practices (cGMPs).
- Same standard operating procedures and controls are used in the original and changed batches.

d. Equipment Changes

Level 1. These include changes from nonautomated or nonmechanical to automated or mechanical equipment and to alternative equipment of the same design and operating principle.

Level 2. These include changes in equipment to different design and operating principles.

There is an equipment addendum for SUPAC-IR with more details.

e. Process Changes

Level 1. These are changes such as rate of mixing, mixing times, operating speeds, and holding times within the approved application ranges.

Level 2. These are process changes as in Level 1 but outside approved application ranges.

Level 3. These are changes in the type of process used in the manufacture of the product, such as a change from wet granulation to direct compression of powder. Level 3 changes are not expected in the SUPAC-SS category (84).

f. Test Documentation

For biopharmaceutical properties in SUPAC, the major decisions in testing concern which situations require bioequivalence studies. Thus the degree of change requires a hierachy of dissolution studies; when the change is considered likely to have an impact on performance, a study of bioequivalence will be expected between the change batch and the original.

Level 1 changes for IR and MR products will not require additional dissolution studies, except for those in the approved application or in the USP for the particular article. In general, Level 3 changes demand a single-dose bioequivalence comparison of original and changed batches.

Dissolution tests in the IR document (15) are described in three "cases" of increasing complexity, A, B, and C.

Case A. This uses USP apparatus 1 at 100 rpm, or apparatus 2 at 50 rpm, with 900 mL of 0.1N hydrochloric acid. The units should meet the Q-value of 85% in 15 min according to the USP table of tolerances for dissolution (80). If changed and original batches meet this standard, it is acceptable.

Case B. This involves a multipoint dissolution profile in the apparatus (1 or 2) and medium in the application or the compendial conditions (if the article is listed), with samples taken at 15, 30, 45, 60, and 120 min or until an asymptote is reached. Profiles of changed and originals are compared using the f_2 equation (see the upcoming section on "Comparison of Profiles").

Case C. Five separate dissolution profiles are obtained in water, 0.1N hydrochloric acid, and in USP buffer media at pH 4.5, 6.5, and 7.5, with sampling times at 15, 30, 45, 60, and 120 min or until an asymptote is reached. The original and changed batch profiles are compared with the f_2 statistic. If it can be justified, a surfactant can be added to the media to generate the profiles.

2. SUPAC-IR

The biopharmaceutical drug classification system (13) is used in SUPAC-IR, and for Level 2 changes in composition for highly soluble, highly permeable drugs the dissolution case A is applied. If this fails, cases B and C dissolution should be performed. For low-permeability drugs with high solubility, a multipoint dissolution (case B) should be applied. For low-solubility, high-permeability drugs, case C should be applied. Comparisons should use the f_2 formula. For Level 3 changes in composition, case B dissolution is applied, but an in vivo bioequivalence study is also required unless there is an in vivo:in vitro correlation.

In SUPAC-IR, changes of site Level 3, batch size Level 2, and process Levels 2 and 3, the case B dissolution profile comparison is suggested. If the changes show similarity of dissolution, bioequivalence comparisons are required only for Level 3 changes in composition and process. As well as preapproval, it is likely that these principles of similarity and use of the biopharmaceutical drug classification system will extend to preapproval changes. Bioequivalence studies may be waived when in vitro:in vivo correlations have been established.

3. SUPAC-MR

The SUPAC-MR application of dissolution requires, for Level 1 change, that dissolution should be compared only by using the method in the approved application or in the USP (if the article is listed). For all Level 2 changes (composition, site, batch size, equipment, and process), extensive dissolution comparison is required. For extended release, for example, in addition to application or compendial release testing, multipoint dissolution profiles should be obtained in three other media. These may include (but are not restricted to) water, 0.1N hydrochloric acid, and pH 4.5 and 6.8 buffer. Sampling times should be appropriate, such as 1, 2, and 4 h and every 2 h until either 80% of the drug is released or an asymptote is reached.

For delayed release (enteric coated) as well as an acid survival (0.1N hydrochloric acid for 2 h), multipoint dissolution profiles are to be obtained at three different agitation speeds in buffer at pH 4.5–7.5.

Bioequivalence studies are required in the absence of an in vitro:in vivo correlation, for Level 3 changes in composition, site, and manufacturing process; this will be accepted with a single-dose study.

4. *SUPAC-SS*

For the SUPAC-SS category of product, dissolution is also considered useful for, and has been described in Section II.C.2.e. Levels 2 and 3 changes in composition or manufacturing site and Level 2 changes in equipment, process, or batch size require comparison of drug release slopes.

5. *Comparison of Profiles*

The comparison of dissolution is considered important not only for any impact on SUPAC decisions, but also for setting specifications as noted in both FDA dissolution guidelines (110) and in the ICH Q6A "specifications" document (28). The guidance on dissolution (101) has a section on comparison of dissolution profiles that describes the similarity factor, f_2, used in some decisions for SUPAC-IR and SUPAC-MR when profiles are being compared. This calculates the percentage difference between the two profiles at each time point and, with an error term, provides a measure of similarity in the percent dissolution. It is a logarithmic reciprocal, square root transformation of the sum of the squared error expressed:

$$f_2 = 50 \log \left\{ \left[1 + \left(\frac{1}{n}\right) \sum_{t=1}^{n} (R_t - T_t) \right]^{-0.5} \times 100 \right\}$$

where n is the number of time points, R_t is the dissolution value of the reference (prechange batch in SUPAC) at time t, and T_t the dissolution value of the test (change) batch at time t. Dissolution profiles of 12 units are obtained, such as with a biobatch (R) and a production lot (T), and the f_2 factor is calculated. Values for f_2 above 50 indicate a sameness of the two curves and suggest that the performance of the two batches will be similar.

The dissolution guidelines discuss other comparison procedures, including model-independent multivariate and model-dependent procedures. These have also been discussed by Polli (85).

C. In Vitro:In Vivo Correlations

Dissolution, as has already been described in several earlier sections, is a key biopharmaceutic test, and it is widely applied as a quality control test for finished products. However, its discriminating power is significantly enhanced if an in vitro:in vivo correlation can be established such that dissolution behavior can predict bioavailability and discriminate between acceptable and inequivalent batches. For highly water-soluble drug classes (high-solubility, high-permeability and high-solubility, low-permeability) when the rate-determining step for absorption is not dissolution, correlations are unlikely. However, for low-solubility, high-permeability drugs correlations are possible. Also, for extended-release drug products, in which the formulation is used to control the release of active drug, correlations are expected. This expectation is formalized in the USP (Chapter 1088) (111) as well as the EU (112) and FDA (SUPAC) modified-release guidelines. It also emerges in the ICH specifications document (28).

1. *Levels of Correlation*

Three correlation levels are defined, in decreasing order of usefulness, for predicting absorption behavior. The relationship in which the in vitro profile reflects the entire plasma drug-concentration profile resulting from the administration of the given drug product is most useful (111).

a. *Level A Correlation*

This represents a point-to-point relationship between dissolution and the in vivo input rate of the drug from the administered drug batch. The curves may be directly superimposable or depend on a constant offset value. There are different means of calculating the in vivo input rate curve, and, increasingly, mathematical deconvolution is being applied. The correlation is validated by in vitro and in vivo testing extremes of the formulation [slow, intermediate (intended for production), and fast dissolution test batches] with dissolution and from pilot bioavailability studies. A solution or immediate-release product is also given. This allows the differential determination of input curves from solution and sustained-release product to compare with the dissolution curve to provide a point-to-point correlation. The two extremes and the production lots can be used to set specifications and also in justifying postapproval changes. Thus, if a change provides a dissolution profile similar to that of the original (within the validated boundaries), the predicted bioavailability will also be expected to be similar.

b. *Level B Correlation*

This uses the statistical moment analysis. The mean in vitro dissolution time is compared to the mean residence time or the mean in vivo dissolution time calculated from time-plasma concentration profiles. However, because the *shape* of the in vivo curve is not defined from mean residence time, it is less reliable for setting specifications.

c. Level C Correlation

Because this relates only one dissolution time point, such as time for 50% to be dissolved, to one bioavailability variable, such as AUC, C_{max}, or t_{max}, it is of limited value. It is a single-point correlation and does not define the complete shape of the plasma level profile and cannot be used in predicting in vivo performance.

V. CONCLUSIONS

The essential features of biopharmaceutical characterization and testing of drug substances and products during development have been described here. From the earliest point after discovery, pharmaceutical development has focused on providing the optimum dose in the optimum time to the patient in a consistent manner. Thus, it is multidisciplinary, including study of the physicochemical attributes of the substance and the likely dosage form, the pharmacokinetic behavior in animals and humans, and the relationships to pharmacologic and clinical effects. The regulatory authorities are required to assess the development as part of safety and efficacy monitoring and to ensure that valid quality tests and specifications are appropriate to provide consistent product performance. Whereas much research has characterized the biopharmaceutics of solid oral dosage forms, the future challenge is to investigate further other types of dosage forms to assist in the management of pre- and postapproval changes.

REFERENCES

1. U.S. Food and Drug Administration, Center for Drugs and Biologics. Guidance for the Format and Content of the Human Pharmacokinetics and Bioavailability Section of an Application, February 1987.
2. Code of Federal Regulations, Title 21. CFR Part 320. Bioavailability and Bioequivalence Requirements. U.S. Government, 1977.
3. JG Wagner. J Pharm Sci 50:359–387, 1961.
4. M Gibaldi. Biopharmaceutics and Clinical Pharmacokinetics. Malvern, PA: Lea & Febiger, 1991, preface.
5. SA Kaplan. In: J Swarbrick, ed. Current Concepts in the Pharmaceutical Sciences: Dosage Form Design and Bioavailability. Philadelphia: Lea & Febiger 1973, pp 1–31.
6. JP Skelly, GA Van Buskirk, DR Savello, GL Amidon, HM Arbit, S Dighe, MB Fawzl, AW Malic, M Malinowski, R Nedich, GE Peck, V Shah, RF Shangraw, JB Schwartz, J Truelove. Pharm Res 10:313–316, 1993.
7. J Wood, J Syarto, H Letterman. J Pharm Sci 54:1068, 1965.
8. Pharmacopeial Forum 24:5617–5619, 1998.
9. M Gibaldi. Biopharmaceutics and Clinical Pharmacokinetics. Malvern PA: Lea & Febiger, 1991, p 40.
10. PA Shore, BB Brodie, CAM Hogben. J Pharmac Exp Ther 119:361–369, 1957.
11. LA Benet. In: JG Hardman, LE Limbird, eds. in chief. Goodman and Gilman's The Pharmacological Basis of Therapeutics. 9th ed. New York: McGraw Hill, 1996, p 6.
12. SA Kaplan, S Cotler. J Pharm Sci 61:1361–1364, 1972.
13. GL Amidon, M. Lennernas, VP Shah, JR Crison. Pharm Res 12:413–420, 1995.
14. H Lennernas, Ö Ahrenstadt, R Hällgren, L Knutson, M Ryde, LK Paalzow. Pharm Res 9:1243–1251, 1992.
15. U.S. FDA. Center for Drug Evaluation and Research. Guidance for Industry Scale-Up and Post Approval Changes Immediate Release Solid and Dosage Forms, November 1995.
16. AS Hussain. APPS/CRS/FDA Workshop on Scientific Foundation and Application for the Biopharmaceutical Classification System and In Vitro In Vivo Correlations. Crystal City, VA, 1997.
17. BH Hirst. Eur J Pharm Sci 1997 5(suppl 2):56–57.
18. H Lennernas. J Pharm Sci 87:403–410, 1998.
19. RM Atkinson, C Bedford, KJ Child, EG Tomich. Nature 193:588–589, 1962.
20. P Finholt, H Kristiansen, O Schmidt, K Wold. Medd Norsk Farm Salsk 28:17–20, 1966.
21. I Gonda, PR Byron. Drug Dev Indust Pharm 4:243–250, 1978.
22. J Shibata, H Kokubu, L Morimoto, K Morisaka, T Ishida, M Inoue. J Pharm Sci 72:1436–1438, 1983.
23. H Ueda, N Nambu, T Nagai. Chem Pharm Bull 32: 244–246, 1984.
24. U Banakar. Pharmaceutical Dissolution Testing. New York: Marcel Dekker, 1992, p 137.
25. J Poole. Drug Inf Bull 3:8–16, 1968.
26. H Meinsardi, E Van den Kleijn, J Van den Meijer, H Van Rees. Epilepsia. 16:353–358, 1975.
27. Food and Drug Administration, Center for Drug Evaluation and Research. Policy Statement for the Development of New Stereoisomeric Drugs, May 1992.
28. International Conference on Harmonization of Technical Requirements for Registration of Pharmaceuticals for Human Use (ICH). Quality Guideline 6A. Specifications: Test Procedures and Acceptance Criteria for New Drug Substances and New Drug Products: Chemical Substances. July 1997.

29. ICH Quality Guideline. Q1 Stability. A. Stability Testing of New Drugs and Products. 1993.
30. HM Abdou. Dissolution, Bioavailability and Bioequivalence. Easton, PA: Mack, 1989, p 57.
31. PJ Atkins, NP Barker, D Mathisen. In: AJ Hickey, ed. Pharmaceutical Inhalation Aerosol Technology. New York: Marcel Dekker, 1992, pp 157–170.
32. Anonymous. Med Lett 23:49, 1981.
33. EL Foltz. Antimicrobial Agents Chemother 14:442–446, 1970.
34. EH Ulm, M Hichens, HJ Gomez, AE Till, E Hand, TC Vassil, J Biollaz, HR Brunner, JL Schelling. Br J Clin Pharmacol 14:357–362, 1982.
35. MA Jacobson, J Gallant, LH Wang, D Coakley, S Weller, D Gary, L Squires, ML Smiley, MR Blum, J Feinberg. Antimicrobial Agents Chemother 38:1534–1540, 1994.
36. J Robinson. Overview of Issues. Proceedings of workshop on in vitro dissolution of immediate-release dosage forms: Development of in vivo relevance and quality control issues. Drug Inform J 30:1030, 1996.
37. E Dahl. Quoted in IJ McGilveray, Drug Inf 30:1029–1037, 1996.
38. ICH, Quality Guideline 4 (Q4), Pharmacopoeial Harmonization, 1997.
39. T Underwood, DE Cadwallader. J Pharm Sci 61:289–241, 1972.
40. M Westerberg, B Jonsson, C Nyström. Int J Pharm 28: 18–23, 1986.
41. U Banakar. Pharmaceutical Dissolution Testing. New York: Marcel Dekker, 1992, p 149.
42. L Bergman, F Bandelin. J Pharm Sci 54:445–449, 1965.
43. GS Rekhi, ND Eddington, MJ Fossier, P Schwartz, LJ Lesko, LL Augsburger. Pharm Dev Technol 2:11–24, 1997.
44. S Reigelman, WJ Crowell. J Amer Pharm Assoc 47: 115–122, 1958.
45. HM Abdou. Dissolution Bioavailability and Bioequivalence. Easton, PA: Mack, 1989, p 82.
46. JE Robinson. In: LF Prescott, WS Nimmo, eds. Rate Control and Drug Therapy. Edinburgh: Churchill Livingstone, 1985, pp 71–82.
47. BW Barry. In: RL Bronaugh, HI Maibach, eds. Percutaneous Absorption: Mechanisms, Methodology, Drug Delivery. New York: Marcel Dekker, 1985, pp 506–507.
48. BW Barry. Dematologic Formulations: Percutaneous Absorption. New York: Marcel Dekker, 1983, pp 295–302.
49. JE Shaw, F Theeuwes. In: LF Prescott, WS Nimmo, eds. Rate Control and Drug Therapy. Edinburgh: Churchill Livingston, 1985, pp 65–70.
50. JP Skelly, GA Van Buskirk, HM Arbit, GL Amidon, LL Augsburger, WH Barr, S Borge, J Clevenger, S Dighe, M Fawzi, D Fox, MA Gonzalez, VA Gray, C Hoiberg, LJ Leeson, L Lesko, H Malinowski, PR Nixon, DM Peace, G Peck, S Porter, J Robinson, DR Savello, P Schwartz, JB Schwartz, VP Shah, R Shangraw, F Theeuwes. Pharm Res 10:1800–1805, 1993.
51. VP Shah. Drug Inf J 30:1085–1089, 1996.
52. G Lukas. Drug Inf J 30:1091–1104, 1996.
53. MS Gordon. Drug Dev Ind Pharm 10:11–29, 1994.
54. J Carless, A Sheak. J Pharm Pharmacol 28:17–18, 1976.
55. HM Abdou. Dissolution, Bioavailability and Bioequivalence. Easton PA: Mack, 1989, p 93.
56. I Ullah, GI Wiley, SN Agharkar. Drug Dev Ind Pharm 18:895–910, 1992.
57. A Johnston, JT Marsden, KK Hla, JA Henry, DW Holt. Brit J Clin Pharmacol 21:331–333, 1986.
58. PJ Neuvonen, PJ Pentikainen, SM Elving. Int J Clin Pharmacol Biopharm 15:84–89.
59. F Bochner. J Neurol Sci 16:48–51, 1972.
60. GK Shiu. Drug Inf J 30:1045–1054, 1996.
61. F Morton. Quoted by IJ McGilveray. Drug Inf J 30: 1029–1037, 1996.
62. Abbot Labs. Quoted in Wall St. Journal, July 28, 1998, p B5.
63. TL Cecil, WL Paul, LT Grady. Pharmacopeial Forum 23:3895–3902, 1997.
64. British Pharmacopoeia, 1993, pp 159–160.
65. United States Pharmacopeia 23, 1995, pp 1796–1798.
66. United States Pharmacopeia 23, 1995, p 803.
67. United States Pharmacopeia 23, 1995, p 9.
68. United States Pharmacopeia 23, 1995, 7th Supplement, pp 3894–3895.
69. British Pharmacopoeia, 1993, p 753.
70. United States Pharmacopeia, 23, 1995, pp 1763–1767.
71. British Pharmacopeia, 1993, p 1091.
72. European Pharmacopoeia. Strasbourg, European Pharmacopoeia Secretariat, Council of Europe, 1997, p 137.
73. JM Aiache, E Beyssac. Pharmeuropa 9:561–565, 1997.
74. British Pharmacopoeia, 1993, Addendum, 1995, pp A416–A419.
75. United States Pharmacopeia, 23, 1995, pp 1790–1791.
76. European Pharmacopoeia, 3rd ed. Strasbourg, European Pharmacopoeia Secretariat, Council of Europe, 1997, p 126.
77. Japanese Pharmacopeia, X ed. (English 1991).
78. European Pharmacopoeia, 3rd ed. Strasbourg, European Pharmacopoeia Secretariat, Council of Europe, 1997, p 127.
79. British Pharmacopoeia, 1993, p. A159.
80. United States Pharmacopeia, 23, 1995, pp Ivi.
81. FDA, Center for Drug Evaluation and Research. Guidance for Industry SUPAC-MR: Modified Release Solid Oral Dosage Forms, 1997.
82. United States Pharmacopeia, 23, 1995, pp 1791–1793.
83. European Pharmacopoeia, 3rd ed. Strasbourg, European Pharmacopoeia Secretariat, 1997, pp 128–131.

84. United States Pharmacopeia, 23, 1995, 7th Supplement (1997), p 3980.
85. JE Polli, G Singh Rekhi, VP Shah. Drug Inf J 30: 1113–1120, 1996.
86. FDA Center for Drug Evaluation and Research. Guidance for Industry. Non Sterile Semi Solid Dosage Forms. Scale-up and Post-Approval Changes. 1997.
87. United States Pharmacopeia, 23, 1995, pp 1796–1799.
88. European Pharmacopeia, 3rd ed. Strasbourg, The European Pharmacopoeia Secretariat, Council of Europe, 1997, pp 131–133.
89. United States Pharmacopeia, 23, 1995, pp 1793–1796.
90. IJ McGilveray. In: PG Welling, FLS Tse, SV Dighe, eds. Pharmaceutical Bioequivalence. New York: Marcel Dekker, 1991, pp 382–383.
91. AB Morrison, D Cook, WGB Casselman. Can Med Assoc J 109:800–802, 1973.
92. Federal Register 21 CFR Part 2, April 28, 1992, pp 17981–18001.
93. Health Protection Branch, Drug Directorate Guidelines, Conduct and Analysis of Bioavailability and Bioequivalence Studies. Part A: Oral Dosage Formations used for Systemic Effects, Ottawa, Health Canada, 1992.
94. SV Dighe, WP Adams. In: PG Welling, FLS Tse, SV Dighe, eds. Pharmaceutical Bioequivalence. New York: Marcel Dekker, 1991, pp 347–380.
95. Proceedings of the FDA, Center for Drug Evaluation and Research, Pharmaceutical Advisory Committee, August 11, 1996.
96. Health Canada, Drugs Programme Guidance. Conduct and Analysis of Bioavailability and Bioequivalence Studies—Part B: Oral Modified Release Formulations. Ottawa, Minister of Public Works and Government Services Canada, 1996, pp 35–130.
97. A Karim. In: KK Midha, T Nagai, eds. Bioavailability, Bioequivalence and Pharmacokinetic Studies. International Conference of the FIP: BioInternational '96. Tokyo, Business Center for Academic Societies Japan, 1996, pp 221–230.
98. Food and Drug Administration, Center for Drug Evaluation and Research. Draft: Guidance for Industry. Food-Effect Bioavailability and Bioequivalence Studies. October 1997.
99. Conference Statement. In: KK Midha, HH Blume, eds. Bioavailability, Bioequivalence and Pharmacokinetics: Proceedings of BioInternational 1992. Bad Homburg, Germany. Stuttgart: Medpharm, 1993, pp 21–23, 209–251.
100. Conference Statement. In: KK Midha, T Nagai, eds. Bioavailability, Bioequivalence and Pharmacokinetic Studies: Proceedings of FIP BioInternational 1996. Tokyo, Business Center for Academic Societies, 1996, pp 5–6.
101. Standards for Comparative bioavailability studies involving drugs with a narrow therapeutic range—oral dosage forms. Therapeutic Products Programme, Ottawa, Health Canada, Draft, May 7, 1997.
102. FDA, Center for Drug Evaluation and Research. Draft Guidance for Industry. In Vivo Bioequivalence Studies Based on Population and Individual Bioequivalence Approaches. October 1997.
103. R Williams. Background to the Workshop. Proceedings of AAPS Workshop on Scientific and Regulatory Issues in Product Quality. Narrow therapeutic Index Drugs/Individual Bioequivalence, Crystal City, VA, 1998.
104. G Ekbohm, H Melander. Biometrics 45:1249–1254, 1989.
105. R Schall, HE Luus. Stat Med 12:1109–1124, 1993.
106. FDA, Center for Drug Evaluation and Research, Guidance for Industry. Topical Dermatologic Steroids: In Vivo Equivalence. June 2, 1995.
107. RB Stoughton. In: HI Maibach, C Surber, eds. Topical Corticosteroids. Basel: Karger, 1992, pp 42–53.
108. FDA, Center for Drug Evaluation and Research. Draft Guidance for Industry. Topical Dermatologic Drug Product NDAs and ANDAs—In Vivo Bioavailability, Bioequivalence, In Vitro Release, and Associated Studies, June 1988.
109. European Economic Community Regulation no. 541/95 Provisions Relating to Variations in Marketing Authorization.
110. FDA Center for Drug Evaluation and Research. Guidance for Industry. Dissolution Testing of Immediate Release Solid Oral Dosage Form, August 1997.
111. United States Pharmacopeia, 23, 1995, pp 1924–1929.
112. Commission of the European Communities. Notes for Guidance. Quality of Prolonged Release and Transdermal Dosage Forms, Section II: Quality of Modified Release and Transdermal Dosage Forms, 1998.

Iain J. McGilveray

Bracketing Design

See also *Stability Matrixing Designs*

Bracketing is a special term used in stability design (see *Stability*). Since the amount of data required to demonstrate the stability characteristics of a product can often be large and expensive to generate, methods to reduce the size of stability studies have been proposed. As defined in the Glossary of the ICH Stability Guidelines Q1A (1), *bracketing* is: "The design of a stability schedule so that at any time point only the samples on the extremes, for example, of container size and/or dosage strengths, are tested." The Glossary further explains that the design assumes that the intermediate condition samples are represented by the extremes and that bracketing may be particularly applicable where strengths are related closely in composition, such as same formulation but different tablet sizes or different amounts filled into capsules. For bracketing across containers, the guideline indicates that bracketing designs may be applicable if the composition of the container and the type of closure are the same. The Food and Drug Administration (FDA) draft stability guidelines (2) provide more detail about when bracketing may be applied without justification and when bracketing must be justified with theory and/or data. The FDA draft guideline provides an example, reproduced in Table 1, of a bracketing design for a case where there are three strengths made from a common granulation to be packaged in three different sizes of HPDE bottles with different counts: 30 counts, C1; 100 counts, C2; and 200 counts, C3.

The FDA draft guidelines also cautions that if the statistical analysis determines that there is dissimilarity between the extremes, the intermediate sizes/strengths should be considered to be no more stable than the least stable extreme. This could be a significant drawback to the use of bracketing depending of course, on the particular situation.

It seem clear that bracketing offers the potential for reducing the number of samples in registration stability studies. The question naturally arises whether bracketing has any application in validation studies and marketed product stability studies. For example, would it be possible to bracket across packages in the validation? Even if the all validation lots must be produced, would it be possible to place only some of the package/strength combinations on stability using bracketing? Since bracketing is a relatively new idea for the design of stability studies, these types of questions will continue to surface and will be discussed over the next few years.

Bracketing seems to be an idea that many stability studies can utilize to good advantage. However, it is also clear that the regulatory agencies worldwide will move forward with some caution, and proposed protocols will need to be cleared almost on a case by case basis at first. The ICH Guidelines and the FDA Draft Stability Guidelines both encourage consultation with the appropriate regulatory body to get agreement about applying bracketing in the stability study before the protocol is executed.

REFERENCES

1. International Conference on Harmonization. "Harmonized Tripartate Guidelines: Stability Testing of New Drug Substances and Products" Q1A (Step5 document effective 1 January 1998).
2. Food and Drug Administration (CDER and CBER). "Guidance for Industry: Stability Testing of Drug Substances and Drug Products," Draft Guidance, June 1998.

John R. Murphy

Table 1 Example of a Bracketing Design

Batch	1									2									3								
Strength	100 mg			200 mg			300 mg			100 mg			200 mg			300 mg			100 mg			200 mg			300 mg		
Container/ Closure	C1	C2	C3	C1	C2	C3	C1	C2	C3	C1	C2	C3	C1	C2	C3	C1	C2	C3	C1	C2	C3	C1	C2	C3	C1	C2	C3
Sample on Stability	X		X				X		X	X		X				X		X	X		X				X		X

C

Cancer Trials

See also *Clinical Trials*

I. INTRODUCTION

Due to the severity of many types of cancer, the critical need for improvement in therapies, and the large amount of research in this area, efficient designs and analyses for cancer clinical trials are of great importance. This review presents the major statistical issues encountered in cancer clinical trials, summarizes the standard approaches, and presents recent proposals regarding the design and analysis of cancer trials. The review is organized by clinical trial phases, followed by a discussion of statistical analysis methods and efficacy endpoints in cancer trials.

II. PHASE I CLINICAL TRIALS

A. Introduction

The primary objectives of phase I cancer trials are to estimate the maximum tolerated dose (MTD) of the compound using a specific schedule and then to select a dose level for subsequent, phase II trials. In addition, the trials provide information regarding the types and severities of toxicities, the pharmacology of the compound, and initial data on efficacy. Because of the toxicities of cancer compounds, participants of phase I cancer trials are typically patients with advanced stages of the disease, patients for whom alternative treatment options have been exhausted, rather than normal volunteers. Also, the dose level for efficacy is often limited by toxicity. The need to avoid toxicity and yet treat patients at a potentially therapeutic dose presents a challenge in the design of phase I trials. Multiple phase I trials are typically completed for a given compound in order to investigate effects from various dose and schedule combinations or to investigate combinations of treatments.

B. The Traditional Design

1. Selection of Dose Levels

The initial dose level for phase I trials is typically 1/10 of the LD10 in mice, in mg/m^2 units, although it may be smaller, based on toxicities observed in other species. This dose is expected to be a conservative dose, not causing any significant toxicity. Additional dose levels, based upon the modified Fibonacci series, are increments of 2.0, 1.67, 1.5, 1.4, 1.33, (and 1.33 thereafter) times the preceding dose (1–3).

2. Dose Escalation Rule

Prior to the trial, specific criteria for determining "significant toxicity" are selected. Typically this is based on toxicity grades, such as toxicity grade 3 or worse. The traditional design (2,4,5) begins by assigning three patients to an initial dose level. If no significant toxicity is observed in any of the three patients treated at a particular dose level, then the next three patients are treated at the next higher dose level. If a toxicity is observed in one of three patients at a particular dose level, then three additional patients are assigned to the same dose level. If none of the additional three patients experience significant toxicity, then the next three patients are assigned to the next higher dose level. Once two patients experience significant toxicities at a particular dose level, the trial ends. That dose level, or the dose level below, is selected as the MTD. Because this is a sequential procedure, the sample size varies. Typically, at most 25 patients are included in a phase I trial, although if the starting dose is too conservative, this number could be higher.

3. *Issues with the Traditional Design*

While the traditional approach has been the standard for many years, it has been increasingly criticized for multiple reasons (6–8). If toxicity in man is proportional to that observed in animal studies, then seven dose escalations are required to reach the MTD using the modified Fibonacci series approach. A review of past trials (7) indicated high variability in the number of dose escalations needed, and several trials requiring at least 12 escalations were noted. The result of a large number of escalations is that a large number of patients will receive a subtherapeutic dose level. In addition, the length of the study and the extended use of limited research resources results in a delay in bringing new compounds to the market and a reduction in the number of new compounds that can be tested in man. The final estimate of the MTD from a traditional design also has poor statistical properties. One has little information regarding the proportion of patients who would have a toxicity at the traditional design estimated MTD. It does not estimate any particular percentile of the dose–toxicity function and is dependent upon the starting dose, the true dose–toxicity function, and the number of dose levels examined. In addition, no accounting for sampling error is taken into account when estimating the MTD. Poor estimates of the MTD can lead to poor selection of doses for patients in subsequent trials, resulting in delays or potential discontinuation of promising compounds. For these reasons, several novel designs for phase I cancer trials have been proposed over the last 12 years.

C. New Designs

1. *Pharmacologically Guided Designs*

In pharmacologically guided designs (6,7,9), dose escalations are based upon pharmacokinetic analyses. The goal of these designs is quickly to escalate patients to achieve a concentration similar to the target concentration based on animal studies. The underlying premises are that toxicity is correlated with concentrations of drug in plasma and that the relationship between toxicity and drug exposure hold across species. A review of previous studies indicated that the relationship between drug concentrations and toxicity from mouse to man is more consistent across compounds than the relationship between doses and toxicity. Retrospective analyses demonstrated a significant savings in numbers of patients and time to complete phase I trials as compared to the traditional approach. To utilize such a procedure, one must have a sensitive drug assay, and interspecies pharmacodynamic differences must be small. Results of trials using pharmacologically guided designs are included in the references (10,11).

2. *Modified Up-and-Down Designs*

Several procedures, based upon the original up-and-down procedure (12), have been proposed (13). These procedures address the issues of treating too many patients at subtherapeutic doses and the lengthy time to complete traditional trials by allowing for more flexibility in the number of patients required prior to dose escalation and in the increments between dose levels and by allowing decreases in dose levels.

Proposals have included two-stage designs. Two-stage designs incorporate an accelerated escalation scheme until the first significant toxicity is observed, and then return to a more conventional escalation scheme. Proposed accelerated escalation schemes include keeping 100% increments between dose levels and/or allowing an increase in dose after a single nontoxic response or after two nontoxic responses. The MTD is defined as a specified percentile of the dose–toxicity function; once the trial is complete, the MTD may be estimated by fitting a logistic dose–toxicity model. Simulation studies have demonstrated that these designs can reduce the number of patients treated at subtherapeutic levels, reduce the time needed to complete trials, and provide improved estimates of the MTD as compared to traditional trials. The main concern with these approaches is the increased risk of treating patients at highly toxic dose levels, although simulation studies suggest this increase is limited.

Recently, a two-stage up-and-down design incorporating data from multiple courses of treatment as well as intrapatient dose escalation has been proposed (3). In this approach, an accelerated escalation in stage I is discontinued when any significant toxicity is observed (any course for any patient) or when the second instance of moderate toxicity is observed. Simulations indicate that this approach does not increase the chance of administering a highly toxic dose, but has lesser gains in reduction of patients at subtherapeutic levels and time to study completion than other up-and-down designs.

3. *Continual Reassessment Method and Modifications*

The continual reassessment method (CRM) (8) utilizes a Bayesian approach to estimate the MTD after each successive patient. The MTD is first defined as a percentile of the dose–toxicity function. In addition, a prior distribution for the MTD and a dose–toxicity model is selected based upon prior data and/or expert opinion (or a noninformative prior may be selected). After each successive patient, an updated estimate of the MTD is obtained by computing the posterior distribution of the MTD. Each patient is then assigned to the dose level closest to the current estimate of the MTD. With this approach, each patient is treated close to the estimate, based on all currently available information, of the MTD. Thus, a reduced number of patients are treated at subtherapeutic levels. In addition, simulations have demonstrated a reduced number of patients needed to complete the trial as well as a more efficient estimate of the MTD (8,14). As with the accelerated up-and-down designs, concerns have been raised that the CRM will result in treating more patients with toxic dose levels. This is especially true with the CRM, because the proposed starting dose is the prior estimate of the MTD, not necessarily a conservative dose level. Another criticism with the CRM is the need to obtain a prior distribution, and the added complexity in computations that make implementation difficult without statistical support.

Several modifications of the CRM have recently been proposed (15–18). These proposals attempt to retain many of the advantages of the CRM while limiting the potential for administering toxic doses. Modifications to the CRM include starting at a conservative dose level as in the traditional approach, limiting dose escalations to those allowed in the traditional approach, always using the closest dose level lower than the current MTD estimate, and using an initial up-and-down procedure prior to the CRM. The references provide one example of a recent completed trial using a modified version of the CRM (19). To address the criticism with the prior distribution and the complexity of the Bayesian computations, recent literature proposes the use of maximum-likelihood estimation (20,21). To obtain the updated estimates of the MTD after each successive patient, maximum-likelihood estimation using a specified dose–toxicity function is performed. However, the use of a seed dataset or some other initial dose escalation is required as unique maximum-likelihood estimates will not exist until certain data conditions are met (22). The advantage of this approach is the ease in computations as compared to the CRM. Drawbacks include the need for a seed dataset, just as the CRM requires the use of a prior distribution.

4. *Other Recent Work*

Other recent proposals include approaches incorporating all levels of observed toxicity grades (rather than dichotemizing into the presence or absence of significant toxicity) (23) and using estimates of parameters of pharmacodynamic models to guide the dose escalation (24–26).

D. Discussion

In summary, phase I cancer trials provide ethical and statistical challenges different than phase I trials in normal volunteers. The traditional method uses a conservative starting dose and escalates using cohorts of three patients at each dose. Recent literature has demonstrated the poor statistical properties of the traditional approach and is full of novel approaches offering varying improvements. The selection of a procedure for a particular trial depends upon many factors, including the compound being studied, the amount and type of available information regarding the toxicities of the compound, the type of toxicities expected, and the patient population for the trial. Due to the small sample sizes, simulation studies are typically required in order to compare competing procedures and to determine the operating characteristics of a procedure. In most of the newer approaches, the MTD is defined as a percentile of the dose–toxicity relationship. Thus, statistically the problem reduces to estimating the percentile of a distribution function. The choice of the specific percentile to be estimated again depends on many factors, although values from .2 to .33 have been cited in the literature.

III. PHASE II CLINICAL TRIALS

A. Introduction

The primary objective of a phase II cancer trial is to determine whether a new experimental treatment is sufficiently active, relative to standard treatments, to warrant the conduct of a large, randomized phase III comparative trial. Phase II cancer studies are important for two reasons. First, they are numerous and, as a result, many cancer patients receive their primary treatment in the context of such studies. Second, they often determine if and what phase III studies are ultimately conducted. Phase II trials are of two basic types. The first type are trials to determine whether a single agent has antitumor activity against a particular type of cancer and provide a rough estimate of the level of that activity. The second type includes phase II trials whose objective is to determine whether the level of activity of a particular combination of active agents is sufficiently promising to warrant a randomized phase III evaluation of the combination against the standard treatment. The first type of phase II trials has received considerable statistical attention in the literature. However, the statistical methods developed for the first type of phase II trials have not been entirely adequate for the second type of phase II trials. Phase II combination trials require explicit consideration of the results of the standard treatment. This comparative nature of phase II trials of combination regimen has often been ignored. For more details, see Ref. 27.

B. Phase II Cancer Clinical Trial Designs

Initiation of phase II clinical trials generally assumes that a "safe" dose regimen has emerged from the phase I clinical trial experience. The selected dose regimen's efficacy must then be tested in phase II patients with advanced disease while further characterizing its safety profile. Therapeutic efficacy often is evaluated primarily on the basis of "regression probability," i.e., the probability that an eligible patient receiving the treatment regimen will experience a tumor regression as defined in the protocol. Historically, phase II study design strategy has been "modality oriented" as well as "disease oriented." With the modality-oriented strategy, patients with different primary tumor types were treated with the same dose regimen. Some of the oldest anticancer agents, such as cyclophosfamide, 5-FU, methotrexate, and adriamycin, were initially developed with this strategy. The most significant issue with the modality-oriented strategy was that factors associated with a given disease, such as cancers of lung, pancreas, head and neck, or breast, introduced bias in the trial results. Increasingly refined understanding of prognostic factors associated with specific types of cancers has rendered the modality-oriented strategy no longer viable and strengthened the case for disease-oriented strategy. With the disease-oriented strategy, the investigator determines several key factors that can affect the outcome of the trial. They include the type of disease to study, where in the course of disease treatment to intervene, and what end result is deemed appropriate for the trial. Prognostic factors that may affect response to treatment are then accounted for in the trial design. Almost all of the statistical methodology put forward for phase II cancer trials has been developed under the disease-oriented strategy.

1. *The Single-Stage Design: The Classical Approach*

The classical approach for designing and conducting phase II cancer trials has been the single-stage design (28). Typically, a total of N evaluable patients are accrued, treated, and observed. Then the number, S, of patients who experience a tumor regression is determined. From the data, one can obtain the point estimate of the incidence of tumor regression probability or tumor response rate, given by the ratio S/N. In addition to the derivation of tumor response rate, it is commonly of interest to formulate a testing procedure to decide whether or not the treatment regimen deserves further investigation. This exercise amounts to specifying in addition to the total number of evaluable patients N, the single-stage design parameters C. The choice of C is made so that if more than C of the N evaluable patients are tumor responders, the new treatment regimen is declared effective, and ineffective otherwise.

2. *The Single-Stage Design: The Bayesian Approach*

Phase II clinical trials, typically single-arm studies with the objective to decide whether a new treatment E is sufficiently promising relative to a standard treatment, S, are inherently comparative even though a standard therapy arm is usually not included. Rarely is the uncertainty about the response rate θ_S of S made explicit, either in planning the trial or interpreting the results. Thall and Simon (29) have put forward some practical Bayesian guidelines for deciding whether the new therapy E is promising relative to the standard therapy S in the setting where patient response is binary and the data are monitored continuously. The rationale for such an approach stems from practical observation. In order to implement a phase II trial, the clinician must specify a single value of the patient's response rate θ_S to S. In many cases, there is uncertainty regarding θ_S. For example, it is not uncommon for a clinician to give a range of values when asked to provide the response rate of the standard therapy S. Under such circumstances, it would be in fact inappropriate to insist on a single value of θ_S for planning purposes, since the resulting design and analysis would treat a variable quantity as if it was a constant. The authors point out that a more realistic approach should explicitly account for the clinician's uncertainty regarding θ_S in the planning of the phase II trial and the interpretation of the results. Briefly, the efficacy of an experimental treatment E is evaluated relative to that of a standard treatment S based on data from an uncontrolled trial of E, and informative prior for θ_S, a targeted improvement for E, and a noninformative prior for θ_E. A noninformative prior for θ_E is appropriate here, since in practice little is known about the efficacy of E prior to the trial, and one wishes to avoid designs that allow strong prior opinion to be used in lieu of empirical evidence. No explicit specification of a loss function is required. The trial continues until E is shown with a high posterior probability to be either promising or not promising, or until a predetermined maximum sample size is reached. The design provides decision boundaries, a probability distribution for the sample size at termination, and operating characteristics under fixed response rate for E. Two extensions of this decision structure were also proposed (29). The first gives criteria from early termination of trials unlikely to yield conclusive results, based on the marginal predictive distribution of the observed response rate. The second allows early termination only if the experimental treatment E is found to be not promising compared to the standard treatment S. For further details on Bayesian approaches in phase II cancer trials, we encourage the reader to consult the references on the topic (29).

3. *Sequential and Multistage Designs*

The rationale for these types of designs stems from ethical concerns. Investigators have become over the years increasingly aware of the need for early termination of phase II trials when early data tend to support the hypothesis of treatment ineffectiveness, and for early reporting of results when early data support the hypothesis of effectiveness of a new treatment. Early termination, with or without a data-driven statistical hypothesis test, has the effect of inflating the false-positive (Type I) and/or false-negative (Type II) error probabilities above prespecified tolerable limits.

a. *Strictly Sequential Designs*

Wald's strictly sequential design (30) is perhaps the most well-known early tool used in designs that permit early termination of a trial while maintaining overall control of Type I and Type II errors. Here, no fixed sample size is specified beforehand. The decision to stop or continue patient accrual is made after each new patient is entered. The method allows for the control of Type I and Type II error probabilities. Drawbacks of strictly sequential design have been discussed by a number of authors (31). They include administrative burdens of maintaining a constant vigil over the data, delays between patient entry and response determination, and delay between response determination and the reporting of response to the trial sponsor.

b. *Multistage Designs*

Multistage designs are a compromise between strictly sequential designs and the single-stage design. A "full" sample size is declared up front, but patients are entered into the trial in batches. After each batch is accrued, a decision is made either to stop the trial or to continue to the next stage on the basis of the number of responding patients observed up to that point. Multistage designs also allow for the control of Type I and Type II error probabilities. The major drawback is mostly the administrative problems associated with monitoring the trial so as to order the incoming observations properly and to minimize the nonrandomness in the sequence of incoming-patient data.

Several articles have appeared in the statistical literature that present a one-sample multiple-testing procedure, developed explicitly to handle phase II cancer trials. The most extensively used multistage designs were put forward by Gehan in the early 1960s (32), Fleming in the early '80s (33), and Simon in the late '80s (34).

Gehan (32) designed a two-stage procedure in which early stopping occurs if no response is observed in the first n_1 patients, where n_1 is chosen to be sufficiently large that such a lack of response would be inconsistent, at a given significance level, with the hypothesis of activity of the new treatment. If any responses are observed in the first n_1 patients, an additional n_2 patients are entered, where the total number of patients $N = n_1 + n_2$ is chosen so that the true antitumor activity of the new treatment can be estimated with approximately a prespecified precision. In practice, however, Gehan's rule has been (and probably still is) greatly misunderstood by clinicians. Most of them believe that the rule is a hypothesis testing procedure rather than an estimation procedure, and that for any specified response rate θ_o the first n_1 patients will be sufficient for a phase II trial. The second sample of n_2 patients is rarely mentioned.

Fleming's one-sample-group sequential design (33) is perhaps the most widely used for phase II trials by cancer investigators. A K-stage multiple testing procedure, the design employs a standard single-stage procedure at the last test, allows for early termination, and essentially preserves the size and power. Simplicity of the single-stage procedure is captured and preserved through the specification of easily calculated acceptance and rejection points for each of the K tests. Simon (34) introduced an optimized version of this type of group sequential design by minimizing expected sample size subject to upper bounds on the error probabilities.

IV. PHASE III CLINICAL TRIALS

Phase III trials are the last step in the planned evaluation of a new treatment. They have one of the following objectives (28,35): (a) to determine the effectiveness of a new treatment relative to placebo (disease natural history), (b) to determine the effectiveness of a new treatment with regard to the best standard therapy (active control), (c) to determine if a new treatment is as effective as the standard treatment but is associated with less severe toxicity. Typical primary endpoints in a phase III cancer trial are response rate, time-to-an-event measure, such as time to disease progression, time to treatment failure, or time to death. Recently, novel endpoints such as clinical benefit (36), as well as quality of life have been explored in phase III cancer trials for advanced, generally symptomatic disease where palliation is deemed a clinically meaningful treatment outcome. However, hard endpoints, such as survival and time to disease progression, remain the most preferred endpoints in phase III cancer trial. As a result, cancer clinical research has been one of the most important breeding grounds for some of the most important and creative statistical developments in the area of survival analysis.

A. Unstratified Designs

With unstratified design, patients are randomly assigned to treatment arms as they become available. As the trial sample size increases, the number of patients per treatment arm becomes approximately the same, but at any point in time, an imbalance between the number of patients per treatment could occur by chance. This is most likely to be the case in small sample trials. Since for the sake of statistical power one would like to ensure approximately equal number of patients per treatment arm, the randomization is "blocked" or "restricted." A block is a sequence of n patients in which each study treatment is allocated an equal number of times. Because of potentially severe toxicities that can be associated with most cancer therapies, treatments are generally not blinded to clinical investigators in order to ensure the best patient care during the course of a phase III cancer trial. This constitutes the major drawback to blocked randomization, because an investigator may figure out the block size and thus know exactly what treatment will be assigned to the last patient(s) of the block and, therefore, can select the patients for a given treatment. This is especially true for single-center studies, where only one or two investigators are enrolling the patients. Simon (37) and Pocock (38) give an elegant discussion of a number of methods to limit this kind of bias, the simplest of which is to vary the dimension of the blocks and to keep investigators blinded to their sizes.

B. Stratified Designs

When factors other than treatment that can influence the response to treatment are known, they must be accounted for when evaluating the meaning of an observed difference between treatment outcomes. The ideal scenario is to account for the possible influence of the prognostic factors by distributing them equally between the treatment groups. The random assignment of patients to treatment arms is therefore performed in each stratum of patients. The most significant drawback is that stratification by more than two factors, or more than two levels, quickly leads to impractical design. Indeed, the number of strata become too large and, as a result, an imbalance between treatment arms with respect to the prognostic factors can occur simply by chance. Excessive stratification in itself is not useful, because, in the long run, it amounts to no stratification and unblocked randomization (37,39,40). Also, the complexity of randomization within a large number of strata is conducive to administrative errors.

C. Adaptive Stratified Designs

In the presence of a large number of strata, the Pocock and Simon (39) "minimization" approach is perhaps the most effective and commonly used way to avoid severe imbalance in the distribution of patients on treatment arms with respect to known prognostic factors. The reader is encouraged to consult the reference for more details on the minimization techniques.

The minimization procedure is not without inconvenience. Indeed, if an investigator is aware of patient treatment assignment with respect to prognostic factors, he can predict the next patient treatment assignment. However, this is likely to be difficult to achieve in the case of multicenter trials with a centralized coordination, where the investigator has no access to information outside his own institution, at least before the end of the trial.

The types of designs just described are by no means the only ones encountered in phase III cancer clinical trials, but they are probably the most frequently used. Factorial and crossover designs are rare but have been used in cancer phase III trials.

V. PHASE II/III DESIGNS WITH MULTIPLE EXPERIMENTAL ARMS

Recently, a number of authors have put forward new designs for the selection of treatments to be tested in randomized clinical trials (41–46). Motivations behind these new designs that interface phase II and phase III cancer trials are ethical, practical, and efficiency-driven in nature. In a program of clinical trials where several experimental treatments are of interest, it may be strategically worthwhile first to identify the single best of the experimental treatments and then to compare it to a standard control therapy. This is particularly desirable if one would be satisfied with one real advance. The goal is to minimize the number of patients expected to be accrued to new regimens that do not offer a tumor-shrinkage or survival benefit over the standard therapy. Examples of clinical settings where such designs are useful include evaluation of combination regimens, because conventional nonrandomized phase II trials of combined modality regimens are often unreliable or difficult to interpret, and evaluation of different dosing regimens of the same drug. Interfacing phase II and phase III trials has an added benefit of reducing the time for protocol development, review, and implementation, among others, and takes advantage of the efficiency of early rejection of treatment regimen(s) without stopping the entire study.

These new designs rely heavily on statistical ranking and selection theory (47,48) for choosing the best among the new regimens. This attractive statistical methodology has not yet been fully leveraged by investigators either in early drug development or in phase II/phase III cancer trial designs. The total sample size needed to select the best among several candidate treatments can be substantially less than that needed to test the classical null hypothesis of equality.

VI. STATISTICAL METHODS AND CANCER TRIAL ENDPOINTS

Tumor response rate remains the most common primary endpoint in phase II cancer trials and is handled with the binomial model (28,33). In phase III trials, however, response rate is increasingly reduced to secondary endpoint status and replaced by time-to-event endpoints, such as time to progressive disease and survival.

The increasing role of time-to-event endpoints as a more reliable tool for assessing efficacy and safety merits of new therapies, especially in frontline settings, has prompted in recent years a prolific surge in the development of new survival analysis methods. These methods nicely complement or refine the popular Kaplan–Meier nonparametric product-limit method (49) of estimating survival probability at a given time point and Cox regression modeling (50,51). Because of the increasing importance of time-to-event endpoints in phase II and phase III cancer trials, it is worthwhile to highlight some of the most recent survival analysis tools in the statistician's toolbox.

"Time to event" data are routinely encountered in clinical trials. They reflect a time measured from randomization or start of therapy to the date of a well-defined, clinically important event. Examples of such events include death, progression of the disease, off study, response (positive or negative), and relapse. The most common time-to-event data are right-censored, because a patient's information may be lost to follow-up before the event of interest occurs, due to a number of reasons, such as dropout, completion of the study, or death from other causes. It is typically assumed that the events must occur in a finite time and censoring occurs at random.

Basic methods used in the analysis of time-to-event data have traditionally consisted of estimation and comparison of survival curves. Although parametric models are available, the nonparametric Kaplan–Meier method for estimating survival distribution function and the Logrank and Wilcoxon tests for comparing two survival distribution functions remain popular. This is the case also for the semiparametric Cox proportional hazard (PH) regression model with one or more fixed cofactors.

The advantages of the Cox PH model are twofold: First, the model does not require an exact parametric specification of the survival distribution. Second, parameters have a rather straightforward and useful interpretation: the risk ratio for an x-unit increase in a given cofactor, all other factors fixed, is given by $\exp(\beta x)$, where β is the regression coefficient for the cofactor of interest in the model. The model has also its disadvantages. It is limited to fixed cofactors. It is limited to univariate time-to-event variables. Finally, the underlying proportional hazard assumption may not be realistic in many cases.

The counting process approach to survival analysis, first formulated by Aalen (52), provided the break needed for extensions of the Cox regression model. Aalen outlined a unified theoretical framework for application of point processes, especially counting processes, for time-to-event analysis. His work precipitated the development and application of important and useful multivariate time-to-event models throughout the 1980s (53–56). It also led to improved computational methods that facilitated the use of time-dependent cofactors as well as time-to-event vectors.

The Anderson–Gill (AG), Prentice–William–Peterson (PWP), and Wei–Lin–Westfield (WLW) models share five nice features: (1) all three models are semiparametric, multiplicative-hazard models (with exponential relative risk), (2) for each model, a common vector parameter β of regression coefficients can be estimated, (3) the risk ratio for an x-unit increase in a given cofactor, all other factors fixed, is given by $\exp(\beta x)$, where β is the regression coefficient for the cofactor of interest in the model, (4) they allow for more than one time-to-event (dependent) variable, and (5) all three models are now fairly easy to implement using readily available software, such as SAS® (57).

There are also subtle but important differences between AG, PWP, and WLW models. First, the AG and PWP models differ in two respects: (1) for the PWP model, patients that have not had the "first" event are not considered at risk for the second and subsequent events, and (2) the PWP model has distinct underlying hazards for each event. Second, the AG and PWP models assume that the distinct events will be interpreted as sequential with independent increments, conditional on cofactors. Finally, the WLW model requires neither a sequential interpretation of the events nor independent increments between the events.

In summary, the AG, PWP, and WLW models are relatively new but immensely useful survival analysis tools in the investigator's toolbox. The assumption of sequential events with independent increments in the AG and PWP models are strong and should be seriously assessed if these models are to be appropriately implemented. The WLW model can be applied to any data for which the assumption of exponential relative risk is reasonable. As straightforward generalizations of the Cox model, all three models are relatively easy to interpret in light of the complexity of the problem. To date, multivariate time-to-event models have been rarely implemented in clinical trials. Yet the three models discussed can be applied relatively easily using the SAS® PHREG procedure (57). They offer a unique exploratory tool for evaluating not only the relationship between a vector of time to events and fixed or time-dependent cofactors, but also the interrelationships between time to events.

VII. CONCLUSIONS

We have shared here our own experience in designing, implementing, analyzing, and reporting cancer clinical trials. We apologize for the omission and superficial treatment of many important areas. We hope that the discussion will point the reader in the right direction for a more comprehensive treatment of the different topics introduced. Our intention has been to focus more on statistical aspects unique to cancer clinical investigation. For readers interested in a much broader perspective, Simon (27) offers a superb account of the state of progress in statistical methodology for clinical trials.

REFERENCES

1. MA Schneiderman. Proc Fifth Berkeley Symp Math Statist Probabil 4:855–866, 1965.
2. NL Geller. Drug Information J 24:341–349, 1990.
3. R Simon, B Freidlin, L Rubinstein, SG Arbuck, J Collins, MC Christian. J Natl Cancer Inst 89(15):1138–1147, 1997.
4. Von Hoff, J Kuh, G Clark. In: ME Buyse, MJ Staquet, RJ Sylvester, eds. Cancer Clinical Trials: Methods and Practice. New York: Oxford University Press, 1984.
5. BE Storer, D DeMets. J Clin Res Drug Devel 1:121–130, 1987.
6. JM Collins, DS Zaharko, RL Dedrick, BA Chabner. Cancer Trt Rep 70:73–80, 1986.
7. J Collins, C Grieshaber, B Chabner. J Natl Cancer Inst 82:1321–1326, 1990.
8. J O'Quigley, M Pepe, L Fisher. Biometrics 46:33–48, 1990.
9. EORTC Pharmacokinetics and Metabolism Group. Eur J Cancer Clin Oncol 23:1083–1087, 1987.
10. BJ Foster, DR Newell, MA Graham, LA Gumbrell, KE Jenns, SB Kaye, AH Calvert. Eur J Cancer 28(2/3): 463–469, 1992.
11. L Gianni, L Vigano, A Surbone, D Ballinari, P Casali, C Tarella, J Collins, G Bonadonna. J Natl Cancer Inst 82(6):469–477, 1990.
12. WJ Dixon, AM Mood. Ann Math Stat 36:800–807, 1948.
13. BE Storer. Biometrics 45:925–937, 1989.
14. S Chevret. Stat in Med 12:1093–1108, 1993.
15. D Faries. J Biopharm Stat 4(2):147–164, 1994.
16. SN Goodman, ML Zahurak, S Piantadosi. Stat in Med 14:1149–1161, 1995.

17. EL Korn, D Midthune, TT Chen, LV Rubinstein, MC Christian, R Simon. Stat in Med 13(18):1799–1806, 1994.
18. S Moller. Stat in Med 14:911–922, 1995.
19. DA Rinaldi, HA Burris, FA Dorr, JR Woodworth, et al. J Oncology 13(11):2842–2850, 1995.
20. J O'Quigley, LZ Shen. Biometrics 52(2):673–684, 1996.
21. JR Murphy, DL Hall. J Biopharm Stat 7(4):635–647, 1997.
22. MJ Silvapulle. J Royal Stat Soc, Ser B 43(3):310–313, 1981.
23. NH Gorden, JKV Willson. Stat in Med 11:2063–2075, 1992.
24. R Mick, MJ Ratain. J Natl Cancer Inst 85:217–223, 1993.
25. MJ Ratain, R Mick, RL Schilsky, et al. J Natl Cancer Inst 85:1637–1643, 1993.
26. S Piantadosi, G Liu. Stat in Med 15:1605–1618, 1996.
27. R Simon. Stat in Med 10:1789–1817, 1991.
28. ME Byse, MJ Staquet, RJ Sylvester. Cancer Clinical Trials: Methods and Practice. Oxford: Oxford University Press, 1984.
29. PF Thall, R Simon. Controlled Clinical Trials 15:463–481, 1994.
30. A Wald. Sequential Analysis. New York: Wiley, 1947.
31. DP Byar, RM Simon, WT Friedewald, JJ Schlesselman, DL DeMets, JH Ellenberg, MH Gail, JH Ware. New Engl J Med 295:74–80, 1976.
32. EA Gehan. J Chonr Dis 13:346–353, 1961.
33. TR Fleming. Biometrics 38:143–151, 1982.
34. R Simon. Controlled Clin Trials 10:1–10, 1989.
35. R Simon. In: VT DeVita, S Hellman, SA Rosenberg, eds. Cancer, Principles and Practice of Oncology. Hagerstown, PA: JB Lippincott, 1982, pp 198–225.
36. JS Andersen, et al. ASCO Program Proceedings, Abstract 1600, 1994.
37. R Simon. Biometrics 35:503–512, 1979.
38. SJ Pocock. Biometrics 35:183–197, 1979.
39. SJ Pocock, R Simon. Biometrics 31:103–115, 1975.
40. M Zelen. In: M Staquet, ed. Cancer Therapy: Prognostic Factors and Criteria of Response. New York: Raven Press, 1975, pp 1–35.
41. PF Thall, R Simon, SS Ellenberg. Biometrika 75:303–310, 1988.
42. PF Thall, R Simon, SS Ellenberg. Biometrics 45:537–548, 1989.
43. DJ Schaid, S Wieand, TM Therneau. Biometrika 77: 507–513, 1990.
44. R Simon, PF Thall, SS Ellenberg. Stat in Med 13:417–429, 1994.
45. PY Liu, S Dahlberg, J Crowley. Biometrics 49:391–398, 1993.
46. TT Chen, R Simon. Biometrics 49:753–761, 1993.
47. JD Gibbons, I Olkin, M Sobel. Selecting and Ordering Populations—A New Statistical Methodology. New York: Wiley, 1977.
48. N Mukhopadyay, KST Solanky. Selection and Ranking Procedures—Second-Order Asymptotics. New York: Marcel Dekker, 1994, V:142.
49. EL Kaplan, P Meier. J Am Statist Assoc 9:457–481, 1958.
50. DR Cox. J R Statist Soc B 34:187–220, 1972.
51. DR Cox. Biometrika 62:269–276, 1975.
52. OO Aalen. PhD dissertation, University of California at Berkeley, 1975.
53. PK Andersen, RD Gill. Ann Stat 10:1100–1120, 1982.
54. RL Prentice, BJ Williams, AV Peterson. Biometrika 68: 372–379, 1981.
55. LJ Wei, DY Lin, L Weissfeld. J Am Statist Assoc 84: 1065–1073, 1989.
56. DY Lin. Stat in Med 13:2233–2247, 1994.
57. SAS. SAS Institute Inc., Cary, NC, 1990.

Clet Niyikiza
Douglas E. Faries

Carcinogenicity Studies of Pharmaceuticals*

See also *Reproductive Studies; Ames Test*

I. INTRODUCTION

The assessment of the risk of exposure to a new drug in humans usually begins with an assessment of the risk of the drug in animals. It is required by law that the sponsor of a new drug conduct nonclinical studies in animals to assess the pharmacological actions, the toxicological effects, and the pharmacokinetic properties of the drug in relation to its proposed therapeutic indications or clinical uses. Studies in animals designed for the assessment of the toxicological effects of the drug include acute, subacute, subchronic, chronic toxicity studies, carcinogenicity studies, reproduction studies, and pharmacokinetic studies. The assessment of the risk of drug exposure in humans includes an important assessment of the risk for carcinogenicity in lifetime tests in mice and rats.

It is recognized that there are distinctive differences in chemical and pharmacological nature between environmental chemicals and human pharmaceuticals. Because of those differences in nature, the design of experiments, methods of analysis, and interpretation of the results used in rodent carcinogenicity studies of pharmaceuticals are not totally the same as those used in carcinogenicity studies of environmental chemicals. For example, it has been recommended in the literature that the highest dose used for carcinogenicity studies of environmental chemicals should be close to the toxicity-based endpoint maximum tolerated dose (MTD). In addition to the traditional approach using toxicity-based endpoints, other criteria have been proposed by the International Conference on Harmonization (ICH) for high dose selection for carcinogenicity studies of human pharmaceuticals. These include approaches using pharmacodynamic endpoints, pharmacokinetic endpoints, saturation of absorption, and maximum feasible dose (1). It is also recognized that knowledge of the pharmacology, pharmacokinetics, and metabolic disposition of the pharmaceuticals in humans may be available and can be used in dose selection, in the determination of target organs, and in assessing the relevance of animal findings to humans.

Traditional long-term animal carcinogenicity studies usually are conducted on both sexes of mice and rats for the majority of those animals' normal lifespans. Treatment on the animals usually lasts for about two years after weaning. A standard, traditional carcinogenicity study without planned interim sacrifices usually uses 50 animals per sex in each of four treatment groups: control, low, medium, and high. To save time and money, the ICH (2) has approved experimental approaches to the evaluation of carcinogenic potential that may obviate the necessity for the routine conduct of two long-term rodent carcinogenicity studies for those pharmaceuticals that need such evaluation. In the document, the ICH made provision for drug sponsors either to continue to conduct two long-term rodent carcinogenicity studies or to use the alternative approach by conducting only one long-term rodent carcinogenicity study (preferably in rats) and a short- or medium-term rodent test system to evaluate the carcinogenic potential of their drug products. The short or medium rodent test systems include the use of models, such as models of initiation-promotion in rodents, or transgenic rodents, or new-born rodents, that provide insight into carcinogenic endpoints in vivo.

The primary purpose of a long-term animal carcinogenicity experiment is to evaluate the oncogenic potential of a new drug when it is administered to animals for the majority of their normal life span. Because of the complicated nature of the rodent carcinogenicity experiment, the statistical analysis in this area involves tests and estimations of other data, such as survival, food consumption, body weight, organ weight, clinical chemistry, and hematology in addition to pathology data. The analyses of the other data provide the information about the design validity of a study, the need for adjustment for mortality differences among treatment groups in the final analysis of tumor data, and the possible use of scientific hypotheses about the mechanisms of carcinogenesis. Because of the limitation of space, this entry concentrates only on the analysis of tumor data, with adjustment for possible differences in survival among treatment groups.

*The views expressed here are those of the author and not necessarily those of the Food and Drug Administration. The techniques or methods of analysis and the interpretation of results described are based on the author's assessment of current literature, his consultations with outside experts, his personal research, and his best scientific judgment, but these are issues where consensus among experts in the evaluation of animal carcinogenicity studies does not always exist.

In a rodent carcinogenicity study of a new drug using a series of increasing dose levels, statistical tests for positive trends in tumor rates are usually of greatest relevance and interest in the evaluation of the oncogenic potential of the drug. However, as is explained in a subsequent section, in some situations pairwise comparisons are considered more appropriate than trend tests for carcinogenicity evaluation based on pharmacological justifications.

The purpose of this entry is to describe some commonly used methods of statistical analysis of tumor data and the interpretation of results of rodent carcinogenicity studies of pharmaceuticals. Section II discusses methods of statistical analysis. Section III discusses how the results should be interpreted. The techniques or methods of analysis, the interpretation of results described in the following sections are based on the author's assessment of current literature, his consultations with outside experts, his internal research, and his best scientific judgment, but these are issues where consensus among experts in the evaluation of animal carcinogenicity studies does not always exist.

II. TESTS FOR TREND AND DIFFERENCES IN TUMOR INCIDENCE

As just mentioned, in a rodent carcinogenicity study of a new drug using a series of increasing dose levels, statistical tests for positive trends or differences in tumor rates are usually of greatest relevance and interest in the evaluation of the oncogenic potential of the drug. There are different methods proposed for analyzing tumor data from a rodent carcinogenicity study. General discussions on various methods of tumor data analysis can be found in the statistical literature (3–14). The methods of analysis discussed in Secs. II.A–II.F were proposed in Peto et al. (12). They have become the most commonly used methods in tumor data analysis. Parts of the texts in Secs. II.A–II.G are from a previously published article by Lin and Ali (11) with or without modifications.

A. Adjustment of Tumor Rates for Intercurrent Mortality

Intercurrent mortality refers to all deaths unrelated to a tumor or class of tumors to be analyzed for evidence of carcinogenicity. Like human beings, older rodents have a many-fold higher probability of developing or dying of tumors than those of younger age. Therefore, in the analysis of tumor data, it is essential to identify and adjust for possible differences in intercurrent mortality (or longevity) among treatment groups to eliminate or reduce biases caused by these differences. It has been pointed out that "the effects of differences in longevity on numbers of tumor-bearing animals can be very substantial, and so, whether or not they [the effects] appear to be [very substantial], they should routinely be corrected when presenting experimental results" (12, p. 323). The following examples clearly demonstrate this important point.

Example 1. Consider an experiment consisting of one control group and one treated group of 100 animals each of a strain of mice (12). A very toxic but not carcinogenic new drug was administered to the animals in the diet for two years. Assume that the spontaneous incidental tumor rates for both groups are 30% at 15 months and 80% at 18 months of age and that the mortality rates at 15 months for the control and the treated groups are 20% and 60%, respectively, due to the toxicity of the drug. The results of the experiment are summarized in Table 1.

If one looks only at the overall tumor incidence rates of the control and the treated groups (70% and 50%, respectively) without considering the significantly higher early deaths in the treated group caused by the toxicity of this new drug, one might misinterpret the apparently significant ($p = 0.002$, one-tailed) decrease in the treated group in this tumor type. The one-tailed p-value is 0.5, however, when the survival-adjusted prevalence method (12) is used.

Example 2. Assume that the design used in this experiment is the same as the one used in the experiment of Example 1. However, assume that the tested new drug in this example induces an incidental tumor that does not directly or indirectly cause animals' deaths, in addition to having severe toxicity, as in the previous example. Assume further that the incidental tumor prevalence rates for the control and treated

Table 1 Effects of Mortality Differences on Tumor Incidence Rates (Example 1)

	Control			Treated		
	T	D	%	T	D	%
15 months	6	20	30	18	60	30
18 months	64	80	80	32	40	80
Totals	70	100	70	50	100	50

T = incidental tumors found at necropsy; D = deaths.

Table 2 Effects of Mortality Differences on Tumor Incidence Rates (Example 2)

	Control			Treated		
	T	D	%	T	D	%
Before 15 months	1	20	5	18	90	20
After 15 months	24	80	30	7	10	70
Totals	25	100	25	25	100	25

T = incidental tumors found at necropsy; D = deaths.

groups are 5% and 20%, respectively, before 15 months of age, and 30% and 70%, respectively, after 15 months of age; and that the mortality rates at 15 months are 20% and 90% for the control and the treated groups, respectively (6). The results of this experiment are summarized in Table 2.

The age-specific tumor incidence rates are significantly higher in the treated group than those in the control group. The survival-adjusted prevalence method yielded a one-tailed *p*-value of 0.003, revealing a clear tumorigenic effect of the new drug. The overall tumor incidence rates, however, are 25% for the two groups. Without adjusting for the significantly higher early mortality in the treated group, the positive finding would be missed.

Before analyzing the tumor data, the intercurrent mortality data should be checked to see if the survival distributions of the treatment groups are significantly different or if there exists a significant trend for survival. The Cox test (6,13,15), the generalized Wilcoxon or Kruskal–Wallis test (13,16,17), and the Tarone trend tests (12,18,19) are routinely used to test for heterogeneity in survival distributions and significant dose–response relationships (trends) in survival.

It is recommended by Peto et al. (12) that, whether or not survival among treatment groups is significantly different, tumor rates should routinely be adjusted for survival when presenting experimental results.

B. Contexts of Observation of Tumor Types

The choice of a survival-adjusted method to analyze tumor data depends on the role that a tumor plays in causing the animal's death. Tumors can be classified as "incidental," "fatal," and "mortality-independent (or observable)" according to the contexts of observation described in Peto et al. (12). Tumors that are not directly or indirectly responsible for the animal's death but are merely observed at the autopsy of the animal after it has died of an unrelated cause are said to have been observed in an *incidental* context. Tumors that kill the animal, either directly or indirectly, are said to have been observed in a *fatal* context. Tumors, such as skin tumors, whose detection occurs at times other than when the animal dies are said to have been observed in a *mortality-independent (or observable)* context. To apply a survival-adjusted method correctly, it is essential that the context of observation of a tumor be determined as accurately as possible. "The distinction between fatal and incidental tumors is important because it is essential to distinguish between a chemical that reduces survival by shortening the time to tumor onset or the time to death following tumor onset (a real carcinogenic effect), and one that also reduces survival, but for which tumors are observed earlier simply because animals are dying of competing causes (a noncarcinogenic effect)" (4).

Different statistical techniques have been proposed for analyzing data on tumors observed in different contexts of observation. For example, the prevalence method, the death-rate method, and the onset-rate method are recommended for analyzing data on tumors observed in incidental, fatal, and mortality-independent contexts of observation, respectively, in Peto et al. (12). In this paper, Peto et al. demonstrate the possible biases resulting from misclassification of incidental tumors as fatal tumors or of fatal tumors as incidental tumors.

C. Tests of Data of Incidental Tumors

The *prevalence* method described in the paper by Peto et al. (12), is routinely used by statisticians in testing for positive trends in prevalence rates of incidental tumors. The method can be described briefly as follows.

The prevalence method focuses on the age-specific tumor prevalence rates to correct for intercurrent mortality differences among treatment groups in the test for positive trends or differences in incidental tumors. The experimental period is partitioned into a set of intervals plus interim (if any) and terminal sacrifices. The incidental tumors are then stratified by those intervals of survival times. The selection of the partitions of the experimental period does not matter very much as long as the intervals "are not so short that the prevalence of incidental tumors in the autopsies they contain is not stable, nor yet so large that the real prevalence in the first half of one interval could differ markedly from the real prevalence in the second half" (12).

In each time interval, for each group, the observed and the expected numbers of animals with a particular tumor type found in necropsies are compared. The expected number is calculated under the null hypothesis that there is no dose-related trend. Finally, the differences between the observed and the expected numbers of animals found with the tumor type after their deaths are combined across all time intervals to yield an overall test statistic using the method described in the paper of Mantel and Haenszel (20).

The following derivation of the Peto prevalence test statistic uses the notations in Table 3. Let the experimental period be partitioned into the following m intervals $I_1, I_2, \ldots, I_m$. As mentioned before, interim (if any) and terminal sacrifices should be treated as separate intervals. The number of autopsied animals expected to have the particular incidental tumor in group i and interval k, under the null hypothesis that there is no treatment effect, is:

$$E_{ik} = O_{.k}P_{ik}$$

The variance-covariance of $(O_{ik} - E_{ik})$ and $(O_{jk} - E_{jk})$ is:

$$V_{ijk} = P_{ik}(\delta_{ij} - P_{jk})$$

where

$$\delta_{ij} = \begin{cases} 1 & \text{if } I = j \\ 0 & \text{otherwise} \end{cases}$$

Define

$$O_i = \sum_k O_{ik}$$

$$E_i = \sum_k E_{ik}$$

$$V_{ij} = \sum_k V_{ijk}$$

The test statistic T for the positive trend in the incidental tumor is defined as:

$$T = \sum_i D_i(O_i - E_i)$$

with estimated variance

$$V(T) = \sum_i \sum_j D_i D_j V_{ij}$$

where D_i is the dose level of the ith group. Under the null hypothesis of equal prevalence rates among the treatment groups, the statistic

$$Z = \frac{T}{[V(T)]^{1/2}}$$

is approximately distributed as a standard normal.

As mentioned earlier, to use the prevalence method, the experimental period should be partitioned into a set of intervals plus interim (if any) and terminal sacrifices. The following partitions (in weeks) are used most often in two-year studies: (a) 0–50, 51–80, 80–104, interim sacrifice (if any), and terminal sacrifice; (b) 0–52, 53–78, 79–92, 93–104, interim sacrifice (if any), and terminal sacrifice (proposed by National Toxicology Program); and (c) partition determined by the "ad hoc runs" procedure described in Peto et al. (12).

The data of liver hepatocellular adenoma of male mice from a carcinogenicity experiment of a new drug are used as an example to explain the prevalence method for testing the positive trend in tumor rates of an incidental tumor. There were four treatment groups in the mouse experiment. The control group had 100 animals; the three treated groups had 50 animals each. The dose levels used were 0, 10, 20, and 40 mg/kg/day for the control, low, medium, and high groups, respectively. The study lasted for 106 weeks. In this example, the study period was partitioned into four intervals: 0–50, 51–80, 81–106, and terminal sacrifice. The numbers of animals died and necropsied and the numbers of necropsied animals with liver hepatocellular adenoma by treatment group in each interval are included in Table 4.

The observed rates and the expected rates of the tumor type calculated under the null hypothesis that there is no trend (or drug induced increase) are shown in Table 5. The expected tumor rates in each interval were calculated in the following way. First, the tumor rate for the interval using data of all treatment groups in the interval was estimated. For example, the estimated tumor rate for the interval 51–80 weeks is 6/74 = 0.0811. Second, the expected rates for individual groups in the interval were calculated by multiplying the numbers of necropsies by the estimated tumor rate. For the interval 50–81 weeks, the expected tumor rates for the control, low, medium, and high groups are 26 × 6/74 = 2.11, 18 × 6/74 = 1.46, 17 × 6/74 = 1.38, and 13 × 6.74 = 1.05, respectively.

The values of the test statistic T and its variance $V(T)$ for the data of the five intervals calculated by the formulas listed earlier are included in Table 6. It is noted that the first interval, 0–50 weeks, did not contribute to the overall test result, since none of the 14 animals that died during the time interval developed liver hepatocellular adenoma. The overall result shows a statistically significant positive trend in tumor rates for this tumor (with one-sided p-value 0.002). Also as mentioned earlier, this method uses normal approximation in the tests for positive trend or difference in tumor prevalence rates. The accuracy of the normal approximation depends on the number of tumor occurrences in each group in each interval, the number of

Table 3 Notations Used in the Derivation of Peto Prevalence Test Statistics

		Group:	0	1	$\cdots$	i	$\cdots$	r	
Interval		Dose:	D_0	D_1	$\cdots$	D_i	$\cdots$	D_r	Sum
I_1	R_1		O_{01}	O_{11}	$\cdots$	O_{i1}	$\cdots$	O_{r1}	$O_{.1}$
			P_{01}	P_{11}	$\cdots$	P_{i1}	$\cdots$	P_{r1}	$P_{.1}$
I_2	R_2		O_{02}	O_{12}	$\cdots$	O_{i2}	$\cdots$	O_{r2}	$O_{.2}$
			P_{02}	P_{12}	$\cdots$	P_{i2}	$\cdots$	P_{r2}	$P_{.2}$
$\vdots$	$\vdots$		$\vdots$	$\vdots$	$\vdots$	$\vdots$	$\vdots$	$\vdots$	$\vdots$
I_k	R_k		O_{0k}	O_{1k}	$\cdots$	O_{ik}	$\cdots$	O_{rk}	$O_{.k}$
			P_{0k}	P_{1k}	$\cdots$	P_{ik}	$\cdots$	P_{rk}	$P_{.k}$
$\vdots$	$\vdots$		$\vdots$	$\vdots$	$\vdots$	$\vdots$	$\vdots$	$\vdots$	$\vdots$
I_m	R_m		O_{0m}	O_{1m}	$\cdots$	O_{im}	$\cdots$	O_{rm}	$O_{.m}$
			P_{0m}	P_{1m}	$\cdots$	P_{im}	$\cdots$	P_{rm}	$P_{.m}$

R_k: number of animals that have not died of the tumor type of interest but come to autopsy in time interval k.
P_{ik}: proportion of R_k in group i.
O_{ik}: observed number of autopsied animals in group i and interval k found to have the incidental tumor type.
$O_{.k} = \Sigma_i O_{ik}$.

intervals used in the partitioning, and the mortality patterns. The approximation may not be stable and reliable when the numbers of tumor occurrences across treatment groups are small. In this situation, an exact permutation trend test based on an extension of the hypergeometric distribution (to be discussed in Sec. II.G) is used to test for the positive trend in tumor prevalence rates.

D. Tests of Data of Fatal Tumors

The death-rate method described in Peto et al. (12) should be utilized to test for the positive trend or difference in tumors observed in a fatal context.

The notations of Sec. II.C, with some modifications, will be used in this section to derive the test statistic of the death-rate method. Now let $t_1 < t_2 < \cdots < t_m$ be the time points when one or more animals died of the fatal tumor of interest. These time points replace the intervals used in the prevalence method. The notations in Table 3 are redefined as follows:

R_k = number of animals of all groups just before t_k

P_{ik} = proportion of R_k in group i (same as in prevalence method)

O_{ik} = observed number of animals in group i dying of the fatal tumor of interest at time t_k

$O_{.k} = \Sigma_i O_{ik}$

As in the prevalence method, the test statistic T for the positive trend in the fatal tumor is defined as:

Table 4 Data on Liver Hepatocellular Adenoma of Male Mice

	Group											
Time interval	Control			Low			Medium			High		
(weeks)	T	N	%	T	N	%	T	N	%	T	N	%
0–50	0	6	0	0	2	0	0	2	0	0	4	0
51–80	1	26	4	1	18	6	3	17	18	1	13	8
81–106	4	37	11	2	14	14	2	14	14	7	19	37
Terminal sacrifice	2	31	6	5	16	31	3	17	18	4	14	29
Total	7	100	7	8	50	16	8	50	16	12	50	24

T = number of necropsies with the particular tumor.
N = number of necropsies during a time interval.
% = percentage of necropsies with the particular tumor.

Table 5 Observed and Expected Tumor Rates for Liver Hepatocellular Adenoma of Male Mice

Time interval (weeks)	Observed and expected rates	Group			
		Control	Low	Medium	High
0–50	Observed:	0	0	0	0
	Expected:	0	0	0	0
51–80	Observed:	1	1	3	1
	Expected:	2.11	1.46	1.38	1.05
81–106	Observed:	4	2	2	7
	Expected:	6.61	2.50	2.50	3.39
Terminal sacrifice	Observed:	2	5	3	4
	Expected:	5.56	2.87	3.05	2.51
Total	Observed:	7	8	8	12
	Expected:	14.28	6.83	6.93	6.95

Expected tumor rates were calculated under the null hypothesis that there is no trend.

$$T = \sum_i D_i(O_i - E_i)$$

with estimated variance

$$V(T) = \sum_i \sum_j D_i D_j V_{ij}$$

where D_i, O_i, E_i, and V_{ij} are defined as in Sec. II.C.

Under the null hypothesis of equal tumor rates across the treatment groups, the statistic

$$Z = \frac{T}{[V(T)]^{1/2}}$$

is distributed approximately as standard normal.

E. Tests of Data of Tumors Observed in Both Incidental and Fatal Contexts

When a tumor is observed in a fatal context for some animals and is also observed in an incidental context for other animals in the experiment, data of the incidental tumor part should be analyzed by the prevalence method, and data of the fatal tumor part by the death-rate method using all animals as at risk of death with the tumor. Results from the different methods can then be combined to yield an overall result. The combined overall result can be obtained simply by adding together either the separate observed frequencies, the expected frequencies, and the variances, or the separate T-statistics and their variances (12).

F. Tests of Data of Mortality-Independent Tumors

Tumors observed in a mortality-independent context, such as skin tumors and mammary gland tumors, that are visible and/or can be detected by palpation in living animals should be analyzed using the onset-rate

Table 6 Test Statistics, Their Variances, z-values, and p-value of Peto Prevalence Analysis of Incidental Tumors for Liver Hepatocellular Adenoma of Male Mice

Time interval (weeks)	T-statistic T	Variance of T-stat $V(T)$	$z = \frac{T}{[V(T)]^{0.5}}$	p-value
0–50	—	—	—	—
51–80	25.6756	1116.583	0.7683	0.2211
81–106	129.2857	3091.314	2.3253	0.0100
Terminal sacrifice	79.7435	2445.855	1.6124	0.0534
Overall Total	234.7048	6653.752	2.8773	0.0020

The z- and p-value columns do not add up to the totals. The z- and p-values for the overall total row were calculated based on the T and $V(T)$ for the row.

method. The onset-rate method for mortality-independent tumors and the death-rate method for fatal tumors are essentially the same in principle except that the endpoint in the onset-rate method is the occurrence of such a tumor (e.g., skin tumor reaching some prespecified size) rather than the time or cause of the animal's death.

In the onset-rate method, all those animals that, although still alive, have developed the particular mortality-independent tumor and hence are no longer at risk for such a tumor are excluded from the calculation of the numbers of animals at risk. The R_k, P_{ik}, and O_{ik} described in Sec. II.D are now redefined as follows for the onset-rate method:

R_k = number of animals alive and free of the mortality-independent tumor of interest in all groups just before t_k

P_{ik} = proportion of R_k in group i (same as in death-rate method)

O_{ik} = observed number of animals in group i found to have developed the mortality-independent tumor of interest at time t_k

The test statistic T and its estimated variance $V(T)$ are the same as those defined in the death-rate method.

G. Exact Tests

As mentioned in previous sections, the prevalence method, the death-rate method, and the onset-rate method use normal approximation in the test for the positive trend in tumor incidence rates. Mortality patterns, the number of intervals used in the partitioning of the study period, and the numbers and patterns of tumor occurrence in each individual interval have effects on the accuracy of the normal approximation. It is also well known that the approximation results may not be stable and reliable and that they tend to underestimate the exact p-values when the total numbers of tumor occurrence across treatment groups are small (21). In this situation, the exact permutation trend test should be used to test for the positive trend (6). The exact permutation trend test is a generalization of the Fisher's exact test to a sequence of $2 \times (r + 1)$ tables. The exact permutation trend test procedure described next is for tumors observed in an incidental context. However, the positive trends in incidence rates of tumors observed in a fatal or in a mortality-independent context can be tested in a similar way. In those cases, the number of $2 \times (r + 1)$ tables will be equal to the number of time points when one or more animals died of a particular fatal tumor or when one or more animals developed a particular mortality-independent tumor. Fairweather et al. (22) contains a discussion on the limitations of applying exact methods to fatal tumors.

The exact method is derived by conditioning on the row and column marginal totals of each of the $2 \times (r + 1)$ tables formed from the partitioned data set of Table 3. Consider the kth interval I_k (in Table 3) and rewrite it as in Table 7. Let the column totals C_{0k}, C_{1k}, . . . , C_{rk} and the row totals $O_{.k}$ and $A_{.k}$ be fixed. Define $P_{ik} = C_{ik}/R_k$. Then the quantities $E_{ik} = O_{.k}P_{ik}$, $V_{ijk} = P_{ik}(\delta_{ij} - P_{jk})$, E_i, and $V(T)$ (defined in Sec. III.C) are all known constants.

Now let y be the observed value of $Y = \sum D_iO_i$, where $O_i = \sum_k O_{ik}$, the total number of tumor-bearing animals with the tumor of interest in treatment group i. Then (under conditioning on the column and row marginal totals in each table) the observed significance level

$$p\text{-value} = P\left[\sum D_iO_i \geq y\right] = P\left(\sum_i D_i \sum_k O_{ik} \geq y\right)$$
$$= P\left(\sum_k \sum_i D_iO_{ik} \geq y\right)$$
$$= P\left(\sum_k Y_k \geq \sum y_k\right) = P(Y \geq y)$$

where $Y = \sum Y_k = \sum_k \sum_i D_iO_{ik}$ and $y = \sum y_k$, the observed value of Y. The p-value $(P(Y \geq y)$ is computed from

Table 7 Data in the kth Time Interval I_k Are Written as a $2(r + 1)$ Table

	Group: 0	1	. . .	i	. . .	r	
	Dose: D_0	D_1	. . .	D_i	. . .	D_r	Total
Number with tumor	O_{0k}	O_{1k}	. . .	O_{ik}	. . .	O_{rk}	$O_{.k}$
Number without tumor	A_{0k}	A_{1k}	. . .	A_{ik}	. . .	A_{rk}	$A_{.k}$
Total	C_{0k}	C_{1k}	. . .	C_{ik}	. . .	C_{rk}	R_k

Table 8 Hypothetical Tumor Data for Exact Permutation Trend Test

Time interval		Dose level 0	1	2	Total
0–50	O:	0	0	0	0
	C:	1	3	3	7
51–80	O:	0	0	0	0
	C:	4	5	7	16
81–104	O:	0	0	2	2
	C:	10	12	15	37
Terminal sacrifice	O:	0	1	0	1
	C:	35	30	25	90

O = observed tumor count; C = number of animals necropsied.

the exact permutational distribution of Y. Given the observed row and column marginal totals in a $2 \times (r + 1)$ table, all possible tables having the same marginal totals can be generated. Let $S_k (k = 1, 2, \ldots, K)$ be the set of all such tables generated from the kth observed table. From a set of K tables taking one from each S_k and assuming independence between the K tables, the foregoing expression for the p-value can now be written as

$$p\text{-value} = \sum [P(Y_1 = y_1) \cdots P(Y_K = y_K)]$$

where $y_k = \Sigma_i D_i O_{ik}$ $(k = 1, 2, \ldots, K)$, the sum is over all sets of K tables such that $y_1 + y_2 + \cdots + y_K \geq y$, the observed value of Y, and $P(Y_k = y_k)$ is the conditional probability given the marginal totals in the kth table; i.e.,

$$P(Y_k = y_k) = \frac{\binom{C_{0k}}{O_{0k}} \binom{C_{1k}}{O_{1k}} \cdots \binom{C_{rk}}{O_{rk}}}{\binom{R_k}{O_{.k}}}$$

The following example illustrates how the probabilities are computed in the exact permutation trend test described earlier. Consider an experiment with three treatment groups (control, low, and high) with dose levels $D_0 = 0$, $D_1 = 1$, and $D_2 = 2$, respectively. Suppose the study period is partitioned into the intervals 0–50, 51–80, and 81–104 weeks and the terminal sacrifice week. Consider a tumor type (classified as incidental) with data as in Table 8. Since all the observed tumor counts (i.e., O's) in the first two time intervals are zero, the data for these intervals will not contribute anything to the test statistic and these intervals may be ignored. The observed subtables formed from the last two intervals are given in Table 9.

Now generate all possible tables from observed subtable 1. Since the marginal totals are fixed, these tables may be generated by distributing the total tumor frequency $O_{.1}$ (=2) among the three treatment groups. Thus, each table will correspond to a configuration of this distribution of $O_{.1}$. The configurations, the values of Y_1, and the $P(Y_1 = y_1)$ are shown in Table 10.

To illustrate the computation of y_1 and $P(Y_1 = y_1)$ consider the last row. Here

$$\begin{aligned} y_1 &= D_0 \times 1 + D_1 \times 1 + D_2 \times 0 \\ &= 0 \times 1 + 1 \times 1 + 2 \times 0 = 1 \end{aligned}$$

and

$$P(Y_1 = 1) = \frac{\binom{10}{1}\binom{12}{1}\binom{15}{0}}{\binom{37}{2}} = \frac{10 \times 12 \times 2}{37 \times 36} = .18018$$

The configurations and probabilities obtained from observed subtable 2 are given in Table 11. Note that the first configuration (0, 0, 2) in Table 10 corresponds to the observed Subtable 1 with a value of $y_1 = (0 \times 0) + (1 \times 0) + (2 \times 2) = 4$ and a probability of .15766, and the second configuration (0, 1, 0) in Table 11 corresponds to the observed Subtable 2 with a value of $y_2 = (0 \times 0) + (1 \times 1) + (0 \times 0) = 1$ and a probability of .33333. Thus, the observed value of $y = y_1 + y_2 = 4 + 1 = 5$. Now, the exact p-value (right-tailed) is calculated as follows:

Table 9 Observed Subtables from the Hypothetical Tumor Data of Table 8

Observed subtable 1					Observed subtable 2				
Dose:	0	1	2	Total	Dose:	0	1	2	Total
O:	0	0	2	$2 = O_{.1}$	O:	0	1	0	$1 = O_{.2}$
A:	10	12	12	$35 = A_{.1}$	A:	35	29	25	$89 = A_{.2}$
C:	10	12	15	$37 = R_1$	C:	35	30	25	$90 = R_2$

Table 10 All Possible Configurations of $O_{.1}$ and the Corresponding Hypergeometric Probabilities

Configuration	y_1	$P(Y_1 = y_1)$
0, 0, 2	4	.15766
0, 2, 0	2	.09910
2, 0, 0	0	.06757
0, 1, 1	3	.27027
1, 0, 1	2	.22523
1, 1, 0	1	.18018

$$
\begin{aligned}
P(Y = Y_1 + Y_2 \geq 5) &= P(Y_1 = 4, Y_2 = 1) \\
&\quad + P(Y_1 = 4, Y_2 = 2) \\
&\quad + P(Y_1 = 3, Y_2 = 2) \\
&= .15766 \times .33333 + .15766 \\
&\quad \times .27778 + .27027 \times .27778 \\
&= .17142
\end{aligned}
$$

For the purpose of comparison it may be noted that the normal approximated p-value for the data set in our example is .0927.

H. Statistical Analysis of Data from Studies with Dual Controls

There are two categories of studies with dual control groups. The first category (Category A) includes studies using an untreated control group and a vehicle control group. The second category (Category B) includes studies that use two identical control groups (23,24).

The main reasons for using an untreated control and a vehicle control in a study in Category A are to check if the vehicle substance has effects, such as spontaneous tumor incidence and pattern, body weight, and food consumption (in dietary studies) on the test animals; and to make sure that the control animals are subjected to the same method and route of exposure (e.g., by gavage or by injection) experienced by the treated animals so that all animals will be under equal physiological responses and stress (i.e., to isolate the treatment effect from other possible effects) (6,25).

Table 11 All Possible Configurations of $O_{.2}$ and the Corresponding Hypergeometric Probabilities

Configurations	y_2	$P(Y_2 = y_2)$
0, 0, 1	2	.27778
0, 1, 0	1	.33333
1, 0, 0	0	.38889

There are arguments for and against using two identical controls in a study in Category B. The arguments for using this type of design are to employ the results from the two identical controls as a quality control mechanism for identifying unsuspected biases in the study (6), and to evaluate the biological significance of increases in tumor incidence in the treated groups (i.e., true increases versus noises). However, as described later, difficulties are encountered in statistical analysis of data from a study in this category. Also, there is a concern about the possible existence of extra-binomial within-study variability between the two identical groups that will result in an inflation of false-positive rates. These difficulties and concerns form the basis for the arguments against the use of designs with two identical controls.

Statisticians and pharmacologists/toxicologists should work together to decide which of the two control groups is appropriate for the analysis of data from a study in Category A. There are situations in which only analyses of data of the vehicle control and the treated groups are meaningful. There are other situations in which three sets of analysis are performed: control 1 versus treated groups, control 2 versus treated groups, and control 1 plus control 2 versus treated groups. Since the concerns about the possible effects of the vehicle substance on the test animals are the reasons for using the vehicle control in addition to the untreated control, it is also of interest to compare the mortality, tumor rates, body weight, and food consumption (in dietary studies) between the two control groups.

Depending on mortality and tumor rates, data from dual control groups may or may not be combined for statistical analysis of data from studies in Category B. If comparisons of the dual controls show no major differences in mortality and tumor rate, then the data of the two controls are combined to form a single control group in subsequent analyses (26). If the data show evidences of major differences in mortality or tumor incidence between the identical controls, then three tests for each tumor/organ combination are carried out: control 1 versus treated groups, control 2 versus treated groups, and control 1 plus control 2 versus treated groups.

There is the question of how to interpret the results of a study in Category B in the second case. Two approaches to the question exist. The first one is that a trend or a difference in tumor rate is considered as significant only if it is significant at the levels of significance described in Sec. III.A in all the three tests. The second one is that the trend or the difference is considered significant as long as any one of the three tests shows a significant result. The first approach may be very conservative, in the sense that the null hypothesis will be rejected much less often than it should be. On the other hand, the second approach may result in a high false-positive error.

Currently there is no information about appropriate levels of significance for the tests of these two approaches in order to maintain the 10% overall false-positive rate, and about the exact consequences of the two approaches in the second practice of dealing with dual controls with differences in tumor incidence or mortality. Unless all three tests yield consistent results —i.e., all are significant or all are not significant from the statistical perspective—the most prudent way of interpreting the test results under this circumstance may be either to regard the study as providing equivocal evidence of carcinogenicity or to consider the study as inadequate for meaningful evaluation (26). Alternatively, from the biological perspective, the dual-control data may be viewed as equivalent to contemporary historical data. In such a biological evaluation of the data, consideration of other appropriate historical control data is essential.

I. Statistical Analysis of Data Without Information About Cause of Death and Without Multiple Sacrifices

As mentioned previously, in the analysis of tumor data it is essential to identify and adjust for possible differences in intercurrent mortality among treatment groups in order to eliminate or reduce biases caused by these differences. However, most tumors, except those, such as skin tumors, that can be detected by palpation and visual inspection, are discovered only at the time of the animal's death. The exact tumor onset times are censored and, therefore, are unidentifiable and unobservable. "Without direct observations of the tumor onset times, the desired survival adjustment usually is accomplished by making assumptions concerning tumor lethality, cause of death, multiple sacrifices, or parametric models" (3).

There is considerable discussion in the literature on statistical tests for positive trend in tumor incidence rate based on those assumptions and models. Some of those proposed statistical tests are not practical, because they are based on either uncommon designs with serial sacrifices or mathematical assumptions that are not realistic and/or cannot be verified. The widely used prevalence method, the death-rate method, and the onset-rate methods for analyzing incidental, fatal, and mortality-independent tumors, respectively, described in previous sections, rely on the information on tumor lethality and cause of death; that is, the choice of a survival-adjusted method to analyze tumor data depends on the role that a tumor plays in causing the animal's death (8,12).

There are situations in which sponsors do not include the tumor-lethality and cause-of-death information in their statistical analyses and in their submitted electronic data sets. Under those situations, statistical reviewers in regulatory agencies either treat all tumors as incidental or, if provided, follow the cause-of-death information determined by reviewing pharmacologists and toxicologists in the agencies. There are consequences in misclassifying tumors and misusing survival-adjusted statistical tests. The use of the prevalence method will reject the null hypothesis of no positive trend less frequently than it should as the lethality of a tumor increases (3,12). This will increase the probability of failing to detect true carcinogens.

The poly-3, and poly-6 (in general, poly-k) tests (3,27,28) have been proposed for testing linear trends in tumor rates. These tests are basically modifications of the survival-unadjusted Cochran–Armitage test (29–31) for linear trend in tumor rate. If the entire study period is considered as one interval, the data of a particular tumor will be in the form of Table 12. The notations in Table 12 to be used to explain these tests are the same as those in Table 7 except that the kth interval is now the entire study period. The second subscript, k (for the kth interval) was dropped from the notations.

The Cochran–Armitage test statistic for linear trend (30) in tumor rate is defined as:

$$\chi^2_{CA} = \frac{R\left\{R\sum O_iD_i - O\sum C_iD_i\right\}^2}{O(R-O)\left\{R\sum C_iD_i^2 - \left(\sum C_iD_i\right)^2\right\}} \quad \text{or}$$

$$= \frac{\left\{\sum D_i(O_i - E_i)\right\}^2}{\sum E_iD_i^2 - \left(\sum E_iD_i\right)^2 \Big/ O}$$

where $O = \Sigma\, O_i$, $A = \Sigma\, A_i$, $R = \Sigma\, C_i$, and $E_i = OC_i/R$. The test statistic χ^2_{CA} is distributed approximately as χ^2 on one degree of freedom. The Cochran–Armitage linear trend test is based on a binomial assumption that

Table 12 Data Using the Entire Study Period as an Interval

	Group: 0	1	...	i	...	r	
	Dose: D_0	D_1	...	D_i	...	D_r	Total
Number with tumor	O_0	O_1	...	O_i	...	O_r	O
Number without tumor	A_0	A_1	...	A_i	...	A_r	A
Total	C_0	C_1	...	C_i	...	C_r	R

all animals in the same treatment group have the same risk of developing the tumor over the duration of the study. However, as mentioned previously, the animal's risk of developing the tumor increases as study time increases. The assumption is no longer valid if some animals die earlier than others. It has been shown that as long as the mortality patterns are similar across treatment groups, the Cochran–Armitage test is still valid, although it might be slightly less efficient than a survival-adjusted test (3). However, if the mortality patterns are different across treatment groups, the Cochran–Armitage test can give very misleading results.

The poly-3 test adjusts the differences in mortality among treatment groups by modifying the number of animals at risk in the denominators in the calculations of overall tumor rates in the Cochran–Armitage test to reflect "less-than-whole-animal contributions for decreased survival" (27). The modification is made by defining a new number of animals at risk for each treatment group. The number of animals at risk for the Ith treatment group C_i^* is defined as

$$C_i^* = \sum w_{ij}$$

where w_{ij} is the weight for the jth animal in the Ith treatment group and the sum is over all animals in the group.

Bailer and Portier (27) proposed applying the weight w_{ij} as follows:

$$w_{ij} = \begin{cases} 1 & \text{to animals dying with the tumor} \\ \left(\dfrac{t_{ij}}{t_{\text{sacr}}}\right)^3 & \text{to animals dying without the tumor} \end{cases}$$

where t_{ij} is the time of death of the jth animal in the Ith treatment group and t_{sacr} is the time of terminal sacrifice.

The power of 3 used in the weighting is from the observation that tumor incidence can be modeled as a polynomial of order 3. Similarly, the poly-6 test (or the general poly-k test) assigns the weight $w_{ij} = (t_{ij}/t_{\text{sacr}})^6$ [or $w_{ij} = (t_{ij}/t_{\text{sacr}})^k$)] to animals dying without the tumor when the tumor incidence is close to a polynomial of order 6 (or order k).

The class of poly-k tests is carried out by replacing the C_i's by the new numbers of animals at risk C_i^*'s in the calculation of the foregoing Cochran–Armitage test statistic.

Bieler and Williams (28) showed that the poly-3 trend test is anticonservative when tumor incidence rates are low and treatment toxicity is high. They proposed a test called the *ratio trend test* that is another modification to the Cochran–Armitage linear trend test. The ratio trend test employs the adjusted quantal response rates calculated in Bailer and Portier (27) and the delta method (32) in the estimation of the variance of the adjusted quantal response rates $p_i^* = O_i/C_i^*$. The ratio trend test treats both the numerator and the denominator of the adjusted quantal response rate as being subject to random variation.

The computational formula for Bieler-Williams ratio trend (modified C-A) test statistic is given as follows:

$$\chi_{\text{BW}}^2 = \frac{\sum m_i p_i^* D_i - \left(\sum m_i D_i\right)\left(\sum m_i p_i^*\right) \Big/ \sum m_i}{\left\{c\left[\sum m_i D_i^2 - \left(\sum m_i D_i\right)^2 \Big/ \sum m_i\right]\right\}^{1/2}}$$

where

$$c = \sum\sum (r_{ij} - r_{i.})2/[R - (r + 1)]$$

$$m_i = C_i^*/C_i$$

$$r_{ij} = y_{ij} - p^* w_{ij}$$

$$r_{i.} = \sum r_{ij}/C_i$$

y_{ij} = tumor response indicator (0 = absent at death, 1 = present at death) for the jth animal in the ith group.

It is noted that the notation c in the above formulas is not the same as C_i, the number of animals in group i, in Table 12, and the notation r for the number of treatment $r + 1$, is not the same as r_{ij} and $r_{i.}$ defined above.

Bieler and Williams (28) further showed that for tumors with low background rates, the ratio trend test yielded actual Type I errors close to the nominal levels used and was observed to be less sensitive than the poly-3 trend test to misspecification of the shape of tumor incidence function and the magnitude of treatment toxicity.

The classes of poly-*k* tests and the ratio trend test adjust for differences in survival, do not need the information on cause of death, and require only a (the terminal) sacrifice. They are also relatively robust to (not affected greatly by) tumor lethality. Results of simulation studies (3,27) show that these tests performed very well under many conditions simulated. Those tests should be used to replace the asymptotic tests that depend on the information on tumor lethality and cause of death when the information is not available.

Theoretically, exact versions of the above tests can be developed for testing data from studies with small numbers of tumor-bearing animals by applying the test procedures to all possible permuted configurations of the outcome. However, because these tests use risk sets based on all animals in each treatment group, the computations involved in the exact tests will be extensive. Therefore, for studies with small numbers of tumor-bearing animals, the current practice of treating them as incidental tumors and applying the exact permutation trend test should continue.

III. INTERPRETATION OF RESULTS OF STATISTICAL TESTS

Interpreting the results of carcinogenicity experiments is a complex process, and there are risks of both false-negative and false-positive results. The relatively small number of animals used and the low tumor incidence rates can cause a carcinogenic drug not to be detected (i.e., a false-negative error is committed). Because of the large number of comparisons involved (usually two species, two sexes, 20–30 tissues examined), there is also a great potential for finding statistically significant positive trends or differences in some tumor types that are due to chance alone (i.e., a false-positive error is committed). Therefore, it is important that an overall evaluation of the carcinogenic potential of a drug take into account the multiplicity of statistical significance of positive trends and differences and that it make use of historical information and other information related to biological relevance (e.g., positive findings at the same site in the other sex and/or in the other species, and evidence of increased preneoplastic lesions at the target organs/tissues).

A. Adjustment for the Effect of Multiple Tests (Control over False-Positive Error)

It is well known that trend tests are more powerful (i.e., with higher probability) than pairwise comparisons for detecting a true carcinogen. Also it is difficult to justify the use of data from only the control and the high-dose groups in a study having four or five treatment groups. Therefore, it is believed that tests for trend, instead of pairwise comparison tests between control and high-dose groups, should be the primary tests in the evaluation of drug-related increases in tumor rate.

Statistical and nonstatistical procedures have been proposed for controlling the overall false-positive rate. Surveys of some of those procedures can be found in Lin and Ali (11) and Fairweather et al. (22). Here, only the statistical decision rules for controlling the overall false-positive rates associated with trend tests and pairwise comparisons in interpreting the final results of carcinogenicity studies are discussed.

In the past, statisticians used the statistical decision rule developed by Haseman (33) in tests for significance of trends in tumor incidence. The decision rule was originally developed for pairwise comparison tests in tumor incidence between the control and the high-dose groups and was derived from results of carcinogenicity studies conducted in the National Toxicology Program (NTP). The decision rule tests the significance differences in tumor incidence between the control and the dose groups at the 0.05 level for rare tumors and at the 0.01 level for common tumors. A tumor type with a background rate of 1% or less is classified as rare by Haseman and as common otherwise. Haseman's original study and a second study using more recent data with higher tumor rates show that the use of this decision rule in the control–high pairwise comparison tests would result in an overall false-positive rate between 7% and 8% and between 10% and 11%, respectively (33,34).

There has been some concern about the possibility of an excessive overall false-positive error when the Haseman decision rule is applied to the trend tests, since data from all treatment groups are used in the tests. Results from recent studies show that this concern is valid. Based on those studies, the overall false-positive error associated with trend tests with the use of the Haseman decision rule just mentioned is about twice as large as that associated with control–high pairwise comparison tests.

Based on recent studies using real historical control data of CD-1 mice and CD rats from Charles River Laboratory and simulation studies conducted by the author and in collaboration with the NTP, a new statistical decision rule for tests for positive trend in tumor incidence has been developed. The new decision rule tests for the positive trend in incidence rates in rare and common tumors at the 0.025 and 0.005 levels of significance, respectively. The new decision rule achieves an overall false-positive rate of around 10% in a standard two-species and two-sex study (35). The 10% overall false-positive rate is seen by regulatory statisticians as appropriate in a new-drug regulatory setting.

Statistical literature concentrates on the methodology of tests for positive trend in tumor rate for the two reasons already mentioned. However, there are situations in which pairwise comparisons between control and individual treated groups may be more appropriate than trend tests based on the following pharmacological consideration: Trend tests assume that a response of the biological system (carcinogenic effect) is related to doses or systemic exposure weights or ranks. The assumption may be true for simple, direct-acting carcinogens in studies not complicated by excessive toxicity. However, there are many cases in which the response to a drug metabolite is mediated through receptor (or enzyme) effects that may be saturated even at the low dose, or is compounded by dose-related toxicity, or is complicated by a myriad of other nonlinear forms. Under those situations, the original decision rule by Haseman (33) should be used in interpreting the results of the pairwise comparison tests. It is suggested that both trend tests and pairwise comparison tests be conducted.

Due to the cost (between $1 million and $2 million) and time (a minimum of 3 years) needed to conduct a standard long-term in vivo carcinogenicity study and to the increased insight into the mechanisms of carcinogenicity due to the advances made in molecular biology, it has been suggested recently to look for alternative in vivo approaches for the assessment of carcinogenicity. The International Conference on Harmonization (ICH) has approved a document entitled "Draft Guideline on Testing for Carcinogenicity of Pharmaceuticals" (2). The guideline outlines experimental approaches to the evaluation of carcinogenic potential that may obviate the necessity for the routine use of two long-term rodent carcinogenicity studies for those pharmaceuticals that need such evaluation. In the document, the ICH made provision for drug sponsors either to continue to conduct two long-term rodent carcinogenicity studies or to use the alternative approach of conducting only one long-term rodent carcinogenicity study and a short- or medium-term rodent test system to evaluate the carcinogenic potential of their drug products. The short- or medium-term rodent test systems include the use of models, such as models of initiation-promotion in rodents, transgenic rodents, or newborn rodents, that provide insight into carcinogenic endpoints in vivo. The current strategy agreed to by the regulatory agencies and associations of pharmaceutical manufacturers of the ICH is to encourage drug sponsors to use the alternative approach to test the carcinogenicity of their products. Results from a separate study by the author using historical control data of CD rats and CD mice (36,37) showed that the use of significance levels 0.05 and 0.01 in tests for positive trend in incidence rates of rare tumors and common tumors, respectively, will result in an overall false-positive rate of around 10% in a study in which only one 2-year rodent bioassay is conducted. The decision rule for testing differences in incidence rates between the control and individual treated groups in a study in which only one 2-year rodent bioassay is conducted is under development and is not available at the moment.

The decision rules for testing positive trend or differences between control and individual treated groups in incidence rates for standard studies using two species and two sexes, and studies following the ICH guideline and using only one 2-year rodent bioassay are summarized in Table 13.

Table 13 Statistical Decision Rules Aiming at Controlling Overall False-Positive Rates Associated with Tests for Positive Trend or with Control–High Pairwise Comparisons in Tumor Incidences to Around 10% in Carcinogenicity Studies of Pharmaceuticals

	Tests for positive trend	Control–high pairwise comparisons
Standard studies with two species and two sexes	Common and rare tumors are tested at 0.005 and 0.025 significance levels, respectively	Common and rare tumors are tested at 0.01 and 0.05 significance levels, respectively (by Haseman's rule)
Alternative ICH Studies with one species and two sexes	Common and rare tumors are tested at 0.01 and 0.05 significance levels, respectively	Under development and not available at the moment

B. Use of Historical Control Data

The concurrent control group is always the one that is most appropriate and important in testing drug-related increases in tumor rates in a carcinogenicity experiment (38). However, if used appropriately, historical control data can be very valuable in the final interpretation of the study results. Due to the large study-to-study variations caused by differences in nomenclature, laboratory, time period, and pathologists, it is extremely important that the historical control data chosen should be comparable with the current study. Appropriate historical control data are useful in classifying and assessing the significance of rare tumors. Since statistically significant findings in rare tumors are less likely to be due to false-positive error, they provide strong evidence of carcinogenic potential. For this reason, rare tumors can be tested with less stringent statistical decision rules. However, as mentioned previously, an overall evaluation of the carcinogenic potential of a drug should take into account the multiplicity of statistical significance of positive trends and differences and make use of historical information and other information related to biological relevance. In cases of marginally significant trends, historical control data can be used to help investigators evaluate if those findings are biologically significant. If the tumor rates in the treated groups in the experiment are within the normal ranges of the historical control data, then the marginally significant findings can be discounted. Historical control data can also be used as a quality control mechanism for a carcinogenicity experiment by assessing the reasonableness of the spontaneous tumor rates in the concurrent control group (38) and for evaluation of disparate findings in dual concurrent controls.

In addition to the informal use of historical control data in the interpretation of statistical testing results just mentioned, there are proposed formal statistical procedures that allow the incorporation of appropriate historical control data in tests for trend in tumor rate. Tarone (39), Hoel (40), Hoel and Yanagawa (41), and Tamura and Young (42,43) proposed some empirical procedures using the beta-binomial distribution to model historical control tumor rates and to derive approximate and exact tests for trend. The results from those studies show that the incorporation of the historical control data improves the power of the tests. The greatest improvement of power is shown in the tests for rare tumors. Dempster et al. (44) proposed a Bayesian procedure to incorporate historical control data in statistical analysis. The procedure uses the assumption that the logits of the historical control tumor rates were normally distributed.

There are some technical issues in the use of these formal statistical procedures. They work well in the situations in which historical data from a large number of studies with relatively large control groups are available for obtaining reliable estimations of the parameters of the prior distributions. The maximum likelihood estimators (MLEs) of the prior parameters were shown to be unstable, and the distributions of the MLEs were skewed to the right, i.e., with bunching above the mean and a long tail below the mean. The skewness is severe for cases in which only historical data of a few small control groups were available. The skewness of the MLEs inflates the Type I error of the tests. Also, these procedures were developed to incorporate historical control data into the Cochran–Armitage test for linear trend in tumor incidence. Since the Cochran–Armitage test is a survival-unadjusted procedure, these procedures cannot be applied to studies with significant differences in survival among treatment groups.

C. Evaluation of Validity of Designs of Negative Studies

In negative or equivocal studies, that is, studies for which statistical tests did not detect any statistically significant positive trend or difference in tumor rate (45), the statistical reviewers should perform a further evaluation of the validity of the design of the experiment to see if there were sufficient numbers of animals living long enough to get adequate exposure to the chemical and to be at risk of forming late-developing tumors and to see if the doses used were adequate to present a reasonable tumor challenge to the tested animals.

As a rule of thumb, a 50% survival rate of the 50 initial animals in any treatment group between weeks 80 and 90 of a 2-year study may be considered a sufficient number and adequate exposure. However, the percentage can be lower or higher if the number of animals used in each treatment/sex group is larger or smaller than 50, but between 20 and 30 animals should still be alive during these weeks.

The adequacy of doses selected and of the animal tumor challenge in long-term carcinogenicity experiments is evaluated by pharmacologists and toxicologists based on the previously mentioned ICH approaches as well as the results of the long-term carcinogenicity experiments. To assist the evaluation, statistical reviewers are often asked to provide analyses of body weight and mortality differences and, occasionally, differences of other data between treated and control groups.

REFERENCES

1. ICH. Guidance for Industry: Dose Selection for Carcinogenicity Studies of Pharmaceuticals, ICH—S1C, 1995.
2. ICH. Guidance for Industry: Draft Guideline on Testing for Carcinogenicity of Pharmaceuticals, ICH—S1B, 1996.
3. Dinse, G. E. A comparison of tumor incidence analyses applicable in single-sacrifice animal experiments. Stat Med 13:689–708, 1994.
4. Dinse, G. E., and J. K. Haseman. Logistic regression analysis of incidental-tumor data from animal carcinogenicity experiments. Fundamental App Toxicol 6:751–770, 1986.
5. Dinse, G. E., and S. W. Lagokos. Regression analysis of tumor prevalence data. J Royal Statistical Soc C 32: 236–248, 1983.
6. Gart, J. J., D. Krewski, P. N. Lee, R. E. Tarone, and J. Wahrendorf. Statistical Methods in Cancer Research, Volume III—The Design and Analysis of Long-Term Animal Experiments. International Agency for Research on Cancer, World Health Organization, 1986.
7. Haseman J. K. Statistical issues in the design, analysis and interpretation of animal carcinogenicity studies. Environmental Health Perspective 58:385–392, 1984.
8. Hoel, D., and H. Walburg. Statistical analysis of survival experiments. J Nat Canc Instit 49:361–372, 1972.
9. Lin, K. K. Peto prevalence method versus regression methods in analyzing incidental tumor data from animal carcinogenicity experiments: an empirical study. 1988 American Statistical Association Annual Meeting Proceedings (Biopharmaceutical Section), New Orleans, LA, 1988.
10. Lin, K. K. CDER/FDA formats for submission of animal carcinogenicity study data. Drug Information J 32: 43–52, 1998.
11. Lin, K. K., and M. W. Ali. Statistical review and evaluation of animal tumorigenicity studies. In C. R. Buncher and J. Y. Tsay, eds. Statistics in the Pharmaceutical Industry, 2nd ed. New York: Marcel Dekker, 1994.
12. Peto, R., M. C. Pike, N. E. Day, R. G. Gray, P. N. Lee, S. Parish, J. Peto, S. Richards, and J. Wahrendorf. Guidelines for simple, sensitive significance tests for carcinogenic effects in long-term animal experiments. In: Long-Term and Short-Term Screening Assays for Carcinogens: A Critical Appraisal, World Health Organization, 1980.
13. Thomas, D. G., N. Breslow, and J. J. Gart. Trend and homogeneity analyses of proportions and life table data. Computer Biomedical Res 10:373–381, 1977.
14. Westfall, P. H., and S. S. Young. Resampling-Based Multiple Testing, Examples and Methods for *P*-Value Adjustment. New York: Wiley, 1993.
15. Cox, D. R. Regression Models and Life Tables (with discussion). J Royal Stat Soc B 34:187–220, 1972.
16. Breslow, N. A generalized Kruskal–Wallis test for comparing *K* samples subject to unequal patterns of censorship. Biometrics 57:579–594, 1970.
17. Gehan, E. A. A generalized Wilcoxon test for comparing *K* samples subject to unequal patterns of censorship. Biometrika:52,203–223, 1965.
18. Cox, D. R. The analysis of exponentially distributed life-times with two types of failures. J Royal Stat Soc B 21:4121–421, 1959.
19. Tarone, R. E. Tests for trend in life table analysis. Biometrika 62:679–682, 1975.
20. Mantel, N., and W. Haenszel. Statistical aspects of the analysis of data from retrospective studies of disease. J Nat Cancer Res 22:719–748, 1959.
21. Ali, M. W. Exact versus asymptotic tests of trend of tumor prevalence in tumorigenicity experiments: A comparison of *P*-values for small frequency of tumors. Drug Information J 24:727–737, 1990.
22. Fairweather, W. R., A. Bhattacharyya, P. P. Ceuppens, G. Heimann, L. A. Hothorn, R. L. Kodell, K. K. Lin, H. Mager, B. J. Middleton, W. Slob, K. A. Soper, N. Stallard, J. Ventre, and J. Wright. Biostatistical methodology in carcinogenicity studies. Drug Information J 32:401–421, 1998.
23. Society for Toxicology. Animal data in hazard evaluation: paths and pitfalls. Fundamental Appl Toxicol 2: 101–107, 1982.
24. Haseman, J. K., J. S. Winbush, and M. W. O'Donnell. Use of dual control groups to estimate false positive rates in laboratory animal carcinogenicity studies. Fundamental Appl Toxicol 7:573–584, 1986.
25. Dayan, A. D. Biological assumptions in analysis of the bioassay. In: Carcinogenicity, The Design, Analysis and Interpretation of Long-Term Animal Studies. H. C. Grice and J. L. Ciminera, eds. New York: Springer-Verlag, 1988.
26. Haseman, J. K., G. Hajian, K. S. Crump, M. R. Selwyn, and K. E. Peace. Dual Control Groups in Rodent Carcinogenicity Studies. In: Statistical Issues in Drug Research and Development. K. E. Peace, ed. New York: Marcel Dekker, 1990.
27. Bailer, A., and C. Portier. Effects of treatment-induced mortality on tests for carcinogenicity in small samples. Biometrics 44:417–431, 1988.
28. Bieler, G. S., and R. L. Williams. Ratio estimates, the delta method, and quantal response tests for increased carcinogenicity. Biometrics 49:793–801, 1993.
29. Cochran, W. (1954). Some methods for strengthening the common χ^2 tests. Biometrics:10,417–451, 1954.
30. Armitage, P. Tests for linear trends in proportions and frequencies. Biometrics 11:375–386, 1955.
31. Armitage, P. Statistical Methods in Medical Research. New York: Wiley, 1971.
32. Woodruff, R. S. A simple method for approximating the variance of a complicated estimate. J Am Statistical Asso 66:411–414, 1971.

33. Haseman, J. K. A reexamination of false-positive rates for carcinogenesis studies. Fundamental Appl Toxicol 3:334–339, 1983.
34. Haseman, J. K. Personal communication to Robert Temple, M.D., CDER, FDA, 1991.
35. Lin, K. K., and M. A. Rahman. Overall false positive rates in tests for linear trend in tumor incidence in animal carcinogenicity studies of new drugs. J Pharmaceut Stat 8:1–22, 1998.
36. Lin, K. K. Control of overall false positive rates in animal carcinogenicity studies of pharmaceuticals. Presented at the 1997 FDA Forum on Regulatory Sciences, December 8–9, 1997, Bethesda, Md, 1997.
37. Lin, K. K., and M. A. Rahman. False Positive Rates in Tests for Trend and Differences in Tumor incidence in Animal Carcinogenicity Studies of Pharmaceuticals under ICH Guideline S1B. Unpublished report, Division of Biometrics 2, Center for Drug Evaluation and Research, Food and Drug Administration, 1998.
38. Haseman, J. K., J. Huff, and G. A. Boorman. Use of Historical control data in carcinogenicity studies in rodents. Toxicological Pathol 12:126–135, 1984.
39. Tarone, R. E. The use of historical control information in testing for a trend in proportions. Biometrics 38: 215–220, 1982.
40. Hoel, D. G. Conditional two-sample tests with historical controls. In: P. K. Sen, ed. Contributions to Statistics, Amsterdam: North-Holland, 1983.
41. Hoel, D. G., and T. Yanagawa. Incorporating historical controls in testing for a trend in proportions. J Am Statistical Asso 81:1095–1099, 1986.
42. Tamura, R. N., and S. S. Young. The incorporation of historical information in tests of proportions: simulation study of Tarone's procedure. Biometrics 42:343–349, 1986.
43. Tamura, R. N., and S. S. Young. A stabilized moment estimator for the beta-binomial distribution. Biometrics 43:813–824, 1987.
44. Dempster, A. P., M. R. Selwyn, and B. J. Weeks. Combining historical and randomized controls for assessing trends in proportions. J Am Statistical Asso 78:221–227, 1983.
45. Chu, K. C., C. Cueto, and J. M. Ward. Factors in the evaluation of 200 national cancer institute carcinogen bioassays. J Toxicol Environmental Health 8:251–280, 1981.

Karl K. Lin

Carry-Forward Analysis

See also *Clinical Trials*

I. INTRODUCTION AND BACKGROUND

In clinical studies, data are often collected over time. The data collected over a period of time from subjects participating in these studies are referred to as *longitudinal* data by statisticians and biometricians. In longitudinal studies, each subject contributes multiple observations to the analysis. These observations are assumed to be correlated within a subject. It is frequently the case in clinical research that subjects drop out of a study before completion. This leads to an analytical challenge usually referred to as the *missing-data problem*. A number of references discussed the issues of missing data (e.g., Rubin, 1976), and dropouts (e.g., Diggle and Kenward, 1994) as well as how to handle these problems (e.g., Little, 1995).

In a report of clinical study results, it is often desirable to perform an "all randomized subjects" analysis. Implementation of this principle becomes a major challenge in data analysis when subjects drop out of the study and no subsequent follow-up information is available. Supposing this principle is interpreted as analyzing all of those subjects who were randomized to the study, imputation techniques can be an easy way to implement it; i.e., for those subjects who did not complete the study, data are inputed so that these subjects can be included in the "all randomized subjects" analysis.

One popular imputation method used in the pharmaceutical industry is to analyze the last observation carry-forward (LOCF) data. The concept of carrying observations forward can best be illustrated by an example. Suppose in a 4-week clinical study, subjects are evaluated at baseline (week 0) and at weeks 1, 2, 3, and 4 after baseline and the clinical objective is to study the drug effect at week 4. Efficacy measurements are collected at these five time points (weeks 0, 1, 2, 3, and 4). In this study, the primary comparison of treatment effect is made by comparing changes-of-efficacy measurements from baseline to week 4. Among all subjects randomized to this study, assume N subjects contribute the baseline measurement and at least one postbaseline measurement. After the study was completed, there were n_1 subjects who dropped out before week 2, an additional n_2 subjects who dropped out before week 3, and finally n_3 more subjects who dropped out before week 4. The week 4 analysis would contain only $N - n_1 - n_2 - n_3 = n$ subjects, if only observed data are analyzed. In order to analyze all N subjects, data collected from the $(N - n)$ subjects who dropped out must be used. One way is to take the last observed value prior to dropout from each of these subjects and treat them as the final data (i.e., carry forward the last observation from each subject who dropped out before the week 4 evaluation).

This concept is very useful in analyzing data at a fixed time point. In this example, if the objective is to compare subjects' responses at week 4, then last observation carry-forward (LOCF) has been a convenient way to include all of the N subjects. In this case, the focus is to analyze data at one or a few selected time points, instead of evaluating the response for the overall time course. It can be viewed as a type of univariate analysis, rather than the repeated-measure analysis. Under LOCF, it is assumed that the distribution of responses after dropping out of the study is not different from the distribution of responses at the last available time point.

Section II discusses some of the regulatory considerations regarding this issue. Section III describes some statistical methods for handling carry-forward analysis. A numerical example is presented in Sec. IV. Section V provides a discussion.

II. REGULATORY REQUIREMENTS

In general, regulatory guidelines do not cover carry-forward analyses specifically. These types of discussions usually appear under the "Intention-to-treat" or other related topics. Recently, the International Conference on Harmonization (ICH) proposed a set of documents in an attempt to set global standards for drug development and approval considerations. These documents are generally referred to as ICH Guidelines. The current version of ICH E9 Guideline (1997) discusses analysis based on "all randomized subjects" and analysis based on "per protocol subjects." In Section 5.2 of ICH E9, this document states:

> The intention-to-treat principle implies that the primary analysis should include all randomized subjects. In practice this ideal may be difficult to achieve.... [A]nalysis sets referred to as "all randomized subjects" may not, in fact, include every subject. For example, it is common practice to exclude from the all randomized set any subjects who failed to take at least one dose of trial medication or any subject without data post randomization. No analysis is complete unless the potential biases arising from these exclusions are addressed and can be reasonably dismissed.

In practice, clinical study results are analyzed following the intention-to-treat principle. It is obvious from the quoted paragraph that the analysis of "all randomized subjects" may not be easily achieved. Therefore, the idea is to define a group of subjects as close to the "all randomized subjects" as possible, and statistical analyses are performed on this group. When this is the case, imputation methods are often used to obtain such a group of subjects. Carry-forward analysis is a natural application of the imputation concept.

Using the example given in the previous section, the best representative group of intention-to-treat subjects is the group of N subjects who contribute the baseline measurement and at least one postbaseline measurement. It is important to note that this group of subjects can be less than the "all randomized subjects," because there may be subjects who were randomized to the study but did not contribute either the baseline or any postbaseline measurements, and those subjects are not included. In analysis of changes from baseline to the week 4 response for these "intention-to-treat" subjects, carry-forward methods can be used to implement this data set.

The ICH E9 (Section 5.2) further states:

> The "per protocol" set of subjects, sometimes described as the "valid cases," the "efficacy" sample or the "evaluable subjects" sample, defines a subset of the data used in the all randomized subjects analysis and is characterized by the following criteria:
>
> (i) the completion of a certain per-specified minimal exposure to the treatment regimen;
> (ii) the availability of measurements of the primary variable(s);
> (iii) the absence of any major protocol violations including the violation of entry criteria where the nature of and reasons for these protocol violations should be defined and documented before breaking the blind.

In the previous example, the primary objective is to study treatment effects after subjects were treated with the study medication for 4 weeks. Hence the group of per protocol subjects is the group of n subjects who completed the 4-week treatment.

ICH E9 further indicates that consistent analytical results across various sets of subjects will support clinical conclusions. In this example, if the intention-to-treat (LOCF), the per protocol subjects (week 4 completer), and other possible analyses demonstrate similar results, then the conclusion would be more robust. In case these results are inconsistent, additional analyses will be needed to investigate the differences further.

In the United States, the Food and Drug Administration (FDA) Guidelines are prepared for diseases or therapeutic areas separately. If carry-forward analysis is relevant to a specific therapeutic area, then the corresponding FDA Guideline may mention it briefly. For example, in the treatment of rheumatoid arthritis, long-term clinical studies typically are carried out, and dropouts are very common in these studies. Hence the FDA Draft Guideline for Rheumatic Diseases (1997) covers this issue. This draft states:

> Methods used to handle dropouts, such as the "LOCF" and "completers" analyses, are not fully satisfactory even though they have often served as the basis for determining that adequate statistical evidence of efficacy has been provided. The LOCF method generally does not preserve the size of the test.

This indicates that clinical conclusions based only on LOCF and completer analysis may not be robust, so additional analyses may also be necessary. The advantages and disadvantages of imputation methods (in particular, LOCF analysis) are discussed in Section V.

III. STATISTICAL METHODS

Statistical methods used to analyze the carry-forward data are the same as those used to analyze the original data. For example, if the proposed analytical method is to apply analysis of variance (ANOVA) on the original data, then ANOVA can be applied to analyze the carry-forward data.

In clinical trials, response measurements are often disease specific. Some of these measurements include blood pressure (for cardiovascular diseases), FEV_1 (forced expiration volume at 1 second, for pulmonary diseases), HbA_{1c} (hemoglobin A_{1c}, for diabetes), number of swollen joints (for rheumatoid arthritis), BPRS (brief psychiatric reading scale, for schizophrenia) are usually treated as continuous variables in data analysis. Hence the carry-forward data from these variables are also analyzed as continuous variables. For these data, *t*-test, ANOVA, ANCOVA (analysis of covariance), other linear models, or nonlinear models are often used to perform the statistical analysis.

In many cases, categorical data such as physician assessment (no symptom, mild, moderate, severe, very severe), pain relief (5-point scale, 0 indicates no pain relief, 4 indicates complete pain relief), number of emetic events in a given time interval, and patient response scale (markedly deteriorated, moderately deteriorated, deteriorated, no change, improved, moderately improved, markedly improved) are used as response variables. Hence, the carry-forward data are categorical. In these cases, methods used to analyze categorical data can also be applied to analyze the carry-forward data. Under certain circumstances, nonparametric techniques are applied for the analysis of the original data; these nonparametric methods can then be applied to analyze the carry-forward data.

Other ways of carrying forward observations are applied in clinical trials, also. Examples are carrying forward the worst case, carrying forward the best case, and carrying forward baseline. In the case of carrying forward the worst case, it is assumed that the subject's condition deteriorated after dropping out, and the worst value the subject ever experienced during the study is used as the analysis value. When the best case is carried forward, it is assumed that the subject dropped out because he or she recovered. In another situation, baseline values may be carried forward for the analysis. Carrying forward baseline requires the assumption that the treatments have no effect on responses for those subjects who dropped out. It is not a common practice to carry forward baselines. One extreme type of imputation in a study comparing a test drug against placebo is to carry forward the best values of placebo-treated subjects and the worst values for test-drug-treated subjects.

Carry-forward analysis can also be viewed as carrying forward the weighted-average responses (see, e.g., Frison and Pocock, 1992), carrying forward the linear trend, or using other modeling approaches to impute data for subjects who dropped out. One simple way of carrying forward weighted-average responses it to compute the mean of all the available observations after baseline for each subject, to treat this mean as the response measurement for the subject, and then to analyze these individual means. Carry-forward linear trend can be implemented by estimating the slope for each subject first and then analyzing these slopes as subject responses. Some of these methods can be illustrated by the numerical example in next section.

Table 1 Simulated Data From Some of These Subjects

Treatment	Subject	BAS	W1	W2	W3	W4	LOCF	WOCF
A	1	18	16	20	16	18	18	18
B	2	13	11	13	9	10	10	10
B	3	15	17	14	11	·	11	17
A	4	13	13	13	15	·	15	15
B	5	19	20	20	18	14	14	14
A	6	7	5	5	8	9	9	9
B	7	12	9	·	·	·	9	9
A	8	12	17	11	9	12	12	12
B	9	12	11	13	14	·	14	14
A	10	14	16	15	10	·	10	16
.	.	.	.	.	.	.	.	.
.	.	.	.	.	.	.	.	.
.	.	.	.	.	.	.	.	.
A	80	15	13	13	14	17	17	17

IV. A NUMERICAL EXAMPLE

Simulated data are used to demonstrate the idea of carry-forward analysis in this numerical example. Suppose in a 4-week clinical trial to study subject responses to a new drug, 80 subjects were randomized to be treated with either the new drug (treatment B) or placebo (treatment A). Each subject visited the clinic five times—baseline (before the subject was dosed with the randomized study medication), 1, 2, 3, and 4 weeks after baseline. For this particular study, the primary response variable is a symptom score, possible values can be between 0 (no symptom) and 20 (worst case), inclusive. Data from some of these subjects are given in Table 1; the number of subjects at each visit by treatment group is summarized in Table 2.

From Table 1 we see that subjects 1, 4, 6, 8, 10, . . . , 80 were randomized to be treated with placebo (treatment A), and subjects 2, 3, 5, 7, 9, . . . were treated with the new drug (treatment B). For subject 1, the baseline value (BAS) was 18; i.e., this subject started with a symptom score of 18 before receiving any dose of the study medication (placebo, for this subject). The symptom improved slightly after the first week of treatment week 1 (W1) value was 16, and then deteriorated to 20 at week 2 (W2), and became 16 at week 3. After 4 weeks of treatment, the symptom score of this subject returned to a value of 18 (W4). In this case, the last observation carried forward (LOCF) is 18, and the worst case carried forward (WOCF) was also 18. Subject 3 visited the clinic for the first 3 weeks of treatment but did not have the week 4 evaluation. Hence the LOCF is to carry forward the last available value: 11 (LOCF = 11), but the worst case this subject experienced was 17 (from week 1), and thus the WOCF is 17. In this example, it is assumed that most subjects dropped out of the study because of lack of disease improvement. Hence carry-forward best values are not included in this example. Baselines are not carried forward here, either.

From Table 2 we see that 40 subjects were randomized to each treatment group ($N = 40$ for both A and B). No subject dropped out before week 1 visit. One and six subjects dropped out after week 1 from treatments A($n_1 = 1$) and B ($n_1 = 6$), respectively. For treatment group A, $n_2 = 7$, $n_3 = 3$, and $n = 29$. For treatment group B, $n_2 = 2$, $n_3 = 4$, and $n = 28$.

The objective of this study is to demonstrate that the new drug improves the disease symptom more than the placebo does. Before entering the study, subjects with

Table 2 Number of Subjects at Each Visit by Treatment Group

Treatment	Baseline	Week 1	Week 2	Week 3	Week 4
A	40	40	39	32	29
B	40	40	34	32	28

the disease under study were randomized at baseline. After 4 weeks of treatment, the improvement can be measured using the difference in symptom score from baseline to week 4. In this example, two-sample t-tests are used to make these comparisons. Table 3 illustrates these analytical results.

Note that in Table 3, the BAS result is the comparison of baseline symptom scores (measured before subjects took any study medication) between treatment groups A and B. DW4 is the comparison of differences from baseline to the week 4 visit; i.e., for each subject, DW4 is obtained by subtracting the baseline score from the corresponding week 4 score. Then the DW4 values are compared between the two treatment groups. DLOCF denotes the difference from baseline to LOCF; DWOCF denotes the difference from baseline to WOCF. In these differences, a negative value indicates an improvement. The results in Table 3 are part of the SAS output from the PROC TTEST based on the assumption that treatments A and B have equal variance for all four variables.

From Table 3, the comparison under BAS (baseline) demonstrates that 40 subjects are randomized to each of treatment A and B ($N = 40$ for TRT = A, and $N = 40$ for TRT = B). Subjects in treatment group A have a baseline mean score of 12.775 with a standard deviation of 3.18. Subjects in treatment group B have a mean score of 11.85 with a standard deviation of 3.33. When comparing these two treatment groups, the t-test value (under the equal-variance assumption) is 1.2696 with 78 degrees of freedom, and the p-value is 0.2080, indicating no significant difference between treatments A and B at baseline.

Comparisons under DW4 indicate that treatment A has 29 subjects completing the week 4 visit ($N = 29$), with a mean improvement of 0.552 units in symptom score and a standard deviation of 2.028. Treatment B has 28 subjects with a mean of 1.786 and a standard deviation of 2.079. Under the equal-variance assumption, the t-test value is 2.2683, which is significant at 0.0273 ($p < 0.05$), with 55 degrees of freedom. This indicates that when the 4-week completers (subjects completed 4 weeks of treatment) are analyzed, the new drug improves disease symptom by 1.786 units, which is significantly more than with the placebo (0.552 units).

DLOCF provides 40 subjects per treatment group because with LOCF, every subject has an improvement score to be analyzed, regardless of whether a subject dropped out or not. Mean improvement from the new drug treatment (B) is 1.65 units, which is significantly more ($p = 0.0099$, assuming equal variance) than the 0.425 units from the placebo-treated subjects. Similarly, DWOCF indicates a mean improvement from the new drug of 1.325, which is significantly more ($p = 0.0209$, assuming equal variance) than the 0.25 from the placebo treatment, based on 40 subjects per treatment group.

Based on the results in Table 3, the 80 subjects entered in the study with comparable baseline disease symptoms ($p = 0.2080$). After 4 weeks, 29 and 28 subjects completed the study treatment of the placebo and the new drug, respectively. By analyzing the differences from baseline to week 4 (DW4), we see that the new drug provides a greater mean improvement than the placebo treatment. Using last observation carry-forward (DLOCF) analysis, the mean improvements from both treatment groups are less than those from the week 4 visit (1.65 vs. 1.786 for treatment B, 0.425 vs. 0.552 for treatment A). This fact indicates that the subjects who dropped out before week 4 tended to experience less improvement than the subjects who completed the 4-week treatment. As expected, worst observation carry forward (DWOCF) provides the least improvement

Table 3 Results Obtained from t-Tests

Variable	TRT	N	Mean	Std dev	Std error	T	DF	Prob > $\|T\|$
BAS	A	40	12.77500000	3.18238340	0.50317900			
	B	40	11.85000000	3.33243578	0.52690436	1.2696	78.0	0.2080
DW4	A	29	−0.55172414	2.02812734	0.37661379			
	B	28	−1.78571429	2.07912273	0.39291726	2.2683	55.0	0.0273
DLOCF	A	40	−0.42500000	1.98568596	0.31396452			
	B	40	−1.65000000	2.15489901	0.34071945	2.6440	78.0	0.0099
DWOCF	A	40	−0.25000000	1.91819894	0.30329388			
	B	40	−1.32500000	2.15296456	0.34041359	2.3578	78.0	0.0209

among these three comparisons (1.325 < 1.65 < 1.786 for treatment B, and 0.25 < 0.425 < 0.552 for treatment A). All three analyses (DW4, DLOCF, DWOCF) provide consistent results that the new drug helps improve the disease symptom more than the placebo does.

In addition to the preceding analyses, these simulated data were used to perform the average change (DAVG) and slope analyses. Average change is computed by taking the mean of all postbaseline measures for each subject first and then subtracting the corresponding baseline of the subject. Let us use subject 1 as an example: This subject has scores 16, 20, 16, and 18 in weeks 1, 2, 3, and 4, respectively. The average of these four values is 17.5. Subtracting baseline 18 from 17.5 gives a DAVG of −0.5 for subject 1. A two-sample t-test is then applied to analyze these DAVG scores between treatment groups A and B.

Slope is estimated from each subject using simple linear regression (using PROC REG in SAS). These individual slopes are then used as the response variable in a two-sample t-test. The results from DAVG and SLOPE are given in Table 4.

The DAVG result in Table 4 indicates that the mean of average improvements are 1.15 for treatment B and 0.381 for treatment A, from 40 subjects of each treatment group. A two-sample t-test under the equal-variance assumption demonstrates a significant difference ($p = 0.0025$) in favor of treatment B. For the SLOPE comparison, an F-test indicates the variance for treatment A is different from the variance for treatment B ($p = 0.006$). Hence an unequal-variance t-test is applied to compare these slopes. From the t-test, the mean slope of −0.5475 for treatment B is significantly different ($p = 0.0145$) from the − 0.075 for treatment A, indicating a more rapid improvement can be obtained from the new drug, as compared to the placebo. Both DAVG results and SLOPE results are consistent with the findings from Table 3.

V. DISCUSSION

As mentioned previously, carry-forward analysis is one of the imputation methods used in analyzing clinical data. The primary advantages of carry-forward analysis include the following: It is convenient to implement for the intention-to-treat population, it is easy to communicate to nonstatisticians, and the results obtained from these analyses can be clearly interpreted. When an experimental drug is filed for regulatory approval, usually a large number of studies using various types of variables are combined to formulate an overall conclusion. In this case, a method applied consistently across different variables and studies can be very useful. Another advantage of carry-forward analysis is that this method can be consistently applied to all types of response variables and to different study designs.

However, there are a few disadvantages in the carry-forward methods. One of them is that the estimates obtained from these results can be biased. When the distribution of responses from subjects who completed the study is different from the distribution from those who dropped out, carry-forward analysis may provide biased estimates. The bias can affect the mean estimate, the variance estimate, or both. Biased estimates from carry-forward analysis may lead to incorrect conclusions from the study. Another disadvantage is that the carry-forward analysis may reduce power. When this is the case, clinical conclusions may also be misleading. Section III mentioned an extreme case—imputing the best values for placebo-treated subjects and imputing worst values for test-drug-treated subjects. Apparently, this method may create more bias.

As discussed previously, in addition to carrying forward the last observation, the worst observation, the best observation, or the baseline, a few other types of imputation can also be considered. The carry-forward weighted-average response and carry-forward linear-trend (slope) methods are illustrated in the numerical example of Sec. IV. Analysis using slope as the response variable is based on the assumption that the rate of change for each subject does not change after dropout. These variables can be analyzed using parametric or nonparametric methods, depending on distributional properties.

In general, imputation methods are frequently used in dealing with missing-data problems. Most of the advantages and disadvantages for imputation methods are similar to those for carry-forward analysis. Since these methods are widely used in application, Rubin (1977) developed multiple-imputation methods to overcome some of the disadvantages from single imputations. The idea of multiple imputation is to impute more than one possible value to each missing data point and to obtain multiple "imputed data sets." Statistical analyses are then performed on these data sets, and results from the multiple analyses are combined to formulate inferences. A similar idea is to perform sensitivity analysis by examining results from many carry-forward analyses. In the numerical example of Sec. IV, analyses performed on the completers (DW4), last observation carry-forward (DLOCF), worst observation carry-forward (DWOCF), average response, and individual slope all demonstrate similar results. This is a case

Table 4 Additional Results from t-Tests

Variable	TRT	N	Mean	Std dev	Std error	T	DF	Prob > $\lvert T\rvert$
DAVG	A	40	−0.38125000	0.92736329	0.14662901			
	B	40	−1.15000000	1.24796130	0.19732001	3.1271	78.0	0.0025
SLOPE	A	40	−0.07500000	0.58474189	0.09245581	2.5152	61.6	0.0145
	B	40	−0.54750000	1.03428100	0.16353418			

For H_0: variances are equal, $F' = 3.13$, DF = (39,39), Prob > F' = 0.0006.

where the clinical conclusion (that the new drug provides more benefit to the subject than the placebo does) can be supported from these analyses. In case different analyses demonstrate different results, further investigation will be needed to identify the reasons for the inconsistencies.

Before choosing appropriate methods to analyze clinical data, the statistician needs to consider the disease under study, the type of design, the study objective, the primary and secondary response variables for analysis, the distributional properties of these response variables, and the clinical interpretation of the results. Subjects often drop out of the study before completion. Under this circumstance, carry-forward methods can be considered for the data analysis. Parametric or nonparametric methods can be used. When parametric methods are used—and in particular, linear or nonlinear models are chosen to perform the data analysis—potential covariates should be considered for inclusion in these models. Sometimes nonparametric analysis may be preferred. Overall, carry-forward analysis is a useful tool in analyzing longitudinal clinical data with dropout problems. However, caution must be taken before choosing these methods and interpreting the analytical results. Sensitivity analysis often will be needed to check the robustness of clinical conclusions made from these analyses.

REFERENCES

P. Diggle and M. G. Kenward. (1994). Informative drop-out in longitudinal data analysis. Journal of Royal Statistical Society, Applied Statistics, 43(1):49–93.

Draft Guidance for Industry—Clinical Development Programs for Drugs, Devices, and Biological Products for the Treatments of Rheumatoid Arthritis (RA). U.S. Department of Human Services, FDA, Jan. 6, 1997.

L. Frison and S. J. Pocock. (1992). Repeated measures in clinical trials: analysis using mean summary statistics and its implications for design. Statistics in Medicine, 11: 1685–1704.

International Conference on Harmonization (ICH) of Technical Requirements for Registration of Pharmaceuticals for Human Use—Draft Consensus Guideline E9. Jan. 16, 1997.

R. J. A. Little. (1995). Modeling the drop-out mechanism in repeated-measure studies. Journal of American Statistical Association, 90:1112–1121.

D. B. Rubin. (1976). Inference and missing data. Biometrika, 63:581–592.

D. B. Rubin. (1977). Formalizing subjective notions about the effect of nonrespondents in sample surveys. Journal of American Statistical Association, 72:538–543.

Naitee Ting

Clinical Endpoint

See also *Clinical Trials*

I. INTRODUCTION

A. Clinical Trial

A *clinical trial* is an experiment performed by a health care organization or professional to evaluate the effect of an intervention or treatment against a control in a clinical environment. It is a prospective study to identify outcome measures that are influenced by the intervention. A clinical trial is designed to maintain health, prevent diseases, or treat diseased subjects. The safety, efficacy, pharmacological, pharmacokinetic, quality-of-life, health economics, or biochemical effects are measured in a clinical trial.

There are two different types of clinical trial, confirmatory and exploratory trials (5). In an *exploratory* trial, the choice of hypothesis is data dependent, even though this study may have clear objectives. These trials explore the doses of subsequent studies and provide a basis for a confirmatory study design. A *confirmatory* trial is a well-controlled study in which the hypothesis of interest is predefined and is intended to provide hard evidence in support of claims that have clinical benefits. A confirmatory trial is less vulnerable to bias and more robust. Atherosclerotic cardiovascular disease (myocardial infarction and cardiovascular death) is the leading cause of death in women in the United States and in most developed countries worldwide. An example of a confirmatory cardiovascular clinical trial would be a multicenter, double-blind, placebo-controlled, randomized, parallel study of 10,000 patients to test whether a new treatment compared with placebo reduces the incidence of the combined endpoint of coronary death and nonfatal myocardial infarction in postmenopausal women at risk for coronary events.

B. Objective

A clinical trial generally consists of two different sets of objectives: primary and secondary. The primary objective is the most important question or the hypothesis an investigator desires to answer definitively at the end of the study and that drives the size and the power of the trial. The hypothesis of interest in a confirmatory trial is directly related to the primary objective. The secondary objective(s) answer other related or unrelated questions based on the efficacy or safety of the intervention. In our earlier example, the primary objective is to compare the effect of the treatment with placebo on the coronary death and nonfatal myocardial infarction. There could be several secondary objectives, for example, to test the effect of the treatment on myocardial revascularization, stroke, all-cause mortality, all-cause hospitalization, breast cancer, fractures, and long-term safety.

C. Outcome Measures

An *outcome measure* is a direct or indirect measurement of a clinical effect in a single, adequate, well-controlled clinical trial (it can be double-blind or open label) to obtain an effective claim regarding the intervention under investigation. The goal of the clinical trial is to assess the effect of the treatment on these outcome measures. In the primary objective there is generally one primary outcome, which is used to test whether the intervention is superior to a comparative agent (placebo control or active) or that it is not clinically inferior (equivalence) to a comparative agent (placebo control or active). Outcome measures should be selected prospectively.

D. Primary Endpoint

The *primary endpoint* is an outcome measure that is related to, and selected based on, the primary objective of the study. The size of the trial is based on the primary objective of the study and the primary endpoints. The primary endpoint in our earlier example is the combined endpoint of coronary death and nonfatal myocardial infarction.

E. Secondary Endpoints

Secondary endpoints are other outcome measures that may or may not be related to the primary objective of the study. Secondary variables are related to secondary objectives. They support the primary variables of the study. Secondary variables can be a "time to an event," "incidence rate of an event," or continuous variables related to efficacy or safety. In the previous example, the secondary endpoints are myocardial revascularization, stroke, all-cause mortality, all-cause hospitalization, breast cancer, fractures, and safety laboratory parameters.

Primary and secondary endpoints are intimately related to the primary and secondary hypotheses. These endpoints can be clinical, surrogate, economic, global, etc. To achieve the study objective, the design should be chosen adequately to reflect the objectives, appropriate primary, secondary, and global variables will be considered that will answer the objectives.

F. Clinical Endpoint

A *clinical endpoint* is a clear, hard, appropriate outcome that can be objectively assessed, not depending on the judgment of an individual, especially the investigator. Nonfatal myocardial infarction is an example of an objective endpoint, in contrast to a subjective endpoint, for example, relief of symptoms or severity of symptoms.

G. Surrogate Endpoint

An endpoint that provides an indirect measurement of a clinical effect when measuring outcome directly is not feasible or practical (due to the requirement of large sample size, long duration, and cost) is called a *surrogate endpoint* (1). Changes induced by a treatment on surrogate variables are expected to reflect changes in a clinical endpoint. Generally these endpoints are continuous response variables that can be substituted for the clinical outcomes. For example, surrogate variables might include biochemical markers of cardiovascular disease such as low-density lipoprotein (LDL) cholesterol, total cholesterol instead of myocardial infarction; spine bone mineral density instead of incidence of vertebral fractures; and CD4 count, viral load effects for the effect on HIV infection.

H. Economic Endpoint

In recent clinical trials, evaluation of a subject's health status data has become increasingly important. Several measurements regarding a subject's use of health care, cost of hospitalization, etc. are considered as economic endpoints. The quality-of-life scores based on a subject's performance, daily activity, mood, etc. (defined by the World Health Organization) are subjective measures of health status. The overall score from these measurements can be analyzed for treatment comparison.

I. Global Assessment Variables

Global assessment variables measure the overall safety and efficacy of an intervention. These are generally based on rating scales. An idea of the risk–benefit profile of an intervention can be assessed from these variables. It helps the investigative physician to balance the safety and efficacy of the intervention and decide on treating subjects by weighing its risk and benefit outcomes.

II. CLINICAL ENDPOINT

Clinical endpoint is a clear, hard, clinically relevant outcome measure that is affected by the intervention under investigation and that can be objectively assessed, not depending on the investigator's judgment. A clinical endpoint is to be selected that can be reasonably and reliably assessed and can answer the primary objective. Sometimes it is difficult to achieve both aims, so in such cases clinical judgment is required. Any definitions used to characterize the primary outcome measure should be explained clearly. Sometimes two different clinically meaningful endpoints can cross-substantiate a claim for the effect of each outcome.

A. Medical Issues

The outcome measures chosen to evaluate the efficacy or the primary objective depends on number of factors, including:

- Knowledge of adverse effects of closely related drugs
- Information from nonclinical or earlier clinical trials
- The types of subjects to be enrolled
- Pharmacodynamic or pharmacokinetic properties of the treatment

The endpoints should be capable of unbiased assessment. To have unbiased assessment of the endpoints, studies should be blinded. In some studies, the intervention is administered by one investigator and the measurements and evaluation of the endpoints is performed by an independent investigator not involved with the study. This is to maintain the trial integrity and to minimize bias introduced by any difference in investigators' judgments. The instrument to assess the primary variable should be reliable and accurate and adequately sensitive to detect any real change for a subject. Clinical endpoints generally provide strong scientific evidence regarding efficacy in a targeted population. Clinical endpoints should be such that they can be assessed on all subjects. The response variable should be measured the same way for all subjects. The assessments of the response variables should be selected as standard, widely used, and generally recognized.

B. Statistical Issues

The sample size and power calculation for a study should be based on the primary endpoint. The endpoint could be incidence rate of an event or time to an event. Since recurrences of the primary event can occur, distribution of the number of events or frequency of episodes can be of interest in this case. Sometimes subject participation ends as soon as the primary event occurs, but sometimes the investigator wants to follow a subject for a subsequent primary variable or other secondary response variables. However, if a secondary variable occurs first, the subject must be followed, because he or she is still at risk for the primary variable. If the secondary variable is a mortality endpoint, then competing risks analyses may be appropriate.

For studies that have long-term duration, it is very important to have complete information on the subjects. Since if the subjects are lost to follow-up, sometimes loss of information occurs in both the treatment and the control groups. All the statistical analyses for primary or secondary variables are usually based on the intent-to-treat principle for all the randomized subjects (3). Sometimes post hoc analyses based on a subset of the randomized population are examined—for example, the effect of an intervention for reducing the rate of occurrence of myocardial infarction in the diabetic population, to test whether the treatment is more effective in the diabetics subset or, irrespective of the population, the treatment effect is the same. Binary data can be analyzed to obtain odds ratio or difference in proportions, and time-to-an-event data can be analyzed using proportional-hazard models, Kaplan–Meier estimates. There are additional statistical issues associated with several clinical endpoints, as shown in the next sections.

C. Combined Endpoints

Sometimes the primary outcome measure occurs very infrequently, so a large sample size or a long study duration is needed to have adequate power to detect a treatment difference. In this situation, combined endpoints (for example, fatal and nonfatal myocardial infarction) are considered, to enable detection of treatment differences with a smaller sample size or shorter study duration. This combined endpoint should have a meaningful clinical interpretation; in our earlier case it is a measure of coronary heart disease. Sometimes it is useful to combine more than one measurement related to primary objective, for example, rating scales in psychiatric disorders. In this case of combined endpoints, multiplicity adjustment to the Type I error is not needed. Different variables of the combined endpoint can also be analyzed separately but may require multiplicity adjustment (see the next section). It is important to remember when answering a question relating to the combined endpoint that only one event per subject should be counted, and also a hierarchy of each component should be placed (for instance, the fatal myocardial infarction will be counted over the nonfatal myocardial infarction). In addition, a combined endpoint could be considered in order to achieve more power to perform subgroup analyses.

D. Multiple Endpoints

Multiple endpoints are different clinical events that reflect a common mechanism of action due to the intervention. Sometimes it is important to have more than one primary variable when the investigator cannot state which of several variables addresses the primary objective of the study. Use of multiple endpoints will result in an increased probability of having false-positive results. If more than one response variable is chosen, a nominal p-value for each variable will be computed. If one of the multiple variables is of most importance and is most impacted by the intervention, then Bonferroni (4) or similar adjustment methods can be used. On the other hand, if the intervention affects all the variables the same way, then adjustment is not necessary, although the effect on the Type II error and the sample size should be evaluated. The O'Brien test (7) of treatment equivalence as a global hypothesis can also be used in that case. Another way to deal with multiple correlated endpoints is to use a weighted average of the variables as a global index. The weight of each variable is the sum of the inverse variance and the covariances of that variable with the others (9). So endpoints that are less correlated with other endpoints have more weights. When all the endpoints result in statistical significance, the study becomes more effective and assertive. For example, the clinical trial for prevention of exacerbations in multiple sclerosis using beta-interferon was based on a study where two completely different endpoints—a rate of exacerbations and an MRI-demonstrated disease activity—were used. Both endpoints were significantly decreased, which resulted in the approval.

III. CLINICAL VERSUS SURROGATE ENDPOINTS

Clinical trials with a clinical endpoint are usually longer in duration, to determine an extended exposure of the treatment. Since the clinical endpoint drives the calculation of the sample size, these trials involve a large number of subjects. On the other hand, surrogate endpoints can replace the true clinical endpoint, since it will often result in a shorter duration and the smaller sample size of the trial. The sample size will be smaller due to the continuous nature of the data. Also, the duration of the trial will be reduced, since the surrogate variables tend to change earlier in the study and the measurements can be taken at several time points.

While choosing the surrogate variables, there are certain important things to consider:

Select the variables that have previously proven in literature to be highly correlated with the clinical endpoint.

Choose the ones that will be accepted by the regulatory authorities and the medical community.

Select the variables that can be reliably and accurately measured.

A lot of invasive procedures should be avoided, since they may result in high discontinuation rates.

According to Prentice (10), a valid surrogate variable should be correlated with the clinical endpoint and should capture fully the treatment's aggregate effect (accounting for all the mechanisms of action) on the clinical endpoint.

A. Statistical Considerations of Surrogate Endpoints

An example of a clinical endpoint in a cardiovascular trial is myocardial infarction; a surrogate variable is LDL cholesterol. Comparisons from the beginning and the end of the study or a certain interval using change or percentage change can be of interest. Comparison of slopes throughout the interval and comparison of treatments based on repeated-measures data are also analyzed.

B. Responder Variable

When a continuous response variable is transformed to a binary or a categorical variable it can be interpreted as a *responder variable*. For example, the original continuous variable percentage change in LDL cholesterol can be modified to a binary variable that takes two values: 1, if percentage change in LDL is less than 0; 0, if percentage change in LDL is greater than or equal to 0. The disadvantage in creating this responder variable is that not all the information is used in the analyses, which results in loss of power.

REFERENCES

1. Armitage P, Colton T, editors-in-chief. Encyclopedia of Biostatistics. Vols 1–6. West Sussex: Wiley, 1998.
2. Friedman LM, Furberg CD, Demets DL. Fundamentals of Clinical Trials. 2nd ed. Littleton, MA: PSG, 1985.
3. Gillings D, Koch G. The application of the principle of intention-to-treat to the analysis of clinical trials. Drug Inform J 25:411–424, 1991.
4. Hochberg Y. A sharper Bonferroni procedure for multiple tests of significance. Biometrika 75:800–802, 1988.
5. ICH Steering committee. ICH Harmonized Tripartite Guideline, Statistical Principles for Clinical Trials. 1998.
6. ICH Steering committee. ICH Harmonized Tripartite Guideline, General Considerations for Clinical Trials. 1997.
7. O'Brien PC. Procedures for comparing samples with multiple endpoints. Biometrics 40:1079–1087, 1984.
8. Pocock SJ. Clinical Trials: A Practical Approach. New York: Wiley, 1983.
9. Pocock SJ, Geller NL, Tsiatis AA. The analysis of multiple endpoints in clinical trials. Biometrics 43:487–498, 1987.
10. Prentice RL. Surrogate endpoints in clinical trials: definition and operational criteria. Statistics Med 8:431–440, 1989.

Sofia Paul

Clinical Pharmacology

See also *Pharmacodynamic Issues; Bioavailability and Bioequivalence*

I. INTRODUCTION AND BACKGROUND

This area of study is fairly broad. The following excerpt from the introduction to *The Pharmacological Basis of Therapeutics* (1) gives a glimpse at this field of study: "The clinician is understandably interested mainly in the effects of drugs in man. This emphasis on *clinical pharmacology* is justified, since the effects of drugs are often characterized by significant interspecies variation, and since they may be further modified by disease." From this quote one may define *clinical pharmacology* as the field of medicine whose main interest is the effects of drugs in humans. Within this definition, the field encompasses a multitude of disciplines.

A different approach for getting at the heart of clinical pharmacology comes from asking what would be the core curriculum for a training program in this field. In a 1997 issue of *Clinical Pharmacology & Therapeutics* (2), knowledge of the following is proposed as a core curriculum in clinical pharmacology and therapeutics for primary care residents:

1. Clinical pharmacokinetics (including drug absorption, distribution, redistribution, clearance, and half-life) and its application to patient-based clinical scenarios.
2. Therapeutic drug monitoring
3. Adverse drug reactions
4. Drug allergy
5. Drug–drug interactions
6. Pharmacokinetics
7. Prescribing for elderly patients
8. Prescribing for pediatric patients
9. Prescribing for pregnant and nursing women
10. Prescribing for patients with renal disease
11. Toxicology
12. Drug regulations applicable to primary care
13. Bioethical issues and principles related to prescribing
14. Therapeutic management of common clinical problems, including both drug and nondrug therapies

This list helps identify areas that need to be studied by clinical pharmacologists before a new chemical entity can be approved as a drug. Clinical pharmacology is often concerned with the safety, pharmacokinetics, and pharmacodynamics of a new compound. (See entries on pharmacokinetics and pharmacodynamics for definitions.) Clinical pharmacology studies are not synonymous with phase I studies. Phase I studies are usually clinical pharmacology studies; however, clinical pharmacology information can often be obtained from studies in other phases of development.

To further classify clinical pharmacology as a discipline, it would be useful to consider the types of clinical pharmacology studies that are performed when a new chemical entity is selected for development. The following list of studies is not meant to be inclusive or a checklist of studies that need to be done. It is meant only to give a flavor of clinical pharmacology research.

A. First Human Dose (Single-Dose Escalation) Studies

These studies are usually open-label, nonrandomized studies performed on healthy volunteers. (If a compound is known to be highly toxic and it is being developed for a life-threatening disease, then often all clinical pharmacology studies are done on patients who are not healthy volunteers.) A "no effect dose," in units of drug per body weight, is determined from toxicology studies. Then a fraction, say, one-tenth, of this dose becomes the highest that will be allowed in the clinical study. The first dose, which is much less than the maximal dose, is administered to subjects. If no adverse effect is observed, on a later date an increased dose is administered to the subject(s). The increased dose is typically restricted according to some accepted ethics-based algorithm. This is continued for each subject until either a drug-related adverse event occurs, a drug-related abnormal laboratory value is observed, or the maximal dose is achieved. The primary objective of these studies is safety; however, blood samples and sometimes urine samples are collected to get a first glimpse at the compound's pharmacokinetic properties. If a surrogate dynamic or efficacy marker is known, then appropriate tests are done to get a first glimpse of whether the compound is doing what it is supposed to do.

B. Multiple-Dose Escalation Studies

These studies are usually open-label, nonrandomized studies performed on healthy volunteers. A maximal dose is chosen from the first study. Then a lower dose is given once or twice a day to each subject for an extended time, which should be sufficient to achieve steady-state systemic concentrations. If no effect is observed, then an increased dose is administered (usually after a washout period). This is continued for each subject until either a drug-related adverse event occurs, a drug-related elevated laboratory value is observed, or the maximal dose is achieved. Again, safety is the primary reason for the study; however, pharmacokinetic and pharmacodynamic information is often obtained.

C. Fed/Fasted Studies

These studies seek to determine if the pharmacokinetic properties of the compound are affected by food. They are usually a randomized two-period crossover design using healthy volunteers. (If a compound has a long half-life, than parallel designed trials may be more appropriate.)

D. Drug Interaction Studies

The goal of these studies is to determine if the pharmacokinetics and the pharmacodynamic properties of the compound are affected by another drug. Similarly, they determine if the pharmacokinetic and pharmacodynamic properties of the other drug are affected by the new compound. Randomized three-period crossover designs with healthy volunteers are the norm.

E. Bioequivalence Studies

These studies seek to determine if the bioavailability of two formulations of the same compound are the same. Again, randomized two-period crossover designs on healthy volunteers are the norm.

F. Special Population Studies

The goal of these studies is to determine if the pharmacokinetic properties and sometimes pharmacodynamic properties are the same for different groups of people. Population traits studied include gender, race, and disease types. Special population studies have interpretation problems similar to those of observational studies. This is because, for instance, one cannot randomize an individual to a gender classification.

G. Absolute Bioavailability Studies

These studies compare the extent of absorption of an oral formulation to the intravenous formulation of a compound. Randomized two-period crossover designs with healthy volunteers are the norm.

H. Dose Proportionality Studies

A range of doses, on separate occasions, is given to each individual to determine if the compound exhibits "linear" pharmacokinetic properties. Some pharmacokinetic variables (for example, the concentration maximum and the area under the concentration time curve) are related to dose in a proportional manner if the compound exhibits "linear" pharmacokinetic properties. Other pharmacokinetic variables (for example, the plasma clearance and the apparent volume of distribution) are independent of dose if the compound exhibits "linear" pharmacokinetic properties. If the rate of elimination and absorption of a compound into the compartments of a pharmacokinetic model are proportional to the amount in the source compartment, then it is said that the compound exhibits linear pharmacokinetic properties. *Linear* in this context is not to be confused with linear models in a statistical context.

I. Population Pharmacokinetic and Pharmacodynamic Work

In phase II and phase III studies, sparse blood samples are collected from many individuals. With the help of nonlinear mixed-effect modeling, the pharmacokinetic and pharmacodynamic properties of a compound can be explored. One thing often done is to relate a pharmacokinetic or pharmacodynamic variable to covariates such as age, disease status, and gender. This work has the same kind of interpretation problems as observational studies.

II. STATISTICAL DESIGN, METHODS, AND ANALYSIS

Although clinical pharmacology studies are broad in range and scope, many of these studies have enough similarities for common statistical themes.

The designs are often sequential or randomized crossover designs. Period and crossover effects may come into play. It should be pointed out, however, that carryover effects should be easy to eliminate in pharmacokinetic studies. If one knows the half-life of a drug, then one should be able to forecast the amount of time necessary to clear almost all of the drug, known as the *washout period*. It is not as easy to eliminate carryover effects if the outcome is pharmacodynamic or safety in nature. Period effects, like carryover effects, are less likely with pharmacokinetic endpoints, but may pose a problem with pharmacodynamic and safety endpoints.

It is not uncommon for the nonrandomized studies discussed to have multiple treatments per subject and repeated measures for each treatment-by-subject combination. The analysis of that data often involves complex mixed-effect models.

The sample sizes are typically small. Sometimes they are chosen for a formal statistical goal; but often they are chosen for nonstatistical reasons, such as cost, or ethical reasons. The small sample size often precludes extensive statistical inferences, particularly where nonsignificant differences exist.

The work is often exploratory in nature. This is especially true for phase I studies. Any clues that could expedite drug development are examined. Undoubtedly, out of the many relationships examined, some will show "statistical significance." These findings should only be considered hints and not conclusive. Also, the traditional 0.05 significance level should be used judiciously, because of the small sample size. A *p*-value of, say, 0.2 may sometimes be giving evidence of an important effect and should not be ignored. Lastly, the goal of a phase I study often is not hypothesis testing but simply the estimation of a pharmacokinetic or pharmacodynamic variable.

In most clinical pharmacology studies, the null hypothesis of no differences, such as "no safety concern" or "no difference in pharmacokinetic properties," is the working hypothesis. Rejecting the no-difference hypothesis may be interpreted negatively for the compound. Yet, it has been mentioned that these studies are often not formally sized. This may encourage undersizing of the studies. There are three options that could be used to fix this undersizing problem.

The first option is to do a proper power analysis, in which the minimal clinically significant effect is elicited from the clinical pharmacologist or pharmacokineticist. One problem with this is that if the drug is in early development, notions of variability may be completely unknown. Another problem is that often there will be multiple significance tests performed. Thus, there is not always an obvious variable to select for sizing purposes.

The second option can be called a field goal approach: Define an interval of clinically irrelevant differences. Construct a confidence interval for the variable. If the confidence interval (football) falls inside the interval (goalposts), then the hypothesis of no difference holds. This is the type of analysis that is currently performed with bioequivalence studies.

The third approach is similar to the field goal approach but depends on the Bayesian paradigm: Use a noninformative prior for the analysis. Set up an interval of clinically irrelevant differences. Calculate the probability that your statistic falls in that interval. If that probability is sufficiently high (higher than a prespecified threshold), then declare that no difference exists.

The difficulty with all three of these methods is getting the clinical pharmacologist or the pharmacokineticist to define clinically irrelevant differences a priori.

III. RECENT DEVELOPMENTS

Two developments in the field are the recently proposed individual and population bioequivalence methods and the recent expectations that population pharmacokinetic and pharmacodynamic work is needed. Discussions of all of these topics can be found as entries in this volume.

REFERENCES

1. LZ Benet, SB Sheiner. In: AC Gilman, LS Goodman, TW Rall, F Murad, eds. The Pharmacological Basis of Therapeutics. 7th ed. New York: Macmillan, 1985, pp 1–2.
2. J Gray, L Lewis, D Nierenberg. Clin Pharmacol Ther 62: 237–240, 1997.

Brian P. Smith

Clinical Trials

See also *Multicenter Trials; International Conference on Harmonization (ICH)*

I. INTRODUCTION

The randomized clinical trial is broadly considered to be the most effective approach to evaluating a new therapeutic modality and to testing an existing therapy in a new setting. In this entry, we provide a broad overview of some general principles that govern the design and conduct of clinical trials. Because they are discussed elsewhere in this volume, we omit such key issues as sample size determination, statistical power, compliance to prescribed therapeutic regimens, and data analytic considerations.

II. DEFINITION

A *clinical trial* is a prospective study in which two or more treatments are compared with one another. All therapeutic arms may be active treatments or, alternatively, a placebo may be used in one group. The most definitive kind of clinical trial is the randomized trial, which requires that group assignment be the result of some random process beyond the control of both the investigator and the subject. A well-designed clinical trial requires a prospectively written protocol that describes in detail the design of the proposed research. The protocol should include a small number of clearly defined hypotheses, specific aims, and outcome measures. It should also provide a detailed delineation of eligibility and exclusion criteria, a description of the proposed interventions, discussions of recruitment and randomization processes, a detailed consideration of how the planned study will avoid potential sources of bias, a discussion of the quality control measures that will minimize the potential for errors in the collection and computerization of data, a justification of the planned sample size, and a presentation of the statistical methods that will be used to evaluate the data.

III. RANDOMIZATION

The purposes of randomization are, first, to reduce the likelihood that there will be selection bias, the bias that may result when the selected sample is unrepresentative of the target population; and, second, to minimize the likelihood that subjective decisions about treatment assignment will yield between-group differences in subject characteristics at baseline. When selection bias yields an unrepresentative sample, it may be difficult to define the population to which study results can be generalized. If one group is older or sicker or otherwise less likely to benefit from treatment, the apparent success of the alternative therapy may reflect nothing more than the positive prognostic status of the sample that has been assigned to that treatment. While carefully conducted nonrandom treatment assignment may yield balance with respect to known risk factors such as age and disease status, there is no way that subjective processes can address the impact of unknown risk factors. Similarly, while covariate adjustment during the analysis stage may account for the effect of known predictors that differ between groups, you cannot adjust for factors you don't know about.

The theoretical benefits of randomization have been demonstrated in a number of studies that have compared results from randomized trials with those that have used nonrandom treatment allocation (1–5). The consistent message is that trials that use nonrandom assignment tend to produce biased overestimates of true therapeutic efficacy. For example, Sacks et al. (1) evaluated 50 randomized clinical trials (RCTs) and 56 trials that used historical controls (HCTs) and found that 44 of the 56 HCT reports (79%) concluded that the therapy was better than the control regimen, while only 10 of the 50 (20%) reports on RCTs reached this conclusion. Chalmers et al. (3) used similar methods in comparing the conclusions of RCTs, nonrandomized trials that used concurrent controls (CCTs) and HCTs in evaluating the efficacy of anticoagulants following acute myocardial infarction. Their conclusion was that efficacy estimates were smallest with RCTs, second smallest with CCTs, and largest with HCTs.

It has recently been argued that it is common for investigators to work actively toward a subversion of the randomization process through a variety of approaches to determining in advance the group to which subjects will be assigned (6). Since inadequately concealed randomization can inflate by 30–40% the odds ratios that estimate treatment effects (7) and can bias results in the direction of unwarranted significance (2), careful attention to the mechanics of randomization is essential. The most commonly used approaches to maximizing the integrity of the process are (a) the use of numbered opaque and sealed envelopes that contain randomization information and that are opened only after eligibility has been established and the subject has given informed consent, and (b) randomization by telephone or fax communication with a coordinating center or some other agent that is physically separated from the investigator.

The literature contains many discussions of the scientific approaches to group assignment (8–11). Regardless of the mechanics of the process itself, randomization is almost always accomplished using random-number generators to make the assignment. While equal allocation is most common, it is not at all unusual to employ randomization schemes that assign different numbers of patients to different groups. The most basic approach to determining the order of treatment assignment is to use simple randomization, a process whose defining characteristic is that each new treatment assignment is random and is fully independent of all preceding assignments. Two disadvantages of this approach are that the laws of chance mean that simple randomization may yield an imbalance in the number of patients assigned to the various study groups, and, alternatively, there may be temporal bias caused by too many subjects being assigned to one group early in the study and too few at later time points.

To address these problems, *blocked* randomization is often used. When blocking is employed, each consecutive group of k subjects is assigned in some fixed proportion to all study groups. In this way, temporal bias is avoided and the target randomization percentages can be guaranteed. When there is concern that investigators in unblinded studies may use information about block sizes to break randomization codes, it may be best to keep investigators uninformed as to the size of blocks or to use varying block sizes. With or without blocking, it is common to use *stratified* randomization. Most frequently, the use of stratification reflects the desire to ensure that there will be between-group balance with respect to some known important risk factor. Thus, if gender is known to be associated with treatment success, stratification by gender can ensure gender balance by using separate blocked randomization schemes for male and female subjects. Other less commonly used randomization schemes include a variety of adaptive methods (10,11) in which nonconstant randomization probabilities depend partly or fully on the degree of between-group balance that has been achieved in previously randomized subjects. With some adaptive procedures, such as the play-the-winner rule (12), randomization probabilities can depend on success rates with previous subjects.

IV. BLINDING

The term *blinding* (or *masking*) refers to the decision to keep subjects and/or investigators uninformed as to treatment assignment as the study progresses. A *single-blind* study is one in which the subject is unaware of his treatment group assignment; a *double-blind* study is one in which neither the subject nor the investigator is provided with this information. In the less commonly used *triple-blind* study, the data and safety monitoring committee is also kept blinded when they evaluate data as the study progresses.

The purpose of blinding is to reduce the likelihood that study assessments will be biased because subject or investigator behavior has been influenced by knowledge of treatment group assignment. Thus, a subject who knows he has been assigned to a usual-care group may be more inclined to drop out or become noncompliant because he would prefer to be in the investigative arm of the trial. Similarly, an investigator who wants to establish the efficacy of a new drug may be influenced in subtle ways when she must make close decisions about ambiguous events if she knows which treatment the subject has received. As is the case with nonrandom or inadequately concealed treatment allocation, the literature suggests that there can be substantial differences between the results of blinded and nonblinded studies. Thus, in a review of 250 controlled clinical trials, Schulz et al. (7) found that the absence of a double blind was associated with a 17% inflation of the odds ratio that describes the benefit of treatment. Colditz and colleagues (4) found, similarly, that RCTs that did not use a double blind were significantly more likely to report a benefit on the new treatment than were double-blind RCTs.

A basic principle of clinical trials conduct is that while blinding is not always feasible, it should not be viewed as an all-or-nothing proposition. Thus, in a medication vs. surgery trial, it is obviously impossible to keep physicians or subjects blinded as to treatment assignment. However, it may be quite reasonable to insist that the individual who interprets the laboratory test or the CT scan that determines therapeutic success is blinded as to the treatment received by the patient. In general, the goal of every trial should be to maximize the level of blinding within the framework of the limitations that are imposed by practical considerations. Whatever the level of blinding, ethical considerations require that there be a mechanism through which blinding codes can be broken when such action is necessitated by concerns about patient safety.

V. SELECTION AND ASSESSMENT OF ENDPOINTS

An endpoint in a clinical trial is a parameter that will be compared across the arms of the study in order to evaluate the success of the study treatment. Endpoints should be prospectively and unambiguously defined quantifiable or categorical measures that correspond precisely to the scientific goals of the trial. Because they are the ultimate determinants of treatment success, every effort should be made to avoid bias in the evaluation of endpoints. Thus, following the suggestion of the previous section, if the blinding of investigators is not feasible in a given trial, procedures that ensure the blinded evaluation of endpoints should be implemented whenever it is practical to do so. When the endpoint is, for example, cardiovascular mortality, committees are often set up to provide an unbiased determination of the true cause of death. When endpoints in multicenter trials involve the evaluation of blood samples or of electrocardiograms, it is common to use a centralized reading center for all clinical sites, to ensure that determinations are made using the same procedures.

Endpoints in clinical trials are often categorized as primary, secondary, and, sometimes, exploratory. A basic principle of trial design is that there should be a small number of *primary* endpoints that are the defining measures of the success of the study treatment. The primary endpoints are the most important measures of treatment success. The emphasis on a small number of such endpoints reflects concern about the possible effect of multiple-comparisons problems, the unavoidable reality that if you have many primary endpoints, the probability that one or more will be significant by chance is increased. If there are many primary endpoints, concerns about multiple comparisons may compromise the most basic conclusions of the study. With one or two primary endpoints, the multiple-comparisons problem does not arise.

In many clinical trials, a decision must be made as to whether the primary endpoints should be "true" endpoints or "surrogate" endpoints. Generally speaking, a *true* endpoint is one that has direct clinical relevance to the patient. Examples include mortality, the remission of a cancer, and the successful treatment of a disease. *Surrogate* endpoints are used as a less rigorous alternative to the clinical endpoint in settings where the clinical endpoint is too difficult, time consuming, or expensive to measure. Such endpoints can only be useful if they are highly associated with the true endpoint of interest. Thus, blood pressure is a potentially viable surrogate for stroke because hypertension is known to be a major risk factor for stroke. For similar reasons, CD4 levels may be a reasonable surrogate for mortality in AIDS patients.

The most important reason for using surrogate endpoints is the potentially large reduction in sample size, trial duration, and cost that they may yield when the true endpoint is infrequent or requires many years to observe. Thus, Wittes and colleagues (13) suggest that a trial whose primary endpoint was a reduction in stroke rate might require 5 years of study and 25,000 patients, whereas the same trial using diastolic blood pressure as a surrogate endpoint could be completed in 1–2 years using 200 patients. Unfortunately, this huge sample size advantage has a large potential downside, because there are many predictors of stroke and reducing blood pressure may leave stroke rates unchanged. In the Cardiac Arrhythmia Suppression Trial (CAST, 14,15), the effect on mortality of three FDA-approved antiarrhythmic drugs were tested in high-risk patients with a history of both myocardial infarction and ventricular ectopy. The logic of CAST was that since these drugs were known to reduce the rate of a major risk factor for cardiac death, they should also reduce death rates. Unfortunately, the exact opposite happened, and two arms of the study were stopped early because of a substantial increase in mortality. Similar problems with surrogate endpoints have been found by meta-analyses that have demonstrated that quinidine is associated with increased mortality even though it can control atrial fibrillation (16) and that lidocaine is associated with both a one-third reduction in ventricular tachycardia and a one-third increase in mortality (17,18).

VI. RECRUITMENT PROCEDURES

A vital component of every clinical trial is the expeditious recruitment of a sufficient number of eligible subjects. Because of this, the protocol writing and planning stages of every trial must include a basic feasibility evaluation: Are we capable of recruiting the required number of subjects? The answer to this practical question can be a driving force in determining eligibility and exclusion criteria. In some cases, it can force a modification of the scientific goals of a study. In the worst case, it can lead to the conclusion that a proposed study should not be attempted. When recruitment is slower than anticipated, the cost of an already expensive process goes up, investigators may lose interest, study quality may deteriorate, and there is an incentive to randomize subjects whose eligibility may be questionable. If delays are sufficiently prolonged, the scientific questions that motivated the study may diminish in importance or become moot. And finally, it is always possible that a study may have to be stopped early because of inadequate recruitment.

In spite of, and perhaps because of, its importance, there is a natural tendency to overestimate recruitment capabilities. This is especially true since no clinic will be invited to participate in a multicenter trial if it cannot justify its ability to enroll some minimum required number of subjects. The tendency toward overestimation is emphasized in a comprehensive review of the recruitment literature by Hunninghake and colleagues (19), who assert that "it is extremely rare for a study to enroll the required number of participants within the time frame originally designated." This observation is highlighted by an evaluation of recruitment success in seven multicenter clinical trials sponsored by the Veterans Administration (20). This review found that one trial was stopped early because of inadequate recruitment, while the remaining six required more time than was initially planned to reach recruitment targets.

The potential magnitude of the recruitment burden is illustrated by large multicenter trials, such as the Lipid Research Clinics Primary Prevention Trial (21) and the Systolic Hypertension in the Elderly Program (22). In both of these studies, it was necessary to screen several hundred thousand subjects because both trials were such that approximately 100 subjects had to be screened in order to find a single subject who was ultimately randomized. While screened to eligible ratios of 100 or more are not typical, these examples provide ample evidence that a realistic and prospective appraisal of recruitment capabilities is vital in every clinical trial. They also emphasize the fact that the mundane details of the recruitment process may sometimes deserve as much attention as do the more glamorous scientific concerns of a trial. The many standard approaches to recruiting subjects for clinical trials include direct contact at clinics; public screenings for conditions such as hypertension, diabetes, and hypercholesteremia; mailings and phone calls to subjects who are known to have a given disease; referrals from private physicians and primary care clinics; retrospective chart reviews, an approach that may be the only option when diseases are rare; and television and radio advertisements. The strengths and weaknesses of these and other approaches to recruitment are discussed by Meinert (23).

VII. LOSS TO FOLLOW-UP

A fundamental objective of every well-conducted RCT is to minimize the rate at which subjects are lost to follow-up. If loss to follow-up occurs in a purely random fashion and to the same degree in each arm of the study, it may be associated with a loss of statistical power because of the reduced sample size, but it is unlikely to be an important source of bias. However, even when the apparent reason for loss to follow-up is benign, the impact on study results may be substantial. For example, patients may terminate their involvement in a study because they have moved out of town or because their level of frailty makes it difficult for them to get to the clinic. While these behaviors are likely to be balanced across groups in a randomized study, they may compromise the generalizability of results by excluding too many of the sickest patients or a preponderance of older patients who have moved away when they retired. More serious consequences may result if the reasons for loss to follow-up are associated with characteristics of the treatment. Thus, the Coronary Drug Project Research Group (24) found that loss to follow-up was greatest in the treatment arm that also had the largest number of drug-related adverse events. In the Lipid Research Clinics Coronary Primary Prevention Trial (25), the somewhat different problem was that treatment side effects led to worse compliance in the treatment group than in the placebo group (26). In a more recent RCT involving the treatment of AIDS patients (27), it was reported that loss to follow-up was significantly associated with several measures of disease progression. The authors concluded that "losses to follow-up probably decreased substantially the observed number of primary endpoints, curtailed the power of the trial to demonstrate any difference between immediate and deferred initiation of antiretroviral therapy, and may have introduced large bias in the estimated hazard ratio for the primary endpoint and its statistical significance."

There are many standard approaches to encouraging clinical trial participants to remain in a study and to attend required follow-up visits. These include providing routine conveniences such as flexible appointment schedules, limiting waiting times, and easy clinic access; regular contacts with subjects between visits; acknowledging special occasions such as patient birthdays and Christmas; payment of parking and travel fees associated with the study and, sometimes, paying the subject for his willingness to participate; and keeping the subject informed of the details of the protocol and of the scientific importance of her contribution as a participant.

It is often the case that loss to follow-up is associated with subjects who relocate. When investigators are not directly informed of the new address, simple steps like contacting friends, relatives, and employers or getting address change information from the post office is likely to be helpful. When the new address is known and when the study is multicenter, it is often feasible to see the subject at another study clinic. Alternatively, subjects who relocate can sometimes be seen when they return for visits to their former city of residence, by enlisting and paying colleagues who live in the city to which a subject moves, or by having study personnel see the subject when they are in the vicinity of a subject's new residence. If a subject is lost to follow-up in a study where the primary endpoint is mortality, the National Death Index can serve as a valuable resource for determining with a high degree of confidence whether or not the subject has died. The fact that these somewhat extreme measures are not uncommon highlights the importance attached to maximizing follow-up rates.

VIII. MULTICENTER TRIALS

Multicenter clinical trials are trials that enroll patients from more than one clinical site. They are generally used because the required sample size is too large for the trial to be feasible using only one clinical site or because the use of a single site would mean that it would take an excessive period of time to complete the study. The primary challenge of multicenter trials is found in the organizational complexity that is necessary to ensure that the protocol is followed in a uniform way at the geographically distant participating clinics. At the top of the trial organization is the study chair, who has the primary executive responsibility for the trial and who is chair of the executive committee. In addition to the study chair, executive committee members will generally include the principal investigator at all resource centers along with other individuals from within and without the trial who have expertise that will facilitate the scientific goals or practical conduct of the trial. The executive committee has overall day-to-day decision-making responsibility for the trial. Its specific responsibilities include maintaining the scientific integrity of the trial; preparing or overseeing the preparation of a manual of procedures and other study documents; appointing subcommittees that are responsible for tasks that include writing manuscripts, verifying endpoints, and codifying publication policies; overseeing the development of study forms; reviewing study manuscripts prior to submission; responding to concerns about clinics that may be having difficulty with enrollment or other trial activities; setting priorities for the study; and having the basic responsibility for the design and conduct of the trial.

The conduct of every multicenter trial is facilitated by one or more resource centers. The one such center that is integral to the functioning of every multicenter trial is the *coordinating center* (CC). The CC has a broad range of responsibilities that cover all phases of the trial. During the protocol development stage, the focus of the CC is on questions of sample size and statistical power, the scientific integrity of the proposed study design, a data analysis plan, and the development of data-processing and quality control procedures. When patients are being recruited and followed up, the CC is focused on such activities as the day-to-day processing and quality control of data; regular communication with the clinical sites so as to ensure the completeness and accuracy of the data they provide; discussing quality control problems with clinical sites and with the executive committee; and preparing reports that measure the progress of the trial through clinic-specific and studywide statistics on recruitment rates, follow-up rates, error rates on forms, missed patient visits, and other measures of study progress. In many multicenter trials, the CC will also be involved in activities that include site visits to the clinics; designing, modifying, and disseminating study forms; and training clinic personnel in assessment procedures. When the enrollment and follow-up of subjects have been completed, the activities of the CC are redirected toward the analysis of data, extensive collaboration in the preparation of manuscripts, and archiving the data generated by the study.

In addition to a coordinating center, most multicenter clinical trials have resource centers that include reading centers and central laboratories. These are centralized facilities whose purpose is to ensure the uniform interpretation of laboratory and clinical data. They tend to be used when the parameter of interest is critical to the trial and when the complexity or lack of standardization of the assessment suggests that local evaluations would be (a) impossible because the expertise or equipment is not available at some sites or (b) difficult to standardize because of the inherent characteristics of the assay.

An additional organizational component of most multicenter clinical trials is the *data and safety monitoring committee* (DSMC). This committee contains outside experts who are recommended by the executive committee and, in general, appointed by the funding agency. The primary purpose of the DSMC is to provide independent oversight and advice regarding all major study activities. It meets periodically with the executive committee to evaluate interim data, to assess the progress of the study as measured in reports prepared by the CC and the reading centers, and to review the ethical conduct of the study. It is frequently responsible for providing external review of the resource center and providing advice as to whether poorly functional clinical centers should be dropped from the study or placed on probation. Following its periodic reviews of interim data, the DSMC may recommend changes in the study protocol, such as an increase in the sample size. In more extreme settings, it may suggest the termination of the study because of treatment side effects or because of inadequate response to study treatment.

IX. INTERIM DATA ANALYSES

A common feature of many clinical trials is the interim data analysis. An *interim* analysis is an evaluation of the data while patient recruitment or follow-up is still progressing. The purpose of these analyses is to determine whether early trends in the data suggest the need to modify the study protocol. Such modifications may take many forms. For example, an interim analysis may suggest that a modification of the dosing schedule is necessary because of excessive toxicity. The Cardiac Arrhythmia Suppression Trial (14,15) provides a more extreme illustration of the potential impact of an interim analysis. In this trial, the interim data resulted in the termination of the encainide and flecainide arms of the study because the data suggested increased mortality with these two drugs. A clinical trial (or an arm of the trial) may also be terminated if an interim analysis suggests that the treatment is ineffective and that there is no hope of ever achieving statistical significance or, as happened with the aspirin arm of the Physicians Health Study (28), that the treatment is so clearly effective that it would be unethical to continue to assign subjects to the placebo group. If the interim data suggest that between-group differences in primary outcome measures are smaller than was anticipated in the original sample size computations, the conclusion may be that the target sample size should be increased. Because interim analyses are concerned with such basic issues as toxicity and the continued use of a placebo when existing data may have established the efficacy of the treatment, such analyses are viewed by many investigators as a being moral imperative.

In a single-center trial, interim analyses are ordinarily conducted by the study statistician. In multicenter trials, these analyses are conducted by the coordinating center, often with active input from the DSMC, the committee responsible for making recommendations in response to the interim report. Results of the interim analysis are presented by the executive committee to the DSMC. However, because of concerns about blinding, it is sometimes the case that these results can be seen only by the DSMC, the principal investigator of the coordinating center, and the representative of the funding agency. Thus, meetings about interim reports are sometimes conducted after the chair of the executive committee and the directors of the reading centers have been asked to leave the room. In single-center trials, the ideal committee structure may not exist, and it may be necessary to devise more ad hoc approaches to ensuring that investigators who recruit or evaluate patients are not aware of the contents of the interim reports.

A statistical concern that arises naturally in trials that contain interim data analyses is the multiple-looks problem. The *multiple-looks* problem reflects the fact that the possible early termination of a study because results from interim looks at the data are significant increases the likelihood that significant results will be achieved when the null hypothesis is true (29,30). Thus, if the Type I error is supposed to be 0.05, and if you may stop the study early because an interim analysis produced a p-value of less than 0.05, the repeated opportunities to claim significance artificially inflate the Type I error. That is, the true probability of claiming significance when the null hypothesis is true is greater than 0.05. In order to address this problem, a variety of "group sequential" testing procedures have been developed. While we do not discuss the details here in this entry, all of these group sequential procedures have a common theme. Using some formal statistical mechanism, the "nominal" p-value that is necessary to establish significance when the various data analyses are conducted is reduced to something less than 0.05. The precise nominal p-values are defined by the desire to ensure that the true Type I error is 0.05. To accomplish this task, the sum of the probabilities of claiming significance at all analyses combined must be exactly 0.05 when the null hypothesis is true. For details on specific group sequential testing procedures, we refer the reader to the methods developed by Pocock (31), O'Brien and Fleming (32), and Lan and DeMets (33). An excellent general review of these methods is provided by Geller and Pocock (34).

X. WHO GETS ANALYZED?

Protocol violations are an unavoidable feature of most clinical trials. Subjects may drop out of the study early, miss follow-up visits, cross over to a different arm of the study, or be otherwise noncompliant with the prescribed therapeutic regimen. The investigator may be the source of the protocol violation when assessment schedules or procedures are not followed and when randomized subjects are ineligible or, as sometimes happens, do not have the disease in question. When protocol violations occur, there can be a powerful incentive to exclude the relevant patients when data are analyzed. After all, the thinking goes, why evaluate a treatment using patients who may have received only a small portion of the prescribed medication?

While this logic has an intuitive appeal, there is ample evidence that in many circumstances, subjects who are excluded from data analyses because of protocol violations will be a source of biased results. The problem with excluding protocol violators is that subjects who violate the protocol may be fundamentally different from those who are fully compliant in the way they respond to treatment, or, in particular, the protocol violation may be partly or entirely a result of the study intervention itself. For these and other reasons, the broadly accepted view is that the intention-to-treat (ITT) principle should govern the primary analyses of clinical trials data unless there is a strong reason to believe that bias will not result from some alternative type of analysis (35–37). When ITT is used for the primary analysis, it is sometimes considered appropriate to perform some exploratory analyses using subgroups, so long as the reader has been clearly cautioned about the potential limitations of violating the ITT principle.

There are many illustrations of the bias that can result from excluding patients from data analyses. For example, in trials aimed at reducing infarct size in patients with acute myocardial infarction, it is essential to include patients in whom an infarct was ruled out because the infarct size was measured as zero. If you exclude such patients, you may be excluding precisely the patients in whom the treatment was most effective, the ones in whom all myocardium was salvaged by the treatment. More typical are the examples of the potential impact of excluding poor compliers. In the Coronary Drug Project (CDP, 38), the mortality rate in compliant clofibrate patients was 15.7% as compared to a much higher 24.6% in noncompliant clofibrate patients. The drug appears to be effective until you discover that compliance to placebo was associated with an even larger apparent "treatment" benefit. The problem is that CDP compliers were fundamentally different from CDP noncompliers, because the compliers were more likely to engage in heart-healthy behaviors or otherwise more likely to have an improved outcome. If the CDP investigators had excluded poor compliers, they would have been analyzing a cohort of patients fundamentally different from the patients who had been randomized in the first place.

REFERENCES

1. Sacks H, Chalmers TC, Smith Jr H. Randomized versus historical controls for clinical trials. Am J Med 72:233–240, 1982.
2. Chalmers TC, Celano P, Sacks HS, Smith Jr H. Bias in treatment assignment in controlled clinical trials. New Engl J Med 309:1358–1361, 1983.
3. Chalmers TC, Matta RJ, Smith Jr H, Kunzler A. Evidence favoring the use of anticoagulants in the hospital phase of acute myocardial infarction. New Engl J Med 297:1091–1096, 1977.
4. Colditz GA, Miller JN, Mosteller F. How study design affects outcomes in comparisons of therapy. I: Medical. Stat Med 8:441–454, 1989.
5. Miller JN, Colditz GA, Mosteller F. How study design affects outcomes in comparisons of therapy. II: Surgical. Stat Med 8:455–466, 1989.
6. Schulz KF. Subverting randomization in controlled trials. JAMA 274:1456–1458, 1995.
7. Schulz KF, Chalmers I, Hayes RJ, Altman DG. Empirical evidence of bias: dimensions of methodological quality associated with estimates of treatment effects in controlled trials. JAMA 273:408–412, 1995.
8. Zelen M. The randomization and stratification of patients to clinical trials. J Chron Dis 27:365–375, 1974.
9. Kalish LA, Begg CB. Treatment allocation methods in clinical trials: a review. Stat Med 4:129–144, 1985.
10. Hoel DG, Sobel M, Weiss GH. A survey of adaptive sampling for clinical trials. In: Elashoff RM, ed. Perspectives in Biometrics. New York: Academic Press, 1975.
11. Signorini DF, Leung O, Simes RJ, Beller E, Gebski VJ, Callaghan T. Dynamic balanced randomization for clinical trials. Stat Med 12:2343–2350, 1993.
12. Zelen M. Play-the-winner rule and the controlled clinical trial. J Am Statistical Assn 64:131–146, 1969.
13. Wittes J, Lakatos E, Probstfield J. Surrogate endpoints in clinical trials: cardiovascular diseases. Stat Med 8: 415–425, 1989.
14. Preliminary report: effect of encainide and flecainide on mortality in a randomized trial of arrhythmia suppression after myocardial infarction. The Cardiac Arrhythmia Suppression Trial (CAST) Investigators. New Engl J Med 321:406–612, 1989.
15. Echt DS, Liebson PR, Mitchell LB, Peters RW, Oblas-Manno D, Barker AH, Arensberg D, Baker A, Friedman L, Greene HL, et al. Mortality and morbidity in patients receiving encainide, flecainide, or placebo. The Cardiac Arrhythmia Suppression Trial. New Engl J Med 324: 781–788, 1991.

16. Coplen SE, Antman EM, Berlin JA, Hewitt P, Chalmers TC. Efficacy and safety of quinidine therapy for maintenance of sinus rhythm after cardioversion. A meta analysis of randomized control trials. Circulation 82: 1106–1116, 1990.
17. Hine LK, Laird N, Hewitt P, Chalmers TC. Meta-analytic evidence against prophylactic use of lidocaine in acute myocardial infarction. Arch Intern Medicine 149:2694–2698, 1989.
18. MacMahon S, Collins R, Peto R, Koster RW, Yusef S. Efects of prophylactic lidocaine in suspected acute myocardial infarction. An overview of results from the randomized controlled trials. JAMA 260:1910–1916, 1988.
19. Hunninghake DB, Darby CA, Progstfield JL. Recruitment experience in clinical trials: literature summary and annotated bibliography. Controlled Clin Trials 8: 6s–30s, 1987.
20. Collins JF, Bingham SF, Weiss DG, Williford WO, Kuhn RM. Some adaptive strategies for inadequate sample acquisition in Veterans Administration Cooperative Clinical Trials. Controlled Clin Trials 1:227–248, 1980.
21. Lipid Research Clinics Program: Recruitment for clinical trials: The Lipid Research Clinics Coronary Primary Prevention Trial experience: Its implication for future trials (Agras WS, Bradford RH and Marshall GD, editors). Circulation 66 (suppl IV):IVI–IV78, 1982.
22. Petrovitch H, Byington R, Bailey G, Borhani P, Carmody S, Goodwin L, Harrington J, Johnson HA, Johnson P, Jones M, et al. Systolic Hypertension in the Elderly Program (SHEP). Part 2: Screening and recruitment. Hypertension 17 (Suppl):II16–II23, 1991.
23. Meinert CL. Clinical Trials: Design, Conduct, and Analysis. New York: Oxford University Press, 1986, pp 149–153.
24. Coronary Drug Project Research Group. Clofibrate and niacin in coronary heart disease. JAMA 231:360–381, 1975.
25. Lipid Research Clinics: The Lipid Research Clinics coronary primary prevention trial results II. The relationship of reduction in incidence of coronary heart disease to cholesterol lowering. JAMA 251:365–374, 1984.
26. Schechtman KB, Gordon ME. The effect of poor compliance and treatment side effects on sample size requirements in randomized clinical trials. J Biopharmaceut Stat 4:223–232, 1994.
27. Ionnidis JPA, Bassett R, Hughes MD, Volberding PA, Sacks HS, Lau J. Predictors and impact of patients lost to follow-up in a long-term randomized trial of immediate versus deferred antiretroviral treatment. J Acquired Immune Deficiency Syndrome Hum Retrovirol 16:22–30, 1997.
28. Preliminary Report: Findings for the aspirin component of the ongoing Physicians' Health Study. N Engl J Med 318:262–264, 1988.
29. Armitage P. Statistical Methods in Medical Research. Oxford: Blackwell, 1971.
30. McPherson K. Statistics: the problem of examining accumulating data more than once. New Engl J Med 290: 501–502, 1974.
31. Pocock SJ. Group sequential methods in the design and analysis of clinical trials. Biometrika 64:191–199, 1977.
32. O'Brien PC, Fleming TR. A multiple testing procedure for clinical trials. Biometrics 35:549–556, 1979.
33. Lan KKG, DeMets DL. Discrete sequential boundaries for clinical trials. Biometrika 70:659–663, 1983.
34. Geller NL, Pocock SJ. Design and analysis of clinical trials with group sequential stopping rules. In: KE Peace, ed. Biopharmaceutical Statistics for Drug Development. New York: Marcel Dekker, 1988, chap 11.
35. Lewis JA, Machin D. Intention-to-treat—who should use ITT? Brit J Cancer 68:647–650, 1993.
36. Peduzzi P, Detre K, Wittes J, Holford T. Intention-to-treat analysis and the problem of crossovers. J Thor Cardiovasc Surg 101:481–487, 1991.
37. Newell DJ. Intention-to-treat analysis: implications for quantitative and qualitative research. Int J Epidemiol 21:837–841, 1992.
38. The Coronary Drug Project Research Group. Influence of adherence to treatment and response of cholesterol on mortality in the Coronary Drug Project. New Engl J Med 303:1038–1041, 1980.

Kenneth B. Schechtman

Clinical Trial Process

See also *Drug Development; Global Database and System; Good Statistics Practice*

I. INTRODUCTION

Regardless of the setting—whether a research group in an academic or healthcare institution, or a research and development organization in a small venture capital–based or large multinational pharmaceutical company—there are forces that demand these organizations, and their members, to respond successfully to the following question: How can more be done with the same resources in shorter time periods with no significant reduction in the quality of the output of the activity? The challenge this question poses to clinical research is now a routine part of the activities in most pharmaceutical companies as the year 2000 approaches. With aggressive commitments from corporate offices to increase dramatically the number of new compounds entering and completing the clinical research and development process, all members of the research team are being affected.

This entry will (a) outline the types of information necessary for finding answers to this vexing question; (b) outline assessments of what was learned during information gathering; (c) outline the importance of monitoring the effects of implementing changes in work processes based on conclusions from the information gathered; and (d) explore the importance of continually looking for answers, acting on what was learned, and monitoring the effects of process changes.

At a basic level, finding out how more can be done with the same resources involves using the steps in the scientific method that are familiar to many readers. That is, gather facts, data, observations, and related information concerning the topic under investigation; articulate a problem based on the what is known; develop a hypothesis surrounding the problem; design an appropriate methodology that will test the hypothesis; execute the methodology to generate data; analyze and evaluate the data to test the hypothesis; and draw appropriate conclusions. As will be discussed later, another parallel to the scientific method is that the critical endpoint is not only the conclusion but the successful application of the conclusion. The application of conclusions may be to provide a practical solution to the problem that was articulated or to use the new data and insights to define the problem better, develop and test a more refined hypothesis, and derive added conclusions to follow along a path of continuous improvement.

II. UNDERSTANDING WORK PROCESSES

In order to answer the question introduced in the preceding section, it is critical to develop a context, by defining what tasks need to be accomplished and what the output is from those tasks. The thoroughness of describing the process, the output(s), and the tasks needed to accomplish them will determine the success of the information-gathering step.

A. Gathering Information

Understanding work processes is not the endpoint but just the beginning. Some of the information needed to understand a work process include the following.

1. Define the Starting Point for the Process

What is the starting point? What are the decisions or events that trigger the initiation of the process? Who makes the decision to start, and how is it communicated? Identify who is responsible for the process or who owns it? What material or information is required to begin the process?

2. Define the End of the Process

What is the ultimate, final output of the completed process? Who is the customer for the final output? How is the customer organizationally related to the person(s) or group(s) responsible for the process? How is the final output delivered? How do you know that the output is ready for delivery? How do you measure and report on the success, or lack of success, of the final output? How is feedback concerning the output received from the customer?

3. Define the Tasks in a Process

Identify and be able to describe each task required to complete a process. Identify who is responsible for a task or who owns it. Is the owner inside or outside of the group responsible for completing the task? What event(s), task(s), information, material(s), or decision(s) are required to begin each task? How are the necessary information, material(s), or decision(s) received? What resources—e.g., persons, equipment, systems—are required to complete each task? For intermediate outputs, to whom is the output delivered? For intermediate outputs, how is the product of the task delivered to the person or group responsible for the next task? Which tasks need to be completed before the next task can begin? Which tasks can be completed in parallel with or independent of other tasks in the process?

4. *Define Intermediate Outputs or Milestones*

Are there any intermediate outputs that must be achieved prior to the final output? What are the intermediate outputs? Who is the customer for (who receives) each intermediate output? How is the customer organizationally related to the persons or groups responsible for the tasks to achieve each output? How is feedback concerning the adequacy of the output received? How do you measure and report on the success, or lack thereof, of each intermediate output? What milestones are meaningful to management?

B. Assessing the Information

In order to understand work processes, the information and data gathered must be assessed, to define or refine the statement of the problem and to develop a statement of hypothesis. The following topics should be considered during work process assessment.

1. *Assess the Starting Point for the Process*

Evaluate the decisions or events that trigger the initiation of the process to determine if there is a predictable pattern. Describe the adequacy, or lack thereof, of the systems for transmitting the decision to start and providing the material or information that is required to begin. Define what information the person(s) or group(s) responsible for the process actually need to start the process. Describe what happens if the decision to start or the material or information that is required to begin is not adequately delivered or communicated. What control or influence, if any, can be exercised by the process owner at the starting point? How can those items that are within the area of authority of the process owner be controlled, influenced, or managed by that person? How can the resources within the area of authority of the process owner be affected by decisions or activities outside the owner's area of authority? For those areas of authority that cannot be controlled or influenced by the process owner, what steps can be taken to manage the impact when those areas do not support, or prove inadequate for, the process?

2. *Assess the End of the Process*

Describe the characteristics of the output that will satisfy the customers' requirements. Describe the adequacy, or lack thereof, of the systems for delivering the output to the customer. Describe the monitoring systems, if any, that can predict if the output will meet the required quality and the delivery schedule. Describe what happens if the output is delivered sooner than scheduled, if delivery is delayed, if the quality is unsatisfactory to the customer or it does not meet the needs of the customer. What control or influence, if any, can be exercised by the process owner at the end of the process? Does the current process work to achieve the desired output, within the desired time schedule, with the systems and resources currently available?

3. *Assess the Tasks Involved in the Process*

Categorize the importance of each task to the completion of the process. Is each moderately or significantly important task well understood? What is needed to start the task? Who provides what is needed? And how? Who executes the task? What resources are needed? How and to whom is the product of the task delivered? Define the tasks on the critical path in the process. What are the dependencies between tasks, e.g., which task(s) have to be completed before another specific task can begin or can be completed in parallel with or independent of other tasks in the process? Describe the adequacy, or lack thereof, of the systems for transmitting the material or information that is required to begin the task. Define what information the person(s) responsible for the task actually need to start the task. Describe what happens if the material or information that is required to begin is not adequately delivered or communicated. What control or influence, if any, can be exercised on the task by the process or task owner? How can those items that are within the area of authority of the task owner be controlled, influenced, or managed by that person to facilitate task completion? How can the resources within the area of authority of the task owner be affected by decisions or activities outside the owner's area of authority? For those areas of authority that cannot be controlled or influenced by the task owner, what steps can be taken to manage the impact when those areas do not support, or prove inadequate for, the process?

4. *Assess the Intermediate Outputs or Milestones*

Describe the requirements the output must achieve. Describe the systems, if any, that can monitor whether the output will meet the required quality and delivery schedule. Describe what happens if the output is delivered sooner than scheduled, if delivery is delayed, or if the quality is unsatisfactory. Do the current tasks work to achieve each of the desired intermediate outputs, within the desired time schedule, with the systems and resources currently available?

C. Prepare a Statement of Problems

Based on a clear understanding of the work process, output(s), and tasks, the next step would be to identify specific problems that could result from work inefficiencies, poor resource utilization, unnecessary or redundant tasks, or low work capacity. Since the objective of these steps is to determine how to do more with the same, avoid focusing on inadequate resources, especially human resources, as a primary problem. Attempt to look beyond this often simplistic statement of the problem. Inadequate resources is a valid problem if you have metrics and benchmarking information to substantiate your conclusions or if the focus of the problem involves equipment or technology that enables increased efficiency or increased work capacity for the available human resources.

The problem list must be prioritized to have the basis for decision making that will manage the limited resources available to explore and/or implement change in the process, output(s), or tasks. Prioritization can be based on one or more of the following parameters: whether the process or task(s) are within the area of authority of the owner (focus on change that is within the area of authority of the task or process owner); the magnitude of the resources and the period of time that will be required to implement the change successfully (look for problems with short-term, fairly easy solutions to build the confidence of the staff, management, and organization while working on other longer-term solutions); the size of the benefit derived from the change to the strategic and/or operational goals of the organization (do not overlook the benefit derived by another part of the organization in addition to the portion implementing the change); objectives and goals established by management (avoid undertaking work process change without having a well-defined, specific set of goals and objectives that have been agreed upon by the necessary levels of management).

From the list of problems, develop a potential solution for the priority problems chosen. The solution should portray the proposed change and the resulting measurable benefit. This could resemble a statement of hypothesis; for example, by developing and implementing a common computer platform, duplicate entry of the same data into two different systems can be eliminated, thus reducing the processing time by X hours or allowing a $Y\%$ increase in the number of forms processed.

III. GATHERING DATA—MONITORING WORK PROCESSES

As with the scientific method, a crucial activity is to test the potential solution, or hypothesis, with a well-designed methodology and to capture data from the endpoints needed to test the solution. As already noted, when stating the solution it is necessary to identify the measurement(s) that can be used to determine if the solution was successful (having preestablished goals for evaluating the solution is critical).

The term *metrics* is often used to refer to the data gathered while monitoring work processes or tasks. Metrics can include such items as the time elapsed for completing a task or process; the number of times a task must be repeated to achieve the standards required for the output; the rate at which errors occur; and objective feedback from the customer of the output.

As is true of clinical research, gathering data solely for the sake of having the data is a poor use of resources. There must be value that can be derived from the data. In the case of evaluating work process, the value could include: providing management with a real-time indication of critical conditions in a process or task; predicting whether the output will meet standards or the delivery schedule; identification of potential problems in a task or process to allow implementation of the required solution before there is a significant delay or loss in quality. An issue that should be addressed when testing a potential solution is to assess the value of the metrics used to evaluate the process, output, or task. Some metrics may not be predictive of or related to the quality or efficiency of the output. Some metrics may be too complex to collect for use in routine assessment of a process.

IV. CONTINUOUS WORK PROCESS IMPROVEMENT

The question introduced at the start of this entry of how more can be done with the same resources in shorter time periods with no significant reduction in the quality of the output of the activity includes an element of continuity. As potential solutions to work process problems are tested, a recurring set of events should be set in place. These events include the ongoing collection of metrics to describe and assess the work process, to identify previously undetected or new problems, to refine insight into previously identified problems, to develop new or alternate solutions and the associated benefits, to implement and test solutions, and to continue monitoring the process through meaningful metrics. Thus, what may have started as a one-time exercise in doing more with the same, or even doing more with less, needs to become a routine part of the culture in which clinical research takes place.

Paul F. Kramer

Confounding and Interaction

See also *Adjustment for Covariates*

I. INTRODUCTION

Confounding and interaction, or effect modification, are most commonly discussed in the epidemiological literature. For epidemiologic studies, which are generally prospective or retrospective, nonrandomized studies, the principle of confounding is particularly critical. Theoretically speaking, confounding is not a concern for large, randomized studies in the pharmaceutical industry. But often randomizations are not effective in achieving balance across all important prognostic factors, particularly when trials are small or when the number of balancing factors is large. In any event, an appreciation for the issues surrounding confounding can aid in the development of informed study designs. An understanding of statistical interaction is relevant to all studies and is critical to performing meaningful analyses of the data once the trial has been conducted.

II. CONFOUNDING

Typically in an epidemiologic study, researchers are interested in exploring the relationship between a disease of interest and an exposure, or potential risk factor for the disease. In very basic terms, then, a confounding factor is something that is related both to the exposure and to the disease, that mixes with the exposure to distort the observed association between disease and exposure (1–3). A confounder may wholly or partially account for an association between the exposure and disease, or it may mask a true underlying association.

Although confounding results in bias, it differs from other biases, such as selection bias and information bias, in that it is not introduced into a study by the investigators or the subjects. Instead, confounding reflects actual underlying associations between the exposure of interest and the confounding factor and between the disease and the confounding factor. Hence, it is a reflection of the causal associations between variables in the study population.

A hypothetical example will help explain this concept and introduce a discussion of the criteria generally used to assess confounding. Suppose a clinical trial is conducted with 200 subjects and that 100 subjects are randomized to receive drug A and 100 are randomized to receive drug B. The primary outcome of the trial is either success or failure. Overall results are contained in Table 1. These results would indicate that drug A is more effective than drug B, since the success rate is 48% for drug A vs. 29% for drug B. Applying a chi square test of no association to these data results in a p-value of 0.006. Results of a more standard epidemiologic analysis give an estimated odds ratio, or ratio of

Table 1 Overall Success and Failure by Treatment

	Success	Failure	Total
Drug A	71	29	100
Drug B	52	48	100

Table 2 Results by Age Category

	Success	Failure	Total
Age <70			
Drug A	63	7	70
Drug B	36	4	40
Age ≥70			
Drug A	8	22	30
Drug B	16	44	60

odds of success for drug A to that for drug B, of 2.26 with a 95% confidence interval of (1.26, 4.05).

However, when these data are tabulated by the level of a potential confounder or prognostic factor, age (<70 vs. ≥70), a different conclusion emerges. Other examples of possible confounders would be smoking status, gender, previous therapy, and functional status at baseline. Table 2 contains these results broken down by age category. For those subjects under 70, the success rate is 90% regardless of treatment, and for those subjects 70 or older, the success rate is 27% regardless of treatment. Accordingly, the estimated odds ratio is 1 in each strata, with 95% confidence intervals of (0.52, 1.94) for those under 70 and (0.37, 2.69) for those aged 70 and older. Hence, the apparent treatment difference in Table 1 is spurious, due to age, which appears to be a confounder.

A. Evaluating Confounding

The general criteria used for assessing confounding (1,3) specify that a confounder

1. Must be associated with the exposure of interest
2. Must be associated with the disease or outcome of interest
3. Must not be a consequence of the exposure or of an intermediate step in the causal pathway of interest

From criterion 1 we see that an extraneous factor can confound the relationship between treatment and outcome only if it is not evenly distributed in the treatment groups. In regard to the example depicted in Tables 1 and 2, criterion 1 requires the proportion of subjects under 70 to differ in the two treatment groups. This criterion is in fact met in this example, since 70% on drug A are under 70 whereas 40% on drug B are in this younger age group.

Similarly, as required by criterion 2, the potential confounder must also be related to the outcome under study. Regardless of how unevenly the potential confounder is distributed in the treatment groups, unless it is also related to outcome there can be no confounding. If the potential confounder is not associated with outcome, there is nothing to mix or add to the association between the treatment and the outcome. In our example, age category, too, is related to outcome, since the success rate among those in the younger age group is 90% as opposed to 27% for those in the older group.

Criterion 3 is most relevant to retrospective epidemiologic studies, where study populations are not followed prospectively. For studies in the pharmaceutical industry, this criterion is equivalent to stating that a confounder may not be a postbaseline factor, or an intermediate result of the treatment under study. In our example, age certainly is not a consequence of treatment, and age category satisfies all criteria of confounding.

Once a particular risk factor satisfies the criteria for being a confounder, the next step is to evaluate the degree of confounding present. One way to do this is to compare the crude and adjusted estimates of association. For example, an estimated odds ratio calculated from the pooled data may be compared to individual odds ratios calculated within each stratum, or level, of the confounder. Alternatively, the crude estimate of association may be compared to a summary estimate of association that ''adjusts'' for the confounder. Methods for finding such adjusted measures of association will be given in Sec. II.C.

Referring back to our example, we can assess the degree of confounding introduced by age group by comparing the crude odds ratio, 2.26, to the stratum-specific odds ratios, which were both 1. For these hypothetical data, then, one would not want to base conclusions on an unadjusted analysis. The disparity between conclusions made based upon pooled data, as in Table 1, and those made based upon stratified data, as in Table 2, is an effect known as *Simpson's paradox* (4). In practice, the amount of confounding requiring adjustment is a subjective judgment on the part of the investigator. There is no generally agreed-upon quantitative procedure for determining whether or not a given variable suspected of confounding should be controlled.

Significance tests should not be used to assess confounding (1,3). In particular, the *p*-values associated with a standard table of baseline characteristics should not be used to determine whether or not a factor is a confounder. The results of these tests are heavily dependent on the size of a study rather than on the degree of confounding present. In addition, for a factor to satisfy the criteria of a confounder it must also be related to outcome. There are no tests able to assess the association of a potential confounder with both treatment group and outcome.

B. Controlling Confounding Through Design

Of course, the best way to control for confounding is through prevention. All potential confounders should be carefully considered at the time the study is designed. The most effective technique employed in the biopharmaceutical industry to control confounding is randomization. The goal of randomization is to ensure that each eligible subject is equally likely to be assigned to each treatment group. But randomizations are not always perfect. For small clinical trials, an imbalance in one or more prognostic factors is likely, and the resulting confounding can be problematic, given that sample sizes have been estimated assuming a large treatment effect. Even for a very large trial, where profound imbalance in prognostic factors is unlikely, perfect balance is also unlikely, and uncontrolled confounding can potentially affect trial results (8).

If a disease outcome under study has known, strong prognostic factors that could act as confounders, then researchers may take additional steps to ensure balance among these factors. One method is through restriction, whereby investigators include only patients with a certain prognostic factor in the trial. Due to the poor generalizability of the results from such a trial, this method may be used only for very small—say, phase II—trials or in studies where the known relationship between the prognostic factor and disease is much stronger than the expected relationship between the experimental treatment and disease. For example, in oncology trials, performance status at baseline is highly predictive of outcome. Hence, most cancer trials include some sort of performance level cutoff as part of its inclusion/exclusion criteria.

A stratified randomization might also be employed, whereby treatment assignments are made separately for each stratum, and balance is preserved within each level of the potential confounder. For example, in a study of obesity, where patterns of weight loss may differ by sex, one might chose to stratify by sex to ensure an equal distribution of males and females within each treatment group. Again, such a procedure is effective only for relatively small number of strata, particularly in multicenter trials. When such a stratified randomization is employed, the model used for analysis should still include the stratification factors.

One common method applied in the design of epidemiologic studies to control confounding is *matching*. In a matched study, the investigator intentionally forces the comparison group to have the same or similar distribution as the index group with regard to a potential confounder or confounders. For example, in a retrospective study, controls, or subjects without disease may be matched to cases or subjects with disease, based upon gender and 5-year age group (e.g., 40–45, 45–50, 50–55). A similar method, called *minimization*, has been proposed for use in clinical trials (5–8). Minimization assigns treatments to subjects sequentially, based upon each subject's individual stratification factors, so that balance may be achieved simultaneously across several prognostic factors. Treatment assignment is made by finding the assignment that "minimizes" imbalance over all prognostic factors considered. This method for assigning treatment is popular for clinical trials in oncology, where there are typically more factors to balance over, such as tumor histology, tumor stage, cell type, prior chemotherapy, and prior radiation (8). This method may also be useful for large, multicenter trials conducted internationally.

The primary disadvantage to such a sequential or dynamic assignment scheme is that randomization lists and prepackaged blinded drugs may not be prepared in advance or may not be allocated in numerical sequence. With the advent of telerandomizations and refined Internet communications, however, this method may become more of a practical reality.

Of course, the randomization employed is only as effective as the final comparison conducted. Only those groups assigned by randomization are truly comparable with regard to all potential confounders. Hence, comparing any other groups at the time of analysis defeats the purpose of the randomization. That is, a true intent-to-treat analysis should be conducted whenever possible. For example, bias may be introduced by comparing only those subjects who have completed a study or those with final observations. In such a situation, where final comparison groups differ from the initial groups randomized, methods adjusting for informed dropouts should be investigated.

C. Controlling Confounding Through Data Analysis

Once a strong confounder has been detected, the simplest way to control for this factor is to perform an analysis within each stratum of the confounder, or by performing a subgroup analysis. By performing such a subgroup analysis, one essentially finds estimates of the association between treatment and outcome, adjusted for the joint effect of the confounders. However, calculating group-specific estimates of association is not the most efficient method of analysis, and often a summary measure across subgroups is required.

Many methods exist for finding point estimates, significance tests, and confidence intervals on summary odds ratios (1). The most common method for conducting a test of association while controlling for other factors is to apply a Mantel–Haenszel chi square test. The Mantel–Haenszel summary odds ratio is a weighted average of the stratum-specific odds ratios, and the Mantel–Haenszel chi square statistic is used to test a null hypothesis of no association (summary odds ratio = 1). When the direction of association varies across strata, the Mantel–Haenszel test has very little power for detecting departures from no association, and opinions vary on the utility of a summary measure of association. One may choose to assess quantitatively the heterogeneity of stratum-specific odds ratios by applying a test. Again, many such tests exist, such as those proposed by Breslow and Day or Woolf (1), though all are unstable when sample sizes within strata are small.

As the number of potential confounders increases or the number of possible levels for a confounder increases, controlling confounding through stratification presents problems unless sample sizes are very large. In this setting, a more unified approach involving the fitting of a linear model with terms for each confounder is preferred. For dichotomous outcomes, such as success rates, a linear logistic model may be used; for continuous outcomes, such as a change from baseline in a laboratory measurement, an ANOVA/ANCOVA model may be employed.

For example, suppose that the outcome under study is a success rate and that x_1 represent treatment ($x_1 = 1$ for drug A, $x_1 = 0$ for drug B) and x_2 through x_p are potential confounders. Performing an analysis that controls for x_2 through x_p involves fitting the linear logistic model:

$$\text{logit}(p) = \beta_0 + \beta_1 x_1 + \beta_2 x_2 + \cdots + \beta_p x_p \qquad (1)$$

where p is the probability of success and the logit (p) is defined as log [$p/(1 - p)$]. If the outcome is continuous, an analysis that controls for potential confounders involves fitting the ANOVA/ANCOVA model:

$$y = \beta_0 + \beta_1 x_1 + \beta_2 x_2 + \cdots + \beta_p x_p \qquad (2)$$

where y is the continuous outcome measure.

Strong confounders will impact the estimate of β_1, and by including them in such a linear model we are able to find an estimate for β_1 that is adjusted for the confounding factors. For example, comparing the estimate of β_1 from the following models (3) and (4) provides an indication of the degree of confounding introduced by factor x_2:

$$\text{logit}(p) = \beta_0 + \beta_1 x_1 \qquad (3)$$

$$\text{logit}(p) = \beta_0 + \beta_1 x_1 + \beta_2 x_2 \qquad (4)$$

Therefore, β_1 as estimated by model (4) may be interpreted as the treatment effect adjusted for factor x_1. In fact, the logistic regression analysis in model (4) is equivalent to the Mantel–Haenszel analysis presented previously.

When controlling for confounding at the time of analysis, it is important to include all strong confounders, but including factors that have little impact on reducing bias will only increase the variability associated with the estimated treatment effect (9). Generally, factors should be included in a model only if they are known from prior studies to be causally related to outcome or if they have been used as a stratifying factor in the design of the trial.

III. INTERACTION

Interaction is a term referring to a change in the magnitude or direction of the association between treatment and outcome according to the level of a third variable. For example, if a drug is more effective in younger patients than in older patients, there would be an interaction between treatment and age. In this situation, the treatment effect is not independent of age. In the epidemiologic literature, this phenomena is often referred to as *effect modification*, because this third variable "modifies the effect" of treatment on outcome. The subtle distinction, however, is that *effect modification* generally implies causality between the effect modifier and outcome, whereas *statistical interaction* dose not (10).

Unlike confounding, interaction is not a bias or something that must be controlled in a study. Instead, interaction is something that should be explored so as to gain a greater understanding of the association between treatment and outcome. Interaction is a natural phenomenon occurring in the underlying population from which study participants are selected and will not vary based upon the design of the study.

A. Evaluating Interaction

An interaction effect is meaningful only within the context of a particular model. Suppose we are analyzing data from a trial with a continuous primary outcome of change from baseline in weight. The variable x_1 corresponds to treatment group (A or B), and x_2 corresponds to gender (male or female). An interaction between treatment and gender may be investigated by fitting the model

$$y = \beta_0 + \beta_1 x_1 + \beta_2 x_2 + \gamma_1(x_1 x_2) \quad (5)$$

where y represents the change from baseline in weight and γ_1 represents the interaction term between treatment and gender. If γ_1 is found to be significantly different from 0, then there is evidence of an interaction between gender and treatment, indicating that the effect of treatment and gender on the mean change from baseline in weight is nonadditive. In other words, the joint effect of treatment and gender is different from the sum of their independent effects.

Often it is helpful to describe such interactions graphically. Plotting mean changes from baseline by treatment and gender can help in assessing the possible interactions present. For example, Figure 1a depicts a situation in which the interaction is simply additive, Figure 1b shows a deviation from an additive interaction in which response to drug B always exceeds that to drug A, and Figure 1c depicts a situation in which response to drug B exceeds that to drug A in males, but response to drug A exceeds that to drug B in females.

Now suppose that our trial has a discrete outcome of success or failure. An interaction between treatment and gender may be investigated by fitting the model

$$\text{logit}(p) = \beta_0 + \beta_1 x_1 + \beta_2 x_2 + \gamma_1(x_1 x_2) \quad (6)$$

where the p is the probability of success and γ_1 again represents the interaction between treatment and gender. If γ_1 is found to be significantly different from zero in this model, we conclude that the relationship between success rate and treatment differs by gender or, using more standard epidemiologic terminology, that the effect of treatment and gender on the log-odds for success is nonmultiplicative. In other words, the joint effect of treatment and gender on the log-odds for success is different from the product of their independent effects.

For pharmaceutical studies, the interaction terms that are of most interest are those involving the treatment group or dose level. For example, in a dose–response analysis, one might want to examine whether or not the dose–response relationship varies by some

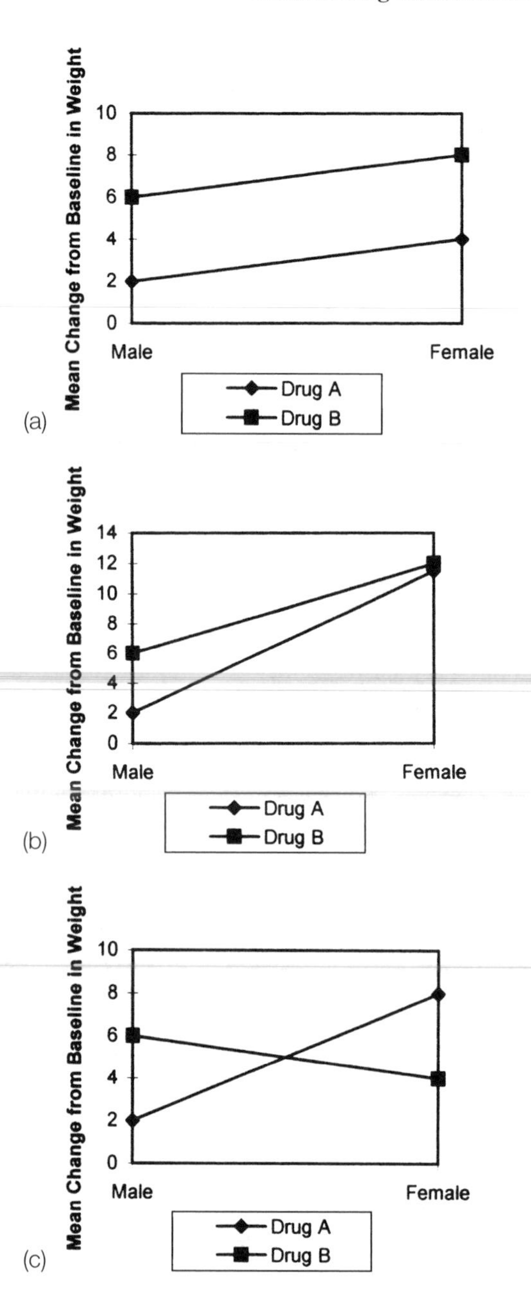

Fig. 1 (a) Additive interaction between treatment and gender. (b) Nonadditive interaction between treatment and gender in which response to drug B always exceeds that to drug A. (c) Nonadditive interaction between treatment and gender in which response "changes direction."

third variable. This could be done by incorporating an interaction term in the model and evaluating its impact on model fit. Or, in a large, multicenter trial, one might want to test for a treatment-by-center effect.

Interaction terms including only prognostic factors are most useful in controlling interrelated confounding effects. In terms of analysis, it is recommended that one first consider and evaluate potential confounders, and then incorporate any interactions present. This order of analysis is recommended because the primary aim of most analyses is to assess the relationship between outcome and treatment, and including interaction terms with treatment while assessing confounding complicates this determination.

IV. REGULATORY REQUIREMENTS

Regulatory guidelines do not specifically address the principle of confounding, but they do discuss the need for randomization in the design of clinical trials. Section 2.3.2 of the current version of the *Guideline on Statistical Principles for Clinical Trials*, proposed by the International Conference on Harmonization (11), discusses the role of randomization in clinical trials. These guidelines cite randomization as one of the critical techniques that should be adopted so as to provide "a sound statistical basis for the quantitative evaluation of the evidence relating to treatment effects." In addition, the ICH recommends stratifying on study center and other important prognostic baseline factors, particularly in small studies. Dynamic allocation procedures are discussed and advocated only after careful consideration of "the complexity of the logistics and potential impact on the analysis."

Section 5.7 of ICH9 addresses the use of covariates, or possible confounders, and interaction terms in the data analysis. The guidelines stress the importance of "prestudy deliberations" concerning the potential covariates affecting outcome and advise that potential confounders be included in the analysis plan whenever possible. They also suggest, "when the potential value of an adjustment is in doubt, it is often advisable to nominate the unadjusted analysis as the one for primary attention, the adjusted analysis being supportive."

In the discussion of interaction, the ICH guidelines recommend prespecification of any planned subgroup or interaction analyses in the analysis plan, adding that "in most cases, however, subgroup or interaction analyses are exploratory and should be clearly identified as such." They go on to caution that "any conclusion of treatment efficacy (or lack thereof) or safety based solely on exploratory subgroup analyses are unlikely to be accepted."

V. DISCUSSION

Confounding and interaction are different yet related phenomena. Confounding is a bias that must be controlled and something that can be addressed both in the design and in the analysis of a study. Interaction, on the other hand, is an inherent quality of the data, something to be explored once the data have been collected. A prognostic factor may be both a confounder and an interaction term, one of these, or neither. Confounding must be assessed qualitatively; interaction may be tested more quantitatively.

When designing a study, the most effective approach to block confounding is to employ randomization. To reduce further the possibility of confounding for known strong potential confounders, study subjects may be restricted by the level of the potential confounder, or a stratified or sequential treatment assignment may be followed.

When analyzing the results of a study, unexpected confounding can be controlled by fitting linear models with terms for such factors as well as any stratification factors used in the design of the study. Such a model may also be used to explore any interactions with treatment, or differences in treatment response based upon the level of another variable.

At some level, the goal of every study is to assess the true association between treatment and outcome. Control of confounding improves the estimate of this underlying association. Identification of treatment interaction terms may affect product labeling and prescribing, in the later phases of research, or may assist in targeting populations for further research, in the earlier stages.

REFERENCES

1. JJ Schlesselman. Case-Control Studies: Design, Conduct, Analysis. New York: Oxford University Press, 1982.
2. OS Miettin. Am J Epidemiology 141:1113–1116, 1995.
3. Z Tongzhang. Principles of Epidemiology: Fundamentals and Study Design. Course notes, Yale University School of Medicine, New Haven, CT, 1993.
4. EH Simpson. J Roy Statist Soc B 13:238–241, 1951.
5. DR Taves. Clin Pharmacol Therap 15:443–453, 1974.
6. SJ Pocock, R Simon. Biometrics 31:103–115, 1975.
7. LS Freedman, SJ White. Biometrics 32:691–694, 1975.
8. T Treasure, KD MacRae. Br Med J 317:362–363, 1998.
9. NE Day, DP Byar, SB Green. Am J Epidemiology 112: 696–706, 1980.
10. N Pearce. Int J Epidemiology 18:976–980, 1989.
11. International Conference on Harmonisation of Technical Requirements for Registration of Pharmaceuticals for Human Use—ICH Harmonised Tripartite Guideline E9: Statistical Principles for Clinical Trials. Feb. 5, 1998.

Catharine B. Stack

Content Uniformity

See also *USP Tests*

Content uniformity is a measure of the variation in the amount of active ingredient from one dosage to the next. In a perfect world, every dosage unit would contain exactly the same amount of drug. However, in the real world of manufacturing tablets, capsules, or other final dosage form, the active drug substance is diluted with excipients to mask taste, to enhance bioavailability, to add bulk, or to promote stability. In such processes, it is not always possible to get absolute homogeneous mixing of the drug with the excipients. Factors such as different densities, different particle sizes, and different particle shapes contribute to different settling tendencies and flow characteristics, which may cause slight variations in the amount of active ingredient present in the dose-sized partitions of the batch. Even if it were possible to get absolute homogeneity in the formulated drug product, uniformity of active ingredient in the final dosage forms would still not result, because it is not possible to fill every capsule or to compress every tablet to contain exactly the same weight of drug product. In order to achieve adequately homogeneous mixtures of bulk drug product, manufacturers validate the mixing process with respect to time and other mixing parameters. However, even with a

Table 1 Requirements of the Three Major Pharmacopeia

	USP	EP	JP
Sample sizes	10 + 20	10 + 20	10 + 20
Definition of M	M = greater of $\bar{X}$ and 100% and lesser of $\bar{X}$ and rubric mean	$M = \bar{X}$	M = 100%[b]
Capsules, etc.	$n1 = 1$ and $n2 = 3$	$n1 = 1$ and $n2 = 3$	—
Tablets, etc.	$n1 = 0$ and $n2 = 1$	$n1 = 0$ and $n2 = 1$	—
Stage 1	Pass: $n \leq n1$ and RSD[a] $\leq 6.0\%$ and $m = 0$ Fail: $n > n2$ or $m > 0$ Stage 2: $n1 < n \leq n2$	Pass: $n \leq n1$ and $m = 0$ Fail: $n > n2$ or $m > 0$ Stage 2: $n1 < n \leq n2$	Pass: $\bar{X} - 2.2s \geq M - 15\%$ and $\bar{X} + 2.2s \leq M + 15\%$ and $m = 0$ Fail: $m > 0$ Stage 2: Neither pass nor fail
Stage 2	Pass: $n \leq n2$ and RSD[a] $\leq 7.8\%$ and $m = 0$ Fail: $n > n2$ or $m > 0$ or RSD $> 7.8\%$	Pass: $n \leq n2$ and $m = 0$ Fail: $n > n2$ or $m > 0$	Pass: $\bar{X} - 1.9s \geq M - 15\%$ and $\bar{X} + 1.9s \leq M + 15\%$ and $m = 0$ Fail: $m > 0$ or $\bar{X} - 1.9s < M - 15\%$ or $\bar{X} + 1.9s > M + 15\%$

Rubric mean = average of potency limits in the monograph.
n = # assays outside 85%–115% of M.
m = # assays outside 75%–125% of M.
[a]RSD = relative standard deviation.
[b]Unless otherwise specified in the individual monograph.

validated process, it is still necessary to demonstrate adequate content uniformity explicitly in the final dosage units on a batch-by-batch basis. This is usually done with a formal dose uniformity test.

All three major pharmacopeias, the United States Pharmacopeia (USP), European Pharmacopeia (EP), and the Japanese Pharmacopeia (JP) have formal criteria for judging the uniformity of dosage units. While the tests are somewhat similar, there are important differences that can lead to cases where a product sample might meet the requirements of one test but not the criteria of the other. Table 1 provides a side-by-side comparison of the requirements of the three pharmacopeia. The reader will note that the wording in Table 1 is different from the descriptions of the test given in the applicable pharmacopeia. This has been done in order to facilitate comparison of the tests. Inspection of the three tests shows that the criteria in both the USP and the JP tests are tied to the mean content, in addition to controlling the allowable variation from the mean, while the EP deals with only the variation in the content of individual units from the mean content.

Although these tests should not be confused with a sampling and acceptance plan for lot release with respect to content uniformity, their performance can be evaluated using operating characteristic (OC) curves, which provide the probability that the test will be passed versus the hypothesized true mean and relative standard deviation (RSD) of the product. Figures 1–3 show the OC curves for the three compendial tests for a range of true means and RSD values. The curves can be generated using Monte Carlo simulation. As illustrated in Fig. 1, when the true mean content is near 100% of label claim, the JP test and both USP tablet and capsule tests behave similarly, while the EP test is different. In Fig. 2, where the true mean is less than 100%, the JP test is less likely to be passed than the USP tests, since it places more emphasis on deviations from 100% of label than the USP. The OC curves of the EP are nearly the same as before, since the operating characteristics are mostly unaffected by changes in the true mean content. Figure 3 illustrates the provision of the USP for when the rubric mean (average of the potency limits) is more than 100%. Here, the JP is much tighter than the USP and EP because deviations from 100% on the high side are treated no differently by the JP than are deviations on the low side of 100%. In these cases, the USP is seen to perform in a middle ground between the JP and the EP. Other cases of interest could be generated and examined in this way through the use of simulations and OC curves.

In the near future, pharmacopeial harmonization will lead to more common global compendial criteria. The tests for uniformity of dosage units is slated for harmonization, and the next few years should see a single compendial test for uniformity of dosage units that is applicable to products in most major markets of the world. It is a natural extension of the concept of content uniformity to ask whether a more simple and straightforward measure of variability, such as RSD, would be more appropriate for judging compliance. Even if the location of the mean content is kept in the equation,

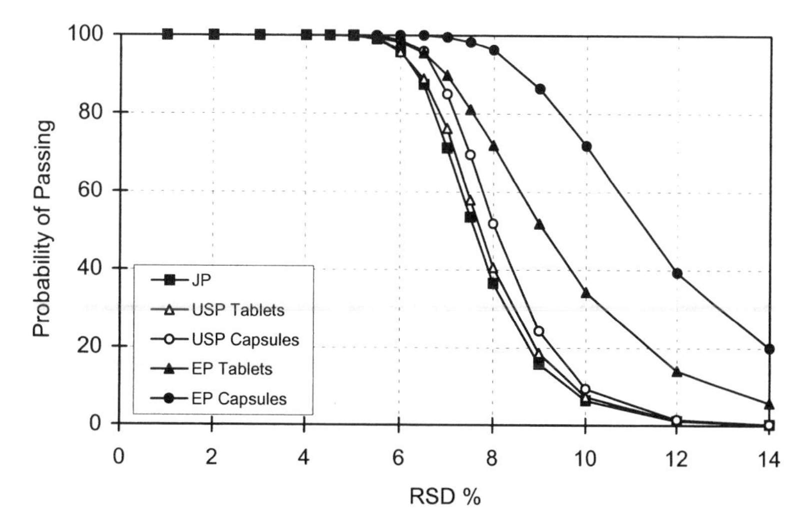

Fig. 1 OC curves for current JP, USP, and EP when rubric mean = 100% and actual mean = 100%.

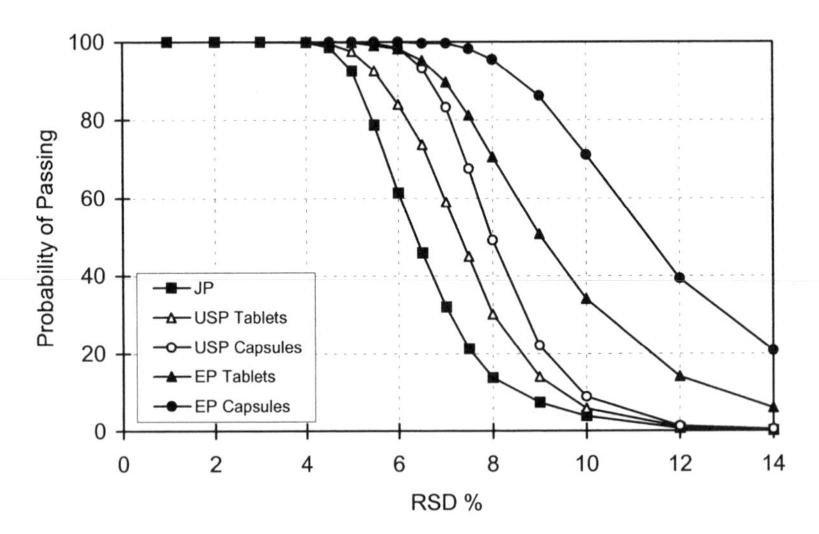

Fig. 2 OC curves for current JP, USP, and EP when rubric mean = 100% and actual mean = 96%.

criteria similar to those in the JP have appeal, since they are analogous to statistical tolerance limits. In such a case, the criterion for passing the test is that there be a given level of confidence that a sufficiently high proportion of the dosage units fall within, say, 85%–115% of label.

The expectation that content uniformity will be maintained throughout the shelf life of the product leads to the question of whether a manufacturer would have product release criteria different from the pharmacopeial tests. By using the OC curves of the compendial tests, one could compare a candidate release criterion to see how it compares to the official test. Another way the OC curves of the compendial tests can be used is to determine the likelihood that the compendial requirements will be met for a product where the mean content and the variation of content is known. Bergum (4) developed methods for determining accep-

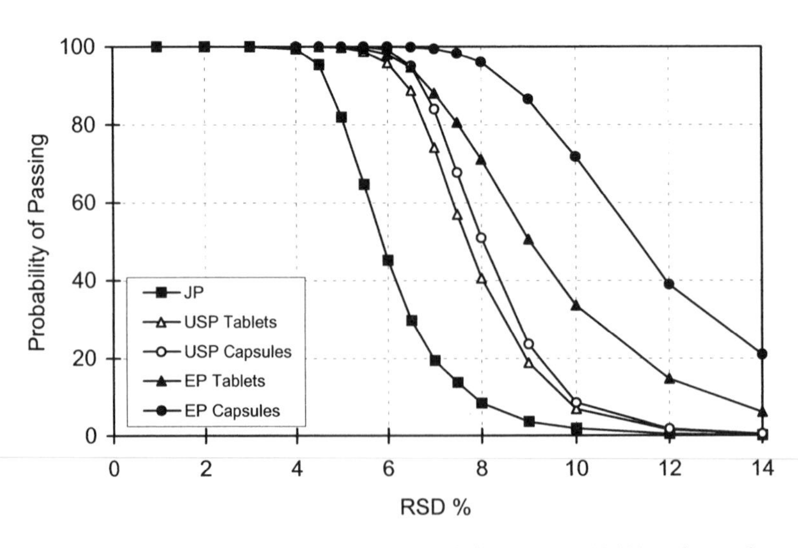

Fig. 3 OC curves for current JP, USP, and EP when rubric mean = 104% and actual mean = 104%.

tance limits for the sample mean and standard deviation that give a specified probability that the USP will be met. Sampson and Breunig (5) and Sampson et al. (6) examined alternate ways to assess content uniformity by exploiting the relationship between weight and content of the dosage unit, and they proposed sampling and acceptance strategies that could be used for product release.

In summary, content uniformity is an important product quality attribute that provides the patient assurance that no matter which tablet or capsule is taken, an acceptable amount of active drug will be delivered. Manufacturers perform validation to demonstrate that the mixing and finishing operations produce uniform dosage units with respect to the content of active ingredient. Where applicable, batches are tested for content uniformity prior to release provide assurance that the product will comply with compendial standards of uniformity. Currently, different compendial requirements are applied in different parts of the world, but international harmonization of the tests for uniformity of dosage units is under way.

REFERENCES

1. United States Pharmacopeia. 23rd ed. Rockville, MD: United States Pharmacopeial Convention, 1994.
2. European Pharmacopeia. 3rd ed. Strasbourg, France: European Dept. for the Quality of Medicines, Council of Europe, 1996.
3. Japanese Pharmacopeia. 13th ed. Tokyo: Society of Japanese Pharmacopeia, 1996.
4. JS Bergum. Drug Devel Indus Pharmacy 16:2153–2166, 1990.
5. CB Sampson, HL Breunig. J Quality Technol 3:170–178, 1971.
6. CB Sampson, HL Comer, DE Broadlick. J Pharm Sci 59:1653–1655, 1970.

John R. Murphy

Contract Research Organization (CRO)

See also *Drug Development*

I. INDUSTRY BACKGROUND

A. Expansion of the Pharmaceutical Industry in the Late 1970s to Early 1990s

Over the years, the pharmaceutical industry has developed a full-scale capability to carry out research, from laboratory discovery to biological screening, animal toxicity studies, formulation, clinical pharmacology, stability, clinical trials, new drug applications (NDAs), and promotional marketing. During the late ’70s to early ’80s, the return on investment was so great that little attention was paid by the industry to the reaction of the consumer. R&D costs for the pharmaceutical industry have been doubling every 5 years since 1970; the average cost of bringing a drug to market was $125 million in 1989 and climbed to $231 million in 1993 (1). “It was not unheard of for a potential shortfall in profitability to be remedied simply by a rise in product prices” (2). Although prescription drug prices have risen an average of 2.3 times faster than consumer prices since 1980 (3), the probability of developing a successful drug is very small. According to the Center for the Study of Drug Development at Tuft’s University, of every 5000 compounds that are synthesized, just five will enter clinical trials and just one will be approved by the FDA. Therefore, the funds from drug development has been decreasing, especially for the high-risk biotechnology products. For biotechnology companies, access to capital was drying up during the early 1990s and was continuing to do so under the impact of health care reform. In 1991, biotech companies raised $6.5 billion; but in 1992, just $5 billion was raised. The value of the industry’s stock fell 6% to $45 billion, a loss of $1.4 billion, while R&D expenses rose 14% to $5.7 billion (4).

B. Pressures on the Pharmaceutical Industry from the Early 1990s

1. *Pricing Pressures.* While drug prices continued to outgrow the consumer price index, the public became discontent with the cost of drugs, hospitalization,

doctors, and insurance. In early 1990, 37 million Americans lacked coverage for health insurance. In the mean time, the influx of generic drugs, therapeutic substitution, and the reduction of tax credits for studies designed for marketing purposes have given the pharmaceutical industry a wake-up call. To counter the criticism of ever-rising health care costs, health care providers and insurance companies have formed their own respective superlarge organizations to consolidate their buying power for pharmaceuticals and health services. These powerful HMOs demand deeper price cuts from the drug companies. In addition, the government demands rebates on sales to Medicare patients, who are the high-volume users of prescription drugs. Congressman Henry Waxman (D, California) stated, "If we are serious about controlling health care costs, we must address the issue of prescription drug prices. Congress, as part of health reform, must find a way to balance profits and price in a way fairer to the American consumer." (5).

2. *R&D Cost Increases Faster Than Price.* Despite the pricing pressure (or perhaps because of it), the pharmaceutical industry has continued to increase spending on research and development. The top 30 companies spent $20 billions on R&D in 1993, representing 18.3% of sales, while 1994 sales grew by only $6 billion (6). R&D costs have been nearly doubling every five years since 1970; and the average cost of bringing a drug to the U.S. market rose from $125 million in 1989 and 59 $231 million in 1993 (just a little over four years). It cost an average of $450 million to develop a new drug in 1997.

C. Alternate Approaches to Pharmaceutical R&D

In order to stay competitive and to maintain a healthy profit margin, pharmaceutical companies have adopted various approaches to respond to the pressures.

1. *Downsizing and early retirement.* As of July 1993, the pharmaceutical industry had already lost 18,000 jobs (according to PMA). Many companies offered early retirement packages to reduce the workforce.
2. *Out-licensing nonmajor products.*
3. *Merger and consolidation.* Numerous pharmaceutical companies merged, to expand their product lines. Once such mergers were consummated, the layoffs and early retirements followed. Companies that merged during this period included: SKF and Beecham, Dow and Merriam, Eastman Kodak and Sterling Drug, AHP and A.H. Robins, Merck and Johnson & Johnson Joint Venture, Hoffman LaRoche and Genentech, Merck and DuPont Joint Venture, Rh'one-Pulenc and Rorer, Bristol-Myers and Squibb, Roche and Syntax, AHP and American Cyanamid.
4. *Comarketing.* The sales forces of different companies are joining together to penetrate the market and to expand market share. For example, Glaxo Holding PLC worked with Roche on Glaxo's Zantac using Roche's 800-strong sales force. Smith Kline Beecham and DuPont comarketed SKB's Tagmet. Beecham and Upjohn joined forces to sell Beecham's heart-attack drug Eminase.
5. *Developing only the drugs with potential major market shares.*
6. *Developing one's own generic business.* Hoechst brought 51% of Copley Pharmaceuticals for $546 million, which is 20 times Copley's 1992 sales of $51.9 million. Merck established a separate generic house, West Point Pharma, to compete with manufacturers of generic drugs.
7. *Developing an OTC market that was not under the price limit.* Warner-Lambert and Wellcome P.L.C. formed a joint venture to comarket OTC drugs.
8. *Price-cut to expand market shares.* Merck's Mevacor is an example of the vulnerability of products that enter crowded therapeutic classes. A 1-month's supply of Merck's Mevacor 20 mg was $60 in April 1994. Sandoz launched Lescol 20 mg with a sale price of $30 per month, which was less expensive than both Merck's Mevacor and BMS Pravachol ($54 per month).
9. *Giving up research.*
10. *Farming out research and development to CROs.* This accelerates R&D without tying up the internal resources of the pharmaceutical companies. To bring a product to market 1 month earlier would generate $10 million in sales if the product had annual sales of $120 million, and $50 million in sales if the product had annual sales of $500 million. It also brings in know-how in a new therapeutic area quickly. The CRO unit can be used as an independent quality control arm.

II. EMERGENCE OF THE CRO INDUSTRY

A. CROs in the Public Market

As R&D costs increased steadily in the late 1970s, a few service organizations emerged to handle specific tasks for the pharmaceutical companies, such as identifying investigator sites, monitoring, data processing, statistical analysis, report writing, and regulatory supports. By the mid-1980s, G.H. Besselaar, for example, had grown to several hundred employees to provide an almost full range of services to the pharmaceutical industry. By the late 1980s and early 1990s, pressure from the health care reform forced pharmaceutical companies to staff at trough and to use CROs to supplement their resource need. The outsourcing of clinical research needs to CROs was established as a branch service industry when G.H. Besselaar was acquired by Corning Glass works in 1989.

B. R&D Spending and Duration of Marketing Exclusivity

Despite cost containment, pharmaceutical companies have continued to increase their spending on R&D, which exceeds $35 billion per year (7). The reasons for the continued elevation of R&D costs are numerous. The number of the NDAs increased from 3233 patients in 1985–88 to 3567 in 1989–92 and to 3936 patients in 1993–96 (8). The number of procedures per patient also increased from 256 in 1992 to 319 in 1993, to 331 in 1994, and to 372 in 1995 (9). Once of the reasons for the increased number of procedures is the globalization of drug development, which in turn increases the complexity of the regulations on clinical trials. The improved efficiency created even more pressures on competitors to market the new drugs faster. Therefore, the number of years of exclusivity enjoyed by the pharmaceutical companies have been reduced drastically (10). Inderal was approved in 1968 and enjoyed marketing exclusively for 10 years. Tagamet was approved in 1977 and had 7 years of exclusivity. The other major drugs' approval dates (duration of exclusivity) were as follows:

Capoten	1980	(5 years)
Prozac	1988	(4 years)
Diflucan	1990	(2 years)
Recombinate	1992	(1 year)
Invirase	1995	(0.25 years)

The pressure from generic-drug producers has never reduced. Brand-name drugs typically lose 70% of their market share at the end of the first year following the introduction of the generic version. Therefore, utilization of CROs becomes a strategic decision.

III. GROWTH OF THE CRO INDUSTRY

A. Increase of Outsourcing by Pharmaceutical and Biotechnology Companies

As of the end of 1997, annual spending on research and development by pharmaceutical and biotechnology companies had exceeded $35 billion. Two-thirds of this spending (approximately $22 billion) was devoted to clinical research. The CRO industry was awarded approximately $2.5 billion and is expected to grow 10% a year.

B. Full-Service CROs Grow Rapidly

Full-service CROs are growing rapidly, due to the following reasons.

1. The incentives for drug firms to outsource clinical testing
2. The continued expansion of biotechnology research
3. The greater clinical testing requirements being imposed on medical device manufacturers
4. Market share gains resulting from industry consolidation
5. The rewards conferred on the first product to market

Pharmaceutical companies are beginning to recognize the advantages of developing broad as well as deep relationships with a limited number of vendors. Contract research organizations able to service the full continuum of research and development needs have the following capabilities:

1. A broad range of therapeutic expertise in designing clinical trials
2. The ability efficiently to collect, analyze, and manipulate data from many patients with various clinical conditions across many geographically dispersed sites
3. The ability to provide a full range of trial management services to customers who wish to nar-

row their relationships to a limited number of CROs for drug development activities
4. The ability to conduct trials at diverse geographical sites, allowing simultaneous filings of registration packages in several major jurisdictions (11)

C. Upstream and Downstream Growth of CROs

By the end of 1994, there were more than 200 DIA-registered CROs in the United States (12), and a similar number of CROs were established in Western Europe. To stay competitive, the wider range of services are an important incentive for the CROs to attract pharmaceutical clients. Starting in 1996, the CRO industry realized that it needs more diversification from the traditional clinical-monitoring, data-processing, statistical analysis, and regulatory support. The leading CROs ventured upstream into central laboratory, drug-packaging, and health-economic services. Starting in 1996, large CROs brought contract sales service organizations. "Although there have been moves to harmonize product registration and requirements, nevertheless it is still a multidomestic market with different medical rules, regulations, laws and medical practices in every country" (13).

Some other CROs acquired organizations that provide prelaunch product strategy services, tactical sales, and reimbursement support services. Some CROs invested in companies that specialized in combinatorial chemistry and functional genomics. To accelerate the recruitment of patients for clinical trials, companies with calling centers became the target of acquisition in 1997 and 1998.

IV. FUTURE OF THE CRO INDUSTRY IN THE PACIFIC RIM

A. Market Size for Pharmaceuticals

The Japanese pharmaceutical market is about $55 billion a year, or 20% of the world market, with an annual growth rate of 10%. "Until recently, there was little recognition of the need for CROs, in Japan" (14). The Ministry of Health and Welfare (MHW) has established a study group (Honma Group) to review the need for CROs in Japan. The Honma Group supports the use of CROs in Japan if they are properly controlled.

Potential total drug consumption in China in the year 2000 is estimated to be $19.2 billion, with an annual growth rate of approximately 13%. Contract research organization groups are being established by various multinational CROs.

The need for CRO support in Taiwan to conduct clinical trials is evidenced by the presence of multinational CROs in recent years. A similar need for CROs is observed in Australia, Southeast Asia, India/Pakistan, Eastern Europe, and Latin America.

B. Critical Issues to Ensure the Quality of Clinical Trials

Critical issues to ensure the quality of clinical trials include the following.

1. *The attitude of medical professionals.* In most Pacific Rim countries (except China), medical professionals enjoy a special social status. The CRO monitors have to deal with the attitude of "I am a doctor and I know what I am doing. Don't come here to tell me what I have to do." Investigators with this type of attitude tend to generate case reports with a higher percentage of errors.
2. *Informed consent.* It will require great effort to educate both investigators and patients that clinical research cannot be done without the proper protection of the rights of the patients. The language of informed consent should be balanced and should be understandable by the local participants.
3. *Transparency of budgeting and payment.* Each budget item should be clearly identified and accountable and should reflect the needs of local trials.
4. *Respect for intellectual property.* Respect for intellectual property will encourage multinational companies to conduct more clinical trials in this region.
5. *Time availability of the investigators and coordinators.* Investigators in some research institutes have very limited time to do research. Their understanding of the efforts required to conduct well-controlled clinical trials is not compatible with the actual time required.
6. *Simplifications of the governmental approval process.* The approval process should be independent of the relationship between the sponsor and governmental officials. The approval process should also be transparent to the public.
7. *Affordable central laboratories.* In the near future, the clinical data from Asia will routinely be part of the multinational submission. The

cost of data generated from the Asia-Pacific Rim should be proportional to the local cost of living.

C. Extension of Traditional CRO Services

Traditionally, pharmaceutical and biotechnology companies treat R&D and sales and marketing as their core value and as proprietary. They guard and manage them with in-house employees. In 1997, a major shift in the FDA policy toward direct marketing by pharmaceutical companies to the potential consumer was announced. This policy shift opens a new avenue for CROs to provide contract sales services (contract sales organizations, CSOs) to pharmaceutical and biotechnology companies. The growth of contract sales is expected to accelerate, and the mergers of CSOs will probably intensify in the coming years.

Although CROs have ventured upstream to acquire discovery and research arms, such as combinatorial chemistry, functional genomics, and special assays, the future of such services is not clear. As site management organizations (SMOs) have continued to grow in number and to consolidate, the experience of mixing CROs and SMOs has not been very impressive.

To accelerate pharmaceutical development, information technology (IT) is critical to the efficient processing of data. More CROs are acquiring firms that provide the technology to enhance integrated information processing, real-time status, information sharing, and global data harmonization. The IT services in this field will grow; however, the development and maintenance cost are high and the success rate is usually low.

V. CONCLUSIONS

The CRO has become an important partner to pharmaceutical and biotechnology companies in the development of new compounds. The CRO has also provided services in drug registration. The total spending on R&D by pharmaceutical and biotechnology companies is estimated to exceed $35 billion. Two-thirds of this is spent on clinical trials, and the CROs have captured only 10% of this. Nineteen of the top 20 pharmaceutical companies are invested in emerging markets, such as China, India, Russia, and Latin America. These markets are accessible to clinical research for fast patient enrollments because of the improvement in adopting the practice of ICH and GCPs. The large CROs are following the path of pharmaceutical companies into the Asia-Pacific market since 1996. The Asia market for CROs will expand rapidly in the next 10 years. Therefore, the potential for further growth for traditional CRO is enormous. Originating in 1997, the CRO has acquired marketing and sales capabilities to extend services downstream. The estimate for worldwide marketing and sales expenses ($36 billion) is very close to worldwide R&D spending. If growth of the traditional CROs is any indication of the potential growth of contract sales and marketing (CSOs), then the growth and consolidation of CSOs will be accelerated. Larger players from other industries may join in to capture the growth. The traditional nature of CROs has changed, and they will evolve into a formidable industry.

REFERENCES

1. S. Thomas. Insight 4:11–14, 1996.
2. J.A. DiMasi, R.W. Hansen, H.G. Grabowski, L. Lasagna. J Health Economics. 10:107–142, 1991
3. P. Sperry. Investor's Business Daily, August 24, 1993.
4. T.Y. Lee. Applied Clinical Trials 3:36–38, 1994.
5. The Star Ledger, February 3, 1994.
6. IMS. Strategic Management Review, 1995.
7. Kendle Internationals Inc. Prospectus, 1997.
8. Boston Consulting Group, 1994.
9. DataEdge, 1997.
10. PhRMA. Facts and Figures, 1997.
11. J.C. Bradford & Co. Healthcare Service Basic Report, 1998.
12. Drug Information Association. Pharmaceutical Contract Support Organizations, 1994.
13. Center Watch Weekly. Volume One, Number 30, March 9, 1998.
14. Stephen Bentley. Clinical Research and CROs in Japan. Applied Clinical Trials; 30–34, April 1997.

T. Y. Lee

Crossover Design

See also *Robust Analysis for Cross-Over Design; Bioavailability and Bioequivalence*

I. INTRODUCTION

"A crossover trial is one in which individual subjects are given sequences of treatments, with the object of studying differences between individual treatments (or subsequences of treatments)" (1, p3). The most studied form of crossover is the so-called AB/BA crossover, in which patients are treated in two periods and are randomized to receive either treatment A followed by treatment B or vice versa.

Example 1. A trial was carried out to compare the effects in Swedish schoolchildren suffering from asthma of a single dose of 200 μg salbutamol and 12 μg formoterol, both given by metered-dose inhaler (2). Fourteen children were randomized in equal numbers to one of two sequences: either formoterol followed by salbutamol or salbutamol followed by formoterol. A washout period of at least 2 days was observed between treatments. The principal outcome measure was peak expiratory flow in 1 second (PEF) (1).

More complicated crossover designs are often used, however, and designs in which patients receive three, four, or five treatments in as many periods are not uncommon.

Example 2. A placebo-controlled trial in migraine was run in Sweden and Finland to estimate and compare the effects of two doses (50 mg and 100 mg) of the potassium salt of diclofenac (3). Each of 72 patients received each treatment once and once only. Patients were randomized to receive one of the six possible sequences of the three treatments. Patients were not to use the first treatment until they had observed a treatment-free and attack-free period of at least 1 week. Thereafter they were to treat the first three attacks with the treatments provided in the order indicated. The main outcome measure was a visual analog pain score 2 hours after treatment (1).

Crossover trials are of greater importance for the biopharmaceutical statistician than is generally the case for fellow medical statisticians working outside of the industry. Some typical applications in drug development are:

- Phase I
  - Pharmacokinetic studies to establish concentration time profiles
  - Bioequivalence (4)
  - Food interaction studies
  - Dose proportionality
  - Dose escalation studies for investigating maximum tolerated dose
- Phase II
  - Studies of pharmacodynamic response
  - Dose finding
  - Parallel assay
- Phase III
  - Specialist scientific studies
  - Determination of individual response
- Phase IV
  - Studies of patient preferences

Example 3. This is an example of a pharmacodynamic parallel assay in asthma (5). Three single doses (6 μg, 12 μg, and 24 μg) for each of two formulations of formoterol to be given by inhalation in the form of a dry powder were compared to placebo. It was not possible to treat patients in more than five periods. Hence, an incomplete blocks design in five periods and 21 sequences was chosen, each patient receiving five of the seven treatments. The trial was carried out in 15 centers in four countries, and 161 patients were recruited. The main outcome variable was the area under the forced expiratory volume in 1 second (FEV_1) curve over 12 hours.

Crossover trials are also used in preclinical work involving animals. This topic is not covered here. Crossover trials are rarely used as pivotal phase III studies in support of a drug application. One reason is the drug regulatory agencies are wary of crossover trials because of the potential danger of carryover. This issue is dealt with in detail in Sec. II. An even more important reason is that a principle advantage of crossover trials is to reduce the number of patients in a trial. However, for most drug developments, more patients are needed to satisfy the regulator as regards the tolerability of the product than are needed to demonstrate its efficacy, even if a parallel group trial is chosen. There is thus less incentive for the sponsor to use crossover trials in phase III. The exception as regards sample size requirements would be trials in which survival is the outcome, but these are, of course, in any case not suitably run as crossovers, nor, indeed, are any trials in which cure or death is the outcome.

In fact, suitable indications for crossovers are chronic diseases in which the patients are relatively stable. Such indications include:

Asthma
Mild or moderate hypertension
Rheumatism
Migraine
Sleep disturbances
Angina
Epilepsy

It is not the disease alone, however, which determines the suitability of a crossover. Also relevant is the nature of the treatment. Quickly reversible treatments are more suitable than those for which the effect is more persistent. Thus, for example, in asthma, beta-agonists are easier to study using crossover trials than are steroids. Other considerations also apply. An important distinction in drug development is between single-dose studies, which often involve pharmacodynamic measures and may study onset and duration of action and multidose studies, in which, more usually, the therapeutic steady-state effect of treatments is studied. Crossover trials may be used for both but are generally more suited to the former.

II. CARRYOVER

The outstanding problem of crossover trials is regarded by many as being carryover. Carryover has been defined as "the persistence (whether physically or in terms of effect) of a treatment applied in one period in a subsequent period of treatment" (1, p8). The simplest example of carryover is where the half-life of a treatment has been underestimated and some relevant concentration of the previous treatment is still present at a time when another is being measured. However, except for bioequivalence studies in which drug concentration itself is the outcome, it is the pharmacodynamic time course rather than the pharmacokinetic course that is relevant. Much play has also been made in the literature of so-called psychological carryover, where perhaps some memory of the therapeutic experience under the previous treatment affects the patient's current judgment. However, it should not be forgotten that in a double-blind trial, to the extent that blinding is successful, the origin of all carryover, however indirect, must ultimately be pharmacological.

Where it occurs, the consequence of carryover is that the trialist will measure the combined (in some cases partial) effects of two or more treatments (S)he may be unaware that this is happening and this may lead to biased assessment; even where detected, the disentangling of individual effects may be difficult. The most appropriate way to deal with carryover is with washout periods. These may be *passive*, in that no treatment is given at all, or *active*. The latter strategy occurs quite commonly in multidose therapeutic trials. Here patients may be switched over almost immediately to the subsequent treatment. The previous treatment will, of course, be being eliminated by the patient's body during the time that the new treatment is being given. Provided that measurement of the effects of the new treatment is delayed until this process is complete, the problem of carryover is dealt with.

If, however, as is sometimes the case in single-dose pharmacodynamic studies, it is desired to study onset of action, then a passive washout is necessary. The issue of modeling carryover is considered in due course later.

III. ANALYSIS USING THE GENERAL LINEAR MODEL

In many cases the primary outcomes for a crossover trial will be continuous. For example, FEV_1 is commonly used in asthma, and diastolic or systolic blood pressure in hypertension. As explained earlier, crossover trials are often more relevant to studying pharmacodynamic outcomes than to studying therapeutic outcomes. In particular, except in rare cases to be discussed later, survival is not a relevant outcome for a crossover trial. A general linear model in which disturbance terms are assumed to be normally distributed will often be a suitable framework for analysis.

Crossover trials share a feature of fractional factorials that can make their representation in a linear model somewhat awkward. Consider an AB/BA crossover trial in two patients only. There are two patients, two periods, and two treatments but only four and not eight observations. A consequence of this is that any two of the three factors patient, period, and treatment determine the third. Thus if patient 1 is allocated to the AB sequence, then if the current observation is the observation in period 2, it must be under treatment B.

If we ignore the problem of carryover, then, following Jones and Kenward (6), a general model for observed values y_{ijk} of a random variable Y_{ijk} from a crossover trial in g sequences, p periods, N patients, and t treatments is as follows:

$$Y_{ijk} = \mu + s_{ik} + \pi_j + \tau_{d[i,j]} + \varepsilon_{ijk} \quad (1)$$

Here, μ is a general mean, s_{ik} is the effect of subject k

in sequence i, $i = 1, 2, \ldots, g$, $k = 1, 2, \ldots, n_i$, π_j is the effect of period j, $j = 1, 2, \ldots, p$, $\tau_{d[i,j]}$ is the direct effect of the treatment administered in period j to subjects (either patients or healthy volunteers) in group i, and ε_{ijk} is a stochastic disturbance term.

A number of points may be noted about this model:

The ε_{ijk} terms are often assumed independent. For designs with three or more periods, however, it may be appropriate to allow for autocorrelation within patients over time.

The model is over parameterized, but it is identifiable contrasts, in particular those of the treatment parameters, that are of interest.

The treatment parameter is indexed by a functional subscript $d[i,j]$ that identifies it by position (sequence and period). This is because, as already discussed, crossover designs are a form of fractional design: there are t treatments, N patients, and p periods but only $N \times p$ observations. To complete the specification, the function $d[i,j]$ must be separately defined, say, by a table that indicates which treatment is given to which sequence group in a given period.

There is no term for carryover in the model. Jones and Kenward propose adding a term $\lambda_{d[i,j-1]}$. This form of carryover has been referred to as *simple* carryover (1) and is appropriate *if* carryover lasts for only one period and depends on engendering treatment only (is *not* modified by the perturbed treatment).

The terms s_k can be treated as fixed or random, depending on approach. This may or may not make a difference to the resulting inference regarding treatment contrasts: whether it does so depends on whether interblock information is present, and this in turn depends on the balance of the design, whether autocorrelation is allowed or not, and whether a carryover term is fitted (7).

Whether or not the *patient* effect is treated as fixed or random, the model does not allow for a true random effect of *treatment*: the possibility that the effect of treatment can vary from patient to patient. The fixed-effect approach is extremely common, but it can be of interest on occasion to use a random-effects approach, in particular in designs where the number of periods exceeds the number of treatments.

A popular approach to analyzing crossover trials within the pharmaceutical industry is to apply ordinary least squares using proc glm® of SAS® and treating the patient effect as fixed. Alternatively, the patient effects are sometimes allowed to be random. Sometimes, where this is done, a factor for the sequence itself is introduced to force the treatment estimate to be a "within-patient" estimate. If this is done, however, between-patient information cannot be recovered.

This issue can be understood simply by considering an AB/BA crossover in which some patients have dropped out at random after the first period. If a fixed effect is included for patients, only patients who have completed both periods will contribute information. However, direct comparison of patients who have received A only with those who have received B only can be made if only the patient effects are allowed to be random (as they must be in any parallel group trial). This will generally contribute a very small amount of information but can in principle be combined with that from patients who have completed both periods, using an appropriate approach to analysis as provided, say, by proc mixed® of SAS®.

This is a simple example where so-called interblock information is available to be recovered. There are more complex cases. Usually, if carryover is introduced into the model, it means that interblock information can in principle be recovered. An exception is given by models in which simple carryover only is allowed for and the design chosen is such that simple carryover is orthogonal to the treatment effect. For example, such a design is the four-period, two-treatment design using the sequences AABB, BBAA, ABBA, and BAAA. For incomplete block designs, such as illustrated by our earlier Example 3, interblock information is available to be recovered whether or not carryover is fitted.

A general feature of conventional approaches to analyzing crossover trials is that the residual degrees of freedom for error can exceed the number of patients (8). Consider, for example, the degrees of freedom for the analysis of variance for a complete blocks design in N patients, p periods, and p treatments. (This sort of design is very common. Our earlier examples 1 and 2 are cases in point, with $N = 14$ and $p = 2$ and $N = 72$ and $p = 3$, respectively.) The degrees of freedom for the analysis of variance may be partitioned as shown in Table 1. Now, in the first example, the degrees of freedom for error are 12 and thus inferior to the number of patients. In the second, however, there are nearly twice as many degrees of freedom for error as patients, and this signals that strong assumptions are involved (8). An alternative approach to using ordinary least squares under such cases can be to reduce the data for a given patient to a contrast of interest and then to analyze these contrasts. A similar problem arises with any crossover trial with more than two periods and

Table 1 Partitioning Degrees of Freedom for Fixed Effects Analyses of a Crossover Trial

	Degrees of freedom		
Source	In general	Example 1	Example 2
Patients	$N - 1$	13	71
Periods	$p - 1$	1	2
Treatments	$p - 1$	1	2
Error	$Np - N - 2p + 2$	12	140
Total	$Np - 1$	27	215

where the same treatment, as may be the case in so-called "*n* of 1 studies," is repeatedly given to the same patient; this summary-measures approach may be particularly valuable (1,8). Alternatively, a true random-effect (multilevel) model can be applied.

IV. MODELING CARRYOVER

There is an enormous literature on adjusting the analysis of crossover trials to allow for the effect of carryover and a corresponding interest in appropriate efficient designs. Most authors have enthusiastically (and usually uncritically) adopted the simple carryover model (6,9–11) whereby carryover depends only on the engendering treatment and lasts for one period only. Of course, for the AB/BA design, there are only two periods and the treatment that follows is completely determined by the treatment that precedes. Hence the simply carryover model is the only one that needs to be considered. Approaches to dealing with carryover in the case of the AB/BA design will be considered in due course; but in any case, those interested in designing supposedly more efficient crossover trials have usually considered more complex designs. For these more complex designs, simple carryover may not be realistic.

Consider, for example, the design in two periods in which patients are randomized in equal numbers to four sequences: AB, BA, AA, and BB (12). We can summarize the responses using eight cell means defined by the product of the four sequences and two periods. If we fit a fixed effect for each patient, any estimate of the treatment effect (A–B) can be expressed as a linear combination of the four contrasts produced by taking the period-2 cell mean from a given sequence from the corresponding period-1 value. If carryover is ignored altogether, the relevant weights for these four contrasts are 1/2, −1/2, 0, and 0. The patient effects are eliminated by virtue of having used within-cell differences, the period effects are eliminated because the second weight is the negative of the first, and it can also be seen by inspection that the treatment difference is appropriately recovered. Let us call this contrast C_1.

In this scheme, the two sequences AA and BB contribute nothing to the estimate of the treatment effect and would appear to be a complete waste of time and resources. If, however, carryover were present and that from A into B were not the same as from B into A, the treatment estimate so formed would have a bias, β. If it could be assumed that the carryover from A into A were the same as from B into B, this bias could be estimated using the weights 0, 0, 1/2, −1/2 on the four cell mean differences. Let us call this contrast C_2. A little reflection shows that this second scheme of weights estimates only a carryover effect; patient, period, and treatment effects are eliminated from it. It thus follows that the difference between the first linear combination and the second, $C_1 - C_2$, provides an unbiased estimate in the presence of *simple* carryover.

This does not mean that such a design is useful. On the contrary it is *not* generally useful. The variance of the estimated treatment effect is four times what it would be if all patients had been allocated to the first two sequences only. This is a very high price to pay for lack of bias, and modern approaches to data analysis recognize the importance of variance bias trade-offs (13). There is, however, a more important criticism, namely, that if carryover occurred, it is extremely unlikely that the carryover from A into A would be the same as from A into B. For example, A might be an active treatment with a response either at steady state (in a multiple trial) or near the top of a dose–response curve. On the other hand, B might be a placebo. The carryover from A into B would thus be more important than from A into A (1,14–16). There are, in fact, occasions where correcting for simple carryover can increase the bias of the resulting treatment estimate (1,16).

When applying the simple carryover model, one can distinguish two rather different cases. The first occurs where the design is such that adjusting for simple carryover has no adverse effect on the variance. This is, for example, the case in the four-period design in two treatments using the four sequences AABB, BBAA, ABBA, and BAAB. Note that in this design A follows A twice and B follows A twice, and now consider the 16 cell means for this design defined by the cross-classification of sequence and period. A natural and fully efficient estimate of the treatment effect A − B is obtained by multiplying each cell mean corresponding to an A by 1/8, each cell mean corresponding to B by −1/8, and forming the weighted sum. But if simple carryover occurs, two of the cell means under A are affected by carryover from a preceding A, and so are two of the cell means by B. The weights associated with these cell means add to zero, and thus simple carryover is eliminated automatically.

Under such circumstances, fitting simple carryover in the model reduces only the error degrees of freedom; since these will generally be many, there is little reason not to do so. The danger here is rather in assuming that *because* simple carryover has been fitted, all forms of carryover have been eliminated.

The second case is where the design is not fully efficient in the presence of simple carryover. Here fitting carryover will increase the variance of the treatment estimate. In some cases it may even increase the bias. It is extremely doubtful whether the simple carryover model is of any value in such cases. A useful review of approaches to modeling carryover in multiperiod designs is given by Matthews (17).

V. TESTING FOR CARRYOVER

Until fairly recently, an alternative to adjusting for carryover was popular with applied statisticians. This was to perform a preliminary test of (simple) carryover and then to use either an adjusted estimate (or test) of the treatment effect if this preliminary test was significant or an unadjusted test if (as would more usually be the case) this was not. This was a particularly popular approach to analyzing the AB/BA design and will now be described in that connection (18,19).

In the presence of carryover, the only unbiased estimate possible from the AB/BA design is that based on first-period data only. Clearly, if the second-period data are discarded, the remaining data have the structure of a parallel group trail and can be analyzed as such. Note that as discussed earlier, the implicit assumption is that the patient effect is random. An estimate produced under such circumstances will be referred to in the discussion that follows as PAR (20).

In the absence of carryover, an unbiased estimate of the treatment effect is provided by weighting the two cell means from the AB sequence with weights 1/2 and −1/2 and the two cell means from the BA sequence with weights −1/2 and 1/2 and then summing. In the discussion that follows, this particular contrast will be referred to as CROS. In the presence of carryover, however, the bias in CROS is $-\lambda/2$, where λ is the difference between the carryover from A into B and from B into A.

An estimate of the *semi*carryover effect, $\lambda/2$, is given by using weights 1/2 for both cell means in the first sequence and −1/2 for both cell means in the second sequence. In other words, it corresponds to the difference between the mean response over both periods in the AB sequence and the corresponding mean response in the BA sequence. Like PAR, this estimate, which will be referred to as CARRY, is a between-patient estimate, and so to base analyses upon it, a random patient effect must be assumed. It is unbiased for carryover, since each sequence mean reflects both treatments and both periods so that the difference between means can reflect only either differences between patients, which are assumed random, or the order in which treatments were administered. The general situation is given in Table 2. Note that these three estimates satisfy the relationship PAR = CROS + CARRY (20).

Table 2 Weights for Estimating Effects for an AB/BA Crossover

		Sequence			
		AB		BA	
		A	B	B	A
	PAR	1	0	−1	0
Treatment	CROS	1/2	−1/2	−1/2	1/2
	CARRY	1/2	1/2	−1/2	−1/2
	PAR − (CROS + CARRY)	0	0	0	0

In 1965, in an influential paper, Grizzle proposed a scheme whereby a preliminary test for CARRY was performed at the 10% level (18). (This higher-than-usual nominal level was suggested in view of the low power of this test). If the result was not significant, CROS would be used as the basis for tests of the treatment effect at the 5% level. If it was significant, PAR would be used, also at the 5% level. This proposal was also clearly described in a further influential paper by Hills and Armitage (19). This scheme was very popular for many years, and some pharmaceutical companies had even written SAS macros to perform this two-stage testing procedure automatically. However, in an extremely important paper published in 1989, Freeman showed that the procedure as a whole does not have the correct implicit nominal size of 5% but, in the case where there is no carryover, has a Type I error rate that lies between 7% and 9.5% (21).

The reason is the extremely high correlation between PAR and CARRY. In fact, if anything, Freeman's paper understates the problem with the two-stage approach, since either the pretest is irrelevant or the conditional Type I error rate of PAR lies between 25% and 50% (1,20). In fact, although general medical statistics textbooks continue to recommend this procedure, none of three monographs devoted to the crossover does so (1,6,9), and its use in drug development and regulation appears to be being abandoned.

It is possible to adjust the two-stage procedure in a number of ways, for example, by carrying out the test using PAR at a lower nominal level; performing this test at the 0.5% rather than the 5% level deals with the problem, for example (22,23). However, under such circumstances any power advantages in the face of carryover disappear when compared to the simpler strategy of always using CROS.

It seems plausible that the problems with the two-stage procedure in the case of the AB/BA designs are likely to be present with strategies involving pretesting for more complex designs. Either the carryover effect will be orthogonal to the treatment effect, in which case the only loss involved in adjusting for carryover would be residual degrees of freedom, or the effects are not orthogonal, in which case similar problems to that with the AB/BA design may occur. In an investigation of the pretesting strategy, Abeyesekera and Curnow came to the conclusion that always adjusting was the best strategy (24). They implicitly assumed, however, that the simple carryover model would be correct. In a more general investigation allowing for the possibility of an alternative steady-state model, whereby carryover from a treatment into itself is zero (14,15), and also allowing for any arbitrary mixture of the steady-state and simple carryover models, Senn and Lambrou came to the opposite conclusion: that never adjusting was the best policy (25). The field remains controversial, and the best advice to the trialist seems to be not to rely on statistical modeling to eliminate carryover.

VI. BAYESIAN APPROACHES

One interpretation of the difficulties that the preceding approaches to carryover have encountered is that they are too extreme. For example, in the case of the AB/BA design, the choice is between an unbiased, inefficient estimator PAR or an efficient but potentially biased estimator CROS. The former estimator would correspond roughly to a Bayesian analysis in which uninformative priors were used for both treatment and carryover effects. The latter would apply in a Bayesian analysis in which a completely informative prior was used for carryover, assigning it the value 0 with probability 1. (The two-stage procedure is, of course, completely incoherent in a Bayesian sense, in that a choice is made between these two extremes on the basis of an amount of information that is quite inadequate to the task at hand (26).) A general advantage of Bayesian approaches, however, is that compromise positions are possible or, to reinterpret in a frequentist framework, that bias–variance trade-offs can be achieved naturally (13).

A series of papers by Grieve starting in 1985 have developed a Bayesian approach to analyzing crossover trials (26–30). (An earlier paper by Selwyn et al. considered Bayesian approaches to bioequivalence (31).) The general difficulty is that of handling the prior for carryover appropriately. Grieve's approach establishes the appropriate posterior distribution under the assumption that carryover is zero and also making no such assumption. As discussed earlier, the former corresponds roughly to the CROS analysis and the latter to the PAR analysis. However, prior odds for these cases are elicited, and these can be updated to produce posterior odds via a Bayes factor. These can be used to mix over the two models and hence produce an integrated posterior distribution for the treatment effect.

VII. OTHER OUTCOMES

The discussion thus far has been in terms of the general linear model. This has relatively more importance for the crossover trial than for parallel group trial, because the former is restricted in use to chronic diseases for which continuous outcome measurements are common. It may be useful, however, to supplement such analyses with nonparametric techniques. A simple approach for the AB/BA design uses the Wlicoxon–Mann–Whitney test (32). A good review of approaches in more complex cases is given by Tudor and Koch (33).

Binary outcomes present particular problems in crossover trials due to the difficulty in handling dependence of observations on the same subject. A useful review of various approaches if given by Kenward and Jones (34), who themselves have developed an approach based on log linear models. The matter is also treated extensively in their book (6). An alternative approach, particularly useful in the more general case where ordered categorical variables are involved, is that of Ezzet and Whitehead. They generalize the proportional-odds model to the crossover context by allowing the true log-odds to have a normal distribution over all patients.

Although survival in the classic sense is not relevant for crossover trials, certain outcome measures may be appropriately analyzed using survival approaches. One example is that of exercise tolerance tests in angina. France et al. (36) show how such data may be analyzed by establishing patient "preferences": if a patient "survived" longer under A than B, then A is said to be *preferred* for that patient. A tie occurs where both measurements are censored. An alternative approach is given by Feingold and Gillespie (37).

VIII. FURTHER READING

The books by Jones and Kenward (6), Ratkowsky et al. (9), and Senn (1) all give very different perspectives on the crossover trial; the last of these deals most closely with the concerns of the biopharmaceutical statistician. The paper by Hills and Armitage (19) is an excellent introduction to the AB/BA design, although, for reasons already explained, its advice as regards carryover is not sound. The encyclopedia articles by Kenward and Jones (38) and Senn (39) may also be useful. A special issue of *Statistical Methods in Medical Research* was devoted to the crossover trial and has articles on the AB/BA design (16), multiperiod designs (17), binary data (34), nonparametric (33), and Bayesian (30) approaches.

REFERENCES

1. SJ Senn. Cross-over Trials in Clinical Research. Chichester, England: Wiley, 1993.
2. V Graff-Lonnevig, L Browaldh. Clin Exper Allergy 20: 429–432, 1990.
3. C Dahlof, R Bjorkman. Cephalalgia 13:117–123, 1993.
4. SC Chow, J Liu. Design and Analysis of Bioavailability and Bioequivalence Studies. New York: Marcel Dekker, 1992.
5. SJ Senn, J Lillienthal, F Patalano, D Till. In: J Vollmar and L Hothorn, eds. Cross-over Trials. Stuttgart: Fischer, 1997, pp 3–26.
6. BJ Jones, MG Kenward. Design and Analysis of Cross-Over Trials. London: Chapman and Hall, 1989.
7. EM Chi. Statist Med 10:115–1122, 1991.
8. SJ Senn, H Hildebrand. Statist Med 10:1361–1374, 1991.
9. DA Ratkowsky, MA Evans, JR Alldredge. Cross-over Experiments Design, Analysis and Application. New York: Marcel Dekker, 1993.
10. RP Kershner, WT Federer. J Am Stat Ass 76:612–619, 1981.
11. EM Lasker, M Meisner, HB Kushner. Biometrics 39: 1089–1091, 1983.
12. LN Balaam. Biometrics 24:61–73, 1968.
13. BP Carlin, TA Louis. Bayes and Empirical Bayes Methods for Data Analysis. London: Chapman and Hall, 1996.
14. JL Fleiss. Biometrics 42:449–450, 1986.
15. JL Fleiss. Cont Clin Trials 10:1121–1130, 1989.
16. SJ Senn. Statist Med 11:715–726, 1992. (Correction, Statist Med, 11:1619, 1992.)
17. J Matthews. Stat Methods Med Res 3:383–405, 1994.
18. JE Grizzle. Biometrics 21:467–480, 1965. (Corrigenda, JE Grizzle. Biometrics 30:727, 1965, and AP Grieve. Biometrics 38:517, 1982.)
19. M Hills, P Armitage. The two-period cross-over trial. Brit J Clin Pharm 8:7–20, 1979.
20. SJ Senn. Stat Methods Med Res 3:303–324, 1994.
21. PR Freeman. Statist Med 8:1421–1432, 1989.
22. SJ Senn. Statist Med 16:2021–2024, 1997.
23. SJ Wang, HMJ Hung. Biometrics 53:1081–1091, 1997.
24. S Abeyasekera, RN Curnow. Biometrics 40:1071–1078, 1984.
25. SJ Senn, D Lambrou. Statist Med 17:2849–2864, 1998.
26. AP Grieve, SJ Senn. J Biopharm Stat 8:191–233, 1998.
27. AP Grieve. Biometrics 41:979–990, 1985.
28. AP Grieve. In: DA Berry, ed. Statistical Methodology in the Pharmaceutical Sciences. New York: Marcel Dekker, 1989, pp 239–270.
29. AP Grieve. Statist Med 13:905–929, 1994.
30. AP Grieve. Stat Methods Med Res 3:407–429, 1994.
31. MR Selwyn, AR Dempster, NR Hall. Biometrics 40: 1103–1108, 1981.
32. GG Koch. Biometrics 28:577–584, 1972.
33. G Tudor, GG Koch. Stat Methods Med Res 3:345–381, 1994.
34. MG Kenward, B Jones. Stat Methods Med Res 3:325–344, 1994.
35. F Ezzet, J Whitehead. A random effects model for ordinal response from a cross-over trial. Statist Med 10: 901–907, 1991.

36. LA France, JA Lewis, RA Kay. Statist Med 10:1099–1161, 1991.
37. M Feingold, BW Gillespie. Statist Med 15:953–967, 1996.
38. MG Kenward, B Jones. In: S Kotz, CB Read, DL Banks, eds. Encyclopedia of Statistics. Update volume 2. New York: Wiley, 1998, pp 167–175.
39. SJ Senn. Cross-over trials. In: P Armitage, T Colton, eds. Encyclopedia in Biostatistics 2: New York: Wiley, 1998, pp 1033–1049.

Stephen Senn

Cutoff Designs*

I. INTRODUCTION

The randomized design is the preferred method for assessing the efficacy of treatments. Randomization of all subjects should be employed whenever possible. Randomization, in principle, serves at least three important purposes: (1) it avoids known and unknown biases on average; (2) it helps convince others that the trial was conducted properly; and (3) it is the basis for the statistical theory that underlies hypothesis tests and confidence intervals (1).

Randomization of all subjects has been criticized, however, because it may raise ethical concerns or practical limitations in certain situations. Ethical tensions may arise, for example, when strong a priori (though inconclusive) information favors the experimental treatment, when the disease is potentially life-threatening, and when randomization does not explicitly incorporate subjects' baseline clinical need or their willingness to incur risk (2,3). Examples that have stirred considerable debate about the ethics of the randomized design include the controversies about the release of drugs for AIDS (4), the availability of drugs for cancer treatment (5), and the use of extracorporeal membrane oxygenation (ECMO) for neonatal intensive care (6,7).

A second potential drawback of the randomized design occurs in instances when randomization is not feasible or practical. Such situations may arise in health services or outcomes research, where, for example, a health education program is to be targeted only to people who need it (8). An evaluation and comparison between managed care and usual care could be made feasible if high users of health care utilization receive managed care only and low users of health care utilization receive usual care only. A study concerned with the effect of a letter as an intervention to control health care costs could be made practical if the letter is sent only to physicians with high billed charges per subscriber, while those with lower billed charges per subscriber don't receive a letter (9). In these contexts, economic constraints and logistical barriers may dictate that an experimental intervention is neither practical nor efficient for those who don't need it or who are not the targeted candidates. Moreover, treatment allocation that reflects actual practice allows for testing the effectiveness of the intervention—its benefit in a real-life setting, as opposed to its efficacy in a controlled setting.

This entry discusses alternative design strategies that are intended to address ethical or practical concerns when it is deemed unethical or infeasible to randomize all subjects to study interventions. These design strategies are called *cutoff* designs, because they involve, at least in part, the assignment of subjects to treatments based on a cutoff score on a quantitative baseline variable that measures clinical need, severity of illness, or some other relevant measure. What follows is an overview of cutoff designs.

II. DESCRIPTION OF THE REGRESSION-DISCONTINUITY DESIGN

The most basic of cutoff designs is the regression-discontinuity design (8,10–13), in which a baseline indicator, for example, severity of illness, can be used to assign subjects to an intervention. All subjects below a cutoff point on the baseline indicator receive one treatment, while all subjects above it receive another treatment. The history of the regression-discontinuity (RD) design is found in the social sciences, specifically in

*This entry is drawn largely from JC Cappelleri. Embedding the regression-discontinuity design within the randomized design. Proceedings of the Biopharmaceutical Session of the American Statistical Association, 1997. ©Joseph C. Cappelleri.

program evaluation. It has been employed to evaluate the effects of compensatory education, being on the dean's list, a criminal justice program, a health education program on serum cholesterol, accelerated math training, and the NIH Career Development Award (14). In these scenarios randomization of subjects was not a viable alternative.

The traditional RD design is a single-cutoff quasi-experimental design that involves no random assignment. The RD design received its name from the "jump," or discontinuity, at the cutoff in the regression line of baseline and outcome (follow-up) scores that occurs when there is a treatment effect. Figure 1 depicts an RD design with a hypothetical 10-point treatment effect (reduction). All subjects with scores above 20 on the baseline assignment indicator are most in need of the intervention and hence are automatically assigned to the test (experimental) treatment, while those with scores of 20 or less (those less in need) are automatically assigned to the control treatment.

As Fig. 1 shows, the outcome scores of the test treatment group (those scoring above the cutoff) are lowered by an average of 10 points from where they would be expected in the absence of a treatment effect. The solid lines show the predicted regression lines for a 10-point effect, and the dashed lines show the expected regression lines for patients in a treatment group if they were given the other intervention instead.

The baseline assignment covariate should be measured on at least an ordinal scale; it is more desirable, though, to have a continuous (ratio-level or interval-level) baseline assignment variable. Baseline and outcome may be the same or different, the cutoff can be placed anywhere along the baseline measure (as long as there are sufficient numbers in the control group), the direction of improvement can be positive or negative for either variable, the treatment groups could have more than two levels, and the response variable can be discrete or continuous.

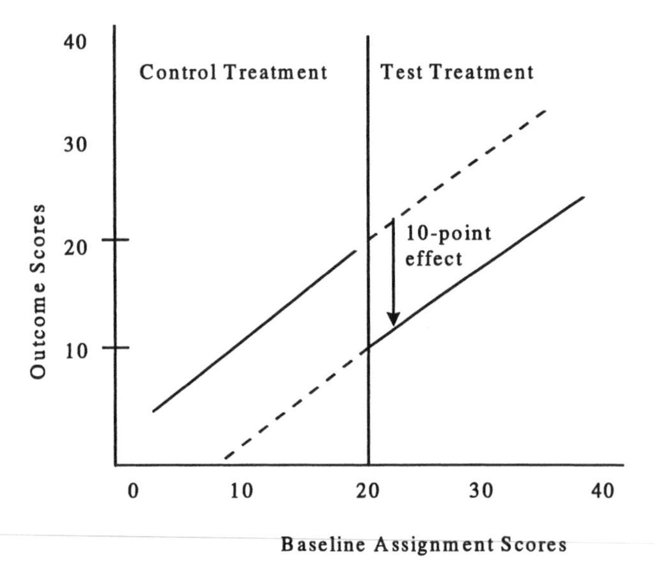

Fig. 1 Regression-discontinuity design with a 10-point treatment effect.

III. VALIDITY OF THE REGRESSION-DISCONTINUITY DESIGN

Under the assumption that the outcome-baseline functional form is correctly specified, the RD design results in an unbiased estimate of treatment effect. An unbiased estimate of treatment effect is obtained because the assignment process is known perfectly and controlled for in the analysis (10). Formal statistical derivations proving this lack of bias are found elsewhere (14–17). Like the randomized experimental (RE) design, the RD design gives a known probability of assignment to treatments. It is imperative, though, that the cutoff assignment rule be followed strictly. If subjects are misclassified, then the treatment effect is likely to be biased.

It can also be demonstrated that the estimate of treatment effect in the RD design, like the RE design, remains unbiased when random measurement error in the observed, fallibly measured baseline covariate is considered (14–17). The reason for this is that once the fallibly measured observed baseline scores are known, treatment assignment is completely determined and hence independent of anything else, including the perfectly measured true baseline scores, in the RD design. Similarly, in the RE design, treatment assignment is completely determined by a randomization scheme and hence independent of anything else.

Regression to the mean, which naturally emanates from random measurement error in the observed baseline covariate, does not therefore affect the estimate of treatment effect in both the RD design and the RE design. Figure 2 graphically shows the impact of regression to the mean, or, equivalently, random measurement error in the observed covariate, in the case of no treatment effect when the same variable is measured at baseline and follow-up. In the absence of a treatment effect, and with no other effects that might change a subject's score at follow-up, the true regression line should be a 45° line beginning at the origin. Regression to the mean causes the fitted regression line to be attentuated by an amount proportional to the reliability

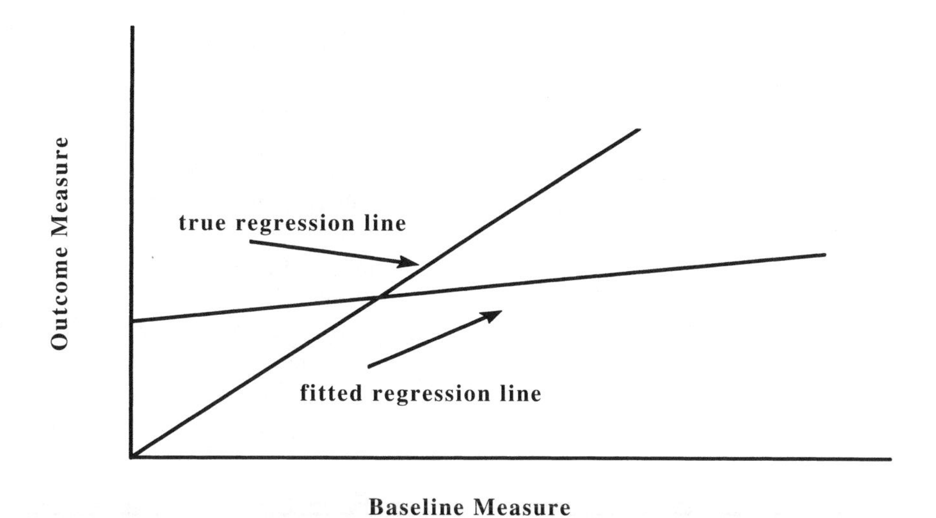

Fig. 2 Regression to the mean: randomized and regression-discontinuity designs.

coefficient of the baseline covariate; therefore, the sample regression coefficient of the baseline covariate on the outcome measure is biased, but the sample regression coefficient of the treatment effect is not (14,15).

The RE design is robust in giving unbiased estimates of treatment effect when the true functional form between the baseline covariate and the outcome measure is not correctly specified. On the other hand, the RD design is not robust here. The most critical step in obtaining an unbiased estimate of effect in the RD design lies in modeling this true functional form correctly. The true functional form, however, is not known in the RD design because of missing data. As shown in Fig. 1, which assumes a linear functional form, the extrapolated regression line of the control group (dashed line, right) if this group's subjects were given the test treatment instead is assumed to continue in the same linear way as its fitted line (solid line, left). The extrapolated regression line of the test-treated group (dashed line, left) if this group's subjects were given the control treatment is assumed to continue in the same way as its fitted line (solid line, right). There is no way to know prospectively whether the form or the slope of the lines in the region of missing data will be the same as that in the region of observed data.

IV. SPECIFYING THE FUNCTIONAL FORM

One suggestion for helping to arrive at the correct functional form is to use a polynomial backward-elimination regression approach (18). Another suggestion uses empirical Bayesian methods to overcome situations when the outcome and baseline relationship may not be linear, as when true baseline scores are not normally distributed (19,20). A third approach, which can be used with either of the other two approaches, is to fit a regression line over a wider range of the baseline-outcome distribution, resulting in less extrapolation and hence a more valid fit. This last approach can be achieved by combining the RD design with the RE design, resulting in a cutoff design with randomization.

V. COMBINING REGRESSION-DISCONTINUITY AND RANDOMIZED EXPERIMENTAL DESIGNS

A regression-discontinuity design can be described as a cutoff design without randomization. This design can also be coupled with a randomized design. For instance, patients who score within the middle range of scores on a baseline severity-of-illness indicator (e.g., those moderately ill) are randomized to either one of two treatments, while patients who score below a given cutoff value on this indicator (e.g., those most ill) are automatically assigned to the novel treatment and patients who score above another, higher cutoff value (e.g., those least ill) are assigned to the control treatment. Another type of cutoff design, for instance, would have subjects below the single cutoff point (e.g., the most ill) randomized to either treatment, while those above it (e.g., the least ill) are automatically assigned to control treatment. These are only two possible design variations that combine cutoff assignment and random assignment. Other variations are mentioned elsewhere (21,22).

Combining the RE design and the RD design may give advantages over either design alone (21–23). Relative to the RE design, this hybrid design may be better suited to address ethical or practical concerns, may result in a larger eligible and diverse sample, and may address better the effectiveness (as opposed to the efficacy) of interventions in particular circumstances. Compared with the RD design, RD-RE design has enhanced validity and improved statistical power.

VI. ILLUSTRATION: COCAINE PROJECT

To illustrate the combined design, we describe a cocaine project, conducted at the University of California at San Francisco, that applied the RD-RE design instead of the completely randomized design, which was considered neither ethical nor feasible (24). The study included about 500 patients with cocaine addiction. The objective of the study was to ascertain whether inpatient (intensive) rehabilitation showed better improvement, and by how much, over outpatient rehabilitation. The baseline assignment covariate was based on a weighted composite of four scales: (1) employment and legal status, (2) family relationship and recovery, (3) alcohol and drug history, and (4) psychological status. Higher scores indicated more clinical need for the more intensive (inpatient) rehabilitation. The primary outcome variable was the same variable measured at follow-up.

Figure 3 portrays how patients may be allocated into inpatient or outpatient rehabilitation in this setting. All patients who score above 60—those most severely ill or most in need—are automatically assigned to inpatient rehabilitation; all patients who score below 40—those least ill or least in need—are automatically assigned to outpatient rehabilitation; and patients who score between 40 and 60, inclusive—those moderately ill or in need—are randomized to either inpatient rehabilitation or outpatient rehabilitation. Note that it is this cutoff interval of randomization that distinguishes the RD-RE design from the RD design, which instead has a cutoff point(s) with no randomization.

Like Fig. 1, Fig. 3 has solid lines representing the predicted regression lines and dashed lines representing the extrapolated regression lines, showing a constant improvement from inpatient rehabilitation over outpatient rehabilitation. An analysis of covariance model, with the baseline assignment measure and the treatment group variable as predictors, would be a correct model

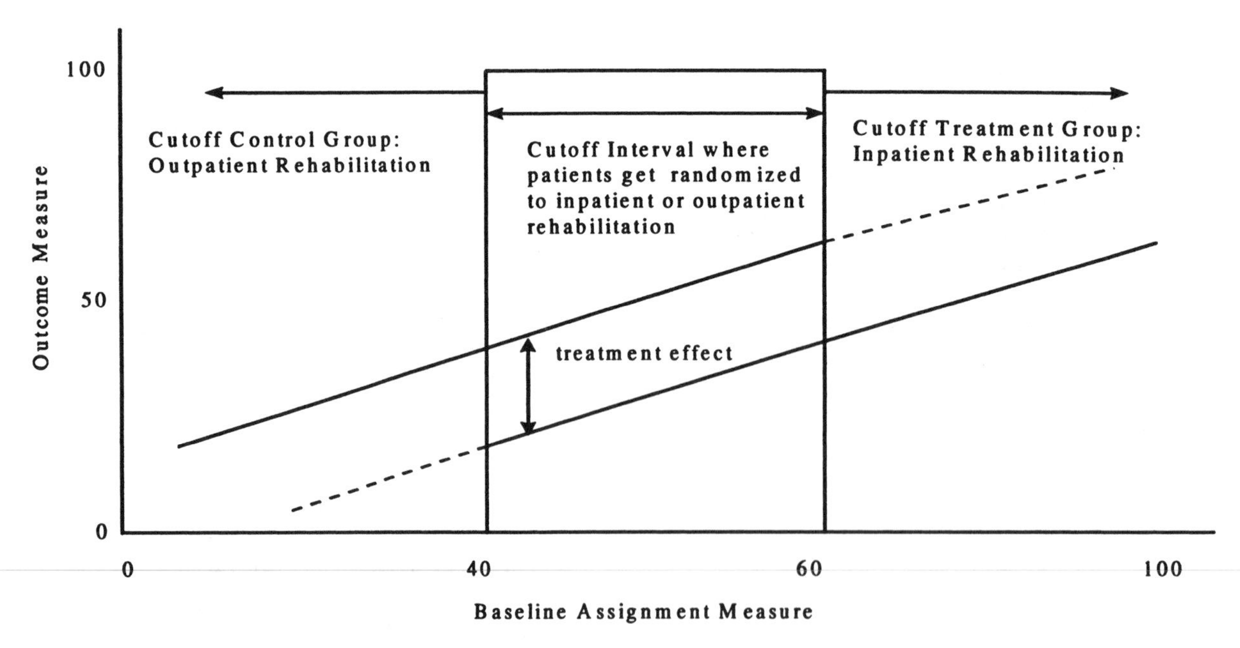

Fig. 3 Illustration of a combined randomized and regression-discontinuity design.

to fit the fitted lines in Figs. 1 and 3. An analysis of variance model, which excludes the baseline assignment variable, should be not fit, for it would result in a biased estimate of treatment effect. While linear relations are highlighted in these two figures, cutoff designs are not restricted to a linear baseline-outcome relationship; higher-order terms (e.g., quadratic or cubic terms), transformations on baseline or outcome variables, and interaction terms may also be fitted.

In a simulation study, several RD-RE design variations, of which the basic design in Fig. 3 is the simplest, were evaluated and compared among themselves, along with the traditional RD design and the traditional RE design (21,22). An unbiased main treatment effect was found for all these designs.

Figure 4 shows one of the more advanced RD-RE designs that may be useful for accommodating varying amounts of resources. One cutoff interval has its bounds at 45 and 55; the other cutoff interval has its bounds at 40 and 60. Both intervals are symmetric around 50. Because the two intervals have different widths, they include different numbers of randomized patients, with the wider interval containing more randomized subjects. As subjects accrue into a study, investigators of a clinical site may favor one interval of randomization over the other in order to address the cost implications of having a shortage or surplus of hospital beds for inpatient rehabilitation. Or one interval may be preferred because it is more commensurate with a hospital's level of resources and expertise with respect to a given treatment.

VII. MODELING AND ANALYZING CUTOFF DESIGNS

The RD-RE combination can be modeled and analyzed with the polynomial backward-elimination approach suggested in Sec. IV for the RD design. Specifically, the initial model equation is

$$
\begin{aligned}
y = {} & \text{bint} + (\text{btrt})*z + (\text{b}x\text{cut})*\text{xcut} \\
& + (\text{b}x\text{cut2})*(x\text{cut})^2 + (\text{b}x\text{cut3})*(x\text{cut})^3 \\
& + (\text{blinint})*(z*x\text{cut}) + (\text{blinquad})*(z*x\text{cut}^2) \\
& + (\text{blincub})*(z*x\text{cut}^3) + \text{error}
\end{aligned}
$$

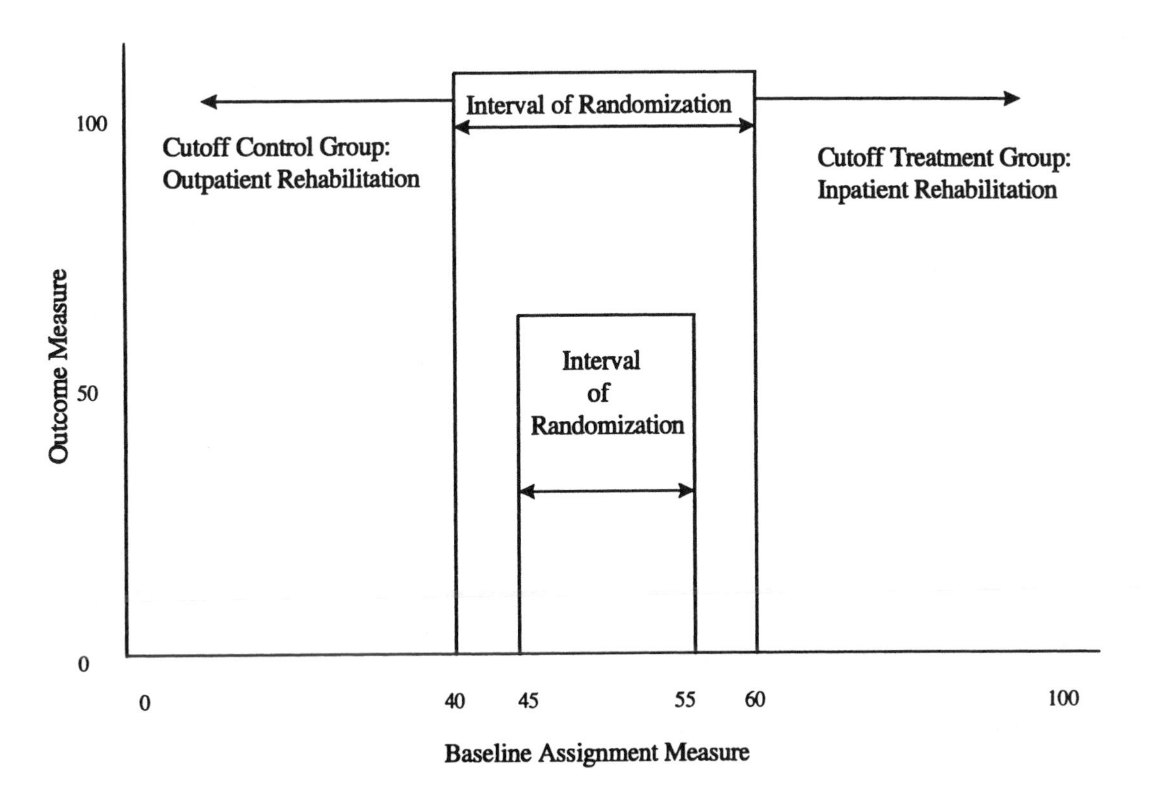

Fig. 4 Randomized and regression-discontinuity design with two cutoff intervals.

where

y = outcome measure
$x$cut = baseline assignment covariate minus a baseline value at which to measure the treatment effect (e.g., the middle value in a cutoff interval in an RD-RE design or the cutoff value itself in an RD design)
z = binary treatment group variable
bint = intercept estimator
btrt = treatment effect estimator
b$x$cut = linear slope estimator
blincut = linear interaction estimator
error = sample regression error term

The other regression coefficients are the coefficients for powers of "$x$cut" higher than 1 and for their corresponding higher-order interactions. The same set of assumptions that apply to linear regression (for continuous responses) and to logistic regression (for discrete responses) also apply here.

The modeling strategy first tests the significance of each regression coefficient separately, beginning with the higher-order interaction terms (i.e., the cubic interaction is tested first, followed by the quadratic interaction, and then linear interaction); interaction terms are tested before main effect terms. All significant terms and their lower-order counterparts are retained. The baseline covariate term and the treatment group variable are always kept in the final model.

VIII. RELATIVE SAMPLE SIZES NEEDED IN CUTOFF DESIGNS

The simulation study mentioned in Sec. VI also showed that, everything else the same, more randomization resulted in lower standard errors of the treatment effect estimate and therefore increased precision. It can be shown that the amount of this precision is completely determined by the multicollinearity or correlation (R) between the baseline assignment covariate and the treatment group variable as expressed by the variance inflation factor (VIF) (15,25):

$$\text{VIF} = \frac{1}{1 - R^2}$$

Suppose that there is a binary treatment group variable and a normally distributed baseline covariate. Table 1 provides the correlation between these two variables (R) and the accompanying variance inflation factor (VIF) in symmetric cutoff designs with varying amounts of randomization and with 50% of the subjects within the interval assigned randomly to either treatment. The VIF can be interpreted as the design effect of how many more subjects are needed in a given cutoff design relative to the completely randomized design in order to achieve the same level of statistical power, everything else the same.

Table 1 shows that, to achieve the same level of statistical power as the RE design, 2.75 times more subjects are needed in an RD design; 2.48 times more subjects are needed in an RD-RE design with 20% of all subjects randomized (i.e., 20% randomization); 1.96 times more subjects are needed in an RD-RE design with 40% randomization; 1.46 times more subjects are needed in an RD-RE design with 60% randomization; and 1.14 times more subjects are needed in an RD-RE design with 80% randomization. Derivations for the efficiency of such a cutoff design using an analogous approach, which gives the same results, are published elsewhere (26).

IX. RECENT CRITIQUES

Cutoff designs are certainly not without limitations. As mentioned earlier, an unbiased estimate of treatment effect requires that the functional relationship between outcome and baseline covariate be correctly modeled. Finklestein et al. (19,20) proposed a mathematical and statistical foundation, illustrated with examples, for how to analyze the RD design and to draw valid statistical conclusions about treatment efficacy. The authors discuss and illustrate their empirical Bayes methodology, which, they mention, can be used in a variety of circumstances, as a way to overcome restrictive assumptions about the functional form between outcome and baseline covariate.

Another reservation with cutoff designs is that they preclude any serious attempt at complete blinding of treatment, making them similar to nonrandomized designs in this regard. A further drawback of cutoff designs is that they are less efficient (precise) than completely randomized designs in terms of their estimates of treatment effects. According to Senn (27), the considerable excess of patients treated with the inferior treatment in cutoff designs (especially the RD design) relative to the RE design is likely to undermine the ethical argument that favors cutoff designs. While it is also true that more patients will receive the superior treatment in cutoff designs, regardless of which treat-

Table 1 Correlations and Variance Inflation Factors for Designs with Varying Amounts of Randomization

Percentage of all subjects within interval of randomization	Correlation coefficient[a]	Variance inflation factor
0 (regression-discontinuity design)	0.79	2.75
20	0.77	2.48
40	0.70	1.96
60	0.56	1.46
80	0.35	1.14
100 (randomized design)	0.00	1.00

[a]Expected correlation between a binary treatment variable and normally distributed baseline covariate.

ment it is, researchers are urged to consider Senn's position (27) before abandoning randomization as a perceived ethical problem in a clinical trial.

X. CONCLUSIONS

Randomization should be employed whenever possible. Cutoff designs should not replace the completely randomized design in the majority of circumstances, usually involving a drug intervention, when no appreciable logistical barriers preclude all subjects from being randomized to interventions. Cutoff designs are an alternative design when circumstances in health services research or outcomes research warrant that randomization of all subjects cannot be undertaken, for whatever reason. Cutoff designs are much more likely to be relevant and appropriate in studies on program evaluation that involve educational or behavioral interventions than in traditional phase 3 studies on drug interventions, but cutoff designs may have potential in phase 2 therapeutic trials as well. When compared with nonrandomized designs, the regression-discontinuity design (a cutoff design with no randomization) is an attractive alternative. When some subjects can be randomized, coupling the regression-discontinuity design with the randomized design is even a more attractive alternative than the regression-discontinuity design.

REFERENCES

1. S Green. Controlled Clin Trials 3:189–198, 1982.
2. KF Schaffner. J Med Philos 11:297–404, 1986.
3. LS Parker, RM Arnold, A Meisel, LA Siminoff, LH Roth. Clin Res 3:537–544, 1990.
4. E Marshall. Science 245:346–347, 1989.
5. JL Marx. Science 245:345–346, 1989.
6. JH Ware. Stat Sci 4:298–340 (with discussion), 1989.
7. RD Truog. Clin Res 38:537–544, 1992.
8. WMK Trochim. In: L Sechrest, P Perrin, J Bunker, eds. Research Methodology: Strengthening Causal Interpretations of Nonexperimental Data. Rockville, MD: Agency for Health Care Policy and Research, U.S. Public Health Service, 1990, pp 119–130.
9. SV Williams. In: L Sechrest, P Perrin, J Bunker, eds. Research Methodology: Strengthening Causal Interpretations of Nonexperimental Data. Rockville, MD: Agency for Health Care Policy and Research, U.S. Public Health Service, 1990, pp 145–149.
10. TD Cook, DT Campbell. Quasi-experimentation: Design and Analysis Issues for Field Settings. Boston: Houghton-Mifflin, 1979, pp 137–147, 202–205.
11. WMK Trochim. Research Design for Program Evaluation: The Regression-Discontinuity Approach. Beverly Hills, CA: Sage Publications, 1984.
12. SL Coyle, RF Boruch, CF Turner, eds. Evaluating AIDS Prevention Programs. Expanded edition. Washington, DC: National Academy Press, 1991, pp 144–159.
13. LB Mohr. Impact Analysis for Program Evaluation. Newbury Park, CA: Sage Publications, 1995, pp 133–155.
14. JC Cappelleri, WMK Trochim, TD Stanley, CS Reichardt. Eval Rev 18:141–152, 1991.
15. AS Goldberger. Selection Bias in Evaluating Treatment Effects: Some Formal Illustrations. Madison, WI: Institute for Research on Poverty, 1972, Discussion paper #123.
16. DB Rubin. J Educ Stat 2:1–26, 1977.
17. CS Reichardt, WMK Trochim, JC Cappelleri. Eval Rev 19:39–63, 1995.
18. JC Cappelleri, WMK Trochim. J Clin Epidemiol 47: 261–270, 1994.
19. MO Finkelstein, B Levin, H Robbins. Am J Public Health 86:691–695, 1996.
20. MO Finkelstein, B Levin, H Robbins. Am J Public Health 86:696–705, 1996.

21. WMK Trochim, JC Cappelleri. Controlled Clin Trials 13:190–212, 1992.
22. JC Cappelleri, WMK Trochim. Med Decis Making 15: 387–394, 1995.
23. RF Boruch. Socio Meth Res 4:31–53, 1975.
24. BE Havassy, BR Wesson, JM Tshann, SM Hall, CJ Henke. Efficacy of cocaine treatment: A Collaborative Study. Grant proposal funded by the National Institute on Drug Abuse (NIDA #DA05582), 1989.
25. JC Cappelleri. Cutoff-based designs in comparison and combination with randomized clinical trials. PhD dissertation, Cornell University, Ithaca, NY, 1991.
26. SJ Senn. Statistical Issues in Drug Development. Chichester, UK: Wiley, 1997, pp 89–92.
27. SJ Senn. Statist Med 15:114–116, 1996.

Joseph C. Cappelleri
William M. K. Trochim

D

Data Monitoring Board (DMB)

I. INTRODUCTION

The use of data monitoring boards (DMBs) in clinical studies has proliferated over the past several years for studies sponsored by the pharmaceutical industry as well as those sponsored by governmental agencies such as the National Institutes of Health (NIH). The purpose of this entry is to present the author's opinion of how this process should work and to describe and discuss issues with respect to the DMB process that arise when interim analyses are conducted.

Data monitoring boards are also referred to in the literature interchangeably as *data safety monitoring boards* (DSMBs), *data safety monitoring committees* (DSMCs), or *data monitoring committees* (DMCs). It our opinion that the word *safety* should be included when the group will be evaluating only safety, but not efficacy, data periodically during the conduct of the study, in order to protect the safety of the patients participating in the clinical study. As will be discussed later, in the pharmaceutical industry there are many phase 2 studies where a DMB is formed primarily to take an early look at efficacy. Hence, we prefer to refer to these groups as DMBs, since the appropriate implication is that they review both efficacy and safety data.

There have been several papers published in recent years on the DMB process. Most have related to large-scale mortality or otherwise significant clinical studies, such as cancer or AIDs trials, sponsored by a governmental agency such as the National Institutes of Health (NIH) (1–6). However, a few have addressed the safety monitoring and DMB processes utilized for studies sponsored by pharmaceutical companies (7–11). Rockhold and Enas (11) point out that data monitoring occurs for all clinical studies in a blinded fashion, whether or not there is an interim analysis or DMB. We would add that in addition to this type of monitoring, someone within the sponsor's organization is often, although not always, unblinded to serious adverse event (SAE) reports. Some regulatory agencies, including the U.S. Food and Drug Administration (FDA), are always notified of and unblinded to SAEs that occur on study drug. This is key safety data, and if the physician who is conducting this monitoring decides that the study drug is not safe, then the study may be terminated early on this basis. Another alternative is for the monitoring physician to request that a DMB be formed and that a full interim analysis of safety data be conducted.

It should also be noted here that most if not all pharmaceutical companies conduct their studies in a triple-blind fashion. Not only are the patient and investigators blinded to the treatment the patient is receiving, but all except a few sponsor personnel are also blinded during the conduct of the study (see Sec. IV for individuals who are typically unblinded). Other than the exceptions, the sponsor does not become unblinded to the data until the data have been cleaned, verified, and validated and the final database is locked. This triple-blinding is standard operating procedure (SOP) to ensure that in particular the sponsor personnel who are in contact with the study sites do not inadvertently share any unblinded information with them. This SOP also ensures that individuals who have authority to make changes to the database, consisting of those who validate the data by checking that source documents, case report forms (CRFs), and the reporting database all agree, remaining blinded until final locking of the database.

Interim analysis (IA) is defined as an analysis of a clinical trial where data are separated into treatment groups that is conducted before the final reporting database is created and locked. Interim data summarized across all treatment groups is not considered IA. Outcome variables can include any data collected in the study, including efficacy, safety, health economics, and pharmacokinetic parameters.

II. REASONS FOR PERFORMING INTERIM ANALYSES AND FOR DATA MONITORING BOARDS

Interim analysis can occur for many reasons. Clearly the most important reason to conduct IA in mortality trials is to protect the safety of the patients participating in the trial. A study in which interim data demonstrate with a high level of confidence that the study drug is unacceptably toxic, or does not have sufficient efficacy, should be stopped early. Much of the literature on DMB processes deal with these types of clinical studies.

However, the large majority of clinical studies conducted do not fall into the category of mortality or irreversible morbidity trials. Although these trials may have IA for the same reason as mortality trials (remember that safety of the patients is always paramount, and hence ongoing safety monitoring is always conducted for all clinical studies), there are several other reasons that are more typical.

First, when the initial phase 2 efficacy study is conducted for a new compound in a particular disorder, the estimates of variability and treatment differences that were used to estimate sample size and power may be far from the truth. An interim analysis in such a study may lead to stopping the study (and other related research activities) due to clear lack of effectiveness of the study compound. Or, if the treatment effect is much more powerful than predicted, the IA can lead to acceleration of the development of the compound, including earlier initiation of phase 3. This is a win–win situation for both the patients with the disorder and the sponsor, in that it shortens the elapsed time to conduct thorough clinical research and development. The International Conference on Harmonization (ICH) Guideline for Statistical Principles for Clinical Trials (12) recognizes this need, where in Section 4.5 they state, "However, it is recognized that drug development plans involve the need for sponsor access to comparative treatment data for a variety of reasons, such as planning other trials."

There are other reasons for conducting IA that deal with activities not related to the conduct of the clinical study. This can include, for example, to initiate the manufacturing of relatively large quantities of study drug material, to commence the building of manufacturing facilities for when the drug is ultimately launched into the marketplace, or to start a 2-year toxicology study.

There are cases in the literature where a sponsor did not follow a DMB process during an IA that led to public criticism from the FDA. One was with respect to the Anturane Reinfarction Trial (ART), where the Task Force of the Working Group on Arrhythmias of the European Society of Cardiology (1) states, "The trial was criticized for not having "stopping rules" and for lack of active involvement of an independent DSMC. Due to these limitations of ART, the FDA disapproved the application for a new indication of sulfinpyrazone. Moreover, FDA officials took the unusual step of publishing their critique of the trial" (13). Another example was in regard to an antisepsis compound, where the FDA was openly critical of the sponsor because they did IA without an independent DMB and changed the primary endpoint as a result of the IA (14).

III. PROTOCOL DESCRIPTION OF INTERIM ANALYSIS PLANS AND THE DATA MONITORING BOARD PROCESS

Interim analysis plans should be stated in the protocol. This includes reasons for conducting the IA, how many there will be, when they will occur (e.g., function of number of patients or calendar time), and how final significance levels will be adjusted, when appropriate. Additionally, any early-stopping rules, even if used only as a guideline, should be described. The protocol should also include a brief description of the DMB membership (e.g., internal, three MDs, two statisticians) and of the DMB process.

Interestingly, much of the literature on DMBs suggests that the popular alpha spending functions, such as Lan and DeMets (15) and O'Brien and Fleming (16), should be utilized only as guidelines, not as rigid stopping rules (17–21). The reasoning given is that there are many other aspects of the study to consider in making such a critical decision. For example, a study might cross the boundary on the primary efficacy outcome but have far worse safety than expected, leading to a poor benefit-to-risk ratio. Or the secondary efficacy variables might not support the primary efficacy variable to the extent that was expected, leading a DMB not to stop the study even when the stopping boundary has been crossed.

Regarding this issue of invoking stopping rules, we also want to add emphatically that taking a statistical penalty for an IA of efficacy is *only* appropriate if the study allows for early termination due to positive interim efficacy results or if a regulatory submission will be based on the interim data when a sufficiently small nominal interim *p*-value is attained (even if the study continues). In the class of all clinical studies, this is by far the minority of the cases. Interim analysis that looks only at safety data, or where efficacy data are analyzed to make decisions external to the ongoing trial, or to stop the study for *lack* of efficacy, do not need to, and in fact should not, make any adjustments to the final nominal alpha level, in order to maintain an overall prespecified Type I error rate. On this topic, the FDA has in recent years insisted on the sponsor's "spending some of the alpha" at the interim, even when both sponsor and FDA acknowledge and agree that the study will not stop early (often this is the case in non-life-threatening disorders where patient exposures are needed to establish safety of the study compound). They do, however, accept a stopping rule whereby the nominal levels at the interims are very small, say 10^{-4}, such that one can still test at the traditional .05 level at the end of the study. Although this essentially ends up in the same place as not adjusting at all, we suggest that one should technically be allowed to test at the interim at a nominal alpha level of 0, which means that the null hypothesis is not rejected under any circumstances at the IA. The final nominal alpha level would then be exactly .05, in order to achieve a .05 Type I error rate.

The process should allow for unplanned IA. Most often when this occurs, it is an early phase 2 study where enrollment was far slower than anticipated. As discussed in the preceding section, such studies may have power computations based on estimates of variability and treatment effect size that are grossly in error, and so an early interim may become beneficial (maybe an IA was not planned in such a case because enrollment was anticipated to be so rapid that by the time an IA could be conducted, the study would have completed).

IV. DEFINITION OF GROUPS OR COMMITTEES INVOLVED WITH THE PROCESS

The various groups involved in the DMB process, and how they interrelate, are defined in the following subsections.

A. Data Monitoring Board (DMB)

The DMB is responsible for interpretation of interim analyses of efficacy and safety data and for producing recommendations that do not, in general, include any data. The responsibilities of the DMB secretary include initiation and documentation of the creation of the DMB membership. The DMB secretary also coordinates the dissemination of materials (e.g., protocol, interim report, or analysis tables) to DMB members, is the scribe at the DMB meeting, and constructs and disseminates minutes to the DMB members for their approval. Typically this person is the project statistician, and is a voting member of the DMB. The DMB can also have a separate chairperson who sets the agenda and leads the DMB meeting. However, in most cases with smaller DMBs, the DMB secretary serves in this dual capacity.

B. Data Monitoring Board Steering Committee (DMBSC)

The DMB makes recommendations to the DMBSC, which then determines whether and how to implement the recommendation. In most cases the DMBSC is comprised of the sponsor's senior research management.

C. Safety Monitoring

Safety monitoring is defined as the process whereby safety data from the ongoing clinical trial is evaluated continuously during the conduct of that trial. This is typically conducted by one or more clinical research physicians (CRPs) from the sponsor. Included are the evaluation of all unexpected SAEs in an unblinded or blinded fashion and reporting them to regulatory agencies around the world. This monitoring can also include blinded tracking of other safety data [e.g., adverse events (AEs)]. In situations where the potential for an important treatment effect becomes apparent, a DMB should be formed to look at the unblinded data (see Ref. 22).

D. Data Coordinating Center (DCC)

This is the group that handles the data, from data entry to generation of data output. This group can be at the sponsor's, or it can be a group not affiliated with the sponsor, such as a clinical research organization (CRO).

E. Data Analysis Group (DAG)

Members of the DAG are a subset of the DCC. Members of the DAG have access to treatment assignments. These members typically include:

- A person who assembles the blinded clinical trial (CT) material, which might be a person in the sponsor's CT material department, where the randomization table can also be produced
- Systems analysts who build and manage the computer database
- Statisticians who are responsible for data analysis and in some cases might be the ones who generate the randomization table

By having access to treatment assignments, the latter two groups can complete the writing of computer programs that will be used for the analyses, results from which will be shared only after the database is locked.

On occasion, other individuals, such as pharmacokineticists, who might be creating population pharmacokinetic models while the data is accruing, also gain access to treatment assignments prior to final database lock; hence, they would then also be considered members of the DAG. Also, in some cases a member of the sponsor's clinical laboratory medicine group has access to the randomization table in order to identify which plasma samples are to be assayed for study drug concentration levels. (Note that it is important that these plasma concentration levels not be included in the database until final data lock, at least not if others outside of the DAG have access, since they would identify the patients who received the study drug.)

F. Project Team

This is the multidisciplinary team within the sponsor's organization; it is responsible for all aspects of research and development of the study compound.

G. Definition of Potential Registration Study

A *potential registration study* is defined as a clinical study in which efficacy is assessed in patients with the target disorder; it will possibly be used for registration or labeling in any country. Usually such studies are designed with at least 80% power, $\alpha = .05$. But potential registration studies can be smaller in size. In early phase 2 studies of a new compound, estimates of power and sample size are often based on estimates derived from other sources; a small study can surprisingly achieve statistical significance. Hence, even small initial phase 2 studies that assess the efficacy of the study drug are often considered as potential registration studies. We advise that the DMB process outlined here be followed for any potential registration study.

H. DMB Administrator

This individual is responsible for the entire DMB process at the company. This individual maintains the company policy and procedures that relate to the IA. She also is one of the approvers of DMB membership, and is invited to DMB meetings and other meetings where DMB recommendations are shared.

I. Relationship Between the Various Committees and Individuals

The reporting structure we propose is relatively simple and straightforward. The DCC is the group that prepares the data for analysis. The DAG, which has access to treatment assignments, conducts the analysis and presents its findings to the DMB. The DMB interprets the results, creating recommendations that it shares with the DMBSC. The DMBSC is responsible for implementation of the recommendations.

Additionally, the physicians who are continuously monitoring safety report their findings to the DMB, and they suggest additional interim analyses whenever they suspect an issue based on their blinded reviews of safety data and unblinded reviews of SAEs. The project team can report to the DMB background information regarding the study compound.

J. Roles and Responsibilities of DMB Members

The role of the DMB needs to be clearly established before the DMB initiates its activities. The key role is to inform the DMB members that under no circumstances are they to discuss the data, their deliberations, or any hint of what the data are showing to anyone outside of the DMB. This point can be drilled home with the use of a letter from a senior member of company management that internal as well as external DMB members must sign, signifying their agreement to abide by this important requirement.

The key responsibilities of the DMB should be to evaluate and interpret the interim data, followed by making its recommendations to the DMBSC. Although the DMB can be made aware of information relating to the trial, such as recruitment, quality of the data, flow of clinical report forms, and adherence to the protocol, we do not believe it is the DMB's responsibility to conduct these tasks. These are better accomplished in an unblinded fashion by the DCC (see Refs. 3, 23, 24 for a differing opinion).

V. FORMATION OF THE DATA MONITORING BOARD

Ideally, the DMB should be formed prior to finalization of the protocol. This gives the board the opportunity to make important contributions to the study design and protocol and in particular to the protocol section describing the interim analyses and DMB process. Of course, an unplanned IA will always result in DMB formation after the study has begun.

A single DMB can serve multiple studies of the same study drug and indication, especially when the DMB will be meeting periodically over several months or years, instead of the situation where the DMB meets only once. As an example, for the AIDS Clinical Trial Group (ACTG), a single DMB served as many as 25 ongoing studies during the course of a year (6). In any case, safety data presented to the DMB should include all available clinical safety data from ongoing and completed studies involving the study compound, which may at least partially be satisfied by providing copies of the clinical investigational brochure (CIB).

A. DMB Membership

The group can have as few as three members and as many as 10 or more, depending on the circumstances. Single interim analyses of the primary efficacy variable in a phase 2 study to make recommendations that do not include the ongoing study (e.g., start phase 3 early) are often entirely internal to the sponsor, and typically involve three to six members. These always include a physician knowledgeable of the disease under study and a statistician (usually the one conducting the interim analysis). In addition, when the study is a potential registration study, a member of senior management is also included. The reason for including this individual on the DMB is that it lessens the need for the full DMBSC to see actual interim data when the DMB makes its recommendations. This person usually is one with knowledge of and responsibility for the particular therapeutic area. Finally, when the DMB recommendation might lead to acceleration or deceleration of manufacturing, toxicology studies, or some other nonclinical aspect of the overall project, the project manager can be included in the DMB, as a representative of these diverse functional areas.

Large-scale multicenter mortality studies typically have DMBs with 8–10 members. In addition to the individuals just specified for smaller DMBs, they might include additional physicians and scientists with particular expertise (e.g., a study of an agent to treat osteoporosis will have at least one physician expert in this disease; and if the drug is believed or known to have positive or negative effects in regards to cancer, then one or more oncologists will also be included). These DMBs might also include an ethicist, who has nothing more than the interests of the participating patients in mind. Finally, these larger DMBs might include members internal and external to the sponsor, a subject discussed in more detail in the following section.

DMBs for potential registration studies should not include individuals who are in contact with the study sites, and they should not include anyone who has the ability to make corrections to the database. This includes clinical research administrators and related personnel from the DCC who check source data to case report forms (CRFs) and who check CRFs to the reporting database. Clearly the perception of bias, and possibly real bias, would be introduced to the data were these individuals to be included on the DMB.

The DMB administrator should be invited to all DMB meetings, as a nonvoting member. The DMB administrator shouldn't be required to attend all DMB meetings, but this is a case-by-case judgment that needs to be made between the DMB administrator, secretary, and chair. The level of involvement of the DMB administrator depends on the experience of the DMB members in the process. The DMB administrator maintains consistency of the process across all therapeutic areas within the company.

B. Internal vs. External Members

We suggest that there are some cases where an internal DMB is appropriate, others where an external DMB should be assembled, and yet others where a mixed DMB is optimal. O'Neill (25) writes:

> Recognizing that independent external data safety monitoring boards (DSMBs) in principle are not necessary for all clinical studies, the scheme depicted [in an accompanying figure] illustrates that any clinical trial evaluating a therapy with a mortality or irreversible morbidity endpoint should probably use externally independent DSMBs. This would include mega-trials, trials with mor-

tality endpoints, and trials of treatments for life-threatening diseases with no alternative therapies. Trials which are exploratory, early phase trials, or which deal with chronic disease, palliative therapy, or non-life-threatening diseases probably can be monitored by designated internal bodies who, although not independent from the sponsor, are insulated from management in decision making and follow guidelines as to who has access to their unblinded data.

Other experts, however, have urged more generally that DMBs be independent not only from those directly involved in conducting the study, but also from those who have any financial interests in the sponsor (1,3,17,20,23,26–29). They point out that this would exclude any sponsor employee from serving on the DMB. Although Harrington et al. (28) suggest that sponsor employees should not be DMB members; they additionally argue in favor of including one or more study investigators on DMBs. Several papers go so far as to suggest that the DCC should be external from the sponsor (1,27,29).

We have several comments and observations regarding this issue. First, "significant" financial interests is difficult to define precisely. Although company employees receive a salary from the sponsor, the external DMB members also receive payment for their time. One could argue that if anyone has a financial incentive that might lead them to bias their decision making, it would be the external members, since their payment is a direct function of how many meetings are held, how long the study is allowed to continue, etc. This is not meant to be accusatory, but to point out that the perception exists. On the contrary, sponsor employees receive their paychecks no matter what happens to the clinical trial, because there are other projects on which to work. (One could argue that employees of small biotech companies, where their companies' survival may depend on the success of the ongoing clinical trial, have the same perception problem that external members have, perhaps to an even greater degree.) The only way fully to satisfy the financial interest issue is for external DMB members to insist on working without compensation for their time or, more realistically, for a governmental agency or other independent group to be responsible for the expenses from external DMB members.

These points have even more relevance to independent external DCCs. These groups, whether commercial or university-based contract research organizations (CROs), are paid large sums of money (often millions of dollars) for their services. We contend that the subtle difference between a sponsor-based DCC and a contract with an external CRO is insignificant.

Our position is that people external to the sponsor should be considered for DMB membership when they have expertise beyond that offered by company employees. This could be a physician who is a global expert in the particular disease or someone expert in the conduct of clinical trials and the ongoing monitoring of safety in those trials. It could be a statistician who is an expert in interim analysis methodology or someone who is expert in the DMB process. However, it will almost always be the case that people who are employed by the sponsor have the greatest prior knowledge of the study drug, including previous preclinical and, most importantly, previous clinical data. Employees of the sponsor often will also have the greatest knowledge of the drug's mechanism of action and, consequently, of what adverse effects the drug might be expected to have. To exclude these individuals from DMBs may be compromising the safety of the patients participating in the clinical trial. Crowley et al. (2) point out the dilemma of wanting to minimize conflicts of interest while maximizing expertise, by stating: "No one with a direct financial conflict of interest can be on the DMC, but we do not want to exclude those with the most knowledge of the trial in order to achieve the appearance of independence." Ellenberg et al. (20) also acknowledge this dilemma by stating that "The pharmaceutical company staff are likely to know more about the properties of the drug than anyone else, and will be knowledgeable about other relevant studies that the company has conducted or is conducting." (See also Ref. 11.)

Another consideration in this issue of internal versus external membership is liability. Ultimately it is the sponsor who is liable for anything that might go wrong in the study. If it is later alleged that a DMB put patients at risk by incorrectly allowing a study to go forward when it turned out the study drug was toxic, the sponsor, not the DMB, would be the one defending itself against lawsuits. For this reason, an internal DMB may be more conservative than an external DMB, leading to earlier termination of a study with a potentially toxic study drug. This point is illustrated by the example of a congestive heart failure trial discussed by Rockhold and Enas (11), in which they state: "Even though the difference was not significant and the external experts recommended continuation of the trial, the company decision was to terminate it."

VI. LEVELS OF BLINDING OF DATA FOR THE DATA MONITORING BOARD

Four strategies for level of unblinding in the data presented to the DMB are as follows:

1. "Partially unblinded," where data are summarized by treatment groups but the treatment groups are identified by blind labels, such as "Group A," "Group B"
2. A variation of the first strategy, in which within a row, or a particular variable, the groups are always ordered from smallest to largest
3. Unblinded at the group level
4. Unblinded at the patient level

As an example of what is meant by strategy 2, a summary of adverse events (AEs) would simply show for each AE the incidence rates, in increasing order. So a single column would not represent a single treatment group. The motivation for choosing this method of blinding is that if one were instead to choose strategy 1 and there was an SAE that occurred in only a single patient, for whom the treatment group assignment is known, then the entire analysis becomes unblinded at the group level. This issue was illustrated by Buyse (3), who stated: "A few groups used to report safety and efficacy analyses 'blindly' (that is, with anonymous codes instead of treatment names), but this policy has fallen out of favor because the blinding tended to be either ineffective or frustrating. Blinding was ineffective when treatments had easily identifiable toxicities or side-effects." However, the method of blinding in strategy 2 does not allow one to interpret data across the many variables, which seriously compromises the ability of the DMB to make informed assessments of the overall treatment effects.

We recommend strategy 3, unblinding at the group level. (See Ref. 3, which concurs.) There is rarely a need for the DMB to be exposed to the randomization code that identifies the treatment received by each patient, which is what occurs with strategy 4. And although strategy 1 has some merits, that is true only if there is little going on in the study, that is, if there are not meaningful or statistically significant treatment differences. As soon as something meaningful is seen, the DMB will logically insist on unblinding the groups (which brings us back to strategy 3), so it can make an intelligent recommendation.

VII. ACTIVITIES PRIOR TO THE DMB MEETING

The IA is conducted by the data analysis group (DAG). The DAG may choose to consult with any DMB members (e.g., the physician) prior to the DMB meeting to resolve any issues that arise. They, or the full DMB, may choose to do additional analyses based on what the interim data indicate or, for that matter, based on what additional outside information becomes available during the course of the study.

The DMB secretary, in concert with the DMB chair and the sponsor's project manager, decides upon the date and location of the meeting. Meetings often are only an hour or two in duration; however, there are cases where the DMB needs a full day to evaluate all of the interim data thoroughly.

At least a week prior to the meeting, the DMB secretary should distribute to the DMB members the study protocol, a list of objectives to be accomplished by the IA, and an agenda. When the IA is complex or when there is a great deal of data to be reviewed, the DMB secretary can distribute to DMB members prior to the meeting a written summary that includes tables of interim results. The DMB members are then expected to review these data before the meeting in order to minimize the time required for data presentation during the meeting.

In the event that interim data are disseminated prior to the DMB meeting, the use of fax or other such nonconfidential methods should be avoided. This is especially true when faxing to a receiving machine that is in an unsecured, open area where others can see it.

VIII. CONDUCT OF THE DMB MEETING

The DMB meeting can optionally begin with an open session. During this session the project team can present background information regarding the study compound. Either the project team or the DMBSC can set the stage for the DMB by clearly delineating the questions and issues the DMB is expected to address and by answering questions from the DMB. This open session can also include a presentation to the DMB by the physician who is monitoring safety in a blinded manner. The IA conducted by the DAG is not shared or discussed during the open session.

At the beginning of the closed session, the DMB secretary or DMB administrator should remind everyone that the results being presented are highly confidential and are not to be shared with anyone outside of the room. Not only should explicit interim data not be shared, but even statements such as ''this is a very viable compound'' or ''the drug is beginning to show serious negative effect on the cardiovascular system'' are not to be shared. It should be explained to the DMB that any leaking of results could lead to a biasing of the future conduct of the study. Or worse, regulatory agencies may not accept data from a study where interim data were released to inappropriate individuals.

All materials either sent to DMB members before the meeting or handed out at the meeting should be collected by the secretary at the end of the meeting. Under special circumstances when this does not occur, the DMB secretary should document which members retained this material and why. Multiple copies should be either destroyed or stored confidentially, maintaining a paper copy or electronic copy for potential future reference.

When it is anticipated that further analyses may be defined by the DMB, the statistician may bring the data on a laptop computer or have some other means to present analyses and summaries during the conduct of the meeting.

The DMB participants must be committed to attending the DMB meeting in its entirety and must not come and go during the meeting. Partial attendance can lead to poorly deliberated recommendations from the DMB. In any case, substitutions should not be allowed without first officially changing the DMB membership.

The goal of the meeting is to construct a recommendation that is then communicated to the DMBSC chair or designee. Agreeing on the final wording of the DMB recommendation may at times take place after the meeting or possibly at a subsequent meeting.

Following are five boilerplate statements that can be used as a guide for writing a DMB recommendation. These are typically relevant to early phase 2 studies in nonmortality disorders. However, the actual recommendation may need to be more specific to the study and drug being evaluated.

Boilerplate Recommendations of Data Monitoring Boards

1. Stop the current study because of extraordinarily positive results. Accelerate other development plans. [This rating should be assigned only if the protocol includes a formal procedure to stop early for positive efficacy.]
2. Accelerate other development plans. Continue the current study, because the prospect for positive results at completion is good.
3. Continue with other developmental activities as planned. Continue the current study; inferences cannot be made before the end of the study [or before the next interim analysis].
4. Decelerate other developmental plans. Continue the current study; however, the prospect for positive results at completion is not good.
5. Stop the current study because of extraordinarily negative results. Rethink other developmental plans.

Brief meeting minutes should be written and kept by the DMB secretary. They should be sent to the DMB members, preferably electronically in such a way that recipients cannot print or forward the document (Lotus Notes, for example, has this capability). These minutes should document who was present, what data were discussed, and what conclusions or recommendations were drawn. These minutes should be kept with a paper copy of the materials disseminated to DMB members, plus any other materials, such as computer printout, that pertain to the interim analysis. Although the minutes generally should not contain any unblinded data, they may contain information or statements regarding the interim results that are unblinding, and hence they are not distributed beyond the DMB.

At the end of the study, when the final reporting database is locked, all of the DMB materials should be archived in the sponsor's project study file.

IX. USE OF TECHNOLOGY TO STREAMLINE THE DMB PROCESS

We suggest that electronic forms be created and then repeatedly used for the following:

1. Creating a DMB and an authorization to conduct the IA
2. Amending the membership of a DMB
3. Authorizing an interim analysis of clinical data
4. Authorizing the final analysis of clinical data
5. Authorizing the replacement of a reporting database

These electronic forms serve as documentation of important aspects of the process. They also eliminate the need to obtain approval signatures physically. Software such as that offered in Lotus Notes allows for a database to be created, enabling tracking of DMBs that exist within the sponsor and much related information. For example, one should track the dates when members have been added to or removed from DMBs. This database should document when interim and final analyses are conducted, along with approvals from the appropriate individuals.

X. COMMUNICATIONS OF THE DMB RECOMMENDATION AND DISSEMINATION OF DATA

Communication of the recommendation should first go to the DMBSC chair or the chair's designee. Approval should be obtained from the DMBSC prior to any further dissemination of recommendations beyond the DMB.

Depending on the DMB recommendation, the DMBSC may then choose from one of the following three choices:

1. Accept the DMB recommendation as written, with no sharing of unblinded data beyond the DMB; forward the DMB recommendation to the project team. This choice is usually appropriate when the DMB recommendation is to continue the study and future research as planned (i.e., do not accelerate or decelerate any other compound-related activities).

2. Instruct the DMB to unblind the DMBSC; this can consist of a written summary, an oral presentation by the DMB, or both. This choice should probably occur when the DMB recommends stopping the study due to lack of efficacy or substantial toxicity. It also may occur when the data are very positive, leading the DMB to recommend initiating phase 3 studies immediately or accelerating other components of research and development. The bottom line is that when DMB recommendations require the sponsor to invest significant financial or human resources faster and sooner than planned, so that accepting the recommendation would take away from other opportunities, they will likely need to see the interim data to make a decision that is balanced among other products in the portfolio. Furthermore, if the DMB recommends stopping the study for positive efficacy, it is prudent first to consult with regulatory agencies for an indication on whether they concur. Pocock (17) points out that the DMBSC may need to see unblinded data before accepting the DMB recommendation:

> Usually, any data monitoring committee meeting has two options: a recommendation that the trial should continue as planned, with no additional details; or a specific recommendation that the trial should be stopped or altered in some way. This second option usually necessitates full or partial release of interim results to the steering committee so that it can rapidly reach a definite decision to concur or (rarely) disagree with the data monitoring committee's recommendation.

3. Instruct the DMB to unblind select members of the project team or external consultants in addition to the DMBSC or even to consult with regulatory agencies. In general, this should occur only when the data are very negative, although it can also occur when the DMB recommends stopping the study early due to very positive results. In the situation where data are very negative, before terminating the study it is worthwhile to consult with those closest to the trial, and others on the project team, before making such a significant decision.

The DMBSC must first give its approval before any interim data are disseminated beyond the DMB, should the need arise. The DMB secretary should maintain a list of who outside of the DMB sees the unblinded data, the date of this unblinding, and a description of what subset of data, if not all of the data, was shared.

Great care needs to be taken in what is shared beyond the DMB and DMBSC, even if no explicit data are included. For example, even a simple identification of a subgroup (e.g., particular investigators) that shows a larger treatment effect should generally not be presented outside of the DMB. This is, not only should the data not be presented, but even words that simply state, for example, that males show a large treatment effect while females do not should not be shared without careful consideration of the implications upon the ongoing study.

Several DMB models that have been published have intermediate groups in between the DMB and the steering committee (the decision-making body) (see Ref 1, for example). We believe these models are inefficient; and more seriously, the intermediary may not accurately convey to the steering committee what the DMB told it. Consequently, we recommend that the DMB communicate directly with the DMBSC.

Regarding the issue of whether the DMBSC should be comprised of individuals from the sponsor, the literature contains several acknowledgments that this is the appropriate composition. For example, the ICH Guidelines (12) corroborate this composition of the DMBSC: "recommend to the sponsor whether to continue, modify, or terminate a trial."

XI. FINAL STUDY REPORT

If the IA includes a stopping rule that lets one stop early and declare effectiveness, a full presentation of the interim analysis results that were discussed at the

DMB meeting needs to be included in the final study report, whether or not the study was altered (e.g., stopped early). Otherwise all that should be required is a description of the IA, including the DMB recommendation, the functional areas represented on the DMB, a description of the data analyzed, the purpose of the IA, etc. However, the interim data and results of the analysis should be archived and should be available if requested.

XII. CONCLUSIONS

Interim analysis is not free; there is a substantial cost. It takes significant resources throughout the medical organization, not just a statistician's time. As suggested by Fleming and DeMets (30), it is important to have data that are current and not to exclude recent data (e.g., within 2 months of the DMB meeting). Because of this need, monitors need to visit the study sites at essentially the same time to get all recent data into the interim database and cleaned in time for the conduct of the IA. Or at times a sponsor might feel the need to include in the analysis some data that have not yet formally been declared "clean,'' in order to satisfy the "current data'' requirement and also to have the greatest statistical power possible at that time. For this reason, IAs should be utilized judiciously.

But on the other hand, use of data monitoring boards in the way described herein maintains the integrity of the data and can allow for more rapid development. In the case where early evidence is overwhelmingly against the study drug, it reduces the number of patients exposed to an ineffective study compound and saves corporate financial and human resources.

REFERENCES

1. Task Force of the Working Group on Arrhythmias of the European Society of Cardiology. The Early Termination of Clinical Trials: Causes, Consequences, and Control, With Special Reference to Trials in the Field of Arrhythmias and Sudden Death. Circulation 89(6): 2892–2906, 1994.
2. J Crowley, S Green, PY Liu, M Wolf. Data Monitoring committees and Early Stopping Guidelines: The Southwest Oncology Group Experience. Stat Med 13:1391–1399, 1994.
3. M Buyse. Interim Analyses, Stopping Rules and Data Monitoring in Clinical Trials in Europe. Stat Med 12: 509–520, 1993.
4. S George. A Survey of Monitoring Practices in Cancer Clinical Trials. Stat Med 12:435–450, 1993.
5. L Friedman. The NHLBI Model: A 25-Year History. Stat Med 12:425–431, 1993.
6. DL DeMets, TR Fleming, RJ Whitley, JF Childress, SS Ellenberg, M Foulkes, KH Mayer, J O'Fallon, RB Pollard, JJ Rahal, M Sande, S Straus, L Walters, P Whitley-Williams. The Data and Safety Monitoring Board and Acquired Immune Deficiency Syndrome (AIDS) Clinical Trials. Controlled Clin Trials 16:408–421, 1995.
7. GG Enas, RL Zerbe. A Paradigm for Interim Analyses in Controlled Clinical Trials. J Clin Res Drug Devel 7: 193–202, 1993.
8. PMA Biostatistics and Medical Ad Hoc Committee on Interim Analysis. Interim Analysis in the Pharmaceutical Industry. Controlled Clin Trials 14:160–173, 1992.
9. GG Enas, BE Dornseif, CB Sampson, FW Rockhold, J Wuu. Monitoring Versus Interim Analysis of Clinical Trials: Perspective from the Pharmaceutical Industry. Controlled Clin Trials 10:57–70, 1989.
10. Committee on Interim Data Review and Analysis (CIDRA). Input into Guidelines on Interim Analyses.
11. FW Rockhold, GG Enas. Data Monitoring and Interim Analyses in the Pharmaceutical Industry: Ethical and Logistical Considerations. Stat Med 12:471–479, 1993.
12. ICH Expert Working Group. ICH Harmonized Tripartite Guideline: Statistical Principles for Clinical Trials 1998.
13. R Temple, GW Pledger. The FDA's Critique of the Anturane Reinfarction Trial. N Engl J Med 303:1488–1492, 1980.
14. R Winslow. Centocor Shares Plummet 41% on FDA Move. Wall Street Journal, April 16, A3, 1992.
15. KKG Lan, DL DeMets. Discrete Sequential Boundaries for Clinical Trials. Biometrika 70:659–663, 1983.
16. PC O'Brien, TR Fleming. A Multiple Testing Procedure for Clinical Trials. Biometrics 35:549–556, 1979.
17. SJ Pocock. When to Stop a Clinical Trial. Brit Med J 305:235–240, 1992.
18. RT O'Neill. Some FDA Perspectives on Data Monitoring in Clinical Trials in Drug Development. Stat Med 12:601–608, 1993.
19. R Simon. Some Practical Aspects of the Interim Monitoring of Clinical Trials. Stat Med 13:1401–1409, 1994.
20. SS Ellenberg, MW Myers, WC Blackwelder, DF Hoth. The Use of External Monitoring Committees in Clinical Trials of the National Institute of Allergy and Infectious Diseases. Stat Med 12:461–467, 1993.
21. JP Pignon, M Tarayre, A Auquier, R Arriagada, T LeChevialier, P Ruffie, A Riviere, I Monnet, P Chomy, C Tuchais. Triangular Test and Randomized Trials: Practical Problems in a Small Cell Lung Cancer Trial. Stat Med 13:1415–1421, 1994.
22. MKB Parmar, D Machin. Monitoring Clinical Trials: Experience of, and Proposals Under Consideration by, the Cancer Therapy Committee of the British Medical Research Council. Stat Med 12:497–504, 1993.

23. J Wittes. Behind Closed Doors: The Data Monitoring Board in Randomized Clinical Trials. Stat Med 12:419–424, 1993.
24. TR Fleming. Data Monitoring Committees and Capturing Relevant Information of High Quality. Stat Med 12: 565–570, 1993.
25. RT O'Neill. Stat Med 13(15):1492–1500, 1994.
26. PW Armstrong, CD Furberg. Clinical Trial Data and Safety Monitoring Boards. Circulation 91:901–904, 1995.
27. JR Hampton, DG Julian. Role of the Pharmaceutical Industry in Major Clinical Trials. Lancet, Nov 28, 1258–1259, 1987.
28. D Harrington, J Crowley, SL George, T Pajak, C Redmond, S Wieand. The Case Against Independent Monitoring Committees. Stat Med 13:1411–1414, 1994.
29. PJH Jongen. Data Handling in Clinical Trials: An Ongoing Debate. Multiple Sclerosis 1:S60–S63, 1995.
30. TR Fleming, DL DeMets. Monitoring of Clinical Trials. Controlled Clinical Trials 14:183–197, 1993.
31. EL Korn, R Simon. Data Monitoring Committees and Problems of Lower-Than-Expected Accrual or Event Rates. Controlled Clin Trials 17:526–535, 1996.
32. M Baum, J Houghton, K Abrams. Early Stopping Rules —Clinical Perspectives and Ethical Considerations. Stat Med 13:1459–1469, 1994.
33. WS Browner. Ethics, Statistics, and Technology Assessment: The Use of a Stopping Rule and an Independent Policy and Data Monitoring Board in a Cohort Study of Perioperative Cardiac Morbidity. Clin Res 39(1):7–12, 1991.
34. TR Fleming, JD Neaton, A Goldman, DL DeMets, C Launer, J Korvick, D Abrams, and the Terry Beirn Community Programs for Clinical Research on AIDS. J Acquired Immune Deficiency Syndromes Hum Retrovirol 10(Suppl 2):S9–S18, 1995.

Walter W. Offen

Dose Proportionality

See also *Bioavailability and Bioequivalence*

I. INTRODUCTION

In drug development, it is essential to manage a safe and effective dosing. While a complete pharmacokinetic (PK) profile for all doses is impossible to establish, prediction of PK effects in a certain dose range can easily be made if the compound possesses the property of dose proportionality (DP). This property, also called *dose independence* or *linear pharmacokinetics*, implies that the rates of absorption, distribution, metabolism, and elimination remain constant over the dose range. With this property, plasma concentration, area under the plasma concentration–time curve (AUC), maximum plasma concentration (C_{max}), and urinary recovery should increase in direct proportion to dose.

Figure 1 depicts a typical pattern of a PK profile with DP. It is assumed that dose 1 < dose 2 < dose 3; the left panel shows the concentration–time curve shifted upward as the dose was increased. Shapes of the three concentration curves at the absorption and the excretion stages are nearly identical. As shown in the middle panel, if the concentrations are divided by the corresponding doses, the difference will be indistinguishable. In the right panel, the three AUC values almost fall on the same line when they are plotted against dose. Similarly, a linear relationship for C_{max} would be observed.

In case of repeated dosing, the concentration after multiple dosing can be well projected if the pharmacokinetics of the drug are time invariant and dose independent. Based upon the superposition principle (1), the concentrations over time can be predicted by laying the profile of the next dose over the residual concentrations resulting from all previous doses. Rowland and Tozer (1) describe the circumstances when DP cannot be established.

For the statistical methods to assess DP, Ezzet and Spiegelhalter (2), Hutchison and Keene (3) provide excellent reviews. They proposed a power model to estimate the degree of proportionality. Besides this approach, two popular models—linear regression and analysis of variance—are also used to assess DP. All of these models will be reviewed and compared in the following section. Some limitations in these models will be examined. In the last section, the design aspects

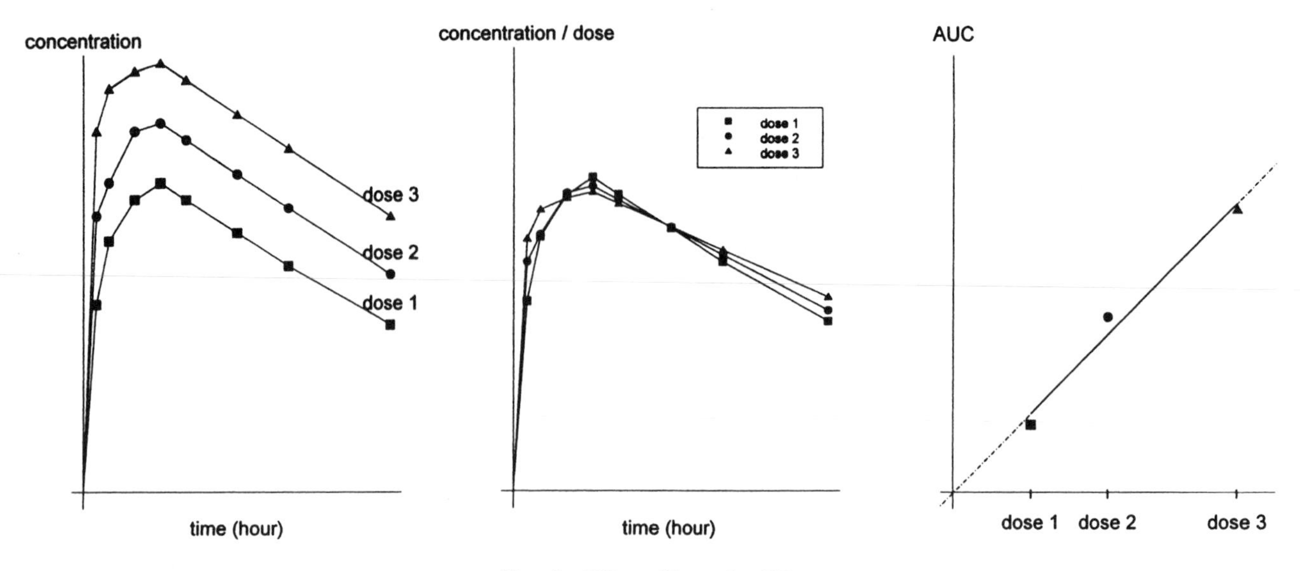

Fig. 1 PK profile under DP.

of the DP study will be discussed. Some interesting works by Ezzet and Spiegelhalter (2) will be illustrated.

II. STATISTICAL MODELS FOR ASSESSING DOSE PROPORTIONALITY

A. Linear Regression

The classical approach to testing DP is to fit a few regression models using the key PK parameters, denoted by Y, as the response and the dose level as the covariate. Since heterogeneity in variances is often found in DP studies, the models are usually fitted with weights equal to the reciprocal of the dose level or with a transformation for Y. The model for initial assessment can be:

$$Y = a + b(\text{dose}) + c(\text{dose})^2$$

When the coefficient for the quadratic term, c, is not significantly different from zero, dose linearity is claimed. The next model of interest is:

$$Y = a + b(\text{dose})$$

If the intercept, a, is not significant, then DP can be declared. In other words, it requires dose linearity and zero intercept to assert DP.

The main drawback of the regression approach is the lack of a measure that can quantify DP. If the measure had existed, we could have derived a confidence interval describing the uncertainty of the estimate. Also, when the quadratic term is significant or the intercept is not close to zero, we are unable to estimate the magnitude of departure from DP.

Besides the classical approach, Haynes and Weiss (4) proposed the following biphasic model to assess DP:

$$Y = \begin{cases} b_1 \text{ dose} & \text{if} \quad 0 < \text{dose} < x_0 \\ b_2 \text{ dose} & \phantom{\text{if}} \quad x_0 < \text{dose} < x_{\max} \end{cases}$$

They modeled DP as a change point problem and considered the slope ratio, b_2/b_1, to be a measure of DP. When the slope ratio is greater than 1, it may reflect capacity-limited elimination and/or first-pass effects. If the ratio is less than 1, it may reflect increased clearance and/or reduced absorption.

B. Analysis of Variance (ANOVA)

The second popular approach to analyze DP studies is to use analysis of variance. Before analysis, correction to the response is made by dividing the response by dose. If the response is proportional to dose, such adjustment will make the corrected response constant. Each dose level is assumed to be associated with a treatment effect. Thus, the basic model can be written as:

$$\frac{Y_j}{(\text{dose})_j} = \mu + \alpha_j$$

where j is the index for dose level, μ denotes the grand effect, and α denotes the treatment effect. If the hypothesis $H_0{:}\alpha_1 = \alpha_2 = \cdots = 0$ is not rejected with an overall F-test, then DP is claimed. Otherwise, statistical methods for multiple comparison can be employed to test if any individuals α_j's are close to zero or to assess the differences among dose levels. Yuh et al. (5) proposed a step-down method using Helmert contrasts to assess DP.

Unlike the regression procedure, the basic model of this approach does not need to assume any relationship among doses. However, because it ignores the ordering of doses, the approach is not as efficient as the regression. Given the estimates of the model, it is difficult to predict the response for the doses not actually studied.

C. Power Model

According to an empirical relationship between AUC or C_{max} and dose, Troy et al. (6) used the following power model to assess DP:

$$Y = a \cdot (\text{dose})^{\beta}$$

In this model, the exponent of dose, β, is a measure of DP. When $\beta = 0$, it implies that the response Y is independent of dose. When $\beta = 1$, DP is declared. The effect of changing dose can be evaluated by:

$$\frac{Y_{j'}}{Y_j} = \left(\frac{(\text{dose})_{j'}}{(\text{dose})_j}\right)^{\beta}$$

Thus, when the dose is doubled, the response is expected to increase by a factor of 2^{β}.

It should be noted that, after log transformation, the model will become linear:

$$\log Y = \alpha + \beta \log(\text{dose}) \quad \text{where } \alpha = \log a$$

The usual linear regression procedures can then be applied in this situation. Because the log transformation tends to stabilize the variance, we do not need to select any weights to correct the heterogeneity in variance. In fact, following FDA guidelines (7), the assumption on the normal error term can well be justified under the transformation.

The primary objective of a DP study is to examine the relationship between the key PK parameters and dose. Rather than test DP solely, it is desirable to estimate the degree of DP or the effects of dose change and to describe the uncertainty. For obtaining confidence intervals, the analysis of the raw ratio in the linear regression model relies on the use of Fieller's method, whereas it is straightforward in the power model. Also, the confidence intervals based on the power model are relatively more narrow than with the ANOVA model, since dose order and values are ignored in the latter.

III. DESIGN OF THE DOSE PROPORTIONALITY STUDY

Recall that the DP study purports to compare doses and to estimate their effects on the key PK parameters. It should not be conducted with any designs where background factors may confound with dose. Usually, Latin squares or a balanced incomplete block design is considered. However, when these crossover designs are used, it is necessary to justify additional sources of variation in models. For example, if the power model is used to fit the PK data, we may consider the following model:

$$\log Y_{ijk} = \alpha_i + \beta \log(\text{dose}_j) + P_k$$

where α_i is the ith subject effect and P_k is the kth period effect.

Dose selection is often dictated by safety concerns, which involve the results of other PK studies. From a statistical perspective, the optimal design for estimating β (or b in linear regression) is to select the lowest and highest doses with equal observations. However, it is unwise to assess DP with only two doses, since the relationship between the PK parameters and dose can be quadratic or in other nonlinear patterns. In practice, at least three doses should be used in a DP study.

When more than three doses are available, one can make the dose selection by comparing the design matrix of our model. To illustrate the idea, Ezzet and Spiegelhalter (2) considered the following model

$$\log \text{AUC}_{ij} = \alpha_i + \beta \log(\text{dose}_j) + \gamma[\log(\text{dose}_j)]^2$$

They specified five doses—25, 50, 100, 200, and 400 mg—and assumed a maximum of five doses per subject. Using the criterion of minimizing the standard error of model estimates, they compared the diagonal element in $(\mathbf{X}'\mathbf{X})^{-1}$ corresponding to β and γ, where $\mathbf{X}$ is the design matrix. For a three-period crossover study, the optimal dose combination was found to be (25, 100, 400). Similar comparisons were also made for crossover design with four and five periods.

REFERENCES

1. M Rowland, T Tozer. Clinical Pharmacokinetics—Concepts and Applications. 3rd ed. Baltimore: Williams & Wilkins, 1995, pp 394–423.
2. F Ezzet, DJ Spiegelhalter. Pharmacokinetic dose proportionality: practical issues on design, sample size and analysis. The Second International Meeting on Statistical Methods in Biopharmacy, Paris, 1993.

3. M Hutchison, O Keene. Assessment of Dose Proportionality. Report from the PSI/PK-UK Joint Working Party, 1994.
4. J Haynes, A Weiss. Modeling pharmacokinetic dose-proportionality data. ASA 1989 Proceedings of the Biopharmaceutical Section, pp 85–89.
5. L Yuh, M Eller, S Ruberg. A stepwise approach for analyzing dose proportionality studies. ASA 1990 Proceedings of the Biopharmaceutical Section, pp 47–50.
6. SM Troy, WH Cevallos, KA Conrad, ST Chiang, JR Latts. The absolute bioavailability and dose proportionality of intravenous and oral dosage regimens of Recainam. J Clin Pharmacol 31:433–439, 1991.
7. Office of Generic Drugs, FDA. Statistical procedures for bioequivalence studies using a standard two-treatment crossover design. 1988.

C. Gordon Law

Dropout*

See also *Patient Compliance*

I. INTRODUCTION

In clinical studies, the length of time for a study can vary tremendously, ranging from a few days or weeks to possibly a few years. For example, in an osteoporosis trial, to illustrate efficacy, subjects must be followed for 2–3 years. During this time, it is often of interest to track subjects periodically in order to characterize their response to treatment throughout the study as well as to obtain subject information in case the subject should discontinue (or, in other words, drop out). Thus, most primary endpoints of interest are collected repeatedly. Studies that follow subjects over periods of time are referred to as *longitudinal studies* and the corresponding data as *longitudinal data*.

It is frequently the case in clinical research that subjects drop out of a study before completion, especially in the case of long trials. Reasons for dropout vary, from relocation to another city, to lack of efficacy, to having a safety concern. For subjects dropping out of the study, responses that were to be captured at time points after which they dropped out are of course lost. Because the objective of most analyses in clinical trials is to estimate or test for differences in treatment effects, the problem then becomes one of estimating or testing for these effects, taking into consideration subject dropout. In this case, there are two sets of variables to be considered in a trial: the clinical response variables (involving the parameters of interest) and the variables describing the dropout process (involving nuisance parameters).

The problem of dropout is a special case of the missing-data problem. In this case, we can assume that observations or responses prior to dropout are not missing (although it is often the case that data are missing before dropout due to subjects' missing visits) and all observations postdropout are missing. Work concerning inference in the case of incomplete data can be found in Rubin (2) and Little and Rubin (3). References dealing with the problem of missing data in longitudinal studies include Laird (4), Wu and Carroll (5), Wu and Bailey (6), Diggle and Kenward (7), and Little (8).

In Sec. II we will explore various relationships between the measurement and dropout processes. Discussion of some general analysis methods currently used in the pharmaceutical industry and the implications of missing data in characterizing the response distribution (for example, estimating treatment effects) will be given in Sec. III. Two parametric methods developed specifically for this problem will be discussed in Sec. IV. The main focus of this entry is on the analysis of continuous data as the outcome variables, with emphasis on a univariate response measurement. Extensions will be mentioned in Sec. V.

II. THE DROPOUT PROCESS

If all subjects remain in the study, the only process of interest is the distribution of responses, in other words, the measurement process. For example, bone mineral density may be the primary response variable of inter-

*Material presented is based in part on the efforts of a working group at Pfizer Central Research, which produced the technical report "Selected Approaches for the Analysis of Longitudinal Data with Dropouts" (1).

est in an osteoporosis trial or glysylated haemoglobin (HbA_{1C}) in a trial for diabetes. However, when subjects drop out we now also need to consider the reasons for dropout, i.e., the dropout process and its relation to the measurement process.

In the first case, methods for analyzing the longitudinal response variables include procedures such as analysis of variance, analysis of covariance, multivariate analysis, and mixed-effect repeated-measures analysis. However, in the case of dropout, many of the current statistical procedures available for analyzing such models are not able to incorporate the missing-data process and thus may lead to biased results.

In general, by the *missing-data mechanism*, we mean the distribution of the variable indicating whether each response is observed or missing. In the case of dropout, this sequence of variables remains constant (say, 1, indicating an observation as being observed) until the subject discontinues, at which point it changes to a new value (say, 0, indicating an observation as being unobserved). Therefore this process can be characterized by the time a subject drops out. The effect this has on analyzing longitudinal data depends on the relationship between the dropout process and the subject response measurements. The reasons for the dropout of a subject may not depend on the response measurements, may depend only on the observed measurements, or may depend on the unobserved measurements. Different terminology is used by various authors. We will follow the terminology given in Diggle and Kenward (7).

When the dropout is independent of the measurement process, then the dropout mechanism is viewed as completely random. For example, suppose a subject moves to a different city for reasons that most probably in this case have nothing to do with the disease or treatment. This subject then drops out of the study completely at random. The case of missing completely at random also applies when the reason for dropout depends only on observable covariables. In other words, given covariables, subjects remaining in the study are a random subset of those dropping out of the study. Thus any analysis that involves conditioning on these covariables (such as in a regression) will provide valid inferences with respect to the whole population being sampled.

When the dropout depends only on the observed cases of the measurement process (and covariables) but does not depend on the unobserved cases—i.e., the subject drops out based only on information collected before the dropout takes place—then the mechanism of dropout is considered random. For instance, a subject who experienced less relief in prior treatment of a migraine headache may be more inclined to terminate a study than those who experienced greater relief. Again, in this case, the distribution of responses (adjusted for prior relief) for unobserved subjects is the same as for observed subjects.

The preceding two types of dropout are referred to as *noninformative* or *ignorable* dropout. When the dropout depends on the unobserved cases of the response measurements, the dropout is said to be *informative* or *nonignorable*. This is the case when dropout cannot be completely characterized by the observed information. Suppose a subject takes the study medication but continues to experience disease deterioration. Furthermore, the subject calls the office, briefly describes the problem, and never returns. This can be considered as a dropout for lack of efficacy, without corresponding response data collected. It is an example of dropout that depends on an unobserved response, or an informative dropout.

III. ANALYSIS POPULATIONS AND ACCOMPANYING METHODS

It is customary to analyze data with dropouts based on specific subgroups of subjects. One simple subgroup is defined as such that data analysis is performed based on those subjects with the complete sequences of data collected over the study period. This is usually referred to as the *completers* subgroup analysis. Those subjects that drop out and therefore do not have data after dropping out will be excluded from the analysis. Inferences based on this population apply to the original population only in the case where subjects who have dropped out (dropped out completely at random) and any covariables affecting the likelihood of dropout are included in the analysis. This case is, in most circumstances, unlikely. Even in the case where missing completely at random holds estimates will be less efficient than when analyzing all the data.

Another approach is to base the analysis on all enrolled subjects, with any unobserved data being filled in (imputed) using some algorithm. Analyses involving all randomized subjects are usually referred to as *intention-to-treat* (ITT) analyses. The simplest imputation technique is to carry forward an observation obtained during the study such that every subject has a complete sequence of data values for every time point. Algorithms for carrying forward an observation include filling in missing values with the last observed response for a subject [last observation carried forward (LOCF)], with the worst observed response, or in some cases with the best observed response. (See the entry on "Last Observation Carried Forward" for more details.)

Obviously there are statistical concerns associated with these approaches. It is possible that subjects who dropped out early were those who did not benefit from the treatment, did not tolerate the treatment well, or experienced serious side effects. Those who completed the study may have either had better health conditions or experienced greater benefit from the treatment. In this case where dropout is not completely at random, a completers subgroup analysis, by selecting a subset of the subgroup based on subjects' final study status, is likely to err toward indicating a fictitious beneficial treatment effect. In the case of an ITT analysis with imputation, the concern is the reduced variability estimates from imputing a single, fixed value (see Ref. 9). Observations to be collected after a subject's dropout would have followed a certain distribution with a certain mean and variance. Imputing a constant value over time would reduce the underlying variance for each subject.

A third approach is to perform statistical analyses based on the observed data, i.e., analyzing all of the subjects with any available data. The linear random-effects modeling approach has been applied successfully to handle longitudinal data with dropout, using all the observed data. Under the linear random-effect model, each subject's set of serial measurements is assumed to follow a linear function to time, with normally distributed errors. It is further assumed that the underlying slope and intercept of this individual linear function are bivariate normal random variables. The treatment effect is determined by the comparison of the slopes.

This procedure produces unbiased estimates in the case of ignorable dropout. However, the problems of bias and inefficiency in estimating treatment effect still exist when the unobserved data after dropping out contains information of the unknown treatment effect parameter; i.e., the dropout process is informative. For example, consider a study in which subjects drop out of the study because of lack of efficacy in the placebo group. In this case, estimates may be biased toward values contributed from subjects remaining in the study, those experiencing greater benefit.

The applicability of the foregoing methods depends on the missing-data process. Because none of the methods mentioned simultaneously consider the dropout process, none are applicable when the dropout mechanism is informative. Methods that model the joint distribution of response and dropout in terms of the marginal distribution of the response and the conditional distribution of dropout given response are called *selection model* methods (8). A variation of this model is the random coefficient-selection model or shared parameter model, where the response and dropout process are conditioned on shared within-subject random effects.

Two such procedures will be described briefly in the next section; Diggle and Kenward (7) and Wu and Bailey (6) (for specific details the reader is referred to the corresponding papers). Diggle and Kenward's approach is an example of the traditional selection model approach. Wu and Bailey's methodology is an approximation of an approach developed by Wu and Carroll (5), which is an example of a shared parameter model.

IV. OTHER METHODS

A. Diggle and Kenward

Diggle and Kenward proposed a likelihood-based method to adjust for informative dropout in analyzing continuous longitudinal data. As mentioned earlier, in order to adjust for dropout, Diggle and Kenward consider a selection models approach to model the joint distribution of response and dropout. Their approach can be broken down into three basic components: an underlying multivariate normal linear model to model the longitudinal response vector; a logit model to model the likelihood of dropping out as a function of the longitudinal responses; and the likelihood function of the observed data, which ties the first two components together. We will describe these components in general next.

The longitudinal responses of individuals across time are assumed to have a multivariate normal distribution. In clinical trials, the primary interest is whether or not there is a treatment effect. The treatment effect as well as other covariates, such as the corresponding measurement time for a particular response, baseline status, and centers are incorporated as parameters in a linear function describing the relationship with the mean response vector. For the covariance structure, Diggle and Kenward allow for three independent variance components: a random intercept component between individuals, a serially correlated within-subject component, and a sampling or measurement error component that is uncorrelated across individuals.

The dropout process is modeled by implementing a linear logistic model. The logit of the probability that the subject, at time d, drops out is modeled as a linear function of the responses prior to dropout and the unobserved response at the time of dropout. Assuming the underlying dropout and response models are correct, submodels can be fit and tested using the likelihood ratio statistic to determine if the dropout process is informative. Using the definitions for the various dropout mechanisms given in Sec. II, the full model corresponds to a case of nonignorable dropout (dropout depends on unobserved data). The reduced model, where the coefficient associated with the unobserved response at the time of dropout is zero, corresponds to the case of ignorable dropout. In this case the dropout process is independent of the missing response. Lastly, completely random dropout corresponds to the model where all coefficients except the intercept are zero. In this case the dropout process is independent of both observed responses and missing responses.

For each subject, the observed data consist of the observed longitudinal responses prior to dropout and the time of dropout. Therefore each subject contributes the following to the likelihood: the marginal distribution of the multivariate linear model corresponding to the responses up to the time prior to dropout, and a sequence of terms corresponding to the probability that a subject drops out at the time he or she did based on the observed responses.

For each subject, the log-likelihood function for (θ, Σ, ϕ) given the observed can be partitioned into three components:

$$L\left(\theta, \sum, \phi\right) = L_1(\theta, \Sigma) + L_2(\phi) + L_3\left(\theta, \sum, \phi\right)$$

The first likelihood component is

$$L_1\left(\theta, \sum\right) = \log\{\phi_{d-1}(\mathbf{y})\} \tag{1}$$

where $\phi_{d-1}(\mathbf{y})$ is the joint marginal multivariate normal density for the first $d - 1$ observed longitudinal responses, where d is the time of dropout for the particular subject.

The second and third components correspond to the probability that a subject drops out at time d given the $d - 1$ observed longitudinal responses. This consists of a sequence of terms corresponding to the probability that a subject doesn't drop out before time d and a term for the probability that a subject drops out at time d. This involves the following two terms:

$$L_2(\phi) = \sum_{j=2}^{d-1} \log\{1 - \Pr(j|y_1, \ldots, y_{j-1}, y_j; \phi)\} \tag{2}$$

and

$$L_3\left(\theta, \sum, \phi\right) = \log(\Pr\left(d|y_1, \ldots, y_{d-1}; \theta, \sum, \phi\right) \tag{3}$$

where $\Pr(j|y_1, \ldots, y_{j-1}, y_j; \phi))$ in Eq. (2) has the following form based on the logistic regression model for modeling the dropout process:

$$\text{PR}(j; \phi)\} = \text{logit}^{-1}\left(\phi_0 + \phi_1 y_j + \sum_{k=2}^{j} \phi_k y_{j+1-k}\right) \tag{4}$$

Only subjects who drop out of the study contribute to Eq. (3). Because the response value y_d at time d (time of dropout) is unobserved, it is not conditioned upon Eq. (3). Equation (3) is evaluated by integrating Eq. (4) for $j = d$ over the univariate normal conditional distribution of y_d given $y_1, \ldots, y_{d-1}$.

Parameters are estimated iteratively via maximum likelihood (ML) using the simplex algorithm of Nelder and Mead (10). Using maximum likelihood, the parameters of the complete-data model and dropout process are estimated simultaneously, and thus parameter estimates corresponding to the response distribution, such as treatment effects, are obtained adjusting for dropout (assuming suitable models for response and dropout have been found). In addition, it is possible to test whether or not the dropout mechanism is ignorable by testing whether coefficients associated with various response measurements given in Eq. (4) are zero.

As mentioned earlier, a generalization of the selection model is the random coefficients-selection model. In this case rather than the marginal response and dropout process, we have the response and dropout process conditional on within-subject random effects. A special case of this is the shared parameter model, where the dropout process possibly depends on unobserved within-subject random effects underlying the response distribution and is thus informative.

B. Wu and Bailey

Wu and Bailey give a methodology to approximate the approach developed by Wu and Carroll. Wu and Carroll use a random-effects linear model and discrete time survival model to model the response and dropout process, respectively, where the likelihood of dropout depends on within-subject random effects. To introduce the concepts, we begin with the random-effects linear model (see Ref. 11).

The random-effects linear model assumes that, for each individual, conditional on the underlying vector of coefficients β_i, the series of measurements $\mathbf{Y}_i$ is independent multivariate normal with mean $\mathbf{K}_i\beta_i$ and covariance matrix $\sigma^2\mathbf{I}$, where $\mathbf{K}_i$ is a $(T \times q)$ within-subject design matrix, σ^2 is an unknown scalar residual variance, and $\mathbf{I}$ is the $(T \times T)$ identity matrix. The β_i are treated as random variables that are independent multivariate normally distributed over subjects with mean $\mathbf{Z}_i\beta$ and covariance matrix Σ_β, where $\mathbf{Z}_i$ is a $(q \times p)$ between-subjects design matrix, β is an unknown $(p \times 1)$ vector of slopes, and Σ_β is an unknown $(q \times q)$ covariance matrix. The linear random-effects model implies that marginally $\mathbf{Y}_i$ are independent multivariate normal with mean $\mathbf{X}_i\beta_i$ and covariance matrix $\mathbf{K}_i\Sigma_\beta\mathbf{K}_i^T + \sigma^2\mathbf{I}$, where $\mathbf{X}_i = \mathbf{K}_i\mathbf{Z}_i$.

Wu and Carroll consider the case where dropout is caused by a subject's death or withdrawal and refer to it as the *right-censoring* process. They assume that the series of measurements follows a linear random-effects model and that the right-censoring process and the observed data share the random-effects parameters β. The corresponding likelihood consists of the joint distribution of the simple least-squares estimates of individual intercept and slope, the random-effects parameters of intercept and slope, and the right-censoring process. To obtain parameter estimates, the unknown within-subject variance and the covariance matrix of the random intercept and slope are substituted with the unbiased estimates of them in the likelihood function, called a pseudo-maximum likelihood, which is then integrated over with respect to the random slope.

Under noninformative censoring, the maximum-likelihood estimates of the random slopes β_i are the weighted least-squares estimates (WLE), which are just a weighted average of the individual simple least-squares estimated slopes, with $\Sigma_\beta + (\mathbf{K}_i^T\mathbf{K}_i)^{-1}\sigma^2\mathbf{I}$, the inverse of the variance of the random slopes, as the weight. In this case, the WLE are consistent and unbiased.

However, when censoring is informative, for example, censoring depends on the individual slope so that subjects who have more serious outcomes (say, steeper rate) are more likely to drop out early, then the WLE are subject to bias. This is due to the fact that weights used in the weighted least-squares estimation are functions of the within-subject design matrix, $\mathbf{K}_i$, which is a function of the number of repeated measures within the subject. Subjects who drop out early have smaller weights than those who drop out late or who complete the study, resulting in estimates that tend to reflect only the treatment effect of those with less serious conditions, which are thus biased. Note that because the individual ordinary least-squares estimates for slope are unbiased, if an unweighted average is used (UWLE) to estimate the random slope effect, then this estimate is unbiased, though it can be inefficient.

Wu and Bailey provided a simple approximate method for the informative censoring problem. They consider that under the random-effects linear model the censoring times and the slopes are related. Instead of modeling the censoring process, they assume that, given the censoring time, the individual least-squares estimated slopes are linear functions of the censoring time with normal errors. Specifically, the conditional linear model assumes that, for each individual, conditional on the censoring time, the random slope is independent univariate normal with mean $\gamma_0 + \gamma_1 E(T_i)$ and conditional variance σ^2_{Ti}, where $E(T_i)$ is the expected value of censoring time for subject i. The conditional variance σ^2_{Ti} is an unknown scalar residual variance, which depends on censoring time and thus on the number of measurements before dropout. Under the random-effects linear model, the σ^2_{Ti} equals $\Sigma_\beta + (\mathbf{K}_i^T\mathbf{K}_i)^{-1}\sigma^2\mathbf{I}$, and when Σ_β and σ^2 are unknown, they can be estimated using the restricted maximum-likelihood method (REML) method.

Two methods to estimate the random effect parameters under the conditional linear model are proposed. These are referred to as the *linear minimum variance unbiased* (LMVUB) and the *linear minimum mean squared error* (LMMSE) estimates, respectively. Because of the good properties associated with the first method, we will discuss it next and refer the reader to Ref. 6 for discussion of the second.

The LMVUB procedure estimates the conditional linear coefficients γ_0 and γ_1 by weighted least squares and then substitutes the estimates of γ_0 and γ_1 into the conditional linear function to obtain the LMVUB. The sample mean censoring time is used as an estimate for $E(T_i)$, the expected value of censoring time for subject i. In practice, the LMVUB estimator can be derived following a sequence of steps. First, the ordinary least-squares estimators of slope for each individual is obtained, and the conditional variances of the estimated conditional slopes given the design matrix of observation times are estimated. Then, weighted analysis of covariance based on these individual slopes, using the inverse of the conditional variances as weights, with dropout times as covariates in the linear model, are used to estimate the group slopes. The difference between the slopes is then tested for significance.

In a simulation study, using a probit model for the right-censoring process, the performance of the LMVUB procedure was competitive when compared to the pseudo-maximum-likelihood estimation of Wu and Carroll. As also shown in a Monte Carlo simulation study by Wang-Clow et al. (12), which compares the performance of several estimators (UWLE, WLE, LMVUB, weighted mean, complete-case, and MLE), when the missing process is informative but the same in both treatment and control groups, the LMVUB is the only unbiased estimator for the difference of the slopes. When the dropout process differs for the two groups, the UWLE has the smallest bias, but it is noncompetitive in terms of mean squared error. The LMVUB has the smallest mean squared error and the lowest bias of all the estimators.

V. DISCUSSION

A key assumption of the dropout model proposed by Diggle and Kenward is that the process depends at most on the responses up to and including the unobserved response at the time of dropout but not the latter unobserved responses. Little (8) mentions that such a dropout model is more plausible when dropout is linked directly to the response being measured (such as withdrawal to avoid a particularly invasive procedure) rather than to some underlying process (such as weight loss due to treatment with a medication).

In the second case, the dependence of dropout on future unobserved responses is a function of how accurately the response variable measures the underlying process. In this case, the framework underlying the random coefficients-selection model may be more satisfied.

An alternative approach to the selection model is the pattern mixture model. In this case, the joint distribution of response and dropout is modeled as the marginal distribution of dropout and the conditional distribution of response given dropout. Because the data are not observed when missingness occurs, parameters underlying the response distribution given dropout are linked by specifying prior information. Little (13) considers this type of model for analyzing incomplete multivariate data.

In this entry we discussed the effect of dropout primarily in analyzing continuous longitudinal data. An example of a shared parameter approach when the longitudinal data are binary is given in Ten Have et al. (14). An example of a random coefficient-selection model approach for ordinal longitudinal data, where dropout is a function of the response variable and within-subject random effects, is given in Sheiner et al. (15). For extensions involving nonparametric analysis of incomplete repeated measures, the reader is referred to Wei and Johnson (16).

Lastly, software for running Diggle and Kenward's procedure using Splus (called "Oswald") can be found on the internet. As far as we know, no available software exists for the Wu and Bailey procedure, but the steps outlined in Sec. IV.B can easily be implemented using SAS®. In addition, single imputation methods, such as LOCF, and multiple imputation methods can be carried out using the software package SOLAS (17).

REFERENCES

1. R Adler, C Balagtas, J Cappelleri, S Cheng, J Finman, G Lan, S Lan, M Li, E Pickering, C Rodenberg, N Ting. Selected Approaches for the Analysis of Longitudinal Data with Dropouts. Biometrics Technical Report, Groton, CT, 1997.
2. DB Rubin. Biometrika 63:581–592, 1976.
3. R Little, DB Rubin. Statistical Analysis of Missing Data. New York: Wiley, 1987.
4. NM Laird. Stat Med 7:305–315, 1988.
5. M Wu, R Carroll. Biomterics 44:175–188, 1988.
6. M Wu, K Bailey. Biometrics 45:939–955, 1989.
7. P Diggle, MG Kenward. Applied Statistics 43(1):49–93, 1994.
8. R Little. J Am Stat Assoc 90:1112–1121, 1995.
9. DB Rubin. Multiple Imputation for Nonresponse in Surveys. New York: Wiley, 1987.
10. JA Nelder, R Mead. Comput J 7:303–313, 1965.
11. NM Laird, JH Ware. Biometrics 38:963–974, 1982.
12. F Wang-Clow, M Lange, NM Laird, JH Ware. Stat Med 14:283–297, 1995.
13. R Little. J Am Stat Assoc 88:125–134, 1993.
14. TR Ten Have, AR Kunselman, EP Pulkstenis, JR Landis. Biometrics 98:367–383, 1998.
15. LB Sheiner, SL Beal, A Dunne. J Am Statis Assoc 92: 1235–1244, 1997.
16. LJ Wei, WE Johnson. Biometrika 72:359–364, 1985.
17. SOLAS for Missing Data Analysis 1.0. Cork, Ireland: Statistical Solutions, Ltd. 1997.

Cindy Rodenberg
Shu-Lin Cheng

Drug Development

See also *Clinical Trial Process*

I. INTRODUCTION AND BACKGROUND

Most of the drugs available in pharmacies started out as a chemical compound or a biologic discovered in laboratories. When first discovered, each new compound or biologic is denoted as a drug candidate. The drug candidate has to demonstrate some activities in the laboratory to indicate that it has the potential of treating certain diseases in humans. Drug development is a process that starts when the drug candidate is first discovered and continues until it is available to be prescribed by physicians to treat patients with the disease. A potential new drug or a drug candidate can generally be classified into one of two categories: a synthetic chemical compound or a biologic. A *compound* is usually a new chemical entity synthesized by scientists from pharmaceutical companies (often referred to as *sponsors*), universities, or research institutes. A *biologic* can be a protein, a part of a protein, DNA, or a different form either extracted from tissues of another live body or cultured by some type of bacteria. In any case, this new compound or biologic will have to go through the drug development process before it can be consumed by the general public.

The drug development process can be broadly classified into two major components: nonclinical development and clinical development. *Nonclinical development* includes all drug testing performed outside of the human body, and *clinical development* is based on experiments conducted in the human body. Nonclinical development can be further divided into pharmacology, toxicology, and formulation. In these processes, experiments are performed in laboratories or pilot plants. Observations from cells, tissues, or animal bodies are collected to derive inferences for potential new drugs. Chemical processes are involved in formulating the new compound into drugs to be delivered into the human body. Clinical development can be further divided into phases 1, 2, 3, and 4. Clinical studies are designed to collect data from normal volunteers or patients who participate in these studies.

Throughout the whole drug development process, two scientific questions are constantly being addressed: Does the drug candidate work? Is it safe? Starting from the laboratory where the compound or biologic is first discovered, the candidate has to go through lots of tests to see if it demonstrates both efficacy (the drug works) and safety. Only candidates that pass all those tests can progress to the next step of development. In the United States, after a drug candidate passes all of the nonclinical tests, an investigational new drug (IND) document is filed with the Food and Drug Administration (FDA). After the IND is approved, clinical trials (tests on human body) can then be performed for the drug candidate. If this drug candidate is shown to be safe and efficacious through phases 1, 2, and 3 clinical trials, the sponsor will then file a new drug application (NDA) with the FDA in the United States. The drug can be made available for general public consumption only if the NDA is approved. Oftentimes the new drug continues to be developed even after it is on the market. These postmarketing studies are known as phase 4 clinical trials.

A great many statistical methods are used in both clinical and nonclinical drug development processes. Details of these processes and statistical applications are discussed in the following sections: Sec. II introduces the regulatory requirements; Secs. III and IV discuss nonclinical and premarketing clinical development, respectively; postmarketing clinical development is presented in Sec. V.

II. REGULATORY REQUIREMENTS

In most modern countries, a drug has to be approved by drug regulatory agencies before it can be marketed. For example, in the United States this agency is the FDA, in Canada it is the Health Protection Branch (HPB) etc. There are guidelines from these agencies to regulate various products and processes in drug development. These guidelines provide a consistent framework for sponsors (pharmaceutical companies) to develop a new drug and for reviewers in these agencies to evaluate documents submitted to them by the sponsors.

The guidelines cover many aspects of drug research and development. For example, the guidance documents from the FDA regulate various steps of new drug development, including: chemistry, pharmacology/toxicology, INDs, clinical/medical, information technology, labeling, generics, advertising, and other areas of drug regulation. Among all of these guidance documents, the one that is most relevant to biopharmaceutical statistics is the "Guideline for the Format and Content of the Clinical and Statistical Sections of an Application" (U.S. FDA, 1988), within the clinical/medical guidelines. This document regulates the format, contents, data sets, and statistical methods used for both efficacy and safety analysis in an NDA report. For details of these documents, interested readers may refer to the FDA guidelines. These guidelines can be ordered from the Superintendent of Documents, U.S Government Printing Office, Washington, DC 20402. They can also be found from the following Web page:

http://www.fda.gov/cder/guidance/index.htm

There has been a continuous effort to harmonize regulation guidelines from various countries. A new organization—the International Conference on Harmonization (ICH)—has arisen, with members from regulatory agencies and drug developers (or sponsors) of various countries (including the United States, Canada, Japan, and many other European and Asian countries), to generate a set of guidelines for drug developers and regulatory agencies to use. As of 1998, a number of ICH guidelines had been developed and finalized; other guidelines were being drafted. The ICH guidelines can be broadly divided into the areas of safety, efficacy, and quality. Statistics is an important component in these guidelines. In particular, ICH Efficacy Guideline E9—"Statistical Principles for Clinical Trials"—was developed specifically for statistical applications in the clinical development of a new drug.

III. NONCLINICAL DEVELOPMENT

A new chemical compound or biologic can be a drug candidate because it demonstrates some desirable pharmacological activities in the laboratory. At such an early stage of drug development, the focus is mainly on cells, tissues, organs, or animal bodies. Experimentation on human beings are performed after the candidate passes these early tests and looks promising. Hence, nonclinical development may also be referred to as *preclinical* development, since these experiments are performed before clinical trials are performed on the human body.

One of the most important properties to be studied at the nonclinical testing stage is whether there is a dose–response relationship for the drug candidate. Figure 1 shows two theoretical curves describing these relationships. The x-axis represents the doses of the drug candidate (can be linear scale or logarithmic scale), and the y-axis represents the responses. Suppose the y-axis allows 0% responses at the bottom and 100% responses (both efficacy response and toxicity response) at the top, then as dose increases, the responses increase. The curve on the left represents the desirable response (potency, efficacy); the curve on the right represents the undesirable response (toxicity). The area in between those two curves would be the doses at which the drug works and is safe. Theoretically speaking, the best drug candidate would be one for which the drug potency (efficacy) increases very rapidly while the toxic response does not increase much as the dose increases. Under this ideal situation, a wide range of doses can be prescribed to treat patients. However, in reality these two curves can be close to each other. When this is the case, there is a narrow range of doses that will be both potent and safe, and dose selection must be very careful. In the process of developing a certain drug candidate, should these two curves become too close to each other, or should the toxic response curve locate to the left of the efficacy curve, then the strategy may be to stop development for this candidate.

At the preclinical stage, the study of drug potency or efficacy is part of pharmacology, whereas the study of toxicity is part of drug safety. In addition to pharmacological activity and toxicity, another important process at this preclinical development stage is the formulation of the compound or biologic into a form that can be consumed by humans (for example, tablet, capsule). This process is generally referred to as *drug formulation*. Nonclinical development processes include pharmacology, toxicology/safety, and drug formulation. These processes are described in subsections III.A, III.B, and III.C, respectively.

A. Pharmacology

Pharmacology is the study of the selective biological activity of chemical substances on living matter (Hubert et al., 1988). A substance has *biological activity* when, in appropriate doses, it causes a cellular response. It is *selective* when the response occurs in some cells and not in others. Therefore, a compound or a biologic has to demonstrate these activities before it can be further developed. In the early stages of drug testing, it is important to differentiate an "active" candidate from an "inactive" candidate. There are screening procedures to select these candidates. Two properties of particular interest are sensitivity and specificity. Given that a compound is active, *sensitivity* is the conditional probability that the screen will classify it as positive. *Specificity* is the conditional probability that the screen will call a compound negative

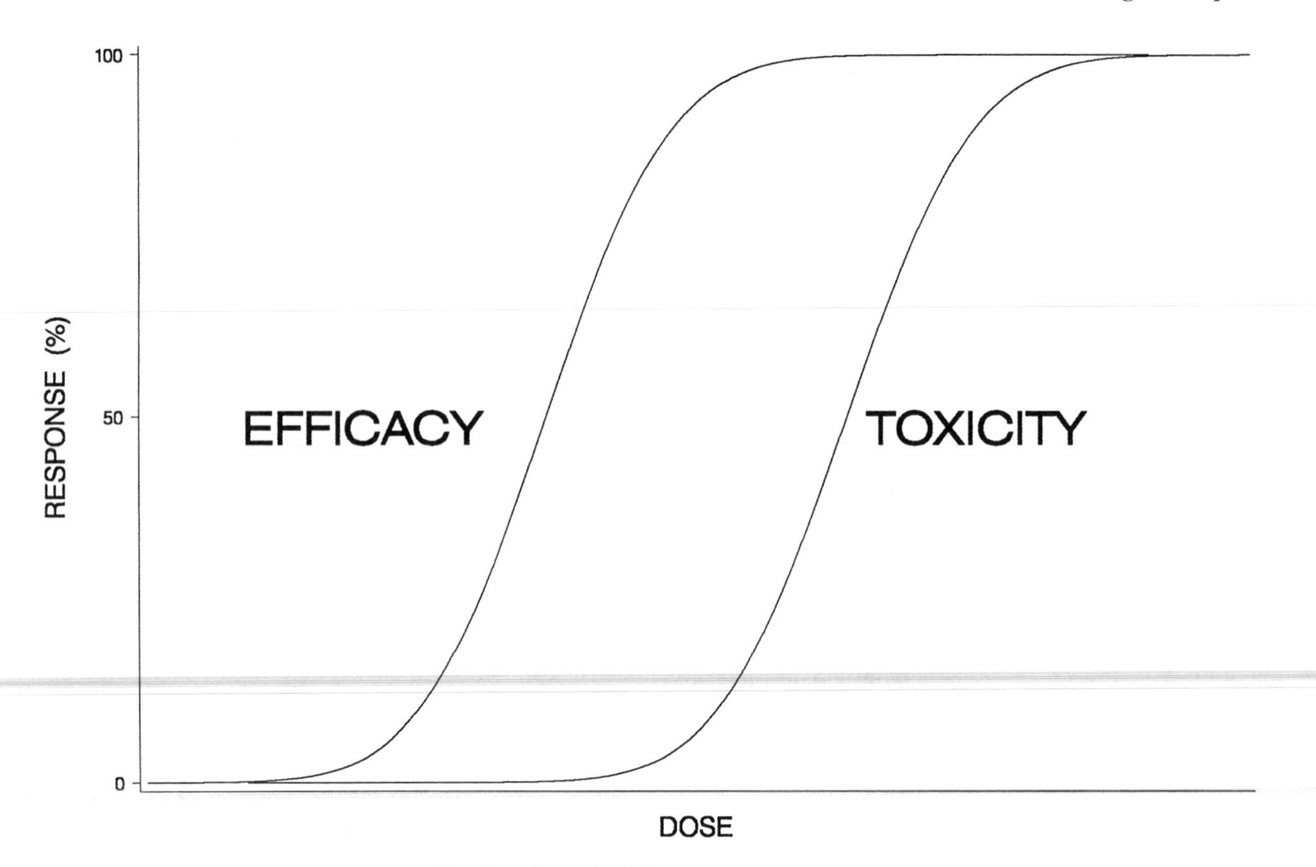

Fig. 1 Theoretical dose–response curves.

given that it is inactive (Redman, 1981). In the ideal case, both of these values should be close to 1.

The level of these pharmacological activities may be viewed as the drug *potency* or *strength*. The estimation of drug potency by the reactions of living organisms or their components is known as *bioassay*. According to Hubert et al. (1988), "the experimental determination of the potency or strength of a chemical or biological substance based on the response observed after its administration to living matter (animal or animal tissue) constitutes a biological assay."

As discussed previously, one of the most important relationships needed to be studied for pharmacological activities is the dose–response relationship. In these experiments, several doses of the drug candidates are selected, and the responses are measured for each corresponding dose. After response data are collected, regression methods may be applied to analyze the results. Regression modeling is a very useful statistical method applied in the analysis of pharmacological activities.

B. Toxicology/Drug Safety

Drug safety is one of the most important concerns throughout all stages of drug development. In the preclinical stage, drug safety needs to be studied in a variety of animals. Studies are designed to observe adverse drug effects or toxic events experienced by animals treated with different doses of the drug candidate. Animals are also exposed to the drug candidate for various lengths of time to see if there are adverse effects caused by cumulative dosing over time. These results are summarized and analyzed using statistical methods. When the results of animal studies indicate potentially serious side effects, drug development is either terminated or suspended pending further investigation of the problem.

Depending on the duration of exposure to the drug candidate, animal toxicity studies are classified as acute, subacute, chronic, or reproductive (Selwyn, 1988). Usually the first few studies are *acute studies*;

i.e., the animal is given one or a few doses of the drug candidate. If only one dose is given, it can also be called *a single-dose study*. Only those drug candidates demonstrated to be safe in single-dose studies progress into *multiple-dose studies*. Acute studies are typically about 2 weeks in duration. Repeat-dose studies of 30–90 days' duration are called *subacute studies*. *Chronic studies* are usually designed with more than 90 days of duration. These studies are conducted in rodents and in at least one nonrodent species. Chronic studies may also be viewed as carcinogenicity studies, since the rodent studies consider tumor incidence as an important endpoint. *Reproductive studies* are carried out to assess the drug's effect on fertility and conception; they can also be used to study the drug's effect on the fetus and developing offspring.

Statistical tools are very useful in designing and analyzing animal toxicity studies. Experimental design techniques are used in designing these studies; hypothesis tests are performed in comparing treatment group differences. Categorical data analysis is used to assess binary responses or count data. Life table techniques and survival analyses are very useful in analyzing chronic and carcinogenicity studies.

C. Drug Formulation Development

As discussed earlier, a potential new drug can be either a chemical compound or a biologic. If the drug candidate is a biologic, then the formulation is typically a solution that contains a high concentration of such a biologic, and the solution is injected into the subject. On the other hand, if the potential drug is a chemical compound, then the formulation can be tablets, capsules, solution, patches, suspension, or other forms. There are many formulation problems that require statistical analyses. The formulation problems that stem from chemical compounds are more likely to involve widely used statistical techniques, but the problems involving formulation of biologics are less likely to succumb to standard statistical solutions. The paradigm of a chemical compound is used here to illustrate some of these formulation-related problems and how statistical tools can be useful.

A drug is the mixture of the synthesized chemical compounds (active ingredients) with other, inactive ingredients designed to improve the absorption of the active ingredients. How the mixture is made depends on the results of a series of experiments, called *optimization experiments*. Usually the optimization experiments are performed under some physical constraints, e.g., the amount of raw materials, the capacity of the container, and the size and shape of the tablets. Some of the samples from the *batch-processed* tablets are stored to study the *stability*, or *shelf life*. These stored samples are exposed to various environmental conditions, e.g., freezing, extra sunlight, very dry or very humid conditions. Part of these stored samples are reanalyzed after half a year, 1 year, and 2 or more years. Such sample properties as potency, dissolution rate, and color, are measured at these time points to study the stability of the drug.

A few branches of statistics can be helpful in designing and analyzing these experiments. Optimization designs, factorial designs, or fractional factorial designs can be used in designing these experiments. After the experiments are over and data are collected, statistical tools are used to analyze these data. Variable selection techniques of multiple regression help select the best subset of independent variables. Analysis of variance (ANOVA) is used to perform data analysis and to test various statistical hypotheses. In cases where multiple dependent variables are to be optimized, multivariate analysis (including MANOVA) may be used to analyze these data. Response surface methodology provides ways to estimate the optimal combination of a set of independent variables. From these results, pharmacists are able to operate under the physical constraints and to select the appropriate materials and processes to produce the tablets for a compound so that the tablets possess some optimal properties.

IV. PREMARKETING CLINICAL DEVELOPMENT

If a compound or a biologic passes the selection process from animal testing and is shown to be safe and efficacious to be tested in human, it progresses into clinical development. As discussed earlier, the major distinction between clinical trials and nonclinical testing is the experimental unit. In clinical trials, the experimental units are human beings, whereas the experimental units in nonclinical testing are nonhuman subjects. Before a drug candidate can be tested in humans, the pharmaceutical company (sponsor) has to submit an IND to the FDA for review. If there is no response from the FDA after 30 days following IND submission, the sponsor can start clinical testing for this drug candidate. This compound or biologic may be referred to as the "new drug."

An IND is a document that contains all the information known about the new drug up to the time the IND is prepared. Generally, the IND includes the name and description of the drug (chemical structure, other ingredients, etc.); how the drug is processed, manufactured, and packaged; information about any preclinical experiences relating to the safety of the drug; marketing information; past experiences or future plans for investigating the drug in foreign countries. In addition, it also contains a description of the clinical development plan. Such a description should contain all of the informational materials supplied to clinical investigators, the educational and scientific background of these investigators, a signed agreement from the investigators, and the initial protocols for clinical investigation.

Clinical development is broadly divided into four phases, namely, phases 1, 2, 3, and 4. Phase 1 trials are designed to study short-term effects of the new drug, e.g., dose range (What range of doses should be tested in humans?), pharmacokinetics (PK; What does a human body do to the drug?), and pharmacodynamics (PD; What does a drug do to the human body?). Phase 2 trials are designed to study the efficacy of the new drug in well-defined subject populations. Dose–response relationships are also studied during phase 2. Phase 3 studies are usually long-term, large-scale studies to confirm findings established from earlier trials. These studies are also used to detect toxic effects caused by cumulative dosing. If a new drug is found to be safe and efficacious during the first three phases of clinical testing, an NDA is filed for regulatory agencies to review (e.g., FDA, in the United States). Once the drug is approved by the FDA, phase 4 (postmarketing) studies are planned and carried out. Many of the phase 4 study designs are dictated by the FDA to examine safety questions; some are used to establish new uses. Section IV.A introduces phase 1 clinical trials; Sec. IV.B introduces phase 2/3 clinical trials. Section IV.C discusses the new drug application.

A. Phase 1 Clinical Trials

In a phase 1 pharmacokinetics (PK) study, the purpose is usually to study the bioavailability of a drug or the bioequivalence among different formulations of the same drug. *Bioavailability* means "the rate and extent to which the active drug ingredient or therapeutic moiety is absorbed and becomes available at the site of drug action" (Chow and Liu, 1992). Experimental units in phase 1 studies are mostly normal volunteers. Subjects recruited for these studies are generally in good health.

A bioavailability or bioequivalence study is carried out by measuring drug concentration in human blood or serum over time and then analyzing summary data from these measurements. Figure 2 presents a drug concentration–time curve; data on this curve are collected at discrete time points. Typical variables used for analysis of PK activities include area under the curve (AUC), maximum concentration (C_{max}), time to maximum concentration (T_{max}), among others. These variables are computed from drug concentration levels, as shown in Fig. 2. Suppose AUC is the variable to be analyzed, then these discretely observed points are connected (for each subject) and the area under the curve is estimated using a trapezoidal rule. For example, AUC up to 24 hours for this curve is computed by adding up the area of the triangle between hour 0 and hour 0.25, the area of the trapezoid between hour 0.25 and hour 0.5, and so on through the area of the triangle between hour 16 and hour 24.

Statistical designs used in phase 1 bioavailability studies are often crossover designs; i.e., a subject is randomized to be treated with drug A first and then treated with drug B after a "washout" period, or randomized to drug B first and then treated with drug A. In some complicated phase 1 studies, two or more treatments may be designed to cross several periods for each subject. The advantages and disadvantages of crossover designs are discussed in Chow and Liu (1992). Response variables, including AUC, T_{max}, and C_{max}, are usually analyzed using ANOVA models. Random-effects and mixed-effects linear/nonlinear models are also commonly used in the analysis of phase 1 clinical studies. In certain designs, model terms considered in these models can be very complicated.

B. Phase 2/3 Clinical Trials

Phase 2/3 trials are designed to study the efficacy and safety of a test drug. Unlike phase 1 studies, subjects recruited in phase 2/3 studies are patients with the disease that the drug is developed for. Response variables considered in phase 2/3 studies are mainly efficacy variables and safety variables. For example, in a trial for the evaluation of hypertension (high blood pressure), the efficacy variables are blood pressure measurements. For an anti-infective trial, the response variables can be the number of subjects cured or the time to cure for each subject. Phase 2/3 studies are designed mostly with parallel treatment groups; i.e., if a patient is randomized to receive treatment A, then this patient would be treated with drug A throughout the whole study.

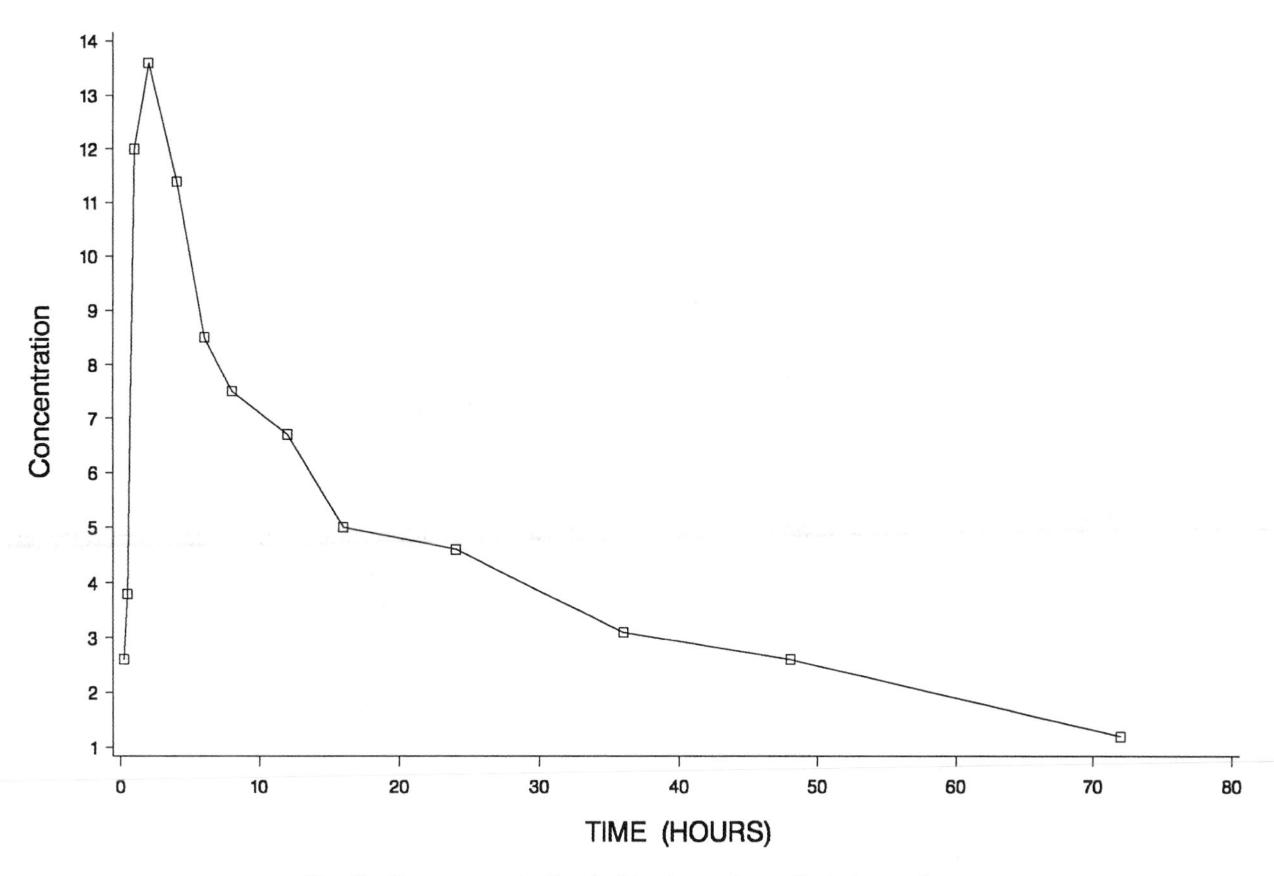

Fig. 2 Drug concentration in blood samples collected over time.

Phase 2 trials are often designed to compare a test drug against placebo. These studies are usually short term (a few weeks) and designed with a small or moderate sample size. Often, phase 2 trials are exploratory trials. One very important type of phase 2 study is the dose–response study. As in the left curve of Fig. 1, drug efficacy may increase as dose increases. In a dose–response study, the following fundamental questions need to be addressed (Ruberg, 1995):

Is there any evidence of a drug effect?
What doses exhibit a response different from the control response?
What is the nature of the dose–response relationship?
What is the optimal dose?

Typical dose–response studies are designed with fixed-dose, parallel treatment groups. For example, in a four-treatment group trial designed to study dose–response relationships, three test doses—low, medium, high—are compared against a placebo. In this case, results may be analyzed using multiple comparison techniques.

Phase 3 trials are long term (can be a few years) and large scale, with less restricted patient populations, and often compared against a known active drug for the disease to be studied. Phase 3 trials tend to be confirmatory trials designed to verify findings established from early studies.

Statistical methods used in phase 2/3 clinical studies can be very different. Statistical analyses are selected based on the distribution of the variables and the objectives of the study. Categorical data analyses are frequently used in analyzing count data (e.g., number of subjects who responded, number with side effects, or number of subjects improved from "severe symptom" to "mild symptom"). Survival analyses are commonly used in analyzing time to an event (such as time to discontinuation of the study medication, time to the first occurrence of a side effect, or time to cure). *T*-tests, regression analyses, analyses of variance (ANOVA), analyses of covariance (ANCOVA), and multivariate analyses (MANOVA) are useful in analyzing continuous data (e.g., blood pressure, grip strength, forced expiration volume, number of painful joints, area under the curve). In many cases, nonparametric analytical methods are selected, because the data do not fit any known parametric distribution well. In other cases, the raw data are transformed (log-transformed, ranked, centralized, combined) before a statistical analysis is performed. A combination of various statistical tools may sometimes be used in a drug development program. Hypotheses tests are often used to compare results obtained from different treatment groups. Point estimates and interval estimates are also frequently used to estimate subject responses to a study medication or to demonstrate equivalence between two treatment groups.

After a clinical study is completed, all of the data collected from the study are to be stored in a database; statistical analyses are performed on data sets generated from the database. A study report is prepared for each completed clinical trial. It is a joint effort to prepare such a study report. Statisticians, data managers, and programmers work together to produce tables, figures, and statistical report. Clinicians and technical writers will then put together clinical interpretations of these results. All of these are incorporated into a study report. Study reports from individual clinical trials are eventually put together as part of an NDA.

C. New Drug Application

When there is sufficient evidence to demonstrate a new drug that works and is safe, a new drug application (NDA) is put together by the sponsor. An NDA is a huge document describing all of the results obtained from both nonclinical experiments and clinical trials. A typical NDA contains sections on proposed drug labeling, pharmacological class, foreign marketing history, chemistry, manufacturing and controls, nonclinical pharmacology and toxicology summary, human pharmacokinetics and bioavailability summary, microbiology summary, clinical data summary, results of statistical analyses, benefit/risk relationship, and so on. If the sponsor intends to market the new drug in other countries, then separate packages will need to be prepared for submission to each corresponding county, too. For example, a new drug submission (NDS) needs to be filed with Canada's regulatory agency, a marketing authorization application (MAA) needs to be filed with the U.K. regulatory agency.

Oftentimes an NDA is filed while the phase 3 studies are ongoing. Sponsors need to be very careful in selecting the data cutoff date, because all of the clinical data in the database up to the cutoff date need to be frozen and stored so that NDA study report tables and figures can be produced from them. The data sets stored in such an NDA database may have to be retrieved and reanalyzed after filing in order to address various queries from regulatory agencies. After these data sets are created and stored, ongoing clinical data can then be entered into the regular database.

An NDA package usually includes not only individual clinical study reports, but also combined study results. These results may be summarized using meta-analyses or pooled-data analyses. Such analyses are performed on efficacy data to produce the "Integrated Summary of Efficacy" and on safety data to produce the "Integrated Summary of Safety." These summaries are important components of any NDA. In some cases, electronic submissions are filed as part of the NDA. Electronic submissions usually include individual clinical data, programs to process these data, and software/hardware to help FDA reviewers review the individual data as well as the whole NDA package.

V. POSTMARKETING CLINICAL DEVELOPMENT

An NDA serves as a landmark of the drug development process. The development process does not stop when an NDA is submitted or approved. But the objectives of the process is changed after the drug is approved and is available on the market. Studies performed after the drug is approved are typically called *postmarketing* studies or phase 4 studies.

One of the major objectives in postmarketing development is to establish a better safety profile for the new drug. Large-scale drug safety surveillance studies are very common in phase 4. Subjects/patients recruited in phases 1, 2, and 3 are typically somewhat restricted (patients would have to be within a certain age range, gender, disease severity, etc.). However, after the new drug is approved and is available to the general patient population, every patient with the indicated disease can be exposed to this drug. Problems related to drug safety that have not been detected from the premarketing studies (phases 1, 2, and 3) may now be observed in this large, general population.

Finally, another type of study frequently found in the postmarketing stage is to use the new drug for additional indications (symptoms or diseases). A drug developed for disease A may also be useful for disease B, but the sponsor may not have sufficient resources (budget, manpower) to develop the drug for both indications at the same time. In this case, the sponsor may decide first to develop the drug for disease A and obtain approval for the drug to be on the market and then to develop it for disease B. Phase 4 studies designed for "new indications" are also very common.

REFERENCES

Chow, S.C., and Liu, J.P. (1992). Design and analysis of Bioavailability and Bioequivalence Studies. New York: Marcel Dekker.

Hubert, J.J., Bohidar, N.R., Peace, K.E. (1988). Assessment of Pharmacological Activity. In: K.E. Peace, ed. Biopharmaceutical Statistics for Drug Development. New York: Marcel Dekker.

Redman, C.E. (1981). Screening Compounds for Clinically Active Drugs. In: C.R. Buncher, J.Y. Tsay, eds. Statistics in the Pharmaceutical Industry. New York: Marcel Dekker.

Ruberg, S.J. (1995). Dose Response Studies, I. Some Design Considerations. J Biopharm Stat 5(1):1–14.

Selwyn, M.R. (1988). Preclinical Safety Assessment. In: K.E. Peace, ed. Biopharmaceutical Statistics for Drug Development. New York: Marcel Dekker.

U.S. FDA. (1988). Guideline for the Format and Content of the Clinical and Statistical Sections of an Application.

Naitee Ting

E–F

Enrichment Design

See also *Titration Design*

I. INTRODUCTION AND DEFINITION

In clinical trials, the target population is defined by the inclusion and exclusion criteria. As a result, a screening period is usually conducted for the purpose of identifying the eligible subjects before actually enrolling them into the study. On the other hand, a placebo run-in period is also usually performed after the screening period and before the double-blind, randomized, active treatment period to assess the patient's compliance and to estimate the potential placebo effect. Both screening and placebo run-in periods can be viewed as sequential screening processes for identification of the patients who meet the inclusion/exclusion and compliance criteria without giving the active treatment under investigation to the patients.

The objective of some trials is to evaluate the pharmaceuticals or treatments based on hard efficacy endpoints such as death, myocardial infarction, and bone fracture. However, these hard endpoints require a long follow-up period to observe. On the other hand, some short-term responses can be obtained much more quickly for some soft endpoints, such as biological markers. In addition, some therapeutic agents are likely to be effective in a population of the patients who may have an underlying disorder that is responsive to manipulation of dose levels of the agents or several agents. Therefore, it is necessary to perform some additional screening processes using the active treatments under evaluation after either the screening period or the placebo run-in period to identify the candidates in whom test agents are likely to be beneficial in the early phase of the trial. This phase of an additional screening process for the same therapeutic agent or test of different agents for identification of patients with drug efficacy is called the *enrichment phase*. The patients with drug efficacy identified in the enrichment phase are then randomized to receive either the efficacious dose (agent) or the matched placebo. This type of design with additional screening processes, with the active treatments evaluated in the study, is called the *enrichment design*. According to this definition, the difference between the designs with either screening or placebo run-in periods and the enrichment designs is that active treatments are used to identify the patients.

II. EXAMPLES

An enrichment design usually consists of at least two phases. The first phase is the enrichment phase, in which a titration design is usually conducted to classify patients into groups according to whether the pharmaceuticals under investigation present some benefits to the patients. The second phase usually is randomized and double-blind, possibly with a placebo concurrent control, to investigate the effectiveness and safety of the test agents in these patients formally and rigorously. The primary efficacy endpoints used for evaluation during the two phases of an enrichment design, depending upon the objectives of the trial, can be the same or different. The concept of enrichment design is illustrated in three clinical trials in the areas of Alzheimer's disease and arrhythmia that have been described in detail in Chow and Liu (1998).

The enrichment design was elected for a clinical trial conducted in the early stage of development of tacrine, with doses of 40 mg and 80 mg four times a day in the treatment of patients with probable Alzheimer's disease. As indicated in Davis et al. (1992), because of the clinical, biochemical, and pathological heterogeneity of the disease and clinical experience, not all patients with probable Alzheimer's disease would respond to any single treatment, and those who did respond might do so only within a limited dose range. This trial consisted of a total of four phases, a 6-week double-blind dose-titration enrichment phase, a 2-week placebo baseline phase, a 6-week randomized, double-blind, placebo-controlled phase, and a 6-week sustained active phase. After the patients met the inclusion and exclusion criteria, they were enrolled into the enrichment phase of the trial, which consists of three titration sequences. Each titration sequence consists of three 2-week dosing periods. The dose in each titration sequence is always titrated up from 40 mg to 80 mg four times a day, with a placebo in dosing periods 1, 2, and 3, respectively, for titration sequences 1, 2, and 3. The patients were randomized into one of the three titration sequences that were conducted in double-blind fashion. The potentially therapeutic responses for each patient at each dose were then assessed at the end of each 2-week dosing period. The "best dose" response for a patient was defined in advance in the protocol as a reduction of at least 4 points from the screening value in a total score on the Alzheimer's Disease Assessment Scale (ADAS, Folstein et al., 1975) and without intolerable side effects. Patients with the identified "best dose" were then entered into a 2-week placebo baseline period, with the hope that this period would be sufficiently long for tacrine to wash out from the body and for patients to return to the screening pretreatment status with comparable efficacy outcomes. At the end of the 2-week placebo baseline phase, the patients with a reduction of at least 4 points in ADAS during the enrichment phase entered the subsequent 6-week randomized, double-blind, parallel-group, placebo-controlled phase. They were randomized either to the active tacrine at their best dose or to the matched placebo. Patients who completed the 6-week double-blind phase then entered into the sustained active treatment phase. As a result, the purpose of selection of an enrichment design with three titration sequences for this study is to identify a group of patients who are likely to respond to tacrine at a certain dose; after a washout period of 2 weeks, these identified patients were randomized to either tacrine at their best dose or to placebo concurrent control in a double-blind phase.

The rationale for selection of the enrichment design in one of the trials during the development of tacrine is to verify whether a short-term response to tacrine has predictability for the long-term efficacy in the prevention of progression of the patients with probable Alzheimer's disease. The same primary efficacy endpoints such as ADAS or the Clinical Global Impression of Change (CGI-C) were used in both the enrichment and the double-blind phases for the evaluation of tacrine's effectiveness. On the other hand, for other therapeutic agents, the real efficacy endpoint is mortality, which requires a long time to observe. Therefore, the short-term efficacy of the agents is assessed by some other objective surrogate endpoints. It is then very important to know whether the short-term efficacy based on surrogate endpoint is predictive of the hard endpoint, such as mortality. As a result, the enrichment design is usually employed for identification of the "short-term" responders at the initial stage followed by the main phase of the long-term study. Examples of this type of trial can be found in the area of arrhythmia, e.g., the Cardiac Arrhythmia Suppression Trial (CAST) (CAST Investigators, 1989; Echt et al., 1991; Ruskin, 1989) and the electrophysiologic study versus electrocardiographic monitoring (ESVEM) (ESVEM Investigators, 1989, 1993; Mason et al., 1993a,b; Ward and Camm, 1993).

CAST is a multicenter, randomized, placebo-controlled study to test the hypothesis of whether the suppression of asymptomatic or mildly symptomatic ventricular arrhythmia after myocardial infarction would reduce the rate of death from arrhythmia. The drugs include two class-IC antiarrhythmic agents—encainide, flecainide; and morcizine with a placebo concurrent control. One of the objectives of the study is to test the predictability of the adequate suppression of ventricular arrhythmia by the active drugs based on ventricular premature contractions (VPC) as recorded by Holter monitor for mortality. As a result, an open-label enrichment design with two titration sequences involved with only active drug was selected for this study. Patients with an ejection fraction of at least 30% were randomly assigned to the two titration sequences: (encainide, morcizine, flecainide) or (flecainide, morcizine, encainide). The reason for the insertion of morcizine in the middle dosing period is its inferior efficacy in suppression of VPC as compared to the other two active agents. Each drug is tested at two dose levels. The doses of encainide were 35 and 50 mg three times a day (tid); of flecainide, 100 and 150 mg twice a day (bid); of morcizine, 200 and 250 mg tid. Because flecainide exhibits negative inotropic properties, it was not administered to the patients with an ejection fraction of less than 30%. The titration sequences for the patients with an ejection fraction of less than 30% were (encainide, morcizine) or (morcizine, encainide). The prespecified criteria for an adequate suppression of ventricular arrhythmia were (a) a reduction of at least 80% in VPC and (b) a reduction of at least 90% in runs of unsustained ventricular tachycardia as measured by 24-hour Holter recording 4 to 10 days after each dose was begun. The titration process for a particular patient was stopped as soon as a drug and a dose were found to yield an adequate suppression. The patients whose arrhythmia were adequately suppressed were then randomized either to the "best drug" identified during the enrichment phase or to placebo for a 3-year long-term follow-up.

The results of CAST indicate that the short-term efficacy measured as the suppression of VPC based on Holter noninvasive ambulatory electrocardiographic monitoring might not be a good predictor for the long-term hard mortality endpoint. Others suggested that the failure to induce ventricular tachycardia or fibrillation by some drug assessed by the invasive electrophysiologic study might be a good independent predictor of recurrence of arrhythmia. Consequently, the ESVEM (electrophysiologic study versus electrocardiographic monitoring) trial was the first large prospective, randomized trial conducted to compare the two methods in the predictability of long-term recurrence of arrhythmia by the short-term efficacy assessed by the two methods (ESVEM investigators, 1989; Mason et al., 1993a,b). Because it requires correlating the difference in recurrence rates of arrhythmia with the short-term efficacy by both methods, an inpatient enrichment phase was elected to identify a group of patients in whom a test drug exhibited a short-term efficacy assessed by either one of the two methods. After the patients had fulfilled entry criteria and 48-hour Holter monitoring and electrophysiologic study criteria, they were randomized to one of the two parallel groups for the two methods to assess the short-term drug efficacy. The first group employed noninvasive ambulatory electrocardiographic monitoring; the second group applied the invasive electrophysiologic study. Within each group, the patients received up to six arrhythmia agents in random order until one drug was predicted to be efficacious or until all drugs had been tried that the patients were eligible to receive. A test drug was identified as effective as assessed by electrocardiographic monitoring during the inpatient enrichment phase if the following predefined criteria for short-term efficacy were met: (a) 70% reduction in mean VPC, (b) 80% reduction in pairs of VPC, (c) 90% reduction in mean ventricular tachycardia count, and (d) absence of any runs of ventricular tachycardia longer than 15 seconds. Drug effectiveness evaluated by the electrophysiologic study during the enrichment phase was defined as failure by the drug to induce a run of ventricular tachycardia longer than 15 seconds with V1V2V3 stimulation at the right ventricular apex. If a drug was proven to be efficacious for a patient during the enrichment phase, then he or she was discharged from the hospital for the long-term follow-up with the drug, and the accuracy of the prediction of efficacy was determined during the long-term follow-up. Patients in whom no drugs were proven to be effective during the enrichment phase were not randomized and were withdrawn from the study, but their vital signs and recurrence of arrhythmia were monitored.

III. DISCUSSION

The main objective of an enrichment design is to screen for possible responders using the active treatments under investigation. However, if placebo is also included in the enrichment phase, then the possible placebo responders will also be identified in the enrichment phase. In addition, as demonstrated in the previous discussion, the design employed in the enrichment phase is usually a dose-titration design with no washout periods between dosing intervals. As a result, the treatment effects, carryover effects, and time effects all confound one another during the process of identification of responders. In CAST or ESVEM, the patients were randomized as soon as the first drug at the first dose was found to yield an adequate suppression of VPC. Another issue for the enrichment design therefore is whether or not the response observed is the best drug at its optimal dose for the responder. On the other hand, due either to lack of randomization or to different methods of randomization for the enrichment phase, statistical methods for analysis based on the data from both the enrichment phases and the double-blind phase of the trial are not fully developed. In summary, enrichment design further restricts the target population into a very small selective group. However, it is not yet always possible to distinguish this small group of patients from the rest with the same ailment in terms of demographic and other prognostic factors for grass-roots clinical practice. Therefore, inference based on the statistical analysis for a trial using an enrichment design remains a challenge for all of us.

REFERENCES

CAST Investigators. (1989). Preliminary report: effect of encainide and flecainide on mortality in a randomized trial of arrhythmia suppression after myocardial infarction. New Eng J Med 321:406–412.

Chow, S.C., and Liu, J.P. (1998). Design and Analysis of Clinical Trials: Concepts and Methodologies. New York: Wiley.

Davis, K.L., Thal, L.J., Gamzu, E.R., Davis, C.S., Woolson, R.F., Gracon, S.I., Drachman, D.A., Schneider, L.S., Whitehouse, P.J., Hoover, T.M., Morris, J.C., Kawas, G.H., Knopman, D.S., Earl, N.L., Jumar, V., Doody, R.S., and the Tacrine Collaborative Study Group. (1992). A double-blind, placebo-controlled multicenter study for Alzheimer's disease. New Eng J Med 327:1253–1259.

Echt, D.S., Liebson, P.R., Mitchell, L.B., Peters, R.W., Obias-Manno, D., Barker, A.H., Arensberg, D., Baker, A., Friedman, L., Greene, H.L., Huther, M.L., Richardson, D.W., and the CAST Investigators. (1991). Mortality and morbidity in patients receiving encainide and flecainide or placebo. New Eng J Med 324:781–788.

ESVEM Investigators. (1989). The ESVEM trial: electrophysiologic study versus electrocardiographic monitoring for selection of antiarrhythmic therapy of ventricular tachyarrhythmias. Circulation 79:1354–1360.

ESVEM Investigators. (1993). Determinants of predicted of efficacy antiarrhythmic drugs in the electrophysiologic study versus electrocardiographic monitoring trials. Circulation 87:323–329.

Folstein, M.F., Folstein, S.E., McHugh, P.R. (1975). "Mini-mental state": a practical method for grading the cognitive state of patients for the clinician. J Psychiatric Rev 12:189–198.

Mason, J.W., for the ESVEM Investigators. (1993a). A comparison of seven antiarrhythmic drugs in patients with ventricular tachyarrhythmias. New Eng J Med 329:452–458.

Mason, J.W., for the ESVEM Investigators. (1993b). A comparison of electrophysiologic testing versus Holter monitoring to predict antiarrhythmic-drug efficacy for ventricular tachyarrhythmias. New Eng J Med 329:445–451.

Ruskin, J.N. (1989). The Cardiac Arrhythmia Suppression Trial (CAST). New Eng J Med 321:386–388.

Ward, D.E., Camm, A.J. (1993). Dangerous ventricular arrhythmias—can we predict drug efficacy? New Eng J Med 329:498–499.

Jen-pei Liu

Equivalence Trials

See also *Bioavailability and Bioequivalence; Therapeutic Equivalence; Individual Bioequivalence*

I. INTRODUCTION

Most of the trials conducted during the development stage of investigational pharmaceuticals before filing the new drug application (NDA) or product license application (PLA) are for demonstration of better efficacy and safety of the test product over the concurrent placebo control. This type of trial is referred to as the *superiority* trial. On the other hand, some trials try to compare the test treatment to an active control, such as some reference treatment or the standard therapy. The purpose of this type of trial is to verify whether the effectiveness or safety of the test treatment is similar to that of the active control. Depending upon their objectives, this type of trial can be further classified into *equivalence*, or *noninferiority*, trials.

After the patent of an innovative product expires, other pharmaceutical companies can manufacture generic copies under the Drug Price Competition and Patent Term Restoration Action passed in 1984 through the abbreviated new drug application (ANDA) based on the pharmacokinetic (PK) responses derived from plasma or blood concentrations collected from bioequivalence studies, such as area under the plasma concentration-time curve (AUC) or the peak concentration (C_{max}). In order to reduce the variation, crossover designs with normal volunteers are usually employed for bioequivalence testing (Federal Register, 1977). Based on the *fundamental bioequivalence assumption* (Chow and Liu, 1992a), if PK parameters of the generic and innovative products are close, within a prespecified allowable limit, it is then assumed that they produce no clinically meaningful difference in effectiveness and safety and hence that the generic drugs and the marketed innovative reference product can be used interchangeably. On the other hand, some drug products do not work through systemic absorption and have negligible plasma levels. These drugs include metered-dose inhalers (MDIs) indicated for the relief of bronchospasm in patients with reversible obstructive airway disease, sucralfate for the treatment of acute duodenal ulcer, and topical and vaginal antifungals. Because these drug products have negligible plasma concentration of the active ingredients, PK responses are no longer adequate for evaluation of bioequivalence between drug products. As a result, the U.S. Food and Drug Administration (FDA) suggested that clinical trials be conducted to demonstrate the therapeutic equivalence based on well-defined clinical endpoints (Huque et al., 1989; Huque and Dubey, 1990). Both clinical equivalence and bioequivalence trials are required to demonstrate similarity based on either clinical endpoints or PK responses; they are referred to as *equivalence trials*.

For many diseases, there exists a standard drug that has been proved to be efficacious with a reasonably tolerable safety profile by more than one adequate well-controlled study. But it is still worthwhile to develop new drugs for the same disease with similar efficacy if they can provide a better safety margin, are easy to administer, are less expensive, require a short duration of treatment, and can substitute for an invasive procedure. As a result, instead of a superior trial, it is sufficient for clinical trials to provide evidence that the new drug is therapeutically no worse than the standard reference treatments with respect to a prespecified limit of no clinical significance (Dunnett and Gent, 1977; Makuch and Johnson, 1990; Durrleman and Simon, 1990; Blackwelder, 1982; Chow and Liu, 1998). This type of trial is called *noninferiority trial*, which will be discussed in more detail in the entry "Therapeutic Equivalence."

II. EQUIVALENCE TRIALS

The concept of equivalence asks for similarity in distributions of efficacy clinical endpoints or PK responses between the test and the reference drug. It is then referred to as *population equivalence*. Under the normality assumption, the distributions of most PK responses or efficacy clinical endpoints or their transformation can be uniquely determined by the first two moments, the average and variability. Hence, the equivalence between the test and reference drug products can be established through the similarity of both the average and the variance. *Average equivalence* requires a similarity in the averages of the distributions between the two drug products. On the other hand, population equivalence, under the normality assumption, asks for equivalence in both average and variability. However, in general, equivalence in the first two moments does not guarantee equivalence between distributions. But the current concept for the assessment of equivalence between a test drug and a reference product focuses only on the comparison of the averages. Only recently has some attention been paid to the equivalence in variabilities for generic products (Liu, 1994; Chow and Liu, 1995). For a complete description of bioequivalence, see Chow and Liu (1992a). Chow and Liu (1998) provide further discussion on therapeutic equivalence.

Failure to reject the null hypothesis of equality does not imply equivalence between the test and reference products (Metzler, 1974; Westlake, 1972; Dunnett and Gent, 1977). Hence we use the average as an example to demonstrate the concept of equivalence. Let μ_T and μ_R be the population averages of the test and reference products, respectively. The concept of equivalence in the averages is that the difference in average is within prespecified allowable limits. In other words, the average of the test product is not too low and is not too high as compared to that of the reference product. As a result, the consumer's risk is the error in declaring equivalence between test and reference products when in fact they are not; and the producer's risk is the error in declaring inequivalence between the two products when in fact they are equivalent. The hypothesis of equivalence in terms of average can therefore be formulated as follows (Chow and Liu, 1992a):

$$H_0: \mu_T - \mu_R \geq L \quad \text{or} \quad \mu_T - \mu_R \leq U$$
$$\text{vs.} \quad H_a: L < \mu_T - \mu_R < U \tag{1}$$

where L and U are some prespecified lower and upper equivalence limits, respectively.

Because the above alternative hypothesis is an interval, it is also called the *interval hypothesis of equivalence*. In addition, both lower and upper equivalence limits are required for the interval hypothesis in Eq. (1); it is also referred to as *two-sided equivalence*. The interval hypothesis in Eq. (1) is the correct formulation of the hypothesis for the assessment of equivalence in such a manner that a Type I error represents the consumer's risk and a Type II error is the producer's risk. The selection of equivalence limits should have clinical or pharmacokinetic justification. For average bioequivalence, the FDA guidance (1992) requires a range of 80 to 125% for the ratio of the average area under curves or the peak concentrations between the test and reference products.

The interval hypothesis in Eq. (1) can be further decomposed into two one-sided hypotheses as:

$$H_{0L}: \mu_T - \mu_R \leq L \quad \text{vs.} \quad H_{aL}: \mu_T - \mu_R > L$$
$$H_0: \mu_T - \mu_R \geq U \quad \text{vs.} \quad H_a: \mu_T - \mu_R < U \qquad (2)$$

A comparison of the interval hypotheses in Eq. (1) and the two one-sided test hypotheses in Eq. (2) reveals that the space defined in the null hypothesis of the interval hypotheses is the union of the space defined in the two one-sided null hypotheses. The space defined by the alternative hypothesis in Eq. (2) is the intersection of the space of the two alternative hypotheses in Eq. (2). From the intersection-union principle (IUT, Berger, 1982), for each of the two one-sided hypotheses, there is a test statistic to verify whether the average for the test product is not too low (high) as compared to that of the reference drug. Therefore, if both null hypotheses of the two one-sided hypotheses are rejected at the prespecified nominal significance level α, then equivalence in averages between the test and reference products is concluded at the α significance level. Because this procedure involves testing two one-sided hypotheses simultaneously, it is called the *two one-sided tests procedure* (Schuirmann, 1987; Chow and Liu, 1992a).

Because equivalence between the two drugs is concluded only if the null hypothesis of both one-sided hypotheses is rejected at the prespecified nominal level of significance, the two one-sided tests procedure can control the consumer's risk at the nominal level. However, if the variability is large or the sample size is too small, the two one-sided tests procedure may be very conservative in concluding equivalence. Hsu et al. (1994); and Brown, Hwang and Munk (1998) proposed other procedures for the interval hypothesis of equivalence that are more powerful than the usual two one-sided tests procedure. However, their procedures are mathematically complicated and lack intuitive interpretation. The open nonconvex shape of their rejection regions implies that any estimate of the mean difference in average might conclude equivalence as long as the standard error of the observed mean difference is sufficiently larger or it increases to infinity.

The hypothesis for the evaluation of equivalence in variability can be similarly formulated (Chow and Liu, 1992a,b) as:

$$H_0: \frac{\sigma_T^2}{\sigma_R^2} \geq \delta_L \quad \text{or} \quad \frac{\sigma_T^2}{\sigma_R^2} \leq \delta_U$$

$$\text{vs.} \quad H_a: \delta_L < \frac{\sigma_T^2}{\sigma_R^2} < \delta_U \qquad (3)$$

where σ_T^2 and σ_R^2 are the variance of the test and reference products, respectively, and $0 < \delta_L < 1 < \delta_U$ are the equivalence limits for the evaluation of equivalence in variability. For bioequivalence testing conducted under crossover design, this is referred to as the *intra-subject variability*. Under a standard two-sequence and two-period (2×2) crossover design, Liu and Chow (1992a) proposed the Pitman–Morgan two one-sided tests procedure for hypothesis (3). Because each subject receives the test or reference product only once under a (2×2) crossover, estimators for σ_T^2, σ_R^2, and σ_T^2/σ_R^2 cannot be obtained without further assumptions. However, for PK responses obtained under replicated crossover design, inference about equivalence based on σ_T^2/σ_R^2, including confidence interval, is quite straightforward (Chow and Liu, 1992a; Liu, 1995a). This discussion about equivalence in average or variability can be applied to the responses obtained from either parallel designs or crossover designs. For bioequivalence trials using either the standard 2×2 or replicated crossover designs, see Chow and Liu (1992a, b). Liu and Chow (1992b), Liu (1995b), and Chow and Liu (1998) provide sample size determination for equivalence in averages.

III. CONFIDENCE INTERVAL APPROACHES

Although the discussion for the assessment of equivalence is in the context of a statistical hypothesis-testing procedure, it also is an estimation problem. As a result, many researchers have advocated the utility of the confidence interval for the assessment of equivalence between the test and reference drugs (Westlake, 1972, 1976; Makuch and Simon, 1978; Chow and Liu, 1992a; Durrleman and Simon, 1990; Jennison and Turnball, 1993). The idea and procedure for the assessment of equivalence are quite straightforward. For two-sided equivalence, if the nominal level is α, then the two drugs are concluded equivalent if the $100(1 - 2\alpha)\%$ confidence interval for the mean difference is completely contained within (L, U). The confidence interval approach is more appealing than the testing procedure because it can provide the magnitude and width of the average difference between the two drugs. In addition, the hypothesis testing and confidence interval approaches in some situations are in fact operationally equivalent (Schuirmann, 1987). For many cases, the pivotal quantity for construction of a confidence interval is the test statistic for the two one-sided test procedures, which follows either a central t-distribution or the standard normal distribution. As a result, for two-sided equivalence problem, the $100(1 - 2\alpha)\%$ confidence interval is totally contained within (L, U) if and only if both one-sided hypotheses in Eq. (2) are rejected at the α significance level.

IV. EQUIVALENCE LIMITS

The decision to assess equivalence either by confidence interval approach or testing procedure depends upon the lower and upper equivalence limits and has nothing to do with whether or not the observed difference is zero. Table 1 illustrates the consistency between the two one-sided tests procedure and the confidence interval approach for two-sided equivalence. As can be seen in the table, although the null hypothesis of equality is not rejected, equivalence between the two drugs cannot be concluded because the 90% confidence interval may be too wide to be contained within the equivalence limits. Another situation is when equivalence between the test and reference drugs is concluded because the confidence interval is very narrow and is totally contained within the limits even though it does not contain 0. In other words, equivalence can be concluded even when the null hypothesis of equality is rejected and it is concluded that there is a difference between the test and reference products. As a result, the length of the equivalence interval, $U - L$ for therapeutic equivalence, in general should be smaller than the difference in averages between the test and concurrent placebo that superior trials try to detect.

Although the equivalence limits for average bioequivalence required by the FDA are well accepted and established, selection of the equivalence limits for assessing similarity is always controversial and remains a challenge in other areas. If the equivalence limits selected are too narrow, it is extremely difficult to conclude equivalence even though the difference between them is of no clinical significance. On the other hand, if the equivalence limits chosen are very wide, then it is easy to declare the two drugs with a clinically meaningful difference to be equivalent.

In general, the equivalence limits depend upon the response of the reference drug. For example, Huque and Dubey (1990) have proposed some equivalence limits for binary data, such as eradication rate for antibiotics. They suggested that the equivalence limits be tighter for higher reference response rates. Suppose that the response rate was 75% from two adequate, well-controlled studies for approval of the reference drug. The equivalence limits suggested by Huque and Dubey (1990) is ±20%. The sample size can be also determined based on ±20% equivalence limits. However, the current equivalence trial yields a reference response rate of 85%. Should the equivalence limits be changed to ±15% as suggested by Huque and Dubey (1990)? Is the sample size calculated with the limits of ±20% large enough to provide sufficient power for the limits of ±15%?

These questions reveal that the equivalence limits based on the reference responses from any previous studies may not be internally valid. Therefore, the equivalence limits chosen may be based on the reference response obtained concurrently within the same equivalence trial for internal validity. In this case, the lower and upper equivalence limits in interval hypothesis (2), however, are no longer fixed, known quantities but random variables. The interval hypothesis should be reformulated as:

$$H_a: L < \frac{\mu_T}{\mu_R} < U \tag{4}$$

where $0 < L < 1 < U$.

The interval hypothesis in Eq. (4) implies that the two drugs are equivalent in average if the ratio of averages is between L and U. Logarithmic transformation of Eq. (4) reformulates the interval hypothesis from a ratio to the difference:

Table 1 Results of Different Procedures for Assessing Equivalence with a Known Symmetric Limit of 0.2

Observed difference	90% CI	H_{0E}	H_{0L}	H_{0U}	Two-sided equivalence
0.00	(−0.10, 0.10)	Fail to reject	Reject	Reject	Yes
0.00	(−0.25, 0.25)	Fail to reject	Fail to reject	Fail to reject	No
0.05	(0.02, 0.08)	Reject	Reject	Reject	Yes
0.10	(−0.02, 0.22)	Fail to reject	Reject	Fail to reject	No

H_{0E} is the null hypothesis for equality.
Source: Chow and Liu (1998).

H_a: $\ln(L) < \ln(\mu_T) - \ln(\mu_R) < \ln(U)$

where ln denotes the natural logarithm and $\ln(L) < 0 < \ln(U)$.

To assess average bioequivalence between a generic and a reference formulation, most regulatory agencies, such as the U.S. FDA and the European Community (EC) request that PK responses such as AUC and $C_{\max}$ be analyzed on the log scale, with $L = 80\%$ and $U = 125\%$ (i.e., $\ln(L) = -0.2231$ and $\ln(U) = 0.2231$).

On the other hand, if data are analyzed on the original scale, the equivalence limits in Eq. (4) depend upon the observed reference mean for internal validity. Suppose that in Eq. (2), $L = f\mu_R$ and $U = g\mu_R$; then the two one-sided hypotheses corresponding to Eq. (2) can be formulated as:

$$H_{0L}: \mu_T - (1 - f)\mu_R \leq 0$$
$$\text{vs.} \quad H_{aL}: \mu_T - (1 - f)\mu_R > 0$$
$$H_{0U}: \mu_T - (1 - g)\mu_R \leq 0$$
$$\text{vs.} \quad H_{aU}: \mu_T - (1 - g)\mu_R > 0 \qquad (5)$$

Consequently, the two one-sided tests procedure can be constructed based on the estimates for $\mu_T - (1 - f)\mu_R$ and $\mu_T - (1 - g)\mu_R$ and their estimated standard errors. Currently, however, the two one-sided tests procedure for average are performed with estimated equivalence limits based on the observed reference mean as the true unknown fixed quantities without taking into consideration the variability of estimated equivalence limits. Liu and Weng (1995) showed that for the data obtained from a two-sequence and two-period crossover design, the current application of the two one-sided tests procedure cannot control the probability of Type I error. When the correlation between the two responses from the same patient approaches 1, the consumer's risk goes to 50%. They proposed a modified two one-sided tests procedure based on the two one-sided hypotheses in Eq. (5), which adequately controls the Type I error rate at the nominal level.

V. POPULATION BIOEQUIVALENCE

Up to the end of 1997, average bioequivalence has been the only evidence required by the regulatory authorities in the world to approve a generic drug product (FDA Guidance, 1992). However, as demonstrated here, average bioequivalence does not guarantee that the generic and reference product can be used interchangeably. In general, interchangeability between the generic and reference products can be classified as prescribability and switchability. *Prescribability* means the physician's choice for prescribing a drug for his or her new patients between an innovative product and a number of its generic copies that have been shown to be bioequivalent to the innovative product (Chow and Liu, 1995). On the other hand, *switchability* is related to the switch from a drug such as an innovative product or a generic drug to an alternative drug, such as a generic copy of a brand name drug or another generic copy of the same innovative drug product within the same patient whose concentration of the drug has been titrated to a steady, efficacious, and safe level. To ensure switchability, the equivalence between drug products must be demonstrated within each individual. This concept is referred to as *individual bioequivalence*.

For the evaluation of population bioequivalence, as mentioned earlier, under the normality assumption, equivalence in both average and variability is required. One approach is to conclude population bioequivalence at the α significance level if both the null hypothesis for average in Eq. (1) and that for variability in Eq. (4) are rejected at the α significance level (Liu and Chow, 1992a). This disaggregate procedure is another application of the union-intersection principle. As a result, although it is a size α test, it can be very conservative. The U.S. FDA issued a draft guidance on December 10, 1997, that requires population bioequivalence for the NDA sponsors who wish to assess bioequivalence during the investigational phase of drug development. The draft guidance suggested using the following aggregate criterion for the assessment of population bioequivalence based on the log scale of AUC or $C_{\max}$:

$$\theta = \frac{(\mu_T - \mu_R)^2 + (\sigma_{TT}^2 - \sigma_{TR}^2)}{\sigma_{TR}^2} \qquad (6)$$

where σ_{TT}^2 is the total variance (sum of intrasubject and intersubject variability) of the test formulation and σ_{TR}^2 is the total variance (sum of intrasubject and intersubject variability) of the reference formulation.

The aggregate criterion in Eq. (6) is scaled by the total reference variability. It is then called the *reference-scaled criterion*. Population bioequivalence is concluded if θ is smaller than some prespecified upper equivalence limit θ_P. When σ_{TR}^2 is smaller than 0.04, the draft guidance proposes that σ_{TR}^2 in the denominator of Eq. (6) be replaced by 0.04. The resulting criterion is then referred to as the *constant-scaled criterion*. The draft FDA guidance suggests that the upper equivalence limit for the aggregate criterion θ be determined by (a) $\mu_T - \mu_R = \ln(1.25)$, (b) $\sigma_{TT}^2 - \sigma_{TR}^2 = 0.02$, and (c) a scale constant of 0.04. It follows that $\theta_P = 1.7448$.

From Eq. (6), the aggregate criterion proposed in the draft FDA guidance is based on the difference in average bioavailability and the difference in total variability scaled by the reference total variability. As a result, crossover design is no longer required for assessing population bioequivalence, and a two-group parallel design with a single PK response from each subject will suffice. Because of a very complicated sampling distribution for estimators of θ, a bootstrap procedure is suggested to obtain the upper 5% confidence bound for θ using the nonparametric percentile method. Population bioequivalence is concluded at the 5% significance level if the upper 5% confidence bound for θ is smaller than θ_P (= 1.7448). The upper 5% confidence bound should be obtained with at least 2000 bootstrap samples. In addition, the selection of the referenced-scaled or constant-scaled criterion for the bootstrap procedure is determined by whether or not the estimated total reference variability from the original data set is greater than 0.04.

The draft guidance recommends that the individual bioequivalence approach be used by sponsors of an ANDA or an abbreviated antibiotic drug application (AADA) to assess bioequivalence between the generic and reference listed drugs. Aggregate referenced-scaled and constant-scaled criteria similar to those for population bioequivalence are also proposed for the evaluation of individual bioequivalence. The criterion for individual bioequivalence is scaled by the reference intrasubject variability. The constant-scaled criterion is selected if the reference intrasubject variability is also smaller than 0.04. Because the reference total variability is larger than the reference intrasubject variability, the draft guidance does not provide a justification for the reason why both reference total and intrasubject variability use the same cutoff point of 0.04 for the determination of the use of their corresponding constant-scaled criteria. There are many other unresolved issues for the aggregate criterion, such as masking effects, unknown bias and power function, inference for second moments, two-stage procedure, and interpretation of results, among others. These issues will be addressed in details in the entry "Individual Bioequivalence."

REFERENCES

Blackwelder, W.C. (1982). "Proving the null hypothesis" in clinical trials. Controlled Clin Trials 3:345–353.

Berger, R.L. (1982). Multiparametric hypothesis testing and acceptance sampling. Technometrics 24:295–300.

Brown, L.D., Hwang, J.T.G., Munk, A. (1998). An unbiased test for the bioequivalence problem. Annals of Statistics 25:2345–2367.

Chow, S.C., Liu, J.P. (1992a). Design and Analysis of Bioavailability and Bioequivalence Studies. New York: Marcel Dekker.

Chow, S.C., Liu, J.P. (1992b). On assessment of bioequivalence with high-order crossover designs. J Biopharmaceut Stat 2:239–256.

Chow, S.C., Liu, J.P. (1995). Current issues in bioequivalence trials. Drug Info J 29:795–804.

Chow, S.C., Liu, J.P. (1998). Design and Analysis of Clinical Trials: Concepts and Methodologies. New York: John Wiley.

Dunnett, C.W., Gent, M. (1977). Significance testing to establish equivalence between treatments with special reference to data in the form of 2×2 table. Biometrics 33: 593–602.

Durrleman, S., Simon, R. (1990). Planning and monitoring of equivalence studies. Biometrics 46:329–336.

FDA Guidance. (1992). Statistical Procedures for Bioequivalence Studies Using a Standard Two-Treatment Crossover Design. Rockville, MD: Center for Drug Evaluation and Research, Food and Drug Administration.

FDA Draft Guidance. (1997). In Vivo Bioequivalence Studies Based on Population and Individual Bioequivalence Approaches. Rockville, MD: Center for Drug Evaluation and Research, Food and Drug Administration.

Federal Register. (1977). 42(5):320.26(b).

Hsu, J.C., Hwang, J.T.G., Liu, H.K., Ruberg, S.J. (1994). Confidence intervals associated with tests for bioequivalence. Biometrika 81:103–114.

Huque, M.F., Dubey, S. (1990). Design and analysis of therapeutic equivalence clinical with a binary clinical endpoint. The 1990 Proceedings of the Biopharmaceutical Section of the American Statistician Association, Alexandria, VA, pp 46–52.

Huque, M.F., Dubey, S., Fredd, S. (1989). Establishing therapeutic equivalence with clinical endpoints. The 1989 Proceedings of the Biopharmaceutical Section of the American Statistician Association, Alexandria, VA, pp 46–52.

Jennison, C., Turnbull, B.W. (1993). Sequential equivalence testing and repeated confidence intervals, with applications to normal and binary responses. Biometrics 49:31–43.

Liu, J.P. (1994). Invited discussion of "Individual bioequivalence: a problem for switchability" by S. Anderson. Biopharmaceut Rep 2:7–9.

Liu, J.P. (1995a). Use of replicated cross-over designs in assessing bioequivalence. Stat Med 14:1067–1078.

Liu, J.P. (1995b). Letter to the Editor on "Sample size for therapeutic equivalence based on confidence interval" by S.C. Lin. Drug Info J 29:1063–1064.

Liu, J.P., Chow, S.C. (1992a). On the assessment of variability in bioavailability/bioequivalence studies. Communic Stat Theory Meth 21:2591–2607.

Liu, J.P., Chow, S.C. (1992b). Sample size determination for the two one-sided tests procedure in bioequivalence. J Pharmacokinetics Biopharmaceutics 20:101–104.

Liu, J.P., Weng, C.S. (1995). Bias of two one-sided tests procedures in assessment of bioequivalence. Stat Med 14: 853–862.

Makuch, R.W., Johnson M. (1990). Active control equivalence studies: planning and interpretation. In: K.E. Peace, ed. Statistical Issues in Drug Research and Development. New York: Marcel Dekker, pp 238–246.

Makuch, R.W., Simon R. (1978). Sample size requirements for evaluating a conservative therapy. Cancer Treatment Rep 6:1037–1040.

Metzler, C.M. (1974). Bioavailability: a problem in equivalence. Biometrics 30:309–317.

Schuirmann, D.J. (1987). A comparison of the two one-sided tests procedure and the power approach for assessing the equivalence of average bioequivalence. J Pharmacokinetics Biopharmaceutics 15:657–680.

Westlake, W.J. (1972). Use of confidence intervals in analysis of comparative bioavailability trials. J Pharmaceutical Sci 61:1340–1341.

Westlake, W.J. (1976). Symmetrical confidence intervals for bioequivalence trials. Biometrics 32:741–744.

Jen-pei Liu

Ethnic Factors

See also *Subgroup Analysis*

I. INTRODUCTION

Can efficacy and safety data about an experimental drug or device from a trial conducted in one region be extrapolated to other regions? Will governmental regulatory bodies accept data collected in a region other than its own in support of drug or device registration? Will the health care environment in one region respond the same way to a new drug or device as in another region? Questions like these often result from the issue of how important ethnic factors are in determining the safety, efficacy, dosage, and therapeutic regimen of a new drug or device.

II. DEFINING ETHNIC FACTORS

Ethnicity is a concept that is defined in many different ways and to meet different objectives. *Ethnicity* may be used to refer to "racial" or "genetic" differences between groups of people. It may be used to refer to behavioral and social differences, such as language, religion, dress, mode of making a living, or kinship patterns. It may also refer to differences in climate and environmental and occupational exposures. The problems inherent in the use of this term are reflected in the current dispute in the United States about how to categorize its citizens into ethnic groups. Should all Spanish-speaking people be labeled "Hispanic"? Such a category would lump together people from as disparate cultural origins as Spain, Argentina, and Cuba. Should the term "African American" apply to a sixth-generation descendant of West Africans brought to North America over 200 years ago as well as to a recent immigrant from Nigeria or South Africa?

In the conduct of clinical trials, the concept of *ethnicity* may also be used in a variety of ways. Ethnic factors may relate to differences in metabolism between groups of people, differences in health care practices, and differences in response to pain, for example. Ethnic factors may have an impact not only on the safety and efficacy of a new drug or device, but also on the acceptance the drug or device is likely to have —how well it will fit in to a particular population's lifestyle, budget, and health care practices.

The International Conference on Harmonization (ICH) has proposed a distinction between what it describes as the physiologic aspects of ethnicity (*intrinsic* ethnic factors) and the cultural and environmental aspects of ethnicity (*extrinsic* ethnic factors). Under *intrinsic* factors, the ICH Guidelines include genetic makeup of a population, similarities in height, body mass, and composition, and organ dysfunction. Under *extrinsic* factors, the ICH Guidelines include environmental factors, such as climate, pollution, and altitude, and cultural or behavioral factors, such as diet, religion, the use of drugs (including tobacco) and alcohol, and indigenous health care systems.

This is a useful way to categorize ethnic factors. However, such a categorization may gloss over the importance of considering the interrelationships between intrinsic and extrinsic ethnic factors. As in many disease syndromes, such as alcoholism, diabetes, heart disease, and mental illnesses, extrinsic and intrinsic factors may have to work together to potentiate the emergence of the disease itself. Genetic predisposition may not inevitably lead to disease, but genetic predisposition plus environmental exposure may make disease development certain.

III. WHY IS IT IMPORTANT TO INCLUDE ETHNIC FACTORS IN DESIGNING SAFE AND EFFECTIVE TRIALS?

There are at least two ways that ethnic factors influence the conduct of clinical trials. One is from the point of view of study design and implementation. The other is from the point of view of the utility of the data collected. Ethnic factors must be considered in the design and implementation of the trial in order to produce useful data for purposes of registration. But ethnic factors must also be considered before the decision to conduct a trial in a foreign region is made. What are some of the reasons for the importance of considering ethnic factors in the conduct of clinical trials?

1. Unless these factors are considered, we will not have an adequate understanding of the relationship among culture, race, environment, and disease in terms of pathobiology, etiology, diagnosis, progression, treatment, outcomes, risk reduction, and drug delivery. These issues will play an important role in the methodological design of a clinical trial.
2. Ethnicity is often a marker of other kinds of distinguishing features among people around the world. Ethnicity may mark disparities in levels of health (malnutrition, hopelessness) or in the transfer of information about health and disease (access to care, types of care available, attitudes toward risk reduction). Ethnic factors may determine whether a drug or device will have market acceptance once it is approved. These are issues to be considered before registration decisions are made.

IV. IMPLICATIONS FOR THOSE CONDUCTING CLINICAL TRIALS IN FOREIGN REGIONS

Let us assume that the decision has been made to conduct a trial in a foreign region. This decision initiates special considerations, responsibilities, and opportunities for the researcher as well as for the region in which the study will be conducted.

1. *Involvement of local researchers and research staff, including social scientists such as anthropologists, sociologists, psychologists, and epidemiologists in the trial design as well as implementation.* Knowledgeable researchers can improve the way trials are conducted in many important ways, including issues of recruitment, retention, adherence, and loss to follow-up. Understanding regional attitudes toward medicine, for example, may improve the design of a trial of a new drug by recognizing potentially confounding variables, such as the use of traditional medicines, diet, and eating and drinking patterns. Providing incentives that are culturally appropriate may improve retention and adherence. Recognizing mobility patterns and anticipating them may result in smaller losses to follow-up.

2. *Designing pharmacokinetic, pharmacodynamic studies tailored to what is known about the population in which the trial will be conducted.* It is important to conduct clinical trials in foreign regions that take into account such ethnic factors as differences in metabolism, dose response, and genetic differences. These factors may affect the levels at which unacceptable toxicity is reached or efficacy rates. Differences in the levels of clearance enzymes, lipids, fibrinogen, and glucose, for example, may have important implications for the extrapolation of data from one region to another. Without accurate information, the human subjects of such trials could be put at unnecessary risk, and the resulting data may be misunderstood and useless.

3. *Designing dose–response and efficacy studies to include well-defined endpoints and disease definitions that are recognized by the local researchers and by the population members.* Ethnic factors may play a role in how *health* and *wellness* are defined. For example, the degree to which the risk of impotence is acceptable may affect the success or failure of a trial of an experimental prostate cancer drug.

4. *Evaluating clinical trials conducted in the population of drugs in the same class as the experimental drug, or similar devices, or of different dosages in a determination of whether or not a new trial must be conducted.* Before a decision to conduct a trial in a foreign region is made, researchers must determine if ethnic factors affect a new drug's safety and efficacy. It may be that trials of drugs in the same class as the experimental drug have shown no outcome or toxicity differences between two regions. It may be that similar dosages and therapeutic regimens have been similar among the two populations. It may not be necessary to

conduct the trial at all. On the other hand, it may be that no trials of similar drugs, dosages, or therapeutic regimens have been conducted, or that previous trials have produced dissimilar results. In this case, if the decision to seek registration in a foreign region has been made, it is important to conduct carefully controlled safety and efficacy trials among that region's population.

V. IMPLICATIONS FOR HUMAN SUBJECTS REVIEW

Clinical trials in foreign regions may pose special challenges for the ethical review boards charged with the protection of human subjects, especially as global development of new drugs and devices expands and matures. The diversity and breadth of clinical trials will increase, and the expertise required of ethical review boards will also expand. What are some of the steps that ethical review boards can take to prepare for these new challenges?

1. *Learn the International Conference on Harmonization's good clinical practices (GCP) with regard to Institutional Review Board (IRB) or Institutional Ethics Committee (IEC) guidance.* The guidance published in the June 10, 1998, Federal Register outlines the important conditions and procedures that are required to conduct ethical reviews of clinical trials.

2. *Require local ethical review of the clinical trial and request documentation of its findings.* It will become even more important than it is today to make sure that clinical trials receive ethical review in the regions in which they are conducted. This may result in a multiplicity of boards reviewing the same trial from quite different perspectives. It will become incumbent upon each board to make sure it maintains open communication with the researcher as well as with the other ethical review boards considering the same trial. Each review board may request documentation of the local region's determination with regard to a particular trial before making its own decision. Not only is requesting documentation helpful to the boards reviewing a study, it makes sure that the population of the region plays a part in determining whether or how the trial is conducted.

3. *Involve experts who can contribute information on ethnic factors in the consideration of a new clinical trial.* Traditionally, ethical review boards have been composed primarily of medical personnel who are experts in certain medical traditions. It will be important to expand the membership of these boards to incorporate less traditional members, including citizens of the region in which the trial will be conducted as well as experts in understanding and evaluating the impact of ethnic factors on the conduct of trials. The education involved in expanding membership is a two-way process. Conventional board members will need to learn more about ethnic factors, and the new members will have to learn about the ethics of research with human subjects.

VI. BARRIERS TO THE ETHICAL CONDUCT OF TRIALS IN FOREIGN POPULATIONS

1. *Time.* In an ever-more-competitive marketplace, speed is of the essence in bringing a drug to approval. Conducting trials in foreign populations will increase the time this process takes.
2. *Budget.* Conducting trials in foreign populations may be even more expensive than conducting trials in the sponsor's home country or in the country that is the primary marketing target.
3. *Communication.* Dealing with researchers and research subjects who are ethnically different from one another and from the sponsor may create problems leading to trials that are conducted poorly, are at variance with the protocol, and result in incomplete or inadequate data. Miscommunication may result not only from language differences but also from religious, political, or other "extrinsic" ethnic factors.
4. *Stigmatization.* Sponsors and researchers may need to work to overcome initial distrust and fear of "foreign" research and of researchers conducting trials in their regions. Education of the region's population and learning from the region may be a slow, and therefore costly, process.

Despite these barriers, the ethical conduct of clinical trials in all regions of the world is now an easier goal to accomplish than ever before. Increasing agreement among different regions on the ethical principles necessary to conduct high-quality trials while protecting human subjects will lead to heightened consistency and cooperation among ethical review boards. The basic ethical principles of respect, beneficence, and justice are grounded in the concept of regional review. Regional ethical review boards best evaluate ethnic factors. Cooperation between these boards and researchers will lead to safer, more efficient clinical drug trials and delivery of high-quality health care to humans everywhere.

Melody H. Lin
Helen McGough

Extra Variation Models

I. INTRODUCTION

Consider a study where experimental units are clusters and each elemental unit within the cluster is classified into two mutually exclusive categories. If we assume elemental units are independent, the binomial model is appropriate for analyzing the data. But clustering introduces a lack of independence, and as a consequence the data usually exhibit larger variances than the variance permitted by the binomial model. This phenomenon is known as *extra variation* or *overdispersion*; in practice, it seems to be the norm and not the exception (1). Extra variation is also observed for count data analyzed under the Poisson model and, on multinomial responses, when the residual variation obtained after fitting a multinomial model may be greater than what can be attributed to the sampling variation assumed by the model. The phenomenon of *underdispersion* is less common, but it can be induced, for example, by stratified sampling. It can also happen when there is competition for a positive response. As discussed in Stigler (2), examples of extra variation arising from a clustered population can be traced in the literature as far back as the year 1876. Early remarks related to extra variation under the Poisson model are found in Student (3).

Examples of binomial (multinomial) models with extra variation may arise (a) in teratological study of a genetic trait that is passed on with a certain probability to offspring of the same mother; (b) in reproductive toxicity experiments where chemicals are administered to litters and responses are measured on individual fetuses; (c) in experiments where individuals receiving treatments are allowed to interact with each other via support groups, as in stop-smoking programs or stop-drinking programs; (d) in household surveys where respondents within a household (or a neighborhood) may be strongly influenced by one or two individuals. Examples of Poisson models with extra variation can be found in the analysis of counts and rates of longitudinal studies. The phenomenon of extra variation is widely recognized by researchers and practitioners alike. A large number of papers in the literature discuss examples, with real data illustrating the extra variation phenomenon and its consequences on erroneous inferences due to inflated Type I error rates. Some relevant examples in the areas of teratology and toxicology are: Williams (4–6), Kupper and Haseman (7), Haseman and Kupper (8), Kupper et al. (9), Shirley and Hickling (10), Paul (11), Pack (12), Lefkopoulou et al. (13), Carr and Portier (14), Donner et al. (15), Morel and Neerchal (16). Other important examples of extra variation are found in Rosner (17) and Donner (18) in ophthalmology; Altham (19) and Cohen (20) in studies of hospitalized sibling pairs; Crowder (21) in the analysis of seed germination; Heckman and Willis (22) in connection with a panel study; Efron (23) in the analysis of toxoplasmosis in 34 cities of El Salvador; Moore (24) in the modeling of chromosome aberration in survivors of the atomic bomb of Hiroshima. This list of examples is by no means exhaustive. For an annotated bibliography of the extra variation literature, we refer the reader to Ashby et al. (25). Breslow (26) provides an interesting example of Poisson data with extra variation.

The phenomenon of binomial outcomes with extra variation is usually characterized in terms of the first two moments. Suppose that T is a binomial random variable representing the number of successes out of m trials. Then,

$$E(T) = m\pi \tag{1}$$

where π represents the probability of success of each Bernoulli trial. Assume now that the Bernoulli trials Y_1, $Y_2, \ldots, Y_m$ that encompass T are positively correlated, so the usual assumption of independence in the binomial model is violated. The case "positively correlated" corresponds to extra variation, which is the focus of this chapter. The occurrence of underdispersion is less common and corresponds to a negative correlation among the Bernoulli outcomes that comprise T. Thus, suppose that for two distinct Bernoulli outcomes the correlation is ρ^2, $0 < \rho^2 < 1$. Then it can be shown that

$$V(T) = m\pi(1 - \pi)\{1 + \rho^2(m - 1)\} \tag{2}$$

A random variable T with first two moments as in Eqs. (1) and (2) is said to have extra variation relative to the binomial model. Among sampling practitioners, the parameter ρ^2 represents the *intracluster correlation* and the factor $\{1 + \rho^2(m - 1)\}$ is the *design effect*.

There are several approaches for modeling data exhibiting extra variation. A widely used method is the one known as quasi-likelihood, where the binomial variance is inflated by a suitable constant. McCullagh and Nelder (1) give details of the method. We discuss this method in Sec. II. Other methods include the use of likelihood models, as we will explain in Sec. III. The generalized estimating equations methodology pro-

Table 1 pH Data

Number of mouth sites with pH ≤ 6.5	0	1	2	3	4
Number of subjects	4	4	6	2	1

posed by Liang and Zeger (27) and by Zeger and Liang (28), which is closely related to the quasi-likelihood method and generalized linear models, is reviewed in Sec. IV. More recently, generalized models that include random effects to model heterogeneity between individuals within a cluster have been proposed. Some references for this method are: McGilchrist (29), Drum and McCullagh (30), Breslow and Clayton (31), Karim and Zeger (32), Zeger and Karim (33), Goldstein (34), Schall (35), and Zeger et al. (36). Random-effect models are reviewed in Sec. V. We will present several examples related to preclinical and clinical experiments from biopharmaceutical studies throughout the remainder of this entry.

II. QUASI-LIKELIHOOD MODELS

In 1974 Wedderburn (37) proposed the use of quasi-likelihood models. A quasi-likelihood function is defined without the knowledge of the likelihood function but through its relation between the mean and the variance of each observation for which the quasi-likelihood function is being constructed. The structure of the first derivative of the quasi-likelihood function (without knowing the function from which this derivative is obtained) resembles the structure of the *score function*, i.e., the derivative of the logarithm of the likelihood function. Parameter estimates are obtained by solving the score function, most times iteratively by methods such as Gauss–Newton. For the one-parameter exponential family, the log likelihood is the same as the quasi-likelihood. Let $T_1, T_2, \ldots, T_n$ be independent random variables such that for each j, $j = 1, 2, \ldots, n$, $E(T_j) = \mu_j$ and $V(T_j) = \phi v(\mu_j)$, where ϕ may be unknown and $v(\mu_j)$ is a known function of μ_j. Usually, μ_j depends on an r-dimensional row vector of known covariates x_j, and an r-dimensional column vector of unknown parameters β^0, throughout a smooth function $g(.)$ so that $g(\mu_j) = x_j\beta^0$. This function is known as the *link function*. The *score function* is

$$U_n = \sum_{j=1}^{n} \frac{T_j - \mu_j}{\phi v(\mu_j)}$$

which under mild regularity conditions has the properties of the log-likelihood derivatives. In the case of either the binomial or Poisson distribution, the constant ϕ is 1. Extra variation corresponds to the case $\phi > 1$. Quasi-likelihoods are used to model extra variation.

We first consider the simple case where $T_1, T_2, \ldots, T_n$ are random variables with extra variation relative to the binomial model so that the mean and the variance of each random variable are as in Eqs. (1) and (2). In this instance, $\mu_j = m\pi$, $v(\mu_j) = m\pi(1 - \pi)$, and $\phi = \{1 + \rho^2(m - 1)\}$ for each j, $j = 1, 2, \ldots, n$. Here, the link function is the identity. Also, there are not covariates associated with the responses. To illustrate the estimation procedure for π and ρ^2, consider the data in Table 1, taken from a clinical trial. The study consisted of $n = 17$ subjects who brushed their teeth for six consecutive weeks with an experimental dental paste. The pH was measured at the end of the study at four sites of each subject's mouth. It is believed that if pH ≤ 6.5, the subject is at higher risk of oral disease. For each site of each subject, we created a binary outcome variable, taking the value of "1" for pH ≤ 6.5 and "0" otherwise. Table 1 provides the distribution of the T_j's (total number of mouth sites with pH ≤ 6.5). The quasi-likelihood estimate of π is

$$\hat{\pi} = \frac{\sum_{j=1}^{n} T_j}{nm} \tag{3}$$

A consistent estimate of ρ^2 is obtained by solving the equation

$$V(\hat{\pi}) = \frac{\pi(1 - \pi)\{1 + \rho^2(m - 1)\}}{nm} \tag{4}$$

with respect to ρ^2. Since a consistent estimator of $V(\hat{\pi})$ is

$$\frac{\sum_{j=1}^{n} (\hat{\pi}_j - \hat{\pi})^2}{n(n - 1)}$$

where $\hat{\pi}_j = T_j/m$, then a consistent estimator of ρ^2 is

$$\hat{\rho}^2 = \frac{1}{m - 1}\left\{\frac{m\sum_{j=1}^{n} (\hat{\pi}_j - \hat{\pi})^2}{(n - 1)\hat{\pi}(1 - \hat{\pi})} - 1\right\}$$

Note that the quantity

$$\frac{m\sum_{j=1}^{n} (\hat{\pi}_j - \hat{\pi})^2}{\hat{\pi}(1 - \hat{\pi})}$$

is the usual Pearson's goodness-of-fit statistic for the

binomial model and the factor $\{1 + \rho^2(m - 1)\}$ is the suitable inflation factor ϕ (the design effect). In this example, the estimates of π and ρ^2 turned out to be 0.3824 and 0.1571, respectively. The inflation factor was $\phi = 1.4713$. Given that the estimated standard error of $\hat{\pi}$ was 0.0715, an approximated 95% confidence interval for π is given by (0.24, 0.52). Since the disease rate (proportion of sites with pH ≤ 6.5) was approximately 71% at baseline, it is believed that the experimental toothpaste was efficacious in reducing the proportion of mouth sites at higher risk of oral disease.

If the T_j's are such that $E(T_j) = m_j\pi$ and $V(T_j) = m_j\pi(1 - \pi)\{1 + \rho^2(m_j - 1)\}$, a consistent estimator of π can be obtained as the ratio estimator $\hat{\pi} = \Sigma_{j=1}^{n} T_j/\Sigma_{j=1}^{n} m_j$. A consistent estimator of $V(\hat{\pi})$ is given by

$$\frac{n \sum_{j=1}^{n} (T_j - m_j\hat{\pi})^2}{\left(\sum_{j=1}^{n} m_j\right)^2 (n - 1)}$$

[see Cochran (38)]. Then a consistent estimator of ρ^2 is obtained by solving

$$\frac{n \sum_{j=1}^{n} (T_j - m_j\hat{\pi})^2}{\left(\sum_{j=1}^{n} m_j\right)^2 (n - 1)} = \frac{\hat{\pi}(1 - \hat{\pi}) \left\{\sum_{j=1}^{n} m_j + \rho^2 \sum_{j=1}^{n} m_j(m_j - 1)\right\}}{\left(\sum_{j=1}^{n} m_j\right)^2}$$

This method for estimating the intracluster correlation can be found in Kish (39). When the m_j's are fairly large, gains in efficiency are obtained when maximum likelihood estimation is used, as we will see in the next section. For a discussion on quasi-likelihoods models we refer the reader to Cox and Snell (40). In general, quasi-likelihood estimates of β^0, the parameters in the link function $g(.)$, are obtained by solving the quasi-likelihood equations

$$U_n = \sum_{j=1}^{n} \frac{T_j - g^{-1}(x_j\beta^0)}{\phi v\{g^{-1}(x_j\beta^0)\}} = 0$$

The use of quasi-likelihood models becomes more complex in the situation we are going to consider next.

Suppose that $T_1, T_2, \ldots, T_n$ are independent random variables with extra variation relative to the binomial model such that $E(T_j) = m_j\pi_j$, $V(T_j) = m_j\pi_j(1 - \pi_j)\{1 + \rho^2(m_j - 1)\}$, and $\ln\{\pi_j/(1 - \pi_j)\} = x_j\beta^0$. This model is fully described in Williams (5). Here, the m_j's and the π_j's may vary among clusters, with the cluster means linked to a set of covariates via the logistic function. Another complexity is that the quasi-likelihood equation

$$U_n = \sum_{j=1}^{n} \frac{T_j - m_j\pi_j}{m_j\pi_j(1 - \pi_j)\{1 + \rho^2(m_j - 1)\}} = 0$$

is nonlinear and needs to be solved iteratively using a procedure such as the Gauss–Newton method. In solving the quasi-likelihood equations we would like to obtain a consistent estimator of β^0 and its asymptotic covariance matrix in a way that the extra variation is properly taken into account. This ensures that the variance estimates are correct and that the Type I error rates are not inflated. Let us review two different approaches. We first consider the case where the estimated covariance matrix of the beta estimates are obtained under the assumption that $\rho^2 = 0$, and the resulting covariance matrix of the beta estimates is inflated by a single constant ϕ. A consistent estimator of this constant (see Morel and Koehler (41) for the calculation of the single inflation constant for binomial and/or multinomial responses) is given by

$$\hat{\phi} = \frac{1}{(n - r)} \sum_{j=1}^{n} \frac{(T_j - m_j\hat{\pi}_j)^2}{m_j\hat{\pi}_j(1 - \hat{\pi}_j)}$$

which is $(n - r)$ times the usual Pearson's goodness-of-fit statistic for the binomial model. Here, $\hat{\pi}_j$ represents the predicted probability under the "naïve" logistic regression, i.e., under the model where the clustering effect has been ignored. The procedure is also known as *iterated reweighted least squares* and is not very sensitive to the choice of weights (see Cox and Snell (42)). The use of either $m_j\pi_j(1 - \pi_j)$ or $m_j\pi_j(1 - \pi_j)\{1 + \rho^2(m_j - 1)\}$ in the score function yields essentially the same estimates of β^0, unless the extra variation factor is very large. The variance estimates are more susceptible to the choice of weights. The second approach follows the estimation methodology proposed by Williams (5): first get a consistent estimator of β^0, say, $\hat{\beta}$, under the assumption that $\rho^2 = 0$, and by holding $\hat{\beta}$ fixed then get a consistent estimator of ρ^2, say, $\hat{\rho}^2$. Then solve iteratively the quasi-likelihood equations for the beta coefficients using the "old" values $\hat{\beta}$ and $\hat{\rho}^2$ (holding $\hat{\rho}^2$ fixed), and get a new estimate of the beta coefficients, say, $\hat{\hat{\beta}}$. Hold $\hat{\hat{\beta}}$

Table 2 Ossification Data*

Group	Observations
Control	8/8, 9/9, 7/9, 0/5, 3/3, 5/8, 9/10, 5/8, 5/8, 1/6, 0/5, 8/8, 9/10, 5/5, 4/7, 9/10, 6/6, 3/5
Sham	8/9, 7/10, 10/10, 1/6, 6/6, 1/9, 8/9, 6/7, 5/5, 7/9, 2/5, 5/6, 2/8, 1/8, 0/2, 7/8, 5/7
PHT	1/9, 4/9, 3/7, 4/7, 0/7, 0/4, 1/8, 1/7, 2/7, 2/8, 1/7, 0/2, 3/10, 3/7, 2/7, 0/8, 0/8, 1/10, 1/1
TCPO	0/5, 7/10, 4/4, 8/11, 6/10, 6/9, 3/4, 2/8, 0/6, 0/9, 3/6, 2/9, 7/9, 1/10, 8/8, 6/9
PHT + TCPO	2/2, 0/7, 1/8, 7/8, 0/10, 0/4, 0/6, 0/7, 6/6, 1/6, 1/7

*Number of fetuses showing ossification/litter size.
PHT: phenytoin; TCPO: trichloropropene oxide.

fixed and get a new estimate of ρ^2, say, $\hat{\hat{\rho}}^2$. In this fashion, perform several steps until the estimators stabilize.

To illustrate these two procedures, consider the ossification data of Table 2, which have previously been analyzed by Morel and Neerchal (16). The data come from a two-way factorial design with 81 pregnant C57BL/6J mice. The purpose of the experiment was to investigate the synergistic effect of phenytoin and trichloropropene oxide. The presence or absence of ossification at the phalanges is considered a measure of the teratogenic effect of the substances under study. At some phalanges the ossification did not occur at all; in some others the ossification was almost complete. The middle third phalanges were selected because they provided about 50% chance of ossification in the control groups. For illustration purposes, we analyzed only the response of the left middle third phalanx. In the experiment each pregnant mouse was randomly allocated to an untreated control group (Control), a sham control group (Sham) receiving gavages of water, and three treated groups receiving daily, by gastric gavages, 60 mg/kg of phenytoin (PHT), 100 mg/kg of trichoropropene oxide (TCPO), and 60 mg/kg of phenytoin and 100 mg/kg of trichoropropene oxide (PHT + TCPO). For illustrative purposes, the two controls were combined and will be referred to as combined controls. The experiment thus can be seen as a 2 × 2 factorial, with TCPO and PHT as the two factors. The levels of TCPO are 0 mg/kg and 100 mg/kg, and the levels of PHT are 0 mg/kg and 60 mg/kg. The data on the left middle third phalanx are shown in Table 2.

Table 3 depicts the beta estimates and their standard errors after fitting the full factorial logistic regression model under no extra variation (fetuses within the litters are wrongly assumed to be independent) and under the two extra variation quasi-likelihood models previously discussed. The no-extra variation model and the single-inflation-factor quasi-likelihood model differ in that the variance-covariance matrix of the latter is 3.27 times the variance-covariance matrix of the former. The estimation procedure proposed by Williams (5) produced results very consistent with those obtained under the single-inflation-factor quasi-likelihood model. We now turn to extra variation Poisson models.

Table 3 Beta Estimates and Standard Errors of Ossification Data for Three Models

	Model					
	Naïve		Quasi-likelihood (single-inflation factor)		Quasi-likelihood (Williams (5) approach)	
Parameter	Estimate	Standard error	Estimate	Standard error	Estimate	Standard error
Intercept	0.83	0.14	0.83	0.25	0.73	0.24
TCPO	−0.85	0.22	−0.85	0.41	−0.75	0.41
PHT	−2.11	0.25	−2.11	0.45	−1.96	0.44
TCPO + PHT	1.05	0.41	1.05	0.74	1.04	0.72
Inflation factor			3.27		3.18	
ρ^2			0.37		0.35	

PHT: phenytoin; TCPO: trichloropropene oxide.

Suppose that $m_1, m_2, \ldots, m_n$ are random counts such that m_j for $j = 1, 2, \ldots, n$ is distributed as a Poisson random variable with mean λ_j. Furthermore, assume that λ_j is a random variable with density function $f(\lambda)$ such that $E_f(\lambda_j) = \mu$ and $V_f(\lambda_j) = \psi\mu^2$. We are assuming $\psi > 0$; otherwise the distribution of the λ_j's would be degenerate at the value of μ. Then the unconditional mean and variance of m_j are $E(m_j) = \mu$ and $V(m_j) = \mu(1 + \psi\mu)$, and the counts m_j's have extra variation relative to the Poisson model. When the prior distribution on the λ_j's is gamma, then the marginal distribution (unconditional distribution) of the m_j's is negative binomial (see Johnson et al. (43)). The assumption of common mean μ can be relaxed and, as in the binomial case, the variance of the Poisson model can be inflated with a constant ϕ ($\phi > 1$) in order to account for extra variation. Thus we can have the counts m_j's such that $E(m_j) = \mu_j$ and $V(m_j) = \phi\mu_j$, where μ_j depends on an r-dimensional row vector of known covariates x_j and an r-dimensional column vector of unknown parameters β^0, throughout a smooth function $g(.)$, $g(\mu_j) = x_j\beta^0$. The natural choice for $g(.)$ is the logarithm function, which is considered the "canonical link" if the m_j's were really distributed as Poisson ($\phi = 1$). Let us consider an example of Poisson counts with extra variation.

In a clinical study of osteoporotic patients at high risk of vertebral fractures, subjects were randomly assigned either to a group receiving active therapy (bisphosphonate) ($n = 28$) or to a placebo group ($n = 25$). The number of incident vertebral fractures (new and worsening) was measured during a period of time that lasted between approximately 14 and 39 months. The data are presented in Table 4.

Let m_{ij} and Y_{ij} represent the number of vertebral fractures and the years of observation of the jth subject in the ith treatment group. Conditioning on Y_{ij}, we fitted the extra variation Poisson model $E(m_{ij}) = \mu_{ij}$, $V(m_{ij}) = \phi\mu_{ij}$, $\phi > 1$, and link function $\ln(\mu_{ij}/Y_{ij}) = \alpha + \tau_i$, where α is an overall mean and τ_i is the effect of the ith treatment group. The estimates are obtained by using iterated reweighted least squares under the assumption the $\phi = 1$; then, as in the binomial case, the usual Pearson's chi square statistic is used to estimate ϕ. Then, the variance-covariance matrix of the parameter estimates is inflated with the estimate of ϕ. The results of this estimation procedure for the vertebral fracture data are presented in Table 5. The estimated fractures-per-year rates are $e^{-0.0449} = 0.96$ for the placebo group and $e^{-0.0449-0.9000} = 0.39$ for the bisphosphonate group, indicating a highly significant reduction (P-value = 0.0034) in the number of vertebral fractures in the bisphosphonate group, compared to placebo.

III. LIKELIHOOD MODELS

In this section we will discuss estimating extra variation models using proper likelihood functions. A likelihood with the first two moments as in Eqs. (1) and (2) can be obtained as the marginal distribution of the total number of successes, where its conditional distribution is binomial and the probability of success varies from cluster to cluster according to a certain distribution. Let $T_1, T_2, \ldots, T_n$ represent independent random variables such that given the jth random variable, $j = 1, 2, \ldots, n$, T_j is distributed binomial (P_j, m). Furthermore, assume that the P_j's are themselves random variables distributed according to a distribution in the interval (0, 1) with density function $f(P)$ (or probability function if the distribution of P is discrete) such that $E_f(P_j) = \pi$ and $V_f(P_j) = \rho^2\pi(1 - \pi)$. Then it can be shown that the unconditional mean and variance of

Table 4 Vertebral Fracture Data*

Group	Observations
Active (Bisphosphonate)	(1, 2.9897), (2, 2.9843), (0, 2.9897), (1, 2.9843), (2, 29459), (0, 3.0445), (1, 3.0582), (0, 2.9897), (1, 1.2567), (0, 2.9678), (3, 2.9596), (1, 2.9843), (0, 2.9925), (1, 1.4346), (1, 3.0116), (0, 2.9733), (2, 3.0637), (1, 2.9706), (0, 1.1882), (0, 2.9733), (1, 1.2676), (2, 2.9541), (0, 2.3874), (0, 1.2868), (0, 1.2813), (2, 2.7625), (2, 2.4504), (4, 2.8747)
Placebo	(0, 1.3771), (11, 3.0418), (0, 2.4312), (3, 3.1650), (1, 3.0034), (2, 2.9843), (0, 2.9651), (2, 1.3087), (5, 3.0363), (8, 3.0691), (2, 2.9706), (2, 3.2553), (4, 2.9624), (0, 3.2060), (1, 2.9596), (7, 3.0144), (0, 2.9569), (2, 2.9843), (0, 1.1937), (3, 3.0062), (1, 2.9870), (4, 3.0089), (1, 1.2621), (1, 2.9788), (5, 2.8556)

*Number of incident vertebral fractures, person-years of observation.

Table 5 Estimates, Standard Errors, Chi Squares, and *P*-Values for Poisson Extra Variation Model of Vertebral Fracture Data

Parameter	Estimate	Standard error	Chi square	*P*-value
Intercept	−0.0449	0.1686	0.0709	0.7901
Bisphosphonate	−0.9000	0.3072	8.5801	0.0034
Scale (ϕ)	1.3591			

each T_j are as in Eqs. (1) and (2). Under a prior $f(P)$, for any $t = 0, 1, 2, \ldots, m$,

$$\Pr(T = t) = E_f\{\Pr(T = t)$$
$$= \int_0^1 \binom{m}{t} P^t(1 - P)^{m-t} f(P)\, dP \tag{5}$$

A distribution of the form (5) is said to be a mixture, with the prior $f(P)$ being the mixing distribution. If $f(P)$ is continuous, the unconditional distribution of T is an infinite mixture. If the prior $f(P)$ is discrete and finite, the integral symbol in (5) changes to a finite summation, and the resulting distribution is a finite mixture. Two examples of mixture distributions used to model extra variation are the beta-binomial distribution originally used by Skellam (44) and the finite-mixture distribution proposed by Morel and Nagaraj (45). We first consider the beta-binomial distribution.

Let $f(P)$ be the density function of a beta distribution with parameters $a = C\pi$ and $b = C(1 - \pi)$, where $C = (1 - \rho^2)/\rho^2$. Note the $E_f(P) = a/(a + b) = \pi$ and $V_f(P) = ab/(a + b + 1)(a + b)^2 = \rho^2\pi(1 - \pi)$. Thus, as noted earlier, the unconditional distribution of T exhibits extra variation. It can be shown that, for any t, $t = 0, 1, 2, \ldots, m$,

$$\Pr(T = t) = \binom{m}{t} \frac{\prod_{k=1}^{t} (C\pi + k - 1) \prod_{k=1}^{m-t} \{C(1 - \pi) + k - 1\}}{\prod_{k=1}^{m} (C + k - 1)} \tag{6}$$

where $C = (1 - \rho^2)/\rho^2$ and a product from $k = 1$ to $k = 0$ is defined as 1. The probability function of Eq. (6) defines the beta-binomial distribution with parameters π, ρ, and m and is denoted as beta-binomial (π, ρ; m).

Morel and Nagaraj (45) proposed a new distribution for modeling extra variation where the unconditional distribution of the total number of successes T can be obtained from placing on P a two-point prior distribution on the points $(1 - \rho)\pi + \rho$ and $(1 - \rho)\pi$ with probabilities π and $(1 - \pi)$, respectively. In this case, the unconditional probability function of T turns out to be, for $t = 0, 1, \ldots, m$,

$$\Pr(T = t) = \pi\Pr(X_1 = t) + (1 - \pi)\Pr(X_2 = t) \tag{7}$$

where $X_1 \sim$ binomial($(1 - \rho)\pi + \rho$; m) and $X_2 \sim$ binomial($(1 - \rho)\pi$; m). Note that this is a mixture of two binomial distributions. If T has the distribution given by Eq. (7), it is denoted as $T \sim$ finite-mixture(π, ρ; m). The finite-mixture distribution results from an effort to model meaningfully the physical mechanism behind the extra variation. It can be derived in the following way.

Let $Y, Y_1^0, \ldots, Y_m^0$ be independent and identically distributed Bernoulli(π) random variables. For each i, $i = 1, \ldots, m$, define Y_i as Y with probability ρ, or as Y_i^0 with probability $(1 - \rho)$, $0 \le \rho \le 1$. Each Y_i can be represented as $Y_i = YI(U_i \le \rho) + Y_i^0 I(U_i > \rho)$, where the U_i's are independent uniform (0, 1) random variables and $I(.)$ denotes an indicator function. In this model for clumped data, the response of any member of the cluster is the same as Y, with probability ρ. Also, for $i \ne j$ it can be shown that $\Pr(Y_i = 1|Y_j = 1) = \rho^2$. That is, the conditional probability that a member of the cluster responds positively, given that one of the others in the cluster has already responded positively, is ρ^2. If we let T be the sum of the Y_i's, then $T = Y \sum_{i=1}^m I(U_i \le \rho) + \sum_{i=1}^m Y_i^0(I(U_i > \rho)$. This leads to the following representation of T:

$$T = (YN) + (X|N) \tag{8}$$

where $N \sim$ binomial(ρ; m), $Y \sim$ Bernoulli(π), N and Y are independent, and $(X|N) \sim$ binomial(π; $m - N$) if $N < m$. It can be shown [see Morel (46)] that the probability function of T in Eq. (8) is given by Eq. (7). Under representation (8), the outcome given by Y is duplicated a random number of times N, $N = 0, 1, \ldots, m$. This is the number of units in the cluster exhibiting the same behavior. The remaining $m - N$ units within the cluster provide independent Bernoulli responses. Contribution of these $m - N$ units to the total count T is represented by $(X|N)$. Figure 1 illustrates the representation of a cluster of size $m = 7$. The cluster is thought of as having two subgroups. One is of size 4 (N; in general $0 \le N \le 7$), and it consists of those elemental units that greatly influence each other

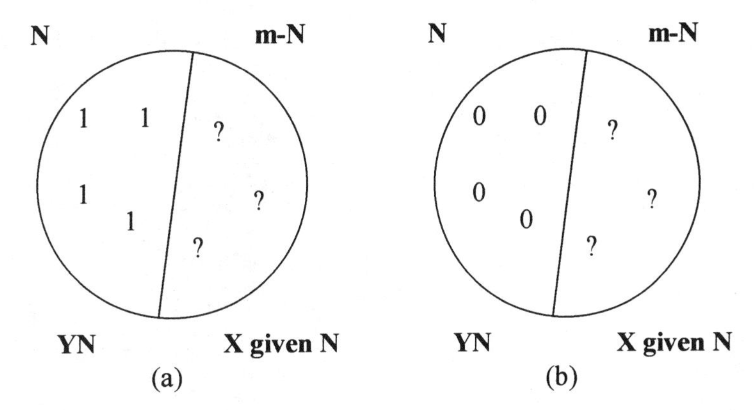

Fig. 1 Two possible outcomes for a cluster of size $m = 7$ under the finite-mixture distribution. Some elemental units provide identical responses, either "1," as in (a), or "0," as in (b). The "?" symbol represents elemental units responding independently, with either a "0" or a "1."

and provide identical responses. The three elemental units forming the second subgroup respond independent of each other and of the first subgroup. Their responses have been shown by a "?" to indicate that they may be either a "0" or a "1." Clearly, the binomial count T in Eq. (8) exhibits extra variation, because the individual respondents (either by consulting each other before responding or by a natural biological inheritance) provide more similar responses than one would expect from independent Bernoulli trials. As far as the authors know, the finite mixture is the only distribution that provides a clear interpretation of the underlying cluster mechanism. To illustrate the fitting of the beta-binomial and finite-mixture distributions we now present an example on chromosome association.

Skellam (44) analyzed data on the secondary association of chromosomes in *Brassica*, a group of Old World temperate zone herbs of the mustard family with beaked cylindrical pods (similar to cabbage). This is the first example in the literature where the beta-binomial distribution was formally used to model extra variation. At meiosis, the special process of cell division, there are three pairs of bivalents. A *bivalent* is a structure consisting of two paired homologous chromosomes. The paired chromosomes in each bivalent may or may not show association. For a nucleus i, $i = 1, 2, \ldots, n$, let T_i represent the total number of pairs of bivalents showing association, $T_i = 0, 1, 2, 3$. If the probability of association is constant for all nuclei and the same for all pairs of bivalents, the binomial distribution would be appropriate for this problem. As we will see next, extra variation is present in this example, and both the beta-binomial and finite-mixture distributions provide reasonable fits to this problem.

Table 6 shows the distribution of the number of chromosome associations for 337 nuclei. We fit three distributions to these data: the binomial, the beta binomial, and the finite mixture. Maximum-likelihood estimates and their asymptotic covariance matrix of the two extra-variation distributions were obtained by using a "modified" Gauss–Newton method. The results of fitting these distributions are given in Table 7.

As we can see in this example, the two extra variation distributions produce almost identical results. Their estimates of π and the one obtained by fitting the binomial distribution differ only in the fourth decimal place. The variance estimates under the beta-binomial and the finite-mixture distributions are 1.17 greater than that obtained by using the binomial distribution. This value is an estimate of the multiplier $\{1 + \rho^2(m - 1)\}$ that appears in Eq. (2). To test for the presence of extra variation, we need to test the hypothesis H_0: $\rho = 0$ versus the alternative hypothesis H_1: $\rho > 0$. Properties of the likelihood ratio test cannot be obtained by the usual asymptotic theory, since the parameter being tested is on the boundary of the parameter space, $0 < \rho < 1$. Self and Liang (47) have shown that in this case (see case 5 in their paper), the asymp-

Table 6 Chromosome Data

Number of associations	0	1	2	3
Frequency	32	103	122	80

Table 7 Results of Fitting Binomial, Beta-Binomial, and Finite-Mixture Distributions to Chromosome Data

	Distribution					
	Binomial		Beta binomial		Finite mixture	
Parameter	Estimate	Standard error	Estimate	Standard error	Estimate	Standard error
π	0.5806	0.0155	0.5807	0.0168	0.5805	0.0168
ρ			0.2944	0.0575	0.2930	0.0578
Log-likelihood function	−440.39		−436.81		−436.88	

totic distribution of the likelihood ratio test is a 50:50 mixture of a point mass at zero and at a chi square distribution with 1 degree of freedom. The likelihood ratio test statistic is 7.16 under the beta-binomial distribution and 7.02 under the finite-mixture distribution with P-values 0.0037 and 0.0040, respectively. These statistics are almost the same, and they strongly support the hypothesis of extra variation.

To conclude this example, we compute the usual Pearson's goodness-of-fit chi square test to see if either distribution is appropriate for the data. Table 8 shows the observed and expected frequencies of the number of associations under the three distributions considered in this example. The binomial distribtuion does not fit the data well, since $\chi^2 = 8.11$, with an associated P-value of 0.017. The beta-binomial and the finite-mixture distributions fit the distribution of the number of associations fairly well, with P-values of the chi square goodness-of-fit tests being 0.383 and 0.348, respectively.

Neerchal and Morel (48) investigated the "relative asymptotic efficiency" of the quasi-likelihood estimator given in Eq. (3) with respect to the maximum-likelihood estimate of π of the beta-binomial and finite-mixture distributions. The relative efficiency of the quasi-likelihood estimator relative to the maximum-likelihood estimator of π under the beta-binomial distribution is

$$\frac{\dfrac{\pi(1-\pi)\{1+\rho^2(m-1)\}}{nm}}{V(\hat{\pi}_{bb})}$$

where $\hat{\pi}_{bb}$ is the maximum-likelihood estimator of π under the beta-binomial distribution and $V(\hat{\pi}_{bb})$ is the asymptotic covariance matrix when the cluster size m is large. The relative asymptotic efficiency with respect to the finite-mixture distribution is defined in a similar way. Table 9 shows the asymptotic relative efficiencies for the combinations $\pi = 0.1, 0.3, 0.5$; $\rho^2 = 0.09, 0.25, 0.49$; and $m = 10, 40$. The results from Table 9 indicate that there is little gain in efficiency in using the beta-binomial distribution over the quasi-likelihood model. The two distributions for modeling extra variation have essentially the same efficiency of the quasi-likelihood estimator when $\rho^2 = 0.09$ and $m = 10$. But as ρ^2 increases and/or the cluster size m gets larger, the gain in efficiency of the finite-mixture distribution over the quasi-likelihood model becomes greater.

Table 8 Chi Square Goodness-of-Fit Tests for Chromosome Data

		Expected frequencies		
Number of associations	Observed frequency	Binomial	Beta binomial	Finite mixture
0	32	24.86	33.97	34.08
1	103	103.24	97.16	96.82
2	122	142.94	127.67	128.24
3	80	65.96	78.20	77.85
χ^2		8.11	0.76	0.88
Degrees of freedom		2	1	1
P-value		0.017	0.383	0.348

Table 9 Asymptotic Relative Efficiency of Quasi-Likelihood Estimator Relative to Beta-Binomial and Finite-Mixture Distributions

ρ^2	m	Distribution	$\pi = 0.1$	$\pi = 0.3$	$\pi = 0.5$
0.09	10	Beta binomial	1.00	1.00	1.00
		Finite mixture	1.03	1.01	1.01
	40	Beta binomial	1.00	1.00	1.00
		Finite mixture	2.33	2.03	1.95
0.25	10	Beta binomial	1.01	1.01	1.01
		Finite mixture	1.25	1.19	1.18
	40	Beta binomial	1.03	1.03	1.03
		Finite mixture	3.85	3.82	3.81
0.49	10	Beta binomial	1.05	1.05	1.05
		Finite mixture	1.46	1.43	1.42
	40	Beta binomial	1.10	1.11	1.12
		Finite mixture	4.05	4.05	4.05

We now revisit the teratology example of Table 2 and fit the beta-binomial and finite-mixture models, assuming a common intralitter correlation ρ^2 and the logistic link with a full factorial design. The results of the fittings are given in Table 10. As we can see from Table 10, the results of the beta-binomial and finite-mixture models are very consistent with each other. The negative coefficients indicate an increased risk of no ossification. Thus, the toxic effect of the chemicals is to impede ossification. If $\hat{\beta}_0$, $\hat{\beta}_1$, $\hat{\beta}_2$, $\hat{\beta}_3$ represent, respectively, the beta estimates for intercept, TCPO, PHT, and TCPO + PHT, then an approximate 95% confidence interval for the odds ratio of PHT versus the combined controls when TCPO = 0 mg/kg is given by $\exp^{(\hat{\beta}_2 \pm 1.96\sqrt{\hat{v}(\hat{\beta}_2)})}$, where $\hat{v}(\hat{\beta}_2)$ represents the estimated asymptotic variance of $\hat{\beta}_2$. Similarly, an approximate 95% confidence interval for the odds ratio of PHT versus the combined controls when TCPO = 100 mg/kg is given by $\exp^{\{(\hat{\beta}_2+\hat{\beta}_3) \pm 1.96\sqrt{\hat{v}(\hat{\beta}_2+\hat{\beta}_3)}\}}$. Table 11 shows the approximate 95% confidence intervals for the two levels of TCPO based on the beta estimates of the binomial, single-inflation-factor, and Williams (5) approach of Table 3 and the beta estimates based on the beta-binomial and finite-mixture models of Table 10. The confidence intervals under the binomial model are much shorter than for all the extra-variation models. This suggests that the level of confidence obtained under the binomial model is below the nominal one. The remaining models, which incorporate the litter effect, suggest that in the presence of TCPO, the chances to impede ossification may be the same for both PHT and the combined control.

IV. GENERALIZED ESTIMATING EQUATIONS MODELS

"Generalized estimating equations," better known as GEE or GEE's, is a methodology adapted by Liang and Zeger (27) and Zeger and Liang (28) for the analysis of longitudinal data when regression is the main focus. The methodology is related to quasi-likelihood models (Wedderburn (27) and McCullagh (49)) and generalized linear models (McCullagh and Nelder (50)). Although the GEE methodology was originally proposed for longitudinal data analysis, it has been applied to almost any type of clustered data. Since we are concerned with extra variation, we restrict ourselves to binary responses or counts. For each subject j, $j = 1, 2, \ldots, n$, let $Y_j = (y_{j1}, y_{j2}, \ldots, y_{jm_j})'$ represent an m_j-dimensional column vector of responses, and let $X_j = (x_{j1}, x_{j2}, \ldots, x_{jm_j})'$ denote an associated $m_j \times k$ matrix of observed covariates, where x_{jl}, $l = 1, 2, \ldots, m_j$ represents a k-dimensional row vector. Let μ_{jl} and $v(\mu_{jl})$ represent, respectively, the mean and the "natural variance" of y_{jl}, the latter being the variance of the y_{jl}'s when they are independent and belong to the exponential family. If the y_{jl}'s were either binomial or Poisson distributed, their corresponding natural variances would be $\pi_{jl}(1 - \pi_{jl})$ or λ_{jl}, respectively. It is further assumed that the μ_{jl}'s are related to the x_{jl}'s throughout a link function $g(.)$ such that $g(\mu_{jl}) = x_{jl}\beta^0$, where β^0 is a k-dimensional column vector of unknown parameters. The canonical links for the binomial and Poisson distributions are the logistic and logarithmic functions, respectively. Other links for the binomial distribution are the "probit" and the "complementary log-log" functions. The identity function may be used as a link function for the Poisson case.

Let A_j represent the $m_j \times m_j$ diagonal matrix of natural variances of the vector Y_j, $A_j = \text{diag}\{v(\mu_{j1}), v(\mu_{j2}), \ldots, v(\mu_{jm_j})\}$. In order to apply the GEE methodology, one needs first to define the correlation structure of $Y_j = (y_{j1}, y_{j2}, \ldots, y_{jm_j})'$ through an $m_j \times m_j$ "working" correlation matrix $R_j(\alpha)$ such that

$$V_j = A_j^{1/2} R_j(\alpha) A_j^{1/2} \phi \tag{9}$$

would be equal to the variance-covariance matrix of Y_j if $R_j(\alpha)$ were the true correlation matrix for the Y_j's. Then define the estimating equations as

$$\sum_{j=1}^{n} D_j' V_j^{-1}(Y_j - \mu_j) = 0 \tag{10}$$

where $\mu_j = (\mu_{jl}, \mu_{jl}, \ldots, \mu_{jm_j})'$, D_j' represents the $k \times m_j$ matrix of first derivatives of μ_j with respect to β^0 and V_j is defined as in Eq. (9).

In order to solve the estimating Eq. (10) for β^0, consistent estimators of ϕ and α (say, $\hat{\phi}$ and $\hat{\alpha}$) are needed. These estimators are functions of the data and of β^0. So they can be acquired by obtaining $\hat{\beta}$, the solution of estimating Eq. (10) assuming that $R_j(\alpha)$ is an identity matrix and that ϕ is equal to 1. Then $\hat{\phi}$ and $\hat{\alpha}$ are plugged into the estimating Eq. (10), which is then solved to obtain an improved version of $\hat{\beta}$. The improved version of $\hat{\beta}$ incorporates the hypothesized covariance matrix structure given in Eq. (9) and can be used to obtain new estimators $\hat{\phi}$ and $\hat{\alpha}$. The procedure continues until no changes are observed on $\hat{\beta}$, $\hat{\phi}$, and $\hat{\alpha}$. Liang and Zeger (27) have shown the consistency of $\hat{\beta}$ under mild regularity conditions. They also showed that a consistent estimator of the variance-covariance matrix of $\hat{\beta}$ is

$$\hat{A} = \left\{ \sum_{j=1}^{n} D_j'(\hat{\beta}) V_j^{-1}(\hat{\beta}, \hat{\alpha}, \hat{\phi}) D_j(\hat{\beta}) \right\}^{-1} \cdot G \left\{ \sum_{j=1}^{n} D_j'(\hat{\beta}) V_j^{-1}(\hat{\beta}, \hat{\alpha}, \hat{\phi}) D_j(\hat{\beta}) \right\}^{-1} \tag{11}$$

where $G = \Sigma_{j=1}^{n} (d_j - \bar{d})(d_j - \bar{d})'$, $d_j = D_j'(\hat{\beta}) V_j^{-1}(\hat{\beta}, \hat{\alpha}, \hat{\phi})\{Y_j - \mu_j(\hat{\beta})\}$, and $\bar{d} = \Sigma_{j=1}^{n} d_j/n$. To take into account the estimation procedure, one should multiply the matrix G by the factor

$$\frac{(n^* - 1)}{(n^* - k)} \frac{n}{(n - 1)}$$

(see Morel (51)), where $n^* = \Sigma_{j=1}^{n} m_j$. If each cluster contains exactly one elemental unit, this factor reduces to $n/(n - k)$, which is the degrees of freedom correction applied to the residual mean square for ordinary least squares in which k parameters are estimated. The quantity $(n^* - 1)/(n^* - k)$ reduces the small sample bias associated with using the estimated link function to calculate residuals.

For large samples, the variance-covariance $\hat{A}$ in Eq. (11) provides correct Type I error rates for arbitrary choice of the working correlation matrix $R_j(\alpha)$. This is one of the most interesting characteristics of GEE: the inferences based on $\hat{\beta}$ are correct as far as one uses the "robust" (also known as "empirical") variance-covariance matrix given in Eq. (11). This is under the assumption that the mean of the model is correct. Liang and Zeger (27) provide results of a simulation study where they show gains in efficiency when the working correlation matrix coincides with the true and unknown

Table 10 Beta Estimates and Standard Errors of Ossification Data for Full Factorial Logistic Regression Model Under Beta-Binomial and Finite-Mixture Models

	Model			
	Beta binomial		Finite mixture	
Parameter	Estimate	Standard error	Estimate	Standard error
Intercept	0.70	0.23	0.64	0.22
TCPO	−0.78	0.40	−0.95	0.37
PHT	−1.69	0.41	−1.53	0.38
TCPO + PHT	0.68	0.68	0.62	0.65
ρ^2	0.34	0.05	0.34	0.04
Log-likelihood	−153.29		−152.53	

PHT: phenytoin; TCPO: trichloropropene oxide.

Table 11 Approximate 95% Confidence Intervals for Odds Ratio of PHT under Binomial, Single-Inflation-Factor, Williams (5), Beta-Binomial, and Finite-Mixture Models

Model	TCPO = 0 mg/kg	TCPO = 100 mg/kg
Binomial	0.07, 0.20	0.18, 0.65
Single inflation factor	0.05, 0.29	0.11, 1.09
Williams (5)	0.03, 0.33	0.13, 1.20
Beta binomial	0.08, 0.41	0.12, 1.07
Finite mixture	0.10, 0.46	0.14, 1.13

PHT: phenytoin; TCPO: trichloropropene oxide.

one. Some examples of working correlation matrices are: independence, exchangeable, m-dependent, first-order autoregressive, and unspecified. For each of these cases, Liang and Zeger (27) provide indications on how to estimate ϕ and α.

Let us now consider the ossification data of Table 2. These data were used to illustrate the fitting of two quasi-likelihood models in Sec. II, and the fitting of two likelihood models in Sec. III. We fit these data using the GEE methodology with two distinct working correlation matrices: independence and exchangeable. The latter covariance matrix structure means that the correlation of two elemental units within the cluster is always the same. The GEE results using the two working correlation matrices are very consistent with each other. They are also consistent with the results of the two quasi-likelihood models of Table 3 and with the two likelihood models of Table 12.

V. RANDOM-EFFECT MODELS

The generalized linear models (GLM) framework provided by McCullagh and Nedler (1) unifies the analysis of continuous and discrete response variables when the responses can be assumed to be independent. In GLM, moments of the response y_{jl}, from the lth elemental unit from the jth cluster, $j = 1, 2, \ldots, n$, $l = 1, 2, \ldots, m_j$ are modeled such that $E(y_{jl}) = \mu_{jl}$ and $V(y_{jl}) = \phi v(\mu_{jl})$, where μ_{jl} depends on a k-dimensional row vector of known covariates x_{jl} and a k-dimensional column vector of unknown paramters β^0 through a link function $g(.)$ such that $g(\mu_{jl}) = x_{jl}\beta^0$. As before, $v(\mu_{jl})$ is a known function that gives the variance of the observation in terms of the mean. As described in Sec. II, GLM formulation is closely related to the quasi-likelihood method of obtaining the estimates. However, the quasi-likelihood methods does not facilitate the simultaneous estimation of the mean (π) and the dispersion (ρ) parameters in the extra variation models. The GLM framework has been extended by Zeger et al. (36) to include random effects. The resulting models, generalized linear mixed models (GLMM), can accommodate the analysis of longitudinal data, where subjects are monitored over a period of time, and the analysis of observations obtained on different subjects belonging to some kind of cluster, like those usually obtained

Table 12 Beta Estimates and Standard Errors of Ossification Data for Full Factorial Logistic Regression Model Using GEE with Independence Working Correlation Matrix and with Exchangeable Working Correlation Matrix

	Model			
	GEE with independence working correlation matrix		GEE with exchangeable working correlation matrix	
Parameter	Estimate	Standard error	Estimate	Standard error
Intercept	0.83	0.25	0.72	0.26
TCPO	−0.85	0.41	−0.74	0.42
PHT	−2.11	0.35	−1.95	0.36
TCPO + PHT	1.05	0.77	1.03	0.77

PHT: phenytoin; TCPO: trichloropropene oxide.

in cluster sampling. As indicated earlier, overdispersed binomial (or multinomial) is usually a result of natural association among observations from individuals belonging to the same cluster. Thus, we include a brief review of GLMM in this entry.

Extension of GLM to obtain GLMM is straightforward and is analogous to the extension of linear models to linear mixed models. A random effect is introduced corresponding to each subject, which describes the deviation of the subject's response from the group average. Following the previous notations, under GLMM the response y_{jl} for the lth elemental unit in the jth cluster is assumed to satisfy

$$g(\mu_{jl}) = x_{jl}\beta^0 + z_{jl}b_j \tag{12}$$

and

$$V(y_{jl}|b_j) = \phi v(\mu_{jl}) \tag{13}$$

where $\mu_{jl} = E(y_{jl}|b_j)$ is an independent observation from a mixture observation. Note that conditional on b_j, Eqs. (12) and (13) describe a GLM with link function $g(.)$ and variance function $v(.)$.

Unconditional moments may be obtained by taking expectations of the conditional moments given by Eqs. (12) and (13) with respect to the distribution b_j's. Thus under GLMM,

$$\begin{aligned}\mu_{jl} &= E(y_{jl}) = E\{E(\mu_{jl}|b_j)\}\\ &= \int g^{-1}(x_{jl}\beta^0 + z_{jl}b_j)\partial F(b_j)\end{aligned} \tag{14}$$

and

$$\begin{aligned}V(y_{jl}) &= E\{V(y_{jl}|b_j)\} + V\{E(y_{jl}|b_j)\}\\ &= \int \phi v\{g^{-1}(x_{jl}\beta^0 + z_{jl}b_j)\}\partial F(b_j)\\ &\quad + z_{jl}z_{jl}^{t}\sigma_b^2\end{aligned} \tag{15}$$

Models described by Eqs. (14) and (15) are referred to as *population average* (PA) models by Zeger et al. (36). They refer to the models given by Eqs. (12) and (13) as *subject specific* (SS) models. Such models are most suitable when response for an individual (rather than description of the population as a whole) is the focus of the study, as in studies of growth curves. Population-average models are most useful in epidemiological studies. Here the focus is on studying the difference between the average response of groups. A salient example given by Zeger et al. (36) is as follows: Suppose x_{jl} indicates whether or not the subject j at time l smokes and y_{jl} is the presence or absence of a respiratory infection. Then the PA model describes the difference in infection rates between smokers and nonsmokers, whereas the SS model estimates the expected change in the probability of infection of a subject given a change in smoking behavior.

Estimation of random GLMM can be done, in principle, by specifying a distribution for the random effects b_j's, obtaining (by integration) the marginal (unconditional) likelihood of the data and then proceeding with likelihood estimation. Even in the case of linear mixed models, closed form for likelihood estimates may not be available. Harville (52) implements this approach to the linear mixed models, assuming a normal distribution for the random effects and a normal distribution for the data conditional on the random effects. Anderson and Aitkin (53) and Im and Gianola (54) use maximum-likelihood estimation in logistic and probit models, where random effects are assumed to follow a normal distribution and the conditional distribution is binomial. In general, being able to obtain likelihood estimates for GLMM depends on whether or not the marginal (unconditional) distribution of the y_{jl}'s can be obtained by integrating with respect to a justifiable distribution of the random effects. Zeger and Karim (33) use the connection between GLMM and the Bayesian framework to give a method of estimation that overcomes the computational limitations using a Gibbs sampling approach. One way of avoiding numerical computation of these integrals is to approximate the integrals by simple expansions or a method such as Laplace's method of approximating integrals. This approach has been explored by Stiratelli et al. (55) and Breslow and Clayton (31). Zeger et al. (36) use the generalized estimating equations approach to estimate GLMM. Schall (35) gives an algorithm for estimation in GLMM that is easy to program. This algorithm also accommodates the specification of just the first two moments of the distribution of the random effects. Drum and McCullagh (30) adapt the residual maximum-likelihood (REML), a method that has been successfully used in linear mixed models, to GLMM with logistic links. Their Table 1 gives a succinct summary of methods used in estimating GLMM.

We conclude the discussion of GLMM with a few general remarks. GLMM methodology provides useful ways of modeling clustered responses, which occur routinely in biopharmaceutical studies. Responses from subjects who belong to the same neighborhood and responses collected from the same subject over a period of time in a longitudinal research are examples of such data. As seen in Sec. III, the extra variation models presented in the entry, namely, the beta-binomial models and finite-mixture models, can, in fact, be considered as random-effect models. However, it should be noted that the models described in Secs. II–V focus on situations where responses from elemental units within the cluster are not available.

VI. CONCLUSIONS

In this entry, we presented a description of the phenomenon known as *extra variation* or *overdispersion*, along with a review of the literature. There are four different sections, each section describing a different approach for modeling extra variation. These sections are: quasi-likelihood models, likelihood models, generalized estimating equations models, and random-effect models. Several examples discussing a variety of real data sets are cited. Quasi-likelihood models are illustrated with data from two clinical studies (pH data of Table 1 and vertebral fracture of Table 4) and one preclinical experiment (ossification data of Table 2). In the section on likelihood models, two distributions for modeling extra variation are presented, namely, the beta-binomial model and the finite-mixture model. An insightful representation of the finite-mixture model is provided (see Figure 1). An example on chromosome association is thoroughly discussed (see Table 6), and the example of ossification data of Table 2 is revisited. The sections on generalized estimating equation models and random-effect models are clearly presented. The ossification data of Table 2 is used one more time to illustrate the use of the generalized estimating equations approach with two distinct working correlation matrices. An epidemiological example is described in the section on random-effect models to clarify the difference between population-average models and subject-specific models.

ACKNOWLEDGMENTS

The authors wish to thank Prof. Raymond Myers, Virginia Polytechnic Institute, Dr. Mohammad Hoseyni, Procter & Gamble Company, and Prof. Bimal K. Sinha, University of Maryland Baltimore County, for reading the manuscript and providing useful suggestions.

REFERENCES

1. P McCullagh, JA Nelder. Generalized Linear Models. 2nd ed. London: Chapman and Hall, 1989, pp 124–128.
2. SM Stigler. The History of Statistics. The Measurements of Uncertainty Before 1990. Cambridge, MA: Belkap Press, 1986, pp 229–238.
3. "Student." Biometrika 12:211–215, 1919.
4. DA Williams. Biometrics 31:949–952, 1975.
5. DA Williams. App Stat 31:144–148, 1982.
6. DA Williams. Biometrics 44:305–308, 1988.
7. LL Kupper, JK Haseman. Biometrics 34:69–76, 1978.
8. JK Haseman, LL Kupper. Biometrics 35:281–293, 1979.
9. LL Kupper, C Portier, MD Hogan, E Yamamoto. Biometrics 42:85–98, 1986.
10. EAC Shirley, C Hickling. Biometrics 37:819–829, 1981.
11. SR Paul. Biometrics 38:361–370, 1986.
12. SE Pack. Biometrics 42:967–972, 1986.
13. M Lefkopoulou, D Moore, L Ryan. J Am Statis Assoc 84:810–815, 1989.
14. G Carr, CJ Portier. Biometrics 49:779–791, 1993.
15. A Donner, M Eliasziw, N Klar. Statistics Med 13:1253–1264, 1994.
16. JG Morel, NK Neerchal. Statistics Med 16:2843–2853, 1997.
17. B Rosner. Biometrics 40:1025–1035, 1984.
18. A Donner. Biometrics 45:605–611, 1989.
19. PME Altham. Biometrika 63:263–269, 1976.
20. JE Cohen. J Am Statis Assoc 71:665–670, 1976.
21. MJ Crowder. Appl Stat 27:34–37, 1978.
22. JJ Heckman, RJ Willis. J Polit Econ 85:27–58, 1977.
23. BE Efron. J Am Statis Assoc 81:709–721, 1986.
24. DF Moore. Appl Stat 36:8–14, 1987.
25. M Ashby, JM Neuhaus, WW Hauck, P Bacchetti, DC Heilbrow, NP Jewell, MR Segal, RE Fusaro. Statistics Med 11:67–99, 1992.
26. NE Breslow. Appl Stat 33:38–44, 1984.
27. KY Liang, SL Zeger. Biometrika 73:13–22, 1986.
28. SL Zeger, KY Liang. Biometrics 42:121–130, 1986.
29. CA McGilchrist. J Roy Statis Soc B 56:61–69, 1994.
30. ML Drum, P McCullagh. Biometrics 49:677–689, 1993.
31. NE Breslow, DG Clayton. J Am Statis Assoc 88:9–25, 1993.
32. MR Karim, SL Zeger. Biometrics 48:631–644, 1992.
33. SL Zeger, MR Karim. J Am Statis Assoc 86:79–86, 1991.
34. H Goldstein. Biometrika 78:45–51, 1991.
35. R Schall. Biometrika 78:719–727, 1991.
36. SL Zeger, KY Liang, PS Albert. Biometrics 44:1049–1060, 1988.
37. RWM Wedderburn. Biometrika 61:439–447, 1974.
38. WJ Cochran. Sampling Techniques. 3rd ed. New York: Wiley, 1977, p 153.
39. L Kish. Survey Sampling. New York: Wiley, 1965, sec 5.4.
40. DR Cox, EJ Snell. Analysis of Binary Data. 2nd ed. New York: Chapman and Hall, 1989, chap 3.
41. JG Morel, KJ Koehler. Appl Statis 44:187–200, 1995.
42. DR Cox, EJ Snell. Analysis of Binary Data. 2nd ed. New York: Chapman and Hall, 1989, p 111.
43. NL Johnson, S Kotz, AW Kemp. Univariate Discrete Distributions. 2nd ed. New York: Wiley, 1992, p 204.

44. JG Skellam. J Roy Statis Soc B 10:257–261, 1948.
45. JG Morel, NK Nagaraj. Biometrika 80:363–371, 1993.
46. JG Morel. Commun Statis-Simulation 21:1255–1268, 1992.
47. SG Self, KY Liang. J Am Statis Assoc 82:605–610, 1987.
48. NK Neerchal, JG Morel. J Am Statis Assoc 93:1078–1087, 1998.
49. P McCullagh. Ann Statis 11:59–67, 1983.
50. P McCullagh, JA Nelder. Generalized Linear Models. London: Chapman and Hall, 1983.
51. JG Morel. Survey Methodol 15:203–223, 1989.
52. DA Harville. J Am Statis Assoc 72:320–340, 1977.
53. DA Anderson, M Aitkin. J Roy Statis Soc B 47:203–210, 1985.
54. S Im, D Gianola. Appl Stat 37:196–204, 1988.
55. R Stiratelli, N Laird, JH Ware. Biometrics 40:961–971, 1984.

Jorge G. Morel
Nagaraj K. Neerchal

Factorial Designs

See also *Screening Design*

I. INTRODUCTION

Factorial design theory is emerging as a fast, effective way of accumulating knowledge in most major sciences. It has been widely used in agricultural research, quality improvement, and cost reduction. Recently we have seen its application growing fast in biology, medical research, and clinical trials. In biological processes, for example, the polymerase chain reaction (PCR) of DNA diagnosis and identification is affected by many factors. In a clinical study, when patients need multiple therapies, such as combinations of several medications, investigators may have to study the toxicity and benefit of different drug combinations (Kesler and Feiden, 1995). A factorial design can be a cost-effective way to study multiple factors in a single study.

Factorial design theory was first introduced by Fisher (1935) in agricultural applications. Bose (1947) established the mathematical foundation of experimental design theory by applying Galois fields and the associated finite geometries. For a symmetric s^n factorial experiment, Rao (1950) showed how to construct an orthogonal array of strength $2d$ [$OA(2d)$] from a single flat design for estimating all effects up to d-factor interactions.

A design is said to be *orthogonal* if and only if the corresponding information matrix is diagonal. Orthogonal designs are optimal designs with respect to many optimality criteria, such as D-, E-, and A-optimality. Using an orthogonal design, the estimators of effects are uncorrelated and the variances of estimates are minimized. Orthogonal arrays play an important role in optimal designs. If a design is an orthogonal array of strength $2d$, it is an optimal design with respect to general criteria [Kiefer (1975) and Cheng (1980)] from which all factorial effects up to d-factor interactions can be estimated, with the assumption that other higher-order interactions are zero.

In practice, it is difficult to use orthogonal arrays of strength $2d$ for $d > 2$. The number of treatments in orthogonal arrays of the strength $2d$ increases exponentially when the number of factors increases. For an experiment with a large number of factors, the number of treatments in an orthogonal array is too large. Often it is not practical to collect all data corresponding to the treatments. In addition, it may be unnecessary to estimate all effects up to d-factor interactions. Prior knowledge may indicate that some higher-order interactions can be ignored. Therefore it is desirable to develop orthogonal designs with a small number of treatments in order to estimate a specified set of higher-order interactions (not necessarily *all* effects up to d-factor interactions).

One effective way to construct factorial designs is to use *parallel flats*. Connor and Young (1961) introduced the class of parallel flats designs (PFD). Srivastava and Throop (1990) showed how to generate an $OA(2d)$ of PFD. The general theory of parallel flats designs is further studied in Srivastava et al. (1984) and Srivastava (1987). In her dissertation, Li (1991) gave a necessary and sufficient condition to generate an orthogonal parallel flats type of design for estimating a specified set of s^n factorial effects. Series of orthogonal designs for 2^n experiments were also presented in Li

(1991). These results were also published in Srivastava and Li (1996). Using representations of complex theory in factorial effects, Liao (1994) extended the foregoing work to mixed $2^n \times 3^m$ factorial experiments.

In this entry, series of orthogonal designs for 2^n, 3^m, and mixed $2^n \times 3^m$ experiments are presented for some specified two-factor interactions by applying theorems in Srivastava and Li (1996) and Liao (1994). It is two-factor interactions that occur most often in practice. Orthogonal designs for higher-factor interactions, such as tree-structured higher-factor interactions, will be considered elsewhere. In addition, applications of mixed orthogonal designs in the PCR process of DNA identification and drug interactions in clinical trials are discussed in Sec. 4. This work may guide applied statisticians to implement optimal designs for practical problems without going through the lengthy and complex mathematics involved in generating orthogonal designs.

II. ORTHOGONAL DESIGNS FOR s^m FACTORIAL EXPERIMENTS

Consider an s^m factorial experiment. Let $\underline{Y} = (y_1, y_2, \ldots, y_N)'$ be an $(N \times 1)$ vector of observations corresponding to N treatment combinations in a design T of size $(m \times N)$. Consider a linear model

$$E(\underline{Y}) = X\underline{F}, \qquad \text{var}(\underline{Y}) = \sigma^2 I_N \tag{1}$$

where $\underline{F}$ is a $(\nu \times 1)$ vector of unknown effect parameters, X is the $(N \times \nu)$ design matrix of rank ν, and I_N is the $(N \times N)$ identity matrix. The elements of $\underline{F}$ form a subset of all factorial effects, assuming other effects not in $\underline{F}$ are zero. The estimate of $\underline{F}$ is given by $\hat{\underline{F}} = (X'X)^{-1}X'\underline{Y}$, and the variance matrix of $\hat{\underline{F}}$ is $\text{var}(\hat{\underline{F}}) = \sigma^2(X'X)^{-1}$. The corresponding information matrix is $M = X'X$. The design T is said to be an orthogonal design if and only if M is diagonal.

Consider the defining vector for any effect in $\underline{F}$. Suppose that $F_1, F_2, \ldots, F_m$ are m factors of interest, and each factor takes three levels. A general factorial effect may be denoted as

$$e = F_1^{\varepsilon_1} \cdots F_m^{\varepsilon_m} \tag{2}$$

where $\varepsilon_i \in GF(3)$ $(i = 1, \ldots, m)$. The vector

$$\underline{e} = (\varepsilon_1, \varepsilon_2, \ldots, \varepsilon_m)' \tag{3}$$

of size $(m \times 1)$ is called the *defining vector* of the factorial effect e. The general mean μ will have the defining vector $\underline{0}$.

Consider the parallel flats type of designs for s^m factorial experiments. Suppose that T_i is an $(m \times s^{m-r})$ matrix, whose columns $\underline{t}$ are the solutions to the linear system

$$A\underline{t} = \underline{c}_i, \qquad (i = 1, \ldots, f) \tag{4}$$

over $GF(3)$, where A is an $(r \times m)$ matrix of rank r and $\underline{t}$ and $\underline{c}_i$ are vectors of the sizes $(m \times 1)$ and $(r \times 1)$, respectively. Let T be the collection of all the T_i; i.e.,

$$T = [T_1\colon T_2\colon \cdots T_i\colon \cdots\colon T_f] \tag{5}$$

Notice that T is a matrix of size $(m \times N)$, where $N = fs^{m-r}$. Let

$$A = [B{:}I_{n-r}] \text{ and } C = (\underline{c}_1, \ldots, \underline{c}_f) \tag{6}$$

The design T is called a parallel flats type of design determined by the pair of matrices (A, C), or given by Eq. (4).

In Srivastava and Li (1996), a simple necessary and sufficient condition on A and C such that T is an orthogonal design is given for a general s^m factorial experiment, where s is a prime or a power of a prime. Theorem 1 provides a way to construct an orthogonal design with a possibly smaller number of treatments.

Theorem 1. (Srivastava and Li, 1996) Consider the linear model of Eq. (1). Let T be determined by (A, C) as in Eqs. (4)–(6). Then T is an orthogonal design for estimating $\underline{F}$ if and only if the following holds.

1. Let e be any effect in $\underline{F}$ with the defining vector $\underline{e}$, if $\underline{e} = \underline{u}'A$, where the vector $\underline{u}(r \times 1)$ is not zero, then $\underline{u}'C(1 \times f)$ must be an $OA(1)$.
2. Let e_1 and e_2 be any two effects in $\underline{F}$ with the defining vectors $\underline{e}_1$ and $\underline{e}_2$, respectively. If $\underline{e}_1 \neq \beta\underline{e}_2$, and $\underline{e}_1 + \gamma\underline{e}_2 = \underline{u}'A$ over $GF(3)$, where β and γ are nonzero in $GF(s)$ and $\underline{u}$ is a nonzero vector of size $(r \times 1)$, then $\underline{u}'C$ must be an $OA(1)$.

Now we apply Theorem 1 to generate orthogonal designs for intermediate two-factor interactions (i.e., a subset of all two-factor interactions). For simplicity, we denote any vector of parameters as follows:

$$\begin{aligned}\underline{F}' &= (\mu, F_1, F_1^2, \ldots, F_i, F_i^2, F_i^{\varepsilon_i}F_j^{\varepsilon_j}, \ldots)\\ &= (0, 1, 1^2, \ldots, i, i^2, i^{\varepsilon_i}j^{\varepsilon_j}, \ldots)\end{aligned}$$

For example,

$$\begin{aligned}\underline{F}' &= (\mu, F_1, F_1^2, F_2, F_2^2, F_1F_2, F_1^2F_2, F_1F_2^2, F_1^2F_2^2)\\ &= (0, 1, 1^2, 2, 2^2, 12, 1^22, 12^2, 1^22^2)\end{aligned}$$

In Secs. 3 and 4, N is the total number of treatments

in a design, n (or m) is the number of factors of interest, and ν is the number of parameters in $\underline{F}$. We assume that n (or m) factors are divided into p subgroups, and each subgroup has m_k ($k = 1, \ldots, p$) factors in it. The two-factor interactions are either within or between some of the subgroups. All the designs of the following parallel flats type satisfy the conditions in Theorem 1.

III. ORTHOGONAL DESIGNS FOR 2^n FACTORIAL EXPERIMENTS

In this section, we provide a class of examples (from 2^n factorial experiments) to illustrate the foregoing theory on orthogonal designs. Given the theory of the last section for any given situation, it is generally relatively simple to obtain A and C matrices by trial and error. Our illustrations include cases where we wish to estimate the general mean, the main effects, and some two-factor interactions, assuming the remaining factorial effects are negligible. Our designs are such that they enable us to estimate the required parameters under the given conditions. However, in many cases, they may turn out to be better than what we claim; i.e., they may enable us to estimate more parameters or to estimate parameters under less restrictive conditions. Illustrations for other cases of the 2^n series and for general s series or mixed cases will be given elsewhere.

In practice, we often do not have to estimate all the two-factor interactions. We may be interested only in estimating some specific type of two-factor interactions plus the general mean and all the main effects, since the remaining effects may be known to be negligible. We shall refer to this as the case of "intermediate resolution." For example, it is very common to see that elderly patients and AIDS patients take combination treatments. Suppose that a study patient population takes four active combination treatments and that each active treatment has two dose levels. To study the effect of interactions within the first three active treatments, a 2^4 factorial experiment given in Case I, Example 1, may be considered.

For a 2^n factorial experiment, orthogonal fractional designs can be used to estimate the factorial effects by applying the classical theory of experimental design. However, the number of treatments is a power of 2. When the number of factors n is large, this generally gives a large number of treatments. In practice, an individual experimental treatment could be very expensive. Therefore one tries to minimize the number of treatments required to estimate effects. Theorem 1 provides a way to construct an orthogonal design with a possibly smaller number of treatments. By applying it, a number of orthogonal designs are presented in this section for 2^n factorial series of intermediate resolution, where n is in the practical range. In the following we discuss the intermediate situation in detail.

Suppose that n factors are divided into p subgroups, where the ith ($i = 1, \ldots, p$) subgroup has n_i factors in it. Let p' be the number of subgroups of interest, where $p' = p$ or $p' = p - 1$, according to whether the pth subgroup is or is not ignored. In other words, if $p' = p - 1$, the factors in the pth subgroup are not involved in the higher-order-factor interactions. Consider the following three cases:

I. The interactions within p' subgroups
II. The interactions that involve factors from different subgroups
III. The interactions that arise within some of the subgroups or between some of the subgroups

We now present the designs for different cases. For convenience, we list the parameters and symbols:

n = number of factors under consideration
p = number of subgroups
p' = number of subgroups actually involved in the two-factor interactions
n_i = number of the factors in the ith ($i = 1, \ldots, p$) subgroup
ν = number of nonnegligible effects
f = number of flats
N = total number of treatments

Let $\tilde{B} = (\underline{b}'_1, \ldots, \underline{b}'_r)$ and $\tilde{C} = (\underline{c}'_1, \ldots, \underline{c}'_r)$, where $\underline{b}'_k$ ($k = 1, \ldots, r$) is the kth row of B in (6) and $\underline{c}'_k$ ($k = 1, \ldots, r$) is the kth row of C. Here, the $\tilde{B}$ is of size $[r(n - r) \times 1]$ and is a "rolled out" form of B; similarly, $\tilde{C}$ is $(rf \times 1)$ and is a "rolled out" form of C. We also denote the parameter vector $\underline{F} = (\mu, F_1, \ldots, F_i, \ldots, F_n, F_1F_2, \ldots, F_iF_j)'$ as $\underline{F} = F(0, 1, \ldots, i, \ldots, n, 1\cdot2, \ldots, i\cdot j)$. For example, $\underline{F} = (\mu, F_1, F_2, F_3, F_1F_2, F_1F_3, F_2F_3)' = F(0, 1, 2, 3, 1\cdot2, 1\cdot3, 2\cdot3)$. Hereafter, all the designs presented are determined by pairs (A, C), where (A, C) satisfy the condition in Theorem 1 and where A and C are defined by the corresponding $\tilde{B}$ and $\tilde{C}$.

A. Case I: Interactions Within Subgroups

Here, the n factors are partitioned into p subgroups, the ith subgroup has n_i factors in it, and the two-factor interactions are within p' subgroups, where $p' = p$ or

$p' = p - 1$. In Table 1, we list the parameters n, ν, N, f, p, p', and $n_1, \ldots, n_p$ for each of 21 designs, where the number of factors n is from 4 to 10. The designs in Table 1 utilize a single flat (i.e., C is a vector, say, $\underline{c}$). Without loss of generality, we just take $\underline{c} = \underline{0}$. So in the following we present only the vector of effects $\underline{F}$, the parameters n, N, ν, and the matrices $\tilde{B}$. The set T of the solutions to the linear system $A\underline{t} = \underline{0}$ is then an orthogonal design for estimating the corresponding vector of effects $\underline{F}$. Next, the 21 designs are presented as Cases 1–21.

1. Consider a 2^4 factorial experimental design (Case 1 in Table 1). We are interested in estimating the general mean, all the main effects, and the two-factor interactions within the subgroup $\{F_1, F_2, F_3\}$. The factor F_4 is not involved in the two-factor interactions. Hence, we have $n = 4$, $\nu = 8$, $p = 2$, $p' = 1$, $(n_1, n_2) = (3, 1)$, which are listed in Table 1. The vector of effects is $\underline{F} = (\mu, F_1, F_2, F_3, F_4, F_1F_2, F_1F_3, F_2F_3)' = F(0, 1, 2, 3, 4, 1\cdot2, 1\cdot3, 2\cdot3)$. The orthogonal design T is the set of solutions to the linear equation $A\underline{t} = 0$ over $GF(2)$, with total number of treatments $N = 8$, where $A = (1, 1, 1, 1)$ and $\tilde{B} = (111)$.

Table 1 Designs for Interactions Within Subgroups

id	n	ν	N	f	p	p'	$(n_1, \ldots, n_p)$
1	4	8	8	1	2	1	(3, 1)
2	5	7	8	1	2	1	(2, 3)
3	6	8	8	1	2	1	(2, 4)
4	6	10	16	1	3	3	(2, 2, 2)
5	6	13	16	1	1	1	(4, 2)
6	6	11	16	1	3	2	(3, 2, 1)
7	7	14	16	1	2	1	(4, 3)
8	7	11	16	1	4	3	(2, 2, 2, 1)
9	7	23	32	1	2	1	(6, 1)
10	8	13	16	1	4	4	(2, 2, 2, 2)
11	8	15	16	1	2	1	(4, 4)
12	8	24	32	1	2	1	(6, 2)
13	8	20	32	1	3	2	(2, 5, 1)
14	9	16	16	1	2	1	(4, 5)
15	9	14	16	1	5	4	(2, 2, 2, 2, 1)
16	9	25	32	1	2	1	(6, 3)
17	9	19	32	1	3	1	(3, 3, 3)
18	10	14	16	1	2	1	(3, 7)
19	10	16	16	1	5	5	(2, 2, 2, 2, 2)
20	10	26	32	1	2	1	(6, 4)
21	10	20	32	1	4	3	(3, 3, 3, 1)

Cases 2–21 in Table 1 are similar to the preceding. We list only the vector of effects $\underline{F}$, parameters n, N, ν, and the $\tilde{B}$ vector for each case.

2. $\underline{F} = F(0, 1, \ldots, 5, 1\cdot2)$, $n = 5$, $N = 8$, $\nu = 7$, $\tilde{B} = (101, 011)$.
3. $\underline{F} = F(0, 1, \ldots, 6, 1\cdot2)$, $n = 6$, $N = 8$, $\nu = 8$, $\tilde{B} = (110, 011, 111)$.
4. $\underline{F} = F(0, 1, \ldots, 6, 1\cdot2, 3\cdot4, 5\cdot6)$, $n = 6$, $N = 16$, $\nu = 10$, $\tilde{B} = (0101, 1010)$.
5. $\underline{F} = F(0, 1, \ldots, 6, 1\cdot2, 1\cdot3, 1\cdot4, 2\cdot3, 2\cdot4, 3\cdot4)$, $n = 6$, $N = 16$, $\nu = 13$, $\tilde{B} = (1110, 0111)$.
6. $\underline{F} = F(0, 1, \ldots, 6, 1\cdot2, 1\cdot3, 2\cdot3, 4\cdot5)$, $n = 6$, $N = 16$, $\nu = 11$, $\tilde{B} = (0011, 1101)$.
7. $\underline{F} = F(0, 1, \ldots, 7, 1\cdot2, 1\cdot3, 1\cdot4, 2\cdot3, 2\cdot4, 3\cdot4)$, $n = 7$, $N = 16$, $\nu = 14$, $\tilde{B} = (0110, 0101, 1111)$.
8. $\underline{F} = F(0, 1, \ldots, 7, 1\cdot2, 3\cdot4, 5\cdot6)$, $n = 7$, $N = 16$, $\nu = 11$, $\tilde{B} = (1100, 0011, 0101)$.
9. $\underline{F} = F(0, 1, \ldots, 7, 1\cdot2, \ldots, 1\cdot6, 2\cdot3, \ldots, 2\cdot6, \ldots, 5\cdot6)$, $n = 7$, $N = 32$, $\nu = 23$, $\tilde{B} = (11001, 00111)$.
10. $\underline{F} = F(0, 1, \ldots, 8, 1\cdot2, 3\cdot4, 5\cdot6, 7\cdot8)$, $n = 8$, $N = 16$, $\nu = 13$, $\tilde{B} = (1100, 0011, 0101, 1011)$.
11. $\underline{F} = F(0, 1, \ldots, 8, 1\cdot2, 1\cdot3, 1\cdot4, 2\cdot3, 2\cdot4, 3\cdot4)$, $n = 8$, $N = 16$, $\nu = 15$, $\tilde{B} = (1100, 1010, 1001, 0111)$.
12. $\underline{F} = F(0, 1, \ldots, 8, 1\cdot2, \ldots, 1\cdot6, 2\cdot3, \ldots, 2\cdot6, \ldots, 5\cdot6)$, $n = 8$, $N = 32$, $\nu = 24$, $\tilde{B} = (11010, 10101, 00011)$.
13. $\underline{F} = F(0, 1, \ldots, 8, 1\cdot2, 3\cdot4, \ldots, 3\cdot7, 4\cdot5, \ldots, 4\cdot7, \ldots, 6\cdot7)$, $n = 8$, $N = 32$, $\nu = 20$, $\tilde{B} = (10001, 11100, 00111)$.
14. $\underline{F} = F(0, 1, \ldots, 9, 1\cdot2, 1\cdot3, 1\cdot4, 2\cdot3, 2\cdot4, 3\cdot4)$, $n = 9$, $N = 16$, $\nu = 16$, $\tilde{B} = (1100, 1010, 1001, 0111, 1111)$.
15. $\underline{F} = F(0, 1, \ldots, 9, 1\cdot2, 3\cdot4, 5\cdot6, 7\cdot8)$, $n = 9$, $N = 16$, $\nu = 14$, $\tilde{B} = (1100, 1001, 1101, 0011, 1111)$.
16. $\underline{F} = F(0, 1, \ldots, 9, 1\cdot2, \ldots, 1\cdot6, 2\cdot3, \ldots, 2\cdot6, \ldots, 5\cdot6)$, $n = 9$, $N = 32$, $\nu = 25$, $\tilde{B} = (10100, 10010, 10001, 00111)$.
17. $\underline{F} = F(0, 1, \ldots, 9, 1\cdot2, 1\cdot3, 2\cdot3, 4\cdot5, 4\cdot6, 5\cdot6, 7\cdot8, 7\cdot9, 8\cdot9)$, $n = 9$, $N = 32$, $\nu = 19$, $\tilde{B} = (10100, 01010, 11001, 10111)$.
18. $\underline{F} = F(0, 1, \ldots, 10, 1\cdot2, 1\cdot3, 2\cdot3)$, $n = 10$, $N = 16$, $\nu = 14$, $\tilde{B} = (1100, 1010, 1001, 0111, 1111, 0110)$.
19. $\underline{F} = F(0, 1, \ldots, 10, 1\cdot2, 3\cdot4, 5\cdot6, 7\cdot8, 9\cdot10)$, $n = 10$, $N = 16$, $\nu = 16$, $\tilde{B} = (1010, 0101, 0110, 1101, 1001, 1110)$.

20. $\underline{F} = F(0, 1, \ldots, 10, 1\cdot 2, \ldots, 1\cdot 6, 2\cdot 3, \ldots, 2\cdot 6, \ldots, 5\cdot 6)$, $n = 10$, $N = 32$, $\nu = 26$, $\tilde{B} = (01100, 01010, 01001, 11110, 10011)$.
21. $\underline{F} = F(0, 1, \ldots, 10, 1\cdot 2, 1\cdot 3, 2\cdot 3, 4\cdot 5, 4\cdot 6, 5\cdot 6, 7\cdot 8, 7\cdot 9, 8\cdot 9)$, $n = 10$, $N = 32$, $\nu = 20$, $\tilde{B} = (11000, 01001, 00111, 00101, 10011)$.

B. Case II: Interactions Between Subgroups

In this section, we consider designs of intermediate resolution, where the two-factor interactions arise between the subgroups. In Table 2, 35 such cases are listed; the parameters n, ν, N, f, p, p', and $n_1, \ldots, n_p$ are given for each case also.

Table 2 Designs for Interactions Between Subgroups

id	n	ν	N	f	p	p'	$(n_1, \ldots, n_p)$
1	4	8	8	1	2	2	(1, 3)
2	5	7	8	1	3	2	(1, 2, 2)
3	6	1	16	1	2	2	(3, 3)
4	6	12	16	1	2	2	(1, 5)
5	7	14	16	1	2	2	(1, 6)
6	7	23	32	1	2	2	(3, 4)
7	8	16	16	1	2	2	(1, 7)
8	8	24	32	1	2	2	(5, 3)
9	6	13	16	4	3	2	(3, 2, 1)
10	9	18	24	12	2	2	(1, 8)
11	10	20	24	12	2	2	(1, 9)
12	11	22	24	12	2	4	(1, 10)
13	11	31	48	12	3	3	(1, 1, 9)
14	12	24	24	12	2	2	(1, 11)
15	12	34	48	12	3	3	(1, 1, 10)
16	13	37	48	12	3	3	(1, 1, 11)
17	14	38	48	12	4	3	(1, 1, 11, 1)
18	17	34	40	20	2	2	(1, 16)
19	18	36	40	20	2	2	(1, 17)
20	18	52	80	20	3	3	(1, 1, 16)
21	19	38	40	20	2	2	(1, 18)
22	19	55	80	20	3	3	(1, 1, 17)
23	20	40	40	20	2	2	(1, 19)
24	20	58	80	20	3	3	(1, 1, 18)
25	21	61	80	20	3	3	(1, 1, 19)
26	22	62	80	20	3	2	(1, 1, 19, 1)
27	21	42	48	24	2	2	(1, 20)
28	22	44	48	24	2	2	(1, 21)
29	22	64	96	24	2	2	(1, 1, 20)
30	23	46	48	24	2	2	(1, 22)
31	23	67	96	24	3	3	(1, 1, 21)
32	24	48	48	24	2	2	(1, 23)
33	24	70	96	24	3	3	(1, 1, 22)
34	25	73	96	24	3	3	(1, 1, 23)
35	26	74	96	24	4	3	(1, 1, 23, 1)

The first 12 designs are given by a single flat. We select $\underline{c} = \underline{0}$. Next, we present the vector of effects $\underline{F}$, the parameters n, N, ν, and the $\tilde{B}$ matrix for each case.

1. $\underline{F} = F(0, 1, 2, 3, 4, 1\cdot 2, 1\cdot 3)$, $n = 4$, $N = 8$, $\nu = 7$, $\tilde{B} = (111)$.
2. $\underline{F} = F(0, 1, \ldots, 5, 1\cdot 2, 1\cdot 3)$, $n = 5$, $N = 8$, $\nu = 8$, $\tilde{B} = (110, 011)$.
3. $\underline{F} = F(0, 1, \ldots, 6, 1\cdot 4, 1\cdot 5, 1\cdot 6, 2\cdot 4, 2\cdot 5, 2\cdot 6, 3\cdot 4, 3\cdot 5, 3\cdot 6)$, $n = 6$, $N = 16$, $\nu = 16$, $\tilde{B} = (1111, 0111)$.
4. $\underline{F} = F(0, 1, \ldots, 6, 1\cdot 2, \ldots, 1\cdot 6)$, $n = 6$, $N = 16$, $\nu = 12$, $\tilde{B} = (1110, 0011)$.
5. $\underline{F} = F(0, 1, \ldots, 7, 1\cdot 2, \ldots, 1\cdot 7)$, $n = 7$, $N = 16$, $\nu = 14$, $\tilde{B} = (1110, 0101, 0011)$.
6. $\underline{F} = F(0, 1, \ldots, 7, 1\cdot 4, 1\cdot 5, 1\cdot 6, 1\cdot 7, 2\cdot 4, 2\cdot 5, 2\cdot 6, 2\cdot 7, 3\cdot 4, 3\cdot 5, 3\cdot 6, 3\cdot 7)$, $n = 7$, $N = 32$, $\nu = 23$, $\tilde{B} = (01111, 11111)$.
7. $\underline{F} = F(0, 1, \ldots, 8, 1\cdot 2, 1\cdot 3, \ldots, 1\cdot 8)$, $n = 8$, $N = 16$, $\nu = 16$, $\tilde{B} = (0111, 1010, 1001, 1011)$.
8. $\underline{F} = F(0, 1, \ldots, 8, 1\cdot 6, 1\cdot 7, 1\cdot 8, 2\cdot 6, 2\cdot 7, 2\cdot 8, 3\cdot 6, 3\cdot 7, 3\cdot 8, 4\cdot 6, 4\cdot 7, 4\cdot 8, 5\cdot 6, 5\cdot 7, 5\cdot 8)$, $n = 8$, $N = 16$, $\nu = 16$, $\tilde{B} = (11000, 01111, 1111)$.

The following orthogonal designs of parallel flats type consist of more than one flat.

9. Consider a 2^6 factorial experiment. We are interested in estimating the general mean, all the main effects, and the two-factor interactions between the subgroup $\{F_1, F_2, F_3\}$ and the subgroup $\{F_4, F_5\}$. The factor F_6 is not involved into the two-factor interactions. So, we have $n = 6$, $p = 3$, $p' = 2$, $\nu = 13$, $(n_1, n_2, n_3) = (3, 2, 1)$, which are listed in Table 2, Case 9. The vector of effects is $\underline{F} = F(0, 1, \ldots, 6, 1\cdot 4, 1\cdot 5, 2\cdot 4, 2\cdot 5, 3\cdot 4, 3\cdot 5)$. The orthogonal design with total number of treatments $N = 16$ is determined by (A, C), where

$$A = \begin{pmatrix} 1 & 0 & 0 & 0 & 0 & 0 \\ 0 & 1 & 0 & 0 & 0 & 0 \\ 0 & 0 & 1 & 0 & 0 & 0 \\ 0 & 0 & 0 & 1 & 1 & 1 \end{pmatrix} \quad \text{and}$$

$$C = \begin{pmatrix} 1 & 0 & 1 & 0 \\ 1 & 1 & 0 & 0 \\ 1 & 0 & 0 & 1 \\ 0 & 1 & 1 & 1 \end{pmatrix}$$

For simplicity, we say that A and C are given

by $\tilde{B} = (00, 00, 00, 11)$ and $\tilde{C} = (1010, 1100, 1001, 0111)$, respectively.

In the following, the C matrices are obtained from the (12×12) Hadamard matrix

$$H_{12\times 12} = \begin{bmatrix} + & + & + & + & + & + & + & + & + & + & + & + \\ + & - & - & + & - & - & - & + & + & + & - & + \\ + & - & + & - & - & - & + & + & + & - & + & - \\ + & + & - & - & - & + & + & + & - & + & - & - \\ + & - & - & - & + & + & + & - & + & - & - & + \\ + & - & - & + & + & + & - & + & - & - & + & - \\ + & - & + & + & + & - & + & - & - & + & - & - \\ + & + & + & + & - & + & - & - & + & - & - & - \\ + & + & + & - & + & - & - & + & - & - & - & + \\ + & + & - & + & - & - & + & - & - & - & + & + \\ + & - & + & - & - & + & - & - & - & + & + & + \\ + & + & - & - & + & - & - & - & + & + & + & - \end{bmatrix}$$

Here, $+$ and $-$ represent $+1$ and -1, respectively. Let $\tilde{H}_{12\times 12}$ be the matrix obtained from $H_{12\times 12}$ by changing all -1's into 0's, and $\tilde{H}_{11\times 12}$ be the matrix obtained by deleting the first row of $\tilde{H}_{12\times 12}$. It is known that $\tilde{H}_{11\times 12}$ and $\tilde{H}_{k\times 12}$ are $OA(2)$, where $\tilde{H}_{k\times 12}$ is the matrix with any k rows ($k \leq 11$) of $\tilde{H}_{11\times 12}$. In the following eight cases, we select C as the matrices $\tilde{H}_{k\times 12}$.

10. $\underline{F} = F(0, \ldots, 9, 1\cdot 2, 1\cdot 3, \ldots, 1\cdot 9)$, $n = 9$, $N = 24$, $\nu = 18$, $\tilde{B} = (1, 0, 0, 0, 0, 0, 0, 0, 0)$, $C = \tilde{H}_{8\times 12}$.
11. $\underline{F} = F(0, \ldots, 10, 1\cdot 2, \ldots, 1\cdot 10)$, $n = 10$, $N = 24$, $\nu = 20$, $\tilde{B} = (1, 0, 0, 0, 0, 0, 0, 0, 0, 0)$, and $C = \tilde{H}_{9\times 12}$.
12. $\underline{F} = F(0, 1, \ldots, 11, 1\cdot 2, 1\cdot 3, \ldots, 1\cdot 11)$, $n = 10$, $N = 24$, $\nu = 22$, $\tilde{B} = (1, 0, 0, 0, 0, 0, 0, 0, 0, 0)$, and $C = \tilde{H}_{10\times 12}$.
13. $\underline{F} = F(0, 1, \ldots, 11, 1\cdot 2, 1\cdot 3, \ldots, 1\cdot 11, 2\cdot 3, \ldots, 2\cdot 11)$, $n = 11$, $N = 48$, $\nu = 31$, $\tilde{B} = (10, 01, 00, 00, 00, 00, 00, 00, 00)$, and $C = \tilde{H}_{9\times 12}$.
14. $\underline{F} = F(0, 1, \ldots, 12, 1\cdot 2, 1\cdot 3, \ldots, 1\cdot 12)$, $n = 12$, $N = 24$, $\nu = 24$, $\tilde{B} = (1, 0, 0, 0, 0, 0, 0, 0, 0, 0, 0)$, and $C = \tilde{H}_{11\times 12}$.
15. $\underline{F} = F(0, 1, \ldots, 12, 1\cdot 2, 1\cdot 3, \ldots, 1\cdot 12, 2\cdot 3, \ldots, 2\cdot 12)$, $n = 12$, $N = 48$, $\nu = 34$, $\tilde{B} = (10, 01, 00, 00, 00, 00, 00, 00, 00, 00)$, and $C = \tilde{H}_{10\times 12}$.
16. $\underline{F} = F(0, 1, \ldots, 13, 1\cdot 2, 1\cdot 3, \ldots, 1\cdot 13, 2\cdot 3, \ldots, 2\cdot 13)$, $n = 13$, $N = 48$, $\nu = 37$, $\tilde{B} = (10, 01, 00, 00, 00, 00, 00, 00, 00, 00, 00)$, and $C = \tilde{H}_{11\times 12}$.
17. $\underline{F} = F(0, 1, \ldots, 14, 1\cdot 2, 1\cdot 3, \ldots, 1\cdot 13, 2\cdot 3, \ldots, 2\cdot 13)$, $n = 14$, $N = 48$, $\nu = 38$, $\tilde{B} = (10, 01, 00, \ldots, 00, 11)$. Here, C is derived from $\tilde{H}_{12\times 12}$ by exchanging the positions of the first row and the last row of $\tilde{H}_{12\times 12}$.

In the following, we use the (20×20) Hadamard matrix

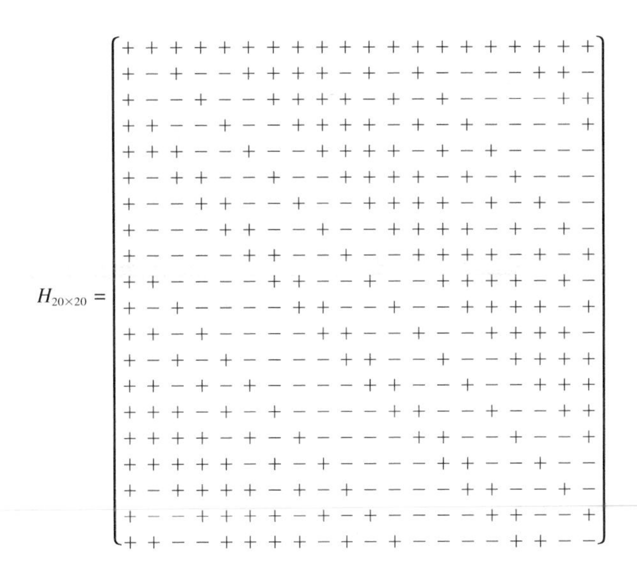

to select the C matrices. Let $\tilde{H}_{20\times 20}$ be the matrix obtained from $H_{20\times 20}$ by changing -1's into 0's, and $\tilde{H}_{19\times 20}$ be the matrix obtained by deleting the first row of $\tilde{H}_{20\times 20}$. It is known that $\tilde{H}_{19\times 20}$ and $\tilde{H}_{k\times 20}$ are $OA(2)$, where $\tilde{H}_{k\times 20}$ is the matrix with any k rows ($k \leq 19$) of $\tilde{H}_{19\times 20}$. In the following cases, we select C as the matrices $\tilde{H}_{k\times 20}$.

18. $\underline{F} = F(0, 1, \ldots, 17, 1\cdot 2, 1\cdot 3, \ldots, 1\cdot 17)$, $n = 17$, $N = 40$, $\nu = 34$, $\tilde{B} = (1, 0, \ldots, 0)$, and $C = \tilde{H}_{16\times 20}$.
19. $\underline{F} = F(0, 1, \ldots, 18, 1\cdot 2, 1\cdot 3, \ldots, 1\cdot 18)$, $n = 18$, $N = 40$, $\nu = 36$, $\tilde{B} = (1, 0, \ldots, 0)$, and $C = \tilde{H}_{17\times 20}$.
20. $\underline{F} = F(0, 1, \ldots, 18, 1\cdot 2, 1\cdot 3, \ldots, 1\cdot 18, 2\cdot 3, \ldots, 2\cdot 18)$, $n = 18$, $N = 80$, $\nu = 52$, $\tilde{B} = (10, 01, 00, \ldots, 00)$, and $C = \tilde{H}_{16\times 20}$.
21. $\underline{F} = F(0, 1, \ldots, 19, 1\cdot 2, 1\cdot 3, \ldots, 1\cdot 19)$, $n = 19$, $N = 40$, $\nu = 38$, $\tilde{B} = (1, 0, \ldots, 0)$, and $C = \tilde{H}_{18\times 20}$.
22. $\underline{F} = F(0, 1, \ldots, 19, 1\cdot 2, 1\cdot 3, \ldots, 1\cdot 19, 2\cdot 3, \ldots, 2\cdot 19)$, $n = 19$, $N = 80$, $\nu = 55$, $\tilde{B} = (10, 01, 00, \ldots, 00)$, and $C = \tilde{H}_{17\times 20}$.
23. $\underline{F} = F(0, 1, \ldots, 20, 1\cdot 2, 1\cdot 3, \ldots, 1\cdot 20)$, $n = 20$, $N = 40$, $\nu = 40$, $\tilde{B} = (1, 0, \ldots, 0)$, and $C = \tilde{H}_{19\times 20}$.
24. $\underline{F} = F(0, 1, \ldots, 20, 1\cdot 2, 1\cdot 3, \ldots, 1\cdot 20,$

$2\cdot3, \ldots, 2\cdot20)$, $n = 20$, $N = 80$, $\nu = 58$, $\tilde{B} = (10, 01, 00, \ldots, 00)$, $C = \tilde{H}_{18\times20}$.

25. $\underline{F} = F(0, 1, \ldots, 21, 1\cdot2, 1\cdot3, \ldots, 1\cdot21, 2\cdot3, \ldots, 2\cdot21)$, $n = 21$, $N = 80$, $\nu = 61$, $\tilde{B} = (10, 01, 00, \ldots, 00)$, and $C = \tilde{H}_{19\times20}$.
26. $\underline{F} = F(0, 1, 22, \ldots, 1\cdot2, 1\cdot3, \ldots, 1\cdot21, 2\cdot3, \ldots, 2\cdot21)$, $n = 22$, $N = 80$, $\nu = 62$, $\tilde{B} = (10, 01, 00, \ldots, 00, 11)$, and C is obtained by exchanging the first row and the last row of $\tilde{H}_{20\times20}$.

In the following, we use the (24×24) Hadamard matrix

$H_{24\times24} =$

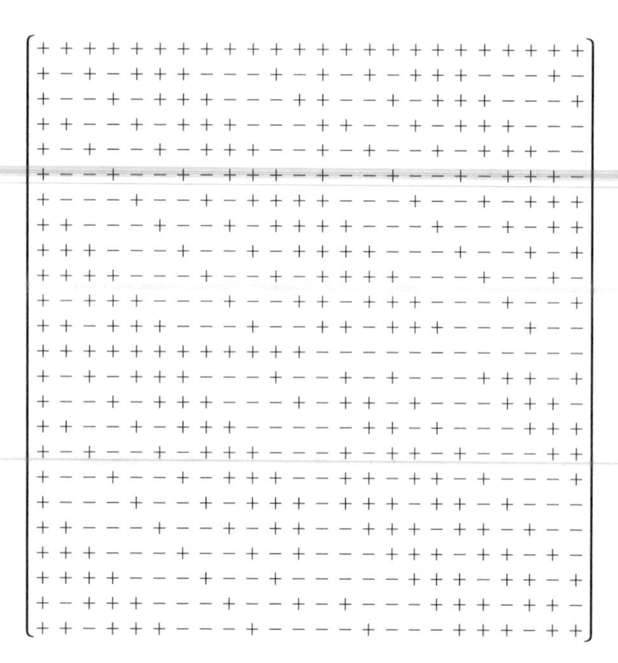

to select C. Let $\tilde{H}_{24\times24}$ be the matrix obtained from $H_{24\times24}$ by changing + into 0, and $\tilde{H}_{23\times24}$ be the matrix obtained by deleting the first row of $\tilde{H}_{24\times24}$. It is known that $\tilde{H}_{23\times24}$ and $\tilde{H}_{k\times24}$ are $OA(2)$, where $\tilde{H}_{k\times24}$ is the matrix with any k rows ($k \leq 23$) of $\tilde{H}_{23\times24}$. In the following cases, we select C as the matrices $\tilde{H}_{k\times24}$.

27. $\underline{F} = F(0, 1, \ldots, 21, 1\cdot2, 1\cdot3, \ldots, 1\cdot21)$, $n = 21$, $N = 48$, $\nu = 32$, $\tilde{B} = (1, 0, \ldots, 0)$, and $C = \tilde{H}_{20\times24}$.
28. $\underline{F} = F(0, 1, \ldots, 22, 1\cdot2, 1\cdot3, \ldots, 1\cdot22)$, $n = 22$, $N = 48$, $\nu = 44$, $\tilde{B} = (1, 0, \ldots, 0)$, and $C = \tilde{H}_{21\times24}$.
29. $\underline{F} = F(0, 1, \ldots, 22, 1\cdot2, 1\cdot3, \ldots, 1\cdot22, 2\cdot3, \ldots, 2\cdot22)$, $n = 22$, $N = 96$, $\nu = 64$, $\tilde{B} = (10, 01, \ldots, 00)$, and $C = \tilde{H}_{20\times24}$.
30. $\underline{F} = F(0, 1, \ldots, 23, 1\cdot2, 1\cdot3, \ldots, 1\cdot23)$, $n = 23$, $N = 48$, $\nu = 46$, $\tilde{B} = (1, 0, \ldots, 0)$, and $C = \tilde{H}_{22\times24}$.
31. $\underline{F} = F(0, 1, \ldots, 23, 1\cdot2, 1\cdot3, \ldots, 1\cdot23, 2\cdot3, \ldots, 2\cdot23)$, $n = 23$, $N = 96$, $\nu = 67$, $\tilde{B} = (10, 01, \ldots, 00)$, and $C = \tilde{H}_{21\times24}$.
32. $\underline{F} = F(0, 1, \ldots, 24, 1\cdot2, 1\cdot3, \ldots, 1\cdot24)$, $n = 24$, $N = 48$, $\nu = 48$, $\tilde{B} = (1, 0, \ldots, 0)$, and $C = \tilde{H}_{23\times24}$.
33. $\underline{F} = F(0, 1, \ldots, 24, 1\cdot2, 1\cdot3, \ldots, 1\cdot24, 2\cdot3, \ldots, 2\cdot24)$, $n = 24$, $N = 96$, $\nu = 70$, $\tilde{B} = (10, 01, \ldots, 00)$, and $C = \tilde{H}_{22\times24}$.
34. $\underline{F} = F(0, 1, \ldots, 25, 1\cdot2, 1\cdot3, \ldots, 1\cdot25, 2\cdot3, \ldots, 2\cdot25)$, $n = 25$, $N = 96$, $\nu = 73$, $\tilde{B} = (10, 01, 00, \ldots, 00)$, and $C = \tilde{H}_{23\times24}$.
35. $\underline{F} = F(0, 1, \ldots, 26, 1\cdot2, 1\cdot3, \ldots, 1\cdot25, 2\cdot3, \ldots, 2\cdot25)$, $n = 26$, $N = 96$, $\nu = 74$, $\tilde{B} = (10, 01, 00, \ldots, 00, 11)$, and $C(24 \times 24)$ is obtained by exchanging the first row and the last row of $\tilde{H}_{24\times24}$.

C. Case III: Interactions Between and Within Subgroups

Consider the situations where the two-factor interactions occur either within subgroups or between subgroups. In Table 3, we list eight such examples. Here, all the factors are divided into three or four subgroups. The two-factor interactions are between the first two subgroups and within the third subgroup. The fourth subgroup (if it exists) is not involved in the two-factor interactions. Since the designs in Table 3 consist of a single flat (i.e., $C = \underline{c} = \underline{0}$), we give the vector of effects $\underline{F}$, parameters n, N, ν, and the corresponding $\tilde{B}$ matrix for each case.

1. $\underline{F} = F(0, 1, \ldots, 6, 1\cdot3, 1\cdot4, 2\cdot3, 2\cdot4, 5\cdot6)$, $n = 6$, $N = 16$, $\nu = 12$, $\tilde{B} = (1110, 1101)$.
2. $\underline{F} = F(0, 1, \ldots, 7, 1\cdot2, 1\cdot3, 1\cdot4, 5\cdot6, 5\cdot7,$

Table 3 Designs for Interactions Within/Between Subgroups

id	n	ν	N	f	p	p'	$(n_1, \ldots, n_p)$
1	6	12	16	1	3	3	(2, 2, 2)
2	7	14	16	1	3	3	(1, 3, 3)
3	7	13	16	1	4	3	(2, 2, 2, 1)
4	8	14	16	1	4	3	(2, 2, 2, 2)
5	8	19	32	1	3	3	(3, 3, 2)
6	9	15	16	1	4	3	(2, 2, 2, 3)
7	9	22	32	1	3	3	(3, 3, 3)
8	10	16	16	1	4	3	(2, 2, 2, 4)

6·7), $n = 7$, $N = 16$, $\nu = 14$, $\tilde{B} = (0111, 1100, 1001)$.

3. $\underline{F} = F(0, 1, \ldots, 7, 1\cdot 3, 1\cdot 4, 2\cdot 3, 2\cdot 4, 5\cdot 6)$, $n = 7$, $N = 16$, $\nu = 13$, $\tilde{B} = (0101, 0011, 1001)$.
4. $\underline{F} = F(0, 1, \ldots, 8, 1\cdot 3, 1\cdot 4, 2\cdot 3, 2\cdot 4, 5\cdot 6)$, $n = 8$, $N = 16$, $\nu = 14$, $\tilde{B} = (0101, 0011, 1010, 1110)$.
5. $\underline{F} = F(0, 1, \ldots, 8, 1\cdot 4, 1\cdot 5, 1\cdot 6, 2\cdot 4, 2\cdot 5, 2\cdot 6, 3\cdot 4, 3\cdot 5, 3\cdot 6, 7\cdot 8)$, $n = 8$, $N = 32$, $\nu = 19$, $\tilde{B} = (11100, 01110, 11101)$.
6. $\underline{F} = F(0, 1, \ldots, 9, 1\cdot 3, 1\cdot 4, 2\cdot 3, 2\cdot 4, 5\cdot 6)$, $n = 9$, $N = 16$, $\nu = 15$, $\tilde{B} = (11100, 01110, 10001, 00111)$.
7. $\underline{F} = F(0, 1, \ldots, 9, 1\cdot 4, 1\cdot 5, 1\cdot 6, 2\cdot 4, 2\cdot 5, 2\cdot 6, 3\cdot 4, 3\cdot 5, 3\cdot 6, 7\cdot 8, 7\cdot 9, 8\cdot 9)$, $n = 9$, $N = 32$, $\nu = 22$, $\tilde{B} = (11100, 00111, 11010, 10001)$.
8. $\underline{F} = F(0, 1, \ldots, 10, 1\cdot 3, 1\cdot 4, 2\cdot 3, 2\cdot 4, 5\cdot 6)$, $n = 10$, $N = 16$, $\nu = 16$, $\tilde{B} = (11000, 10100, 10010, 10001, 00111)$.

IV. ORTHOGONAL DESIGNS FOR 3^m FACTORIAL EXPERIMENTS

In the following examples, N is the total number of treatments in a design, m is the number of factors of interest, and ν is the number of parameters in $\underline{F}$. We assume that m factors are divided into p subgroups, and each subgroup has m_k ($k = 1, \ldots, p$) factors in it. The two-factor interactions are either within or between some of the subgroups. All of the following designs of parallel flats type satisfy the conditions in Theorem 1.

Case 1. Consider a 3^4 factorial experiment. The factorial effects given in the vector

$$\begin{aligned}\underline{F} = (&\mu, F_1, F_2, F_3, F_4, F_1^2, F_2^2, F_3^2, F_4^2, F_1F_2, F_1F_3, \\ &F_2F_3, F_1F_2^2, F_1F_3^2, F_2F_3^3, F_1^2F_2, F_1^2F_3, F_2^2F_3, \\ &F_1^2F_2^2, F_1^2F_3^2, F_2^2F_3^2) \\ = (&0, 1, 2, 3, 4, 1^2, 2^2, 3^2, 4^2, 12, 13, 23, 12^2, 13^2, \\ &23^2, 1^22, 1^23, 2^23, 1^22^2, 1^23^2, 2^23^2)\end{aligned}$$

have the property that the two-factor interactions occur within the subgroup $\{1, 2, 3\}$. Consider the design T given by the linear equation

$$A\underline{t} = t_1 + t_2 + t_3 + t_4 = C$$

over $GF(3)$, where $A = (1\ 1\ 1\ 1)$, and $C = \underline{0}$. By Theorem 1, the design

$$T = \begin{pmatrix} 0&0&0&0&0&0&0&0&0&1&1&1&1&1&1&1&1&1&2&2&2&2&2&2&2&2&2 \\ 0&0&0&1&1&1&2&2&2&0&0&0&1&1&1&2&2&2&0&0&0&1&1&1&2&2&2 \\ 0&1&2&0&1&2&0&1&2&0&1&2&0&1&2&0&1&2&0&1&2&0&1&2&0&1&2 \\ 0&2&1&2&1&0&1&0&2&2&1&0&1&0&2&0&2&1&1&0&2&0&2&1&2&1&0 \end{pmatrix}$$

is an orthogonal design for estimating $\underline{F}$. Here, $m = 4$, $\nu = 21$, $p = 2$, $m_1 = 3$, $m_2 = 1$, $N = 27$.

In the following examples of orthogonal designs, we only give matrices A, C, and $\underline{F}$.

Case 2. Consider a 3^5 factorial experiment. The factorial effects given in the vector

$$\begin{aligned}\underline{F}' = (&0, 1, \ldots, 5, 1^2, \ldots, 5^2, 12, 13, 23, 12^2, 13^2, \\ &23^2, 1^22, 1^23, 2^23, 1^22^2, 1^23^2, 2^23^2)\end{aligned}$$

are such that the two-factor interactions occur within the subgroup $\{1, 2, 3\}$. The design T given by $A\underline{t} = C$, where

$$A = \begin{pmatrix} 1 & 0 & 0 & 1 & 1 \\ 0 & 1 & 0 & 1 & 1 \end{pmatrix}, \qquad C = \underline{0}$$

is an orthogonal design for estimating F. Here, $m = 5$, $\nu = 23$, $p = 2$, $m_1 = 3$, $m_2 = 2$, $N = 27$.

Case 3. Consider a 3^6 factorial experiment. The factorial effects given in the vector

$$\begin{aligned}\underline{F}' = (&0, 1, \ldots, 6, 1^2, \ldots, 6^2, 12, 13, 23, 12^2, 13^2, \\ &23^2, 1^22, 1^23, 2^23, 1^22^2, 1^23^2, 2^23^3)\end{aligned}$$

are such that the two-factor interactions occur within the subgroup $\{1, 2, 3\}$. The orthogonal design T given by $A\underline{t} = \underline{0}$, where

$$A = \begin{pmatrix} 1 & 0 & 0 & 0 & 1 & 1 \\ 0 & 1 & 0 & 1 & 1 & 0 \\ 0 & 0 & 1 & 1 & 0 & 1 \end{pmatrix}, \qquad C = \underline{0}$$

is an orthogonal design for estimating $\underline{F}$. Here, $m = 6$, $\nu = 25$, $p = 2$, $m_1 = 3$, $m_2 = 3$, $N = 27$.

In the following examples, the two-factor interactions occur between subgroups.

Case 4. Consider a 3^4 factorial experiment. The factorial effects given in the vector

$$\begin{aligned}\underline{F}' = (&0, 1, 2, 3, 4, 1^2, 2^2, 3^2, 4^2, 14, 24, 34, 14^2, 24^2, \\ &34^2, 1^24, 2^24, 3^24, 1^24^2, 2^24^2, 3^24^2)\end{aligned}$$

are such that the two-factor interactions are between the subgroups $\{1, 2, 3\}$ and $\{4\}$. By Theorem 1, the design T given by the solution to the linear equation $A\underline{t} = t_1 + t_2 + t_3 + t_4 = C$, where $A = (1\ 1\ 1\ 1)$ and $C = \underline{0}$, is an orthogonal design for estimating $\underline{F}$. Here $m = 4$, $\nu = 21$, $p = 2$, $m_1 = 3$, $m_2 = 1$, $N = 27$.

Case 5. Consider a 3^5 factorial experiment. The factorial effects given in the vector

$$\underline{F}' = (0, 1, \ldots, 5, 1^2, \ldots, 5^2, 15, 25, 35, 45, 15^2, 25^2, 35^2, 45^2, 1^25, 2^25, 3^25, 4^25, 1^25^2, 2^25^2, 3^25^2, 4^25^2)$$

are such that the two-factor interactions are between the subgroups $\{1, 2, 3, 4\}$ and $\{5\}$. The design T given by $A\underline{t} = C$ is an orthogonal design for estimating $\underline{F}$, where

$$A = \begin{pmatrix} 1 & 1 & 1 & 1 & 0 \\ 0 & 1 & 1 & 1 & 1 \end{pmatrix}, \qquad C = \underline{0}$$

Here, $m = 5$, $\nu = 27$, $p = 2$, $m_1 = 4$, $m_2 = 1$, $N = 27$. The number of the parameters in $\underline{F}$ is equal to the number of treatments in T. The orthogonal design T is a saturated orthogonal design.

Case 6. Consider a 3^6 factorial experiment. The factorial effects given in the vector

$$\underline{F}' = (0, 1, \ldots, 6, 1^2, \ldots, 6^2, 14, 24, 34, 14^2, 24^2, 34^2, 1^24, 2^24, 3^24, 1^24^2, 2^24^2, 3^24^2)$$

are such that the two-factor interactions are between the subgroups $\{1, 2, 3\}$ and $\{4\}$. The subgroup $\{5, 6\}$ is not involved in the two-factor interactions. By Theorem 1, the design T given by $A\underline{t} = C$ is an orthogonal design for estimating $\underline{F}$, where

$$A \begin{pmatrix} 1 & 0 & 1 & 0 & 1 & 0 \\ 0 & 1 & 0 & 0 & 1 & 1 \\ 1 & 1 & 1 & 1 & 1 & 0 \end{pmatrix}, \qquad C = \underline{0}$$

Here, $m = 6$, $\nu = 25$, $p = 3$, $m_1 = 3$, $m_2 = 1$, $m_3 = 2$, $N = 27$.

Case 7. Consider a 3^7 factorial experiment. The factorial effects given in the vector

$$\underline{F}' = (0, 1, \ldots, 7, 1^2, \ldots, 7^2, 14, 24, 34, 14^2, 24^2, 34^2, 1^24, 2^24, 3^24, 1^24^2, 2^24^2, 3^24^2)$$

are such that the two-factor interactions are between the subgroups $\{1, 2, 3\}$ and $\{4\}$. The subgroup $\{5, 6, 7\}$ is not involved in the two-factor interactions. The design T given by $A\underline{t} = C$ is an orthogonal design for estimating $\underline{F}$, where

$$A = \begin{pmatrix} 1 & 1 & 1 & 0 & 0 & 0 & 0 \\ 1 & 0 & 1 & 0 & 0 & 0 & 1 \\ 0 & 0 & 0 & 1 & 0 & 1 & 1 \\ 0 & 0 & 0 & 0 & 1 & 1 & 1 \end{pmatrix}, \qquad C = \underline{0}$$

Here, $m = 7$, $\nu = 25$, $p = 3$, $m_1 = 3$, $m_2 = 1$, $m_3 = 3$, $N = 27$.

Bose and Bush (1952) gave an orthogonal array $H(7 \times 18)$ of strength 2 as follows:

$$H = \begin{pmatrix} 0&1&2&0&1&2&0&1&2&0&1&2&0&1&2&0&1&2 \\ 0&1&2&0&1&2&1&2&0&2&0&1&1&2&0&2&0&1 \\ 0&1&2&1&2&0&0&1&2&2&0&1&2&0&1&1&2&0 \\ 0&1&2&2&0&1&2&0&1&0&1&2&1&2&0&1&2&0 \\ 0&1&2&1&2&0&2&0&1&1&2&0&0&1&2&2&0&1 \\ 0&1&2&2&0&1&1&2&0&1&2&0&2&0&1&0&1&2 \\ 0&0&0&0&0&0&1&1&1&1&1&1&2&2&2&1&2&2 \end{pmatrix} \tag{7}$$

Notice that any submatrix $\hat{H}(2 \times 18)$ of H is $OA(2)$. Therefore, a linear combination of any two rows of H over $GF(3)$ is an $OA(1)$. Now we use the matrix H to construct orthogonal designs.

Case 8. Consider a 3^4 factorial experiment. The factorial effects given in the vector

$$\underline{F}' = (0, 1, 2, 3, 4, 1^2, 2^2, 3^2, 4^2, 12, 13, 23, 12^2, 13^2, 23^2, 1^22, 1^23, 2^23, 1^22^2, 1^23^2, 2^23^3)$$

are such that the two-factor interactions occur within the subgroup $\{1, 2, 3\}$. An orthogonal design for estimating $\underline{F}$ is given by $A\underline{t} = C_{3\times18}$, where

$$A = \begin{pmatrix} 1 & 0 & 0 & 1 \\ 0 & 1 & 0 & 0 \\ 0 & 0 & 1 & 0 \end{pmatrix}$$

and $C_{3\times18}$ consists of the first three rows of H in Eq. (7). Here, $m = 4$, $\nu = 21$, $p = 2$, $m_1 = 3$, $m_2 = 1$, $N = 3 \times 18 = 54$.

Case 9. Consider a 3^5 factorial experiment. The factorial effects given in the vector

$$\underline{F}' = (0, 1, \ldots, 5, 1^2, \ldots, 5^2, 15, \ldots, 45, 15^2, \ldots, 45^2, 1^25, \ldots, 4^25, 1^25^2, \ldots, 4^25^2)$$

are such that the two-factor interactions occur between the subgroups $\{1, 2, 3, 4\}$ and $\{5\}$. An orthogonal design for estimating $\underline{F}$ is given by $A\underline{t} = C_{4\times18}$, where

$$A = \begin{pmatrix} 1 & 0 & 0 & 0 & 0 \\ 0 & 1 & 0 & 0 & 0 \\ 0 & 0 & 1 & 0 & 0 \\ 0 & 0 & 0 & 1 & 1 \end{pmatrix}$$

and $C_{4\times18}$ consists of the first four rows of H in Eq. (7). Here, $m = 5$, $\nu = 27$, $p = 2$, $m_1 = 4$, $m_2 = 1$, $N = 3 \times 18 = 54$.

Case 10. Consider a 3^6 factorial experiment. The factorial effects given in the vector

$$\underline{F}' = (0, 1, \ldots, 6, 1^2, \ldots, 6^2, 16, \ldots, 56, 16^2, \ldots, 56^2, 1^26, \ldots, 5^26, 1^26^2, \ldots, 5^26^2)$$

are such that the two-factor interactions occur between the subgroups {1, 2, 3, 4, 5} and {6}. An orthogonal design for estimating $\underline{F}$ is given by $A\underline{t} = C_{5\times 18}$, where

$$A = \begin{pmatrix} 1 & 0 & 0 & 0 & 0 & 0 \\ 0 & 1 & 0 & 0 & 0 & 0 \\ 0 & 0 & 1 & 0 & 0 & 0 \\ 0 & 0 & 0 & 1 & 0 & 0 \\ 0 & 0 & 0 & 0 & 1 & 1 \end{pmatrix}$$

and $C_{5\times 18}$ consists of the first five rows of H in Eq. (7). Here, $m = 6$, $\nu = 33$, $p = 2$, $m_1 = 5$, $m_2 = 1$, $N = 3 \times 18 = 54$.

Case 11. Consider a 3^7 factorial experiment. The factorial effects given in the vector

$$\underline{F}' = (0, 1, \ldots, 7, 1^2, \ldots, 7^2, 17, \ldots, 67, 17^2, \ldots, 67^2, 1^27, \ldots, 6^27, 1^27^2, \ldots, 6^27^2)$$

have the property that the two-factor interactions occur between the subgroups {1, . . . , 6} and {7}. An orthogonal design for estimating $\underline{F}$ is given by $A\underline{t} = C_{6\times 18}$, where

$$A = \begin{pmatrix} 1 & 0 & 0 & 0 & 0 & 0 & 0 \\ 0 & 1 & 0 & 0 & 0 & 0 & 0 \\ 0 & 0 & 1 & 0 & 0 & 0 & 0 \\ 0 & 0 & 0 & 1 & 0 & 0 & 0 \\ 0 & 0 & 0 & 0 & 1 & 0 & 0 \\ 0 & 0 & 0 & 0 & 0 & 1 & 1 \end{pmatrix}$$

and $C_{6\times 18}$ consists of the first six rows of H in Eq. (7). Here, $m = 7$, $\nu = 39$, $p = 2$, $m_1 = 6$, $m_2 = 1$, $N = 3 \times 18 = 54$.

Case 12. Consider a 3^8 factorial experiment. The factorial effects given in the vector

$$\underline{F}' = (0, 1, \ldots, 8, 1^2, \ldots, 8^2, 18, \ldots, 78, 18^2, \ldots, 78^2, 1^28, \ldots, 7^28, 1^28^2, \ldots, 7^28^2)$$

have the property that the two-factor interactions occur between the subgroups {1, . . . , 7} and {8}. An orthogonal design for estimating $\underline{F}$ is given by $A\underline{t} = C_{7\times 18}$, where

$$A = \begin{pmatrix} 1 & 0 & 0 & 0 & 0 & 0 & 0 & 0 \\ 0 & 1 & 0 & 0 & 0 & 0 & 0 & 0 \\ 0 & 0 & 1 & 0 & 0 & 0 & 0 & 0 \\ 0 & 0 & 0 & 1 & 0 & 0 & 0 & 0 \\ 0 & 0 & 0 & 0 & 1 & 0 & 0 & 0 \\ 0 & 0 & 0 & 0 & 0 & 1 & 0 & 0 \\ 0 & 0 & 0 & 0 & 0 & 0 & 1 & 1 \end{pmatrix}$$

and $C_{7\times 18} = H$ given by Eq. (7). Here, $m = 8$, $\nu = 45$, $p = 2$, $m_1 = 7$, $m_2 = 1$, $N = 3 \times 18 = 54$.

V. ORTHOGONAL DESIGNS FOR MIXED $2^n \times 3^m$ FACTORIAL EXPERIMENTS

The mixed $2^n \times 3^m$ factorial experiments arise very commonly in practice. It is useful to be able to generate an orthogonal design to estimate any specified set of mixed factorial effects. Consider the parallel flats type of designs for the mixed factorial experiment. Let T_{1i} be the set of solutions to the linear system

$$A_1\underline{t}_1 = \underline{c}_{1i}, \qquad (i = 1, 2, \ldots, f) \tag{8}$$

over $GF(2)$. Let T_{2i} be the set of solutions to the linear system

$$A_2\underline{t}_2 = \underline{c}_{2i}, \qquad (i = 1, 2, \ldots, f) \tag{9}$$

over $GF(3)$. The PFD design T for $2^n \times 3^m$ with $N = f \times 2^{n-r_1} \times 3^{m-r_2}$ treatments is defined as follows:

$$T = T_{11} \otimes_f T_{21} + \cdots + T_{1f} \otimes_f T_{2f} \tag{10}$$

where $\otimes_f$ is the product defined in Connor and Young (1961), shown as in Cases 1–4 of this section.

The extension of Theorem 1 to the mixed case was given in Liao (1994). Now we review the results. First consider the following definition.

Definition 1. Let Q be a $2 \times f$ array whose first row has the elements from the set {0, 1}, and whose second row has the elements from the set {0, 1, 2}. Let k_{ij} be the number of column vectors $(i, j)'$ of Q ($i = 0, 1$; $j = 0, 1, 2$).

1. Q is said to have the property O_1 if $k_{00} - k_{10} = k_{01} - k_{11} = k_{02} - k_{12}$.
2. Q is said to have property O_2 if $k_{00} = k_{10}$, $k_{01} = k_{11}$ and $k_{02} = k_{12}$.
3. Q is said to have property O_3 if $k_{00} = k_{01} = k_{02}$ and $k_{10} = k_{11} = k_{12}$.
4. Q is said to have property O_4 if $k_{00} = k_{01} = k_{02} = k_{10} = k_{11} = k_{12}$; i.e., Q is an $OA(2)$.

Notice that the model of Eq. (1) and the defining vectors can easily be extended to the mixed $2^n \times 3^m$ factorial experiment. We use the same notation as in Sec. II for the mixed experiments.

Let $\underline{F}$ be a set of factorial effects in a $2^n \times 3^m$ experiment. Suppose that e_1 and e_2 are any effects in $\underline{F}$, and $\underline{e}_1$ and $\underline{e}_2$ are the corresponding defining vectors. Let $\underline{e}_1' = (\underline{e}_{11}', \underline{e}_{21}')$ and $\underline{e}_2' = (\underline{e}_{12}', \underline{e}_{22}')$, where $\underline{e}_{11}'(1 \times n)$ and $\underline{e}_{12}(1 \times n)$ are over $GF(2)$, $\underline{e}_{21}'(1 \times m)$ and $\underline{e}_{22}(1 \times m)$ are over $GF(3)$. Then we define the operation $\oplus$ as

$$\underline{e} \oplus \gamma\underline{e}_2 = (\underline{e}_{11} + \gamma\underline{e}_{12}, \underline{e}_{21} + \gamma\underline{e}_{22})$$

Define

$$\underline{F}^2 = \{\underline{e}_1 \oplus \gamma\underline{e}_2 \mid \text{for } e_1, e_2 \in \underline{F}, \gamma = 1, 2\}$$

Theorem 2. (Liao, 1994) In a $2^n \times 3^m$ experiment, let design T be described as in Eq. (10), and let M be the information matrix for estimating $\underline{F}$ by using T. Then T is orthogonal if and only if the following conditions hold.

1. For $(\underline{e}_1, \underline{0}) \in \underline{F}^2$, if there exists a nonzero vector $\underline{u}_1'(1 \times r_1)$ over $GF(2)$ such that $\underline{e}_1 = \underline{u}_1'A_1$, then $\underline{u}_1'C_1$ must be an $OA(1)$, where $C_1 = (\underline{c}_{11}, \underline{c}_{12}, \ldots, \underline{c}_{1f})$.
2. For $(\underline{0}, \underline{e}_2) \in \underline{F}^2$, if there exists a nonzero vector $\underline{u}_2'(1 \times r_2)$ over $GF(3)$ such that $\underline{e}_2 = \underline{u}_2'A_2$, then $\underline{u}_2'C_2$ must be an $OA(1)$, where $C_2 = (\underline{c}_{21}, \underline{c}_{22}, \ldots, \underline{c}_{2f})$.
3. Suppose $(\underline{e}_1, \underline{e}_2) \in \underline{F}^2$, but $(\underline{e}_1, \underline{0}), (\underline{0}, \underline{e}_2) \notin \underline{F}^2$. If there exist a nonzero vector $\underline{u}_1'(1 \times r_1)$ over $GF(2)$ and a nonzero vector $\underline{u}_2'(1 \times r_2)$ over $GF(3)$ such that $\underline{e}_1 = \underline{u}_1'A_1$ and $\underline{e}_2 = \underline{u}_2'A_2$, then the matrix $\begin{bmatrix} \underline{u}_1' & C_1 \\ \underline{u}_2' & C_2 \end{bmatrix}$ must have property O_1.
4. Suppose $(\underline{e}_1', e_2), (\underline{e}_1, \underline{0}) \in \underline{F}^2$, but $(\underline{0}, \underline{e}_2) \notin \underline{F}^2$. If there exist a nonzero vector $\underline{u}_1'(1 \times r_1)$ over $GF(2)$ and a nonzero vector $\underline{u}_2'(1 \times r_2)$ over $GF(3)$ such that $\underline{e}_1 = \underline{u}_1'A_1$ and $\underline{e}_2 = \underline{u}_2'A_2$, then $\begin{bmatrix} \underline{u}_1' & C_1 \\ \underline{u}_2' & C_2 \end{bmatrix}$ must have property O_2.
5. Suppose $(\underline{e}_1, \underline{e}_2), (\underline{0}, \underline{e}_2) \in \underline{F}^2$, but $(\underline{e}_1, \underline{0}) \notin \underline{F}^2$. If there exist a nonzero vector $\underline{u}_1'(1 \times r_1)$ over $GF(2)$ and a nonzero vector $\underline{u}_2'(1 \times r_2)$ over $GF(3)$ such that $\underline{e}_1 = \underline{u}_2'A_2$, then $\begin{bmatrix} \underline{u}_1' & C_1 \\ \underline{u}_2' & C_2 \end{bmatrix}$ must have property O_3.
6. Suppose $(\underline{e}_1, \underline{e}_2), (\underline{e}_1, \underline{0}), (\underline{0}, \underline{e}_2) \in \underline{F}^2$. If there exist a nonzero vector $\underline{u}_1'(1 \times r_1)$ over $GF(2)$ and a nonzero vector $\underline{u}_2'(1 \times r_2)$ over $GF(3)$ such that $\underline{e}_1 = \underline{u}_1'A_1$ and $\underline{e}_2 = \underline{u}_2'A_2$, then $\begin{bmatrix} \underline{u}_1' & C_1 \\ \underline{u}_2' & C_2 \end{bmatrix}$ must have property O_4.

Now we apply Theorem 2 to generate the mixed factorial experiments.

Case 1. Consider a $2^3 \times 3$ factorial experiment. There are four factors F_1, F_2, F_3, and F_4, where F_1, F_2, and F_3 take two levels $\{0, 1\}$ and F_4 takes three levels $\{0, 1, 2\}$. The factorial effects of interest are given in the vector

$$\underline{F}' = (\mu, F_1, F_2, F_3, F_4, F_4^2, F_1F_4, F_2F_4, F_3F_4, F_1F_4^2, F_2F_4^2, F_3F_4^2)$$
$$= (0, 1, 2, 3, 4, 4^2, 14, 24, 34, 14^2, 24^2, 34^2)$$

The two-factor interactions in $\underline{F}$ are between the subgroups $\{1, 2, 3\}$ and $\{4\}$. Let $T_1 = [T_{11}:T_{12}:T_{13}]$, and T_{1i} $(i = 1, 2, 3)$ with size (3×4) be the solutions to the linear $A_1\underline{t}_1 = \underline{c}_{1i}$, where $\underline{t}_1' = (t_1, t_2, t_3)$, $A_1 = (111)$, and $\underline{c}_{1i} = 0$. Let $T_2 = [T_{21}:T_{22}:T_{23}]$ and let T_{2i} $(i = 1, 2, 3)$ with size (1×3) be the solutions to the linear $A_2\underline{t} = \underline{c}_{2i}$, where $\underline{t}_2' = (t_4)$, $A_2 = (1)$, and $\underline{c}_{2i} = 0$. Here $n = 3$, $m = 1, f = 3$. Clearly T_1 and T_2 are orthogonal designs for estimating the main effects of a 2^3 and a 3^1 factorial, respectively. The linear system

$$A_1\underline{t}_1 = \underline{c}_{1i}$$
$$A_2\underline{t}_2 = \underline{c}_{2i}, \qquad \text{for } (i = 1, 2, 3)$$

satisfies the conditions in Theorem 2. Therefore the design

$$T = T_{11} \otimes_f T_{21} + T_{12} \otimes_f T_{22} + T_{13} \otimes_f T_{23}$$
$$= \begin{pmatrix} 0&1&1&0&0&1&1&0&0&1&1&0 \\ 0&1&0&1&0&1&0&1&0&1&0&1 \\ 0&0&1&1&0&0&1&1&0&0&1&1 \\ 0&0&0&0&1&1&1&1&2&2&2&2 \end{pmatrix}$$

obtained as in Eq. (10) is an orthogonal design for estimating $\underline{F}$. Here number of treatments is $N = 3 \times 4 = 12$, the number of parameters in $\underline{F}$ is $\nu = 12 = N$. The design T is a saturated orthogonal design.

Case 2. Consider a $2^4 \times 3$ factorial experiment. The factorial effects of interest given in the vector $\underline{F}$ as

$$\underline{F}' = (0, 1, 2, 3, 4, 5, 5^2, 12, 13, 23, 15, 25, 35, 45, 15^2, 25^2, 35^2, 45^2)$$

are such that the two-factor interactions are either within the subgroup $\{1, 2, 3\}$ or are between the subgroups $\{1, 2, 3, 4\}$ and $\{5\}$. Let $T_1 = [T_{11}:T_{12}:T_{13}]$ and let T_{1i} $(i = 1, 2, 3)$ with size (4×8) be the solutions to the linear $A_1\underline{t}_1 = \underline{c}_{1i}$, where $A_1 = (1111)$ and $\underline{c}_i = 0$. Let $T_2 = [T_{21}:T_{22}:T_{23}]$, where T_{2i} $(i = 1, 2, 3)$ with size (1×3) are the solutions to the linear $A_2\underline{t}_2 = \underline{c}_{2i}$, where $\underline{t}_2' = (t_4)$, $A_2 = (1)$, and $\underline{c}_{2i} = 0$. Here $n = 4$, $m = 1$, $f = 3$. The linear system

$$A_1\underline{t}_1 = \underline{c}_{1i}$$
$$A_2\underline{t}_2 = \underline{c}_{2i}, \qquad \text{for } (i = 1, 2, 3)$$

satisfies the conditions in Theorem 2. Therefore the design

$$T = T_{11} \otimes_f T_{21} + T_{12} \otimes_f T_{22} + T_{13} \otimes_f T_{23}$$

$$= \begin{pmatrix} 0&1&1&1&0&0&0&1&0&1&1&1&0&0&0&1&0&1&1&1&0&0&0&1 \\ 0&1&0&0&1&1&0&1&0&1&0&0&1&1&0&1&0&1&0&0&1&1&0&1 \\ 0&0&1&0&1&0&1&1&0&0&1&0&1&0&1&1&0&0&1&0&1&0&1&1 \\ 0&0&0&1&0&1&1&1&0&0&0&1&0&1&1&1&0&0&0&1&0&1&1&1 \\ 0&0&0&0&0&0&0&0&1&1&1&1&1&1&1&1&2&2&2&2&2&2&2&2 \end{pmatrix}$$

obtained as in Eq. (10) is an orthogonal design for estimating $\underline{F}$. Here number of treatments is $N = 3 \times 8 = 24$, and the number of parameters in $\underline{F}$ is $\nu = 18$.

Now consider orthogonal arrays for mixed designs. An $(n + m) \times N$ matrix T is said to be an $OA((n + m), N, 2^n \times 3^m, d)$, if:

1. Its first n rows have elements from the set $\{0, 1\}$ and the last m rows have elements from the set $\{0, 1, 2\}$.
2. Any $d \times N$ submatrix T_0 of T has the property that all possible columns vectors of T_0 occur in T_0 with the same frequency.

It is easy to see that the following matrix H is an $OA(5, 12, 2^4 \times 3, 2)$:

$$H = \begin{pmatrix} 0&0&1&1&0&0&1&1&0&0&1&1 \\ 0&1&0&1&0&1&0&1&0&1&0&1 \\ 0&0&1&1&1&1&0&0&1&0&0&1 \\ 0&1&1&0&1&0&1&0&0&1&0&1 \\ 0&0&0&0&1&1&1&1&2&2&2&2 \end{pmatrix}$$

Now we use the matrix H to generate the following orthogonal designs.

Case 3. Consider a $2^5 \times 3$ factorial experiment. The factorial effects given in the vector

$$\underline{F}' = (0, 1, \ldots, 5, 6, 6^2, 16, 26, 36, 46, 56, 16^2, 26^2, 36^2, 46^2, 56^2)$$

have the property that the two-factor interactions are between the subgroups $\{1, 2, 3, 4, 5\}$ and $\{6\}$. Let C_1 (4×12) be the first four rows of H, and let $T_1 = [T_{1,1}\!: \cdots T_{1,12}]$, $T_{1,i}$ $(i = 1, \ldots, 12)$ with size (5×2) be the solutions to the linear $A_1\underline{t}_1 = \underline{c}_{1,i}$, where $\underline{c}_{1,i}$ is the ith column of C_1, and

$$A_1 = \begin{pmatrix} 1&0&0&0&1 \\ 0&1&0&0&0 \\ 0&0&1&0&0 \\ 0&0&0&1&0 \end{pmatrix}$$

Let $C_2(1 \times 12)$ be the last row of H. Let $T_2 = [T_{2,1}\!: \cdots T_{2,12}]$, and let $T_{1,i}$ $(i = 1, \ldots, 12)$ with size (1×1) be the solutions to the linear $A_2\underline{t}_2 = \underline{c}_{2,i}$ over $GF(3)$, where $A_2 = (1)$, $\underline{t}_2 = t_6$, $\underline{c}_{2,i}$ is the ith column of C_2. The linear system

$$A_1\underline{t}_1 = \underline{c}_{1i}$$

$$A_2\underline{t}_2 = \underline{c}_{2i}, \qquad \text{for } (i = 1, \ldots, 12)$$

satisfies the conditions in Theorem 2. Therefore the design

$$T = T_{1,1} \otimes_f T_{2,1} + \cdots + T_{1,12} \otimes_f T_{2,12}$$

$$= \begin{pmatrix} 0&1&0&1&0&1&0&1&0&1&0&1&0&1&0&1&0&1&0&1&0&1&0&1 \\ 0&0&1&1&0&0&1&1&0&0&1&1&0&0&1&1&0&0&1&1&0&0&1&1 \\ 0&0&0&0&1&1&1&1&1&1&1&1&0&0&0&0&1&1&0&0&0&0&1&1 \\ 0&0&1&1&1&1&0&0&1&1&0&0&1&1&0&0&0&0&1&1&0&0&1&1 \\ 0&1&0&1&0&1&0&1&0&1&0&1&0&1&0&1&0&1&0&1&0&1&0&1 \\ 0&0&0&0&0&0&0&0&1&1&1&1&1&1&1&1&2&2&2&2&2&2&2&2 \end{pmatrix}$$

obtained as in Eq. (10) is an orthogonal design for estimating $\underline{F}$. Here the number of treatments is $N = 12 \times 2^{5-4} \times 3^{1-1} = 24$, and the number of parameters in $\underline{F}$ is $\nu = 18$.

Case 4. Consider a $2^4 \times 3^2$ factorial experiment. The factorial effects of interest given in the vector

$$\underline{F}' = (0, 1, \ldots, 5, 6, 5^2, 6^2, 16, 26, 36, 46, 56, 16^2, 26^2, 36^2, 46^2, 56^2, 5^26, 5^26^2)$$

are such that the two-factor interactions are between the subgroups $\{1, 2, 3, 4, 5\}$ and $\{6\}$. Take $T_1 = H$. Therefore the design

$$T = T_1 \otimes_f (0, 1, 2)'$$

$$= H \otimes_f (9, 1, 2)'$$

$$= \begin{pmatrix} 0&0&1&1&0&0&1&1&0&0&1&1&0&0&1&1&0&0&1&1&0&0&1&1&0&0&1&1&0&0&1&1&0&0&1&1 \\ 0&1&0&1&0&1&0&1&0&1&0&1&0&1&0&1&0&1&0&1&0&1&0&1&0&1&0&1&0&1&0&1&0&1&0&1 \\ 0&0&1&1&1&1&0&0&1&0&0&1&0&0&1&1&1&1&0&0&1&0&0&1&0&0&1&1&1&1&0&0&1&0&0&1 \\ 0&1&1&0&1&0&1&0&0&1&0&1&0&1&1&0&1&0&1&0&0&1&0&1&0&1&1&0&1&0&1&0&0&1&0&1 \\ 0&0&0&0&1&1&1&1&2&2&2&2&0&0&0&0&1&1&1&1&2&2&2&2&0&0&0&0&1&1&1&1&2&2&2&2 \\ 0&0&0&0&0&0&0&0&0&0&0&0&1&1&1&1&1&1&1&1&1&1&1&1&2&2&2&2&2&2&2&2&2&2&2&2 \end{pmatrix}$$

is an orthogonal design for estimating $\underline{F}$. Here the number of treatments is $N = 12 \times 3 = 36$, and the number of parameters in $\underline{F}$ is $\nu = 21$.

VI. APPLICATION OF MIXED FACTORIAL DESIGN IN BIOMEDICAL RESEARCH

In this section, we consider two examples in the biomedical research area, in which factorial experiments are widely used for screening the important factors or studying the higher-factor interactions.

A. Example 1: Allele-Specific Polymerase Chain Reaction for Screening Point Mutation

Polymerase chain reaction (PCR) is a powerful biological technique to facilitate molecular analysis, including the DNA diagnosis and identification of a human being's genetic disease. Many components affect a PCR process. In addition, there may be complex interactions among the components of PCR. Factorial designs are considered as cost-effective techniques to achieve optimization of PCR.

In an experiment we consider, the PCR yield is determined by two procedures, DNA extraction and PCR conditions. The five most important factors (or components) are identified in the PCR array as follows:

1. F_1 = autoclave, the degree of autoclave blood spots prior to DNA extraction
2. F_2 = DNA extraction, the methods of extracting human genetic DNA from dried blood spots
3. F_3 = number of blood spots in each DNA extraction tube
4. F_4 = template DNA, the volume of template DNA used in the PCR
5. F_5 = magnesium concentration

The first four factors, F_1 to F_4, take on two levels, coded as $\{0, 1\}$. The last factor, F_5 (magnesium), takes on three levels, coded as $\{0, 1, 2\}$. Prior experience indicates that there may be interactions among autoclave, DNA extraction, and number of blood spots in each of the DNA levels. Also, magnesium is very active, and it may interact with other factors. Therefore we need to estimate the two-factor interactions within the subgroup $\{F_1, F_2, F_3\}$ and the two-factor interactions between the subgroups $\{F_1, F_2, F_3, F_4\}$ and $\{F_5\}$, besides the general mean and all the main effects. The parameter vector of interest can be written as

$$\underline{F}' = (\mu, F_1, F_2, F_3, F_4, F_5, F_5^2, F_1F_2, F_1F_3, F_2F_3, F_1F_5, F_2F_5, F_3F_5, F_4F_5, F_1F_5^2, F_2F_5^2, 2F_3F_5^2, F_4F_5^2)$$

By Sec. V, Case 2, there exists an orthogonal design with 24 treatment combinations to estimate the parameter vector $\underline{F}$. The orthogonal design is presented as follows

Treatment index	1	2	3	4	5	6	7	8	9	10	11	12	13	14	15	16	17	18	19	20	21	22	23	24
Autoclave	0	1	1	1	0	0	0	1	0	1	1	1	0	0	0	1	0	1	1	1	0	0	0	1
DNA extraction	0	1	0	0	1	1	0	1	0	1	0	0	1	1	0	1	0	1	0	0	1	1	0	1
No. blood spots	0	0	1	0	1	0	1	1	0	0	1	0	1	0	1	1	0	0	1	0	1	0	1	1
Template DNA	0	0	0	1	0	1	1	1	0	0	0	1	0	1	1	1	0	0	0	1	0	1	1	1
Magnesium	0	0	0	1	0	1	1	1	1	1	1	1	1	1	1	1	2	2	2	2	2	2	2	2

Here the 24 columns correspond to the 24 treatments of different factorial combinations, which can give optimal estimates of 18 parameters in $\underline{F}$.

B. Example 2: Drug Interaction in a Clinical Study

It is very common for patients with life-threatening diseases, such as AIDS and cancer, to undergo multiple therapies. Suppose, in a clinical study, we want to study the drug interactions among patients in different demographic groups. The following factors affect the outcome of therapy:

1. F_1 = gender of patients, with two levels (male = 0, female = 1)
2. F_2 = drug A, with two doses, denoted as 0 and 1
3. F_3 = drug B, with two doses, denoted as 0 and 1
4. F_4 = drug C, with two doses, denoted as 0 and 1
5. F_5 = severity of the disease, with three levels (mild = 0, moderate = 1, severe = 2)
6. F_6 = newly approved drug D, with three doses, denoted as (0, 1, 2)

The objective is to study the synergistic or antagonistic effects between newly approved treatment D and existing treatments A, B, and C in male and female patients with different levels of severity of the disease. Based on previous experiments, it is believed that interactions occur between D and A, B, C. The three-factor interactions can be ignored. We are interested in estimating all the main effects for each factor and two-factor interactions between drug D and the rest of the factors. Therefore, the 21 factor effects of interest are given in the vector

$$\underline{F}' = (\mu, F_1, \ldots, F_6, F_5^2, F_6^2, F_1F_6, F_2F_6, F_3F_6, F_4F_6, F_5F_6, F_1F_6^2, F_2F_6^2, F_3F_6^2, F_4F_6^2, F_5F_6^2, F_4^2F_6, F_5^2F_6^2)$$

By Case 4 in Sec. V, the following design is an orthogonal design for estimating $\underline{F}$:

Index	1	2	3	4	5	6	7	8	9	10	11	12	13	14	15	16	17	18
Gender	0	0	1	1	0	0	1	1	0	0	1	1	0	0	1	1	0	0
Drug A	0	1	0	1	0	1	0	1	0	1	0	1	0	1	0	1	0	1
Drug B	0	0	1	1	1	1	0	0	1	0	0	1	0	0	1	1	1	1
Drug C	0	1	1	0	1	0	1	0	0	1	0	1	0	1	1	0	1	0
Severity	0	0	0	0	1	1	1	1	2	2	2	2	0	0	0	0	1	1
Drug D	0	0	0	0	0	0	0	0	0	0	0	0	1	1	1	1	1	1

Index	19	20	21	22	23	24	25	26	27	28	29	30	31	32	33	34	35	36
Gender	1	1	0	0	1	1	0	0	1	1	0	0	1	1	0	0	1	1
Drug A	0	1	0	1	0	1	0	1	0	1	0	1	0	1	0	1	0	1
Drug B	0	0	1	0	0	1	0	0	1	1	1	1	0	0	1	0	0	1
Drug C	1	0	0	1	0	1	0	1	1	0	1	0	1	0	0	1	0	1
Severity	1	1	2	2	2	2	0	0	0	0	1	1	1	1	2	2	2	2
Drug D	1	1	1	1	1	1	2	2	2	2	2	2	2	2	2	2	2	2

where the 36 columns of factor level combinations are the treatments in the orthogonal design.

REFERENCES

Bose, R. C. (1947). Mathematical theory of the symmetrical factorial design. Sankhya 8:107–166.

Bose, R. C., and Bush, K. A. (1952). Orthogonal arrays of strength two and three. Ann Math Stat 23:502–524.

Bose, R. C., and Srivastava, J. N. (1964). Analysis of irregular factorial fractions. Sankhya A 26:117–144.

Cheng, C. S. (1980). Optimality of some weighing and 2^n fractional factorial designs. Ann Statist 8:437–446.

Connor, W. S. (1960). Construction of fractional factorial designs of the mixed 2^m3^n. Contribution to Probability and Statistics. Stanford, CA: Stanford University Press.

Connor, W. S., and Young, S. (1961). Fractional Factorial Designs for Experiments with Factors at Two and Three Levels. Applied Mathematics Series—National Bureau of Standards. Washington, DC: U.S. Government Printing Office.

Finney, D. J. (1945). The fractional replication of factorial experiments. Ann Eugenics 11:341–353.

Fisher, R. A. (1935). The Design of Experiments. London: Oliver and Boyd.

Kesler, David A., and Feiden, Karyn L. (1995). Faster evaluation of vital drugs. Scientific American March 1995: 48–54.

Kiefer, J. (1975). Construction and optimality of generalized Youden designs. In: J. N. Srivastava, ed. A Survey of Statistical Design and Linear Models. Amsterdam: North-Holland, pp 331–351.

Li, J. (1991). Sequential and optimal single stage factorial designs, with industrial applications. Ph.D. Thesis, Colorado State University, Department of Statistics, Fort Collins, Colorado, USA.

Liao, C. T. (1994). Fractional factorial designs for estimating location effects and screening dispersion effects. PhD dissertation, Colorado State University, Fort Collins, CO.

Rao, C. R. (1947). Factorial arrangements derivable from combinatorial arrangements of arrays. Suppl J Royal Stat Soc 9:128–139.

Rao, C. R. (1950). The theory of fractional replications in factorial experiments. Sankhya 10:81–86.

Srivastava, J. N. (1987). Advances in the general theory of factorial designs on parallel pencils in Euclidean n-space. Utilitas Mathematica 32:75–94.

Srivastava, J. N. (1990). Modern factorial design theory for experimenters and statisticians. In: S. Ghosh, ed. Statistical Design and Analysis in Industrial Experiments. New York: Marcel Dekker, pp 311–401.

Srivastava, J. N., Anderson, D. A., and Mardekian, J. (1984). Theory of factorial designs of parallel flat type. I. The coefficient matrix. J Statis Planning Inference 9:229–252.

Srivastava, J. N., and Li, J. (1996). Orthogonal designs of parallel flats type. J Statis Planning Inference 53:261–283.

Srivastava, J. N., and Throop, D. (1990). Orthogonal arrays obtainable as solutions to linear equations over finite fields. Linear Algebra and Its Applications, Vol. 127. City: Publisher.

Junfang Li

Food and Drug Administration

See also *Drug Development; Biologics*

I. INTRODUCTION

The Food and Drug Administration (FDA) is a United States government agency under the Department of Health and Human Services (DHHS) that plays a complex and important role in the lives of all Americans as well as in the lives of consumers of many products grown, processed, and/or manufactured in the United States. As mandated by federal law, the FDA helps ensure that the food we eat is safe and wholesome, the cosmetics we use are pure and safe, and medicines and medical products are safe and effective for uses indicated in the product labels. The FDA is also responsible for overseeing veterinary medical and food products.

The FDA mission, formalized in the FDA Modernization Act of 1997, is as follows:

1. To promote the public health by promptly and efficiently reviewing clinical research and taking appropriate action on the marketing of regulated products in a timely manner;
2. With respect to such products, protect the public health by ensuring that foods are safe, wholesome, sanitary, and properly labeled; human and veterinary drugs are safe and effective; there is reasonable assurance of the safety and effectiveness of devices intended for human use; cosmetics are safe and properly labeled; and public health and safety are protected from electronic product radiation;
3. Participate through appropriate processes with representatives of other countries to reduce the burden of regulation, harmonize regulatory requirements, and achieve appropriate reciprocal arrangements; and
4. As determined to be appropriate by the Secretary [of DHHS], carry out paragraphs (1) through (3) in consultation with experts in science, medicine, and public health, and in cooperation with consumers, users, manufacturers, importers, packers, distributors and retailers of regulated products.

To accomplish these goals, the FDA is divided into six centers, along with a network of regional field offices (see Fig. 1). Three centers focus on human medical products: the Center for Biologics Evaluation and Research (CBER), the Center for Devices and Radiological Health (CDRH), and the Center for Drug Evaluation and Research (CDER). The Center for Food Safety and Applied Nutrition (CFSAN) is responsible for overseeing food and food products; the Center for Veterinary Medicine (CVM) enables the marketing of effective and safe veterinary medical products; and the National Center for Toxicological Research (NCTR) is a basic research facility that primarily studies problems related to the biological mechanisms underlying the toxicity of products faced by all of the centers with regulatory responsibilities.

The goal of this entry is to provide an overview of the FDA while discussing the role of statistics in drug regulation. While this discussion will be far from comprehensive, we will attempt to provide practical commentary regarding the interaction between the FDA and its customers. We will begin by discussing the origin of the FDA and important historical events that shaped its current structure and purpose.

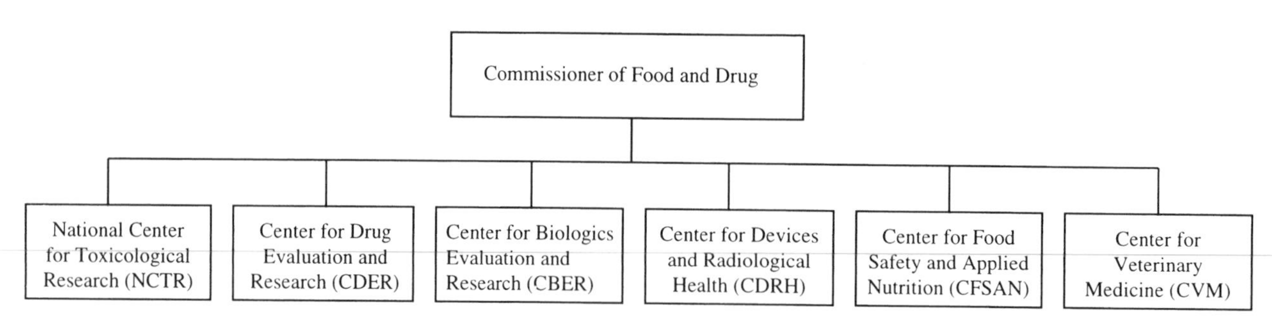

Fig. 1 Organizational structure of the Food and Drug Administration.

II. HISTORICAL DEVELOPMENT

The organization we know today has evolved via a long and interesting series of events, beginning in 1785 when the state of Massachusetts enacted the first food adulteration law in the United States. The FDA itself had its origins in the establishment of the Bureau of Chemistry in the Department of Agriculture in 1862 by President Abraham Lincoln. A national food and drug law was first recommended in 1880, but one was not enacted until 1906, following a flurry of congressional activity between 1879 and 1906 resulting in over 100 bills relating to the safety of food and drug products.

Three landmarks of food and drug law are the Biologics Control Act of 1902 and the Food and Drugs Act and the Meat Inspection Act, both of 1906. The first established the basis for the regulation of sera, vaccines, and related products and represents the origin of biologics regulation in the United States. The second prohibited interstate commerce in misbranded and adulterated food and drugs and provided the basis for regulation in these areas. The third dealt with sanitation in meat-packing plants. These laws were all motivated by tragic events and/or disclosures of clearly dangerous conditions and procedures in the preparation of food and drugs marketed to the U.S. population.

In 1938 the Federal Food, Drug and Cosmetic (FDC) Act was passed. This act substantially extended the responsibilities of the executive branch with regard to the regulation of these products. Cosmetics and therapeutic devices became subject to safety requirements; drugs were required to be shown safe prior to marketing; inspections of manufacturing facilities were authorized; and a variety of standards were established for foods. Shortly after this legislation was enacted, the regulatory authority for food and drugs was moved from the Department of Agriculture to the Federal Security Agency, and the first Commissioner of Food and Drugs, Walter G. Campbell, was appointed in 1940. (In 1953 the Federal Security Agency became the Department of Health, Education and Welfare, the predecessor to the current DHHS.)

In 1962, the FDC Act was amended to require for the first time that drugs be demonstrated effective as well as safe prior to being approved for marketing. Provisions strengthening drug safety requirements were also included. These legislative actions were motivated largely by the discovery that thalidomide, a sleeping pill marketed widely outside the United States was responsible for an epidemic of birth defects.

More recent milestones in food and drug regulation include the transfer of the Bureau of Radiological Health to the FDA in 1971; the transfer of responsibility for the regulation of biologics from the NIH to the FDA in 1972 and the initiation of over-the-counter drug review that same year; the passage of medical device amendments in 1976, formalizing and strengthening regulation in this growing area; the Orphan Drug Act of 1983, promoting the development of drugs needed to treat rare diseases; and the Prescription Drug User Fee Act of 1992, requiring manufacturers of drugs and biologics to pay fees related to the submission of marketing applications. As a result of these acts, the FDA is authorized by Congress to monitor activity surrounding $1 trillion worth of goods annually, approximately one-quarter of the gross national product. Each of the six centers of the FDA plays an important role in helping to accomplish this enormous task.

III. FDA CENTERS AND THE PRODUCT DEVELOPMENT PROCESS

Four of the FDA's six centers—CBER, CDER, CDRH, and CVM—focus on biopharmaceutical/biomedical products. We will focus most of our discussion on these four centers. The primary tasks of these centers relate to oversight of product development and evaluation of licensing applications for new products (or new uses of old products). Regulatory requirements for drugs and biologics can be found in the Code of Federal Regulations, Title 21 (Food and Drugs), Chapter 1 (The Food and Drug Administration), Subchapter D (Parts 300–499); requirements specific to veterinary products in Subchapter E (Parts 500–599); requirements specific to biologics in Subchapter F (Parts 600–699); and requirements for medical devices and radiological health in Subchapters H–J (Parts 800–1050).

For drugs and biologics intended for human use, the regulatory cycle begins when the product developer, or sponsor, intends to begin the first studies in humans. At that time, an investigational new drug (IND) application is submitted to the Agency. Sufficient preclinical data must be provided in this application to support the safety and the rationale for the use of the product. For medical devices, such applications are required only for specific classes of devices that pose "substantial risk." In these cases, an investigational device exemption (IDE) application will be submitted. Reviewers for the FDA (representing a variety of disciplines depending on the product) will consider the application and determine whether clinical investigations may proceed.

Plans for each clinical study to be carried out must be submitted to the agency prior to initiation if the studies are to be carried out in the United States. Review of these studies by the FDA serves several functions. First, this review provides an additional check on safety—assurance that the study is unlikely to result in harm to participants. Second, FDA reviewers will assess whether the study design is likely to achieve its goals, and they may make suggestions to the sponsor regarding the approach taken. Interaction between the sponsor and the FDA at early stages of drug development can be extremely helpful in ensuring that the development strategy laid out by the sponsor will ultimately provide the persuasive data regarding safety and efficacy that will be necessary to gain marketing approval.

The culmination of the sponsor's evaluation of the product is the submission of a dossier to the appropriate agency: for pharmaceuticals, a new drug application (NDA) to CDER; for biologics, a biologics licensing application (BLA) to CBER; and for medical devices, a premarket assessment (PMA) to CDRH. This comprehensive package must summarize the efficacy and safety of the investigational drug, biologic, or device as demonstrated by clinical and other studies. It must also contain adequate evidence that the product will have the effect that the sponsor claims under the conditions of use (505(d) FDA Act). Necessary components of a submission package have been clearly defined by the appropriate center of the FDA and can be found easily on the Internet (http://www.fda.gov). In the case of an NDA, the dossier must contain an application summary, copies of study protocol(s), study reports, integrated summaries of efficacy and safety and proposed product labeling (package insert).

In addition to reviewing data from studies evaluating product, the FDA engages in a wide variety of oversight activities. Product manufacturing processes are held to established standards, and manufacturing facilities are inspected by FDA staff at regular intervals to ensure compliance with the regulations. FDA inspectors also visit clinical sites where studies are carried out to document that correct and appropriate data were recorded, that all patients treated with investigational product were accounted for in study reports, and that study procedures were followed according to the study protocol. The records of Institutional Review Boards (IRBs) are also subject to scrutiny by FDA inspectors.

The FDA's responsibility for ensuring product safety does not end with the decision that a product may be marketed. Manufacturers are required to report adverse experiences associated with their products to the FDA within timelines dependent on the type of experience. These reports, together with others received directly from health care providers and consumers, are reviewed by the FDA to determine if changes are needed in the way the product is labeled, marketed, or prescribed.

IV. STATISTICS AT THE FDA

Each of the FDA's six centers has its own statistical unit. In 1998, the number of statisticians employed by the FDA was approximately 150. FDA statisticians participate in all aspects of regulatory review, assessing laboratory and preclinical data as well as data from pharmacokinetic studies, early clinical assessments, and large-scale clinical trials. They are also involved in setting and evaluating standards for testing and quality control, assessing data from postmarketing safety surveillance programs, collaborating with other FDA scientists on research projects, evaluating newly proposed statistical methods potentially applicable to regulatory settings, conducting methodological research, providing training to other scientific staff, and contributing to policy development.

The Center for Drug Evaluation and Research has the largest cadre of statisticians, divided among three divisions and a staff devoted to special projects. Each of the three divisions includes about 20–30 statisticians and works with a particular subset of medical review divisions. For example, the Division of Biometrics I evaluates drugs for cancer, heart, kidney, and neurological/mental disorders. All other centers have a single group of statisticians, ranging in approximate size from 5 to 20. Most FDA statisticians have doctoral degrees in statistics or biostatistics.

Probably the largest fraction of statistical effort at the FDA goes into the review of licensing applications for new products. The data from studies that the sponsor believes support the efficacy and safety of the product are generally provided in electronic format, facilitating the FDA statistical analysis. The FDA statistical reviewers will assess the analyses presented by the sponsor and may perform supplementary analyses to assist in interpreting the data presented. Some FDA statisticians attend the advisory committee meetings at which the application is discussed and may make a presentation in cases where statistical issues need to be highlighted. The statistical assessment of the data in the application is key to the decision to approve or not to approve a product for marketing.

A major ongoing challenge for FDA statisticians is the evaluation of the applicability of new statistical approaches to the regulatory setting. For example, there has been increased interest in applying Bayesian methodology to clinical trials of investigational products (1). Statisticians at the Center for Devices and Radiological Health are actively evaluating incorporation of Bayesian methodology in the clinical development process. Because so many new devices differ in only minimal ways from earlier versions, the formal inclusion of information on the effectiveness and safety of these earlier versions in the evaluation of new devices is appealing. As reported by the February 1998 International Medical Device Regulatory Monitor, "The FDA is preparing to issue a guidance on an advanced statistical methodology, known as the Bayesian model, that incorporates historical information into medical device clinical trial data'' (2). For a discussion of the use of Bayesian methodology in medical device development, see Berry (3).

V. FDA RELATIONSHIP TO SPONSORS

The relationship formed between the FDA and a sponsor is extremely important. Each drug, device, or biologic that a sponsor decides to research and submit for regulatory approval represents an enormous and risky investment. Only one in approximately 10,000 compounds tested ever makes it to market. Additionally, of those that do eventually make it to market, the cost of testing and satisfying the regulators worldwide has been estimated as $528 million to $792 million by the Association of British Pharmaceutical Industry (ABPI). Due to the costly nature of the registration process, the importance of good communication between the sponsor and the FDA cannot be overemphasized.

Each center of the FDA has an office dedicated to communication with sponsors and other interested parties. These offices work to ensure that all of the FDA's customers (including representatives of the pharmaceutical industry, health care professionals, government officials in other agencies, the media, and consumers) have easy and open access to information and are fully informed about the drug regulation process.

The FDA encourages communication with the sponsor throughout the review process and specifically advocates meetings prior to development of an IND application and immediately preceding submission of a marketing application. In planning a meeting with the FDA, the sponsor should take great care in designing the agenda. As noted by Waymack and Rutan (4), sponsors should carefully consider the attendees and the content of this meeting to ensure that the relevant issues can be fully and efficiently addressed.

One important component of the review process is the presentation to one of the FDA's advisory committees by the sponsor and the FDA review team of their respective assessments of the data contained in a marketing application. These committees consist of a panel of experts, external to the FDA and independent of the sponsor, that generally include physicians, statisticians, other scientists and health professionals, and a lay consumer representative. The committees are governed by the Federal Advisory Committee Act of 1972 and later amendments. The primary purposes of FDA advisory committees are to review available data (efficacy and safety) for medical products under investigation for human use and to address issues as requested by the FDA to assist in labeling decisions. Advisory committees may also be asked to comment on methodological issues germane to many applications. The FDA is not required to follow the recommendations of advisory committees but does act in accordance with their recommendations in most instances. While the size of the committee can range between 10 and 15 members, the panel can be supplemented with additional experts on an ad hoc basis when necessary for particular applications. A thorough discussion of the FDA's use of advisory committees can be found in an Institute of Medicine report that was issued in 1992 (5) and formed the basis for many recent changes in operating procedures. For a practical article related to FDA advisory committees, see, for example, Farley (6).

VI. FDA GUIDANCE DOCUMENTS

A list of guidelines that have been prepared by the FDA to aid sponsors in their pursuit of regulatory product approval is maintained by each FDA center. It is important to note that guidance documents differ from regulations, in that sponsors are not required to abide by practices described in guidance documents; rather, these documents represent what the FDA considers acceptable approaches to regulated research, although there may be other approaches that are also acceptable. While guidance is provided by the FDA on many aspects of design and analysis, the responsibility for adequate studies and credible analyses ultimately lies with the sponsor. A list of all guidelines that apply to the FDA's regulations across all centers, as well as the corresponding documents, can be found at the FDA's website (http://www.fda.gov/opacom/morechoices/industry/guidedc.htm).

An important ongoing initiative of the FDA is participation in the International Conference on Harmonization (ICH). The ICH comprises a joint effort of regulatory authorities and the pharmaceutical industry in the United States, Europe, and Japan to standardize regulatory requirements and establish standards for pharmaceutical research to the extent possible among these three regions, thereby reducing inefficiencies and ensuring a consistently high quality of research for manufacturers in an increasingly global industry. The ICH has developed documents in the areas of quality, safety, and efficacy, and for some cross-cutting topics as well. A description of the ICH, a listing and brief description of all documents issues, and access to the full documents themselves can be found at the ICH website (www.ifpma.org/ich1.html).

An ICH guideline entitled "Statistical Principles for Clinical Trials" has recently been issued by the FDA as a guidance document (http://www.fda.gov/ohrms/dockets/98fr/091698c.txt). This document discusses approaches to trial design and analysis in terms of the prevention of bias and the preservation of the desired error probabilities. This document should help ensure that adequate attention is given to statistical considerations for studies of new pharmaceutical products performed in the three ICH regions, and should facilitate the acceptability of registration data collected for marketing authorization in one region by regulatory authorities in the others.

VII. COMMON PROBLEMS AND CURRENT CONTROVERSIES

As methods for clinical research become more sophisticated, the approaches to evaluating investigational medical products also evolve. In many cases, FDA regulations and guidelines may be updated to account for new information. For example, guidance documents were updated in the late 1980s to account for methods of interim analysis that were becoming widely used in trials with mortality or major morbidity endpoints. In addition, patient groups in recent years have raised concerns about FDA regulations being so focused on the protection of study subjects that many individuals were being denied access to clinical trials. These policies had the effect (perhaps inappropriately) of restricting access to investigational therapy of some patient groups, as well as reducing the generalizability of the data generated. As a result, guidance documents have been made available to stress the importance that all relevant subgroup populations be included in clinical trials.

Despite the ready availability of FDA regulations and guidelines, as described earlier, and the continual reassessment of optimal research approaches, certain problems regularly arise. These issues include the size of studies, the appropriate control group, the appropriate population(s) to study, and the primary and secondary study endpoints that will define product efficacy. It is important to demonstrate efficacy and safety in a way that not only meets the requirements of regulatory agencies but also documents and motivates use from both a clinical and a commercial perspective.

A. Study Size

Study size is determined by the desired error probabilities and the estimated effects of the investigational and control treatments. Companies wish to perform the smallest possible studies that will still be adequate to demonstrate efficacy and safety. In some cases, FDA reviewers believe that the company is overly optimistic about the expected effects of the new treatment and suggest that a study be enlarged to be able to detect more modest but still clinically important effects. In some instances, the study size proposed may be adequate to assess the primary endpoint but not important secondary endpoints or specific safety concerns.

B. Control Group

There are a number of aspects to the selection of a control group: chronology (historical vs. concurrent controls); treatment assignment method (randomized vs. systematic); type of control treatment (active vs. inactive) and approach to blinding (double-blind, single-blind, open label). The randomized, double-blind placebo-controlled trial is generally considered the optimal approach but is not feasible in all instances. For example, active controls rather than placebo controls would be used when available active treatment is known to prevent or delay death or major morbidity. Random allocation of patients between treatment and control is considered essential, in most instances, to reliable interpretation of observed treatment effects. However, historical controls have been accepted as a basis for approval in some instances when the target patient population is extremely small and/or the effect of the new product is so dramatic as to make selection bias an unlikely explanation for the outcome.

C. Study Population

The definition of the study population is extremely important. Clearly, one wishes to study a new treatment in a population in which it is most likely to prove effective. Thus, individuals whose disease is so minimal that improvements would be difficult to document and individuals whose disease is so advanced that no treatment is likely to be effective are generally not considered for studies of new products. On the other hand, it is important to study the treatment in the population for which it is likely to be prescribed, should the study be positive and the product marketed. Attention is given at the design stage to ensure inclusion of the elderly, women, and children in studies of treatments likely to be used in these populations.

D. Primary and Secondary Endpoints

The *primary* endpoint for the trial must provide an adequate basis for assessing the clinical value of the treatment. Much discussion between the FDA and sponsors centers around the choice of clinical trial endpoints. Endpoints may be measured as continuous, binary, or multinomial variables, as time to an event of interest, as counts of events over a specified time interval, and as the slope of a regression line, just to give a few examples. In some cases, a *composite* endpoint—counting any of two or more outcomes as a primary endpoint—may be used. Such endpoints may be particularly relevant when treating conditions that have multiple manifestations, any of which might be improved by an effective treatment. The choice of endpoint will depend on the particular circumstances of the trial and must be clearly specified in the study protocol prior to initiating the study.

Secondary endpoints are those that are important indicators of disease status and thus may provide supportive evidence of treatment efficacy or safety. The number of secondary endpoints specified should be limited to those that are truly important to patients and to physicians managing patients with the disease in question and likely to reflect the effects of the treatment being evaluated. Secondary endpoints generally cannot provide the basis for marketing approval in the absence of effects on the primary endpoint [see Fisher (7) and Moye (8) for recent extensive discussion of this issue] but can provide useful supportive data as well as further understanding of the product's mechanisms of activity.

In trials of therapies that are potentially life-saving or have other potentially major benefits, "*surrogate*" endpoints—laboratory or early clinical measures that are believed to be predictive of the clinical outcome of interest—have been accepted as the basis for product approvals, with the condition that adequate studies documenting effects on the primary clinical outcome are ongoing. Because rapid approval based on surrogate endpoints implies shorter-term studies, there is a risk that longer-term safety issues might not be identified in such studies and that surrogates might not accurately predict the true risk–benefit ratio of a product (9,10). For this reason, action based on surrogate endpoints is limited to settings of serious disease where agents with improved efficacy are urgently needed.

Many other topics that frequently are debated in regulatory meetings include washout periods in crossover designs, the need for and/or adequacy of blinding, treatment of dropouts and noncompliers in study analyses, the need for statistical adjustment for multiple comparisons, procedures for carrying out interim analyses, definition of an equivalence margin in active control trials, and the necessary extent of follow-up for safety and efficacy assessments. Some of these issues are discussed by Lynch and Lachenbruch (11) in the context of trials for biologics, but most of the discussion is relevant to clinical trials more generally. An excellent source of insight on current clinical trial controversies is Senn (12).

VIII. CHALLENGES FOR THE FUTURE

In a time of rapid pharmaceutical development and the opening of innovative research made possible by the explosion in biotechnology, the FDA is facing a number of difficult but exciting challenges. Perhaps the most obvious of these is the need to review an increasing number of new products rapidly so that new products that meet standards of safety and effectiveness can be made available to the public as quickly as possible. The Prescription Drug User Fee Act (PDUFA) of 1992, which has been extended through September 2002, commits the FDA to faster review times for most new drugs and biologics applications. It was anticipated that this act would enable the new product application review to be cut from 15.5 to 10 months. As shown in a progress report to Congress in 1998, FDA has been successful in hiring and training new reviewers and is ahead of schedule in meeting the escalating PDUFA goals.

Other challenges relate to the application of new biological knowledge to the development of new products. Gene therapy is in its infancy, as are therapies involving the manipulation of a patient's own cells and then replacing them. Both of these approaches are showing the potential for improving clinical status in disease settings with currently little or no available satisfactory treatment. Gene therapy is one of a number of promising approaches for which theoretical safety concerns will need to be carefully studied. Other products, such as thalidomide (now approved for treatment of leprosy and AIDS) and xenotransplants (transplants of animal tissues to humans; not yet approved but under study), raise even stronger safety concerns. The task of balancing the risks of dramatically new treatment approaches carrying major risks with the potential benefits of such treatments for extremely ill patients will be confronted more frequently in coming years.

REFERENCES

1. DJ Spiegelhalter, LS Freedman, MKB Parmar. JRSS A 157:357–416, 1994.
2. FDA officials endorse statistical model that uses historical data in clinical trials. In: International Medical Device Regulatory Monitor. Lanham, MD: Newsletter Services Inc. 6(2):2, 1998.
3. D Berry. Using a Bayesian Approach in Medical Device Development. Duke University, Durham, NC, 1997. Available at: http://www.stat.duke.edu/people/personal/db.html
4. JP Waymack, R Rutan. Appl Clin Trials 28–34, 1996.
5. Institute of Medicine. Food and Drug Administration Advisory Committees. Washington, DC: National Academy Press, 1992.
6. D Farley. FDA Consumer. 2nd ed. 1995, pp 31–35.
7. LD Fisher. Contr Clin Trials 20:16–39, 1999.
8. LA Moye. Contr Clin Trials 20:40–49, 1999.
9. R Temple. A regulatory authority's opinion about surrogate endpoints. In: WS Nimmo, GT Tucker, eds. Clinical Measurement in Drug Evaluation. New York: Wiley, 1995, pp 3–22.
10. TR Fleming, DL DeMets. Ann Internal Med 125:605–613, 1996.
11. CJ Lynch, PA Lachenbruch. Drug Inf J 30:921–32, 1996.
12. Senn S. Statistical Issues in Drug Development. New York: Wiley, 1997.

Stacy R. David
Susan S. Ellenberg

G

Global Database and System

See also *Good Statistics Practice; Clinical Trial Process*

I. INTRODUCTION AND BACKGROUND

Standardization is a generic solution applied in business re-engineering. Analysis of the clinical data management, statistical analysis, and reporting processes has identified several elements where standardization would result in time savings.

Analysis has identified the database as the true center of the clinical development process. In the database, the "upstream" of clinical research raw data obtained in clinical trials, transformed in analyzable variables, meets the "downstream"—statistical analysis and reporting, and clinical reporting. Standardization of this element will yield large time-saving benefits for the development process and higher-quality databases and reports; in addition, it will make more effective use of available resources.

At the same time, a standardized database will allow for different modes of data entry, e.g., centralized as well as geographically distributed. It will also take advantage of technological changes such as electronic data capture versus manual data entry methods and provide customized analyses to meet different regulatory preferences in different countries and efficiently to answer additional regulatory requests during the review process. Standardized databases also provide the opportunity to harmonize and streamline the systems for clinical data management and statistical analysis. Applying such standardized systems will definitely result in consistent and comprehensive results and high-quality reporting, independent of where the system is applied in a decentralized approach of processing clinical trials.

This entry will give an overview of the requirements leading to such a standardized environment. It will briefly touch upon the system architecture and the methodology applied to support the development, deployment, and maintenance of such a validated system. A section is devoted to the data components of such a standardized database that are based upon a simple implementation of data-warehouse techniques. The last sections will briefly lay out the transition into a standardized approach to processing clinical data and the critical success factors involved.

II. REQUIREMENTS ANALYSIS

With the guidelines of the International Conference on Harmonization (ICH) of Technical Requirements for Registration of Pharmaceuticals for Human Use (1) in place, regulatory submissions to different authorities, such as the Department of Health and Human Services, the Food and Drug Administration (FDA), and the European Agency for the Evaluation of Medicinal Products (EMEA), can be based on studies conducted all over the world. Consequently, studies conducted in the United States are used for submissions to the EMEA, and studies conducted in the rest of the world can be used in a new drug application (NDA) submitted to the FDA.

Managing, analyzing, and reporting individual trials can be done independently. However, consolidating the data and the results, for instance, into integrated summaries for safety and efficacy, can become a very challenging task when the databases of the individual trials were designed and constructed without having this final global objective of integration in mind. Therefore it is necessary to define and construct similar databases for each trial in terms of structure and content that are based on similar protocols and case report forms (CRFs).

In a decentralized environment for processing clinical data, the likelihood of duplicating work is always present. Standardized databases and systems, however, provide the opportunity to leverage resources globally by reusing database designs and programs from one trial to another. Appropriate and to-the-point communication is an important and necessary proactive tool in leveraging efforts around the globe.

Standardized approaches in clinical databases and systems, leveraging resources by reusing designs and programs, and standardizing the processes of data management and statistical analysis are solutions for shortening the time to the market and for producing higher-quality databases and reports. However, these solutions don't guarantee these important results when the approach is not global and applied in all departments involved in clinical trials. Proactive management and careful planning are key factors in achieving those quality and time objectives.

Many regulatory authorities, such as the FDA and the EMEA, and other organizations, such as ICH and the International Organization on Standardization (ISO), are also imposing more strict regulations in the area of computerized systems (2). These regulations require that a pharmaceutical company be able to show at any time that the applied computer systems produce reliable, accurate, and complete data and results. All requirements mentioned in this paragraph will help in validating the programs and systems used for clinical data processing; however, the challenge of building and, especially, of maintaining a validated system is still there and must be a goal for all levels and functions in the different departments involved.

III. SYSTEM ARCHITECTURE

The architecture of a clinical system needs to be considered from the perspectives of data management, statistics, and system development. The foundation of the architecture is the physical computing environment. It is essential to separate the computing environment into three distinct environments—development, testing, and production—so that each function can be performed without adversely affecting other functions.

The development environment needs to be a very open one in which the developers have few restrictions. The developers should be able to try out as many alternatives as possible for designs and programs, some of which may even affect other developers in a negative way, such as bad performance or disk congestion. Ultimately one developer could even cause the computer system to crash. However, by applying a split environment, such a tryout should not affect the production environment or the user acceptance environment, since this environment will be separate.

User acceptance is a formal stage during development and testing of a computer system, and the results of the acceptance test are part of the validation package. Therefore, the environment for this function needs to be more controlled and restricted as compared to the development environment and should be set up according to formal installation guidelines. At the same time, a user, being very familiar with real data, expects copies of real data or simulated data in the test environment. The user can then use and manipulate these data to cover as many test scenarios as possible.

The actual processing of the clinical trial data is performed in a secured, validated production environment that is separated from the development and testing functions. The entire physical computing environment as part of the system architecture is depicted in Fig. 1.

The next building block in this computer environment is the data. The *database* containing the data refers in a broad sense to a "knowledge database" of a clinical project for developing a compound of a certain indication and formulation. The limits are:

Upstream: the construction of a normalized, relational, "raw" database the design for which is based on the layout of the case record forms (CRFs). This database contains the actual data collected using the CRF, diaries, quality-of-life questionnaires, etc.

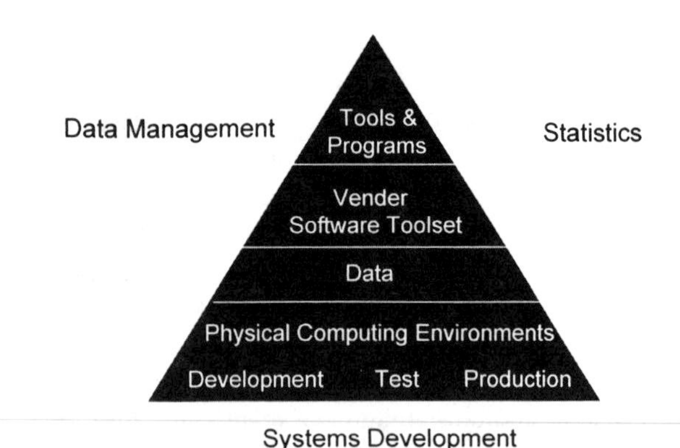

Fig. 1 The system architecture supports system and trial specific development in a validated environment.

Downstream: datasets containing the data for the tables and figures that format the output of the statistical analysis, and which are used for the clinical trial report and its addenda.

Another level in the hierarchy of the system architecture consists of the basic software for storing, retrieving, and analyzing data. A relational database management system (RDBMS), such as oracle®, is most often used for storing data in a normalized and structured way. Clinical data management systems installed upon such as RDBMS are used for entering, cleaning, and reviewing the data. Examples are Clintrial™, Recorder™, and Oracle Clinical™. Statistical analysis of the clinical data is supported mostly by a separate package, such as SAS® or SPlus™. However, this package is closely integrated with the RDBMS containing the raw data.

A modular approach to developing systems will result in a very maintainable system. Therefore the system can be split into a number of tools and a number of subsystems. These tools and subsystems form the last level of the infrastructure hierarchy.

A *subsystem* will have a very specific function supporting a specific task in the process. An example of such a system is the data display subsystem that provides computer-based CRF printouts. These are used for evaluating the trial database; in addition, they are included in submissions to regulatory agencies. Another example of a subsystem is the efficacy system that analyzes all types of efficacy data: discrete, ordinal, and continuous.

Tools, on the other hand, don't support a specific task as a subsystem does, but is a piece of code with specific input and output that can be used for building the subsystems. In other words, the tools are the building blocks for subsystems. Tools can have different objectives, such as data handling, graphic representation, reporting, and applied statistics. Examples are tools to merge two or more datasets, to create summary reports, to output to a word processor, and to calculate the *p*-value according to a specific statistical test.

IV. DATA COMPONENTS

The global database contains many types of data and is used for many purposes. It will be much easier to manage the complex data by separating them into logical components. One such way is to separate them into the following five components (see Fig. 2).

1. The first component contains all data collected in a clinical trial—whether they be paper CRFs, ancillary data on tape, or other paper- or computer-based data. These *assorted* data are named *A* database.
2. The second component is the *base* physical CRF database, the *B* database. This database is usually a relational database that contains raw CRF data and other ancillary data and that is organized in a normalized structure most suited for data entry.

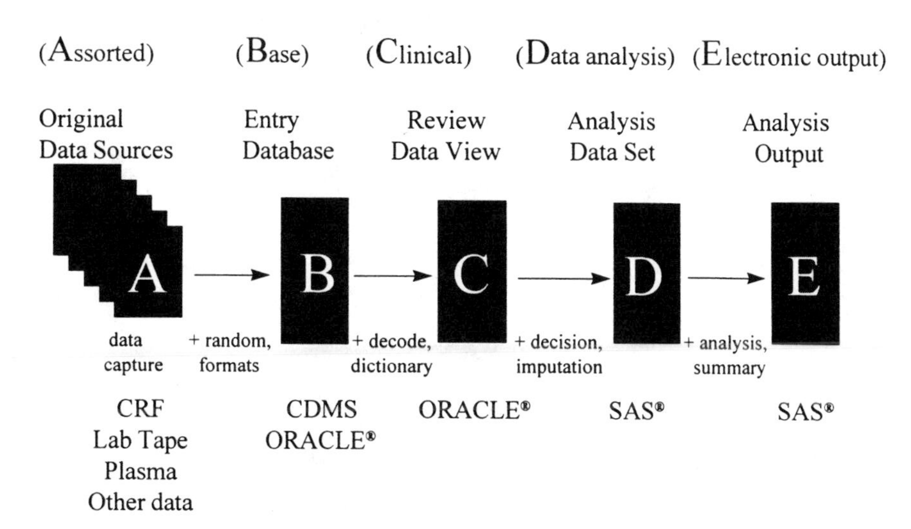

Fig. 2 The different database components enable an efficient workflow by mirroring the processes in data management and biostatistics.

3. The third component, the *C* database, is intended for *clinical* usage. The *C* database structure provides a view into the raw CRF data in the *B* database, but it is organized in a denormalized structure most suited for retrieval and data review. Dictionaries are linked to the *C* views.
4. The fourth component, the *D* datasets, is designed for *data analysis*. The *D* database is constructed by adding calculated values, derived variables, and analysis decisions to the *B* database. Statisticians provide the requirements and specifications of the *D* database. The *D* datasets are usually in SAS format. They are organized on a subject level and include separate datasets for demographics, adverse events, efficacy, etc.
5. The fifth component contains the *electronic outputs* of the statistical summary and analysis, the *E* datasets. The statistical subsystem generates statistical summary and analysis outputs, i.e., summary tables, listings, and graphics.

V. SYSTEM DEVELOPMENT

A. Methodology

One of the most commonly applied development and qualification methodologies is the V-model (3). This model is very simple and straightforward and is schematically described in Fig. 3.

1. The first step is to specify the system *concept*, including the scope of the project, a preliminary project schedule, and a first assessment of costs and resources needed.
2. The next step is to define and formalize the user and functional requirements of the system (4). The *system requirement specification* (SRS) describes each of the essential requirements of the software and the external interfaces. One or more reviews will be performed to verify its compliance with stated user requirements.
3. After approval of the SRS, the design specifications are formalized in the *system design description* (SDD) for the system architecture, software components, interfaces, and data. One or more reviews will be performed to ensure that the SDD for a given module achieves the objectives provided in the applicable SRS.
4. The system is then built during step 4, according to programming guidelines and naming conventions. Every developer performs construction testing that is freeform testing; however, some documentation is maintained. The objective is to ensure that the programming was done properly and that individual modules meet their design specifications.
5. During step 5, technical testers will perform system verification, in which formal testing is performed, independent of the developer, in a separate test environment. The objective is to

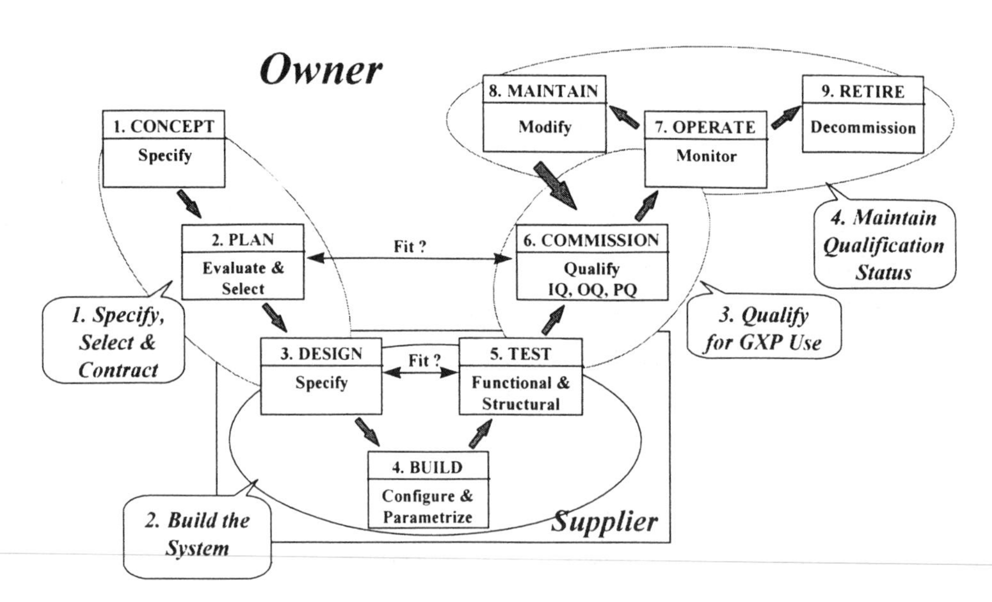

Fig. 3 The V-model is an accepted development methodology with validation components built in.

ensure that the programming of modules and interfaces was done properly and that the resulting product conforms to design specifications.

6. System qualification ("user acceptance testing") during step 6 is formal testing performed in an operating environment that is equivalent to the one used for normal operations. The objective is to ensure that the integrated system performs as expected under normal working conditions. System qualification consists of installation qualification (IQ), operational qualification (OQ), and performance qualification (PQ).

B. Validation

In the computer business many definitions and approaches for the validation of computer systems are available. The definition applied for the qualification of the system is the following:

> Validation of a system is to establish documented evidence that provides a high degree of assurance that a specific process will consistently produce results meeting its predetermined specifications and quality attributes.

The implementation of this validation approach resulted in three formal stages of validation: installation qualification, operational qualification, and performance qualification. Every qualification phase consists of three steps: defining the plan for qualifying the system, defining and executing the test scripts, and writing a summary report of the qualification. These steps are formally approved and signed off by the system owners. This process is repeated for every qualification.

The IQ will demonstrate that the computerized system installed is the system designed to meet the requirements. The IQ will contain the evidence that the computerized system has been installed in an appropriate environment and that all system documentation is in place to start the use and maintenance of the system.

The OQ will produce the evidence that the system meets all its requirements by means of functional testing of the system under test conditions. The OQ will demonstrate that the system performs as specified according to predefined test scripts that are executed in a separate test environment but within the boundaries of normal usage of the system.

The PQ provides evidence that the system functions as intended by the user, using the defined procedures. The PQ will guarantee that the necessary standard operating procedures are in place, users are properly trained, errors are tracked and managed, changes to the system are controlled, etc.

VI. MAINTENANCE

A. Change Management and Training

Systems to support clinical data management and biostatistics are very complex and not straightforward to use. Change management of people and customized training sessions are key solutions to a successful implementation and delivery of such a system.

A formal definition of *change management* is:

> The process of aligning an organization's people and culture with changes in strategy, structure, and systems.

A formal change management approach will clearly define the customers and all other people that are directly or indirectly involved in the project and how they will be affected by the implementation of the system. By laying out this structure, it becomes obvious who can help in implementing the system and who might be obstructive. The change management approach must be supported by a good communication plan. Having all these parts of a change management procedure defined, the actual change management can start during the development of the system, way before the actual implementation.

Adequate training sessions are part of the implementation of a good change management procedure and of performance qualification. The audience for the training for this system is very heterogeneous, for instance, technical and nontechnical people, or people from different functional areas, such as data management, biostatistics, and medical writing. Also, every new hire needs training, though more personalized, before he can start working with the system.

The training therefore is divided into three parts. The first part is an introduction to the new system. The second part is to explain and clearly define the rationale behind the data management databases and systems and the biostatistics databases and systems. Newly introduced definitions, structures, techniques, and tests, among other items, are explained in detail, along with why a certain solution is chosen. The third part of the training package consists of technical usage of the system and is focused primarily on technical people such as database administrators, analysts, and programmers.

B. Change Control

Research in general and clinical research in particular are very dynamic and change very frequently. Clinical trials within the same development plan will vary; clinical trials can have a standard part, e.g., safety, and will have a very specific part as well, e.g., efficacy. Many changes and enhancements will be requested of a system supporting the process of analyzing clinical data. A *change request* is a request to modify one or more existing deliverables, such as documents or system components, for reasons other than nonconformance, and which have a direct impact on the configuration of the system.

A good clinical practices (GCP) system must remain in a validated state during its operational use. All changes to the system have to be recorded and evaluated for their impact on the validation status. If necessary, the system owner will perform a repeat of validation activities or have them performed. Therefore it is crucial to have a good change control approach in place to track, resolve, and implement changes to the system. The main objectives of the change control procedure that is applied is to collect all bugs, anomalies, and enhancements, to track these requests from originator to implementation of the change, and to document changes in code, documentation, procedures, etc. The impact of such a system change on cleaned and closed databases or on reported trials needs to be carefully assessed, since clinical development programs generally span many years.

VII. TRANSITION

The introduction of a new clinical database and system is always complicated by the question of how to handle the ongoing trials, existing databases, and legacy systems. It is not possible to leave all current databases and systems to run their own course, since clinical programs usually last many years. It is also not possible to move all current databases and systems into the new one, due to incompatibility and high cost. Therefore, decisions regarding the transition level for the current database and system need to be made. The transition level may depend on the CRFs' layout, programming status, and submission status. One other factor determining transition level is the balancing of the benefits with the cost in resources and budget. After detailed inventory, the transition level may range from no conversion to full conversion. For example, we may opt not to convert projects without a plan for further development, and convert only integrated safety-summary-related safety data for projects that are at the submission stage. A detailed transition plan, including time schedule, resource needs, cost, and transition level, will provide a guide as to when, what, and how to transition all existing databases and systems.

VIII. CRITICAL SUCCESS FACTORS

The successful development and implementation of a new global system require both good system development methodology and good process. In previous sections, the importance of training and transition when implementing a new system was discussed. Throughout the life of the system, all involved parties need to co-own the system by being actively involved in the design, planning, development, implementation, and maintenance of the system. The involved parties include users and customers from all relevant sites of the company. Good project management in planning, resource, and communication are critical in ensuring the success of system development.

The main objective of a global clinical system is to shorten the drug development time while improving the quality of clinical data, analysis, summary, and reports. A good system alone will not reach this goal. It is necessary to review, modify, or even re-engineer the analysis process, from finalized protocol and CRFs to the generation of individual and integrated reports. The analysis process conducive to up-front planning, good communication and teamwork, and global collaboration will make optimal use of the global system. Coupling the new process with the new global clinical database and system will achieve the benefits of improved submission quality, synergy among sites, shared resources among sites, shortened submission preparation time, shortened response time for regulatory agency inquiries, and an enabling of the electronic data review internally and by the regulatory agencies.

REFERENCES

1. International Conference on Harmonization of Technical Requirements for Registration of Pharmaceuticals for Human Use. E6: Good Clinical Practice: Consolidated Guidelines; E8: General Considerations for Clinical Trials; E9: Statistical Considerations in the Design of Clinical Trials; E3: Clinical Trial Reports, Structure and Content.
2. Department of Health and Human Services, Food and Drug Administration. 21 CFR Part 11, Electronic rec-

ords, Electronic signatures; Guidance for industry, computerized systems used in clinical trials, draft guidance.
3. Teri Stokes, Ronald C. Branning, Kenneth G. Chapman, Heinrich Hambloch, Anthony J. Trill. Good Computer Validation Practices, Common Sense Implementation. Interpharm Press.
4. Institute of Electrical and Electronic Engineers. ANSI/IEEE Std 1058.1—1987, IEEE Standard for Software Project Management Plans; ANSI/IEEE Std 730.1—1989, IEEE Standard for Software Quality Assurance Plans; ANSI/IEEE Std 830–1984, IEEE Guide to Software Requirements Specifications; ANSI/IEEE Std 1016–1987, IEEE Recommended Practice for Software Design Descriptions; ANSI/IEEE Std 1012–1986, IEEE Standard for Software Verification and Validation Plans.

Alice T. M. Hsuan
Patrick Genyn

Good Programming Practice

See also *Good Statistics Practice*

I. INTRODUCTION

The operation of a biostatistics group in the drug development process is analogous to the process of building a house. Data coordinators prepare building blocks and lay a solid foundation. Statisticians are the architects with overall responsibility for the conception and realization of the project. Programmers are the construction engineers who execute the statisticians' plan and transform building materials into a house. In other words, a synergy of efforts among data coordinators, statisticians and programmers is necessary. Data coordinators ensure the integrity of the data. Statisticians do detailed analysis of the drug, choose appropriate statistical methods, write specifications, and analyze results. Programmers implement statistical rules to convert clinical data into useful information for assessing the safety and efficacy of a drug.

This entry presents ideas on the fitness, robustness, remodeling friendliness, extendibility, and economy of program construction. The importance of documenting, organizing, validating, and archiving programs, as well as the need to keep updated and to have a cooperative spirit in a shared project, are highlighted.

II. QUALITY IS IN THE EYE OF BEHOLDER?

The goal of good programming practices is quality work. *Quality* is an elusive term. Quality as practiced may sometimes be different from quality as idealized in theoretical discussions. There are constraints to balance among available resources, budgets, schedules, and programming requirements. Quality needs to be defined within business concerns, corporate and industry culture, on the basis of the expected life span of the programming code and, sometimes, on a project-by-project basis.

In general, quality for programming in the pharmaceutical industry refers to *fitness* (whether the programs meet functional requirements and are suitable for the purpose for which they are intended), *robustness* (accurate and reliable under different programming conditions and data scenarios), *remodeling friendliness* (easy to read, understand, and maintain), *extendibility* (reusable, automated to handle multiple studies of the same drug), and *economy* (on time and within budget).

Within the stringent requirements, this entry emphasizes programming practices that build in quality to do things right the first time. The SAS® System* is the most commonly used software in clinical trials, so "SASese" is inevitably spoken. However, the general principles discussed within the context of programming in SAS can be applied regardless of programming technique, tool, or language selected. Analogies and simple programming examples are given to make the concepts behind the principles more concrete.

---

*SAS® is a registered trademark or trademark of SAS Institute Inc. in the USA and other countries. Other brand and product names are registered trademarks or trademarks of their respective companies.

III. DESIGNING

Just as building a house needs blueprints, designing is an important prerequisite for good programming practice before actual coding starts. Winograd et al. stressed the importance of bringing design to programming (1). Their book is based on a workshop on software design. Two software developers at the workshop listed the components in the development of designs as understanding, abstracting, structuring, representing, and detailing. In clinical trials, the crucial role of design as a counterpart to programming is very similar.

Understanding involves learning the nature of the drug and the protocol as well as comprehending the analysis specification. What are the rules? Are the rules appropriate, feasible, sufficient or contradictory? Understanding also means having a good grasp of the data structure, formats, and categories. Are the data standardized across studies? Do the data match the case report form or the protocol requirements? *Abstracting* means sketching out the main elements for programming. Are all the requirements covered? Can a consistent vision of all the elements in the whole project be built in, from investigational new drug (IND) to new drug application (NDA) or CANDA? *Structuring* entails the ordering of relationships among the elements. Is there logical, functional, sequential, or procedural coherence? Is the layout elegantly clear? *Representing* refers to representing the elements in computer language. What are the steps or procedures selected? Are the selections relevant? Will they work in various situations? Do the methodologies conform with FDA requirements? *Detailing* is the programming approach. Is the approach lean, simple, and straightforward? How can the programs be written so that they are easy to debug, modify, maintain, and enhance? What are the goals of a specific program? Are the goals to build applications or systems where continuous fine-tuning and adding extra features are needed? Or are the goals to write good enough programs for basic operation? What is seen as indispensable in some situations can be unnecessary overhead in other situations.

The process of designing fosters communication among programmers, statisticians, clinicians, and data coordinators, and encourages different ways of looking at problems. Good designs clarify a program and convey the content. Good programmers are designer-programmers. The design approach selected has to be tailored to project needs. Design deals with the issue of whether the programs are fitting and appropriate to address the statistical questions of the drug under investigation.

IV. CODING

In addition to good designs, quality engineering is needed. Good programming practice as defined and illustrated herein are organized around the goals of robustness, remodeling friendliness, extendibility, and economy. Not all of these goals are completely compatible. For example, making a program more robust may add complexity to a program. The pragmatic challenge lies in prioritizing and finding a happy medium among these quality goals.

A. Robustness

Robustness is the most essential characteristic in quality program construction. Robustness encompasses accuracy, reliability, thoroughness, and reproducibility. As a general rule, making a program robust means writing solid code and having a good understanding of how to prevent bugs effectively and how to detect bugs easily or automatically (2).

Here is an example of built-in robustness. If you need to create a response flag that indicates an average increase of 1% from baseline over 3 years at each of the scheduled 6-month intervals, you can code the program as follows:

```
if month=6 and pchange=.50 then response=1;
    else if month=12 and pchange=1.0 then response=1;
    else if month=18 and pchange=1.5 then response=1;
    else if month=24 and pchange=2.0 then response=1;
    else if month=30 and pchange=2.5 then response=1;
    else if month=36 and pchange=3.0 then response=1;
    else response=0;
```

If patients do not follow the study protocol and come in at unscheduled time points or after 3 years, or if there are missing data, the preceding code will not work properly. A more robust way to write the code is:

```
if pchange=month/12 then response=1;
    else if pchange^=. then response=0;
```

Not all the programming code can be made robust and elegant simultaneously as effortlessly as this example. Sometimes it depends on the level of overengineering expended to check in every conceivable way and at every conceivable place.

What follows is a discussion of the various aspects of strengthening robustness.

Syntax errors, uninitialized variables, automatic data type conversions, division by zero, unknown formats, undeclared array elements, and many-to-many merges are some of the ordinary types of bugs that usually exist in the early stage of programming development. They are the easiest to locate and fix. Compilers usually give error or warning messages on these kinds of errors.

Logic errors are a less obvious kind of bugs that inhibits the proper function of a program. They are less obvious because the program is syntactically correct but may not give the correct result. An example of logic error is referring to the first or the last observation that has just been deleted in the same step. It is very important to test programming code in small chunks. It is also important to know the assumptions behind the design of the program, and to build in statements to output warning messages if assumptions are violated. Bugs multiply quickly. Any bug found needs to be fixed immediately, especially at a step that supplies data to another step.

A common type of bug is introduced by data. A program may not have accounted for all the data scenarios or for unexpected data errors. For example, the studies of a drug are conducted in both the United States and Europe, and a programmer writes code to convert the height values into the same unit for analysis. If the programmer expects only inches and centimeters in the conversion, the program will miss the height values that are in millimeters or have missing units. One way to avoid this kind of bug is to do a frequency check on the unit categories and to write the program based on the frequency results. But the studies may not have been completed when the program is written and new data may create bugs in the program. A more reliable way is to build routines into the program to flag unexpected data values and to output warning messages, or to capture those values and print them out. It also helps to draw a Venn diagram or to sketch out a list to see if all possible combinations of conditions are covered.

Missing values are a type of bug most often neglected by programmers. A simple example is to write a condition for test results of less than 5, to indicate significant improvement; patients with missing test results end up in the "significant improvement" category:

```
if score<5 then improve='SIG';
    else if score>=5 then improve='NOT SIG';
```

This code is syntactically correct, but there are two other problems besides assigning missing values to the wrong category. First, the length of the variable *improve* is not defined. Its length is defined by the first IF statement. Thus, the value of *improve* in the second IF statement is truncated to 3. Second, invalid scores, such as negative scores or zeroes, are not accounted for. One way to prevent the bugs is to change the code to the following:

```
length improve $7;
if score=. then improve ' ';
  else if 1<=score<5 then improve='SIG';
  else if score>=5 then improve='NOT SIG';
  else put 'DATA ERROR:' 'PATIENT=' patid 'SCORE='
      score;
```

One other kind of data-related bug comes from merging or concatenating data. It is necessary to examine database structures before combining the data, to sort the data and join them by key variables, to keep only the variables in each data set that are necessary in order to prevent variables of the same name from being overwritten in merging, to add debug code to output warning if the data cannot be combined properly, and to verify the number of observations before and after the data are combined.

In clinical trials, because NDAs and CANDAs require the integration of information from several studies of a drug under investigation, maintaining internal consistency is a very important aspect of ensuring robustness. Consistency is needed across all studies of the same drug in naming variables, in choosing data type, in formatting and labeling variables, in assigning the values of categorical variables, and in structuring derived data. Here are reasons why consistency is important:

1. If a character variable is assigned different length in different studies, the values of the character variable are truncated when data from the studies are concatenated.
2. If the same variable is given different names in different studies, proper integration of data is not possible until the variable is renamed to the same name in all studies.
3. The effects of inconsistency are subtle; usually there is no error message. It is necessary to adopt a uniform approach for all studies to rule out unexpected errors.

Numeric representation errors, also known as *rounding errors*, are one type of bug that can creep into programming code without notice. Fractional values and values that do not have enough storage space have the potential for numeric representation problems. Numeric representation errors propagate. The more computation the values go through, the larger the errors will become. An example of the occurrence of numeric representation errors in programming laboratory data is in comparing laboratory values to normal ranges, where both the laboratory values and the normal ranges are fractional values. Here are some major ways to minimize numeric representation problems in SAS software:

1. Only use the LENGTH statement or the ATTRIB statement to truncate values when disk space is limited.
2. Use integers if possible.
3. Use the ROUND function or the FUZZ function on fractional values before doing a comparison on the values.

Related to numeric representation errors is mixed variable types that result in automatic data type conversions. Avoid comparing numeric variables with character variables. The comparison figures out the conversion type, converts one type into another, does rounding, and determines the answer.

Hard-coding is a cardinal sin. *Hard-coding* is usually an attempt to fix data errors programmatically. Hard-coding disrupts the integrity of data and the replication of results. It also undermines the robustness of a program. Data errors are best handled at the source, that is, in the database.

Using available formats, functions, and routines from software is a way to prevent bugs. The software is usually tested thoroughly before it is released. In SAS software, for example, outputting *p*-values directly from a procedure into a data set is easier and less error-prone than writing a program to extract the *p*-values. There is also less chance for error if you use formats such as PVALUE and PERCENT than if you build the formats with FORMAT statements. The PVALUE format automatically transforms values of less than zero to the standard SAS output *p*-value format. The PERCENT format can automatically take care of rounding.

B. Remodeling Friendliness

Change is a fact of life in programming for clinical trials. Therefore, making a program easy to understand, easy to use, easy to modify, and easy to debug is vital. Some requests for changes are made for cosmetic reasons. Most cosmetic changes are like repainting a house: they are simple and rarely affect original program design. Requirement or analysis changes are usually structural changes; the sections that are intricately interwoven may need major rework. Changes also arise from error correction.

No matter how big or small changes are, each change requires review of the changed code, testing of the impact of the changes on the whole program, and revalidation and sometimes rewriting of all the results. It is better to funnel all the change requests through lead statisticians or programming managers who are familiar with the intent and the budget of the studies than to make changes on demand. The leads will coordinate the changes where ideas for changes contradict among clinicians, data coordinators, statisticians, medical writers, and the regulatory group. The leads will also separate the wheat from the chaff change requests and will estimate and make sure everyone knows the cost of each change.

To insulate a program from the effects of remodeling, the following are some guidelines to write or revise programs by anticipating and documenting changes:

1. Write simple and straightforward code in a clear layout, and have sufficient comments so that any programmer can understand and modify the code with little chance of introducing bugs.
2. Comply with your company's programming standards and naming conventions; they are tools to help communicate the nature of a program or a variable. For example, all program names that start with the letter *m* are macro programs, or all variables that start with an underscore are array variables.
3. Keep the initialization and the reference of variables close together so that they are easy to locate.
4. Do not nest too deep, and cut down interrelated parts to minimize the impact of changes.
5. Global macro variables are convenient to access, but the values of the global macro variables may be inadvertently changed in a program, whereas local variables that are confined to a small section of code are easier to maintain.
6. Modularize parts that are likely to change.
7. When there are extensive programming structural changes, it is sometimes easier to rewrite a program than to modify an old one.
8. Avoid tricks; the time spent writing and documenting tricky code is better spent in writing plain code that is easy to maintain.
9. Make changes that are compatible with the existing program design.

10. Adhere to your company's system for documenting and tracking changes.

C. Extendibility

In clinical trials, multiple studies are performed on the same drug. A major goal in programming is to attain extendibility. Extendibility here means that the programs are flexible in accommodating varying number of treatments, visits, hours, etc. and are reusable for many studies. A basic example of extendibility is to make study number a macro variable so that programs can be used for any study with the same data structure and analysis requirements by reassigning the value of the macro variable.

The advantages of extendibility are that reusing the code makes the output identical and that the value of the code increases if the code is reusable for multiple studies of one or more drugs.

Some of the techniques for building extendible programs are: look for patterns in data and analysis; design easily shared code; make use of macros and parameter-passing mechanism; avoid hard-coding; modularize parts that are different and make it easy to isolate and replace a section of code; use local variables when possible so that it is easy to pull out some parts of the code and plug them into another program.

There is a trade-off between generality and specificity. The more general and flexible the design is, the harder it is to write the program and to detect bugs, and sometimes the harder it is to use or to modify. Usually, it is easier to write general programs for safety analyses than for efficacy analyses, because safety analyses tend to be similar, except for the number of treatment groups or visit numbers, while efficacy analyses tend to be substantively different.

D. Economy

Conventional wisdom in programming defines *economy* as leanness of programming code. Each and every piece of code is important and cannot be taken away. In a more practical sense from the viewpoint of research on drugs, *economy* means completing the project and being able to submit NDAs on time and within budget.

Economy is often discussed in terms of computer efficiency and human efficiency. Computer efficiency means minimal use of system resources, such as execution time, input-output time, memory, and data storage. Human efficiency means less programming time spent in designing, coding, validating, documenting, and maintaining.

Some techniques for ensuring computer efficiency are: order IF conditions by frequency of occurrence; eliminate unnecessary input and output operations; minimize macro reference; compile frequently used macros; use random access (WHERE condition) instead of sequential access (IF condition); use the fewest array dimensions possible; automate repetitive tasks.

Here is an example of human efficiency in creating a variable called *block* where patients 1–10 are in block 1, patients 11–20 are in block 2, and patients 21–30 are in block 3:

```
block=1*(1<=patient<=10)+2*(11<=patient<=20)
    +3*(21<=patient<=30);
```

There is only one statement to type. The trade-offs are that it is hard to maintain and may lack robustness. The code is fine in this case, because randomization schemes for assigning patients to blocks are usually established at the beginning of a study and seldom change.

In clinical trials, good programming design, careful algorithm selection, the preventing of errors, and the writing of clean code that is easy to maintain contribute to economy. Having a few dedicated programmers on a project, and not rotating programmers in and out of projects, saves time and increases quality, since each additional person added to a project has a learning curve. In addition, economy means joint efforts to increase productivity. Clean and complete data and less rework from rule changes are important. Locate data problems and give them to data coordinators promptly to rule out programming errors. Work with statisticians in early stages to reduce the amount of time spent in requirement changes or debugging. If each and every person makes an effort to do the job right the first time, economy is built into the process. It also spares the medical writers from having to rewrite results due to changes or errors.

Economy is integral to overall program quality, but it is not the most critical condition. Priorities need to be established if economy conflicts with other goals, especially robustness.

V. DOCUMENTATION AND ORGANIZATION

Documentation and organization are controversial topics, because they are products of programming style.

Rigid standards cannot be imposed without contention; the best way is to work out a team style. Style is not just a matter of aesthetics; good style communicates content. The essence of good style is discussed here.

Documentation in programming is generally known as *comments*. Effective comments are not the comments that simply repeat the code; they should explain the objectives at a higher level of abstraction than the code (3). Code-level comments can be replaced by good programming style, which includes meaningful variable names, descriptive labels, easily understood logic, well-thought-out design, clear layout, and naming conventions that distinguish among constants, temporary variables, local or global macro variables, loop index variables, and array variables. Good style documents code without extensive comments. The best documentation is self-documenting code. Comments are expensive, not only to write but also to maintain. Therefore, comments need to be accurate, brief, and to the point.

Organization refers to the formatting of the code. Good organization provides clues to the logic of a program. Some of the techniques of good organization include: group related topics together; use alignment, white space, blank lines, indentation, blocks; use parentheses to clarify how expressions are evaluated; use a minimum of nesting; be consistent throughout the whole project; make the code easy to read and maintain. The purpose of organization is to make the visual structure match the logical structure of the code.

VI. VALIDATION

The Food and Drug Administration's (FDA's) definition of process validation (4) is: "Establishing documented evidence that provides a high degree of assurance that a specific process will consistently produce a product meeting its predetermined specifications and quality attributes."

Every company needs to establish a checklist of validation procedures that reflect the essence of the FDA's requirements. Validation is making sure that the code does what it purports to do and that it is stable and reliable. The emphasis is on documentation. If there is no written document, there is no proof of validation. The validation procedures have to be understood and followed.

Validation is costly and potentially endless. Good programming practices at every development step is essential to ensure high quality, regardless of the amount of final testing and inspection. It is hard to define what constitutes meaningful validation, but here are some ways to produce bug-free code:

1. Design programs to include debugging aids during the development phase and to exclude debugging aids during the production phase.
2. Test every step of programming logic.
3. Try out various data scenarios.
4. Falsify data errors to see if the code breaks down.
5. Devise a group review and change control process.
6. Review initial program development and changes thoroughly.
7. Conduct independent testing.

Most of the testing procedures can prove only the presence, but not the absence, of bugs. Doing a combination of various validation approaches increases the chances of finding bugs. If budget and resources are limited, fixing problems is more important than fixing symptoms. All the test results, the comments of table and listing review from statisticians, data coordinators, clinicians, medical writers, and the regulatory group need to be kept as proofs of validation.

VII. SECURITY PLAN

Many factors can contaminate a program. A computer virus can alter the behavior of a program. A person can accidentally delete or modify the wrong program. Power failure may cause the loss of information. A disk may crash. A hard drive may be accidentally reformatted.

Most companies have standard operating procedures (SOPs) for storage, archival, and retrieval. The keys are to have appropriate antivirus software and procedures in place, to have password authentication or audit trail controls, to backup periodically, to have off-site storage, to test the backup recovery procedures, and to have contingency plans for conducting the studies in the event of disasters.

VIII. KEEPING UPDATED

Programming is a fast-paced occupation. The SAS software and other application software frequently come out with updated versions with new and improved features. Some time ago, it was a tedious job to extract *p*-values or least squares means differences; now it is possible to output those values into a data set directly from SAS procedures. Keeping updated with new features helps to make programs more efficient.

It is also necessary to keep up with FDA requirements. Software and hardware changes can result in a change in the requirements, such as electronic versus paper submissions and the use of Adobe Acrobat reader software.

The Windows point-and-click or drag-and-drop environment makes programming easier and provides multiple options to perform the same task. At the same time, programming can be more error-prone in such environment. A single movement of the mouse button can delete a whole directory. It is crucial to select the correct variables and the correct directory and to define standards common to all programmers. For dynamic reports or graphs, make sure they always reflect the most current data available. Here are the important steps for building an application in the new environment:

1. Determine the variables and valid ranges for variable values.
2. Determine the necessary sequences.
3. Group the sequences according to function.
4. Use descriptive labels, icons, or pushbuttons.
5. Maintain internal consistency in the interface design, and work with users to make the interface user-friendly.
6. Make sure numeric and logical operations are correct.
7. Test the individual sequences.
8. Link the sequences, and test the entire application.
9. Add code to capture errors and make a graceful exit instead of letting the application crash.
10. Document the application.

IX. PROGRAMMING ETIQUETTE ON A SHARED PROJECT

Beyond good coding habits, the remaining programming virtue is attitude. Programmers usually share different aspects or different studies of programming a clinical trial. A cooperative and understanding attitude facilitates the sound and speedy completion of a project.

Standards and conventions are communication tools. Abide by the standards and conventions so the whole team speaks a common language. Write readable, easily shared, and reusable code whenever possible, keeping in mind that others may have to maintain, modify, or validate the code. Clean up test programs, or stack the most recent test on top of the previous ones in the same program and comment out the previous tests. Putting all test programs in one file not only makes the programming area cleaner, it also makes it easy to locate previous tests.

Make the project a good learning experience for everyone. Be tolerant of individual approaches, keep an open mind, and learn from one another. To err is human. We learn not only from good programming tips and techniques but also from making mistakes.

X. CONCLUSION

Since less is more, my one-sentence conclusion is: In programming, the chain is only as strong as its weakest link, so make good programming practice a habit.

REFERENCES

1. Winograd, Terry; Bennett, John; De Young, Laura; Hartfield, Bradley. Bringing Design to Software. New York: ACM Press, 1996, pp xiii–xxv, 46–47.
2. Maguire, Stephen A. Debugging the Development Process. Redmond, WA: Microsoft Press, 1994, pp 31–32.
3. McConnell, Steven C. Code Complete: A Practical Handbook of Software Construction. Redmond, WA: Microsoft Press, 1993, pp 453–491.
4. Stokes, Teri; Branning, Ronald C.; Chapman, Kenneth G.; Anthony J. Good Computer Validation Practices: Common Sense Implementation. Hambloch, Heinrich; Trill, Anthony J. IL: Interpharm Press, 1994, p 76.

Aileen L. Yam

Good Statistics Practice

See also *Clinical Trial Process; Good Programming Practices*

I. DEFINITION

Good statistics practice (GSP) in drug research and development is defined as a set of statistical principles for the best biopharmaceutical practices in design, conduct, analysis, reporting, and evaluation of studies at various stages of drug research and development (Chow, 1997). The purpose of GSP is not only to minimize bias but also to minimize variability that may occur before, during, and after the conduct of the studies. More importantly, GSP provides a valid and fair assessment of the drug product under study. The concept of GSP in drug research and development can be seen in many regulatory requirements, standards/specifications, and guidelines/guidances set by most health authorities, such as the United States Food and Drug Administration (FDA). For example, the United States regulatory requirements for drug research and development are codified in the United States Code of Federal Regulations (CFR), while the United States Pharmacopeia and National Formulary (USP/NF) includes standard procedures, test and sampling plans, and acceptance criteria and specifications of many pharmaceutical compounds. In addition, the FDA also develops a number of guidelines and guidances to assist the sponsors in drug research and development. These guidelines and guidances are considered gold standards for achieving good laboratory practice (GLP), good clinical practice (GCP), current good manufacturing practice (cGMP), and good regulatory practice (GRP). The concept of GSP is well outlined in the guideline on *Statistical Principles for Clinical Trials* recently issued by the International Conference on Harmonization (ICH, 1997). As a result, GSP not only provides accuracy and reliability of the results derived from the studies but also ensures the validity and integrity of the studies.

II. ROLE OF STATISTICS IN DRUG DEVELOPMENT

In the research and development of a drug product, statistics are necessarily applied at various critical stages to meet regulatory requirements for the effectiveness, safety, identity, strength, quality, purity, and stability of the drug product under investigation. These critical stages include pre-IND (investigational new drug application), IND, NDA (new drug application) and post-NDA. The role of statistics at these critical stages are briefly described later.

At the very early stages of pre-IND, pharmaceutical scientists may have to screen thousands of potential compounds in order to identify a few promising compounds. An appropriate use of statistics with efficient screening and/or optimal designs will assist pharmaceutical scientists to identify the promising compounds within a relatively short time frame and cost effectively.

As indicated by the FDA, an IND should contain information regarding chemistry, manufacturing, and controls (CMC) of the drug substance and drug product to ensure the identity, strength, quality, and purity of the investigational drug. In addition, the sponsors are required to provide adequate information about pharmacological studies for absorption, distribution, metabolism, and excretion (ADME) and acute, subacute, and chronic toxicological studies and reproductive tests in various animal species to show that the investigational drug is reasonably safe to be evaluated in clinical trials in humans. At this stage, statistics are usually applied to (a) validate a developed analytical method, (b) establish a drug expiration dating period through stability studies, and (c) assess toxicity through animal studies. Statistics are necessarily applied to meet standards of accuracy and reliability.

Before the drug can be approved, the FDA requires that substantial evidence of the effectiveness and safety of the drug be provided in the Technical Section of Statistics of an NDA submission. Since the validity of statistical inference regarding the effectiveness and safety of the drug is always a concern, it is suggested that a careful review be performed to ensure an accurate and reliable assessment of the drug product. In addition, to have a fair assessment, the FDA also establishes advisory committees, each consisting of clinical, pharmacological, and statistical experts and one advocate (not employed by the FDA) in designated drug classes and subspecialities, to provide a second but independent review of the submission. The responsibility of the statistical expert is not only to ensure that a valid design is used but also to evaluate whether statistical methods used are appropriate for addressing the scientific and medical questions regarding the effectiveness and safety of the drug.

After the drug is approved, the FDA also requires that the drug product be tested for its identity, strength, quality, and purity before it can be released for use. For this purpose, the current good manufacturing practice is necessarily implemented to (a) validate the manufacturing process, (b) monitor the performance of the manufacturing process, and (c) provide quality assurance of the final product. At each stage of the manufacturing process, the FDA requires that sampling plans, acceptance criteria and valid statistical analyses be performed for the intended tests, such as potency, content uniformity, and dissolution (USP/NF, 1995). For each test, sampling plan, acceptance criteria, and valid statistical analysis are crucial for determining whether the drug product passes the test based on the results from a representative sample.

III. STATISTICAL PRINCIPLES

In this section, we will discuss some key statistical principles in the design and analysis of studies that may be encountered at various stages of drug development.

A. Bias and Variability

For approval of a drug product, regulatory agencies usually require that the results of the studies conducted at various stages of drug research and development be accurate and reliable to provide a valid and fair assessment of the treatment effect. The accuracy and reliability are usually referred to as the *closeness* and the *degree of closeness* of the results to the true value (i.e., true treatment effect). Any deviation from the true value is considered a bias, which may be due to selection, observation, or statistical procedures. Pharmaceutical scientists would make any attempt to avoid bias whenever possible to ensure that the collected results are accurate.

The reliability of a study is an assessment of the precision of the study, which measures the degree of the closeness of the results to the true value. The reliability reflects the ability to repeat or reproduce similar outcomes in the targeted population. The higher precision a study is, the more likely it is that the results would be reproducible. The precision of a study can be characterized by the variability incurred during the conduct of the study.

In practice, since studies are usually planned, designed, executed, analyzed, and reported by a team that consists of pharmaceutical scientists from different disciplines, bias and variability inevitably occur. It is suggested that possible sources of bias and variability be identified at the planning stage of the study, not only to reduce the bias but also to minimize the variability.

B. Confounding and Interaction

In drug research and development, there are many sources of variation that have an impact on the evaluation of the treatment. If these variations are not identified and properly controlled, then they may be mixed up with the treatment effect that the studies are intended to demonstrate. In this case, the treatment is said to be confounded with the effects due to these variations. To provide a better understanding, consider the following example. Last winter, Dr. Smith noticed that the temperature at the emergency room was relatively low, which had caused some discomfort among medical personnel and patients. Dr. Smith suspected that the heating system might not function properly and decided to improve it. As a result, the temperature of the emergency room has been raised to a comfortable level this winter. However, this winter is not as cold as last winter. Therefore, it is not clear whether the improvement of emergency room temperature was due to the improvement of the heating system or the effect of a warmer winter.

The statistical interaction is to investigate whether the joint contribution of two or more factors is the same as the sum of the contributions from each factor when considered alone. If an interaction between factors exists, an overall assessment cannot be made. For example, suppose that a placebo-controlled clinical trial was conducted at two study centers to assess the effectiveness and safety of a newly developed drug product. Suppose that the results turned out that the drug is efficacious (better than placebo) at one study center and inefficacious (worse than placebo) at the other study center. As a result, a significant interaction between treatment and study center occurred. In this case, an overall assessment of the effectiveness of the drug product can be made.

In practice, it is suggested that possible confounding factors be identified and properly controlled at the planning stage of the studies. When significant interactions among factors are observed, subgroup analyses may be necessary for a careful evaluation of the treatment effect.

C. Type I Error, Significance Level and Power

In statistical analysis, two different kind of mistakes are commonly encountered when performing hypotheses testing. For example, suppose that a physician is to determine whether or not one of his/her patients is still alive. If the patient is dead, then the physician may remove his/her life-support equipment for other patients who need it. Therefore, the null hypothesis of interest is that the patient is still alive, while the alternative hypothesis is that the patient is dead. Under these hypotheses, the physician may make two mistakes, which are: (a) he/she concludes that the patient is dead when in fact the patient is still alive, and (b) he/she claims that the patient is still alive when in fact the patient is dead. The first kind of mistake is usually referred to as a Type I error; the latter is the so-called Type II error. Since a Type I error is usually considered more important or serious, we would like to limit the probability of committing this kind of error to an acceptable level. This acceptable level of probability of committing a Type I error is known as the *significance level*. As a result, if the probability of observing a Type I error based on the data is less than the significance level, we conclude that a statistically significant result is observed. A statistically significant result suggests that the null hypothesis be rejected, in favor of the alternative hypothesis. The probability of observing a Type I error is usually referred to as the p-value of the test. On the other hand, the probability of committing a Type II error subtracted from 1 is called the *power* of the test. In our example, the power of the test is the probability of correctly concluding the death of the patient when the patient is dead.

For the pharmaceutical application, suppose that a pharmaceutical company is interested in demonstrating that a newly developed drug is efficacious. The null hypothesis is often chosen as the drug is inefficacious, versus the alternative hypothesis that the drug is efficacious. The objective is to reject the null hypothesis in favor of the alternative hypothesis and consequently to conclude that the drug is efficacious. Under the null hypothesis, a Type I error is made if we conclude that the drug is efficacious when in fact it is not. This error is also know as *consumer's risk*. Similarly, a Type II error is committed if we conclude that the drug is inefficacious when in fact it is efficacious. This error is sometimes called *producer's risk*. The power is then considered to be the probability of correctly concluding that the drug is efficacious when in fact it is. For assessment of drug effectiveness and safety, a sufficient sample size is often selected to have a desired power with a prespecified significance level. The purpose is to control both Type I (significance level) and Type II (power) errors.

D. Randomization

Statistical inference on a parameter of interest of a population under study is usually derived under the probability structure of the parameter. The probability structure depends upon the randomization method employed in sampling. The failure of the randomization will have a negative impact on the validity of the probability structure. Consequently, the validity, accuracy, and reliability of the resulting statistical inference of the parameter is questionable. Therefore, it is suggested the randomization be performed using an appropriate randomization method under a valid randomization model according to the study design to ensure the validity, accuracy, and reliability of the derived statistical inference.

E. Sample Size Determination/Justification

One of the major objectives of most studies during drug research and development is to determine whether the drug is effective and safe. During the planning stage of a study, the following questions are of particular interest to the pharmaceutical scientists: (a) How many subjects are needed in order to have a desired power for detecting a meaningful difference? (b) What is the trade-off if only a small number of subjects is available for the study due to limited budget and/or some scientific considerations? To address these questions, a statistical evaluation for sample size determination/justification is often employed. Sample size determination usually involves the calculation of sample size for some desired statistical properties, such as power or precision; sample size justification is to provide statistical justification for a selected sample size, which is often a small number.

For a given study, sample size can be determined/justified based on some criteria of a Type I error (a desired precision) or a type II error (a desired power). The disadvantage for sample size determination/justification based on the criteria of precision is that it may have a small chance of detecting a true difference. As a result, sample size determination/justification based on the criteria of power becomes the most commonly used method. Sample size is selected to have a desired power for detection of a meaningful difference at a prespecified level of significance.

In practice, however, it is not uncommon to observe discrepancies among study objective (hypotheses), study design, statistical analysis (test statistic), and sample size calculation. These inconsistences often result in (a) the wrong test for the right hypotheses, (b) the right test for the wrong hypotheses, (c) the wrong test for the wrong hypotheses, or (d) the right test for the right hypotheses with insufficient power. Therefore, before the sample size can be determined, it is suggested that the following be carefully considered: (a) The study objective or the hypotheses of interest should be clearly stated. (b) A valid design with appropriate statistical tests should be used. (c) Sample size should be determined based on the test for the hypotheses of interest.

F. Statistical Difference and Scientific Difference

A *statistical difference* is defined as a difference that is unlikely to occur by chance alone, while a *scientific difference* is a difference that is considered to be of scientific importance. A statistical difference is also referred to as a *statistically significant difference*. The difference between the concepts of *statistical difference* and *scientific difference* is that statistical difference involves chance (probability) while scientific difference does not. When we claim there is a statistical difference, the difference is reproducible with a high probability.

When conducting a study, there are basically four possible outcomes. The result may show that (a) the difference is both statistically and scientifically significant, (b) there is a statistically significant difference yet the difference is not scientifically significant, (c) the difference is scientifically significant yet it is not statistically significant, and (d) the difference is neither statistically significant nor scientifically significant.

If the difference is both statistically and scientifically significant or it is neither statistically or scientifically significant, then there is no confusion. However, in many cases, a statistically significant difference does not agree with the scientifically significant difference. This inconsistency has created confusion/arguments among pharmaceutical scientists and biostatisticians. The inconsistence may be due to large variability and/or insufficient sample size.

G. One-Sided Test Versus Two-Sided Test

For the evaluation of a drug product, the null hypothesis of interest is often that there is no difference. The alternative hypothesis is usually that there is a difference. The statistical test for this setting is called a *two-sided test*. In some cases, the pharmaceutical scientist may test the null hypothesis of no difference against the alternative hypothesis that the drug is superior to the placebo. The statistical test for this setting is known as a *one-sided test*.

For a given study, if a two-sided test is employed at the significance level of 5%, then the level of proof required is 1 out of 40. In other words, at the 5% level of significance, there is 2.5% chance (or 1 out of 40) that we may reject the null hypothesis of no difference in the positive direction and conclude that the drug is effective at one side. On the other hand, if a one-sided test is used, the level of proof required is 1 out of 20. It turns out that a one-sided test allows more ineffective drugs to be approved because of chance as compared to the two-sided test. It should be noted that when testing at the 5% level of significance with 80% power, the sample size required increases by 27% for a two-sided test as compared to a one-sided test. As a result, there is a substantial cost saving if a one-sided test is used.

However, agreement is not universal among the regulatory, academia, and the pharmaceutical industry as to whether a one-sided test or a two-sided test should be used. The U.S. Food and Drug Administration tends to oppose the use of a one-sided test, though this position has been challenged by several pharmaceutical companies on Drug Efficacy Study Implementation (DESI) drugs at the Administrative Hearing. Dubey (1991) pointed out that several viewpoints that favor the use of one-sided tests were discussed in an administrative hearing. These points indicated that a one-sided test is appropriate in the following situations: (a) where there is truly concern with outcomes in one tail only and (b) where it is completely inconceivable that the results could go in the opposite direction.

IV. IMPLEMENTATION OF GSP

The implementation of GSP in drug research and development is a team project that requires mutual communication, confidence, respect, and cooperation among statisticians, pharmaceutical scientists in the related areas, and regulatory agents. The implementation of GSP involves some key factors that have an impact on the success of GSP. These factors include (a) regulatory requirements for statistics, (b) the dissemination of the concept of statistics, (c) an appropriate use of statistics, (d) an effective communication and flexibility, and (e) statistical training. These factors are briefly described next.

In the drug development and approval process, regulatory requirements for statistics are the key to the implementation of GSP. They not only enforce the use of statistics but also establish standards for statistical evaluation of the drug products under investigation. An unbiased statistical evaluation helps pharmaceutical scientists and regulatory agents in determining (a) whether the drug product has the claimed effectiveness and safety for the intended disease, and (b) whether the drug product possesses good drug characteristics, such as the proper identity, strength, quality, purity, and stability.

In addition to regulatory requirements, it is always helpful to disseminate the concept of statistical principles described earlier whenever possible. It is important for pharmaceutical scientists and regulatory agents to recognize that (a) a valid statistical inference is necessary to provide a fair assessment with certain assurance regarding the uncertainty of the drug product under investigation; (b) an invalid design and analysis may result in a misleading or wrong conclusion about the drug product; (c) a larger sample size is often required to increase the statistical power and precision of the studies. The dissemination of the concept of statistics is critical to establish the pharmaceutical scientists' and regulatory agents' brief in statistics for scientific excellence.

One of the commonly encountered problems in drug research and development is the misuse or sometimes the abuse of statistics in some studies. The misuse or abuse of statistics is critical, which may result in either having the right question with the wrong answer or having the right answer for the wrong question. For example, for a given study, suppose that a right set of hypotheses (the right question) is established to reflect the study objective. A misused statistical test may provide a misleading or wrong answer to the right question. On the other hand, in many clinical trials, point hypotheses for equality (the wrong question) are often wrongly used for establishment of equivalency. In this case, we have the right answer (for equality) for the wrong question. As a result, it is recommended that appropriate statistical methods be chosen to reflect the design that should be able to address the scientific or medical questions regarding the intended study objectives for implementation of GSP.

Communication and flexibility are important factors to the success of GSP. Inefficient communication between statisticians and pharmaceutical scientists or regulatory agents may result in a misunderstanding of the intended study objectives and consequently in an invalid design and/or inappropriate statistical methods. Thus, effective communications among statisticians, pharmaceutical scientists, and regulatory agents is essential for the implementation of GSP. In addition, in many studies, the assumption of a statistical design or model may not be met, due to the nature of the drug product under investigation, the experimental environment, and/or other causes related/unrelated to the studies. In this case, the traditional approach of doing everything by the book does not help. In practice, since a concern from a pharmaceutical scientist or the regulatory agent may translate into a constraint for a valid statistical design and appropriate statistical analysis, it is suggested that a flexible yet innovative solution be developed under the constraints for the implementation of GSP.

Since regulatory requirements for the drug development and approval process vary from drug to drug and from country to country, various designs and/or statistical methods are often required for a valid assessment of a drug product. Therefore, it is suggested that statistical continued/advanced education and training programs be routinely held for both statisticians and nonstatisticians, including pharmaceutical scientists and regulatory agents. The purpose of such a continued/advanced education and/or training program is threefold. First, it enhances communications within the statistical community. Statisticians can certainly benefit from such a training and/or educational program by acquiring more practical experience and knowledge. In addition, it provides the opportunity to share/exchange information, ideas, and/or concepts regarding drug development between professional societies. Finally, it identifies critical practical and/or regulatory issues that are commonly encountered in the drug development and regulatory approval process. A panel discussion from different disciplines may result in some consensus to resolve the issues, which helps in establishing standards of statistical principles for the implementation of GSP.

V. CONCLUSIONS

During the development and regulatory approval process, good pharmaceutical practices are necessarily implemented to (a) ensure the effectiveness and safety of the drug product under investigation before approval, and (b) ensure that the drug product possesses good drug characteristics, such as the proper identity, strength, quality, purity, and stability, in compliance with the standards as specified in the USP/NF after regulatory approval. These good pharmaceutical practices include GLP for animal studies, GCP for clinical development, cGMP for chemistry, manufacturing, and controls (CMC), and GRP for regulatory review and approval process. In essence, GSP is the foundation of GLP, GCP, cGMP, and GRP. The implementation of GSP is a team project that involves statisticians, pharmaceutical scientists, and regulatory agents as well. The success of GSP depends upon mutual communication, confidence, respect, and cooperation among statisticians, pharmaceutical scientists, and regulatory agents.

REFERENCES

Chow, S.-C. (1997). Good statistics practice in the drug development and regulatory approval process. Drug Inf J 31: 1157–1166.

Dubey, S.D. (1991). Some thoughts on the one-sided and two-sided tests. J Biopharm Stat 1:139–150.

ICH. (1997). Statistical Principles for Clinical Trials. International Conference on Harmonization.

USP/NF. (1995). The United States Pharmacopeia XXIII and the National Formulary XVIII. United States Pharmacopeial Convention, Inc., Rockville, MD.

Shein-Chung Chow

Group Sequential Methods

I. INTRODUCTION

The group sequential procedure (design, monitoring, and analysis) is the most commonly applied form of sequential procedure to facilitate the conduct of interim analysis in the biopharmaceutical field. A primary motivation of a group sequential analysis in general is conserving experimental materials and/or research effort. An important biopharmaceutical application in this respect was in drug screening (1). Recently, active developments of the group sequential procedure have been in clinical trials, especially trials with mortality or irreversible morbidity as the primary endpoint. In these clinical trials a further consideration is an ethical obligation to terminate the study at the earliest opportunity to prevent administration of a less favorable therapy to additional patients. Conducting the clinical trial in stages, with group sizes predetermined by a certain schedule, provides logistical feasibility over fully sequential methods, since these experiments or clinical trials generally will involve a lengthy time to respond. Yet the advantage of reducing the average sample number is largely retained (2). Many different methods of group sequential procedure are available: different stopping boundaries, different Type I error spending/use functions, conditional power or predictive power approach, Bayesian methods, etc. The logistics of a group sequential trial are, however, complicated and require careful planning and conducting. A review article is given in Ref. 3.

II. REGULATORY REQUIREMENTS

The recent International Conference on Harmonization (ICH) Guidelines (February 1998) (4) discuss regulatory requirements on interim analyses in general (E3, Section 11.4.2.3) and group sequential procedures in specific regarding data monitoring and early stopping (E9, Sections 3.4 and 4.1–4.6). The following is a summary.

All interim analyses should be carefully planned in advance and described in the protocol. Any unplanned interim analysis should be avoided if at all possible. In the event that an unplanned interim analysis was conducted, a protocol amendment describing the interim analysis should be completed prior to unblinded access to treatment comparison data, and the study report should explain why it was necessary and the degree to which the blindness had to be broken, and provide an assessment of the potential magnitude of bias introduced and of the impact on the interpretation of the results.

The schedule of analyses, or at least the considerations that will govern its generation, should be stated in the protocol or amendment before the time of the first interim analysis; the stopping guidelines and their properties should also be clearly stated there.

The procedure plan should be written or approved by the data monitoring committee (DMC), when the trial has one.

Any changes to the trial and any consequent changes to the statistical procedures should be specified in an amendment to the protocol at the earliest opportunity.

The procedures selected should always ensure that the overall probability of Type I error is controlled.

The execution of an interim analysis must be a completely confidential process when unblinded data and results are potentially involved. All staff involved in the conduct of the trial should remain blind to the results of such analyses. Investigators should be informed only about the decision to continue or discontinue the trial or to implement modifications to trial procedures.

Any interim analysis planned for administrative purposes only should also be described in the protocol and subsequently reported. The specific purposes should be clearly stated and should specifically exclude any possibility of early stopping. The blind should not be broken.

III. STATISTICAL METHODS

A. Fundamentals

We start by considering the design of a clinical trial with a plan for monitoring the data at calendar time $ct_1, ct_2, \ldots, ct_K$ for some fixed $K > 1$. For example, a trial monitoring committee might plan to meet every March 31 and September 30 for the first 2 years and then every June 15 for the rest of the trial duration. (The Data Monitoring Committee will be discussed later.) Corresponding to the ct_i, $i = 1, \ldots, K$, are the information time (or *information fraction*) $t_1, t_2, \ldots, t_K = 1$. (The *information time/fraction* and the *maximum duration trial* versus *maximum information trial* will be discussed later.) At information time t_i, $i = 1, \ldots, K$, standardized Z-statistics, $Z_{t1}, Z_{t2}, \ldots, Z_{tK}$ are calculated based on the accumulated data for testing hypotheses about the treatment effect represented by a parameter θ. A focal statistical problem of group sequential procedures is to find critical values to satisfy certain desirable operational characteristics. These critical values are the so-called *group sequential boundaries*. For example, for a one-sided hypothesis with the decision options being "either reject H_0: $\theta = 0$ or continue at the interim stages," and "reject or accept H_0 at the final stage," the problem is to find values $b_1, b_2, \ldots, b_K$ such that the overall Type I error α is maintained at a prespecified level:

$$P_{H_0}(\text{reject } H_0) = P_{H_0}(Z_{t1} \geq b_1, \text{ or } Z_{t2} \geq b_2, \text{ or } \ldots . Z_{tK} \geq b_K) = \alpha \quad (1)$$

(Different forms of one-side hypotheses are given in Ref. 3.) For a corresponding two-sided hypothesis test, a symmetric boundary would replace α by $\alpha/2$ in Eq. (1) and then apply the one-sided boundary symmetrically. (Asymmetric boundaries and interim analyses with chance to accept H_0 at an early stage will be discussed later.) In the following we first review several commonly cited group sequential boundaries.

B. Equally Spaced Group Sequential Boundaries

Suppose that the (information) times are equally spaced: $t_1 = 1/K$, $t_2 = 2/K$, $\ldots$, $t_K = K/K = 1$.

Pocock (5): Find $b_i = c$ for $i = 1, \ldots, K$ such that

$$P_{H_0}(Z_{t1} \geq c, \text{ or } Z_{t2} \geq c, \text{ or } \ldots Z_{tK} \geq c) = \alpha \quad (2)$$

For example, if set $\alpha = 0.025$ and $K = 5$, then $c = 2.413$. The boundaries are (2.413, 2413, 2.413, 2.413, 2.413).

O'Brien–Fleming (6): Find c such that

$$P_{H_0}(Z_{t1}\sqrt{t_1} \geq c, \text{ or } Z_{t2}\sqrt{t_2} \geq c, \text{ or } \ldots Z_{tK}\sqrt{t_K} \geq c) = \alpha \quad (3)$$

For $\alpha = 0.025$ and $K = 5$, $c = 2.04$. Hence $b_i = 2.04/\sqrt{t_i} = 2.04/\sqrt{i/5}$. The boundaries are (4.562, 3.226, 2.634, 2.281, 2.040).

Haybittle–Peto (7,8): Find $b_i = c$ for $i = 1, \ldots, K - 1$ such that

$$P_{H_0}(Z_{t1} \geq c, \text{ or } Z_{t2} \geq c, \text{ or } \ldots Z_{tK} \geq z_{l-\alpha}) = \alpha \quad (4)$$

For example, if set $\alpha = 0.025$ and $K = 5$, then $c = 3.291$. The boundaries are (3.291, 3.291, 3.291, 3.291, 1.960).

The preceding boundaries, for the two-sided case, are not "closed" at the final analyses. Whitehead (9,10) proposed "triangular" closed boundaries, which plots the unstandardized test statistic versus its variance.

The calculations of the boundaries require iterative numerical integrations (11,12). Commercial computer packages available for obtaining the boundaries for a variety of K and α include EaSt (13) and S-plus (14) for the Pocock and O'Brien–Fleming, and PEST (15) for the Whitehead boundaries.

Choice of the boundaries in the trial design depends on the intention of early stopping of a trial. O'Brien–Fleming boundaries are the most stringent of the three at early times, while Pocock boundaries stay constant throughout. Haybittle–Peto boundaries are similar to Pocock at interim stages, but set the final analysis resembling the fixed-sample-size design.

At the design stage it may be reasonable to plan for fixed-K analyses at equally spaced information times. But when the trial is ongoing (i.e., when monitoring the study), changes often occur. Interim analyses may well not be on the preplanned schedule, and the frequency of the analyses, K, may also change. This requires a more flexible procedure. The Slud–Wei (16) and Lan–DeMets (17) Type I error spending/use function approach meets this need.

C. Type I Error Spending/Use Function Approach

The idea of the Type I error spending/use function approach can be illustrated as follows. Untie the fixed K in expression (1) and decompose the rejection region,

$$R = \{Z_{t1} \geq b_1, \text{ or } Z_{t2} \geq b_2, \text{ or } Z_{t3} \geq b_3, \ldots\}$$

into disjoint regions:

$$R_1 = \{Z_{t1} \geq b_1\}$$

$$\{Z_{t1} \geq b_1, \text{ or } Z_{t2} \geq b_2\} = \{Z_{t1} \geq b_1\} \cup \{Z_{t1} < b_1, Z_{t2} \geq b_2\} = R_1 \cup R_2$$

$$\{Z_{t1} \geq b_1, \text{ or } Z_{t2} \geq b_2, \text{ or } Z_{t3} \geq b_3\} = (R_1 \cup R_2) \cup \{Z_{t1} < b_1, Z_{t2} < b_2, Z_{t3} \geq b_3\} = (R_1 \cup R_2) \cup R_3$$

and so on. Hence,

$$P_{H_0}(R) = P_{H_0}(R_1 \cup R_2 \cup R \cdots) = P_{H_0}(R_1) + P_{H_0}(R_2) + P_{H_0}(R_3) + \cdots = \alpha$$

At t_1 we specify a value of $\alpha(t_1) = P_{H_0}(R_1)$ and solve for b_1. At t_2, specify

$$\alpha(t_2) = P_{H_0}(R_1 \cup R_2) = P_{H_0}(R_1) + P_{H_0}(R_2) = \alpha(t_1) + P_{H_0}(R_2)$$

This leads to $P_{H_0}(R_2) = \alpha(t_2) - \alpha(t_1)$, and we can solve for b_2, since b_1 is already known from the first stage. Similarly at t_3, specify

$$\alpha(t_3) = P_{H_0}(R_1 \cup R_2 \cup R_3) = (P_{H_0}(R_1) + P_{H_0}(R_2)) + P_{H_0}(R_3) = \alpha(t_2) + P_{H_0}(R_3)$$

Thus $P_{H_0}(R_3) = \alpha(t_3) - \alpha(t_2)$. We then solve for b_3, since b_1 and b_2 have already been solved in the previous steps. The process continues.

Note that, when solving for b_2, we only need the joint distribution of (Z_{t1}, Z_{t2}). For a partial sum process with independent increments (see later discussion), where we can write $Z_{ti} = S(t_i)/\sqrt{I(t_i)}$, with the numerator being the cumulative sum of the observations (i.i.d. variates with mean θ) and the denominator being the square root of the cumulative statistical information in estimating θ at time t_i, the correlation is

$$\rho(t_1, t_2) = \text{corr} = \text{cov}(Z_{t1}, Z_{t2}) = \text{cov}\left(\frac{S(t_1)}{\sqrt{I(t_1)}}, \frac{S(t_2)}{\sqrt{I(t_2)}}\right)$$

$$= \text{cov}\left(\frac{S(t_1)}{\sqrt{I(t_1)}}, \frac{S(t_1) + (S(t_2) - S(t_1))}{\sqrt{I(t_2)}}\right)$$

$$= \frac{1}{\sqrt{I(t_1)I(t_2)}} \text{var}(S(t_1)) = \sqrt{\frac{I(t_1)}{I(t_2)}}$$

Other time-point boundaries follow similarly. [When information is proportional to the sample size, then $\rho(t_1, t_2) = (n_1/n_2)^{1/2} = (t_1/t_2)^{1/2}$.] Therefore, the joint distribution involves the ratio of information only at the current stage and before, not at the later time-points. This indicated the flexibility of this approach for monitoring trials, since there is no requirement for a given K, the total number of analyses, or for the "equally spaced t_i" restriction. All that is needed is $\alpha(t_1) < \alpha(t_2) < \cdots < \alpha(1) = \alpha$, a strictly increasing function $\alpha(t)$, known as the "Type I error (α) spending/use function," prespecified in the study design protocol.

Example: Two-Stage Design (One Interim and One Final Analysis) Symmetrical Boundaries. Specify the overall Type I error $\alpha = 0.05$ (two-sided test) and the spending function as $\alpha(t) = \alpha t$. Suppose we conduct an interim analysis when $t_1 = 1/2$, then $\alpha(t_1 = 1/2) = 0.025$. The corresponding boundary point at $t_1 = 1/2$ is $b_1 = 2.244$. This leaves $\alpha - \alpha(t_1 = 1/2) = 0.025$ to be spent at the final stage $t_2 = 1$. Solving $P\{|Z_{t1}| < 2.244, |Z_{t2}| \geq b_2\} = 0.025$ given that the correlation of the bivariate normal distribution is $t_1/t_2 = 1/2$, we obtain $b_2 = 2.126$. Notice that at the final stage, $\alpha_2 \equiv P(|Z| > 2.126) = 0.034$. That is, with this linear/uniform spending function, if an interim testing is to be conducted in the middle of the trial, the p-value at the final stage has to be less than 0.034, instead of the 0.05 level, to be significant.

A natural question is: "What kind of α-spending function describes the Pocock boundary and the O'Brien–Fleming boundary?"

Example: $K = 5$, $\alpha = 0.025$, Pocock boundary: (2.413, 2.413, 2.413, 2.413, 2.413).

1. $P_{H_0}(Z_{t1} \geq 2.413) = 0.0079 \Rightarrow \alpha(t_1 = 1/5) = 0.0079$
2. $P_{H_0}(Z_{t1} \geq 2.413 \text{ or } Z_{t2} \geq 2.413)$
   $= P_{H_0}(Z_{t1} \geq 2.413) + P_{H_0}(Z_{t1} < 2.413, Z_{t2} \geq 2.413)$
   $= 0.0079 + 0.0059$
   $= .0138 \Rightarrow \alpha(t_2 = 2/5) = 0.0138$
3. $P_{H_0}(Z_{t1} \geq 2.413, \text{ or } Z_{t2} \geq 2.413, \text{ or } Z_{t3} \geq 2.413)$
   $= P_{H_0}(Z_{t1} \geq 2.413 \text{ or } Z_{t2} \geq 2.413) + P_{H_0}((Z_{t1} < 2.413, Z_{t2} < 2.413), Z_{t3} \geq 2.413)$
   $= 0.0138 + 0.0045$
   $= 0.0183 \Rightarrow \alpha(t_3 = 3/5) = 0.0183$
4. $P_{H_0}(Z_{t1} \geq 2.413, \text{ or } Z_{t2} \geq 2.413, \text{ or } Z_{t3} \geq 2.413, \text{ or } Z_{t4} \geq 2.413)$
   $= \cdots$ (similar steps)
   $= 0.0183 + 0.0036$
   $= 0.0219 \Rightarrow \alpha(t_4 = 4/5) = 0.0219$
5. $P_{H_0}(Z_{t1} \geq 2.413, \text{ or } Z_{t2} \geq 2.413, \text{ or } Z_{t3} \geq 2.413, \text{ or } Z_{t4} \geq 2.413, \text{ or } Z_{t5} \geq 2.413)$
   $= \cdots$ (similar steps)
   $= 0.0219 + 0.0031$
   $= 0.0250 \Rightarrow \alpha(t_5 = 1) = 0.0250$

Example: $K = 5$, $\alpha = 0.025$, O'Brien–Fleming Boundary: (4.56, 3.23, 2.63, 2.28, 2.04). Following similar steps as previously, we can obtain the (discrete) Type I error spending function $\alpha(t_1 = 1/5) = 0.0000$, $\alpha(t_2 = 2/5) = 0.0006$, $\alpha(t_3 = 3/5) = 0.0045$, $\alpha(t_4 = 4/5) = 0.0128$, and $\alpha(t_5 = 1) = 0.0250$.

Removing the restriction of a fixed K and equally spaced $t_i = i/K$, Lan and DeMets (17) gave the continuous α-spending function for the general Pocock boundary as

$$\alpha_{\text{Pocock}}(t) = \alpha \ln[1 + (e - 1)t]$$

and for the general O'Brien–Fleming boundary as

$$\alpha_{\text{OB-F}}(t) = 2\left[1 - \Phi\left(\frac{z_{\alpha/2}}{\sqrt{t}}\right)\right]$$

where $\Phi(.)$ is the c.d.f. of the standard normal variate.

Other Type I error rate spending functions have also been proposed in the literature. Lan and DeMets (17) and Kim and DeMets (18) studied certain members of the "power family" $\alpha(t) = \alpha t^{\theta}$, $\theta > 0$. The landmark clinical trial "Simvastatin Scandinavian Survival Study," or the "4S" (19), used a member of the "truncated exponential distribution family" of α-spending functions proposed by Hwang et al. (20):

$$\alpha(\gamma, t) = \begin{cases} \alpha\left[\dfrac{1 - e^{-\gamma t}}{1 - e^{\gamma}}\right], & \gamma \neq 0 \quad \text{for } 0 \leq t \leq 1 \\ \alpha t, & \gamma = 0 \end{cases} \tag{5}$$

As noted in Hwang et al. (20), the continuous Pocock boundary is included in Eq. (5), with $\gamma = 1$; the continuous O'Brien–Fleming boundary is also a member of Eq. (5), with approximately $\gamma = -4$ or -5; the "power family" member $\alpha(t) = \alpha t^{3/2}$ is closely approximated by $\gamma = -1$, and $\alpha(t) = \alpha t^2$ is approximated by $\gamma = -2$.

Group sequential methods provide a chance for early rejection of H_0 and for stopping a trial in such a way that the overall Type I error rate is maintained. The trade-off is some loss of power, since the tests at early stages involve smaller sample sizes, and at later stages the rejection regions are tighter (both implying lower power). To have the same power as the fixed-size design, a group sequential trial would require a larger *maximum* sample size, but the *average* sample size would still be smaller (under H_A). The maximum sample size factor is a function of the Type I and II errors, the number of analyses, and the way of spending α.

Some authors (18,21) tried to search for "optimal boundaries"—that minimize *average* sample size or the expected first boundary crossing time under a given H_A and power. However, in practice the choice of boundaries is usually based more on clinical decisions and the nature of the trial than on the statistical properties. For example, with either the "power family" or the "truncated exponential family" one can choose a member [e.g., $\gamma = 0$ in Eq. (5)] for uniform spending of the Type I error. Choosing a member (e.g., $\gamma > 0$, usually $1 \leq \gamma \leq 4$) for a concave spending of the Type I error is suitable for short-term trials with immediate response, such as single-dose analgesic studies using time to pain relief as the endpoint, or when accelerated early stopping is desirable in the early phases of drug development. A negative γ in Eq. (5) (usually $-5 \leq \gamma \leq -1$, including the O'Brien–Fleming boundary) is for convex spending, which is suitable for a large mortality trial where the patient recruitment is slow and staggered and the follow-up period is long-term so that very early stopping with limited samples is not much encouraged. As was pointed out by Kim and DeMets (18), the more convex the Type I error spending function is, the more powerful the group sequential test is.

D. Partial Sum Process with Independent Increments

Most of the group sequential methods were developed earlier using immediate response, either continuous or binary, where cumulative sums of the independent random variables from time to time are easily seen to have independent increments. The usefulness of this ''partial sum with independent increments'' property was seen in the previous section.

An important development was provided by Tsiatis (22,23), who showed that the log-rank statistic computed over time behaves much like a partial sum of independent normal random variables. This result extended the use of group sequential methods to clinical trials with survival data. (See the entry on *Survival Data* for the log-rank test.) Recently, Jennison and Turnbull (24) provided a unified theory that explains the ''independent increments'' structure commonly seen in group sequential test statistics. Scharfstein et al. (25) demonstrated that all sequentially computed Wald statistics based on efficient estimators [e.g., maximum likelihood estimators (MLEs)] of the parameter of interest will, under mild regularity conditions, have the asymptotic multivariate normal distribution similar to the earlier setup. Hence, the group sequential procedure extends to more complicated situations, such as proportional-hazards model (26) and correlated observations, including longitudinal data with random-effects model (27,28) or with distribution-free analyses (29).

E. Information Time/Fraction and Maximum-Duration Trial versus Maximum-Information Trial

Taking statistical information as the inverse of the variance of the parameter estimate, Lan and Zucker (30) defined the *information time/fraction* as the amount of information accrued by calendar time divided by the total information at the scheduled end of the trial. We have seen that information time played an essential role in the Type I error spending-function approach. Depending on the statistics used [see Lan et al. (31) for a general discussion], a complete unit information can be approximated by either a ''patient'' (for comparing means) or an ''event'' (for comparing survival distributions). In either case, the total information must be known. If not known, as is often the case, then the information time/fraction can only be estimated.

For example, in the time-to-event case, Tsiatis (22) showed that the variance of the log-rank statistics, when calculated over time, grows proportional to the number of events observed. Hence the information time is equal to the proportion of the maximum number of events expected by the end of a study, with the numerator being the observed number of events at the (calendar) time of interim analysis. For a maximum-information trial, the maximum number of events expected by the end of a study is chosen in advance to achieve a desired power, given other design parameters. However, for a maximum-duration trial, i.e., when the maximum trial duration is fixed, the maximum information is random. In such a trial, the denominator of the information time can be estimated under either the null or the alternative hypothesis, thus leading to two information time scales.

A natural compromise to overcome this difficulty of uncertain information time scale is first to choose an estimate of the maximum information, either ''under the null'' or ''under the alternative'' hypothesis. Then set the information time to be 1 if the proportion exceeds 1 or if the current analysis is the last analysis and the proportion has not yet reached 1. [See, e.g., Kim et al. (32).] The consequence of this compromise is that the Type I error spending function will be altered from the original one that was prespecified in the design.

The following simple example (32) illustrates the preceding discussion.

Example: Over-Running and Under-Running. Suppose $K = 2$ analyses were planned for a trial where the one-sided significance level of 0.05 was specified in the protocol. At the design stage, the uniform (linear) Type I error spending function, $\alpha(t) = 0.05t$, was chosen for monitoring. The expected total number of events is 200 under the null hypothesis, but is 100 under the alternative. Suppose that there are 50 events at the first analysis.

If we have chosen the information time scale based on the null hypothesis, then $t_1 = 50/200 = 0.25$, and $\alpha(t_1) = 0.0125$. The group sequential boundary b_1 such that $P\{Z_{t1} \geq b_1\} = 0.0125$ is $b_1 = 2.24$. Suppose at the final analysis, the number of events is truly 200, as expected under the null hypothesis, the correlation between Z_{t1} and Z_{t2} is $(t_1/t_2)^{1/2} = (0.25)^{1/2} = 0.5$. Thus, the group sequential boundary b_2 such that $P\{Z_{t1} < 2.24, Z_{t2} \geq b_2\} = 0.05 - 0.0125 = 0.0375$ is $b_2 = 1.74$.

However, if the realized number of events at the final analysis turns out to be 100 instead (i.e., an under-running situation), then the "true" t_1 should be 50/100 = 0.5 and we should have spent $\alpha(t_1) = 0.025$. But we cannot go back in time, since we have already adopted $b_1 = 2.24$ for the test at the first analysis. What we can do is to recognize that (a) at $t_1 = 0.5$ (not 0.25), we used $\alpha(t_1) = 0.0125$, and (b) the correlation between Z_{t1} and Z_{t2} is $(t_1/t_2)^{1/2} = (0.5)^{1/2}$. Thus, the group sequential boundary b_2 such that $P\{Z_{t1} < 2.24, Z_{t2} \geq b_2\} = 0.05 - 0.0125 = 0.0375$ is $b_2 = 1.70$.

From the fact that $\alpha(0.5) = 0.0125 < 0.025$ [see point (a)], we see that the Type I error spending function was no longer the uniform (linear) spending function. Instead, it is a convex function, running under the uniform (linear) spending function.

The over-running case is the opposite, when 100 events under the alternative hypothesis was used to estimate t_1, but it turned out that 200 events occurred at the final analysis. The consequence is that the linear spending function is altered to a concave function running over it.

F. Other Group Sequential Procedures: Stochastic Curtailment and Bayesian Methods

Extending the repeated significance testing procedure, a host of other group sequential procedures, frequentist or Bayesian, have been proposed in the literature. These procedures do not necessarily specify formal boundaries. They can be outlined by the following principal steps at each interim analysis: (a) Divide the outcome space (the treatment–difference scale) into a region where the test treatment is considered superior and a region where the control is considered superior. (b) Guide the trial to stop if the "chance" is minimal that the control is superior (in which case we would use the test treatment) or the same is true for the test treatment (in which case we would use the control treatment). The key is that this "chance" may be measured by the conditional (33), posterior (34,35), or predictive (36) probability of what would be the final outcome given the data at the interim stage. We then set to find the regions in step (a) by requiring certain prespecified operating characteristic. The latter two (posterior and predictive) approaches are considered preferable by Bayesian statisticians (37).

IV. OTHER MONITORING MATTERS

A. Data Monitoring Committee (DMC)

In order to ensure objectivity and to safeguard the blinding of the study, trials that use a group sequential design often have a data monitoring committee (DMC) to evaluate the interim analyses and relevant external information. The ICH guidelines (E9, Sections 4.5 and 4.6) also call for an external independent DMC (IDMC) to be formed for monitoring comparisons of efficacy and/or safety outcomes for trials that have major public health significance. The composition of a DMC membership usually involves disciplines in the following areas: clinical, laboratory, epidemiology, biostatistics, data management, and ethics (38). The organization structure for a DMC to function varies from trial to trial. For example, government-sponsored collaborative trials usually have different reporting structure from industry-sponsored trials. More references on the DMC issues can be found in Friedman and DeMets (39), Fleming and DeMets (40), DeMets et al. (41), and Armstrong and Furberg (42).

When a DMC reviews interim data, not only the outcome measures (primary and secondary) are involved, but also the recruitment, baseline variables, the adverse effects, and compliance. Early termination of a trial is a serious matter, and considerations of such cover a wide range of issues of baseline comparability, quality of the data, internal and external consistency of the results, risk and benefit ratio, length of the follow-up, public impact, and, of course, the probability of Type I and Type II errors. Group sequential boundaries often serve as an important "guideline," not an absolute "rule," in the consideration.

B. Confidence Intervals Following a Group Sequential Test

In addition to hypothesis testing, the calculation of a confidence interval for the difference of treatment effects, θ, is often an important part of data analysis. A problem analogous to inflation of the Type I error arises when estimating the population parameters upon termination of a group sequential procedure, since the study is stopped preferentially when extreme data have been observed. Thus the usual fixed-sample estimators are biased toward the extremes.

The duality of hypothesis testing and confidence interval estimation, however, is helpful for the problem. Basically, the confidence interval of θ involves inversion of the acceptance region of the test statistic, which, in turn, in the sequential testing setting, involves the boundary crossed, the stopping stage ($i = M$), and the cumulative/partial sum of the observations at the stopping stage ($S_i = S_M$). Since the acceptance region is based on "nonextreme" results for the sufficient statistic (i, S_i), clearly an ordering on the outcome space of (i, S_i) is necessary. However, the best choice for such an ordering is not as clear, as pointed out by Emerson and Fleming (43), since the densities for the group sequential statistics lack a monotone likelihood ratio, so the theory regarding the uniformly most powerful tests and the uniformly most accurate confidence bounds does not apply. For the binomial parameter, Jennison and Turnbull (44) proposed a procedure based on an intuitive outcome space ordering, in which results corresponding to earlier termination are regarded more extreme than those that terminate later. This approach was then adopted to the normal mean case by Tsiatis et al. (45). Chang and O'Brien (46) investigated an ordering based on the likelihood ratio test for the binomial case; Chang (47) then applied the same approach to the normal mean case. In both cases the likelihood-ratio ordering of the sample space was shown to give shorter confidence intervals, but Emerson and Fleming (43) reported that there were occasions on which the confidence set based on the likelihood ratio was not an interval. Emerson and Fleming (43) proposed the ordering based on sample mean. Confidence intervals based on the sample mean ordering agrees with the test results for all practical group sequential test designs, in the sense that a $(1 - 2\alpha)$ confidence interval will not include hypotheses that are rejected by a one-sided level-α group sequential test. For random group sizes, Emerson and Fleming also recommended the approximate confidence intervals based on the sample mean ordering as well as the approximation to Whitehead's (48) bias-adjusted mean method.

V. OTHER RECENT DEVELOPMENTS

A. Early Acceptance/Rejection of the Null Hypothesis, Type I and Type II Error Spending-Function Approaches, and Group Sequential Equivalence Trials

Group sequential methods allowing early acceptance of the null hypothesis for futility is as important and as popular as that for early rejection of the null in the sense of ethics and economics. DeMets and Ware (49,50) have proposed modifications to the Pocock and the O'Brien and Fleming designs to test one-sided hypotheses with allowance for early stopping in favor of the null. Gould and Pecore (51) proposed a procedure that takes the cost function into account and pointed out that, although these procedures are less powerful than the ones that allow early rejection only, the saving of cost can be much greater. Lan and Friedman (52), when emphasizing the one-sided null hypothesis as being a "harmful" direction for the test treatment, suggested a combination of Type I error spending function for the upper boundary and a conditional probability (stochastic curtailing) approach for the lower boundary. An extension of the idea of the flexible Type I error spending function is naturally the use of both Type I error and Type II error spending functions in constructing the rejection and acceptance boundaries [Pampallona and Tsiatis (53); Chang et al. (54)]. Recently the idea of both acceptance and rejection boundaries has also been extended to group sequential equivalence trials by Whitehead (55).

B. Sequential Monitoring Trials with Multiple (Survival) Endpoints

In addition to the interim analyses with repeated measurements previously mentioned (Sec. III.D), design of group sequential clinical trials with multiple endpoints in general or in survival outcomes has also been considered by authors. Tang et al. (56) addressed multivariate normal distribution based on O'Brien's global test (57). Previously generated group sequential boundaries can be used for the combined test statistic. Lin (58) proposed a sequential nonparametric testing for detecting the stochastic ordering of two multivariate distributions, with emphasis on multiple time-to-event endpoints, based on a weighted sum of linear rank statistics. Stopping boundaries were computed that preserve the overall significance level and may lead to earlier trial termination than procedures based on univariate survival endpoints. Williams (59) extended the method of Lin by developing approaches for constructing repeated joint confidence regions for the hazard ratios, in the spirit of Jennison and Turnbull (60,61), based on inverting linear rank statistics in terms of the parameter of interest. Technically, in all procedures the basic building blocks are the same as that laid out earlier in Sec. III, namely, finding the joint distribution (asymptotically, if necessary) of the test statistic over time by forming the test statistic at each time as a sum of i.i.d. random variables. The property of partial sum with independent increment structure can then be utilized to fit in the flexible Type I (and Type II) spending function framework. Of course, as in the nonsequential case, for the procedure to be meaningful the endpoints being combined need to be biologically related and to demonstrate treatment effects in the same direction.

C. Monitoring Design Specifications to Extend a Trial

The traditional group sequential methods, as just reviewed, have been focused on early termination of a trial, for ethical and economic reasons. Recently, considerable interest has also been expressed in possibly extending a trial beyond its originally planned sample size and/or duration, based on interim data. This occurs when the assumed values for some parameters used for the initial sample size and/or duration calculation are uncertain, as is often the case for a new treatment studied for an unfamiliar disease. For example, the within-group variance may be underestimated initially, or treatment effects are less, or the patient accrual rate is slower than anticipated. Allowing for an upward adjustment of the sample size or study duration can prevent a trial from being inconclusive due to underpower after investing considerable resources.

Most of the methods in this area suggest a two-stage design due to practical logistics involved in extending a trial. Wittes and Brittain (62) termed the first stage/phase an *internal pilot study*. For the univariate continuous case, Shih (63) and Gould and Shih (64) proposed a procedure to estimate the within-group variance without breaking the treatment codes and to update the sample size needed to secure the planned power of the trial. Shih and Gould (65) extended the technique to the case of repeated measures where the key response is the slope. Shih and Long (66) examined the robustness of the procedure with unequal variances and center effects present in the trial. For the binary case, Gould (67) suggested using the pooled event rate, and Shih and Zhao (68) proposed a simple design with a dummy stratification, to re-estimate the sample size without breaking the treatment code at the interim stage. To keep the treatment code masked is essential for maintaining the integrity of a trial where no IDMC is employed. Since these procedures only increase sample size and/or duration when necessary (otherwise they keep the original size and/or duration intact) through updating nuisance parameters without unveiling the treatment effects at the interim stage, the Type I error probability is affected minimally, if at all. Gould and Shih (69) recently further examined strategies to combine blinded sample size re-estimation with the traditional group sequential designs.

When an IDMC is in place and unblinding is not an issue, Andersen (70), Henderson et al. (71), and Proschan and Hunsberger (72) proposed the use of conditional power for updating the design specifications and extending a trial. For more complex group sequential designs (e.g., multiple-arm and factorial with arbitrary survival curves), Halpern and Brown (73) and Natarajan et al. (74) developed simulation-based programs. Recently, Scharfstein and Tsiatis (75) proposed using the interim data to revise the initial guesses in the computer simulation and updating the study design to attain the maximum information.

REFERENCES

1. JR Schultz, FR Nichol, GL Elfring, SD Weed. Multiple stage procedure for drug screening. Biometrics 29:293–300, 1973.
2. GL Elfring, JR Schultz. Group sequential designs for clinical trials. Biometrics 29:471–477, 1973.
3. C Jennison, BW Turnbull. Statistical approaches to interim monitoring of medical trials: a review and commentary. Statistical Sci 5:299–317, 1990.
4. ICH Guidelines. Web site: www/fda.gov/cder/guidance/guidance.htm.
5. SJ Pocock. Group sequential methods in the design and analysis of clinical trials. Biometrika 64:191–199, 1977.
6. PC O'Brien, TR Fleming. A multiple testing procedure for clinical trials. Biometrics, 35:549–556, 1979.
7. JL Haybittle. Repeated assessment of results in clinical trials of cancer treatment. Br J Radiol 44:793–797, 1971.
8. R Peto, MC Pike, P Armitage, NE Breslow, DR Cox, SV Howard, N Mantel, CK McPherson, J Peto, PG Smith. Design and analysis of randomized clinical trials requiring prolonged observation of each patient. Br J Cancer 35:585–611, 1976.
9. J Whitehead, I Stratton. Group sequential clinical trials with triangular continuation regions. Biometrics 39: 227–236, 1983.
10. J Whitehead, H Brunier. The double triangular test: a sequential test for the two-sided alternative with early stopping under the null hypothesis. Sequential Anal 9: 117–136, 1990.
11. RC Milton. Computer evaluation of the multivariate normal integral. Technometrics 14:881–889, 1972.
12. MJ Schervish. Multivariate normal probabilities with error bound. Appl Stat 33:81–94, 1984.
13. EaSt: A software package for the design and interim monitoring of group sequential clinical trials. Cambridge, MA: Cytel Software Corporation, 1992.

14. S+SeqStat: Reference Manual. AG Bruce and S Emerson, MathSoft, Inc., Research Report No. 51, 1997.
15. PEST 3.0 Operating Manual. H Brunier and J Whitehead, Reading University, Reading, UK, 1993.
16. ER Slud, LJ Wei. Two-sample repeated significance tests based on the modified Wilcoxon statistic. J Am Statistical Asso 77:862–868, 1982.
17. KKG Lan, DL DeMets. Discrete sequential boundaries for clinical trials. Biometrika 70:659–663, 1983.
18. K Kim, DL DeMets. Design and analysis of group sequential tests based on type-I error spending rate function. Biometrika 74:149–154, 1987.
19. The Scandinavian Simvastatin Survival Study Group. Design and baseline results of the Scandinavian Simvastatin Survival Study of patients with stable angina and/or previous myocardial infarction. Am J Cardiol 71:393–400, 1993.
20. IK Hwang, WJ Shih, JS deCani. Group sequential designs using a family of type I error probability spending functions. Stat Med 9:1439–1445, 1990.
21. SK Wang, AA Tsiatis. Approximately optimal one-parameter boundaries for group sequential trials. Biometrics 43:193–199, 1987.
22. AA Tsiatis. The asymptotic joint distribution of the efficient scores test for the proportional hazards model calculated over time. Biometrika 68:311–315, 1981.
23. AA Tsiatis. Repeated significance testing for a general class of statistics used in censored survival analysis. J Am Stat Asso 77:855–861, 1982.
24. C Jennison, BW Turnbull. Group-sequential analysis incorporating covariates information. J Am Stat Asso 92: 1330–1341, 1997.
25. DO Scharfstein, AA Tsiatis, JM Robbins. Semiparametric efficiency and its implication on the design and analysis of group sequential studies. J Am Stat Asso 92:1342–1350, 1997.
26. T Sellke, D Siegmund. Sequential analysis of proportional hazards model. Biometrika 70:315–326, 1983.
27. LJ Wei, JQ Su, JM Lachin. Interim analyses with repeated measurements in a sequential clinical trial. Biometrika 77:359–364, 1990.
28. JW Lee, DL DeMets. Sequential comparison of changes with repeated measurements data. J Am Stat Asso 86: 757–762, 1991.
29. JM Lachin. Group sequential monitoring of distribution-free analyses of repeated measures. Stat Med 16: 653–668, 1997.
30. KKG Lan, D Zucker. Sequential monitoring for clinical trials: the role of information and Brownian motion. Stat Med 12:753–765, 1993.
31. KKG Lan, DM Reboussin, DL DeMets. Information and information fractions for designing sequential monitoring of clinical trials. Communications Stat (A)—Theory Methods 23:403–420, 1994.
32. KM Kim, H Boucher, AA Tsiatis. Design and analysis of group sequential log-rank tests in maximum duration versus information trials. Technical report, Department of Biostatistics, Harvard School of Public Health, 1993.
33. KKG Lan, R Simon, M Halperin. Stochastically curtailed tests in long term clinical trials. Communications Stat—Sequential Anal 1:207–219, 1982.
34. LS Freedman, DJ Spiegelhalter. Comparison of Bayesian with group sequential methods for monitoring clinical trials. Controlled Clin Trials 10:357–367, 1989.
35. PM Fayers, D Ashby, MKB Parmar. Bayesian data monitoring in clinical trials. Stat Med 16:1413–1430, 1997.
36. DJ Spiegelhalter, LS Freedman, PR Blackburn. Monitoring clinical trials: conditional or predictive power? Controlled Clin Trials 7:8–17, 1986.
37. A Berry. Interim analysis in clinical trials: classical vs. Bayesian approaches. Stat Med 4:521–526, 1985.
38. DL DeMets. Data monitoring and sequential analysis—an academic perspective. J AIDS 3(suppl 2):S124–S133, 1990.
39. L Friedman, DL DeMets. The data monitoring committee: how it operates and why. IRB 3:6–8, 1981.
40. TR Fleming, DL DeMets. Monitoring of clinical trials: issues and recommendations. Controlled Clin Trials 14: 183–197, 1993.
41. DL DeMets, SS Ellenberg, TR Fleming, JF Childress, KH Mayer, RB Pollard, JJ Rahal, L Walters, J O'Fallon, P Whitley-Williams, S Straus, M Sande, RJ Whitley. The data and safety monitoring board and acquired immune deficiency syndrome (AIDS) clinical trials. Controlled Clin Trials 16:408–421, 1995.
42. PW Armstrong, CD Furberg. Clinical trials data and safety monitoring boards: the search for a constitution. Circulation 1:Sess:6, 1994.
43. SS Emerson, TR Fleming. Parameter estimation following group sequential hypothesis testing. Biometrika 77: 875–892, 1990.
44. C Jennison, BW Turnbull. Confidence intervals for a binomial parameter following a multistage test with application to MIL-STD 105D and medical trials. Technometrics 25:49–58, 1983.
45. AA Tsiatis, GL Rosner, CR Mehta. Exact confidence intervals following a group sequential test. Biometrics 40:797–803, 1984.
46. MN Chang, PC O'Brien. Confidence intervals following group sequential tests. Controlled Clin Trials 7:18–26, 1986.
47. MN Chang. Confidence intervals for a normal mean following a group sequential test. Biometrics 45:247–254, 1989.
48. J Whitehead. On the bias of maximum likelihood estimation following a sequential test. Biometrika 73:573–581, 1986.
49. DL DeMets, JH Ware. Group sequential methods for clinical trials with a one-sided hypothesis. Biometrika 67:651–660, 1980.

50. DL DeMets, JH Ware. Asymmetric group sequential boundaries for monitoring clinical trials. Biometrika 69: 661–663, 1982.
51. AL Gould, VJ Pecore. Group sequential methods for clinical trials allowing early acceptance of H_0 and incorporating costs. Biometrika 69:75–80, 1982.
52. KKG Lan, L Friedman. Monitoring boundaries for adverse effects in long-term clinical trials. Controlled Clin Trials 7:1–7, 1986.
53. S Pampallona, AA Tsiatis. Group sequential designs for one-sided and two-sided hypothesis testing with provision for early stopping in favor of the null hypothesis. J Statistical Planning Inference 42:19–35, 1994.
54. MN Chang, IK Hwang, WJ Shih. Group sequential designs using both type I and type II error probability spending functions. Communication Stat (A)—Theory Methods 27:1323–1339, 1998.
55. J Whitehead. Sequential designs for equivalence studies. Stat Med 15:2703–2715, 1996.
56. DI Tang, C Gnecco, NL Geller. Design of group sequential clinical trials with multiple endpoints. J Am Statistical Asso 77:862–868, 1989.
57. PC O'Brien. Procedure for comparing samples with multiple endpoints. Biometrics 40:1079–1087, 1984.
58. DY Lin. Nonparametric sequential testing in clinical trials with incomplete multivariate observations. Biometrika 78:123–131, 1991.
59. P Williams. Sequential monitoring of clinical trials with multiple survival endpoints. Stat Med 15:2341–2357, 1996.
60. C Jennison, BW Turnbull. Repeated confidence intervals for group sequential clinical trials. Controlled Clin Trials 5:33–45, 1984.
61. C Jennison, BW Turnbull. Interim analyses; the repeated confidence interval approach (with discussion). J Roy Statistical Soc B 51:305–361, 1989.
62. J Wittes, E Brittain. The role of internal pilot studies in increasing the efficiency of clinical trials. Stat Med 9: 65–72, 1990.
63. WJ Shih. Sample size reestimation in clinical trials. In: KE Peace, ed. Biopharmaceutical Sequential Statistical Applications. New York: Marcel Dekker, 1992, pp 285–301.
64. AL Gould, WJ Shih. Sample size re-estimation without unblinding for normally distributed outcomes with unknown variance. Communication Stat (A)—Theory Methods 21:2833–2853, 1992.
65. WJ Shih, AL Gould. Re-evaluating design specifications of longitudinal clinical trials without unblinding when the key response is rate of change. Stat Med 14: 2239–2248, 1995.
66. WJ Shih, J Long. Blinded sample size re-estimation with unequal variances and center effects in clinical trials. Communication Stat (A)—Theory Methods 27: 395–408, 1998.
67. AL Gould. Interim analyses for monitoring clinical trials that do not materially affect the type I error rate. Stat Med 11:55–66, 1992.
68. WJ Shih, PL Zhao. Design for sample size re-estimation with interim data for double-blind clinical trials with binary outcomes. Stat Med 16:1913–1923, 1997.
69. AL Gould, WJ Shih. Modifying the design of ongoing trials without unblinding. Stat Med 17:89–100, 1998.
70. PK Andersen. Conditional power calculation as an aid in the decision whether to consider a clinical trial. Controlled Clin Trials 8:67–74, 1987.
71. GWG Henderson, SG Fisher, L Weber, KE Hammermeister, G Sethi. Conditional power for arbitrary survival curves to decide whether to extend a clinical trial. Controlled Clin Trials 12:304–313, 1991.
72. MA Proschan, SA Hunsberger. Design extension of studies based on conditional power. Biometrics 51: 1315–1324, 1995.
73. J Halpern, BW Brown. A computer program for designing clinical trials with arbitrary survival curves and group sequential testing. Stat Med 14:109–122, 1993.
74. R Natarajan, BW Turnbull, EH Slate, LC Clark. A computer program for sample size and power calculation in the design of multiple-arm and factorial clinical trials with survival time endpoints. Computer Meth Programs Biomed 49:137–147, 1996.
75. DO Scharfstein, AA Tsiatis. The use of simulation and bootstrap in information-based group sequential studies. Stat Med 17:75–87, 1998.

Weichung Joseph Shih

I

Individual Bioequivalence

See also *Bioavailability and Bioequivalence; Equivalence Trials*

I. INTRODUCTION

Current practice for the assessment of bioequivalence is based on the *fundamental bioequivalence assumption* that when two formulations of the same drug product or two drug products (e.g., a brand-name drug and its generic copy) are equivalent in the rate and extent of drug absorption, it is assumed that they will reach the same therapeutic effect or that they are therapeutically equivalent (Chow and Liu, 1992). Pharmacokinetic (PK) responses such as area under the blood concentration-time curve (AUC) and maximum concentration (C_{max}) are usually considered to assess the rate and extent of drug absorption. The current regulation of the United States Food and Drug Administration (FDA) requires that the evidence of bioequivalence in average bioavailabilities in terms of some primary PK responses such as AUC and C_{max} between the two formulations of the same drug product or the two drug products be provided (FDA, 1992). This type of bioequivalence is usually referred to as *average bioequivalence* (ABE). The FDA indicates that a generic drug product can be used as a substitution for a brand-name drug if it has been shown to be bioequivalent to the brand-name drug under the ABE criterion. However, the FDA does not indicate that two generic drug products of the same brand-name drug can be used interchangeably, even when they are shown to be bioequivalent to the same brand-name drug. In addition, the FDA does not require that bioequivalence among generic copies of the same brand-name drug be provided.

In the medical community, as more generic drug products become available in the marketplace, it is of great concern whether a number of generic drug products of the same brand-name drug can be used safely and interchangeably. Basically, drug interchangeability can be classified as drug prescribability or drug switchability. *Drug prescribability* is defined as the physician's choice for prescribing an appropriate drug product for his/her new patients between a brand-name drug product and a number of generic drug products of the brand-name drug product that have been shown to be bioequivalent to the brand-name drug product. The underlying assumption of drug prescribability is that the brand-name drug product and its generic copies can be used interchangeably in terms of the efficacy and safety of the drug product. Under current practice, the FDA only requires that evidence of equivalence in average bioavailabilities be provided; the bioequivalence assessment does not take into account equivalence in variability of bioavailability. A relatively large intrasubject variability of a test drug product (e.g., a generic drug product) as compared to that of the reference drug product (e.g., its brand-name drug product) may present a safety concern. To overcome this disadvantage, in addition to providing evidence of ABE, it is recommended that bioequivalence in variability of bioavailabilities between drug products be established. This type of bioequivalence is called *population bioequivalence* (PBE). In practice, although PBE is often considered for assessment of drug prescribability, it does not fully address drug switchability, due to the possible existence of subject-by-formulation interaction.

Drug switchability is related to the switch from a drug product (e.g., a brand-name drug product) to an alternative drug product (e.g., a generic copy of the brand-name drug product) within the same subject, whose concentration of the drug product has been titrated to a steady, efficacious, and safe level. As a result, drug switchability is considered more critical than drug prescribability in the study of drug interchangeability for patients who have been on medication for a while. To ensure drug switchability, it is recommended that bioequivalence be assessed within individual subjects. This type of bioequivalence is known as *individual bioequivalence* (IBE). The concept of IBE has attracted the FDA's attention since introduced by Anderson and Hauck (1990), which has led to a serious consideration for change in the regulatory requirement for assessment of bioequivalence. In this entry, we will focus on the review of a draft guidance on *In Vivo Bioequivalence Studies Based on Population and Individual Bioequivalence Approaches*, which was recently distributed by the FDA for public comments (FDA, 1997).

In the next section, the limitations of ABE for the assessment of drug interchangeability is outlined. The concepts, decision rules, and statistical methods for evaluation of PBE and IBE as described in the FDA draft guidance are given in Sec. III. In Sec. IV, we provide a comprehensive review of the FDA draft guidance. A brief conclusion is given in the last section.

II. LIMITATIONS OF AVERAGE BIOEQUIVALENCE

Under current FDA regulation, two formulations of the same drug or two drug products are claimed bioequivalent if the ratio of the means of the primary PK responses, such as AUC and C_{max}, between the two formulations of the same drug or the two drug products is within (80%, 125%) with 90% assurance (FDA, 1992). A generic drug product can serve as the substitution of its brand-name drug product if it has been shown to be bioequivalent to the brand-name drug. The FDA, however, does not indicate that one generic drug can be substituted by another generic drug, even though both of the generic drugs have been shown to be bioequivalent to the same brand-name drug. Bioequivalence between generic copies of the same brand-name drug is *not* required. As more generic drugs become available in the marketplace, it is very likely that a patient may switch from one generic drug to another. Therefore, an interesting question to physicians and patients is whether the brand-name drug and its generic copies can be used safely and interchangeably.

Chen (1997) pointed out that the current ABE approach for bioequivalence assessment has limitations for addressing drug interchangeability, especially for drug switchability. These limitations include: (a) ABE focuses only on the comparison of population average between the test and reference drug products; (b) the distribution of the primary pharmacokinetic response of interest is not taken into account, nor is the intrasubject variance of the drug products under study; and (c) ABE ignores the subject-by-formulation interaction, which has an impact on drug switchability. As a result, Chen (1997) suggested that current regulation of ABE be switched to the approach of IBE to overcome these disadvantages.

Many practitioners, however, indicated that the impact of the foregoing limitations on safety might be negligible, because little clinical evidence was observed in the past decade to indicate that the generic drugs approved based on ABE were unsafe. In addition, if the limitations of ABE do have an impact on safety, it is recommended that the decision rule (i.e., the one-fits-all bioequivalence limits) of ABE be modified before switching to a totally new and yet much more complicated approach of IBE. For example, if the safety of the drug product under study is a great concern, we may lower the upper bioequivalence limit from 125% to 115%, say. Similarly, we may adjust the lower bioequivalence limit to ensure the efficacy of the drug. This flexibility reflects not only the nature of the drug product under study but also allows a more accurate and reliable assessment of bioequivalence between drug products. The adjustment for bioequivalence limits of ABE should be flexible according to drug classification, which may depend upon either (a) the intrasubject variability and the therapeutic window of the drug product or (b) the population difference ratio (PDR) and/or the individual difference ratio (IDR) as specified in the draft guidance.

Chow and Liu (1997) proposed to perform a meta-analysis for an overview of ABE. The proposed meta-analysis provides an assessment of bioequivalence among generic copies of an innovator drug that can be used as a tool to monitoring the performance of the approved generic copies of the innovator drug. In addition, it provides more accurate estimates of inter- and intrasubject variabilities of the drug product.

III. DRUG INTERCHANGEABILITY

As indicated earlier, drug interchangeability can be classified as drug prescribability or drug switchability.

It is recommended that PBE and IBE be used to assess drug prescribability and drug switchability, respectively. More specifically, the FDA draft guidance recommends that PBE be applied to new formulations, additional strength, or new dosage forms in NDAs (new drug applications), while IBE should be considered for ANDA (abbreviated new drug application) or AADA (abbreviated antibiotic drug application) for generic drugs. In what follows, we will provide a brief overview of the concepts, decision rules, and statistical methods of PBE and IBE for assessment of drug interchangeability separately.

A. Population Bioequivalence

Statistically speaking, drug prescribability can be addressed through PBE by examining the closeness between distributions of the primary pharmacokinetic responses for the rate and extent of absorption. Under the normality assumption, PBE can be established by demonstrating equivalence in both the average and variability of bioavailability. As a result, we may consider a disaggregate criterion for the assessment of PBE. In other words, we assess bioequivalence in terms of the average bioavailability and variability of bioavailability. For the assessment of bioequivalence in average bioavailability, statistical methods such as the confidence interval approach and Schuirmann's two one-sided tests procedure are well established. For the assessment of bioequivalence in the variability of bioavailability, Liu and Chow (1992) proposed two parametric and nonparametric one-sided Pitman–Morgan test procedures for assessing bioequivalence in intrasubject variabilities. It, however, should be noted that there is no discussion on the decision rule for bioequivalence in the variability of bioavailability in the FDA 1992 guidance and 1997 draft guidance as well. In addition, it is not clear whether testing for bioequivalence in average bioavailability and testing for bioequivalence in the variability of bioavailability should be done separately (in sequential order) or simultaneously. In either case, it is suggested that an appropriate α-level be adjusted in order to have a predetermined overall Type I error rate.

To address drug prescribability, the FDA proposed the following aggregated, scaled, moment-based one-sided criterion:

$$\text{PBC} = \frac{(\mu_T - \mu_R)^2 + (\sigma_{TT}^2 - \sigma_{TR}^2)}{\max(\sigma_{TR}^2, \sigma_{T0}^2)} \leq \theta_P$$

where μ_T and μ_R are the mean of the test drug product and the reference drug product, respectively, σ_{TT}^2 and σ_{TR}^2 are the total variance of the test drug product and the reference drug product, respectively, σ_{T0}^2 is a constant that can be adjusted to control the probability of passing PBE, and θ_P is the bioequivalence limit. The numerator on the left-hand side of the criterion is the sum of the squared difference of the traditional population averages and the difference in total variance between the test and reference drug products, which measures the similarity for the marginal population distribution between the test and reference drug products. The denominator on the left-hand side of the criterion is a scaled factor that depends upon the variability of the drug class of the reference drug product. The guidance suggests that θ_P be chosen as

$$\theta_P = \frac{(\ln 1.25)^2 + \varepsilon_P}{\sigma_{T0}^2}$$

where ε_P is guided by the consideration of the variability term $\sigma_{TT}^2 - \sigma_{TR}^2$ added to the ABE criterion. As suggested by the guidance, it may be appropriate that $\varepsilon_P = 0.02$. For the determination of σ_{T0}^2, the guidance recommends the use of the population difference ratio (PDR), which is defined as

$$\begin{aligned}
\text{PDR} &= \left[\frac{E(T - R)^2}{E(R - R')^2}\right]^{1/2} \\
&= \left[\frac{(\mu_T - \mu_R)^2 + \sigma_{TT}^2 + \sigma_{TR}^2}{2\sigma_{TR}^2}\right]^{1/2} \\
&= \left[\frac{\text{PBC}}{2} + 1\right]^{1/2}
\end{aligned}$$

Therefore, assuming that the maximum allowable PDR is 1.25, substitution of $(\ln 1.25)^2/\sigma_{T0}^2$ for PBC without adjustment of the variance term approximately yields $\sigma_{T0} = 0.2$.

The draft guidance suggests that a mixed-effects model in conjunction with the restricted maximum-likelihood (REML) method be used to estimate total variances of σ_{TT}^2 and σ_{TR}^2. An intuitive statistical test can then be obtained by simply replacing the unknown parameters with their corresponding estimates. However, the exact statistical properties of the resultant test are unknown. In this case, the draft guidance recommends that the bootstrap method be employed to obtain the confidence interval or confidence bound of the test. If the upper 95% confidence bound is less than θ_P, we conclude PBE.

B. Individual Bioequivalence

The individual bioequivalence is motivated by the 75/75 rule, which claims bioequivalence if at least 75%

of individual subject ratios (i.e., relative individual bioavailability of the generic drug product to the innovator drug product) are within (75%, 125%) limits. Along this line, Anderson and Hauck (1990) first proposed the concept of testing for individual equivalence ratios (TIER). The idea is to test individual bioequivalence based on the dichotomization of continuous PK metrics by calculating the p-value for at least the observed number of subjects who fall within bioequivalence limits with the minimum proportion of the population in which the two drug products must be equivalent in order to claim individual bioequivalence.

It should be noted that no universal definition of IBE exists that is uniformly accepted by researchers from the regulatory agency, academia, and the pharmaceutical industry. For example, IBE may be established based on the comparison between distributions within each subject, or it could be based on the distribution of the difference or ratio within each subject (Liu and Chow, 1997). Similar to the concept of disaggregate criteria for PBE, in addition to average bioavailability and variability of bioavailability, we may also consider assessment for the variability due to subject-by-formulation interaction. In this case, IBE can be assessed by means of a union-intersection test approach, which concludes IBE if and only if all of the three hypotheses are rejected at the α-level of significance. Most current methods for assessment of IBE, however, are derived from the distribution of either the difference or the ratio within each subject. Under this setting, IBE can be classified as probability-based and moment-based according to different criteria for bioequivalence (e.g., Anderson and Hauck, 1990; Esinhart and Chinchilli, 1994; Sheiner, 1992; Schall and Luus, 1993; Holder and Hsuan, 1993).

To address drug switchability, the FDA proposed the following aggregated, scaled, moment-based one-sided criterion:

$$\mathrm{IBC} = \frac{(\mu_T - \mu_R)^2 + \sigma_D^2 + (\sigma_{WT}^2 - \sigma_{WR}^2)}{\max(\sigma_{WR}^2, \sigma_{W0}^2)} \le \theta_I$$

where σ_{WR}^2 and σ_{WR}^2 are the within subject variance for the test drug product and the reference drug product, respectively, σ_D^2 is the variance due to subject-by-formulation interaction, σ_{W0}^2 is a constant that can be adjusted to control the probability of passing IBE, and θ_I is the bioequivalence limit. The guidance suggests that θ_I be chosen as follows:

$$\theta_I = \frac{(\ln 1.25)^2 + \varepsilon_I}{\sigma_{W0}^2}$$

where ε_I is the variance allowance factor, which can be adjusted for control sample size. As indicated by the guidance, ε_I may be fixed between 0.04 and 0.05. For the determination of σ_{W0}^2, the guidance recommends the use of the individual difference ratio (IDR), which is defined as

$$\begin{aligned}\mathrm{IDR} &= \left[\frac{E(T - R)^2}{E(R - R')^2}\right]^{1/2} \\ &= \left[\frac{(\mu_T - \mu_R)^2 + \sigma_D^2 + (\sigma_{WT}^2 + \sigma_{WR}^2)}{2\sigma_{WR}^2}\right]^{1/2} \\ &= \left[\frac{\mathrm{IBC}}{2} + 1\right]^{1/2}\end{aligned}$$

Therefore, assuming that the maximum allowable IDR is 1.25, substitution of $(\ln 1.25)^2/\sigma_{W0}^2$ for IBC without adjustment of the variance term approximately yields $\sigma_{W0} = 0.2$.

The draft guidance suggests that a mixed-effects model in conjunction with the restricted maximum-likelihood (REML) method be used to estimate variance components of σ_D^2, σ_{WT}^2, and σ_{WR}^2. An intuitive statistical test can then be obtained by simply replacing the unknown parameters with their corresponding estimates. However, the exact statistical properties of the resultant test are unknown. In this case, the draft guidance recommends that the bootstrap method be employed to obtain the confidence interval or confidence bound of the test. If the upper 95% confidence bound is less than θ_I, we conclude IBE.

IV. REVIEW OF THE US FDA DRAFT GUIDANCE

As stated in the FDA draft guidance, the guidance is intended to address drug interchangeability. When the draft guidance is finalized, it is intended to replace the current regulation of ABE. The draft guidance claims that the usual ABE cannot address drug interchangeability. However, the draft guidance not only failed to provide any clinical evidences regarding the limitations of ABE in drug interchangeability, but also made no attempts to revise/fix the current regulations for ABE. In what follows, we provide a comprehensive review of the FDA draft guidance from both scientific/statistical and practical points of view.

A. Aggregate vs. Disaggregate

The FDA draft guidance recommends aggregate criteria as described earlier for assessment of both PBE and IBE. The PBE criterion accounts for the average of bioavailability and the variability of bioavailability; in addition, the IBE criterion takes into account the variability due to subject-by-formulation interaction. Under the proposed aggregate criteria, however, it is not clear whether the IBE criterion is superior to the ABE criterion or the PBE criterion for the assessment of drug interchangeability. In other words, it is not clear whether or not IBE implies PBE and PBE implies ABE under aggregate criteria. Hence, the question of particular interest to pharmaceutical scientists is *whether the proposed aggregate PBE or IBE criterion can really address drug drug interchangeability (i.e., prescribability and switchability).*

Liu and Chow (1997) suggested that disaggregate criteria be implemented for the assessment of drug interchangeability. The concept of disaggregate criteria for the assessment of PBE and IBE is described later. In addition to ABE, we may consider the following hypotheses testing for assessing bioequivalence in the variability of bioavailabilities:

$$H_0: \frac{\sigma_{WT}^2}{\sigma_{WR}^2} \geq \Delta_v$$

vs.

$$H_\alpha: \frac{\sigma_{WT}^2}{\sigma_{WR}^2} < \Delta_v$$

where Δ_v is the bioequivalence limit for the ratio of intrasubject variabilities. We conclude PBE if the 100 $(1 - \alpha)$% upper confidence limit for $\sigma_{WT}^2/\sigma_{WR}^2$ is less than Δ_v. For assessing IBE, we further consider the following hypotheses:

$$H_0: \sigma_D^2 \geq \Delta_s$$

vs.

$$H_\alpha: \sigma_D^2 < \Delta_s$$

where Δ_s is an acceptable limit for variability due to subject-by-formulation interaction. We conclude IBE if both the $100(1 - \alpha)$% upper confidence limit for $\sigma_{WT}^2/\sigma_{WR}^2$ is less than Δ_v and the $100(1 - \alpha)$% upper confidence limit for σ_D^2 is less than Δ_s. Under the foregoing disaggregate criteria, it is clear that IBE implies PBE and PBE implies ABE.

In practice, it is then of interest to examine the relative merits and disadvantages of the FDA proposed aggregate criteria and the disaggregate criteria described earlier for the assessment of drug interchangeability. In addition, it is also of interest to compare the aggregate and disaggregate criteria of PBE and IBE with the current ABE criterion in terms of the consistencies and inconsistencies in concluding bioequivalence for regulatory approval.

B. Interpretation of the Criteria

The draft guidance suggests that a logarithmic transformation be performed for pharmacokinetic metrics, except for T_{max} and the degree of fluctuation. For ABE, the interpretation on both the original and the log-scale is straightforward and easily understood, because the difference in arithmetic means on the log-scale is the ratio of geometric means on the original scale. For log transformation, two drug products are concluded ABE if the ratio of the average bioavailabilities on the original scale is between 80% and 125%. The statement is the same as saying that the two drug products are concluded ABE if the difference between the arithmetic means on the log-scale is between -0.2231 [= ln(0.8)] and 0.2231 [= ln(1.25)]. However, the interpretation of the aggregate criteria for both PBE and IBE is not straightforward and clear on the original scale. For example, the exponentials of the subject-by-formulation interaction and intrasubject variability on the log-scale do not correspond to the subject-by-formulation and intrasubject variability on the original scale. In addition, the guidance needs to provide an explanation of the criteria on the original scale for PBE on page 5 and for IBE on page 6 and their corresponding bioequivalence limits θ_P and θ_I.

C. Masking Effect

The goal for evaluating bioequivalence is to assess the similarity of the distributions of the PK metrics obtained either from the population or from individuals in the population. However, under aggregate criteria, different combinations of values for the components of the aggregate criterion can yield the same value. In other words, bioequivalence can be reached by two totally different distributions of PK metrics. This is another artifact of the aggregate criteria. At the 1996 Advisory Committee meeting, it was reported that the data sets from the FDA's files showed that a 14% increase in the average (ABE allow only 80–125%) is offset by a 48% in the variability, and the test passes IBE but fails ABE.

D. Power and Sample Size Determination

For the proposed aggregated criterion, it is desirable to have sufficient statistical power to declare IBE (or PBE) if the value of the aggregated criterion is small. On the other hand, we would not want to declare IBE or PBE if the value is large. In other words, a desirable property for assessment of bioequivalence is that the power function of the statistical procedure is a monotone decreasing function. However, since different combinations of values of the components in the aggregated criteria may reach the same value, the power function for any statistical procedure based on the proposed aggregated criteria is not a monotone decreasing function. Experience for implementing the aggregate criteria in the regulatory approval of generic drugs is lacking.

Another major concern is how the proposed criteria for PBE and IBE will affect the sample size determination based on power analysis. Unlike ABE, there exists no closed form for the power function of the proposed statistical procedure for IBE. As a result, there exists no closed form for sample size determination. The sample size can be determined based only on the bootstrap method through a Monte Carlo simulation study.

E. Statistical Procedures

The proposed criterion for IBE is a nonlinear function of μ_T, μ_R, σ_D^2, σ_{WT}^2, and σ_{WR}^2. To provide a valid statistical assessment of IBE according to the proposed criterion, an appropriate statistical test is necessarily derived. The statistical test will involve the estimation of μ_T, μ_R, σ_D^2, σ_{WT}^2, and σ_{WR}^2. An intuitive test can be obtained by simply replacing μ_T, μ_R, σ_D^2, σ_{WT}^2, and σ_{WR}^2 with their corresponding estimates. Although unbiased estimates of μ_T, μ_R, σ_D^2, σ_{WT}^2, and σ_{WR}^2 can be obtained by the method of the analysis of variance under a mixed model, the statistical properties of the resultant test are still unknown. Besides, since the test is a nonlinear function of μ_T, μ_R, σ_D^2, σ_{WT}^2, and σ_{WR}^2, the nonlinear function of the best estimates of μ_T, μ_R, σ_D^2, σ_{WT}^2, and σ_{WR}^2 may not preserve the property of the best estimate for θ_I. As a result, the draft guidance recommends that the bootstrap method be used with at least 2000 samples stratified by sequence to characterize the statistical properties of the statistical test. The method of restricted maximum likelihood (REML) is recommended for estimation of each variance component using a mixed-effects model as described in the draft guidance. Based on the resultant estimator of θ_I, the guidance suggests that a nonparametric bootstrap percentile procedure be applied to obtain the 95% upper confidence limit for θ_I. The guidance failed to provide a scientific or statistical justification as to why a nonparametric bootstrap method rather than a parametric bootstrap method was recommended, because the REML is a parametric procedure. In addition, dropouts and/or missing values were briefly mentioned, and yet no statistical methodology was recommended in the draft guidance.

The criteria for both PBE and IBE are functions of the second moments of the distribution. For ABE, the criterion is based on the average, which is the first moment. As a result, statistical inference for ABE is quite robust against normality. On the contrary, the statistical inference for IBE or PBE aggregate second-moment criteria will be very sensitive to any mild violation of normality. In other words, we need to check the normality assumption before we apply the bootstrap methods as suggested in the draft guidance.

Note that the estimators suggested in the draft guidance for either PBE or IBE are biased for the corresponding aggregate criteria. As a result, the point estimator or 95% confidence interval obtained from the bootstrap method as suggested by the draft guidance is not an estimator or 95% upper confidence interval for the aggregate criteria for IBE or PBE.

F. Two-Stage Test Procedure

To apply the proposed criteria for the assessment of PBE or IBE, the draft guidance suggests that the constant scale be used if the observed estimator of σ_{TR} or σ_{WR} is smaller than σ_{T0} or σ_{W0}. However, statistically, the observed estimator of σ_{TR} or σ_{WR} being smaller than σ_{T0} or σ_{W0} does not mean that σ_{TR} or σ_{WR} is smaller than σ_{T0} or σ_{W0}. A test on the null hypothesis that σ_{TR} or σ_{WR} is smaller than σ_{T0} or σ_{W0} is necessarily performed. As a result, the proposed statistical procedure for the assessment of PBE or IBE becomes a two-stage test procedure. It is then recommended that the overall Type I error rate and the calculation of power be adjusted accordingly.

G. Study Design

The draft guidance recommends that two replicated designs, i.e., (TRTR,RTRT) and (TRT,RTR), be used for the assessment of individual bioequivalence without any scientific and/or statistical justification. It is not clear whether the two replicated crossover designs the optimal design in terms of power in 2×4 and 2×3 replicated crossover designs with respect to the aggregate criterion? In addition, the draft guidance indicates that the 2×4 design is recommended, while the 2×3 design is preferred as an alternative. Several questions are raised. First, it is not clear what the relative efficiency of the two designs is if the total number of observations is to be fixed. Second, it is not clear how these two designs compare to other 2×4 and

2×3 replicated designs, such as (TRRT,RTTR) and (TTRR,RRTT) designs and (TRR,RTT) and (TTR, RRT) designs. Finally, it may be of interest to study the relative merits and disadvantages of these two designs as compared to other designs, such as latin square designs and four-sequence and four-period designs.

Other issues regarding the proposed replicated designs include: (a) it will take longer time to complete; (b) the subject's compliance may be a concern; (c) it is likely to have a higher dropout rate and missing values, especially in 2×4 designs; and (d) there is little in the literature on statistical methods dealing with dropouts and missing values in a replicated crossover design setting.

H. Outlier Detection

The procedure suggested for detection of outliers is not appropriate for the standard 2×2 crossover designs or for the 2×3 or the 2×4 replicated crossover designs, because the observed PK metrics from the same subject are correlated. For a valid statistical assessment, the procedures proposed by Chow and Tse (1990) and Liu and Weng (1992) should be used. These proposed statistical procedures for outlier detection in bioequivalence studies were derived under crossover designs that incorporate the correlations within the same subject. The draft guidance provides little or no discussion regarding the treatment of identified outliers.

V. CONCLUSIONS

The rationale for switching from ABE to PBE or IBE lacks (a) convincing clinical evidence regarding the limitations of ABE for addressing drug interchangeability, and (b) a valid scientific/statistical justification of PBE and IBE. It is then recommended that clinical evidence for the limitations of ABE and scientific/statistical justification of PBE or IBE be carefully evaluated before the proposed PBE and IBE criteria and the corresponding approaches be implemented as standard requirement for the assessment of bioequivalence. To provide convincing clinical evidence regarding the limitations of ABE for addressing drug interchangeability, it would be helpful to study the incidence rates concerning the failure or critical safety issue of approved generic drug products in the past decade. A high incidence rate may indicate that there is a deficiency in the current regulation of ABE. If the incidence rate is relatively low and considered clinically acceptable, then there is no need to switch from ABE to PBE or IBE, especially when clinical performance is unknown. On the other hand, to provide scientific/statistical justification for the use of PBE or IBE, it is suggested that the probabilities of consistencies and inconsistencies for bioequivalence assessment based on ABE, PBE, and IBE be evaluated (Liu and Chow, 1997). If the probability of consistency between ABE and PBE or IBE is high and the probability of inconsistency is small, then there is no need to switch from ABE to PBE or IBE, because ABE, PBE, and IBE will basically reach the same conclusion regarding bioequivalence. If there is a relatively high probability of inconsistency between ABE and PBE or IBE with a relatively low probability of consistency between ABE and PBE or IBE, then the impact on clinical performance for the switch from ABE to PBE or IBE is necessarily provided. In this case, instead of switching from ABE to PBE or IBE, we might tighten the bioequivalence limits for ABE in order to achieve the same result as IBE.

In addition to efficacy and safety, the identity, strength, quality, purity, and stability of approved generic drug products are important drug characteristics that have an impact on average bioavailability, on the variability of bioavailability, and on variability due to subject-by-formulation interaction and consequently to drug interchangeability. As a result, it is suggested that the regulatory requirements for postapproval equivalence in laboratory development and the manufacturing process be established regardless of any switch from ABE to PBE or IBE.

More information regarding the relative merits and disadvantages of ABE, PBE, and IBE can be found in a special issue of *Clinical Pharmacology Therapy and Toxicology on Bioequivalence Assessment: Methods and Applications* (Steinjans and Schulz, 1992), Chow and Liu (1995), a special issue of *Drug Information Journal* on *Bioavailability and Bioequivalence* (Chow, 1995), the proceedings of FIP Bio-International '96 *Bioavailability, Bioequivalence and Pharmacokinetics Studies* (Midha and Nagai, 1997), a special issue of *Journal of Biopharmaceutical Statistics* on *Bioavailability/Bioequivalence* (Chow, 1997), Liu and Chow (1997), and a special issue of *Statistics in Medicine* on *Individual Bioequivalence* (Chow and Liu, 1999).

REFERENCES

Anderson, S. and Hauck, W. W. (1990). Consideration of individual bioequivalence. J Pharmacokinet Biopharmac 18:259–273.

Chen, M. L. (1997). Individual bioequivalence—a regulatory update. J Biopharmaceut Stat 7:5–11.
Chow, S. C. (1995). Bioavailability and Bioequivalence. Special issue of Drug Inform J 29(3).
Chow, S. C. (1997). Bioavailability and Bioequivalence. Special issue of J Biopharmaceut Stat 7(1).
Chow, S. C. and Liu, J. P. (1992). Design and Analysis of Bioavailability and Bioequivalence. New York: Marcel Dekker.
Chow, S. C. and Liu, J. P. (1995). Current issues in bioequivalence trials. Drug Inform J 29:795–804.
Chow, S. C. and Liu, J. P. (1997). Meta-analysis for bioequivalence review. J Biopharmaceut Stat 7:97–111.
Chow, S. C. and Liu, J. P. (1999). Individual Bioequivalence. Special issue of Stat Med. In press.
Chow, S. C. and Tse, S. K. (1990). Outliers detection in bioavailability/bioequivalence. Stat Med 9:549–558.
Esinhart, J. D. and Chinchilli, V. M. (1994). Extension to the use of tolerance intervals for assessment of individual bioequivalence. J Biopharmaceut Stat 4:39–52.
FDA. (1992). Guidance on Statistical Procedures for Bioequivalence Studies Using a Standard Two-Treatment Crossover Design. Rockville, MD: Food and Drug Administration.
FDA. (1997). Draft Guidance on In Vivo Bioequivalence Studies on Population and Individual Bioequivalence. Rockville, MD: Food and Drug Administration.
Holder, D. J., and Hsuan, F. (1993). Moment-based criteria for determining bioequivalence. Biometrika 80:835–846.
Liu, J. P. and Chow, S. C. (1992). On the assessment of variability in bioavailability/bioequivalence studies. Communications Stat Theory Meth 21:2591–2608.
Liu, J. P. and Chow, S. C. (1997). Some thoughts on individual bioequivalence. J Biopharmaceut Stat 7:41–48.
Liu, J. P. and Weng, C. S. (1992). Detection of outlying data in bioavailability/bioequivalence studies. Stat Med 10: 1375–1389.
Midha, K. K. and Nagai, T. (1997). Bioavailability, Bioequivalence and Pharmacokinetic Studies. Proceedings of the FIP Bio-International '96, Business Center for Academic Societies Japan, Tokyo, Japan.
Schall, R. and Luus, R. E. (1993). On population and individual bioequivalence. Stat Med 12:1109–1124.
Sheiner, L. B. (1992). Bioequivalence revisited. Stat Med 11: 1777–1788.
Steinijans, V. W. and Schulz, H.-U. (1992). Bioequivalence Assessment: Methods and Applications. Special issue of Clin Pharmacol Therapy Toxicol 30:Suppl. 1.

Shein-Chung Chow
Jen-pei Liu

Integrated Summary Report

See also *Clinical Trials*

I. INTRODUCTION

There are many reasons to integrate and summarize all the data from a clinical trial program. Each clinical trial in the program is unique in its objective and design. Some are small safety studies among normal volunteers, while others are efficacy trials in a large patient population. The primary reason to create an integrated summary is to compare and contrast all the various study results and arrive at one consolidated review of the benefit/risk profile. A second and important reason is to reach a defensible statistical conclusion through exploring the integrated data that no competing alternative hypothesis exists that can reasonably account for the observed findings. Third, pooling the data from various studies enables the examination of trends in rare subgroups of patients, such as the elderly, those with differing disease states (mild versus severe), and those with comorbidities at baseline. Last, providing such a summary in the new drug application is required by the FDA and other international authorities.

An integrated summary report is a compilation of all the evidence for efficacy and safety from the data collected in all completed clinical trials. There are two integrated summary reports: the integrated summary of efficacy (ISE) and the integrated summary of safety (ISS). Both are required for all new drug applications in the United States.

The analysis approach for the ISE is substantially different from that for the ISS. The ISS summarizes all data collected from the clinical trial program, including normal volunteers and patients. The ISE includes data only from those clinical trials that present some evidence of efficacy, either through surrogate markers or through well-designed tests for efficacy. The ISE requires a more prospective approach to analysis, whereas the ISS approach is more sensitive, because it allows a retrospective search through all the data to expose possibly rare but important side effects. As a result, pooling data across studies is required in the ISS to obtain more reliable estimates of the incidence of rare adverse events. A pooled analysis in the ISE is not required, but it can be useful for estimating a more precise treatment effect. However, pooling is generally not useful for substantiating evidence of efficacy; without side by side study results each showing evidence of efficacy separately. Therefore, this entry has two sections. The first section is devoted to the ISE; the second pertains to the ISS.

There are a number of areas of specific interest that arise within the integrated summary report. A description of the demographic and baseline clinical features of the population treated during the course of the clinical trial program is necessary. Key questions concerning efficacy need to be addressed by considering the results of relevant trials and highlighting the degree to which they reinforce or contradict each other. For example, when is it useful to pool trials to obtain a more precise estimate of the treatment effect? All the safety information available from the entire database of all studies should be summarized. How can the data be thoroughly searched so that any potential safety concern is identified (2)?

II. INTEGRATED SUMMARY OF EFFICACY

A. Dose Rationale

One of the most important sections of the ISE is that on dose selection. Efficacy information at all doses studied should be analyzed is such a way that a dose or dosing interval can be recommended. Information that is needed to support a new drug application should identify an appropriate starting dose, as well as how to adjust dosage to the needs of a particular patient (2). The maximum dosage beyond which there is no likely benefit or that would produce unacceptable side effects should also be identified. It is important to identify the lowest dose exhibiting a clinically important effect. To know for certain that one has identified the minimally effective dose, it is often necessary to study a dose that has no clinical benefit compared to placebo. This section of the ISE should contain the information from the dose response studies, that is, controlled information on the minimum, maximum, and average dose from the response curve for efficacy. It should also contain any other information from other doses. Safety needs to be brought into the discussion, since choosing the dose is always a benefit/risk decision.

In order to provide proper dose–response information, there should be at least one prospective dose–response study in the ISE that explores more than three doses. In principle, being able to detect a statistically significant difference using pairwise comparisons between doses is not necessary in such a trial, if a statistically significant trend (upward slope) across doses can be established using all doses. The more doses explored in this study, the better the dose–response curve will be characterized. It should be demonstrated, however, that the dose to be recommended has a statistically significant and clinically meaningful effect. This can be done in the dose–response trials or in other adequate and well-controlled trials.

The entire database should be examined extensively for possible dose–response effects. Limitations of study design should be considered. For example, the titration-scheme designs should not be pooled with fixed-dose designs. Many trials will titrate to the desired response. Weaker responders would then receive the higher doses. This may lead to an inverted U-shaped dose–response curve. Despite the different designs, all clinical data should be examined for dose–response information.

The entire database should be examined to explore possible differences in subgroups of patients based on baseline differences. Age, gender, and race should always be examined. Other important covariates may be baseline disease severity and baseline comorbidities. Height, weight, renal function, and lean body mass are important for looking at the dose–response data as a function of body size and metabolism.

B. Efficacy Data

The International Conference on Harmonization guideline (1) says that individual clinical trials to demonstrate efficacy must be performed and large enough to satisfy their objectives. Additional valuable information may also be gained by summarizing a series of clinical trials that address essentially identical key efficacy questions. This can be done in the ISE. The main results of such a set of studies should be presented in an identical form to permit comparison across studies, focusing on estimates plus confidence limits. The use of meta-analysis to combine these estimates of treatment effect across studies of the same dose regimen is often useful, because it allows a more precise overall estimate of the treatment effect at those doses. However, only under exceptional circumstances would a meta-analysis be the most appropriate way to demonstrate efficacy via an overall hypothesis test.

Figure 1 shows the clinical trials that are summarized in the ISE. In this program, phase IIb and phase III studies are the focus of the ISE. These studies should be planned while taking into account that they will be combined and presented together for an overall estimate of the benefit of the therapy. They should have similar endpoints (primary and secondary), similar populations, and some overlap in the dose regimens. Phase IIB is the study that demonstrates "proof of principle," usually through a statistically significant increase in efficacy with an increase in dose. There should be two phase III studies: one to demonstrate a statistically significant and clinically meaningful benefit, and the other to replicate that demonstration.

C. Meta-Analysis

In pharmaceutical development, it is known at the time of positive phase II results that a new drug application is likely to be filed. It is at this time that the researchers should start planning for the ISE. In the planning of the adequate and well-controlled trials, one should incorporate any needs that the ISE plan may require. It is at this time that the decision to do meta-analyses should be made in a prospective manner. The individual studies can then be designed with pooling as a goal. Which studies will be pooled and which studies will not can be decided prior to obtaining the results. Then pooling will not be influenced by the results. The decision to undertake a meta-analysis after the data from all the efficacy studies have been reviewed cannot be made to save any negative or borderline results that were obtained in the individual studies. In such a case, the researchers could pick and choose studies so that the results favor the treatment being proposed. A simple principle is to perform a meta-analysis in a manner that is consistent with the conduct of a single well-designed randomized clinical trial (3).

Flather and others (4) address why we do meta-analysis: "Even large randomized clinical trials may not answer specific questions reliably because of weaknesses in their design or, more commonly, because they are not large enough to detect the moderate but medically important treatment effects that can be expected realistically." The primary reason for performing meta-analysis is to obtain more reliable estimates of the treatment effects. Secondary reasons for performing meta-analysis are (a) to summarize formally the available information on a particular treatment, and (b) to generate hypotheses that may be tested in future trials. For example, in the development of a treatment for ischemic stroke, each individual study could be powered for functional status, while the meta-analysis could be powered for mortality (which would generally take larger numbers). In this case, the randomized controlled trials are powered for differences in endpoints that are more sensitive (possibly, surrogates), while meta-analysis is powered for the less sensitive clinical endpoints. The pooling may allow detection of clinically significant changes on clinical endpoints due to the increase in sample size. This is similar to the optimal information size discussed in Pogue and Yusuf (5).

In pooled analysis, all appropriate clinical trials should be included to avoid bias. Studies should be excluded if not of a consistent, similar design: eligibility criteria, trial treatment regimens, concomitant medications, definitions of outcome events, or length of follow-up. Studies with incongruent results should not be pooled but should be examined for an explanation of their differences. Given the retrospective nature of most meta-analyses and the multiple comparisons that are carried out, many researchers believe that

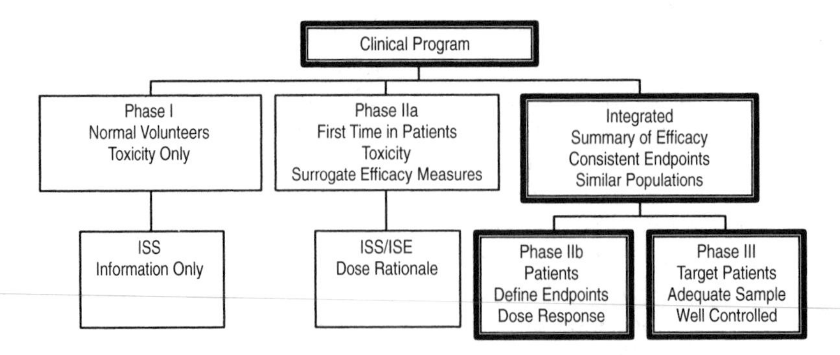

Fig. 1 The integration of efficacy information to support a marketing application.

a p-value of 0.05 is not stringent enough. Levels of 0.01 or even 0.001 should be routinely considered (6).

After the decision to do phase III studies, develop a prospective protocol for your meta-analysis defining specific hypotheses (primary and secondary) to be tested:

Frame a question that is medically relevant and biologically sensible.
Define the study population and relevant subgroups.
Define treatment regimens to be included.
Define outcomes of interest.
Identify trials that will be pooled in a prospective manner.
Use only properly randomized trials.
Include all patients randomized in each trial.
Prespecify valid statistical methods to combine data.
Determine the primary endpoint as well as the effect size necessary to be clinically meaningful.

The ISE should have a description of the entire efficacy database's demographics and baseline disease characteristics. Subgroup analyses for age, race, gender, and other characteristics should be looked at by pooling data over different studies of the same dose regimen. The intent-to-treat population should be the focus; however, if there is a large number of discontinued patients or major protocol violators, the results from a "clean" subpopulation, valuable for efficacy, should be presented as well.

The pooled results should always be presented in combination with the individual study results. This allows the examination of the degree of heterogeneity between the studies (7)

III. INTEGRATED SUMMARY OF SAFETY

The summary of safety data is an important and very large part of a new drug application. It is important to investigate the safety data thoroughly through pooling to search for rare adverse events. For example, if an adverse-event rate is truly only 1%, then more than 300 patients are needed to observe it with 95% confidence. This is from "the rule of three," which states: If none of n patients shows the event of interest, then the upper 95% confidence limit is approximately $3/n$. So if the event were truly rare, say, 0.01%, then 30,000 patients would have to be observed, which is larger than most randomized clinical trials. It is also important to estimate accurately the expected rate of any common adverse event associated with the use of the drug.

The ICH guideline (1) states that all safety data need to be examined thoroughly to uncover any indications of potential toxicity, and to follow up any indications by searching for an associated supportive pattern of observations. The combination of the safety data from all human exposure to a drug should provide the main source of information, because its larger sample size provides the best chance of detecting the rarer adverse events and, perhaps, of estimating their approximate incidence. However, incidence data are difficult to evaluate without a natural comparator group; so a section on the examination of safety from the controlled studies alone is particularly useful. The results from the controlled studies should be combined separately for placebo and active controlled studies.

The components of an ISS should include the extent of exposure, characteristics of the population, deaths, and dropout analyses for serious or potentially serious adverse events, rates of common adverse events, as well as clinical laboratory results. All indications of potential toxicity arising from statistical exploration of the data should be reported. The evaluation of the reality of these potential adverse effects should take into account the issue of multiplicity arising from the numerous statistical comparisons made. The evaluations should make use of survival analysis methods to exploit the potential relationship of adverse-events incidence to duration of exposure and follow-up. The risk associated with identified adverse events should be appropriately quantified to allow a proper assessment of the risk/benefit relationships.

There are at least three reasons to pool data across studies for the safety summary. One goal is to improve the precision of the incidence estimates. This is especially important for rare events. A second goal is to improve the statistical power for detecting any risk factors that may be associated with the use of the drug and the adverse effect. A final goal is to generate hypotheses about risk that may be explored in future studies (8).

It is appropriate to consider carefully the pooling strategy. It is more appropriate to pool studies of a similar design. One needs to consider the patient population, dosing regimens, duration of exposure, and study methods prior to pooling. Furthermore, pooling may not give a meaningful incidence rate if the incidence rates in the individual studies differ dramatically. The differences may be important predictors of the event and should not be ignored. An example of this may be a drug for which phototoxicity was observed. Outpatient studies may reflect the adverse event, while inpatient studies would not. Therefore, they should not be combined. If the incidence rates are comparable, then pooling to get a more precise estimate is appropriate (8).

Once the pooling has been decided and appropriate patient samples are obtained for estimating incidence, it is important to consider important predictive factors, such as dose, plasma level, duration of treatment, concomitant medications, age, sex, and concurrent illness, among others (8). These factors should be looked at in combination, as well. For example, elderly patients with severe disease may be more prone to a particular side effect, especially when treated simultaneously with a particular concomitant medication. These three-way interactions are difficult to detect and require large sample sizes, and it is exactly the type of information that the clinician is interested in.

IV. CONCLUSIONS

During the design of a clinical program, careful attention should be paid to the uniform definition and collection of measurements that will facilitate subsequent interpretation of the series of trials, particularly if the data are likely to be combined across trials. A common dictionary for recording medication details, medical history, and adverse events should be selected. A common definition of primary and secondary variables is nearly always worthwhile for replication of results. The manner in which key variables are collected, the timing of assessments, the handling of protocol violations, and the definition of prognostic factors should all be kept compatible. To change these items from trial to trial makes it difficult to summarize them into a consistent story. Changes should be made only when there is a compelling reason to do so (1).

Integrated summaries are the ultimate conclusion for the benefit/risk ratio of the drug therapy in humans. These are the end product for a typically long period during which clinical studies are performed. Therefore, it is important to plan the summaries early in the clinical development program. In this way, each clinical study can be planned so that it will fit clearly into the ISE and/or the ISS and will thus be easier to design. If we "Begin with the end in mind" (9), the plan for the integrated summaries should be written in advance and should evolve as new results are obtained.

REFERENCES

1. International Conference on Harmonization; Dose Response Information to Support Drug Registration; Guideline; Availability. Notice, Federal Register (IV), Vol. 59, No. 216, 1994.
2. International Conference on Harmonization; Statistical Principles for Clinical Trials. Draft 2C, 13 November 1996.
3. Pogue, J. and Yusuf, S. Overcoming the limitations of current meta-analysis of randomised controlled trials. Lancet 351:915–916, 1998.
4. Flather, M. D., et al. Strengths and limitations of meta-analysis: larger studies may be more reliable. Controlled Clin Trials 18:568–579, 1997.
5. Pogue, J. and Yusuf, S. Cumulating evidence from randomized trials: Utilizing sequential monitoring boundaries for cumulative meta-analysis. Controlled Clin Trials 18:580–593, 1997.
6. Furberg, C. D. and Morgan, T. M. Lessons from overviews of cardiovascular trials. Stat. Med. 6:295–303, 1987.
7. Egger, M., Smith, G. D., and Phillips, A. N. Meta-analysis: principles and procedures. BMJ 315:1533–1537, 1997.
8. Draft Guidance: Conducting a Clinical Safety Review of a New Product Application and Preparing a Report on the Review. U.S. Dept. of Health and Human Services, FDA, CDER, 11/96.
9. Covey, S. R. The Seven Habits of Highly Effective People. Provo, Utah: Franklin-Covey, 1986.

Laura J. Meyerson

Intention-to-Treat Analyses

See also *International Conferences on Harmonization (ICH); Clinical Trials*

I. INTRODUCTION

Randomization of therapy ensures that a therapy assignment is based on chance alone. This means that the baseline characteristics (measured and unmeasured) of the therapy groups should be comparable, with any differences due to chance. Thus, differences in trial outcomes among the groups should be due only to therapy assignment. Most statistical tests of significance require randomization as a fundamental principle. Therefore, randomization is the foundation for statistical inference of prospective clinical trials comparing therapy regimens (1).

Intention to treat (ITT) is a principle used for primary analyses in most randomized clinical trials testing new therapeutic agents. In a randomized clinical trial, subjects are randomly assigned to different therapies or regimens. The ITT principle requires that any comparison among therapy groups in a randomized clinical trial is based on results for all subjects in the therapy groups to which they were randomly assigned (2). This principle helps maintain the benefits of randomization. An alternative is to analyze only a subset of the subjects, such as those deemed to be compliant. However, the subset of compliant subjects may be different from the entire randomized population in ways that can lead to incorrect conclusions. The Coronary Drug Project provides an example of the possible bias introduced when looking only at the compliant (3).

If the randomized subjects in a clinical trial follow the therapy regimen assigned and complete the procedures specified in the protocol exactly, then the ITT analysis is the natural analysis for looking at all data from all randomized subjects. Unfortunately, in most randomized clinical trials, a combination of predictable and unpredictable circumstances leads to a subset of subjects who are unable to follow the intended protocol exactly. These circumstances include: subjects entering the trial when ineligible, subjects who do not follow their assigned therapy regimen, subjects erroneously given or assigned the incorrect therapy, subjects who refuse trial procedures, subjects taking disallowed concomitant medication, and subjects who terminate trial participation prior to completion of the trial. These factors (along with others) lead to subsets of randomized subjects at the end of the trial who may no longer be comparable or representative of the population under study.

II. REGULATORY REQUIREMENTS AND GUIDELINES

The International Conference on Harmonization (ICH) Guidelines on Statistical Principles for Clinical Trials (4) contain extensive comments on ITT principles in Section 5.2, including the following discussion:

> The intention-to-treat principle implies that the primary analysis should include all randomized subjects. Compliance with this principle would necessitate complete follow-up of all randomized subjects for study outcomes. In practice this ideal may be difficult to achieve, for reasons to be described. In this document the term "full analysis set" is used to describe the analysis set which is as complete as possible and as close as possible to the intention-to-treat ideal of including all randomized subjects. Preservation of the initial randomization in analysis is important in preventing bias and in providing a secure foundation for statistical tests. In many clinical trials the use of the full analysis set provides a conservative strategy. Under many circumstances it may also provide estimates of treatment effects which are more likely to mirror those observed in subsequent practice.
>
> There are a limited number of circumstances that might lead to excluding randomized subjects from the full analysis set, including the failure to satisfy major entry criteria (eligibility violations), the failure to take at least one dose of trial medication and the lack of any data postrandomization. Such exclusions should always be justified. Subjects who fail to satisfy an entry criterion may be excluded from the analysis without the possibility of introducing bias only under the following circumstances:
>
> 1. the entry criterion was measured prior to randomization;
> 2. the detection of the relevant eligibility violations can be made completely objectively;

3. all subjects receive equal scrutiny for eligibility violations; (This may be difficult to ensure in an open-label study, or even in a double-blind study if the data are unblinded prior to this scrutiny, emphasizing the importance of the blind review.)
4. all detected violations of the particular entry criterion are excluded.

In some situations, it may be reasonable to eliminate from the set of all randomized subjects any subject who took no trial medication. The intention-to-treat principle would be preserved despite the exclusion of these patients provided, for example, that the decision of whether or not to begin treatment could not be influenced by knowledge of the assigned treatment. In other situations it may be necessary to eliminate from the set of all randomized subjects any subject without data postrandomization. No analysis is complete unless the potential biases arising from these specific exclusions, or any others, are addressed.

When the full analysis set of subjects is used, violations of the protocol that occur after randomization may have an impact on the data and conclusions, particularly if their occurrence is related to treatment assignment. In most respects it is appropriate to include the data from such subjects in the analysis, consistent with the intention-to-treat principle. Special problems arise in connection with subjects withdrawn from treatment after receiving one or more doses who provide no data after this point, and subjects otherwise lost to follow-up, because failure to include these subjects in the full analysis set may seriously undermine the approach. Measurements of primary variables made at the time of the loss to follow-up of a subject for any reason, or subsequently collected in accordance with the intended schedule of assessments in the protocol, are valuable in this context; subsequent collection is especially important in studies where the primary variable is mortality or serious morbidity. The intention to collect data in this way should be described in the protocol. Imputation techniques, ranging from the carrying forward of the last observation to the use of complex mathematical models, may also be used in an attempt to compensate for missing data. Other methods employed to ensure the availability of measurements of primary variables for every subject in the full analysis set may require some assumptions about this subject's outcomes or a simpler choice of outcome (e.g., success/failure). The use of any of these strategies should be described and justified in the statistical section of the protocol and the assumptions underlying any mathematical models employed should be clearly explained. It is also important to demonstrate the robustness of the corresponding results of analysis especially when the strategy in question could itself lead to biased estimates of treatment effects.

Because of the unpredictability of some problems, it may sometimes be preferable to defer detailed consideration of the manner of dealing with irregularities until the blind review of the data at the end of the trial, and, if so, this should be stated in the protocol.

The Center for Drug Evaluation and Research (CDER) of the Food and Drug Administration (FDA) has released a Guideline for the Format and Content of the Clinical and Statistical Sections of an Application (5) that discusses ITT analyses. The subsection on data sets analyzed (in the efficacy results section) contains the following:

> Exactly which patients are included in the effectiveness analysis should be precisely defined, e.g., all patients with any effectiveness observation or with a certain minimum number of observations, only patients completing the trial, all patients with an observation during a particular time window, only patients with a specified degree of compliance, etc. It should be clear, if not defined in the study protocol, when, *relative* to study completion, and how, inclusion/exclusion criteria were developed. As a general rule, even if the applicant's preferred analysis is based on a reduced subset of the patients with data, there should be an additional "intent-to-treat" analysis using all randomized patients.

The subsection on demographic and baseline features of individual patients and comparability of treatment groups contains the following statement:

> If the data sets in the "intent-to-treat" analysis and the applicant's preferred analysis are substantially different, comparability of treatment groups should be examined and any other reasons for such differences should be discussed.

In the section on statistical/analytical issues, the subsection on handling of dropouts or missing data states:

> The results of a clinical trial should be assessed not only for the subset of patients who completed the study, but also for the entire patient population randomized (the intent-to-treat analysis). Several factors need to be considered and compared for the treatment groups in analyzing the effects of dropouts: the reasons for the dropouts, the time to dropout, and the proportion of dropouts among treatment groups at various time points.
>
> Procedures for dealing with missing data, e.g., use of imputed data, should be described. Detailed explanation should be provided as to how such imputations were done and what underlying assumptions were made.

The FDA also provides guidance for some therapeutic areas that refer to preferences regarding ITT analyses. For example, the draft guidelines for preclinical and clinical evaluation of agents used in the prevention or treatment of postmenopausal osteoporosis (6) states that "Plans for imputing missing data (both fracture follow-up assessment and BMD data) should be explained in the protocol, and one or more supplementary intent-to-treat analyses using imputed data should be performed.'' Such guidelines are an important resource for studies that might support the registration of a therapy being studied.

III. STATISTICAL ISSUES

When designing a study, sample size calculations should consider all aspects of the primary analyses. When ITT principles are used, calculation of sample size should include assumptions about the effect of factors such as noncompliance and crossover to other active therapies that are available. These factors will generally attenuate the assumed therapy effect and lead to the need for more subjects (7).

IV. SCIENTIFIC ISSUES

It should be the goal of any randomized clinical trial to develop a protocol that can be reasonably followed by at least a large proportion of the subjects in all therapy groups (8). However, it is fairly common to see a degree of partial or complete noncompliance with the assigned therapy regimen (it is important to note that compliance may involve more than simply taking the medication, but also following specific instructions related to use [e.g., timing of multiple doses, relationship to food intake, or the need to remain upright]). In particular, subjects may not be able to tolerate side effects perceived to be related to therapy, and may stop taking medication or switch to other therapies. Subjects may also become noncompliant because of a perceived lack of efficacy of the therapy. It is also possible for a subject to be given a different therapy than that assigned, due to error; thus, subjects may take their study medication but still be noncompliant with the *assigned* therapy. In extreme cases, noncompliance may make study inference impossible.

In cases where subjects do not follow the protocol, a natural question is whether such subjects should be excluded from analysis because they do not seem to provide information relevant to the therapy regimens being studied. The ITT principle does not allow comparisons that exclude randomized subjects, regardless of their compliance with the therapy regimens and procedures of the trial, although practically such exclusions may be necessary in cases such as subjects missing all postbaseline measurements. This ensures that the randomization is protected (i.e., all groups have comparable baseline characteristics and that any differences besides therapy are due to chance), and precludes the possibility of bias due to selectively excluding subjects from therapy groups, which may lead to systematic differences among the groups that are attributed to factors other than therapy assignment.

In studies with an inactive control (e.g. placebo) or where the therapy under study is more effective than the control, the ITT analysis will usually lead to attenuated estimates of the true therapy, particularly when noncompliance is limited to subjects taking nothing or other therapies under study instead of the assigned therapy. In particular, ITT analyses will generally underestimate the effects of an efficacious therapy regimen. The ITT analysis will also generally underestimate any effects of a therapy on side effects. When active alternatives are available and are used by trial participants, effects on estimates of efficacy or side effect profile are generally not predictable and depend on factors such as the effectiveness of the alternative therapy and the characteristics of the noncompliant trial participants. It should also be noted that ITT analyses can increase variability, which might mask differences in equivalence trials.

It is intuitive to want to compare therapy groups based on the subjects who complete and are fully compliant with the protocol (completers or evaluable analysis) (10). If subjects are removed from analysis in a completely random fashion, excluding them should have no effect on estimates of response or therapy effect. However, excluding nonrandomly selected subsets of subjects from an analysis may introduce bias in the results, even if membership in the subset to be excluded is equally distributed among the therapy groups, because the subset may have different characteristics than the randomized population. For example, it has been demonstrated that subjects compliant with study medication may respond better than noncompliant subjects in both treated and placebo groups (3). In the case where the subjects removed from analysis are not representative of the randomized populations, but are similar among therapy arms, estimates of therapy effects will usually be unbiased, but inference will no longer be generalizable to the intended study population.

Once therapy groups are compared based on subsets of subjects not solely defined by the randomization, it is impossible to show that the subset of subjects from which the inferences are being drawn are comparable. This leads to the potential introduction of bias and limits the generalizability of the trial (11). An exception to this rule is removing subjects found to be ineligible based on objective data that was known or could be determined in the same manner for all subjects prior to randomization (12). However, if the subject is found to be ineligible based on postrandomization data, it is possible that therapy assignment could have affected the availability of the information and, thus, that subject should be included in the ITT analysis.

Intention-to-treat analyses are often considered to be more generalizable to clinical practice. In particular, when subjects in a therapy group stop taking drug or are not fully compliant with the regimen, the response rate of the therapy declines, as it would in clinical practice. In contrast, with most objective measures, not taking study medication should have minimal effect on placebo control groups, provided that the subjects are followed to study completion. However, this advantage of generalizability can be lost when subjects are allowed to switch to or add on other effective therapies. In this situation, active therapy arms may still be reflective of what happens to subjects who initiate therapy in clinical practice, and should not have a large effect on the therapy arm if the other therapies are similar in efficacy to that being tested. However, the placebo group will generally have a higher observed response rate than what would be seen with a group treated with placebo alone. Therefore, effective therapies will be penalized by the additional active medication use in the placebo arm. This can lead to the need for larger studies to compensate for the attenuated therapy effect due to bias in the placebo arm. It should also be noted that such crossovers may lead to overestimates of the therapy effect when other medications available are more effective than the test therapy, particularly when more subjects in the test arm take the additional agent. Careful consideration should be given to minimizing the use of additional active agents in placebo-controlled clinical trials when alternative therapies (or the therapy being studied) are available.

Finally, it should be noted that some subjects end up with missing data even when efforts are made to follow all randomized subjects. For example, subjects may terminate participation in the study prior to completion, miss or refuse procedures, or move away from the clinical center. When no postbaseline information is collected for a randomized subject, the subject may be excluded or some value indicating ''failure'' may be imputed. In most cases, only some of the data are missing for a subject. Even in a strict ITT analysis, data will need to be imputed for these subjects. Many methods exist for imputing missing data (see the entry on *Missing Data*). Common methods include carrying forward the last value for subjects leaving the study, and treating those subjects as failures (13). Other authors have proposed methods for imputing endpoints based on earlier data obtained in the study (14,15). The primary method for imputation should be documented prior to unblinded data analysis. Several different imputation methods can be used to confirm the primary results, evaluate the sensitivity of the primary method used, and examine the robustness of the therapy effect.

In conclusion, statisticians generally agree that following the ITT principle reduces bias and is appropriate for most trials. However, there are many important decisions that need to be considered during the planning phase of data analysis, including the strictness of the application of the principle; handling of ineligible subjects who were erroneously randomized; biasing factors, such as noncompliance, concomitant medication use, and crossover to active medications; methods for imputing missing data; and the effect of the analysis on safety or other endpoints where equivalence might be expected. Karl Peace has written that a goal of clinical research in the development of new drugs is to design protocols and conduct clinical trials so that the evaluable-only and ITT analyses are the same (11).

REFERENCES

1. DP Byar, RM Simon, WT Friedewald, JJ Schlesselman, DL Demets, JH Ellenberg, MH Gail, JH Ware. N Engl J Med 295:74–80, 1976.

2. D Gillings, G Koch. Drug Inform J 25:411–424, 1991.
3. Coronary Drug Project Research Group. N Engl J Med 303:1038–1041, 1980.
4. International Conference on Harmonization (ICH). Guidelines on Statistical Principles for Clinical Trials (E9).
5. FDA. Guideline for the Format and Content of the Clinical and Statistical Sections of an Application.
6. FDA. Draft Guidelines for Preclinical and Clinical Evaluation of Agents Used in the Prevention or Treatment of Postmenopausal Osteoporosis.
7. YJ Lee, JH Ellenberg, DG Hirtz, KB Nelson. Stat Med 10:1595–1605, 1991.
8. LM Freidman, CD Furberg, DL Demets. Fundamentals of Clinical Trials. St. Louis: Mosby-Year Book, 1996.
9. JH Ellenberg. In: P Armitage and T Colton, eds. Encyclopedia of Biostatistics. Chichester, Eng: Wiley, 1998, pp 2056–2060.
10. JH Ellenberg. Drug Inform J 30:535–544, 1996.
11. L Fisher, DO Dixon, J Herson, R Frankowski, MS Hearron, KE Peace. In: K Peace, ed. Statistical Issues in Drug Research and Development. New York: Marcel Dekker, 1990, pp 331–350.
12. Gail MH. Cancer Treatment 69:1107–1113, 1985.
13. Gould AL. Biometrics 36:721–727, 1980.
14. RJA Little, DB Rubin. Statistical Analysis with Missing Data. New York:Wiley, 1987.
15. B Efron. J Am Statistician 89:463–474, 1994.

Ronald K. Knickerbocker

International Conference on Harmonization (ICH)

See also *Food and Drug Administration*

In drug research and development, health authorities in different countries have very similar though different requirements for approval of the commercial use of drug products. As a result, pharmaceutical companies may repeatedly have to conduct similar studies or prepare different documents of the same pharmaceutical product for regulatory submissions. Therefore, there is a strong industrial and regulatory interest and need in the international harmonization of requirements for drug research and development so that information generated in one country or area would be acceptable to other countries or areas. Consequently, the International Conference on Harmonization (ICH) of Technical Requirements for the Registration of Pharmaceuticals for Human Use was organized to provide an opportunity for important initiatives to be developed by regulatory authorities as well as industry associations for the promotion of international harmonization of regulatory requirements.

Currently, however, ICH is concerned only with tripartite harmonization of technical requirements for the registration of pharmaceutical products among three regions: the European Union, Japan, and the United States. Basically, the organization of the ICH consists of two representatives (one from a health authority and one from the pharmaceutical industry) from each of these three regions. As a result, the ICH organization consists of six parties: the European Commission of the European Union, the European Federation of Pharmaceutical Industries' Associations (EFPIA), the Japanese Ministry of Health and Welfare (MHW), the Japanese Pharmaceutical Manufacturers Association (JPMA), the United States Food and Drug Administration (FDA), and the Pharmaceutical Research and Manufacturers of America (PhRMA). The ICH Steering Committee was established in April 1990 (1) to determine policies and procedures, (2) to select topics, (3) to monitor progress, and (4) to oversee the preparation of biannual conferences. Each of the six parties has two seats on the ICH Steering Committee. The ICH Steering Committee also includes several observers from the World Health Organization, the Canadian Health Protection Branch, and the European Free Trade Area, which have one seat each on the committee. In addition, two seats on the ICH Steering Committee are given to the International Federation of Pharmaceutical Manufacturers Association (IFPMA), which represents research-based pharmaceutical industries from 56 countries outside the regions covered by the ICH. Note that the IFPMA has responsibility to run the ICH Secretariat at Geneva, Switzerland, and to coordinate the preparation of documentation.

In order to harmonize technical procedures, the ICH has issued a number of guidelines and draft guidelines. After the ICH Steering Committee selects a topic, the ICH guidelines initiated by a concept paper and go through a five-step review process, described in Table 1. Table 2 provides a list of currently available ICH guidelines or draft guidelines pertaining to clinical trials. As an example, Table 3 gives the table of contents for the ICH draft guideline on *General Considerations for Clinical Trials*. In addition, Tables 4, 5, and 6 give the tables of contents for the ICH guidelines for *Good Clinical Practices: Consolidated Guidelines*, for *Structure and Content of Clinical Study Report*, and for *Statistical Principles for Clinical Trials*, respectively. From these tables, it can be seen that these guidelines are not only for harmonization of design, conduct, analysis, and report for a single clinical trial but also for consensus in protecting and maintaining the scientific integrity of the entire clinical development plan of a pharmaceutical entity.

Jen-pei Liu
Shein-Chung Chow

Table 1 Review Steps for the ICH Guidelines

Step 1
1. Harmonize topic identified
2. Expert working group (EWG) formed
3. A topic leader and a deputy selected by each party
4. Rapporteur for EWG selected
5. Other parties represented on EWG as appropriate
6. A guideline, policy statement, "points to consider" produced
7. Scientific issues agreed on
8. Sign-off and submission to ICH steering committee
Step 2
1. Review of ICH document by steering committee
2. Sign-off by all six parties
3. Formal consultation in accordance with regional requirements
Step 3
1. Regulatory rapporteur appointed
2. Collection and review of comments across all three regions
3. Step 2 draft revised
4. Sign-off by EWG regulatory members
Step 4
1. Forwarded to steering committee
2. Review and sign-off by three regulatory members of ICH
3. Recommended to regulatory bodies for adoption
Step 5
1. Recommendations adopted by regulatory agencies
2. Incorporation into domestic regulations and guidelines

Table 2 ICH Clinical Guidelines and Draft Guidelines

Guidelines
Safety
1. S1A: The Need for Long-Term Rodent Carcinogenicity Studies of Pharmaceuticals
2. S1B: Testing for Carcinogenicity of Pharmaceuticals
3. S1C: Dose Selection for Carcinogenicity Studies of Pharmaceuticals
4. S1C(R): Guidance on Dose Selection for Carcinogenicity Studies of Pharmaceuticals: Addendum on a Limit Dose and Related Notes
5. S2A: Specific Aspects of Regulatory Genotoxicity Tests for Pharmaceuticals
6. S2B: Genotoxicity: A Standard Battery for Genotoxicity Testing of Pharmaceuticals
7. S3A: Toxicokinetics: The Assessment of Systemic Exposure in Toxicity Studies
8. S3B: Pharmacokinetics: Guidance for Repeated Dose Tissue Distribution Studies
9. S5A: Detection of Toxicity to Reproduction for Medicinal Products
10. S5B: Detection of Toxicity to Reproduction for Medicinal Products: Addendum on Toxicity to Male Fertility
11. S6: Preclinical Safety Evaluation of Biotechnology—Derived Pharmaceuticals
Joint Safety/Efficacy
1. M3: Nonclinical Safety Studies for the Conduct of Human Clinical Trials for Pharmaceuticals
Efficacy
1. E1A: The Extent of Population Exposure to Assess Clinical Safety for Drugs Intended for Long-Term Treatment of Non-Life-Threatening Conditions
2. E2A: Clinical Safety Data Management: Definitions and Standards for Expedited Reporting
3. E2B: Data Elements for Transmission of Individual Case Report Forms
4. E2C: Clinical Safety Data Management: Periodic Safety Update Reports for Marketed Drugs
5. E3: Structure and Content of Clinical Studies
6. E4: Dose–Response Information to Support Drug Registration
7. E5: Ethnic Factors in the Acceptability of Foreign Clinical Data
8. E6: Good Clinical Practice: Consolidated Guideline
9. E7: Studies in Support of Special Populations: Geriatrics
10. E8: General Considerations for Clinical Trials
11. Choice of Control Group in Clinical Trials
12. Electronic Standards for the Transfer of Regulatory Information
Quality
1. Q1A: Stability Testing of New Drug Substances and Products
2. Q1B: Photostability Testing of New Drug Substances and Products
3. Q1C: Stability Testing for New Dosage Forms
4. Q2A: Test on Validation of Analytical Procedures
5. Q2B: Validation of Analytical Procedures: Methodology
6. Q3A: Impurities in New Drug Substances
7. Q3B: Impurities in New Drug Products
8. Q3C: Impurities: Residual Solvents
9. Q5B: Quality of Biotechnology Products: Analysis of the Expression Construct in Cells Used for Production of r-DNA Derived Protein Products
10. Q5C: Quality of Biotechnology Products: Stability Testing of Biotechnological/Biological Products
11. Q5D: Quality of Biotechnological/Biological Products: Derivation and Characterization of Cell Substrates Used for Production of Biotechnological/Biological Products
Draft Guidance
Safety
1. S4A: Duration of Chronic Toxicity in Animals (Rodent and Nonrodent Toxicity Testing)
Efficacy
1. E9: Statistical Principles for Clinical Trials
2. Timing of Nonclinical Studies for the Conduct of Human Clinical Trials for Pharmaceuticals
3. Standardization of Medical Terminology for Regulatory Purposes
Quality
1. Q5A: Viral Safety Evaluation of Biotechnology/Products Derived from Cell Lines of Human or Animal Origin
2. Q6A: Specifications: Test Procedures and Acceptance Criteria for New Drug Substances and New Drug Product: Chemical Substances
3. Q6B: Test Procedures and Acceptance Criteria for Biotechnological/Biological Products

Table 3 Table of Contents for Draft Guideline on General Considerations for Clinical Trials

1. Objectives of This Document
2. General Principles
   - 2.1 Protection of clinical trial subjects
   - 2.2 Scientific approach in design and analysis
3. Development Methodology
   - 3.1 Considerations for development
     - 3.1.1 Nonclinical studies
     - 3.1.2 Quality of investigational medicinal products
     - 3.1.3 Phases of clinical development
     - 3.1.4 Special considerations
   - 3.2 Considerations for individual clinical trials
     - 3.2.1 Objectives
     - 3.2.2 Design
     - 3.2.3 Conduct
     - 3.2.4 Analysis
     - 3.2.5 Reporting

Table 4 Table of Contents for ICH Guideline on Good Clinical Practice: Consolidated Guideline

Introduction
1. Glossary
2. The Principles of ICH GCP
3. The Institutional Review Board/Independent Ethnic Committee (IRB/IEC)
   - 3.1 Responsibilities
   - 3.2 Composition, Functions, and Operations
   - 3.3 Procedures
   - 3.4 Records
4. Investigators
   - 4.1 Investigator's Qualifications and Agreements
   - 4.2 Adequate Resources
   - 4.3 Medical Care of Trial Subjects
   - 4.4 Communication with IRB/IEC
   - 4.5 Compliance with Protocol
   - 4.6 Investigational Products
   - 4.7 Randomization Procedures and Unblinding
   - 4.8 Informed Consent of Trial Subjects
   - 4.9 Records and Reports
   - 4.10 Progress Reports
   - 4.11 Safety Reporting
   - 4.12 Premature Termination or Suspension of a Trial
   - 4.13 Final Report(s) by Investigator/Institution
5. Sponsor
   - 5.1 Quality Assurance and Quality Control
   - 5.2 Contract Research Organization
   - 5.3 Medical Expertise
   - 5.4 Trial Design
   - 5.5 Trial Management, Data Handling, Recordkeeping and Independent Data Monitoring Committee
   - 5.6 Investigator Selection
   - 5.7 Allocation of Duties and Functions
   - 5.8 Compensation to Subjects and Investigators
   - 5.9 Financing
   - 5.10 Notification/Submission to Regulatory Authority(ies)
   - 5.11 Confirmation Review by IRB/IEC
   - 5.12 Information on Investigational Product(s)
   - 5.13 Manufacturing, Packaging, Labeling, Coding Investigation Product(s)
   - 5.14 Supplying and Handling Investigational Product(s)
   - 5.15 Record Access
   - 5.16 Safety Information
   - 5.17 Adverse Drug Reaction Reporting
   - 5.18 Monitoring
   - 5.19 Audit
   - 5.20 Noncompliance
   - 5.21 Premature Termination or Suspension of a Trial
   - 5.22 Clinical Trial/Study Reports
   - 5.23 Multicenter Trials
6. Clinical Trial Protocol and Protocol Amendment(s)
   - 6.1 General Information
   - 6.2 Background Information
   - 6.3 Trial Objectives and Purpose
   - 6.4 Trial Design
   - 6.5 Selection and Withdrawal of Subjects
   - 6.6 Treatment of Subjects
   - 6.7 Assessment of Efficacy
   - 6.8 Assessment of Safety
   - 6.9 Statistics
   - 6.10 Direct Access to Source Data/Documents
   - 6.11 Quality Control and Quality Assurance
   - 6.12 Ethics
   - 6.13 Data Handling and Recordkeeping
   - 6.14 Financing and Insurance
   - 6.15 Publication
   - 6.16 Supplements
7. Investigator's Brochure
   - 7.1 Introduction
   - 7.2 General Considerations
   - 7.3 Contents of the Investigator's Brochure
   - 7.4 Appendix 1
   - 7.5 Appendix 2
8. Essential Documents for the Conduct of a Clinical Trial
   - 8.1 Introduction
   - 8.2 Before the Clinical Phase of the Trial Commences
   - 8.3 During the Clinical Conduct of the Trial
   - 8.4 After Completion or Termination of the Trial

Table 5 Table of Contents for ICH Guideline on Structure and Contents of Clinical Study Reports

	Introduction to the Guideline
1.	Title page
2.	Synopsis
3.	Table of Contents for the Individual Clinical Study Report
4.	List of Abbreviations and Definition of Terms
5.	Ethics
6.	Investigators and Study Administrative Structure
7.	Introduction
8.	Study Objectives
9.	Investigational Plan
10.	Study Patients
11.	Efficacy Evaluation
12.	Safety Evaluation
13.	Discussion and Overall Conclusions
14.	Tables, Figures, Graphs Referred to But Not Included in the Text
15.	Reference List
16.	Appendices

Table 6 Table of Contents for ICH Guideline on Statistical Principles for Clinical Trials

I.	Introduction	
	1.1	Background and Purposes
	1.2	Scope and Direction
II.	Considerations for Overall Clinical Development	
	2.1	Study Context
	2.2	Study Scope
	2.3	Design Techniques to Avoid Bias
III.	Study Design Considerations	
	3.1	Study Configuration
	3.2	Multicenter Trials
	3.3	Type of Comparisons
	3.4	Group Sequential Designs
	3.5	Sample Size
	3.6	Data Capture and Processing
IV.	Study Conduct	
	4.1	Study Monitoring
	4.2	Changes in Inclusion and Exclusion Criteria
	4.3	Accrual Rates
	4.4	Sample Size Adjustment
	4.5	Interim Analysis and Early Stopping
	4.6	Role of Independent Data Monitoring Committee (IDMC)
V.	Data Analysis	
	5.1	Prespecified Analysis Plan
	5.2	Analysis Sets
	5.3	Missing Values and Outliers
	5.4	Data Transformation/Modification
	5.5	Estimation, Confidence Interval and Hypothesis Testing
	5.6	Adjustment of Type I Error Rate and Confidence Levels
	5.7	Subgroup, Interactions, and Covariates
	5.8	Integrity of Data and Computer Software
VI.	Evaluation of Safety and Tolerability	
	6.1	Scope of Evaluation
	6.2	Choice of Variables and Data Collection
	6.3	Set of Subjects to Be Evaluated and Presentation of Data
	6.4	Statistical Evaluation
	6.5	Single Study versus Integrated Summary
VII.	Reporting	
	7.1	Evaluation and Reporting
	7.2	Summarizing the Clinical Database
Annex 1	Glossary	

IVRS

See also *Clinical Trial Process*

I. INTRODUCTION

The rapid development of computer telephony technology in the past decade has enabled automatic interchange of information between a computer and a push-button tone phone. It has had many applications in industry. The interactive voice response system (IVRS), for example, has applications ranging from widely used banking services to clinical trials.

The IVRS uses a tone phone as an input device and responds with a computer-controlled, high-quality digitized human voice. The multilingual computer response capability enables IVRS to function worldwide 24 hours a day and to collect real-time online quality data. All these features make IVRS an ideal tool for patient randomization and drug management for both national and international clinical trials.

This entry describes the application of IVRS to patient randomization and drug management in clinical trials. We start with a description of the traditional randomization and drug management procedure, follow with a description of the application of IVRS to mapping patient randomization to the drug management system, and end with a description of a stand-alone patient randomization system, including a minimization procedure.

For our discussion, *randomization* is defined as a process by which each subject has an equal chance or a prespecified probability of being assigned to one of the treatments in a study. This process generates a randomization schedule containing patient numbers and the corresponding treatment assignments.

The drug management system involves several activities: packaging of clinical supplies, shipping them from a warehouse to study sites, and dispensing them to patients by a third party. Unless otherwise specified, a double-blind randomized study is assumed. The terms *drug* and *treatment* are used interchangeably.

II. TRADITIONAL APPROACH TO RANDOMIZATION AND DRUG MANAGEMENT

In a double-blind randomized clinical trial, much pre-study preparation needs to be completed before a site can enroll and randomize the first eligible patient. One such preparation involves the design and generation of a randomization schedule and a plan to manage and ensure an adequate drug supply throughout the study. The traditional approach, which links a patient number to the clinical supply containing the treatment for that patient, is described as follows:

1. The statistician generates the randomization schedule according to the specifications in the final protocol. The end product is a schedule containing patient identification numbers and the corresponding treatment assignments. The total number of patients generated in the randomization schedule depends on the sample size required for the study and the projected number of dropouts.
2. A copy of the validated randomization schedule is sent to the Clinical Supply Group.
3. The Clinical Supply Group prepares and packages the drug supply according to the randomization schedule. In general, an extra drug supply of at least 20% is recommended to cover any potential drug wastage during the study.
4. Individual patient kits, each of which contains the treatment assigned by the randomization process, are prepared. Each kit is labeled with a patient number only. The treatment code is either printed on the blinded portion of a two-part label or kept in a sealed envelope and secured by a third party at the site or at a central location. In either case, the seal or the blind for a given patient can be broken only in case of a medical emergency, and for that patient only.
5. A prespecified, fixed quantity of drug supply is shipped to a given site before the first patient is randomized. If a permuted-block randomization is used, the drug supply sent to a site needs to be in multiples of block size. A permuted block of size 4 in a two-treatment randomization results in two repeats of the two treatments in any random order.
6. When the first patient becomes eligible at a site, the kit labeled Patient Number 1 is dispensed by the pharmacist or designee at that site. The next patient receives the kit labeled Patient Number 2. This process continues until the last

patient has been randomized at the site or until study termination.

7. If the number of patients enrolled at a site exceeds the drug supply, then an additional drug supply needs to be shipped in a timely fashion before the next patient is ready for randomization. On the other hand, if a site enrolls fewer patients than the drug supply, then the remaining drugs will be wasted.

The process works as follows. Assume a study has a randomized, double-blind, parallel-group design. Two treatments, A and B, are to be evaluated in a total of 16 patients at two study sites. A permuted-block design with a block size of 4 is used. That is, for every four patients randomized, two are assigned to treatment A and two to treatment B. According to these specifications, a randomization schedule is generated (see Table 1).

The Clinical Supply Group, upon receiving the randomization schedule, packages eight kits containing drug A, labeled Patient Number 1, 4, 5, 8, 9, 11, 14, or 15, and the other eight kits containing drug B, labeled Patient Number 2, 3, 6, 7, 10, 12, 13, or 16. Since each center plans to randomize eight patients, the kits labeled Patient Numbers 1 through 8 are shipped to site 1 and the kits labeled Patient Numbers 9 through 16 are sent to site 2. Note that the number of patients enrolled at each site has to be in multiples of the block size to maintain treatment balance.

The first patient randomized at site 1 is dispensed the kit labeled Patient Number 1 and therefore receives treatment A. The first patient randomized at site 2 is dispensed the kit labeled Patient Number 9 and therefore receives treatment A. The second patient randomized at sites 1 (Patient Number 2) and 2 (Patient Number 10) receive treatment B. This process continues until the enrollment is complete or the study is terminated for other reasons.

Table 1 Randomization Schedule

Patient	Treatment	Patient	Treatment
1	A	9	A
2	B	10	B
3	B	11	A
4	A	12	B
5	A	13	B
6	B	14	A
7	B	15	A
8	A	16	B

In this example, each patient in the study has a unique patient number. Alternatively, each site can number the patient starting with 1 according to the chronological order of entry. In this case, the unique patient number for the study consists of the site number and the patient number. Applying this method, Patient Numbers 9 through 16 at site 2 in the example would now become Patient Numbers 201 through 208. The leading digit "2" identifies the site number and the second and third digits are for numbering the patients within a site. Thus, the patients at site 1 will now have Patient Numbers 101 through 108. In this numbering system, it is important to have both the site number and the patient number preprinted on the label of the drug supply.

III. PROBLEMS ASSOCIATED WITH THE TRADITIONAL APPROACH

A number of problems arise with the traditional approach to randomization and drug management. One concerns the wastage of drugs as a result of the inability to share drug supplies between sites; another is the inability to ensure treatment balances across sites.

In our example, suppose that only two patients were randomized, one from each site. Only two kits would be used. Since 14 of the total 16 kits would not be used, the overall wastage for the study would be 87.5% (14/16). According to the randomization schedule, both patients enrolled would be assigned treatment A. Ideally, when a study enrolls two patients, it is preferable to have both treatments represented. In order to accommodate this type of treatment balance across sites, a central randomization is needed. Under the traditional approach, it would be impossible to balance the treatment codes across all sites. Since the rate of enrollment in clinical trials varies (some sites enroll patients rapidly whereas others may enroll a few or none at all), the traditional approach does not have a mechanism to allow for adjusting the drug supply when a site does not perform (i.e., drug wastage) or a site overenrolls (i.e., drug shortage).

These problems are readily resolved by using IVRS. The wastage problem can be overcome by assigning patient numbers independent of the supply kits and dynamically matching them through IVRS at the time of randomization. IVRS, which is accessible to all sites at all times, is ideal for providing central randomization to check treatment balance across sites. The following section describes the use of IVRS in drug management and patient randomization.

IV. HOW DOES THE INTERACTIVE VOICE RESPONSE SYSTEM WORK?

The interactive voice response system has the capability of linking information from different tables stored in the relational database and performing queries using common identifiers. With this capability, a randomization schedule such as Table 1 can be separated into many different tables and connected by IVRS at the time of randomization. This mapping process allows IVRS the flexibility to perform many special functions unmatched by the traditional randomization approach. The following section explains some of these features and the connection of tables by IVRS.

A. Drug Supply and Packaging

As already mentioned, if the drug supply can be packaged separately, independent of the patient numbers and the randomization schedule, with only the minimal quantity shipped to a site when the site is ready to randomize the first patient, the savings in the drug supply can be enormous. The Clinical Supply Group can package the drug as soon as the total supply of each drug type is known.

Let us applying this approach to our earlier example, in which 16 kits need to be packaged. The Clinical Supply Group, without the knowledge of the randomization schedule, can package the drug in two groups: eight kits for treatment A and eight for treatment B. The kits can be labeled with any numbering system unrelated to the patient number. The simplest method is to label the eight kits containing treatment A with consecutive numbers d01 through d08 and the eight kits containing treatment B with numbers d09 through d16. However, this method, although straightforward, risks revealing the treatment codes through clever guesswork.

An alternative approach is to scramble the kit numbers before treatment assignment. One such sequence for 16 kits might be d02, d05, d10, d06, d13, d09, d12, d08, d16, d011, d07, d03, d014, d04, d15, and d01. The next step is to assign the first eight numbers to kits containing treatment A and the last eight to kits containing treatment B. After the assignment, the kit numbers within each treatment group can be sorted in an ascending order for easy packaging and assembling. The resulting kit numbers are displayed in Table 2.

Table 2 Kit Numbers

Treatment A	Treatment B
d02	d01
d05	d03
d06	d04
d08	d07
d09	d11
d10	d14
d12	d15
d13	d16

After the kit numbers have been assigned, the Clinical Supply Group can package the drugs accordingly and store them in a warehouse ready for shipment. At this point, the kits are not associated with any patient number or any study site. When a site is anticipating the first eligible patient, a minimal quantity containing equal numbers of treatment A and treatment B is shipped to the site. The minimal quantity of an initial shipment depends on a number of factors: it must take into account the enrollment rate at each site, and, preferably, it must be in multiples of the block size in the case of a permuted-block design. If a site can enroll four patients a day, the initial shipment has to be eight kits or more to ensure an adequate drug supply. In our example, the enrollment is assumed to be one patient per week at all sites; therefore, an initial shipment of four kits, two treatment A and two treatment B, is acceptable. The first site to request shipment receives kit numbers d01, d02, d03, and d05, and the next site receives kit numbers d04, d06, d07, and d08; the contents, however, remain blinded to the site. In theory, an initial shipment of two, one for each treatment, is sufficient; however, this is not recommended in practice because of the potential risk of revealing the drug codes by the staff at the site.

As soon as the file containing kit numbers is available, as in Table 2, it is loaded into the IVRS relational database, which is accessible through a tone phone. When a site is ready for randomization, IVRS notifies the warehouse to ship the initial supply to the site. The kit numbers and the associated site number are recorded in the IVRS database. The system is then ready to manage (i.e., dispense or resupply) the drug supply.

When the drug supply is low, IVRS will trigger the warehouse to resupply the site. For a slow-enrolling site, the trigger point may be one of each treatment type remaining at the site. For a fast-enrolling site, the trigger point depends on the enrollment rate. For example, if a site enrolls four patients a day, the trigger point has to be at least four kits remaining at a site, two of each drug type. The quantity of resupply also depends on the enrollment rate and how close it is to the study completion. When the study approaches its completion, no more resupply is needed. The trigger point and the quantity of resupply must be established when the IVRS system is being developed for the study.

B. Randomization Schedule

In the traditional approach, the randomization schedule (see Table 1) is used for packaging, shipping, and dispensing; a patient number is associated with a kit. With IVRS, the same schedule is used, but the process is more flexible and has additional features.

One advantage of using IVRS is its ability to separate patient numbers from the randomization codes and to match them dynamically at the time of randomization. The randomization codes serve as a bridge between the treatment assignment and any patient numbering system. This bridge is formed at the time of randomization. For example, Table 3 uses the same randomization schedule as Table 1, but the patient numbers are now randomization codes and these codes are not associated with patient numbers.

Assume that the patients are sequentially numbered at each site, starting with 1. When the first patient at site 1 (Patient 101) is ready for randomization, the site calls IVRS, which would then search through Table 3 stored in the IVRS database for the next available code and the treatment it is linked to. In this case, code 1 is available; hence, Patient 101 is assigned to treatment A. If the next patient is from site 2 (Patient 201), this patient is linked to code 2 by IVRS and treatment B is assigned. Should the study be terminated at this point, there will be two patients from the two sites, one assigned to treatment A and the other to treatment B.

Table 3 Randomization Codes

Code	Treatment	Code	Treatment
1	A	09	A
2	B	10	B
3	B	11	A
4	A	12	B
5	A	13	B
6	B	14	A
7	B	15	A
8	A	16	B

Thus, a central randomization is achieved. In this type of randomization, all the randomization codes are shared by all patients from all sites. The end result is a central randomization, with a balanced treatment across all sites but not necessarily within one site.

If, however, the study is designed to have treatment balance within a given center, then the IVRS database needs to be structured so that codes 1 through 8 are reserved for site 1 and codes 9 through 16 for site 2. With this new structure, the second patient in our example (Patient 201) would be linked to code 9, receiving treatment A, instead of code 2, receiving treatment B.

C. Mapping by Interactive Voice Response System

After the clinical supply list (Table 2) and the random assignment of treatment (Table 3) are separated and available, they are stored in the IVRS relational database. The IVRS can query the database and connect these tables at the time of randomization and respond to the caller with real-time randomization and drug assignment. This mapping process is described as follows:

1. To ensure system security, a caller from any site must first pass stringent system and study-specific security measures (e.g., site identification, password, caller identification) to continue the dialog with IVRS. For reference, a sample IVRS script is given in the appendix to this entry.
2. Through the tone phone keypad, the caller enters any prespecified study and patient information, which may include site and patient numbers, patient eligibility, and other patient characteristics, into the IVRS database.
3. As soon as the eligibility is verified, IVRS searches the randomization schedule (Table 3) for the next available randomization code and the corresponding treatment.
4. Once the treatment is identified, IVRS then searches the drug supply list (Table 2) for the next available kit number containing the treatment at that site.
5. The IVRS, through the voice response system, informs the caller which kit number to dispense, without revealing the treatment code.
6. The randomization procedure is complete when the caller confirms the kit number by means of the tone phone. The entire phone call takes no more than two minutes.

7. The IVRS stores the site, patient, and kit numbers as well as the treatment assignment in the database.
8. The IVRS checks the drug supply at the site to assess the need for resupply.

Let us now illustrate by extending our earlier example. Figure 1 provides a graphic presentation of the mapping process for this example. Let us assume the study calls for a central randomization and a patient numbering system by site. The randomization schedule in Table 3 and the kit list in Table 2 are stored in the IVRS database. Because of a slow enrollment rate, an initial shipment of four kits is provided to each site: kits d01, d02, d03, and d05 are sent to site 1, and d04, d06, d07, and d08 are sent to site 2. The trigger point for resupply is set at the time when two kits (one of each treatment) still remain at the site.

If the first caller is from site 1 (Patient Number 101), IVRS matches this patient to the first available randomization code in Table 3, i.e., code 1, which is linked to treatment A. The IVRS then searches Table 2 for the next available kit containing treatment A in site 1; that is d02. The IVRS tells the caller to dispense kit d02 without revealing its contents. The double-blind status of the study is therefore maintained. After confirmation that kit d02 has been dispensed, IVRS checks the trigger point for resupply. Because the site still has three kits, IVRS will not generate a resupply notice.

Applying the same approach, if the next patient comes from site 2 (Patient Number 201), IVRS maps this patient to code 2 in Table 3 and hence assigns treatment B and kit d04. Again, no replacement kit is needed.

If the study is terminated after two patients have been randomized, a maximum of eight kits would have been shipped to the sites and only two used. The number of kits wasted would be six, which would be much lower than the 14 kits under the traditional approach in the earlier example. The percentage of wastage would be 37.5% (6/16) versus 87.5% (14/16). In addition to saving drug supply, the treatment would be balanced (one treatment A and one treatment B) by the central randomization capability of IVRS even though only two patients were enrolled. In traditional randomization, treatment balance across sites can happen only by chance, not by design.

In all of the examples given, the size of the study (i.e., number of patients and sites) is kept small for easy illustration and understanding. In real studies, the numbers of sites and patients are much larger, which generally would lead to a much greater saving in drug supply.

V. PATIENT RANDOMIZATION USING THE INTERACTIVE VOICE RESPONSE SYSTEM

In addition to the dynamic mapping of patient number, randomization code, drug assignment, and drug supply, IVRS also performs stand-alone patient randomization for various randomization methods. The most commonly used methods in clinical trials include nonstratified and stratified randomization, and minimization. The implementation of these methods using the IVRS is described next.

A. Nonstratified Randomization

Whether it is a permuted-block or a simple randomization, the implementation is straightforward in a nonstratified randomization because it involves only one randomization schedule. The IVRS, upon verification of the caller and the collection of patient eligibility, connects the patient to the next available treatment code in the randomization schedule.

This type of randomization is commonly used in randomized large-scale, open-label studies. Large-scale studies would have a greater chance of achieving treatment balance within a stratum because of sufficient number of patients enrolled in each stratum. In open-label trials, treatment assignment is known to both patient and investigator; therefore, it is likely to be subject to selection bias (e.g., the investigator may choose to assign the active treatment rather than a placebo to a more severely ill patient). The IVRS is especially valuable in these studies, because the treatment is assigned at the moment when the phone call is completed and therefore the selection bias is avoided.

B. Stratified Randomization

One of our earlier examples involved treatment balance within each study site. Study site is considered a stratification factor in this case because of potential "institution effect" in carrying out the trials. Many clinical trials, especially in cancer research, involve multiple prognostic factors that could affect treatment outcome. Traditionally, treatment balance within a stratum is achieved by generating a permuted-block randomization schedule for each stratum. All such schedules are stored in the IVRS database. The IVRS, upon collecting patient information from the caller, first matches the stratum and then searches for the next available randomization code in that stratum.

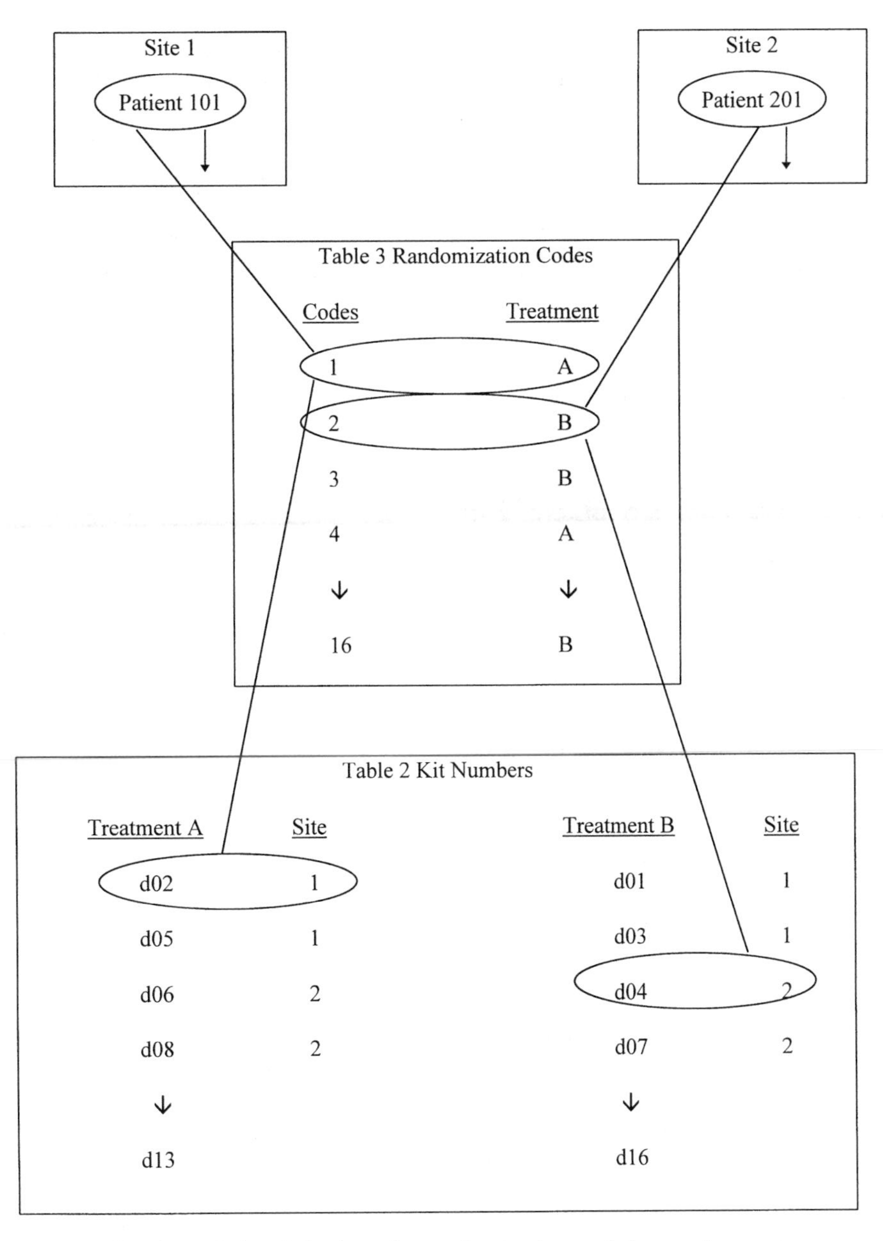

Fig. 1 Relational database for randomization and drug assignment.

For example, in the treatment of stress incontinence, the baseline incontinence episode frequency (IEF) and gender are thought to have impact on the treatment outcome; therefore, treatment balance within each stratum defined by baseline IEF and gender is deemed necessary. Since there are three levels of baseline IEF (mild, moderate, severe) for each gender, a total of six baseline strata is considered for randomization. A randomization schedule is generated for each of the six strata, and all six schedules are stored in the IVRS database. When a site calls to randomize a patient, IVRS first collects the stratification information through the tone phone. If the patient is female with moderate baseline IEF, IVRS will search for the next available randomization code in the randomization schedule for the stratum of female with moderate baseline IEF, and will assign the corresponding treatment to that patient.

One difficulty of using this procedure is that the number of strata increases geometrically with the number of prognostic factors and their levels. For example, three prognostic factors, each with four levels, will generate 64 strata ($4 \times 4 \times 4$). As the number of strata increases, the ratio between the total number of patients and the number of strata becomes smaller. In other words, the number of patients in each stratum could be so small that in most strata the first permuted block of treatments would not be completely assigned. The result would be a considerable imbalance of treatment assignments across all stratification factors.

Even though this procedure has traditionally been used, it does have limitations in balancing treatments simultaneously across several prognostic factors. A minimization procedure that overcomes these difficulties has since been proposed by Taves (1974) and expanded by Pocock and Simon (1975). The implementation of a minimization procedure by IVRS is described and illustrated in an example in the next section.

C. Minimization

Minimization is a method of assigning patients to treatments in order to minimize the differences between the treatment groups in the number of patients as well as in patient characteristics and prognostic factors. Although the minimization method is mathematically straightforward, it had not been widely used for a decade following its inception because of the difficulties in implementing it on a routine basis without an interactive computer. Therefore, the introduction of IVRS has made this dynamic procedure a popular one for clinical trials in recent years, because it assigns treatment on an ongoing basis and does not require a prespecified randomization schedule. This procedure is best illustrated by the following example.

Suppose that a trial for controlling stress incontinence is conducted in five centers and that there are two treatment groups (active treatment and placebo) to be balanced with regard to baseline IEF, gender, and centers. A total of 93 patients has already been entered. The distribution of these patients by the three stratification factors is given in Table 4. This table is stored in the IVRS relational database. The IVRS collects from the caller the stratification factors associated with the next patient: Patient 94 is from site 1 and is a male with severe baseline IEF. For each treatment group, IVRS adds up the number of patients who had already been randomized and who shared the same baseline characteristics as Patient 94 and assigns the patient to the group having the lower total (Reed, 1988). In this example, the sum for the active group is 39 (24 + 8 + 7) and for the placebo group is 40 (25 + 7 + 8). Therefore, Patient 94 is assigned to the active group, thus minimizing the imbalance in the number of patients between the two treatment groups across all the stratification factors. Note that the marginal total is used for decision making in this example simply for easy illustration; other criteria for decision making can be found in other references (Pocock and Simon, 1975; Taves, 1974).

In this example, Patient 94 is assigned to the active drug with a probability of 1. Alternatively, a less "deterministic" approach can be applied, which assigns the new patient to the treatment with a lower total with a probability of less than 1 (e.g., 0.75 or 0.95). For this type of randomization, IVRS uses a uniform random-number generator to perform the task. Even though the random numbers generated by a computer are in fact pseudo-random numbers, they are able to serve the purpose of randomization without compromising the integrity of the process. Let us extend our current example and assume that the probability is set at 0.80. A list of uniform random numbers, ranging from 0.00 to 0.99, is generated. The IVRS searches the list for the next random number and assigns Patient 94 to active treatment if the next available random number is between 0.00 and 0.79, inclusively, and to placebo otherwise.

As a general practice, minimization should not be applied until a given number of patients has been randomized using a permuted-block design so that the risk of revealing the treatment codes of the first few enrolled can be avoided.

VI. OTHER IVRS APPLICATIONS

There are several additional capabilities of IVRS besides randomization and drug management. One is the collection of real-time quality data in clinical trials. For example, the patient diary in clinical trials has been a cumbersome, hand-written process subject to human error and missing data. Using IVRS, quality data can be obtained directly from the patient. The IVRS can also call and remind the patient to enter the diary data so that missing data can be avoided, and a high patient compliance can also be expected. This has proven to be a major cost savings because of the ease of data processing as well as the accelerated availability of quality data.

In addition, in long-term-safety studies, IVRS can collect and provide real-time crucial safety information, such as serious adverse events and mortality data. The fast and accurate information provided by IVRS has proven to be valuable for monitoring the safety of clinical trials. Other operational aspects of clinical trials,

Table 4 Number of Patients by Stratification Factors for the First 93 Patients

Stratification factor	Active	Placebo
Sex		
Male[a]	24	25
Female	23	21
Baseline IEF		
Mild	19	20
Moderate	20	19
Severe[a]	8	7
Site		
1[a]	7	8
2	9	8
3	13	12
4	14	12
5	4	6

[a]Strata for Patient 94.

such as patient enrollment and withdrawal status, can also be collected through IVRS. This type of real-time data is very useful for strategic planning (e.g., enrollment may need to increase if the dropout rate is higher than expected in the protocol).

VII. CONCLUSIONS

The interactive voice response system (IVRS) is an application based on the rapidly growing computer telephony technology. Its multilingual capability and its ability to function without human intervention provide services at any time and place. The computer-based technology makes worldwide access possible, making it therefore an ideal tool for central randomization and drug management in clinical research. This entry has described the application of IVRS in randomization and drug management. One of the great advantages of using IVRS in managing the clinical drug supply is the reduction of drug wastage. Our example demonstrated a reduction in drug wastage from 87.5% under the traditional approach to 37.5% using IVRS. In addition to saving drug supply, IVRS has the ability to randomize centrally the patients from all site so that treatment balance can be achieved across all sites for the entire study. The IVRS technology can also provide clinical researchers with a new and cost-effective method for collecting real-time quality data and for providing critical monitoring information any time and any place a push-button tone phone is available.

APPENDIX: SAMPLE IVRS SCRIPT FOR RANDOMIZATION AND DRUG MANAGEMENT

A. General

1.1 IVRS: Please enter your identification number.
1.2 CALLER: XXXXXXX
1.3 IVRS: Please enter your access code.
1.4 CALLER: XXXX

Language is determined by IVRS from identification number and access code.

1.5 IVRS: Welcome to the Covance automated randomization and drug assignment system for the BONE study. For computer-automated help at any time, please press the "star'' key, or to speak with the help desk, please press 0.

Upon leaving the system:

1.6 IVRS: Thank you for using the Covance automated randomization and drug assignment system for the BONE study. Good-bye.

These prompts are played for every caller. They are used to determine who the caller is and verify security clearance to main menu.

B. Main Menu

These prompts provide the caller with a list of system options.

2.1 IVRS: To screen a patient, press 1.
2.2 IVRS: To record a patient screen failure, press 2.
2.3 IVRS: To randomize a patient, press 3.
2.4 IVRS: To assign drug, press 4.
2.5 IVRS: To record a patient discontinuation, press 5.
2.6 IVRS: To confirm a drug shipment, press 6.
2.7 IVRS: To review patient information, press 7.
2.8 IVRS: To exit, press 9.

REFERENCES

SJ Pocock, R Simon. Sequential treatment assignment with balancing for prognostic factors in the controlled clinical trial. Biometrics 31:103–115, 1975.

JV Reed, EA Wickham. Practical experience of minimization in clinical Trials. Pharmaceutical Med 3:349–359, 1988.

DR Taves. Minimization: A new method of assigning patients to treatment and control groups. Clin Pharmacol Therapeutics 15:443–453, 1974.

Mon-Gy Chen

L–M

Lilly Reference Ranges

See also *Integrated Summary Report*

I. INTRODUCTION

The Lilly reference ranges are sets of reference intervals for routine clinical laboratory analytes measured in clinical trials. The Lilly reference ranges were designed specifically for use in clinical trials. There are 40 published sets of adult reference intervals for 39 routine clinical laboratory analytes (1).

In clinical laboratory medicine, reference ranges provide a point of comparison. The interpretation of specific observed laboratory test results is a comparative decision-making process. "Health-associated" reference values are most frequently used to screen patients for the presence of disease, to confirm a diagnosis, and to monitor disease progression. However, the Lilly reference ranges are used for very different purposes in the clinical trials setting. For this reason, there are sharp contrasts between "health-associated" reference values and the Lilly reference ranges.

In a parallel fashion, Thompson et al. suggested that three major decisions could be made using observed routine clinical laboratory results in clinical trials (2). First, patients need to be screened for inclusion in clinical trials. Second, patients need to be protected during clinical trials. Third, drugs' effects need to be assessed during and after the clinical trial.

The two most striking differences between "health-associated" reference intervals and the Lilly reference ranges are (1) the former is narrower and the latter is wider, and (2) the reference population of the former is healthy participants and that of the latter is clinical trial participants. However sharp these contrasts are, reference intervals share a similar protocol for their definition and determination (3).

II. DETERMINATION OF REFERENCE INTERVALS

A. Methods

Thompson et al. published the methodology for determination of Lilly reference ranges in 1987 (4). Briefly, the reference sample was selected by predetermining inclusion and exclusion criteria for all patients entering a clinical trial. The 39 most frequently sampled routine analytes were selected for definition. The succeeding steps are: Obtain written consent from study participants. Prepare and collect specimens properly and consistently. Inspect reference value data to evaluate the distribution. Identify possible outliers. Select methods of estimating reference limits. Analyze reference values. Subcategorize populations for separate reference intervals. These steps are discussed in detail later.

Also in 1987, the approved recommendations on the theory of reference values that provided definitions, principles, and procedures for the determination and use of reference values was published by the Expert Panel on the Theory of Reference Values (EPTRV) of the International Federation of Clinical Chemistry and the Standing Committee on Reference Values of the International Council for Standardization in Hematology. These definitions, principles, and procedures provided a rational approach and sound basis for the determination of reference values (5–10).

In 1995, approved guidelines for the definition and determination of reference ranges were published by the National Committee for Clinical Laboratory Standards (NCCLS) (11). The work of the EPTRV proved to be a most useful basis for the development of the NCCLS guidelines. It was the intent of the subcommittee to publish guidelines for determining "health-associated" reference values. Basically the a posteriori method as described by the NCCLS protocol guideline was the methodology employed to construct the Lilly reference ranges.

B. Selection of Reference Sample Group

The reference population is a group consisting of all possible reference individuals. The reference sample group is composed of an adequate number of reference individuals selected on the basis of well-defined criteria. The researchers wish to draw inferences from the sample and generalize them to the population.

During the selection of reference individuals, careful consideration should be given to the reference population about which the study is attempting to draw an inference. Unlike, "health-associated" reference intervals, which are very useful diagnostically, Lilly reference ranges are used for a very different purpose. Clinical trial researchers dealing with routine clinical laboratory measurements are most interested in testing hypotheses concerning the safety of an experimental compound relative to a control compound. The population about which the researchers are attempting to draw an inference is clinical trial participants. This population generally exhibits greater variability in selected routine laboratory parameters than noted in populations of disease-free individuals. Therefore, Thompson et al. (4) included all clinical trial participants, even if they were not subsequently randomized, with an available reference value in the reference sample group.

Reference values are defined to be those values, or test results, obtained from a reference sample group. The *empirical distribution of reference values* is defined to be the reference distribution, and the summary statistics from the reference distribution used to describe the reference distribution is called a *reference limit*. These reference limits are commonly order statistics. The interval between, and including, the two reference limits is the *reference interval*. Few reference intervals should be defined as reference ranges. The term *range* should be reserved for describing a set of values defined by the actual minimal and maximal measured values. Lilly reference ranges are defined as such because no trimming or censoring of the distribution occurred and because the minimal and maximal values are available for use.

Potential reference values were only those values for reference individuals that were sampled during qualification for enrollment into a clinical trial. These values were the screening and baseline values for the clinical trial that the reference individuals were screened for. Analytes that were used as efficacy variables and not as routine laboratory safety variables were excluded as reference values. For example, patients in diabetic studies did not contribute glucose values.

Patients were then to be partitioned into 32 subclasses dichotomizing gender, race, age, and smoking and drinking status. Results for glucose values were to be partitioned into fasting and nonfasting status.

The a posteriori sampling process was used for the Lilly reference ranges, since the process of exclusion and partitioning took place after analyte testing.

C. Selection of Sample Size and Statistical Methods for Estimating Reference Limits

In general, reference limits can be determined using parametric or nonparametric methods. Lilly reference ranges were determined nonparametrically. There were several considerations for this decision. Nonparametric estimation does not require any assumptions about the distribution for a given analyte. Also, it is a methodology that could be applied consistently for all analytes. It is simple to implement and is readily understandable. The disadvantage of using nonparametric estimation is that when the reference distribution is normal-Gaussian, the order statistics are slightly biased and more variable. However, large sample sizes were anticipated for the Lilly reference ranges, and few distributions were thought to be normally distributed.

The reference value data were inspected to evaluate the reference distribution for each analyte and to identify possible outliers. Because an important implicit assumption is that all routine safety reference values come from the same distribution, none were excluded from the reference distribution or subsequent calculations of the Lilly reference ranges, regardless of how aberrant they appeared. However, the primary Lilly reference ranges was the 98% interval defined by the first and ninety-ninth percentiles. The ninety-fifth and ninetieth intervals were also calculated.

D. Delta Limits

All patients who had repeated measurements during their qualification period were included in the determination of the delta values. These patients had both screening and baseline values. These samples were drawn at an interval of 1 or 2 weeks. This interval was typically associated with the placebo baseline or washout period for their clinical trial.

The difference between the screening and baseline

reference values was calculated. A distribution of these differences was defined to be the *delta distribution*. These differences express the within-patient variability. The same methods for the determination of the reference intervals were applied to these difference values for the creation of the *delta limits*. Delta limits can be very useful to the researcher. They provide a guideline for the amount of change that can be expected in a given analyte. This guideline can then be used in the interpretation of the observed change due to an experimental compound for a given patient. Procedures for validating the reference ranges and delta limits as well as for calculating the tolerance intervals were also given by Thompson et al. (4).

III. RESULTS

The Lilly reference ranges are based on a sample of 20,102 patients with at least one analyte result recorded. These patients had blood and urine tests during qualification for entry into 50 clinical trials. A total of 1121 investigators studied 11 indications throughout the United States and Europe. Each of these patients had nonmissing values for gender, race, and age.

This reference population appeared comparable to the United States population, based, as reported in the national almanac, on gender, race, age, smoking, and ethanol use. Smokers and imbibers are defined to be those patients who at the screening visit report that they currently smoke or take alcohol. Patients whose baseline value for smoking or ethanol use was missing were deleted from the partitioned interval calculation. Table 1 contains the final sample sizes for each of the five categorized groups. The NCCLS subcommittee recommends a sample size of 120 reference subjects per analyte.

Each of the five categorized groups were further partitioned into eight groups defined by gender, race, and age. The 98% reference intervals for 39 are shown in Appendix 1 for the total sample. The 98% intervals for 39 analytes for the remaining 32 demographic groups are contained in Appendixes 2–5.

In addition, the results of two blood samples drawn 1–4 weeks apart was examined in 5802 patients taking placebo at the onset of a clinical trial, to calculate delta limits. Not all 20,102 patients had a second blood draw, because some failed to meet inclusion criteria, did not take placebo as directed, or did not return for their study. For the patients who did return, their delta values, defined to be the difference between the two blood draws, were included in the reference sample group. The reference distribution was found to be symmetric

Table 1 Sample Sizes by Category

Smoker	Imbiber	Sample size	Appendix
Total	Total	20,102	1
Yes	Yes	2,974	2
Yes	No	2,037	3
No	Yes	4,265	4
No	No	4,801	5

about zero. Delta limits were not as influenced by demographic factors as were reference ranges, so one pair of delta limits was used to describe an analyte. These delta limits are contained in Appendix 1.

For the Lilly reference range data of 20,102 patients, the Brown–Mood multisample procedure was used to compare analytically the impact of these five demographic characteristics on the extreme tails. The Brown–Mood procedure is a nonparametric procedure that answers the question: Do patient demographics influence the reference distribution in the tails? A yes answer supports the need for partitioning. This procedure was also applied to the delta limits.

The observed significance levels, p-values, of the Brown–Mood test for the reference ranges are given in Appendix 6. These p-values are derived from the test of the hypothesis for equality between a specific demographic category in tail proportions. This tests whether the proportion of patients with values below the pooled low reference limit is the same for both strata in a demographic category and whether the proportion of patients with values above the pooled high reference limit is the same for both strata in a demographic category defined by each of the five demographic variables. Missing p-values occur when there are no observations below (above) the reference limit. The observed significance levels, p-values, of the Brown–Mood test for the delta limits are not reported.

Contained in Appendix 7 are the estimates of the difference in the low (high) reference limit between the two strata defined by each of the five demographic variables relative to the low (high) reference limit calculated from the pooled sampled group. For example, the low AST values for females and males are 9 and 10 U/L, respectively. The low AST value for the entire population is 9 U/L. The entry for the low value of AST for gender is $(9 - 10)/9 * 100 = -11.1\%$. The sign of each table entry indicates the direction of the difference between reference limits. Missing values are shown for those estimates where the Brown–Mood test was not statistically significant at the 0.05 level.

APPENDIX 1 Eli Lilly and Co. Adult Reference Ranges in Clinical Trials for Routine Laboratory Test Results (n = 20,102)*

		Male				Female					
		Caucasian		Non-caucasian		Caucasian		Non-caucasian			
		Age		Age		Age		Age		Delta limits	
Analyte	Units	<50	≥50	<50	≥50	<50	≥50	<50	≥50	Low	High
Liver											
ALT	U/L	7–125	5–95	5–121	4–84	5–80	5–80	4–82	4–82	−27	28
AST	U/L	11–79	10–78	11–122	8–84	9–62	9–67	9–66	9–65	−25	22
GGT	U/L	6–194	6–193	7–320	5–274	5–127	5–201	5–161	5–190	−30	29
Alk. phosphatase	U/L	30–153	28–163	32–151	27–174	28–131	31–168	31–138	33–174	−44	28
Bilirubin	μmol/L	3–31	3–29	3–29	3–22	3–22	3–19	2–21	2–19	−10	10
Muscle											
Creatine kinase	U/L	17–640	11–426	27–1105	15–940	19–265	0–267	20–453	10–463	−277	236
Kidney											
Creatinine	μmol/L	62–141	71–168	71–168	71–186	53–124	53–150	53–124	53–150	−35	35
Urea	mmol/L	2.5–8.9	2.9–11.1	2.1–8.9	2.1–12.5	1.8–7.9	2.5–11.4	1.8–7.5	2.5–10.7	−3.2	3.2
Uric acid	μmol/L	167–547	161–583	184–595	208–625	89–470	113–535	107–482	143–619	−125	119
Phosphate (i)	mmol/L	0.72–1.61	0.68–1.49	0.77–1.65	0.68–1.58	0.74–1.65	0.74–1.58	0.74–1.52	0.74–1.61	−0.45	0.42
Calcium	mmol/L	2.07–2.64	2.02–2.62	2.05–2.64	2.05–2.64	2.02–2.62	2.05–2.64	2.02–2.64	2.02–2.69	−0.30	0.30
Electrolytes											
Sodium	mmol/L	135–147	133–148	135–150	134–149	135–147	134–150	135–150	134–152	−8	8
Potassium	mmol/L	3.5–5.5	3.3–5.4	3.0–5.5	3.0–5.4	3.4–5.3	3.3–5.5	3.1–5.3	3.0–5.3	−1.1	1.0
Chloride	mmol/L	96–113	95–112	96–110	95–112	96–114	94–113	98–112	93–111	−9	8
Bicarbonate	mmol/L	21–35	21–35	21–36	21–36	20–33	21–36	20–35	21–36	−7	8

Nutritional											
Glucose, fasting	mmol/L	3.7–10.8	3.9–14.2	3.8–9.5	3.9–16.8	3.7–9.2	3.7–15.0	3.8–15.3	3.9–20.3	−3.1	3.2
Glucose, nonfasting	mmol/L	3.4–9.8	3.8–14.4	3.4–10.8	4.2–20.3	3.6–11.2	4.3–15.2	3.8–14.7	4.2–18.4	−4.8	4.2
Albumin	g/L	36–52	32–50	34–53	32–49	34–50	32–49	33–49	32–50	−7	6
Protein	g/L	62–83	60–82	63–89	61–87	61–82	60–82	63–86	63–86	−11	10
Cholesterol	mmol/L	2.84–8.56	3.41–8.51	2.77–8.46	2.84–8.95	3.10–7.73	3.65–9.10	3.10–7.73	3.78–9.36	−1.78	1.73
HDL cholesterol	mmol/L	0.44–2.02	0.47–2.17	0.47–2.48	0.41–2.48	0.59–2.64	0.57–2.56	0.52–2.51	0.52–2.64	−0.85	0.72
LDL cholesterol	mmol/L	1.50–6.13	1.68–6.08	1.22–6.26	1.42–6.43	1.37–5.46	1.78–6.23	1.16–5.82	1.81–6.813	−1.97	1.78
Triglycerides	mmol/L	0.50–9.53	0.53–8.69	0.45–8.99	0.51–5.05	0.38–6.93	0.58–6.21	0.44–5.59	0.52–4.41	−3.01	2.88
Erythrocytes											
Hemoglobin (Fe)	mmol/L	7.90–11.29	7.10–11.36	6.83–11.23	6.60–11.00	6.80–10.18	6.60–10.36	6.14–10.10	6.21–10.10	−1.20	1.20
Erythrocytes	TI/L	4.2–6.2	3.9–6.2	3.9–6.4	3.7–6.4	3.7–5.5	3.6–5.7	3.6–5.7	3.5–5.9	−0.7	0.7
Hematocrit	1	0.38–0.55	0.35–0.55	0.35–0.55	0.33–0.54	0.33–0.49	0.33–0.51	0.31–0.49	0.33–0.49	−0.07	0.06
MCV	fL	78–105	77–107	71–104	74–104	76–103	77–105	67–102	73–102	−9	8
MCH (Fe)	fmol	1.60–2.17	1.60–2.20	1.40–2.17	1.40–2.11	1.50–2.10	1.60–2.11	1.30–2.05	1.40–2.05	−0.19	0.20
MCHC (Fe)	mmol/L	19–23	19–23	18–23	18–22	19–23	19–23	18–22	18–22	−2	2
Leukocytes											
Leukocytes	GI/L	3.6–13.8	3.4–12.8	3.0–13.0	2.9–11.6	3.5–13.4	3.6–12.4	3.2–13.6	3.1–11.4	−4.0	3.7
Bands	GI/L	0–0.62	0–0.54	0–0.27	0–0.32	0–0.75	0–0.55	0–0.37	0–0.20	−0.29	0.26
Neutrophils	GI/L	1.83–9.29	1.82–8.74	1.17–9.24	1.29–7.81	1.70–8.99	1.79–8.52	1.19–8.66	1.31–7.87	−3.38	3.18
Lymphocytes	GI/L	0.88–4.49	0.81–4.32	0.93–4.95	0.82–4.28	0.97–4.20	0.90–4.30	0.94–4.51	0.95–4.47	−1.57	1.54
Monocytes	GI/L	0–0.89	0–0.96	0–0.81	0–0.85	0–0.82	0–0.85	0–0.83	0–0.75	−0.44	0.46
Eosinophils	GI/L	0–0.65	0–0.66	0–0.74	0–0.59	0–0.66	0–0.58	0–0.62	0–0.55	−0.38	0.37
Basophils	GI/L	0–0.18	0–0.18	0–0.15	0–0.14	0–0.17	0–0.17	0–0.16	0–0.17	−0.13	0.13
Platelets	GI/L	136–425	130–483	131–486	123–518	151–501	141–513	144–541	114–559	−93	107
Urine											
Specific gravity		1.006–1.037	1.006–1.035	1.007–1.036	1.006–1.036	1.004–1.036	1.005–1.036	1.005–1.038	1.006–1.037	−0.018	0.017
pH		5.0–8.0	5.0–8.0	5.0–8.0	5.0–8.0	5.0–8.0	5.0–8.0	5.0–8.0	5.0–8.0	−2.0	2.0

*The number of adult patients with at least one analyte result is 20,102.

APPENDIX 2 Eli Lilly and Co. Adult Reference Ranges in Clinical Trials for Routine Laboratory Test Results, Smoker/Imbiber (n = 2974)*

		Male				Female					
		Caucasian		Non-caucasian		Caucasian		Non-caucasian			
		Age		Age		Age		Age		Delta limits	
Analyte	Units	<50	≥50	<50	≥50	<50	≥50	<50	≥50	Low	High
Liver											
ALT	U/L	5–143	6–114	6–109	3–77	5–63	7–62	4–112	6–57	−27	28
AST	U/L	11–111	9–104	11–140	11–114	9–55	10–80	8–81	9–62	−25	22
GGT	U/L	8–230	5–167	7–390	6–460	6–168	5–148	7–111	7–98	−30	29
Alk. phosphatase	U/L	31–144	32–137	35–140	36–155	27–136	36–142	35–145	31–235	−44	28
Bilirubin	μmol/L	3–29	3–26	3–29	3–22	3–19	2–17	2–17	2–19	−10	10
Muscle											
Creatine kinase	U/L	16–615	12–474	27–2770	18–600	20–218	0–308	26–316	0–389	−277	236
Kidney											
Creatinine	μmol/L	62–133	71–141	71–168	62–186	53–124	53–115	53–115	53–159	−35	35
Urea	mmol/L	2.1–8.9	2.5–9.6	2.1–8.6	1.4–12.5	1.4–7.5	1.8–8.6	2.0–7.9	2.1–8.2	−3.2	3.2
Uric acid	μmol/L	172–547	167–553	178–607	220–571	95–440	143–476	143–476	143–648	−125	119
Phosphate (i)	mmol/L	0.74–1.58	0.74–1.45	0.77–1.61	0.71–1.71	0.81–1.65	0.84–1.58	0.81–1.65	0.61–1.71	−0.45	0.42
Calcium	mmol/L	2.07–2.62	2.05–2.57	2.05–2.64	2.10–2.69	2.02–2.62	2.05–2.64	2.07–2.62	1.90–2.60	−0.30	0.30
Electrolytes											
Sodium	mmol/L	135–147	134–146	136–149	135–153	135–150	134–151	135–150	128–149	−8	8
Potassium	mmol/L	3.6–5.7	3.4–5.4	3.5–5.3	2.8–5.4	3.5–5.3	3.5–5.5	3.4–6.8	2.9–4.7	−1.1	1.0
Chloride	mmol/L	97–115	96–112	96–110	97–114	98–116	95–117	99–112	90–109	−9	9
Bicarbonate	mmol/L	20–34	21–35	21–32	22–38	19–32	22–36	19–35	20–35	−7	8

Nutritional											
Glucose, fasting	mmol/L	3.6–9.9	3.8–12.1	3.6–8.4	4.1–15.1	3.7–8.0	3.8–8.2	3.9–7.8	4.4–23.3	−3.1	3.2
Glucose, nonfasting	mmol/L	2.2–9.8	2.8–11.9	4.1–8.0	4.2–14.5	0.5–11.2	4.4–13.7	0.9–7.4	4.6–18.4	−4.8	4.2
Albumin	g/L	35–52	33–50	33–51	33–51	34–51	35–50	33–53	32–52	−7	6
Protein	g/L	62–84	61–83	63–88	59–87	60–82	61–80	64–93	60–88	−11	10
Cholesterol	mmol/L	2.97–8.84	3.59–8.56	2.61–8.67	2.92–8.61	3.03–8.25	3.70–9.00	3.54–8.84	3.82–11.2	−1.78	1.73
HDL cholesterol	mmol/L	0.52–2.02	0.42–2.17	0.59–2.64	0.44–2.48	0.70–2.38	0.33–3.10	0.52–2.35	0.08–2.64	−0.85	0.72
LDL cholesterol	mmol/L	1.45–6.15	1.63–6.10	0.75–6.80	1.42–7.16	1.27–5.46	1.60–5.59	1.06–6.13	2.15–6.15	−1.97	1.78
Triglycerides	mmol/L	0.58–8.24	0.52–8.74	0.53–7.95	0.51–4.14	0.42–5.61	0.51–9.04	0.41–7.69	0.44–3.51	−3.01	2.88
Erythrocytes											
Hemoglobin (Fe)	mmol/L	8.0–11.61	7.60–11.79	6.80–11.42	6.95–11.48	7.02–10.40	7.50–10.98	6.39–10.30	5.80–11.11	−1.20	120
Erythrocytes	TI/L	4.2–6.1	3.6–6.1	3.6–6.2	3.4–6.9	3.7–5.5	4.0–5.9	3.7–5.6	3.6–6.0	−0.7	0.7
Hematocrit	1	0.39–0.58	0.37–0.56	0.33–0.56	0.34–0.57	0.34–0.50	0.35–0.53	0.33–0.52	0.31–0.53	−0.07	0.06
MCV	fL	78–109	80–111	72–109	75–107	76–105	79–107	71–102	78–104	−9	8
MCH (Fe)	fmol	1.60–2.23	1.68–2.40	1.40–2.23	1.49–2.17	1.60–2.20	1.70–2.17	1.37–2.05	1.49–2.11	−0.19	0.20
MCHC (Fe)	mmol/L	19–23	19–23	10–23	19–23	19–23	19–22	19–23	19–22	−2	2
Leukocytes											
Leukocytes	GI/L	3.9–14.3	3.6–12.9	3.0–12.9	3.2–11.5	3.6–14.6	4.4–13.3	3.3–14.4	3.1–11.4	−4.0	3.7
Bands	GI/L	0–0.77	0–0.78	0–0.44	0–0.19	0–1.15	0–1.71	0–0.72	0–0.14	−0.29	0.26
Neutrophils	GI/L	1.96–9.91	1.84–9.05	0.90–9.15	1.44–8.01	1.71–10.34	2.26–8.79	1.41–8.66	1.60–8.59	3.39	3.18
Lymphocytes	GI/L	0.89–4.80	0.05–4.15	0.93–4.66	0.75–4.65	0.97–4.47	0.90–4.79	1.01–4.88	0.78–4.79	−1.57	1.54
Monocytes	GI/L	0–0.95	0–1.13	0–0.81	0–1.00	0–0.83	0–0.93	0–0.84	0–0.66	−0.44	0.46
Eosinophils	GI/L	0–0.77	0–0/66	0–0.61	0–0.83	0–0.69	0–0.49	0–0.47	0–0.54	0.38	0.37
Basophils	GI/L	0–0.18	0–0.18	0–0.14	0–0.16	0–0.18	0–0.19	0–0.18	0–0.18	0.13	0.13
Platelets	GI/L	142–424	149–432	148–377	105–574	159–490	121–475	109–489	114–450	−93	107
Urine											
Specific gravity		1.006–1.035	1.005–1.034	1.007–1.036	1.005–1.037	1.004–1.035	1.004–1.036	1.005–1.034	1.005–1.038	−0.018	0.017
pH		5–9	5–8	5–9	5–7	5–8	5–8	5–7	5–7	−2.0	2.0

*The number of adult patients with at least one analyte result is 2974.

APPENDIX 3 Eli Lilly and Co. Adult Reference Ranges in Clinical Trials for Routine Laboratory Test Results, Smoker/Nondrinker (n = 2037)*

		Male				Female					
		Caucasian		Non-caucasian		Caucasian		Non-caucasian			
		Age		Age		Age		Age		Delta limits	
Analyte	Units	<50	≥50	<50	≥50	<50	≥50	<50	≥50	Low	High
Liver											
ALT	U/L	5–93	4–72	5–95	3–96	4–63	5–61	3–228	4–92	−27	28
AST	U/L	10–60	9–48	11–60	7–84	8–55	10–60	9–139	10–71	−25	22
GGT	U/L	8–194	6–94	8–207	6–101	4–252	7–189	5–181	5–190	−30	29
Alk. phosphatase	U/L	34–138	38–166	36–144	27–194	30–133	39–211	32–121	39–174	−44	28
Bilirubin	μmol/L	3–24	3–21	3–22	3–21	3–17	3–15	2–22	3–15	−10	10
Muscle											
Creatine kinase	U/L	18–670	15–281	30–1481	17–788	0–218	0–297	15–246	0–328	−277	236
Kidney											
Creatinine	μmol/L	53–141	62–150	53–141	71–194	53–115	53–133	53–124	62–124	−35	35
Urea	mmol/L	2.5–9.6	2.5–10.7	2.1–9.3	2.5–12.9	1.8–7.5	2.1–9.6	1.8–7.1	2.5–9.3	−12	3.2
Uric acid	μmol/L	167–512	119–541	208–571	190–630	83–470	119–547	50–494	155–529	−125	119
Phosphate (i)	mmol/L	0.68–1.58	0.68–1.49	0.81–1.58	0.61–1.58	0.77–1.58	0.80–1.52	0.77–1.49	0.81–1.52	−0.45	0.42
Calcium	mmol/L	2.10–2.67	2.02–2.64	2.05–2.70	2.10–2.72	2.05–2.60	2.10–2.64	2.05–2.64	1.67–2.64	−0.30	0.30
Electrolytes											
Sodium	mmol/L	135–148	134–146	135–148	125–149	135–150	134–153	134–151	135–149	−8	8
Potassium	mmol/L	3.3–5.3	3.1–5.2	3.0–5.3	3.1–5.2	3.4–5.4	2.8–5.7	3.2–4.9	3.1–5.3	−1.1	1.0
Chloride	mmol/L	94–108	96–112	98–109	94–111	96–113	93–112	99–112	95–107	−9	8
Bicarbonate	mmol/L	19–36	22–34	21–35	22–34	19–31	21–34	21–35	22–34	−7	8

Nutritional											
Glucose, fasting	mmol/L	3.4–8.3	4.2–15.4	2.8–18.9	3.7–25.1	3.6–9.4	3.7–14.4	4.1–10.3	3.9–20.3	−3.1	3.2
Glucose, non-fasting	mmol/L	4.2–9.0	4.3–22.2	0.7–16.5	4.0–16.7	3.0–21.6	4.5–18.8	3.8–10.7	4.8–13.1	−4.8	4.2
Albumin	g/L	34–53	32–49	36–51	34–47	34–51	33–49	35–49	33–50	7	6
Protein	g/L	61–83	61–81	63–86	63–89	61–81	59–81	62–84	62–85	−11	10
Cholesterol	mmol/L	2.87–8.77	3.54–8.56	2.74–8.74	2.43–8.90	3.21–7.60	4.03–9.39	3.28–7.32	3.70–9.31	−1.78	1.73
HDL cholesterol	mmol/L	0.26–1.78	0.36–2.02	0.52–1.76	0.41–2.80	0.47–2.22	0.54–2.30	0.54–2.30	0.67–2.69	−0.85	0.72
LDL cholesterol	mmol/L	1.81–5.72	1.94–6.71	1.22–5.99	0.70–5.38	1.89–8.04	1.97–6.15	1.16–5.48	2.25–6.83	−1.97	1.78
Triglycerides	mmol/L	0.72–12.78	0.63–9.03	0.43–3.62	0.45–11.49	0.53–6.25	0.87–15.72	0.53–3.82	0.75–4.30	−3.01	2.88
Erythrocytes											
Hemoglobin (Fe)	mmol/L	8.10–11.61	7.00–11.36	7.57–11.23	6.27–11.61	7.10–10.60	6.90–10.43	6.45–10.80	6.60–10.18	−1.20	1.20
Erythrocytes	TI/L	4.3–6.4	3.9–6.2	4.0–6.4	3.9–6.6	3.6–5.7	3.8–5.9	3.5–6.0	3.8–5.8	−0.7	0.7
Hematocrit	1	0.29–0.56	0.35–0.55	0.37–0.56	0.33–0.56	0.34–0.52	0.36–0.52	0.32–0.54	0.33–0.50	−0.07	0.06
MCV	fL	78–104	75–106	74–99	73–107	78–105	81–105	70–104	79–103	−9	8
MCH (Fe)	fmol	1.60–2.10	1.40–2.11	1.60–2.05	1.30–2.10	1.60–2.20	1.70–2.11	1.46–2.17	1.49–2.20	−0.19	0.20
MCHC (Fe)	mmol/L	19–23	19–23	18–22	18–22	19–23	19–22	18–22	18–22	2	2
Leukocytes											
Leukocytes	GI/L	4.1–14.9	3.5–13.5	3.1–15.5	2.9–13.1	4.3–14.1	3.9–13.6	3.5–14.8	3.1–12.5	−4.0	3.7
Bands	GI/L	0–1.03	0–0.71	0–0.54	0–0.38	0–0.78	0–0.58	0–0.39	0–0.20	−0.29	0.26
Neutrophils	GI/L	2.18–10.35	2.05–9.05	1.21–9.66	1.25–9.96	1.96–9.84	1.95–9.80	1.31–10.50	1.85–7.60	−3.38	3.18
Lymphocytes	GI/L	1.09–4.84	1.02–4.54	1.06–6.01	0.84–4.37	1.22–4.71	1.06–4.75	1.22–4.87	0.43–4.72	−1.57	1.54
Monocytes	GI/L	0–0.93	0–0.92	0–1.04	0–0.85	0–0.85	0–0.87	0–0.93	0–0.75	−0.44	0.46
Eosinophils	GI/L	0–0.78	0–0.71	0–0.86	0–0.60	0–0.84	0–0.59	0–0.63	0–0.58	−0.38	0.37
Basophils	GI/L	0–0.23	0–0.19	0–0.14	0–0.14	0–0.19	0–0.21	0–0.15	0–0.15	−0.13	0.13
Platelets		136–457	149–586	116–394	158–486	154–499	127–476	184–550	101–480	−93	107
Urine											
Specific gravity		1.006–1.036	1.006–1.034	1.005–1.036	1.006–1.036	1.005–1.036	1.005–1.043	1.005–1.039	1.005–1.032	−0.018	0.017
pH		5–7	5–9	5–8	5–7	5–7	5–8	5–7	5–7	−2.0	2.0

*The number of adult patients with at least one analyte result is 2037.

APPENDIX 4 Eli Lilly and Co. Adult Reference Ranges in Clinical Trials for Routine Laboratory Test Results, Nonsmoker/Imbiber (n = 4265)*

		Male				Female					
		Caucasian		Non-caucasian		Caucasian		Non-caucasian			
		Age		Age		Age		Age		Delta limits	
Analyte	Units	<50	≥50	<50	≥50	<50	≥50	<50	≥50	Low	High
Liver											
ALT	U/L	9–102	8–100	6–125	7–101	6–81	7–90	5–52	2–114	−27	28
AST	U/L	11–67	11–83	12–131	12–132	10–70	11–87	8–52	12–106	−25	22
GGT	U/L	6–160	6–200	7–207	9–152	5–117	5–201	5–121	5–113	−30	29
Alk. phosphatase	U/L	25–125	22–124	33–119	22–132	25–115	31–158	29–131	32–192	−44	28
Bilirubin	μmol/L	3–36	3–31	3–29	3–27	3–22	3–22	3–22	3–15	−10	10
Muscle											
Creatine kinase	U/L	20–563	0–427	35–1070	17–2124	25–273	11–275	34–420	10–361	−277	236
Kidney											
Creatinine	μmol/L	71–141	71–159	71–177	71–150	53–124	53–133	53–133	44–133	−35	35
Urea	mmol/L	2.5–9.6	3.2–10.4	2.5–9.3	2.1–10.4	1.8–7.5	2.5–10.4	1.8–7.9	2.9–10.4	−3.2	3.2
Uric acid	μmol/L	178–583	178–583	202–607	190–613	89–446	113–547	131–541	83–613	−125	119
Phosphate (i)	mmol/L	0.71–1.52	0.68–1.49	0.77–1.58	0.65–1.45	0.74–1.58	0.74–1.61	0.74–1.55	0.90–1.68	−0.45	0.42
Calcium	mmol/L	2.10–2.62	2.05–2.62	2.10–2.69	2.10–2.69	2.02–2.62	2.00–2.59	2.05–2.59	1.90–2.64	−0.30	0.30
Electrolytes											
Sodium	mmol/L	134–147	135–148	135–150	133–148	135–146	134–147	136–145	137–153	−8	8
Potassium	mmol/L	3.5–5.5	3.3–5.4	3.0–5.5	3.0–6.0	3.2–5.2	3.3–5.3	3.3–5.9	3.1–5.1	−1.1	1.0
Chloride	mmol/L	96–112	97–111	97–110	97–110	95–114	95–113	90–110	90–115	−9	8
Bicarbonate	mmol/L	22–35	22–35	23–36	20–44	21–35	22–36	22–33	22–35	−7	8

Nutritional											
Glucose, fasting	mmol/L	3.7–8.8	4.2–12.5	4.1–8.7	2.9–12.4	3.8–8.1	3.7–18.0	0.9–16.3	2.6–16.6	−3.1	3.2
Glucose, nonfasting	mmol/L	3.4–7.7	1.7–12.1	3.5–13.7	4.8–10.5	4.7–7.7	4.6–13.7	4.8–6.3	4.7–12.0	−4.8	4.2
Albumin	g/L	37–53	36–50	37–53	32–50	34–50	34–50	34–48	33–49	−7	6
Protein	g/L	62–83	61–82	64–86	61–83	62–82	62–83	63–88	59–82	−11	10
Cholesterol	mmol/L	2.95–8.48	3.49–8.43	3.23–8.46	3.41–8.87	3.21–7.47	3.93–9.13	3.59–7.37	3.67–9.36	−1.78	1.73
HDL cholesterol	mmol/L	0.47–2.17	0.47–2.22	0.44–3.03	0.28–2.53	0.72–2.95	0.65–2.74	0.75–2.87	0.78–2.61	−0.85	0.72
LDL cholesterol	mmol/L	1.85–5.96	1.68–6.08	1.55–6.26	1.50–6.49	1.37–5.43	1.84–6.54	1.68–4.76	1.24–7.68	−1.97	1.78
Triglycerides	mmol/L	0.47–10.21	0.53–7.81	0.65–8.99	0.54–8.79	0.29–4.44	0.53–4.90	0.38–6.42	0.52–6.51	−3.01	2.88
Erythrocytes											
Hemoglobin (Fe)	mmol/L	7.88–10.98	7.32–11.23	7.50–11.11	5.83–10.67	6.83–9.70	7.00–10.18	6.30–9.62	6.33–10.05	−1.20	1.20
Erythrocytes	TI/L	4.2–6.0	4.0–6.1	4.4–6.4	3.2–6.4	3.7–5.4	3.7–5.6	3.8–5.6	3.5–5.7	−0.7	0.7
Hematocrit	1	0.38–0.53	0.36–0.54	0.38–0.54	0.29–0.52	0.33–0.48	0.34–0.50	0.32–0.47	0.33–0.49	−0.07	−0.06
MCV	fL	77–102	77–105	74–104	69–102	78–101	78–103	65–100	71–103	9	8
MCH (Fe)	fmol	1.60–2.11	1.60–2.20	1.40–2.11	1.31–2.05	1.60–2.10	1.60–2.11	1.20–2.00	1.43–2.17	−0.19	0.20
MCHC (Fe)	mmol/L	19–23	19–23	19–23	10–22	19–23	19–23	18–22	18–24	−2	2
Leukocytes											
Leukocytes	GI/L	3.4–10.8	3.3–11.8	2.9–9.4	2.9–10.2	3.4–11.8	3.3–10.8	3.0–12.5	3.0–9.4	−4.0	3.7
Bands	GI/L	0–0.29	0–0.26	0–0.16	0–0.20	0–0.34	0–0.21	0–0.08	0–0.30	−0.29	0.26
Neutrophils	GI/L	1.72–7.67	1.63–7.79	1.08–6.12	0.75–6.28	1.67–8.38	1.39–8.36	1.09–8.54	0.87–5.45	−3.38	3.18
Lymphocytes	GI/L	0.87–3.76	0.75–4.42	0.96–4.10	0.82–4.17	0.91–3.65	0.85–3.64	0.94–3.95	0.95–3.96	−1.57	1.54
Monocytes	GI/L	0–0.79	0–0.96	0–0.75	0–0.72	0–0.76	0–0.78	0–0.78	0–1.00	−0.44	0.46
Eosinophils	GI/L	0–0.53	0–0.68	0–0.59	0–0.50	0–0.50	0–0.59	0–0.48	0–0.54	−0.38	0.37
Basophils	GI/L	0–0.15	0–0.17	0–0.15	0–0.19	0–0.17	0–0.17	0–0.20	0–0.19	−0.13	0.13
Platelets	GI/L	147–399	131–429	152–508	123–397	168–474	155–514	165–448	145–409	−93	107
Urine											
Specific gravity		1.006–1.037	1.005–1.036	1.007–1.035	1.005–1.036	1.004–1.036	1.005–1.035	1.006–1.036	1.006–1.034	−0.018	0.017
pH		5–8	5–8	5–8	5–9	5–8	5–8	5–7	5–7	−2.0	2.0

*The number of adult patients with at least one analyte result is 4265.

APPENDIX 5 Eli Lilly and Co. Adult Reference Ranges in Clinical Trials for Routine Laboratory Test Results, Nonsmoker/Nondrinker (n = 4801)*

		Male				Female					
		Caucasian		Non-caucasian		Caucasian		Non-caucasian			
		Age		Age		Age		Age		Delta limits	
Analyte	Units	<50	≥50	<50	≥50	<50	≥50	<50	≥50	Low	High
Liver											
ALT	U/L	9–97	6–79	4–109	6–68	5–94	5–77	4–89	4–82	−27	28
AST	U/L	11–67	10–55	11–88	9–60	10–64	9–62	8–64	10–65	−25	22
GGT	U/L	6–117	5–190	7–148	6–135	5–137	5–148	6–130	5–123	−30	29
Alk. phosphatase	U/L	29–136	26–162	29–149	29–146	29–136	23–156	24–138	34–171	−44	28
Bilirubin	μmol/L	5–32	3–36	3–29	3–24	3–22	3–19	2–19	2–21	−10	10
Muscle											
Creatine kinase	U/L	17–667	15–419	15–717	14–1043	10–269	0–235	13–407	11–610	−277	236
Kidney											
Creatinine	μmol/L	62–141	62–168	71–159	71–194	53–124	53–150	53–133	44–150	−35	35
Urea	mmol/L	2.5–9.3	2.9–10.7	1.8–8.6	2.5–11.4	1.8–8.6	2.5–11.1	1.8–7.5	2.9–10.7	−3.2	3.2
Uric acid	μmol/L	155–547	178–601	184–589	244–601	95–500	113–517	125–476	155–619	−125	119
Phosphate (i)	mmol/L	0.81–1.52	0.68–1.45	0.77–1.68	0.68–1.55	0.77–1.55	0.74–1.58	0.77–1.52	0.84–1.58	−0.45	0.42
Calcium	mmol/L	2.07–2.64	2.05–2.59	2.05–2.64	2.05–2.59	2.02–2.59	2.05–2.64	2.02–2.64	2.10–2.69	−0.30	0.30
Electrolytes											
Sodium	mmol/L	132–148	134–148	135–157	135–150	129–145	134–150	135–150	135–152	−8	8
Potassium	mmol/L	3.5–5.5	3.3–5.5	2.8–7.4	2.8–5.4	3.3–5.3	3.4–5.5	3.0–5.0	3–5.2	−1.1	1.0
Chloride	mmol/L	96–112	96–111	93–112	96–111	96–110	95–112	97–110	95–111	−9	8
Bicarbonate	mmol/L	22–36	22–35	21–36	21–36	20–34	21–35	19–36	21–36	−7	8

Nutritional											
Glucose, fasting	mmol/L	3.6–13.2	3.8–16.6	3.9–9.5	3.6–17.5	3.8–12.1	3.8–15.2	3.8–16.0	2.7–21.7	−3.1	3.2
Glucose, nonfasting	mmol/L	4.3–12.6	4.7–14.4	3.4–8.3	4.5–17.3	4.3–9.8	3.5–29.2	4.1–15.8	3.9–22.8	−4.8	4.2
Albumin	g/L	37–51	34–49	34–53	30–48	34–50	33–49	32–49	33–49	−7	6
Protein	g/L	63–84	60–82	61–88	59–86	61–83	62–83	64–85	63–86	−11	10
Cholesterol	mmol/L	3.08–7.81	3.44–8.51	2.79–8.24	3.21–9.05	3.18–8.33	3.70–8.82	3.36–8.51	3.80–9.31	−1.78	1.73
HDL cholesterol	mmol/L	0.42–2.07	0.55–2.12	0.28–2.43	0.36–2.16	0.49–2.22	0.54–2.53	0.47–2.25	0.52–2.61	−0.85	0.72
LDL cholesterol	mmol/L	1.48–5.77	1.84–5.48	1.42–6.30	1.32–6.43	0.91–5.28	1.63–6.23	1.66–5.82	2.20–6.57	−1.97	1.78
Triglycerides	mmol/L	0.54–11.02	0.52–8.85	0.44–16.81	0.50–4.19	0.40–9.08	0.67–7.77	0.55–5.59	0.52–4.44	−3.01	2.88
Erythrocytes											
Hemoglobin (Fe)	mmol/L	7.80–11.30	7.50–11.20	6.30–11.00	6.80–10.92	6.64–9.90	6.90–10.24	5.00–10.43	6.30–9.99	−1.20	1.20
Erythrocytes	Tl/L	4.2–6.3	4.0–6.2	4.0–6.4	3.7–6.2	3.7–5.6	3.7–5.6	3.6–5.9	3.6–5.8	−0.7	0.7
Hematocrit	l	0.37–0.54	0.36–0.54	0.33–0.54	0.33–0.54	0.33–0.48	0.34–0.51	0.27–0.51	0.33–0.49	−0.07	0.06
MCV	fL	77–99	78–103	68–100	74–103	75–101	76–101	66–100	72–100	−9	8
MCH (Fe)	fmol	1.60–2.05	1.60–2.11	1.40–2.00	1.40–2.10	1.49–2.05	1.55–2.10	1.24–2.05	1.40–2.05	−0.19	0.20
MCHC (Fe)	mmol/L	19–23	10–23	18–23	18–22	19–23	19–23	18–22	18–22	−2	2
Leukocytes											
Leukocytes	Gl/L	3.6–11.5	3.6–12.6	3.1–12.8	2.8–10.9	2.4–13.0	3.7–11.8	3.0–11.5	3.0–10.6	−4.0	3.7
Bands	Gl/L	9–0.32	0–0.38	0–0.15	0–0.28	0–0.57	0–0.40	0–0.20	0–0.20	−0.29	0.26
Neutrophils	Gl/L	1.86–7.31	1.89–8.61	1.19–9.06	1.33–7.50	1.63–8.78	1.88–7.83	1.13–8.24	1.33–7.31	−3.38	3.18
Lymphocytes	Gl/L	0.80–3.92	0.88–4.29	0.84–4.26	0.75–4.15	1.03–4.12	0.96–4.17	0.84–4.03	1.02–4.14	−1.57	1.54
Monocytes	Gl/L	0–0.85	0–0.90	0–0.78	0–0.82	0–0.80	0–0.88	0–0.83	0–0.71	−0.44	0.46
Eosinophils	Gl/L	0–0.48	0–0.64	0–0.68	0–0.54	0–0.62	0–0.59	0–0.81	0–0.55	−0.38	0.37
Basophils	Gl/L	0–0.19	0–0.16	0–0.14	0–0.13	0–0.16	0–0.16	0–0.16	0–0.14	−0.13	0.13
Platelets	Gl/L	152–240	127–483	156–476	150–443	160–530	157–486	164–541	149–559	−93	107
Urine											
Specific gravity		1.006–1.037	1.006–1.036	1.011–1.040	1.006–1.036	1.005–1.037	1.005–1.036	1.006–1.040	1.006–1.038	0.018	0.017
pH		5–8	5–8	5–7	5–8	5–8	5–8	5–8	5–8	−2.0	2.0

*The number of adult patients with at least one analyte result is 4801.

APPENDIX 6 Eli Lilly and Co. Effect of Demographic Characteristics on Reference Ranges, *p*-Values

	Gender		Race		Age		Smoking		Drinking	
Analyte	Low	High	Low	High	Low	High	Low	High	Low	High
Liver										
ALT	0.099	<0.001	<0.001	0.580	0.289	0.002	0.035	0.886	<0.001	0.003
AST	0.021	<0.001	0.254	0.011	0.777	0.824	0.165	0.010	0.053	<0.001
GGT	<0.001	<0.001	0.647	<0.001	0.122	0.098	0.525	<0.001	0.140	0.012
Alk. phosphatase	0.961	0.078	0.098	0.542	0.893	<0.001	0.006	0.308	0.141	<0.001
Bilirubin	<0.001	<0.001	0.004	0.197	0.258	0.036	0.006	<0.001	0.377	0.020
Muscle										
Creatine kinase	0.031	<0.001	0.164	<0.001	<0.001	0.004	0.375	0.744	0.003	0.741
Kidney										
Creatinine	<0.001	<0.001	0.706	<0.001	<0.001	<0.001	0.462	0.110	0.607	0.018
Urea	<0.001	0.005	0.281	0.946	<0.001	<0.001	<0.001	0.018	0.476	0.001
Uric acid	<0.001	<0.001	<0.001	<0.001	<0.001	<0.001	0.435	0.061	0.846	0.180
Phosphate (i)	0.015	0.109	0.275	0.587	0.55	<0.001	0.029	0.196	0.808	0.465
Calcium	0.179	0.360	0.129	0.004	0.268	0.135	0.502	0.815	0.999	0.040
Electrolytes										
Sodium	0.495	0.013	0.060	0.032	0.289	0.424	0.143	0.527	0.939	0.200
Potassium	0.465	0.265	<0.001	0.626	0.002	0.826	0.574	0.710	<0.001	0.178
Chloride	0.117	<0.001	0.639	0.004	0.006	0.004	0.727	<0.001	0.201	<0.001
Bicarbonate	0.304	0.623	0.380	0.103	0.265	0.573	0.089	0.006	0.755	0.980

Nutritional										
Glucose										
Fasting	0.335	0.350	0.355	<0.001	0.372	<0.001	0.399	0.019	0.636	<0.001
Nonfasting	0.194	0.578	0.610	0.031	0.046	0.002	0.975	0.850	0.068	0.017
Albumin	0.516	<0.001	0.142	0.664	<0.001	<0.001	0.372	0.402	0.030	0.013
Protein	0.944	0.124	0.025	<0.001	<0.001	0.034	0.028	0.047	0.363	0.141
Cholesterol	<0.001	0.832	0.030	0.221	<0.001	<0.001	<0.001	0.171	0.650	0.692
HDL cholesterol	0.006	<0.001	0.676	0.007	0.963	0.948	0.381	0.087	0.040	0.030
LDL cholesterol	0.670	0.881	0.071	0.009	<0.001	0.059	0.074	0.695	0.279	0.677
Triglycerides	<0.001	<0.001	0.520	0.261	<0.001	0.592	0.624	0.532	0.004	0.464
Erythrocytes										
Hemoglobin	<0.001	<0.001	<0.001	0.092	0.208	0.443	0.017	<0.001	0.002	0.035
Erythrocytes	<0.001	<0.001	0.023	<0.001	<0.001	0.187	0.719	0.021	0.550	0.037
Hematocrit	<0.001	<0.001	<0.001	0.343	0.871	0.011	0.019	<0.001	0.028	0.014
MCV	0.086	<0.001	<0.001	0.012	0.057	<0.001	0.176	<0.001	0.002	<0.001
MCH	0.003	<0.001	<0.001	0.082	0.601	0.008	0.483	<0.001	<0.001	<0.001
MCHC	0.076	0.698	<0.001	0.127	0.019	0.373	0.167	0.792	<0.001	0.200
Leukocytes										
Leukocytes	0.092	0.781	<0.001	0.095	0.929	<0.001	0.017	<0.001	0.912	0.439
Neutrophils	0.083	0.083	<0.001	0.205	0.112	0.015	<0.001	<0.001	0.605	0.491
Lymphocytes	0.002	0.162	0.938	0.012	0.014	0.681	0.275	<0.001	0.188	0.070
Monocytes	—	<0.001	—	0.009	—	0.110	—	<0.001	—	0.664
Eosinophils	—	0.162		0.783		0.079		<0.001		0.230
Basophils	—	0.391		<0.001		0.963		<0.001		0.295
Urine										
Specific gravity	<0.001	0.149	0.007	0.904	0.104	0.303	0.911	0.103	0.067	0.036
pH	—	<0.001	—	0.239	—	<0.001	—	0.011	—	0.167

APPENDIX 7 Eli Lilly and Co. Percent Difference Between Reference Values for Each Demographic Group Relative to Pooled Reference Values

Analyte	Gender (female–male)		Race (caucasian–non-caucasian)		Age (young–mature)		Smoking (nonsmoker–smoker)		Drinking (imbiber–nondrinker)	
	Low	High	Low	High	Low	High	Low	High	Low	High
Liver										
ALT	—	−27.1	20.0	—	—	13.5	−12.3	—	34.0	13.7
AST	−11.1[a]	−31.5	—	−28.2	—	—	—	17.6		34.8
GGT	−18.7	−25.0	—	−43.1	—	—	—	39.6	—	22.0
Alk. phosphatase	—	—	—	—	—	−17.6	17.9	—	—	−12.4
Bilirubin	0.0[b]	−31.0	0.0	—	—	5.2	0.0	−13.2	—	11.5
Muscle										
Creatine kinase	−23.1	−70.0	—	−73.3	92.3	26.7	—	—	33.3	—
Kidney										
Creatinine	−17.0	−23.3	—	−12.0	−0.1	−17.9	—	—	—	−5.9
Urea	−35.0	−6.7	—	—	−35.0	−27.2	−14.3	−10.7	—	−8.0
Uric acid	−58.4	−11.8	−26.5	−10.8	−26.5	−9.7	—	−3.3	—	—
Phosphate (i)	4.2	—	—	—	—	3.8	4.1	—	—	—
Calcium	—	—	—	−0.2	—	—	—	—	—	−0.9
Electrolytes										
Sodium	—	1.3	—	−1.3	0.7	—	—	—	—	—
Potassium	—	—	9.4	—	6.3	—	—	—	6.3	—
Chloride	—	1.8	—	1.8	1.0	1.8	—	1.8	—	2.7
Bicarbonate	—	—	—	—	—	—	—	−2.9	—	—

Nutritional										
Glucose										
Fasting	—	—	—	−29.6	—	−45.1	—	−21.9	—	−37.3
Nonfasting	—	—	—	−20.4	−13.2	−50.4	—	—	—	−27.6
Albumin	—	−2.0	—	—	6.1	2.0	—	—	2.9	2.0
Protein	—	—	−1.6	−4.8	1.6	0.0	−1.6	0.0	—	—
Cholesterol	5.9	—	3.3	—	−11.0	−9.1	−8.3	—	—	—
HDL cholesterol	21.0	16.0	—	−6.3	—	—	—	—	10.5	8.5
LDL cholesterol	—	—	15.4	−6.7	−21.5	—	—	—	—	—
Triglycerides	−19.6	−34.7	—	—	−21.9	—	—	—	−17.3	—
Erythrocytes										
Hemoglobin	−10.2	−9.5	9.7	—	—	—	4.6	3.4	5.3	1.0
Erythrocytes	−7.0	−9.5	0.0	−3.3	3.0	—	—	1.7	—	−1.7
Hematocrit	−8.8	−9.3	5.9	—	—	0.0	0.0	3.7	0.0	1.9
MCV	—	−1.9	8.0	1.9	—	−1.9	—	4.8	2.7	2.9
MCH	−6.6	−4.7	13.3	—	—	−2.9	—	4.1	8.3	4.6
MCHC	—	—	5.3	—	0.0	—	—	—	2.0	—
Leukocytes										
Leukocytes	—	—	−15.2	—	—	−7.9	−7.3	−16.6	—	—
Neutrophils	−5.7	—	−38.3	—	—	−6.1	−13.3	−17.9	—	—
Lymphocytes	11.1	—	—	−6.9	8.9	—	—	18.4	—	—
Monocytes	—	−9.3	—	7.4	—	—	—	10.5	—	—
Eosinophils	—	—	—	—	—	—	—	19.7	—	—
Basophils	—	—	—	17.6	—	—	—	11.8	—	—
Urine										
Specific gravity	−0.1	—	−0.1	—	—	—	—	—	—	0.0
pH	—	0.0	—	—	—	0.0	—	0.0	—	—

[a]Example: [(9 − 10)/9] * 100 = −11.1.
[b]Example: [(3 − 3)/3] * 100 = 0.0.

IV. RECENT DEVELOPMENTS

Authors have suggested that for the purposes of clinical trials, disease-specific reference intervals would be helpful. Voss et al. have constructed these reference intervals for use in clinical trials specific to schizophrenic disorders (12).

REFERENCES

1. Wilson MG, Enas GG. Reference values and analysis of routine clinical laboratory data in clinical trials. Joint Meeting of the Society of Clinical Trials and the International Society for Clinical Biostatistics, Brussels, Belgium, July 8–12, 1991.
2. Thompson WL, Brunelle RL, Enas GG, Simpson PJ. Routine laboratory tests in clinical trials. In: AE Cato, ed. Clinical Trials and Tribulations. New York: Marcel Dekker, 1988, pp. 1–54.
3. Wilson MG, Enas GG. What analyses should be performed on test results. Drug Information Association Workshop on the Integrated Safety Summary, Bethesda, MD, April 27, 1990.
4. Thompson WL, Brunelle RL, Enas GG, Simpson PJ. Routine laboratory tests in clinical trials. J Clin Research Drug Devel 1:95–119, 1987.
5. Solberg HE. Approved recommendation (1986) on the theory of reference values. Part 1. The concept of reference values. Clin Chim Acta 167:111–118, 1987.
6. PetitClerc C, Solberg HE. Approved recommendation (1987) on the theory of reference values. Part 2. Selection of individuals for the production of reference values. J Clin Chem Clin Biochem 25:639–644, 1987.
7. Solberg HE, PetitClerc C. Approved recommendation (1988) on the theory of reference values. Part 3. Preparation individuals and collection of specimens for the production of reference values. Clin Chim Acta 177: S1–S12, 1988.
8. Solberg HE, Stamm D. Approved recommendation on the theory of reference values. Part 4. Control of analytical variation in the production, transfer, and application of reference values. Eur J Clin Chem Clin Biochem 29:531–535, 1991.
9. Solberg HE. Approved recommendations (1987) on the theory of reference values. Part 5. Statistical treatment of collected reference values. Determination of reference limits. J Clin Chem Clin Biochem 25:645–656, 1987.
10. Dybkaer R, Solberg HE. Approved recommendation (1987) on the theory of reference values. Part 6. Presentation of observed values related to reference values. J Clin Chem Clin Biochem 25:657–662, 1987.
11. Sasse EA, Aziz KJ, Harris EK, Krishnamurthy S, Lee HT, Ruland A, Seamonds B. How to Define and Determine Reference Intervals in the Clinical Laboratory: Approved Guideline. Villanova, PA: National Committee for Clinical Laboratory Standards, Publication C28-A, 1995.
12. Voss S, MFPHM; Sanger T, Beasley Jr C. Haematology Reference Ranges in a Schizophrenic Population. Work in progress.

Michael G. Wilson

Meta-Analysis of Therapeutic Trials

See also *Statistical Significance; Power*

I. INTRODUCTION

How can large amounts of independent quantitative information on the same question come together in a coherent and meaningful manner? Many researchers have relied on meta-analysis to achieve such syntheses of evidence. Traditional journal review articles and textbook chapters, upon which most clinicians have traditionally relied for therapeutic guidance, generally tend to make selective and subjective appraisals of the evidence and usually do not provide a quantitative synthesis of the data (1). While it can be applied to data of any sort, meta-analysis has been commonly used to combine results from different studies to draw conclusions about treatments. *Meta-analysis* may be defined as the statistical analysis of data from multiple studies that can synthesize and summarize results and evaluate quantitatively sources of heterogeneity and bias.

The advent of this research method in the biopharmaceutical sciences has paralleled the explosion in the number of randomized trials being conducted and the increasing need to provide an unbiased quantitative synthesis in the era of "evidence-based" medicine. Meta-analysis thus far has used mostly data from completed trials, but there is an increasing momentum for using the same principles of data synthesis in the prospective design of sets of clinical trials addressing a similar question (2,3). Meta-analysis has also been applied to epidemiologic studies to derive better risk estimates and to evaluations of diagnostic tests to obtain more reliable estimates of test performance. The basic statistical principles can be expanded to apply to diverse types of data. In this entry, however, we focus primarily on the meta-analysis of randomized therapeutic trials that combines aggregated estimates of treatment effect.

We provide a review of the application of meta-analysis as a reference to aid pharmaceutical scientists, regulatory reviewers, and biostatisticians who are engaged in pharmaceutical research and development. Section II highlights the benefits and limitations of meta-analysis. Section III discusses common statistical methods for pooling discrete data from two-by-two contingency tables. Section IV covers the basic steps to perform and evaluate a meta-analysis of randomized therapeutic trials. Section V illustrates several applications on how meta-analysis has helped to determine optimal therapeutic use and to enhance the understanding of the efficacy, safety, and cost-effectiveness of different pharmaceuticals. Section VI highlights a few of the recent major methodological developments.

II. BENEFITS AND LIMITATIONS

A. Benefits

Meta-analysis offers several benefits. It may be used to address uncertainty and heterogeneity when results of studies disagree, to increase statistical power for primary outcomes and subgroups, to improve estimates of treatment effect, and to lead to new knowledge and formulate new questions (4). It is also a powerful tool for exploring sources of heterogeneity and bias in clinical research. The rapidly increasing volume of research, often with discrepant findings, has led to an increased need for meta-analysis. We identified 225 published meta-analyses of randomized controlled trials in 1993, compared with 7 in 1980. At least 500 meta-analyses of randomized trials were published in 1996 and 1997.

Pooling results from studies with inadequate sample sizes may lead to a more precise and statistically significant estimate of a treatment effect. Applying regression methods to combined (pooled) results from studies may help to explain heterogeneity of treatment effects and differences across studies; as such, enhanced understanding may be gained and research hypotheses may be generated and subsequently evaluated. A comprehensive, well-done meta-analysis may help the researcher to know the strengths and weaknesses of the studies and therefore may aid in the design and analysis of future studies, which may bring about improvements in the quality of research. Finally, when performed prospectively, a meta-analysis may allow advance planning in the collection and analysis of large-scale evidence.

B. Limitations

Several criticisms have been leveled at meta-analysis (5,6). The major limitation of retrospective meta-analysis is that it is based only on what studies are available. One offshoot of this is the "apples and oranges" phenomenon pertaining to the external validity (generalizability) of the process: Combining different studies may lead to uncertainty as to which study or to what specific combination of studies and populations the results apply. Furthermore, the studies available for pooling may vary in design, quality, outcome measures, or populations studied, calling into the question whether it is logical and useful to combine their results. To address this issue, the researcher should prepare a protocol that includes well-defined criteria and objectives for including the studies in a meta-analysis, as well as plans for subgroup analyses and regression analyses that may examine differences (heterogeneity) among studies.

Publication bias is a second limitation of using only available studies. Publication bias is induced when the decision to publish is influenced by whether the study result of an experimental treatment is positive (i.e., favorable and statistically significant) or negative (i.e., statistically unfavorable or not statistically significant). Theoretical and empirical evidence suggests that, if the treatment effects of the studies included in the meta-analysis are found to be related to the sample size or the variance of the treatment effects, this association may suggest publication bias (7), because small negative studies are more likely to be unpublished than are large ones. On the other hand, it can be argued that unpublished data have not been evaluated by peer review and therefore may be of dubious quality and should not be trusted. Although a source of concern, if there are only a few unpublished studies and each has a small sample size, their omission may be less likely to cause a substantial problem because they may then have minimal impact on the overall result. Nevertheless, there is evidence that even large efficacy trials may suffer from "publication lag" (8), a term coined to reflect the situation when negative efficacy trials appear in the literature with a time lag compared with their positive counterparts that are completed, submitted, and published more promptly. This may result in larger treatment effects in meta-analyses of the early evidence.

Publication bias may be detected by formal statistical tests and a "funnel plot," in which a measure of precision (e.g., sample size or the reciprocal of the variance) is plotted against the treatment effect (7). One analytic way to identify the impact of potential publication bias is the "fail-safe N" approach (9), which estimates the number of additional studies with null results required to reverse statistically significant findings of a meta-analysis. Although no satisfactory solution exists yet for correcting publication bias, research registries (10,11) and other reasonable means (e.g., following up on published abstracts) can be used to track down unpublished studies. Prospective trial registries is the best way to circumvent publication bias and publication lag, because, as new trials become registered before they are started, the decision to register a trial cannot be affected by the result of the study, which is not yet known.

A final criticism against meta-analysis centers on its internal validity, in that it cannot correct the qualitative flaws of the studies that it includes and thus carries along the biases inherent in each included trial. Since a careful meta-analysis identifies the quality defects of the individual trials, however, the identification of bias becomes one of the prime benefits of a systematic meta-analysis. There are several meticulous scales for assessing the quality of randomized trials in meta-analyses, such as the one developed by Chalmers et al. (12). A comprehensive bibliography of quality assessment methods has been published (13). So far, there is little evidence that the magnitude of the treatment effect corresponds to the quality of the studies. Nevertheless, by focusing attention on the quality of clinical trials, meta-analysis has probably provided a powerful stimulus for improving their conduct and their reporting (14).

III. METHODS OF ANALYSIS FOR TWO-BY-TWO TABLES

This section emphasizes the main methods of meta-analysis that have been used when a study's results come from a two-by-two contingency table, with two treatment groups and a binary outcome, a situation common in randomized therapeutic trials. A compendium of methods for combining continuous data, as well as discrete data, can be found in *The Handbook of Research Synthesis* (15), which is the most comprehensive book to date on statistical, as well as nonstatistical, matters pertaining to meta-analysis.

It is not advisable simply to lump the results of different studies together as if only one large study had been performed. Not only may the results of larger studies unduly overwhelm the results of smaller studies, but the overall results may be distorted. As shown in Table 1, the risk ratio (i.e., relative risk) for each study is the same (0.50), yet simply adding their results gives a misleading value (0.39), often referred to as *Simpson's paradox*. Correct methods of pooling would, of course, give a pooled risk ratio of 0.50.

A. Fixed-Effects and Random-Effects Models

Two types of models have been employed to adjust appropriately the potential confounding effect of study.

Table 1 Fictional Results of Two Randomized Controlled Trials

	Treatment group			Control group			
Trial	Deaths	Number	Risk	Deaths	Number	Risk	Risk ratio
A	20	100	20%	40	100	40%	20/40 = 0.50
B	50	500	10%	20	100	20%	10/20 = 0.50
Total	70	600	11.7%	60	200	30%	11.7/30 = 0.39 11.7/30 ≠ 0.50

In the fixed-effects model, which leads to inferences only about the studies assembled, the true treatment effect is assumed to be the same across all studies, and the only variability of treatment effect considered is within each study (i.e., within-study variability). In the random-effects model, which leads to inference about all studies from a hypothetical population, each study can have a different effect that is randomly drawn from a normal distribution and is positioned about a central value; in this model, not only within-study variability but also variability of treatment effects across studies (i.e., among-study variability) is considered.

The fixed-effects model weights each study by its sampling variability of treatment effect, which incorporates the number of events and sample size in each treatment group of a study. Along with weighting by this sampling variation, the random-effects model also weights each study by an overall estimate of the variability of true treatment effects across studies. Compared with the fixed-effects model, the random-effects model tends to give wider confidence intervals, which makes the statistical significance of its results more conservative than those in the fixed-effects model.

In practice, important differences in the results between these two statistical models occur infrequently. The random-effects model should be used when heterogeneity of treatment effect is present—that is, when there is evidence that there is no single effect of treatment across studies (see the next section, Sec. III.B).

For combining discrete data from two-by-two tables, fixed-effects models that have been used to combine estimates of treatment effect across studies include (but are not limited to) the following: directly weighted pooled risk differences and pooled risk ratios (16); Mantel–Haenszel procedure for pooling odds ratios (16) and pooling risk ratios (16); and Peto's method for pooling odds ratio (17). Analogous random-effects models include the DerSimonian–Laird method for pooling risk differences (18), pooling odds ratios (19), and pooling risk ratios (20), which add to the fixed-effects models an extra variance component of between-study variability of treatment effects.

Table 2 compartmentalizes these methods for pooling results from different studies. Programs, publicly available without charge, to analyze such data include Meta-Analyst (21), EasyMA (22), and Revman (23). We recommend that the user of such programs be fully familiar with the limitations of different formulas for calculating treatment effects and variances. For example, the Peto method may be biased if the allocation ratio of patients in the study arms is far from 1 (24).

If person-time were used instead of number of persons in the denominator of a rate in a study, formulas for computing the incidence rate ratio and incidence rate difference of a study—along with how to pool them with a fixed-effects model—can be found in Rothman (16). Corresponding methods of pooling for a random-effects model are the same as those for a

Table 2 Statistical Methods for Pooling Treatment Effects from Two-by-Two Contingency Tables

	Risk difference	Odds ratio	Risk ratio
Fixed-effects model	Directly weighted	Mantel–Haenszel with RBG Peto	Mantel–Haenszel with GR Directly weighted
Random-effects model	D&L	D&L	D&L

D&L: DerSimonian & Laird (Ref. 18).
RBG: Robins–Breslow–Greenland variance estimator (see Ref. 16, p. 220).
GR: Greenland–Robins variance estimator (see Ref. 16, p. 216).

fixed-effects model, but they add between-study variability of treatment effects (20). For continuous data, pooled effect sizes exist for fixed-effects models (25) and random-effect models (26).

We delineate the steps and formulas to derive the pooled risk ratio, a common measure used to combine the results of prospective trials, in Table 3 with the directly weighted fixed-effects approach (16) and in Table 4 with the DerSimonian–Laird (random-effects) approach (20). The directly weighted fixed-effects approach is essentially the DerSimonian–Laird method with no interstudy variability of treatment effect (i.e, $D = t^2 = 0$ in Table 4).

B. Assessing Homogeneity of Treatment Effects

A chi square test of homogeneity is used to test whether there is evidence that the treatment effects are different across studies (27). Step 1 of Table 4 shows a chi-square statistic to assess whether the risk ratios are different across studies. The issue is not whether differences exist, for the population treatment effects of different studies are not likely to be identical, but whether they can be reasonably ignored. Given the low power of the chi-square test of homogeneity, as an arbitrary rule of thumb, *P*-values at or below 0.1 can

Table 3 Steps in Computing an Estimate and 95% Confidence Interval of a Pooled Fixed-Effects Risk Ratio

1. Compute natural logarithm of the risk ratio for the ith study ($i = 1, \ldots, n$ total studies):

$$\ln(\mathrm{RR}_i) = \ln\left(\frac{\mathrm{TR}_i}{\mathrm{CR}_i}\right)$$

2. Compute its standard error:

$$\mathrm{se}_i = \left[\frac{b_i}{a_i(a_i + b_i)} + \frac{d_i}{c_i(c_i + d_i)}\right]^{1/2}$$

3. Compute the fixed-effects weight for each study:

$$w_i = \frac{1}{(\mathrm{se}_i)^2}$$

4. Compute the pooled natural logarithm of the fixed-effects risk ratio:

$$\ln(\mathrm{RRf}) = \frac{\sum w_i \ln(\mathrm{RR}_i)}{\sum w_i}$$

5. Compute its antilogarithm to obtain the pooled fixed-effects risk ratio:

$$\mathrm{RRf} = e^{\ln(\mathrm{RRf})}$$

6. Compute the corresponding 95% confidence interval:

$$e^{\ln(\mathrm{RRf}) \pm 1.96/\sqrt{\Sigma w_i}}$$

a_i = number of treated patients with the event of interest in the ith study ($i = 1, \ldots, n$ studies)
b_i = number of treated patients without the event of interest in the ith study
c_i = number of control patients with the event of interest in the ith study
d_i = number of control patients without the event of interest in the ith study
$\mathrm{TR}_i = a_i/(a_i + b_i)$ = treatment rate in the ith study
$\mathrm{CR}_i = c_i/(c_i + d_i)$ = control rate in the ith study

Table 4 Steps in Calculating an Estimate and 95% Confidence Interval of a Pooled Random-Effects Risk Ratio

1. Calculate the test of homogeneity statistic for the underlying risk ratios:

$$Q = \sum w_i[\ln(\mathrm{RR}_i) - \ln(\mathrm{RRf})]^2$$

Note: The null hypothesis is that the n underlying risk ratios are equal. For hypothesis testing, this test statistic (Q) is compared with a given percentile (for example, 90th) of a chi square distribution with $n - 1$ degrees of freedom.

2. Calculate the interstudy variability of treatment effects:

$$D = \frac{Q - (n - 1)}{\sum w_i - \dfrac{\sum w_i^2}{\sum w_i}}$$

$$t^2 = \begin{cases} D & \text{if } D > 0 \\ 0 & \text{if } D \leq 0 \end{cases}$$

3. Calculate the random-effects weight for each study:

$$w_i^* = \left[t^2 + \left(\frac{1}{w_i}\right)\right]^{-1}$$

4. Calculate the pooled natural logarithm of the random-effects risk ratio:

$$\ln(\mathrm{RR}^*) = \frac{\sum w_i^* \ln(\mathrm{RR}_i)}{\sum w_i^*}$$

5. Calculate its antilogarithm to obtain the pooled random-effects risk ratio:

$$\mathrm{RR}^* = e^{\ln(\mathrm{RR}^*)}$$

6. Calculate the corresponding 95% confidence interval:

$$e^{\ln(\mathrm{RR}^*) \pm 1.96/\sqrt{\Sigma w_i^*}}$$

Calculation and definition of terms not appearing here appear in Table 3.

usually be taken to mean that the differences should not be ignored. The Galbraith plot (28), a plot of the z-score of treatment effect versus the square root of the weight for each trial, is a convenient graphical technique to examine heterogeneity of effects. The L'Abbe plot (29), a plot of the treatment rate against the control rate, is another graphical technique to explore such heterogeneity.

C. Meta-Regression Analysis

Meta-regression analysis (30,31) is being increasingly used as part of a meta-analysis to describe how a treatment effect may vary with patient and study characteristics. Meta-regressions have been applied to several clinical areas (examples can be found in Refs. 32–36). In a meta-regression, the unit of analysis is the individual study in the meta-analysis. Aggregate quantities reported in the studies, such as the mean dosage, mean age, mean duration of disease, could be extracted and used as independent variables. The treatment effect, for example, the risk difference, can be the dependent variable. Taking the natural logarithm of a treatment effect measured on a multiplicative scale, such as the risk ratio and odds ratio, is suggested to help ensure that the dependent variable becomes normally distributed in order to make valid statistical inferences. It is more desirable to weight each study in the meta-regression proportional to its precision (e.g., the inverse for the variance of a study's treatment effect) rather than weighting the studies equally. When output from standard statistical packages, like the Statistical Analysis System (SAS), is used to obtain the standard error of a regression coefficient, this standard error should be adjusted by dividing it by the square root of the mean squared error (30). While potentially informative, the results of meta-regression based on covariates with aggregate numbers (e.g., group averages) do not necessarily prove causality and should be interpreted as cautiously as other ecological investigations when making inferences to individual patients (37,38).

IV. BASIC STEPS IN PERFORMING AND EVALUATING A META-ANALYSIS

A properly conducted meta-analysis can be most demanding and requires the utmost advanced planning and care in its conduct. Figure 1 depicts the basic steps to perform a retrospective meta-analysis.

Quantitative methods give an aura of credence to any analysis. A meta-analysis, similar to any clinical

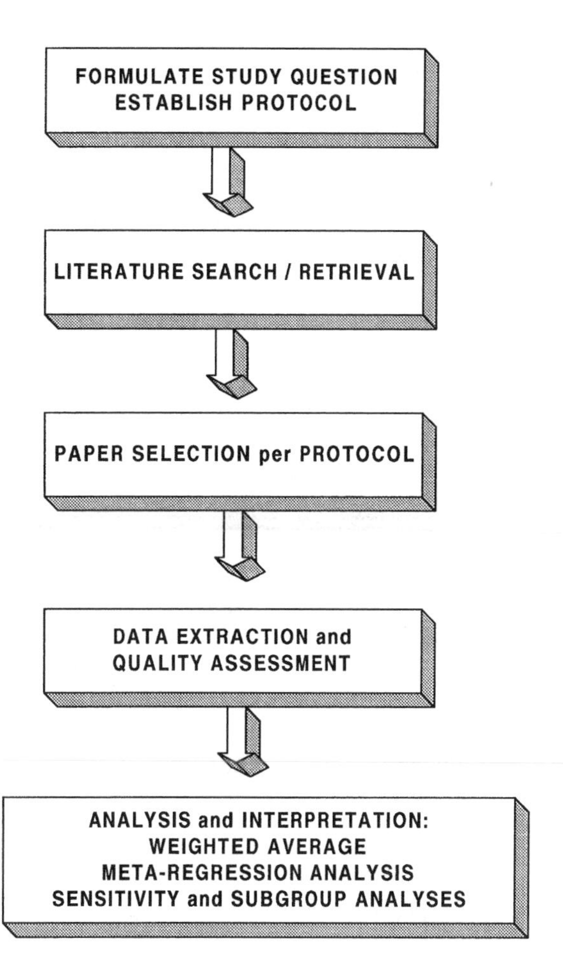

Fig. 1 Basic steps in performing a meta-analysis.

study, must be read carefully, recognizing its strengths and limitations. Guidelines are available (39–41) on how to assess the quality of a meta-analysis. In general, the following questions may be posed.

A. *Objectives:* Were the objectives clearly stated, clinically useful, and well-formulated? Was the design of the meta-analysis addressing a well-specified population and clinical question?
B. *Identification and selection of studies:* How were the studies selected? Were the inclusion criteria explicit? Which studies were excluded? Why? Were unpublished papers sought and their impact examined? How many sources of data were screened (e.g., electronic databases, meeting presentations, communications with experts and pharmaceutical sponsors)?
C. *Information about the studies:* Was information given on the specific study design, patient and disease characteristics, each treatment being compared, and study duration? What precau-

tions were taken to avoid bias? What were the results of the individual studies? Were the clinical outcomes of the studies well defined and useful? What criteria were chosen to appraise the quality of the studies?

D. *Methods of analysis:* What method was employed to combine the results? Was the method technically correct? Were confidence intervals, as well as *P*-values, given? Was a fixed-effects model or a random-effects model used? Was there justification to combine the trials? Was a test of homogeneity of treatment effect performed? Were subgroups analyses defined a priori? Was sensitivity analysis used? Were meta-regressions performed to explore whether treatment effect was influenced by potential modifier variables? Was it considered whether the baseline disease risk of the patients affected the magnitude of the treatment effect?

E. *Results and conclusions:* Were the results interpreted cautiously? Were the conclusions justified by the data? What was the generalizability (external validity) of the included studies and the meta-analysis? Were limitations of the meta-analysis mentioned?

V. APPLICATIONS

Numerous applications appear in the literature on how the results of meta-analyses of pharmaceuticals have helped to determine optimal therapeutic use and to enhance the understanding of pharmaceuticals with respect to their efficacy, safety, and cost-effectiveness. We highlight some of these applications, classifying them by a methodological framework.

A. Cumulative Meta-Analysis vs. Expert Opinion

Cumulative meta-analysis is a method of updating previous meta-analyses with the appearance of new studies (42). The method allows the monitoring of developing trends of therapeutic efficacy. When performed routinely, the earliest time when statistical significance is reached (by whatever criterion is chosen) can be identified.

The first placebo-controlled randomized study of a thrombolytic drug in the treatment of acute myocardial infarction was reported in 1959. A cumulative meta-analysis on this treatment (Fig. 2) could have found that in 1973, after eight studies that included 2432 patients, a statistically significant reduction in overall mortality of about 25% had been achieved with thrombolytic drugs relative to placebo (Mantel–Haenszel odds ratio = 0.74; 95% confidence interval, 0.59–0.92) (43). Yet 62 more placebo-controlled trials were reported subsequently, involving an additional 45,000 patients. The Food and Drug Administration (FDA) did not approve a thrombolytic drug for this indication until 1988, 15 years after cumulative meta-analysis indicated that the treatment could have been found effective in saving lives. Compared with the results of the cumulative meta-analysis, expert opinions lagged substantially in recognizing the efficacy of the treatment and did not begin to recommend it until around the time of FDA approval.

B. Meta-Analysis in Cost-Effectiveness Analysis

Pooled estimates from a meta-analysis could serve as input into a decision analysis or a cost-effectiveness analysis. For example, the cost-effectiveness of enoxaparin, a low-molecular-weight heparin derivative, was compared with that of low-dose warfarin in the prevention of deep-vein thrombosis (DVT) after total hip replacement (44). To determine the proportion of DVT with enoxaparin and warfarin prophylaxis, the authors pooled the proportions from the available studies to arrive at the efficacy of the drugs for input into determining that the incremental cost-effectiveness of enoxaparin relative to warfarin was estimated as $29,120 per life-year gained (in Canadian dollars).

C. Meta-Analysis for Optimal Dosing

A meta-analysis of 35 randomized placebo-controlled trials was undertaken to study the efficacy and safety of different aspirin dosages for patients at high risk of vascular diseases (34). Dose–response meta-regressions indicated no evidence that lower doses of aspirin were more efficacious or less efficacious in preventing overall mortality and vascular events. There was evidence, however, that lower doses of aspirin significantly reduced gastrointestinal symptoms. For a daily dosage increase of 100 milligrams of aspirin, the (unadjusted) odds ratio of gastrointestinal symptoms increased by 4.5% on average (99% confidence interval, 1.83% to 7.25%).

Aminoglycosides are one of the most commonly used antibiotics for serious infections. There are several agents in this class with very similar mechanism of action and toxicity profile. The drugs have traditionally

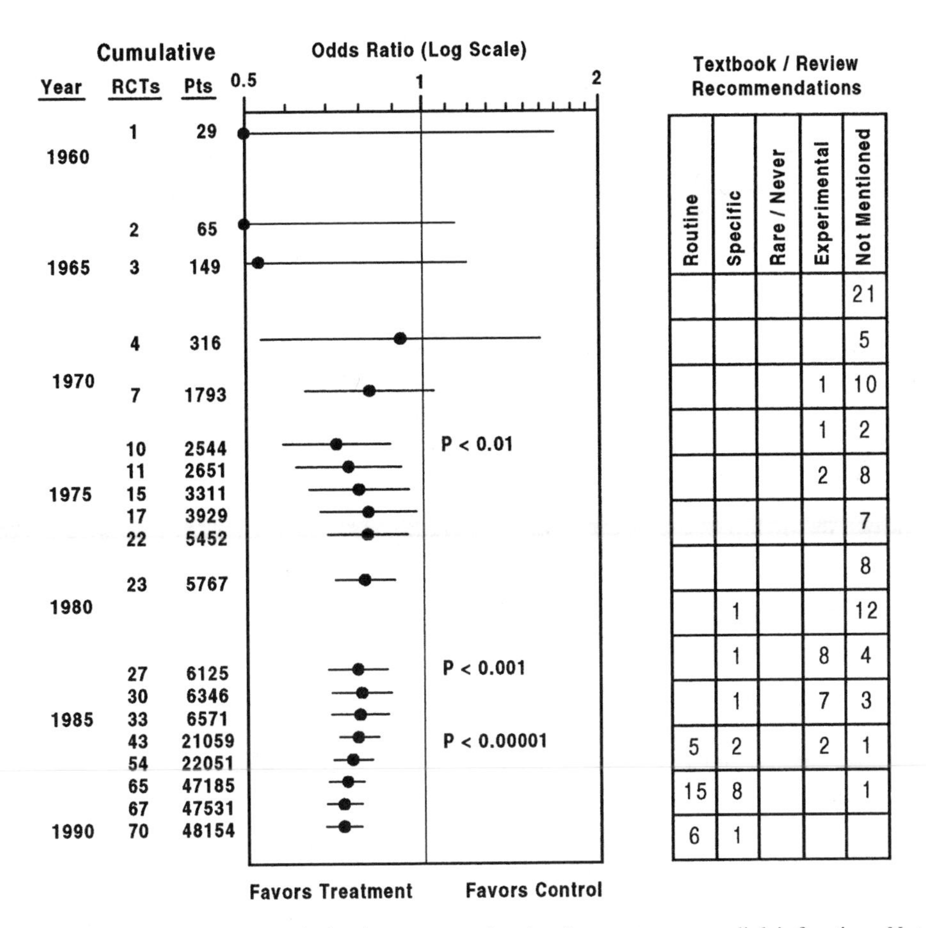

Fig. 2 Cumulative meta-analysis of thrombolytic therapy vs. placebo for acute myocardial infarction. *Note:* The cumulative meta-analyses by year of publication of randomized controlled trials (RCTs) are presented on the left. The cumulative number of trials and patients (Pts) are also presented. On the right, the recommendations of the clinical expert reviewers are presented in 2-year segments, except for the entry in 1966, which represents all previous years. (*Source:* Ref. 43. Copyright 1992, American Medical Association.)

been administered with multiple doses a day, although experimental evidence would suggest that giving the whole daily dose in a single daily administration should be as efficacious and possibly less toxic. Twenty-one trials were used to compare single versus multiple daily doses of aminoglycosides. A meta-analysis of these trials convincingly showed that single daily doses were about 25% less nephrotoxic than and at least as efficacious as multiple daily doses (35). These findings are expected to result in better outcomes, substantial cost savings, and increased convenience.

Many randomized controlled trials addressed the efficacy of many different regimens in the prophylaxis of *Pneumocystis carinii* infection among patients with advanced human immunodeficiency virus infection. A meta-analysis of 35 randomized trials showed the marked superiority of trimethorpim-sulfamethoxazole regimens over aerosolized pentamidine and dapsone (36). The meta-analysis predicted a 43% reduction in the odds of discontinuations of prophylaxis due to side effects when trimethoprim-sulfamethoxazole was used at a lower dose (one double-strength tablet three times a week) instead of the regular dose (one double-strength tablet daily), with no substantial loss of efficacy.

D. Meta-Analysis for Bioequivalence

Under current FDA regulation, a patient may switch from the brand-name drug to a generic drug if they are shown to be bioequivalent. Bioequivalence between generic copies of the brand-name drug, however, is not required by the FDA. Chow and Liu (45) proposed a methodology to provide an overview on the assessment of (average) bioequivalence among approved generic copies of an expired brand-name drug. Their methodology was intended to help guide the FDA and others in evaluating the bioequivalence among the approved generic copies. More work is needed in this area.

In general, meta-analysis has been gaining recognition by the FDA. Although it is frequently used in issuing warnings and treatment guidelines, large and randomized clinical trials continue to be relied on and are necessary for approval, according to FDA spokesman Don McLearn (S.P. Hoffert, personal communication, 1997).

E. Drug Planning and Development

Perhaps the most fruitful applications of meta-analysis in the years to come may stem from its adoption in the prospective planning of drug and therapeutic strategy development. Current drug development typically is trying to meet the regulatory mandates for showing treatment efficacy and safety, and is also largely dictated by piece meal planning and clinical opportunities for the drug. Prospective meta-analysis offers a framework for building all the trials in the development of a drug within a single matrix, with explicit quantitative plans for the analysis and interpretation of the complete accumulated evidence. This approach will largely dissolve the threat of publication bias and will give full transparency and credibility to the results of developmental trials seen in their totality. On an even larger scale, prospective meta-analysis matrices may incorporate different drugs for the same disease or indications. Such prospective meta-analyses are one of the most appealing challenges of current therapeutics.

VI. SOME RECENT DEVELOPMENTS

There has been a multitude of recent developments in meta-analysis, and the field has been a fertile area for statistical and epidemiological research. We highlight three major developments here.

A. Control Rate

For a dichotomous outcome, the control rate is the proportion of patients in the control group with the event of interest. The control rate in a clinical trial has been gaining acceptance as an available summary measure of patient or study differences that may help to explain the heterogeneity of treatment effects across studies (46–48). The control rate is a summary measure that is always reported. Variation in control rates across studies may reflect different patient populations, underlying baseline risk of patients, length of study follow-up, and treatment delivery. As a readily available, simple measure to explore treatment effect heterogeneity, the control rate may help delineate when one needs further research on risk factors.

The size of the treatment effects of studies in a meta-analysis may tend to vary with the size of the control rates in these studies. Several methodological advancements have been made in this area (49–53). For example, in a recent empirical investigation that analyzed 112 meta-analyses, Schmid et al. (53) used a hierarchical model to find that the control rate was significantly associated with the risk difference in 35 meta-analyses (31%), with the risk ratio in 15 instances (13%), and with the odds ratio in 16 (14%). The impact of differing patient risks on the results of clinical trials, along with the advantages and limitations of using the control rate as a measure of baseline risk, have been discussed (54,55).

B. Large Trials and Meta-Analyses

The extent of concordance between the results of large trials and the results of meta-analyses of smaller trials on the same topic has stirred considerable debate. Three different groups of investigators evaluated and compared these results. Villar et al. (56) found moderate agreement, beyond chance, between the results of large trials and those of smaller trials. Cappelleri et al. (57) concluded that the results of the two methods usually agree, but discrepancies do occur; however, clinically important discrepancies without an obvious explanation were unusual. LeLorier et al. (58) concluded that up to 35% of the time a meta-analysis failed to predict accurately the results of a subsequent, large trial.

In their evaluation of the three protocols, Ioannidis et al. (59) concluded that a comparison of large trials with meta-analyses may reach different conclusions, depending on the design used, the studies selected, and the method of analysis employed. When a robust approach was used that accounts for both the magnitude and the uncertainty of treatment effect, the estimate of disagreements was fairly similar in the three protocols (10–23%). Future clinical trials may benefit in addressing sources of heterogeneity of treatment effect between large trials themselves and even within a large trial. Similarly, future meta-analyses may benefit in addressing sources of heterogeneity rather than trying always to fit a common estimate among diverse studies.

C. Meta-Analyses of Individual Patient Data

Most meta-analysis in the past have been conducted using summary data on treatments and subgroups from clinical trials. Meta-analyses of individual patient data (MIPD) offer a distinct advantage of performing more detailed analyses using all the data on each individual patient participating in a clinical trial to be included in meta-analysis. This may allow, with the use of regression analyses, for detailed time-to-event analysis, the development of more accurate predictive models for the risk of the studied outcome, and the quantification of different treatment effects for different subgroups. Furthermore, the process of performing MIPD may also improve the quality of trials by revealing additional issues that cannot be gleaned from published reports. Stewart and Clarke (60) provided a description of the organizational steps required for performing a MIPD.

Overall, the results of MIPD against meta-analyses on group data should give similar results when the same data are used. Discrepancies may arise when different data are used by the MIPD and the meta-analysis of group data (such as updated information beyond the main trial follow-up in the MIPD). Disadvantages of MIPD include the fact that they require more effort and coordination on an international level, and the potential for retrieval bias when individual patient data can be retrieved only for a selected group of the pertinent trials.

REFERENCES

1. C Mulrow. Ann Intern Med 106:485–488, 1987.
2. A Laupacis, SJ Connolly, M Gent, RS Roberts, J Cairns, C Joyner. Ann Intern Med 115:818–822, 1991.
3. WG Henderson, T Moritz, S Goldman, J Copeland, G Sethi. Controlled Clin Trials 16:331–341, 1995.
4. HS Sacks, J Berrier, D Reitman, VA Ancona-Berk, TC Chalmers. N Engl J Med 316:450–455, 1987.
5. SB Thacker. JAMA 259:1685–1689, 1988.
6. SN Goodman. Ann Intern Med 114:244–246, 1991.
7. CB Begg. In: H Cooper, LV Hedges, eds. The Handbook of Research Synthesis. New York: Russell Sage Foundation, 1994, pp 399–409.
8. JPA Ioannidis. JAMA 279:281–286, 1998.
9. R Rosenthal. Psychol Bull 86:638–641, 1979.
10. PJ Easterbrook. Stat Med 11:345–359, 1992.
11. K Dickersin. In: H Cooper, LV Hedges, eds. The Handbook of Research Synthesis. New York: Russell Sage Foundation, 1994, pp 71–83.
12. TC Chalmers, H Smith, B Blackburn, B Silverman, B Schroeder, D Reitman, A Ambroz. Controlled Clin Trials 2:31–49, 1981.
13. D Moher, AR Jadad, G Nichol, M Penman, P Tugwell, S Walsh. Controlled Clin Trials 16:62–73, 1995.
14. C Begg, M Cho, S Eastwood, R Horton, D Moher, I Olkin, R Pitkin, D Rennie, KF Schulz, D Simel, DF Stroup. JAMA 276:637–639, 1996.
15. H Cooper, LV Hedges, eds. The Handbook of Research Synthesis. New York: Russell Sage Foundation, 1994.
16. KJ Rothman. Modern Epidemiology. Boston: Little, Brown, 1986, pp 177–236.
17. S Yusuf, R Peto, J Lewis, R Collins, P Sleight. Prog Cardiovasc Med 27:335–371, 1985.
18. R DerSimonian, N Laird. Controlled Clin Trials 7:177–188, 1986.
19. JL Fleiss, AJ Gross. J Clin Epidemiol 44:127–139, 1991.
20. JPA Ioannidis, JC Cappelleri, J Lau, PR Skolnik, B Melville, TC Chalmers, HS Sacks. Ann Intern Med 122:856–866, 1995.
21. Lau J. Meta-Analyst. Version 0.991. Boston, MA: New England Medical Center, 1997.
22. M Cucherat, J-P Boissel, A Leizorovicz, MC Haugh. Comp Methods Programs Biomed 53:187–190, 1997.
23. CD Mulrow, AD Oxman. Analysing and Presenting Results. Cochrane Collaboration Handbook; Section 8. In: The Cochrane Library [database on disk and CDROM]. Oxford, UK: Update Software, 1997.
24. S Greenland, A Salvan. Stat Med 9:247–252, 1990.
25. LV Hedges. In: H Cooper, LV Hedges, eds. The Handbook of Research Synthesis. New York: Russell Sage Foundation, 1994, pp 285–299.
26. SW Raudenbush. In: H Cooper, LV Hedges, eds. The Handbook of Research Synthesis. New York: Russell Sage Foundation, 1994, pp 301–321.
27. JL Fleiss. Statistical Methods for Rates and Proportion. 2nd ed. New York: Wiley, 1982, pp 161–165.
28. RF Galbraith. Stat Med 11:141–158, 1992.
29. KAL'Abbe, AS Detsky, K O'Rourke. Ann Intern Med 107:224–233,1987.
30. S Greenland. Epidemiol Rev 9:1–30, 1987.
31. JA Berlin, EM Antman. Online J Curr Clin Trials, Doc No 134, 1994, June 4.
32. BL Kasiske, RSN Kalil, JZ Ma, M Liao, WF Keane. Ann Intern Med 118:129–138, 1993.
33. SG Thompson. Statist Meth Med Res 2:173–192, 1993.
34. JC Cappelleri, J Lau, B Kupelnick, TC Chalmers. Online J Curr Clin Trials, Doc No 174, 1995, March 14.
35. M Barza, JPA Ioannidis, JC Cappelleri, J Lau. BMJ 312:338–345, 1996.
36. JPA Ioannidis, JC Cappelleri, PR Skolnick, J Lau, HS Sacks. Arch Intern Med 156:177–188, 1996.
37. LI Langbein, AJ Lichtman. Ecological Inference. Newbury Park, CA: Sage Publications, 1978.
38. H Morgenstern. Am J Public Health 72:1336–1344, 1982.
39. JH Abramson. Making Sense of Data. 2nd ed. New York: Oxford University Press, 1994, pp 318–389.

40. CD Mulrow, AD Oxman, eds. Cochrane Collaboration Handbook [updated September 1997]. In: The Cochrane Library [database on disk and CDROM]. Oxford, UK: The Cochrane Collaboration, 1997.
41. NL Geller, M Proschan. J Biopharm Stat 6:377–394, 1996.
42. J Lau, EM Antman, J Jimenez-Silva, B Kupelnick, F Mosteller, TC Chalmers. N Engl J Med 327:248–254, 1992.
43. EM Antman, J Lau, B Kupelnick, F Mosteller, TC Chalmers. JAMA 268:240–248, 1992.
44. BJ O'Brien, DR Anderson, R Goeree. Can Med Assoc J 150:1083–1090, 1994.
45. SC Chow, JP Liu. J Biopharm Stat 7:97–111, 1997.
46. GD Smith, F Song, RA Sheldon. BMJ 306:1367–1373,1993.
47. JP Boissel, JP Collet, M Lievre, P Girard. J Cardiovasc Pharmacol 22:356–363, 1993.
48. M Rotwell. Lancet 345:1616–1619, 1995.
49. SJ Sharp, SG Thompson, DG Altman. BMJ 313:735–738.
50. M McIntosh. Stat Med 15:1713–1728, 1996.
51. SG Thompson, TC Smith, SJ Sharp. Stat Med 16:2741–2758, 1997.
52. SD Walter. Stat Med 16:2883–2900,1997.
53. CH Schmid, J Lau, M McIntosh, JC Cappelleri. Stat Med 17:1923–1942, 1998.
54. J Lau, JPA Ioannidis, CH Schmid. Lancet 351:123–127, 1997.
55. JPA Ioannidis, Lau J. J Clin Epidemiol 50:1089–1098, 1997.
56. J Villar, G Carroli, JM Belizan. Lancet 345:772–776, 1995.
57. JC Cappelleri, JPA Ioannidis, CH Schmid, SD de Ferranti, M Aubert, TC Chalmers, J Lau. JAMA 276: 1332–1338, 1996.
58. J LeLorier, G Gregoire, A Benhaddad, J Lapierre, F Derderian. N Engl J Med 337:536–542, 1997.
59. JPA Ioannidis, JC Cappelleri, J Lau. JAMA 279:1089–1093, 1998.
60. LA Stewart, MJ Clarke. Stat Med 14:2057–2079, 1995.

Joseph C. Cappelleri
John P. A. Ioannidis
Joseph Lau

Mixed Effects Models

See also *Clinical Trials*

I. INTRODUCTION

Mixed effect models have been the subject of active theoretical research for over two decades. The increasing availability of software for performing the intensive calculations required to implement mixed effect model methods has made it practical to implement these methods in a wide variety of applications. One area of application for which mixed effect models have been particularly well suited is that of repeated measurement trials, in which a subject receives a treatment or a sequence of treatments repeatedly over a period of time. Typical examples of repeated measurement trials include longitudinal trials of the effect of long-term administration of a treatment and pharmacokinetic trials in which sequences of blood samples are obtained to study the time course of concentration of a drug in the blood following one or more applications of the drug. Another area of application for mixed effect models is in accounting for excess variation attributable to intraclass correlation arising from clustering. Typical examples include members of a family, students within a class and classes within a school, patients within a center contained in a multicenter trial, etc. Mixed effect models express an observed response explicitly as a function of fixed effects such as administered treatment, gender, and age, and of random effects such as measurement error, biological variation within a subject, or population variation among subjects. Mixed effect models may be linear, when the response is expressed as the sum of fixed and random effects, or nonlinear, when the response is a nonlinear function of the fixed and random effects. The application of mixed effect model methods requires assumptions about the distributions of the random effects. The usual assumption is that the effects are normally distributed, although other distributional assumptions may be made depending on the application. This entry briefly outlines linear and nonlinear mixed effect models. More comprehensive treatments of the issues may be found in recent books by Bryk and Raudenbush (1), Khuri, Mathew and Sinha (2), Rao and Kleffe (3), Searle (4), Searle, Casella, and McCullogh (5), which deal primarily with linear mixed effect models; Gallant (6), Carroll, Rupert and Stefanski (7), Davidian and Giltinan (8) and Vonesh and Chinchilli (9), which address nonlinear models; and Andersen (10) and McCullagh and Nelder (11), which address mixed effect model methods appropriate when the data are discrete, such as categories, rather than continuous. Random effects can be introduced into survival type models to account for missing covariates and clustering (12–29).

II. LINEAR MODELS

A. Background

Conventional general fixed effect linear models can be written as

$$\mathbf{y} = \mathbf{X}\boldsymbol{\beta} + \boldsymbol{\varepsilon} \tag{1}$$

where $\mathbf{y}$ denotes a vector of observations, $\boldsymbol{\beta}$ denotes a vector of unknown constants (parameters), $\mathbf{X}$ denotes a matrix of (known) covariates or indicator variables, and $\boldsymbol{\varepsilon}$ denotes a vector of unknown residual errors. The elements of $\boldsymbol{\varepsilon}$ ordinarily are assumed to be independently distributed with zero mean and constant variance σ^2. Circumstances in which a more complex covariance structure for $\boldsymbol{\varepsilon}$ would be appropriate usually can be approached effectively using mixed model methods as described below. Methods for estimating the elements of $\boldsymbol{\beta}$ under a variety of assumptions about the distribution of the elements of $\boldsymbol{\varepsilon}$ are well known (4). Nonlinear fixed effect models usually replace the $\mathbf{X}\boldsymbol{\beta}$ term in Eq. (1) with a nonlinear function $f(\mathbf{X}, \boldsymbol{\beta})$ of parameters and covariates, although they also can be formulated in other ways such as through link functions (11).

In applications such as longitudinal studies when the aim is to study the evolution of response over time, the observation vector $\mathbf{y}$ may consist of sequences of values, each sequence corresponding to a different subject or patient. At least two sources of variability arise in such cases. One source of variability corresponds to variation among a subject's sequence of observations beyond what the fixed effects would account for. Another source of variability is the variation between the observations on different subjects, which often will be greater than the variability among the observations made on the same subject. The idea that the observations could be functions of various kinds of random effects as well as fixed effects underlies mixed effect models.

Random effects can be incorporated into linear models by generalizing model (1) slightly, to express a vector of observations as

$$\mathbf{y} = \mathbf{X}\boldsymbol{\beta} + \mathbf{Z}\boldsymbol{\delta} + \boldsymbol{\varepsilon} \tag{2}$$

The essential difference is the addition of the term $\mathbf{Z}\boldsymbol{\delta}$, where $\mathbf{Z}$ is a matrix of known covariates and $\boldsymbol{\delta}$ denotes a vector of random effects, analogous to $\boldsymbol{\varepsilon}$. Expression (2) describes a particularly simple way to incorporate random effects. Random effects actually can be incorporated into nonlinear models in a number of ways, outlined in Sec. III. Ordinarily, the elements of $\boldsymbol{\varepsilon}$ are assumed independently distributed with zero mean and constant variance σ^2, although more complex covariance structures (e.g., autoregressive) could be assumed. The elements of $\boldsymbol{\delta}$ also are assumed to have zero mean and to be uncorrelated with the elements of $\boldsymbol{\varepsilon}$, but may have a complex covariance structure, with covariance matrix $\mathbf{D}$. More usually, especially in practice, $\mathbf{Z}\boldsymbol{\delta}$ is expressed as

$$\mathbf{Z}\boldsymbol{\delta} = \sum_{i=1}^{q} \mathbf{Z}_i\boldsymbol{\delta}_i$$

where $\boldsymbol{\delta}$ has m_i elements and covariance matrix $\sigma_i^2\mathbf{I}_{m_i}$ where $\mathbf{I}_{m_i}$ denotes an $m_i \times m_i$ identity matrix, and the elements $\boldsymbol{\delta}_i$ and $\boldsymbol{\delta}_j$ are uncorrelated if $i \neq j$. Then the covariance matrix $\mathbf{D}$ is block-diagonal with the matrices $\sigma_i^2\mathbf{I}_{m_i}$, $i = 1, \ldots, q$ on its diagonal and the covariance matrix of the elements of the observation vector $\mathbf{y}$ can be written as

$$\mathbf{V} = \mathbf{V}(\mathbf{y}) = \sum_{i=1}^{q} \sigma_i^2\mathbf{Z}_i\mathbf{Z}_i^T + \sigma^2\mathbf{I}_n. \tag{3}$$

B. Estimation

The objective of analyses of mixed effect models is inference about the fixed effects $\boldsymbol{\beta}$ and estimation of σ^2 and the convariance matrix $\mathbf{D}$ (usually the variance components σ_i^2, $i = 1, \ldots, q$). This commonly is accomplished by maximizing the likelihood of the data under a set of distributional assumptions about the random effects such as multivariate normality. The computations generally are not explicit, but instead entail successively more accurate iterative approximations to solutions of the likelihood equations or estimating equations (30–32). The iterative nature of the computations means that a number of numerical analysis issues need to be considered such as convergence (at all, global vs local), dependence on starting values, potential singularity of $\mathbf{V}$, algorithm used for calculations, effect of boundaries (e.g., variance components must be positive), and computing cost. (See, e.g., Searle, Casella, and McCullogh (5), Chapter 8.)

1. Maximum Likelihood Estimation

Estimation can be carried out under the assumption of multivariate normality of the observations by conventional maximum likelihood (ML) or by restricted maximum likelihood (REML). Conventional maximum likelihood estimates the fixed effect parameters $\boldsymbol{\beta}$ and the variance components σ_i^2 $(i = 1, \ldots, q)$ as those values that maximize the logarithm of the multivariate normal likelihood

$$L = L(\boldsymbol{\beta}, \mathbf{V}|\mathbf{y}) = -\frac{1}{2}\log|\mathbf{V}| - \frac{1}{2}(\mathbf{y} - \mathbf{X}\boldsymbol{\beta})^T\mathbf{V}^{-1}(\mathbf{y} - \mathbf{X}\boldsymbol{\beta}). \tag{4}$$

These usually are taken as the values for which the first derivatives of L equal zero; it is necessary to verify that these values really do maximize expression (4) and that the estimates of the variance components are positive. The derivative of L with respect to $\boldsymbol{\beta}$ is

$$L_{\boldsymbol{\beta}} = \mathbf{X}^T\mathbf{V}^{-1}\mathbf{y} - \mathbf{X}^T\mathbf{V}^{-1}\mathbf{X}\boldsymbol{\beta} \tag{5}$$

and the derivative with respect to σ_i^2, $i = 0, 1, \ldots, q$, is

$$L_{\sigma_i^2} = \frac{1}{2}((\mathbf{y} - \mathbf{X}\boldsymbol{\beta})^T\mathbf{V}^{-1}\mathbf{Z}_i\mathbf{Z}_i^T\mathbf{V}^{-1}(\mathbf{y} - \mathbf{X}\boldsymbol{\beta}) - tr(\mathbf{V}^{-1}\mathbf{Z}_i\mathbf{Z}_i^T) \tag{6}$$

with the convention that $\sigma_0^2 = \sigma^2$ as defined in Eq. (3) and $\mathbf{Z}_0 = \mathbf{I}_n$. The estimates of the variance components derived by maximizing the log likelihood in Eq. (4) via the first derivatives provided Eq. (5) and (6) do not reflect the degrees of freedom involved in estimating $\boldsymbol{\beta}$.

If $\hat{\mathbf{V}} = \mathbf{V}(\hat{\boldsymbol{\theta}})$ denotes the estimated covariance matrix of the observations using the ML estimator $\boldsymbol{\theta}$ of variance components, then the maximum likelihood estimator of $\mathbf{X}\boldsymbol{\theta}$ is

$$MLE(\mathbf{X}\hat{\boldsymbol{\beta}}) = \mathbf{X}(\mathbf{X}^T\hat{\mathbf{V}}^{-1}\mathbf{X})^-\mathbf{X}^T\hat{\mathbf{V}}^{-1}\mathbf{y}, \tag{7}$$

and its asymptotic covariance matrix is

$$\mathbf{V}(MLE(\mathbf{X}\hat{\boldsymbol{\beta}})) = \mathbf{X}(\mathbf{X}^T\hat{\mathbf{V}} - 1\mathbf{X})^-\mathbf{X}^T. \tag{8}$$

2. REML Estimation

Estimates of the variance components that do reflect the degrees of freedom involved in estimating $\boldsymbol{\beta}$ are obtained using restricted (or residual) maximum likelihood, usually abbreviated as REML. REML estimates of the variance components are based on residuals after fitting the fixed effects by ordinary least squares. Thus, if $\mathbf{U}^-$ denotes a generalized inverse of a matrix $\mathbf{U}$ ($\mathbf{U}^-$ is any matrix satisfying $\mathbf{U}\mathbf{U}^-\mathbf{U} = \mathbf{U}$), then $\mathbf{r} = (\mathbf{I} - \mathbf{X}(\mathbf{X}^T\mathbf{X})^-\mathbf{X}^T)\mathbf{y}$ is the vector of residuals corresponding to the observation vector $\mathbf{y}$. The expectation of $\mathbf{r}$ is zero regardless of the value of $\mathbf{V}$. The matrix $\mathbf{I} - \mathbf{X}(\mathbf{X}^T\mathbf{X})^-\mathbf{X}^T$ has $n - p$ linearly independent rows, where p = the rank of $\mathbf{X}$. Let M denote an $(n - p) \times n$ matrix whose rows are linearly independent vectors in the row space spanned by the rows of $\mathbf{I} - \mathbf{X}(\mathbf{X}^T\mathbf{X})^-\mathbf{X}^T$. $\mathbf{M}$ could be any $n - p$ linearly independent rows of $\mathbf{I} - \mathbf{X}(\mathbf{X}^T\mathbf{X})^-\mathbf{X}^T$. If $\mathbf{w} = \mathbf{M}\mathbf{y}$, then $\mathbf{w}$ represents an observation from a multivariate normal distribution with zero mean and nonsingular covariance matrix $\mathbf{U} = \mathbf{M}\mathbf{V}\mathbf{M}^T$. The logarithm of the likelihood under multivariate normality is

$$L = -\frac{1}{2}(\log|\mathbf{U}| + \mathbf{w}^T\mathbf{U}^{-1}\mathbf{w}) \tag{9}$$

and the derivative with respect to σ_i^2, $i = 1, \ldots, q$, is

$$L_{\sigma_i^2} = \frac{1}{2}(\mathbf{w}^T\mathbf{U}^{-1}\mathbf{Z}_i\mathbf{Z}_i^T\mathbf{U}^{-1}\mathbf{w} - tr(\mathbf{U}^{-1}\mathbf{Z}_i\mathbf{Z}_i^T)) \tag{10}$$

The variance components can be estimated from these expressions with attention to the same computational considerations, namely that the estimates really do maximize expression (9) and the estimates are positive (or at least non-negative).

Estimates of the fixed effects and the asymptotic covariance matrix of the estimates are given by Eq. (7) and (8) with $\hat{\mathbf{V}}$ calculated using the REML estimates of the variance components.

3. Other Estimation Procedures

The parameters can be estimated by means other than maximum likelihood and REML. Extended generalized least squares (EGLS) can be used when the observations at least conceptually represent series of observations on individuals (33). Write the *jth* observation on individual i as

$$y_{ij} = f(x_{ij}, \boldsymbol{\beta}_i) + \sigma g(f(x_{ij}, \boldsymbol{\beta}_i), \boldsymbol{\theta})\varepsilon_{ij} \tag{11}$$

where the parameter vector $\boldsymbol{\beta}_i$ for individual i is drawn from a distribution (usually assumed p-variate normal) with mean $\boldsymbol{\beta}$ and covariance matrix Σ. The within-individual errors ε_{ij} are assumed independently and identically distributed with zero mean and unit variance. The scale parameter σ and the variance function $g(\ )$ allow for heterogeneity due to intra-individual correlation. An analogue to REML estimation can be applied to EGLS estimators as well. The general formulation provides a unified basis for a number of variations on this theme (34–40).

The analyses also can be carried out using Bayesian (41–52) or Empirical Bayesian (53) methods, which start with a likelihood for the observations based on a model such as Eq. (2) or (11) and then append expressions for prior distributions of the various parameters to obtain an expression for the joint distribution of the data and the parameters. Integrating with respect to the model (likelihood) parameters yields a marginal distribution for the observations that depends at most on the (known) parameters of the prior distributions. Dividing the joint distribution of the data and the model parameters by this marginal distribution gives the posterior distribution of the model parameters. Inferences about the parameters and about functions of the parameters (for example tolerance limits or predictive limits for future observations) are based on the posterior distribution. The difficult part of the computation is the integration to produce the marginal distribution of the observations. One can partly avoid the problem by approximating the mean of the posterior distribution by its mode (50). Current practice employs Markov Chain Monte Carlo techniques (43,48) that sample from conditional posterior distributions in such a way as to generate samples from the full joint posterior distribution. These can be summarized by appropriate techniques to yield summary statistics and descriptions of the posterior distributions (54,55). Markov Chain Monte Carlo methods are powerful tools for analyzing mixed models, both linear and nonlinear. However, the results they provide need to be checked carefully to assure that the assumptions on which they are based are met, notably that the realizations from the sampling process comprise independent (or at least uncorrelated) stationary series.

III. NONLINEAR MODELS

As noted in Sec. I.A above, the essential difference between nonlinear and linear models is the expression of the expectation of the observation vector as a nonlinear function $f(\ )$ of the covariates and parameters instead of a linear function. The functional form of f may be obtained from previous studies, derived from theory based on fundamental physical or biological principles, or represent empirical description of observations, e.g., from a graphical display of data. The functional form of f ordinarily is assumed known. Analyses can be carried out without assuming f is known by using a spline approximation to f or using generalized additive models (56).

Random effects can enter the model nonlinearly as well as linearly. This occurs, for example, in generalized linear models and in survival analyses, where the random effects scale the hazard function. Inference in these cases can be considerably more complicated than when the effects enter linearly and standard methods often cannot be applied. Two approaches often used to deal with this problem are extended general linear models as described in Sec. II.B.3 where parameters are estimated for each individual in the sample and then combined, and approximation of the nonlinear model by a first-order Taylor series expansion that effectively linearizes it (38,57,58).

IV. SOFTWARE

The computations required for the analysis of mixed models range from intensive to very intensive. Standard software packages usually include some facility for analyzing linear mixed models for continuous, normally distributed data. Some of these packages are commercial, some are available free of charge. Packages also are available to carry out calculations for nonlinear models, e.g., logistic mixed models, ordinal regression, Poisson regression, etc. The availability of diagnostic procedures to test the validity of the results of compute-intensive methods is an important consideration in the choice of a software package.

REFERENCES

1. AS Bryk, SW Raudenbush. Hierarchical Linear Models: Applications and Data Analysis Methods. Thousand Oaks: Sage Publications, 1992.
2. AI Khuri, T Mathew, BK Sinha. Statistical Tests in Mixed Linear Models. New York: John Wiley, 1998.
3. CR Rao, J Kleffe. Estimation of Variance Components and Applications. New York: Elsevier-Science, 1988.
4. SR Searle. Linear Models. New York: John Wiley, 1997.
5. SR Searle, G Casella, CE McCullogh. Variance Components. New York: John Wiley, 1992.
6. AR Gallant. Nonlinear Models. Northampton: Edward Elgar, 1997.

7. RJ Carroll, D Ruppert, LA Stefanski. Measurement Error in Nonlinear Models, CRC Press, Boca Raton, 1995.
8. M Davidian, DM Giltinan. Nonlinear Models for Repeated Measurement Data. London: Chapman and Hall, 1995.
9. EF Vonesh, VM Chinchilli. Linear and Nonlinear Models for the Analysis of Repeated Measurements. New York: Marcel Dekker, 1997.
10. EB Andersen. Introduction to the Statistical Analysis of Categorical Data Analysis. New York: Springer Verlag, 1997.
11. P McCullagh, JA Nelder. Generalized Linear Models. London: Chapman and Hall, 1989.
12. V Ducrocq, G Casella. Genetics Selection 28:505–529, 1996.
13. P Hougaard. Lifetime Data Anal 1:255–273, 1995.
14. N Keiding, PK Andersen, JP Klein. Stat Med 16:215–224, 1997.
15. TM King, TH Beaty, KY Liang. Genetic Epidemiology 13:139–158, 1996.
16. KF Lam, AC Kuk. J Am Statistical Asso 92:985–990, 1997.
17. KY Liang, SG Self, RK Bandeen, SL Zeger. Lifetime Data Anal 1:403–415, 1995.
18. C Morris, C Christiansen. Lifetime Data Anal 1:347–359, 1995.
19. JH Petersen. Biometrics 54:646–661, 1998.
20. A Pickles, R Crouchley. Stat Med 14:1447–1461, 1995.
21. DJ Sargent. Biometrics 54:1486–1497.
22. TH Scheike, TK Jensen. Biometrics 53:318–329, 1997.
23. D Sinha. Biometrics 54:1463–1474, 1998.
24. SG Walker, BK Mallick. J Royal Statistical Society, Series B 59:845–860, 1997.
25. XN Xue, R Brookmeyer. Stat Med 16:1983–1993, 1997.
26. XN Xue. Biometrics 54:1631–1637, 1998.
27. KW Yau, CA McGilchrist. Stat Med 17:1201–1213, 1998.
28. HB Yue, KS Chan. Biometrics 53:785–793, 1997.
29. PH Zahl. Stat Med 16:1573–1585, 1997.
30. RT Burnett, WH Ross, D Krewski. Environmetrics 6: 85–99, 1995.
31. MG Kenward, E Lesaffre, G Molenberghs. Biometrics 50:945–953, 1994.
32. SL Zeger, KY Liang. Stat Med 11:1825–1839, 1992.
33. M Davidian, DM Giltinan. Biometrics 49:59–73, 1993.
34. RJ Carroll, D Ruppert. Transformation and Weighting in Regression. Chapman and Hall, London, 1988.
35. SL Beal, LB Sheiner. Technometrics 30:327–338, 1988.
36. CC Peck, SL Beal, LB Sheiner, AI Nichols. J Pharmacokinetics and Biopharmaceutics 11:303–319, 1984.
37. SL Zeger, KY Liang, PS Albert. Biometrics 44:1049–1060, 1988.
38. MJ Lindstrom, DM Bates. Biometrics 46:673–687, 1990.
39. A Racine-Poon. Biometrics 41:1015–1023, 1985.
40. JL Steimer, A Mallet, JL Golmard, JF Boisvieux. Drug Metabolism Reviews 15:265–292, 1984.
41. DA Berry, DK Stangl. Bayesian Biostatistics. New York: Marcel Dekker, 1996.
42. LD Bromeling. Bayesian Analysis of Linear Models. New York: Marcel Dekker, 1984.
43. BP Carlin, TA Louis, Bayes and Empirical Bayes Methods for Data Analysis. London: Chapman and Hall, 1996.
44. A Chaturvedi. J Statistical Planning And Inference 50: 175–186, 1996.
45. MK Cowles, BP Carlin, JE Connett. J Am Statistical Asso 91:86–98, 1996.
46. AE Gelfand, SE Hills, A Racine-Poon, AM Smith. J Am Statistical Asso 85:972–985, 1990.
47. A Gelman, JB Carlin, HS Stern, DB Rubin. Bayesian Data Analysis. London: Chapman and Hall, 1995.
48. WR Gilks, S Richardson, DJ Spiegelhalter. Markov Chain Monte Carlo in Practice. London: Chapman and Hall, 1996.
49. DA Harville, AG Zimmermann. J Statistical Computation And Simulation 54:211–229, 1996.
50. DV Lindley, AFM Smith. J Royal Statistical Society Series B 34: pp 1–42, 1972.
51. R Simon, LS Freedman. Biometrics 53:456–464, 1997.
52. MA Waclawiw, KY Liang. J Am Statistical Asso 88: 171–178, 1993.
53. JS Maritz, T Lwin. Empirical Bayes Methods. Chapman and Hall, London, 1989.
54. NG Best, MK Cowles, SK Vines. CODA: Convergence diagnosis and output analysis software for Gibbs sampling output, version 0.4. MRC Biostatistics Unit, Cambridge, 1997.
55. DJ Spiegelhalter, A Thomas, NG Best, WR Gilks. BUGS: Bayesian inference using Gibbs sampling. MRC Biostatistics Unit, Cambridge, 1997.
56. TJ Hastie, RJ Tibshirani. Generalized Additive Models. London: Chapman and Hall, 1990.
57. AJ Boeckmann, LB Sheiner, SL Beal. NONMEM Users guide. NONMEM Project Group C255, University of San Francisco, 1992.
58. EF Vonesh, RL Carter. Biometrics 48:1–17, 1992.

Donghui Zhang
A. Lawrence Gould

Multicenter Trials

See also *Clinical Trials; Data Monitoring Board*

I. INTRODUCTION AND BACKGROUND

A multicenter clinical trial is a clinical trial (see *Clinical Trials* entry) involving more than one clinical center (i.e., study site, investigator, field site, clinic) that is conducted to evaluate the effectiveness of a therapeutic agent according to a common protocol. Multicenter trials have been utilized in clinical medicine since the 1940s, as documented by Bradford Hill (1) in his seminal work on the statistical aspects of trials of antihistamines, cortisone, and streptomycin. Multicenter trials are employed primarily for two reasons: (a) the use of multiple centers allows adequate and/or rapid accrual of patients, and (b) the involvement of multiple centers enhances the generalizability of the results due to the anticipated heterogeneity in patient populations and center practices. This latter rationale is attractive because it more closely resembles how the therapeutic agent will be utilized when commercially available. Other reasons for utilizing multicenter trials are to introduce state-of-the-art therapies to the medical community as well as to solicit a wider range of clinical opinions concerning the utility of the therapeutic agent.

Many of the aspects of multicenter trials run by the pharmaceutical industry are similar to those of multicenter trials run by collaborative efforts sponsored by government agencies (e.g., National Institutes of Health). However, there are several important differences. Pharmaceutical industry trials evaluate new therapeutic agents for regulatory approval in a specific patient population. Often, these multicenter trials are used in phase 3 of drug development (see *Drug Development* entry), sometimes as the pivotal studies in the new drug application (see *NDA* entry) to the Food and Drug Administration (see *FDA* entry). Collaborative trials address outstanding medical questions for approved therapeutic agents in a broader patient population; the results of these trials profoundly affect current medical practice (2). In addition, because pharmaceutical industry trials evaluate new therapeutic agents, there is a need for timely safety monitoring and periodic safety reporting to regulatory agencies during the trial. Collaborative trials are not subject to such regulatory oversight, since they evaluate approved therapeutic agents about which the safety profile is known. See "Multicenter Trials" in the *Encyclopedia of Biostatistics* (3) for a more complete description of collaborative multicenter trials. The following discusses aspects of multicenter trials in the pharmaceutical industry.

By its nature, a multicenter clinical trial is organizationally complex, requiring careful protocol development (see *Protocol Development* entry) and implementation as well as clear organizational structure and decision-making.

A. Protocol Development and Implementation

The protocol should be clearly written so that it can be easily understood and uniformly implemented across clinical centers. In particular, entry and therapy/patient discontinuation criteria should be rigorously standardized using verifiable decision rules. The pharmaceutical company (i.e. drug sponsor) should hold investigator meetings and training sessions for study coordinators to assure common understanding of evaluation procedures, especially for the efficacy endpoint (entries on *Clinical Endpoint* and *Surrogate Endpoint*). Manuals of standard operating procedures should be created to enhance uniformity and protocol adherence further. Ongoing monitoring by the sponsor of each clinical center's compliance to the protocol should occur throughout the trial. The foregoing concerns are critically important to the success of the study.

Many of the data-collection and data-management (see entry on *Data Management*) issues associated with multicenter trials are similar to those associated with single-center clinical trials. For example, the data collection tool (see entry on *Case Report Forms*) should be designed to collect essential data in a consistent manner. Also, there should be a quality control process in place to ensure that the collected data match the patient's medical records.

Several new data-management issues are introduced by multicenter trials compared to single-center trials. For example, a central resource center can be used to perform the evaluation of important objective safety and efficacy endpoints as well as entry criteria. Use of a central resource center ensures standardized, blinded review of these critical data. The logistics of sample collection and flow from the clinical centers to the central resource centers (e.g., collection of blood sent for evaluation to a central laboratory) and the subsequent transfer of the data to the data-management center should be clearly spelled out, facilitated, and monitored by the sponsor.

It is the sponsor's responsibility to have in place a process for monitoring the study for emerging safety issues and for communicating any issues in a timely manner to the regulatory agencies and, if necessary, to the clinical investigators.

B. Organizational Structure and Decision Making

The organizational structure of and decision-making process for the multicenter trial should be determined and agreed to prior to study initiation. In particular, the roles and responsibilities of each committee should be defined. Usually, there is a study chair (e.g., the principle investigator) and a committee of investigators. The sponsor facilitates communication between investigators by conducting investigator meetings as well as frequent mailings. Often, it is the role of the investigator committee, in consultation with the sponsor, to plan publications and to resolve authorship.

The study coordinating center is responsible for managing the clinical centers (e.g., monitoring patient accrual, ensuring protocol compliance) and the data-coordinating center is responsible for managing the data and analysis. These coordinating centers may be a contracted research organization (see entry on *CROs*), an academic center, or medical/data-management/statistical personnel of the sponsor. Also, as already mentioned, there may be several additional resource centers, such as central laboratories and evaluation centers, which are usually external to the sponsor so as to maintain perceived integrity of the results.

Often, Data Safety Monitoring Boards (see entry on *DSMBs*) are used to monitor accumulating safety data. In addition, the DSMB may perform an interim analysis (see entry on *Interim Analysis*) of efficacy endpoints, if specified in the protocol. The membership and operation of the DSMBs are kept separate from the study coordinating center so as to minimize risks to study integrity (4).

II. REGULATORY REQUIREMENTS

There are no specific regulatory requirements beyond those noted for clinical trials (5).

III. STATISTICAL DESIGN AND ANALYSIS

A. Statistical Design

One drawback to the use of multicenter trials is the presence of additional sources of variability not present in single-center trials. In particular, a major source of variability is the clinical center. Clinical centers often vary on factors that may be related to either the therapy or clinical endpoint: characteristics of the patients enrolled at the center, specific interests/skills of the center and physician (e.g., family practice versus referral center), and implementation of the protocol (e.g., how the physician evaluates the patient) (6). To minimize this variability, the design of multicenter trials usually considers the clinical center as a block. Prospective stratification by clinical center with randomization is generally preferred to achieve balance among therapy groups. However, there are occasions where stratification is not possible or is not necessary. For example, in mortality or cancer trials (see entry on *Cancer Trials*), there may be many centers, each with very few patients, precluding stratification by center. Also, other factors may be felt to be more predictive of the clinical outcome than clinical center. In these cases, randomization may be performed using a stratification scheme administered by a central system (e.g., interactive voice randomization system—see entry on *IVRS*) that stratifies on other important factors besides center (7).

One should consider the number of centers and the number of patients allocated per center in the design. The guidance from the International Conference on Harmonization (see entry on *ICH*) on Statistical Principles for Clinical Trials (8) recommends that one try to avoid excessive variation in the number of patients per center and to avoid centers with very few patients. Such variation, if present, complicates the interpretation of the therapy-by-center interaction (to be discussed in the Statistical Analysis section that follows).

The sample size determination (see entry on *Sample Size Determination*) should be specified in the protocol with enough detail so that an independent reviewer (e.g., regulatory reviewer) can verify the calculations. The sample size estimated should take into account the design of the trial (e.g., comparative versus equivalence, parallel versus crossover), anticipated noncompliance, possibly varying hazards, and the multicenter nature of the study (i.e., the variability may be larger than that seen in single-center trials) (9). Centralized readings of objective safety and efficacy endpoints may allow a reduction in variability. Sample size and power calculations for the multicenter trial are based on the assumption that the therapy differences from each center are estimating the same quantity. Often clinical trials are not designed to consider the presence of therapy-by-center interaction (10), leaving it to the statistical analysis to explore the robustness of the results to this assumption.

B. Statistical Analysis

The statistical model for the analysis should be specified in the protocol. If the trial has been stratified on center, this effect should be included in the model to account for variability associated with center as well as to ensure that the appropriate error term is used (11). It is generally accepted in the pharmaceutical industry that center is a fixed effect (VA Uthoff, personal communication, 1986). This assumption limits the generalizability of the trial results to the centers employed in the trial. Alternatively, the center effect could be considered a random effect so that inference can be more generalizable. However, random-effects models often require greater sample sizes due to reduced degrees of freedom in the error term. In addition, there is concern that centers do not represent a random sampling of all investigative centers but, rather, are selected on the basis of past performance and the current set of entry criteria.

The protocol should state a priori how therapy-by-center interaction (see entry on *Treatment-by-Site Interaction*) is to be evaluated. In particular, the protocol should state the level of significance to use, whether interaction will be evaluated for all safety and efficacy measures and time points, how to pool centers with few patients, and what will be done if interaction is present.

If therapy-by-center interaction is not significant, the protocol should specify how inference is to be made; e.g., will the full or the reduced model be used? The reduced model is a parsimonious model having greater power for the detection of therapy effects, whereas the full model provides unbiased estimates (i.e., one can never be sure that the interactions are really zero). In the latter case, there is controversy whether sequential or partial sums of squares should be used for inference (12). The choice ultimately depends on the hypothesis to be tested.

If the therapy-by-center interaction is significant, the interpretation of the main effect is controversial. As already mentioned, a critical assumption for multicenter trials is that the therapy effect is consistent from center to center. When this is not valid, the nature of the interaction should be explored in an attempt to characterize it. Generally, there are two types of interaction: quantitative and qualitative (13). *Quantitative* interaction refers to therapy differences that are in the same direction but vary in magnitude across the centers; *qualitative* interaction refers to therapy differences that vary in direction across the centers. Graphical presentation of the therapy effect for each center, as well as identification of the centers having the greatest statistical contribution to the interaction (14), is useful to gain insight into the cause of the interaction.

IV. SCIENTIFIC ISSUES

One rationale for the use of multicenter trials is increased heterogeneity in patients and centers, enhancing the generalizability of the trial conclusions. However, there is a cost associated with multicenter trials. Greater heterogeneity can imply greater variability, leading to compromised power and larger sample sizes for the detection of therapy effects. This increased variability may be reduced through the use of a central reading of objective endpoints.

Often it is difficult to define *center*. Is it the region, which may have several investigative centers, is it the hospital/practice, which may have several investigators, or is it the investigator? The driving principle is to define *center* in such a way that one achieves homogeneity in the important factors that affect the outcome variable and therapy.

V. RECENT DEVELOPMENTS

As mentioned in the statistical design section, there are vague notions concerning the optimal number of centers and number of patients per center as well as concerning the effect of differential sample sizes among centers on the analysis. Shao and Chow (15) design a multicenter trial using a two-stage sampling plan normally used for USP tests (see entry on *USP Tests*). They show that this plan is optimal in minimizing the variance of the associated statistical testing procedures. Using this sampling plan, one can select the number of centers and number of patients per center to maintain the optimality criterion. Salsburg (1991, personal communication) addresses the issue of minimum sample size per center from the standpoint of the power of the tests for interaction in a two-way fixed-effects ANOVA. He concludes that, when there is no therapy-by-center interaction or when therapy allocations are balanced within center, differential sample sizes among centers have no effect on alpha level or on power for hypothesis tests.

Recent work (16) considers an alternative to sequential or partial sums of squares for inference. They propose weighting of the therapy effect from each center, depending on the variability observed within each center. Centers are grouped by the observed variability, and more variable centers are given less weight, and vice versa. The authors show that this procedure provides a reasonable estimate of the therapy effect in the presence of heteroscedasticity and "modest" therapy-by-center interaction.

Nonparametric statistical tests (specifically, extended Mantel–Haenszel tests) for analyzing data from multicenter trials have proliferated in recent years. Simulation studies have shown that these tests have excellent performance, even under optimal conditions for normal-theory tests (17).

In an attempt to bring speed and uniformity to the drug decision-making process, Enas et al. (18) present a collection of standard statistical and graphical techniques for analyzing multicenter clinical trials. These techniques, easily automated, facilitate better understanding of the efficacy findings. In addition, strategies evaluating the robustness of inference to varying assumptions are also presented. These strategies are consistent with the use of sensitivity analysis to evaluate robustness as promulgated by Jones (19).

The FDA Guidelines for the Clinical and Statistical Sections of an NDA (20) state that the approval of a new therapeutic agent should be "supported by more than one well-controlled trial. . . . This interpretation is consistent with the general scientific demand for replicability." However, when a multicenter trial is viewed as a collection of single-center trials using a common protocol, replication of the therapy effect within the multicenter trial is feasible. Huster and Louv (21) present a minimax statistic that assesses whether the therapy effect has been replicated in a multicenter trial. They describe the operating characteristics of this statistic under the null and several alternative hypotheses. They also observe that, as described by Salsburg earlier, the power for the detection of a therapy effect is not affected by differential sample sizes.

REFERENCES

1. B Hill. Statistical Methods in Clinical and Preventive Medicine. New York: Oxford University Press, 1962.
2. J Herson. Stat Med 12:555–564, 1993.
3. J Bryant, W Cronin, S Wieand. In: P Armitage and T Colton, eds. Encyclopedia of Biostatistics: Multicenter Trials. New York: Wiley, 1997, pp 2710–2716.
4. WJ Huster, AS Shah, W Dere, GV Kaiser, RD DiMarchi. Harmonization of Global Requirements: Case Study of EVISTA. Drug Information Association Meetings, Boston, Massachusetts, 1998.
5. International Conference on Harmonization Harmonized Tripartite Guideline. Guidelines for Good Clinical Practice (E6). 1996.
6. JL Fleiss. The Design and Analysis of Clinical Experiments: Multicenter Trials. New York: Wiley, 1986, pp 176–180.
7. JM Lachin. Controlled Clin Trials 2:93–113, 1981.
8. International Conference on Harmonization Harmonized Tripartite Guideline. Statistical Principles for Clinical Trials (E9). 1998.
9. M Halperin, D DeMets, J Ware. Stat Med 9:881–892, 1990.
10. VM Chinchilli, EB Bortey. J Biopharmaceut Stat 1:67–80, 1991.
11. JL Fleiss. Controlled Clin Trials 7:267–275, 1986.
12. FM Speed, RR Hocking, OD Hackney. JASA 73:105–112, 1978.
13. DP Byar. Stat Med 4:255–263, 1985.
14. DS Hwang, E Barry, H Hamot. Treatment-by-center interaction: resolving the question by presenting evidence. Proceedings of Biopharmaceutical Section for the ASA Meetings in Toronto, Canada, 1983, pp 40–44.
15. J Shao, S-C Chow. Stat Med 12:1999–2008, 1993.
16. JL Ciminera, JF Heyse, HH Nguyen, JW Tukey. Stat Med 12:1033–1045, 1993.
17. CS Davis, Y Chung. Biometrics 51:1163–1174, 1995.
18. GG Enas, TM Sanger, WJ Huster. Biopharmaceut Rep 2:1–12, 1993.
19. DR Jones. Drug Inform J 27:833–836, 1993.
20. Food and Drug Administration Guideline for the Format and Content of the Clinical and Statistical Sections of New Drug Applications, 1988.
21. WJ Huster, WC Louv. J Biopharmaceut Stat 2:219–238, 1992.

William J. Huster

Multiple Comparisons

See also *Multiple Endpoints*

I. INTRODUCTION

Generally, a phase II/III trial involves collecting data for evidence to show the efficacy and safety of an experimental drug. The integrity of this evidence relies heavily on such factors as the design, the patients recruited, the overall conduct of the study, and the appropriateness of the endpoints and the analysis. Descriptions of these factors must be clearly specified a priori in the protocol, as required by the regulatory guidelines.

Efficacy and safety claims of an experimental drug are never based on one endpoint, but on several endpoints. For example, in a trial involving patients with schizophrenia, a set of efficacy endpoints may include more than one derived measure from the Positive and Negative Symptom Scales (PANSS) and Clinical Global Assessment Scales (CGI-severity and CGI-improvement). Also, for most of the trials, a safety claim is based on adverse events, vital signs, laboratory measurements, and electrocardiogram. In addition to several endpoints, most of the phase II/III trials are typically comparative trials—comparison against placebo or an active therapy, or dose–response trials that involve more than one dose of the experimental drug. There is also a time factor that gets into the overall equation; viz., patient data for various endpoints and doses are collected across several time points in a longitudinal fashion. It is now not difficult to imagine the dimensionality of the accumulated data collected with one overall objective of showing that the drug is either efficacious or safe or both.

Statisticians play an important role in assuring their clients, clinicians, that the design and the methods chosen are appropriate and most sensitive to the primary questions that are asked in the protocol. If the trial is an exploratory one, then it is typical to present and interpret the data using data summaries such as means and standard deviations, without getting into any inferential procedures. For a confirmatory trial, final conclusion about the success or failure of the study is generally based on statistical inference and a presentation of the overall p-value. With the unavoidable multiplicity in a trial such as just described, if the overall conclusion is based on several comparisons, it can then result in presentation of several p-values from various statistical tests. There is always some chance that not all these tests would result in a statistical significance, typically considered as $p \leq 0.05$.

Suppose, in a confirmatory pivotal trial with three doses of an experimental drug and a placebo with one primary endpoint, that out of three pairwise comparisons with placebo, only two yield significant results. For this example, suppose we would like to make an overall claim that the trial has proven the efficacy of the drug. Regulators [such as the Food and Drug Administration (FDA)] may not accept as evidence of efficacy, especially if it is not specified a priori, which comparison is more important than the other. From a regulatory perspective, they are not comfortable with an inefficacious drug reaching the market, as compared to an efficacious drug not reaching the market. That is, a false-positive result is more of an issue to the regulators; this idea will be covered further in the next section.

II. ERROR RATES AND NEED FOR MULTIPLE COMPARISONS

When there is more than one treatment, an overall F-test can be performed to see whether there is any difference in the treatments. If the overall F-test is significant, then individual comparisons between the treatments, generally pairwise comparisons, are done to investigate which treatment is different from what. At this point, it must be made clear that considering these pairwise comparisons only if the overall F-test is significant might be erroneous (4). Suppose in a trial involving three doses of an experimental drug (low, medium, high) and a placebo that there are six possible pairwise comparisons in this trial (or experiment). In general, if there are k treatments, then there are $k(k - 1)/2$ possible pairwise comparisons.

A *comparisonwise error rate* is a Type I error rate for each comparison; that is, it is the probability of erroneously rejecting the null hypothesis between treatments involved in the comparison. On the other hand, an *experimentwise error rate* (or *familywise error rate*) is the error rate associated with one or more Type I errors for all comparisons included in the experiment. When multiple statistical tests are performed using the same data set, the *experimentwise error rate* can be much larger than the significance level associated with each test itself. For illustration, assume that $P\{\text{Type I}\} = \alpha$ for each test or comparison. Then, for k comparisons,

comparisonwise error rate (CWE) = α

experimentwise error rate (EWE) = $1 - (1 - \alpha)^k$

To illustrate the inflation of the *experimentwise error rate*, these rates can be expressed in numbers as (with $\alpha = 0.05$):

k	CWE	EWE
1	0.05	0.05
5	0.05	0.25
10	0.05	0.40
20	0.05	0.64
50	0.05	0.92

As can be seen in the table, even with $k = 5$ comparisons, the EWE increases to 25%, which may make the overall conclusion based on these $k = 5$ comparisons questionable. There are multiple comparison methods available for controlling these error rates, and their descriptions will be provided next.

III. DESCRIPTION

A. All Pairwise Comparisons

Assume that there are k treatment means (or LSMEANS in SAS™ terminology), $\bar{y}_1, \bar{y}_2, \ldots, \bar{y}_k$, and let s^2 be an estimate of the variance, typically obtained from an analysis-of-variance table. In the case of an unbalanced one-way model, with n_i observations for treatment i,

$$\bar{y}_i = \frac{\sum y_{ij}}{n_i}$$

$$s^2 = \frac{\sum \sum (y_{ij} - \bar{y}_i)^2}{\sum (n_i - 1)} \qquad (1)$$

where $\nu = \Sigma\,(n_i - 1)$ is the degrees of freedom associated with the estimate s^2.

1. *Fisher's Least Significant Difference (LSD)*

Fisher's LSD method is a two-step process. As indicated in the previous section, one starts with an overall F-test for testing the equality of treatment means, H_0: $\mu_1 = \mu_2 = \cdots = \mu_k$. If H_0 is not rejected, then no further steps are taken. Otherwise, this method concludes that the treatment means μ_i and μ_j are different, for every $i \neq j$, if

$$|\bar{y}_i - \bar{y}_j| > t_{\alpha/2}(\nu)[s^2(n_i^{-1} + n_j^{-1})]^{-1/2} \qquad (2)$$

where $t_{\alpha/2}(\nu)$ denotes a critical value for the t-distribution having ν degrees of freedom and an upper-tail probability of $\alpha/2$.

This two-step Fisher's LSD procedure, also known as the *protected LSD method*, controls the comparisonwise error rate at the level α is the overall null hypothesis is true. If the overall null hypothesis is false, then it does not control the comparisonwise error rate unless $k \leq 3$. The unprotected LSD procedure is used with a significance level that is chosen to control the experimentwise error rate; that is, the rate used to assert the simultaneous correctenss of the comparisons is of no consequence. If a moderate or large number of comparisons ($k > 3$) is to be made, then neither of the LSD procedures is recommended, and the Bonferroni method can be considered as an alternative.

2. *Bonferroni Method*

The *Bonferroni method* uses the inequality (2) with a comparisonwise error rate of α/m, where α is the desired experimentwise error rate and m is the number of pairwise comparisons that are to be made within the experiment. This procedure is not recommended if there is a large number of pairwise comparisons, which can make the comparisonwise error rate too small to be of any value as α/m gets small when m is increased. In such situations, multiple range tests provide some alternatives, some of which are discussed next.

3. *Tukey's Multiple Range Test*

The method suggested by Tukey controls the experimentwise error rate incurred in testing for treatment differences, especially when the averages involved in those differences are based on the same number of observations, that is, when they are taken in pairs. Using this test, we can declare the two means $\bar{y}_i$ and $\bar{y}_j$ to be significantly different if

$$|\bar{y}_i - \bar{y}_j| > q(\alpha, k, \nu)\left\{s^2\,\frac{(n_i^{-1} + n_j^{-1})}{2}\right\}^{1/2} \qquad (3)$$

where $q(\alpha, k, \nu)$ is the "studentized range statistic," k is the number of averages being compared, s^2 is the mean squared error (MSE) from the ANOVA model with ν degrees of freedom, and α is the experimentwise error rate. Tables of critical values for the studentized range statistic are widely available.

This multiple range test suggested by Tukey can also be used to construct simultaneous confidence intervals on all pairs of mean differences $\mu_i - \mu_j$, and it has been shown that the confidence intervals have a confidence coefficient $100(1 - \alpha)\%$ collectively; that is,

$$P\left\{\mu_i - \mu_j \in \bar{y}_i - \bar{y}_j \pm |q| \cdot \left\{s^2 \frac{(n_i^{-1} + n_j^{-1})}{2}\right\}^{1/2} \text{ for all } i \neq j\right\} = 1 - \alpha \quad (4)$$

4. *Duncan's Multiple Range Test*

For comparison of pairs of treatment means, Duncan has developed a test, called a *multiple range test*; its comparisonwise error rate is a function of the number of comparisons. It is a good compromise between Fisher's LSD and Tukey's multiple range test, but it does not control the comparisonwise error rate. To use this test, first the treatment means, obtained from a balanced design with a sample size n, are arranged in ascending order. The criterion is to conclude that the largest and smallest of the treatment means, $\bar{y}_i$ and $\bar{y}_j$, are significantly different if

$$|\bar{y}_i - \bar{y}_j| > q(\alpha_p, p, \nu)\left\{\frac{\text{MSE}}{n}\right\}^{1/2} \quad (5)$$

where p = number of averages, $q(\alpha_p, p, \nu)$ is the critical value from the "studentized range statistic" with an experimentwise error rate of α_p, and MSE is the mean squared error with ν degrees of freedom. When α is the comparisonwise error rate, then α_p is related to α as $\alpha_p = 1 - (1 - \alpha)^{p-1}$.

5. *Other Tests*

There are other tests available for pairwise comparisons. These are not discussed here, but detailed discussions on these procedures are available in the literature (4).

B. Multiple Comparisons with a Control

In many clinical studies, when there are several treatment groups (e.g., multiple doses of an experimental drug) and a control group, the primary objective specified in the protocol may dictate comparison of the treatment group against the control group, but within-treatment-group comparisons may not be required. For example, if the interest is only to show that the new therapy is "better" than no therapy (placebo), then there is no need to make comparisons between the dose groups of the new therapy. Comparisons of such types are known as *multiple comparisons with a control*. *Dunnett's test* is the most popularly used multiple comparison procedure for comparison against a control.

1. *Dunnett's Test*

Suppose the $k - 1$ treatment groups are given as μ_i, $i = 1, 2, \ldots, k - 1$, and the control group is denoted as μ_k. Further, the treatment groups are modeled in terms of a balanced one-way model as

$$y_{ij} = \mu_i + \varepsilon_{ij}, \quad i = 1, 2, \ldots, k; j = 1, 2, \ldots, n \quad (6)$$

Assume that the error terms ε_{ij} are normally distributed with mean 0 and unknown variance σ^2. Equation (1) can now be used to estimate μ_i and σ^2 for this balanced model.

Dunnett's method provides both one-sided and two-sided simultaneous confidence intervals for $\mu_i - \mu_k$ (4). For the one-sided alternatives, the lower bound of the confidence interval is given as

$$\mu_i - \mu_k > \hat{\mu}_i - \hat{\mu}_k - T\hat{\sigma}\sqrt{\frac{2}{n}}, \quad \text{for } i = 1, 2, \ldots, k - 1 \quad (7)$$

where $T = T_{k-1,\nu,\{\rho ij\}}(\alpha)$ is solved using the following equation:

$$\int_0^{\infty}\int_{-\infty}^{\infty} [\Phi(z + \sqrt{2}Tu)]^{k-1}\, d\Phi(z)\gamma(u)\, du = 1 - \alpha \quad (8)$$

In Eq. (8), γ is the density of $\hat{\sigma}/\sigma$, Φ is the distribution function of the standard normal, and T depends only on α, k, $\nu = k(n - 1)$, and $\rho_{ij} = 1/2$ for all $1 \leq i \neq j \leq k - 1$. Also, ρ_{ij} is the correlation coefficient of the estimators $\delta_i = \hat{\mu}_i - \hat{\mu}_k$ $(1 \leq i \leq k - 1)$. It is widely known (2) that $T = T_{k-1,\nu,\{\rho ij\}}(\alpha)$ are the critical points from the distribution of the random variable max T_i, where $T_1, T_2, \ldots, T_k$ have a multivariate t-distribution with ν degrees of freedom and correlation matrix $\{\rho_{ij}\}$; its tabulated values are widely available for the case $\rho_{ij} = \rho$.

For two-sided simultaneous confidence intervals, $\hat{\mu}_i - \hat{\mu}_k \pm |h|\hat{\sigma}\sqrt{2n}$, the solution to $|h|$ is obtained through the equation

$$\int_0^\infty \int_{-\infty}^\infty [\Phi(z + \sqrt{2}|h|t) - \Phi(z - \sqrt{2}|h|t)]^{k-1} \cdot d\Phi(z)\gamma(t)\, dt = 1 - \alpha$$

In other words, $|h|$ are critical points from the distribution of max $|T_i|$. For the unbalanced case, tabulation of the critical values may be impractical in many cases. But for some special correlation structures, the integration in Eq. (8) can be simplified to produce iterative solutions. For example, in the case of an unbalanced one-way design, $\rho_{ij} = \lambda_i\lambda_j(1 \leq i \neq j \leq k - 1)$, where $\lambda_i = (1 + n_k/n_i)^{-1/2}$ for $i = 1, 2, \ldots, k - 1$; tables are available for this case (2).

2. *Other Tests*

Other procedures, such as step-up and step-down procedures, available for multiple comparisons against a control (4) are not discussed here.

C. Multiple Comparisons in Dose-Finding Studies

Dose–response studies are performed to evaluate the relationship between the effect (both efficacy and safety) and the dose and to evaluate the recommended dose regimen and frequency. In such studies, several doses of the experimental drug are compared against the zero-dose (or placebo) control. If there is no monotonicity assumed in the effect (that is, group means are not ordered), then Dunnett's test is recommended to compare the group means against placebo. But if the group means are expected to follow an increasing or decreasing order, then there are several procedures available for comparing the doses against placebo. One of the main benefits of some of these procedures is the control of familywise error rate.

1. *Closed Test Procedures*

Suppose there is a family of hypotheses denoted by $\{H_i, 1 \leq i \leq k\}$ and its closure is represented by taking all nonempty intersections $H_P = \cap_{j \in P} H_j$ for $P \subseteq \{1, 2, \ldots, k\}$. A *closed testing method* (5) asserts that any hypothesis H_P is rejected if and only if every H_Q is rejected by its associated α-level test for all $Q \supseteq P$, assuming that an α-level test for each hypothesis H_P is available. For tests of this type, Marcus et al. (5) showed that it controls the Type I familywise error rate. In order to use this method, if we have α-level tests individually for each H_i, then they can be applied in a step-down manner. A simple example is provided by a modified Dunnett's test.

Suppose, in a dose response study with several doses of an experimental drug and a placebo, that it is assumed that the primary objective is to make one-sided comparisons with placebo. For this example, consider a family of hypotheses: $\{H_i\colon \mu_i - \mu_k \leq 0\ (1 \leq i \leq k - 1)\}$ against upper one-sided alternatives, where placebo is the kth group. Also, assume that the treatment groups have the same size n and that the number of patients in placebo is n_k, and let $\rho = n/(n + n_k)$. The step-down procedure is carried out as follows:

1. Calculate T_i, the t-statistics for $1 \leq i \leq k - 1$, and let the ordered t-statistics be $T_{(1)} \leq T_{(2)} \leq \cdots \leq T_{(k-1)}$ with their corresponding hypotheses denoted as $H_{(1)}, H_{(2)}, \ldots, H_{(k-1)}$.
2. Reject $H_{(j)}$ if $T_{(i)} > T_{i,\nu,\rho}(\alpha)$ for $i = k - 1, k - 2, \ldots, j$. If $H_{(j)}$ is not rejected, then conclude that $H_{(j-1)}, \ldots, H_{(1)}$ are also to be retained.

This step-down version of Dunnett's test is more powerful than the test given in Eq. (7), which is based on a single critical value (5).

2. *Other Tests*

Other multiple test procedures for dose-finding studies are described in Ref. 6.

IV. DESIGN ISSUES

A. Active Control Studies

In many areas of clinical research (e.g., anti-infectives), standard therapies (positive or active controls) are commonly available in the market. Inclusion of one or more doses of the positive control would serve as reference standard for efficacy comparisons with the experimental drug, while comparison with placebo would validate the clinical trial. In other words, it would serve to differentiate between an "ineffective" drug versus an "ineffective" study.

When a trial involves one or more doses of an experimental drug, placebo, and one or more doses of an active control, the first dilemma always is to establish a family of hypotheses a priori in the protocol. Based on the study design and various underlying hypotheses, strategies should be explored for testing various hypotheses. One such set of hypotheses, {drug vs. placebo} and {positive control vs. placebo}, would help to conclude whether both the drug and the positive control are superior to placebo. For each one of these hypotheses, multiple comparison procedures would then have to be proposed in the protocol, depending on the objective of either controlling *comparisonwise* or *familywise* error rate (1).

B. Sample Size

As study designs in most of the clinical trials involve an ANOVA or an ANCOVA model, it is not uncommon to perform sample size calculations based on an overall *F*-test, which may not be appropriate if the primary objective involves multiple comparisons. When multiple comparisons are involved, another approach used is to adjust the Type I error level with a Bonferroni correction. Again, this may lead to more patients than what is actually required. Hsu (3) suggests a confidence interval approach as follows: Given a confidence interval approach with level $1 - \alpha$, do the sample size computations so that with a prespecified power $1 - \beta$ ($<1 - \alpha$) the confidence intervals will cover the true parameter value and be sufficiently narrow. Technical details for exact sample size computations are given in Appendix C of Hsu's excellent textbook (4). Certain computer software, including SAS™ procedures, is also available for sample size computations.

REFERENCES

1. D'Agostino, R. B., Massaro. J., Kwan, H., and Cabral, H. (1993). Strategies for dealing with multiple treatment comparisons in confirmatory clinical trials. Drug Information Journal, 27:625–641.
2. Hochberg, Y., and Tamhane, A. C. (1987). Multiple Comparison Procedures. Wiley & Sons, New York.
3. Hsu, J. C. (1988). Sample size computation for designing multiple comparison experiments. Computational Statistics and Data Analysis, 7:79–91.
4. Hsu, J. C. (1996). Multiple Comparisons—Theory and Methods. Chapman & Hall, London.
5. Marcus, R., Peritz, E., and Gabriel, K. R. (1976). On closed testing procedures with special reference to ordered analysis of variance. Biometrika, 63:655–660.
6. Tamhane, A. C., Hochberg, Y., and Dunnett, C. W. (1996). Multiple test procedures for dose finding. Biometrics, 52:21–37.

Mani Y. Lakshminarayanan

Multiple-Dose Bioequivalence Studies

See also *Bioavailability and Bioequivalence; Dose Proportionality*

I. INTRODUCTION

The purpose of a bioequivalence study is to determine if two products, containing the same amount of a therapeutic agent, are sufficiently similar in their rate and extent of drug absorption that they can be used interchangeably in a patient to produce the same therapeutic outcome.

Methods currently required by the U.S. Food and Drug Administration (FDA) generally involve a single measurement of metrics of the rate and extent of absorption (AUC, C_{max}, T_{max}) in 16–32 normal, healthy volunteers. This procedure is referred to as *average bioequivalence*, because only the means of the two products are compared statistically. There are several assumptions inherent in this method:

1. The sample of individuals is homogeneous.
2. The volunteer study population is representative of the target patient populations.
3. For each metric, the variances of the test and reference products are the same.

A modification of average bioequivalence is *population bioequivalence*, in which estimated means and variances are compared. Generally, it is recognized that population bioequivalence is adequate to show that the test product is safe and effective in the population and is, therefore, *prescribable*. However, this process does not examine differences between the test and reference products in individual subjects (or patients). Therefore, it may not be a good indicator of whether the test and reference products can be used interchangeably in individual subjects. This has been referred to as the *switchability* of products (Sheiner 1992). Anderson and Hauck (1990) proposed an early approach for assessing the switchability, or *individual bioequivalence*, of test and reference products.

To assess better the interchangeability or switchability of products in individual subjects, there is increasing scientific (Sheiner 1992, Wellek 1997, Marzo and Balant 1995) and regulatory (Chen 1997, U.S. Department of Health and Human Services 1997) interest in developing new statistical procedures that allow the assessment of the following questions, which are not adequately examined by current methods for population bioequivalence:

1. Do the intrasubject variances of the test and reference products differ?
2. Are there subsets of the study population (and, thus, possibly the patient population) that show greater differences between test and reference products?
3. Can these differences be detected statistically as a subject × treatment interaction?

These questions can be answered only if replicate measures in the same individual are available. There are basically two experimental approaches to obtaining replicate measures: replicate designs in single-dose studies, and repeated-measurements designs in multiple-dose studies.

A replicate design in a *single-dose study* refers to the situation in which a subject receives each of the test and reference products in a single-dose administration, which event is replicated on more than one occasion. Thus, a replicate design in a single-dose study provides replicate observations of the important metrics.

Subjects in a *multiple-dose study* are administered each formulation on repeated regular intervals in order to attain steady-state conditions. Blood sampling is performed after the steady-state conditions supposedly have occurred. However, multiple-dose studies provide the opportunity to conduct blood sampling on repeated occasions for each subject within each period of the crossover design. For example, if subjects in a multiple-dose study are to be administered a formulation on each of d consecutive days, and it takes $d - q + 1$ days to attain steady-state conditions, then this multiple-dose study can provide multiple observations of the important metrics on q days of that period (days $d - q + 1$ through d).

The focus of the following statistical section is the assessment of average bioequivalence. Although multiple-dose studies provide an appropriate framework for the assessment of individual bioequivalence, as just discussed, the various statistical approaches that have been proposed are still under consideration (U.S. Department of Health and Human Services 1997).

II. STATISTICAL METHODS

We consider a general $s \times p$ crossover design in which subjects are randomized to s sequences containing test (T) and reference (R) formulations administered over p periods. Within sequence i, $i = 1, \ldots, s$, the test formulation appears in p_{Ti} periods and the reference formulation appears in p_{Ri} periods, with $p_{Ti} + p_{Ri} = p$. Suppose that measurements are taken on q occasions within each period after steady-state conditions have been attained. Let

$$\mathbf{Y}_{Tijk} = [Y_{Tijk1} \ldots Y_{Tijkq}]' \quad \text{and} \quad \mathbf{Y}_{Rijl} = [Y_{Rijl1} \ldots Y_{Rijlq}]' \tag{1}$$

denote the q-vector of observations for the kth occurrence of the test formulation and the lth occurrence of the reference formulation, respectively, for the jth subject within the ith sequence, $i = 1, \ldots, s$, $j = 1, \ldots, n_i$, $k = 1, \ldots, p_{Ti}$, and $l = 1, \ldots, p_{Ri}$. For example, the $\mathbf{Y}_{Tijk}$ and $\mathbf{Y}_{Rijl}$ could denote measurements of AUC or C_{max} on the natural log scale. We use the natural log transformation because there is reason to believe that AUC and C_{max} usually are log-normally distributed (Westlake 1988).

The model we assume is

$$Y_{Tijkm} = \mu_T + \mu^*_{Tm} + U_{Tij} + \gamma_{Tikm} + E_{Tijkm} \tag{2a}$$

$$Y_{Rijlm} = \mu_R + \mu^*_{Rm} + U_{Rij} + \gamma_{Rilm} + E_{Rijlm} \tag{2b}$$

for $i = 1, \ldots, s$, $j = 1, \ldots, n_i$, $k = 1, \ldots, p_{iT}$, $l = 1, \ldots, p_{iR}$ and $m = 1, \ldots, q$, where

μ_T = population mean for formulation T

μ_R = population mean for formulation R

μ^*_{Tm} = deviation from population mean at occurrence m for formulation T

μ^*_{Rm} = deviation from population mean at occurrence m for formulation R

γ_{Tikm} = a nuisance parameter (occurrence × sequence nested within formulation T)

γ_{Rilm} = a nuisance parameter (occurrence × sequence nested within formulation R)

$\mathbf{U}_{ij} = [U_{Tij} U_{Rij}]'$ = bivariate random subject effect

$\mathbf{E}_{Tijk} = [E_{Tijk1} \cdots E_{Tijkq}]' = q$-vector of random errors under formulation T

$\mathbf{E}_{Rijl} = [E_{Rijl1} \cdots E_{Rijlq}]' = q$-vector of random errors under formulation R

Constraints on the nuisance parameters are necessary to prevent overparameterization of the model with respect to the fixed location parameters;

$$\sum_{m=1}^{q} \mu^*_{Tm} = 0 \quad \text{and} \quad \sum_{m=1}^{q} \mu^*_{Rm} = 0 \tag{3a}$$

$$\sum_{i=1}^{s} \sum_{k=1}^{p_{Ti}} \gamma_{Tikm} = 0 \quad \text{and} \quad \sum_{i=1}^{s} \sum_{l=1}^{p_{Ri}} \gamma_{Rilm} = 0 \quad \text{for each } m = 1, \ldots, q \tag{3b}$$

With the constraints in place, the model in Eq. (2) is said to be *saturated*, because there are a total of spq unrestricted location parameters, which corresponds exactly to the number of cells in the $s \times p$ crossover design with q occurrences within each of the p periods. It is not necessary to include all of the nuisance parameters in the model and force it to be saturated. Although a saturated model is conservative, it protects against confounding of the formulation means with any excluded nuisance parameters.

The nuisance parameters $\boldsymbol{\mu}^*_T = [\mu^*_{T1} \cdots \mu^*_{Tq}]'$ and $\boldsymbol{\mu}^*_R = [\mu^*_{R1} \cdots \mu^*_{Rq}]'$ are of special interest because they represent deviations from test and reference formulation means, respectively, over the q occurrences within each period. If steady-state conditions have not been achieved, then it is anticipated that the estimates of $\boldsymbol{\mu}^*_T$ and $\boldsymbol{\mu}^*_R$ will differ significantly from zero and the variability of the estimated μ_T and μ_R will be relatively large. In such a situation, there will be decreased statistical power for bioequivalence testing.

In addition, we assume that carryover effects do not exist in the proposed model. This is a reasonable assumption in bioequivalence trials (Westlake 1988) if the washout period is of adequate length, but may not be so for other types of crossover trials. If the existence of carryover effects is suspected, then the model can be modified to include them.

The distributional assumptions for the jth subject within the ith sequence, $i = 1, \ldots, s$, and $j = 1, \ldots, n_i$ are as follows:

$$\mathbf{U}_{ij} = \begin{bmatrix} U_{Tij} \\ U_{Rij} \end{bmatrix} \sim \mathcal{N}_2\left(\begin{bmatrix} 0 \\ 0 \end{bmatrix}, \boldsymbol{\Omega} = \begin{bmatrix} \omega_{TT} & \omega_{TR} \\ \omega_{TR} & \omega_{RR} \end{bmatrix}\right) \tag{4}$$

where $\boldsymbol{\Omega}$ is an unknown positive definite matrix;

$$\mathbf{E}_{Tijk} \sim \mathcal{N}_m(\mathbf{0}, \sigma^2_T \mathbf{I}_m) \quad \text{and} \quad \mathbf{E}_{Rijl} \sim \mathcal{N}_m(\mathbf{0}, \sigma^2_R \mathbf{I}_m) \tag{5}$$

where σ^2_T and σ^2_R are unknown positive scalars and $\mathbf{I}_m$ is the $m \times m$ identity matrix. Further, we assume that the random subject effects and the random errors are mutually independent.

The model and assumptions described in Eqs. (1)–(5) are a generalization of those presented by Chinchilli et al. (1994) for multiple-dose studies with a 2×2 crossover design and Chinchilli and Esinhart (1996) for single-dose studies with an $s \times p$ crossover design. Other researchers have considered similar types of models with the distributional assumptions in Eqs. (4)–(5) for the case $q = 1$, such as Ekbohm and Melander (1989) and Sheiner (1992). The distributional assumptions in Eqs. (4)–(5) differ from those typically assumed for a crossover experiment, such as univariate random effects, i.e., $U_{Tij} = U_{Rij}$ for $i = 1, 2$ and $j = 1, \ldots, n_i$, and homogeneous intrasubject variances, i.e., $\sigma^2_T = \sigma^2_R$, such as the books by Jones and Kenward (1989) and Ratkowsky et al. (1993).

The model and assumptions in Eqs. (1)–(5) fall into the general realm of a mixed-effects linear model. Therefore, standard statistical software packages can be invoked to perform a likelihood analysis of a particular data set. A special case of interest is a design that is uniform within sequences, defined by Laska et al. (1983) as the situation in which the test formulation appears the same number of times within each sequence ($p_{T1} = \cdots = p_{Ts}$) and the reference formulation appears the same number of times within each sequence ($p_{R1} = \cdots = p_{Rs}$). If the crossover design is uniform within sequences and the model is saturated in terms of the location parameters, then, as Chinchilli and Esinhart (1996) showed, the maximum likelihood (ML) and restricted maximum likelihood (REML) estimates of the location parameters and the variance-covariance components have a closed-form solution. Under these circumstances, if $\delta = \mu_T - \mu_R$, then

$$\hat{\delta} = \bar{Y}_{T\ldots} - \bar{Y}_{R\ldots} \tag{6}$$

is the ML and REML estimator of the formulation difference. The estimator in Eq. (6) is intuitive, because it simply involves averaging the response over all appropriate observations. The variance expression for the estimator in Eq. (6) is

$$\text{var}(\hat{\delta}) = \frac{1}{s^2}\left(\frac{1}{n_1} + \cdots + \frac{1}{n_s}\right) \cdot \left(\omega_{TT} + \omega_{RR} - 2\omega_{TR} + \frac{1}{p_T q}\sigma^2_T + \frac{1}{p_R q}\sigma^2_R\right) \tag{7}$$

The variance formula in Eq. (7) is useful for sample

size calculations, as described in Chinchilli et al. (1994) for the 2×2 crossover design.

The typical approach for assessing bioequivalence with respect to average bioavailability based on pharmacokinetic endpoints such as AUC and C_{max} is the two one-sided testing approach introduced by Schuirmann (1987). Alternatively, researchers often calculate the $100(1 - 2\alpha)\%$ confidence interval for $\mu_T - \mu_R$, denoted as $[\hat{\delta}_\alpha, \hat{\delta}_{1-\alpha}]$, exponentiate the endpoints of this confidence interval, and ascertain whether this exponentiated confidence interval lies within (0.80, 1.25). The null hypothesis of bioinequivalence at the α significance level is rejected in favor of the alternative hypothesis of bioequivalence if the exponentiated $100(1 - 2\alpha)\%$ confidence interval lies entirely within (0.80, 1.25). However, Berger and Hsu (1996) have demonstrated that the $100(1 - 2\alpha)\%$ confidence interval does not correspond with the bioequivalence testing problem. They argue that the appropriate confidence set is the $100(1 - \alpha)\%$ confidence interval constructed as $[\min(0, \hat{\delta}_\alpha), \max(0, \hat{\delta}_{1-\alpha})]$.

III. RESULTS AND DISCUSSION

A multiple-dose study was conducted by the Department of Pharmacy and Pharmaceutics at Virginia Commonwealth University, supported by Mova Pharmaceutical Company, in order to assess the average bioequivalence of test and reference formulations of levothyroxine sodium. The study consisted of 24 healthy male volunteers who were randomized to receive either sequence TR or sequence RT. Blood samples were drawn from each subject on days 41 and 43 of steady state within each period of formulation administration, so $q = 2$ in the notation of our model in Eqs. (1)–(2). For purposes of illustration, we focus on the analysis of area under the curve (AUC) only.

We applied the natural log transformation to the data and proceeded with the statistical model and analysis as described earlier. The SAS code for the analysis using PROC MIXED is as follows:

```
PROC MIXED;
CLASS subject formulat;
MODEL log_auc=mu_t mu_r mustar_t mustar_r
  gamma_t1 gamma_t2 gamma_r1 gamma_r2 /
  NOINT SOLUTION;
RANDOM mu_t mu_r / SUBJECT=subject
  TYPE=UN G;
REPEATED / SUBJECT=subject TYPE=SIM R
  GROUP=formulat;
ESTIMATE 'formulation diff' mu_t 1 mu_r - 1
  /DF=22;
```

The uppercase components of the SAS code represent statements and keywords, whereas the lowercase components represent user-provided variable names and options. Variables corresponding to the two formulation means and the six nuisance parameters (mustar_t, mustar_r, gamma_t1, gamma_t2, gamma_r1, gamma_r2) are created using IF-THEN statements within an SAS DATA step. The RANDOM and REPEATED statements are constructed in a manner to provide the inter- and intrasubject variance components, as described in the model and assumptions of the previous section. The ESTIMATE statement yields the estimated formulation difference and its standard error, which are needed to construct the 95% confidence interval for assessing average bioequivalence.

The estimated test and reference formulation means for the logarithm of the AUC are 6.892 and 6.886, respectively. None of the six estimated nuisance effects are significantly different from zero ($p > 0.15$). The estimated variance components are as follows:

Parameter	Estimate
ω_{TT}	0.0518
ω_{RR}	0.0422
ω_{TR}	0.0400
σ_T^2	0.0035
σ_R^2	0.0039

With respect to the assessment of average bioequivalence, the ratio of the geometric means on the original scale is 1.001, and the 95% confidence interval using the Berger and Hsu (1996) approach is (0.96, 1.05). Therefore, average bioequivalence is concluded for AUC. As an aside, the intrasubject variances of the two formulations, 0.0035 and 0.0039, appear to be very similar.

REFERENCES

Anderson S, Hauck WW. (1990). Consideration of individual bioequivalence. J Pharmacokinetics Biopharmaceutics 18: 259–273.

Berger RL, Hsu JC, (1996). Bioequivalence trials, intersection-union tests and equivalence confidence sets. Statistical Sci 11:283–319.

Chen ML. (1997). Individual bioequivalence—a regulatory update. J Biopharmaceutical Stat 7:5–11.

Chinchilli VM, Esinhart JD. (1996). Design and analysis of intra-subject variability in crossover experiments. Stat Med 15:1619–1634.
Chinchilli VM, Esinhart JD, Barr WH. (1994). Analysis of multiple-dose bioequivalence studies. J Biopharmaceutical Stat 4:423–435.
Ekbohm G, Melander H. (1989). The subject-by-formulation interaction as a criterion of interchangeability of drugs. Biometrics 45:1249–1254.
Jones B, Kenward MG. (1989). Design and Analysis of Cross-Over Trials. New York: Chapman and Hall.
Laska EM, Meisner M, Kushner HB. (1983). Optimal crossover designs in the presence of carryover effects. Biometrics 39:1087–1091.
Marzo A, Balant LP. (1995). Bioequivalence. An updated reappraisal addressed to applications of interchangeable multi-source pharmaceutical products. Arzneimittelforschung 45:109–115.
Ratkowsky DA, Evans MA, Alldredge JR. (1993). Cross-Over Experiments: Design, Analysis, and Application. New York: Marcel Dekker.
Schuirmann DJ. (1987). A comparison of the two one-sided tests procedure and the power approach for assessing the equivalence of average bioavailability. J Pharmacokinetics Biopharmaceutics 15:657–680.
Sheiner LB. (1992). Bioequivalence revisited. Stat Med 11: 1777–1788.
U.S. Department of Health and Human Services, Food and Drug Administration, Center for Drug Evaluation and Research. (1997). Guidance for Industry: In Vivo Bioequivalence Studies Based on Population and Individual Bioequivalence Approaches (available Internet site is http: //www.fds.gov.cder/guidance/index.htm).
Wellek S. (1997). A comment on so-called individual criteria of bioequivalence. J Biopharmaceutical Stat 7:17–21.
Westlake WJ. (1988). Bioavailability and bioequivalence of pharmaceutical formulations. In: KE Peace, ed. Biopharmaceutical Statistics for Drug Development. New York: Marcel Dekker, pp 329–352.

Vernon M. Chinchilli
William H. Barr

Multiple Endpoints

See also *Multiple Comparison*

I. INTRODUCTION

A common problem arising in medical research is that of comparing some groups of patients (for instance, the arms of a clinical trial) based on multiple outcome measures named *endpoints*, all of which must be considered of primary importance and have no hierarchical structure. The aim of this entry is to give a synthetic account of some actually known methods of analysis of multiple endpoints. Each method is described with its merits and with the situations where it is most appropriate. The procedures of application will be detailed wherever possible. However, sometimes the reader will be referred to the original article for the steps of computation. Most of the published material on the topic concerns the comparison of two groups. So in most of this entry, only the comparison of two arms of a study will be considered. They will be identified as experimental and control. Consistent with the informative nature of this entry, mathematical formalism will be kept to a minimum.

In the analysis of a multiple endpoint study two types of questions are possible:

1. Are there any endpoints for which the experimental treatment is more effective than the control (or vice versa)? Which are these endpoints?
2. Does the combined evidence of all endpoints support the global superiority of the experimental treatment, even if no clear conclusions can be drawn for any determined endpoint?

By answering the first question one often answers the second too. Indeed, if one proves that the experimental treatment is more effective than the control for determined endpoints, and no less effective for the others, one has also proved that the treatment is globally superior to the control. The reverse is not true.

II. NOTATION

C is the control treatment, T is the experimental treatment.

k is the number of endpoints.

n_T and n_C are the numbers of subjects, respectively, in the T and C samples.

S_i or s_i is a test statistic that compares the effect of T and C on the ith endpoint. (Uppercase indicates a random variable, lowercase an observed value.)

P_i (or p_i) is the unadjusted P-value of the univariate test on the ith endpoint. (Uppercase indicates a random variable, lowercase an observed value). P_i doesn't take into account the multiplicity of endpoints. Therefore P_i (or p_i) is called *raw* or *nominal*.

P_{ia} (or p_{ia}) is the P-value of the test on the ith endpoint adjusted by taking into account the multiplicity of endpoints. Therefore P_{ia} (or p_{ia}) is called *adjusted*.

S or s is a test statistic that compares the effect of T and C simultaneously on all endpoints.

P or p is the P-value of the global comparison, testing the global difference of the effect of T and C on all endpoints, without identifying the endpoints where the difference is "significant."

H_0 is the global null hypothesis. It states that the difference of effect of T and C is zero on all endpoints.

H_{0i} is the null hypothesis that the effect of T and C on the ith endpoint is the same.

$\sim H_{0i}$ is an alternative hypothesis to H_{0i} and to H_0. It states that the effect of T and C on the ith endpoint is not the same; it is the complement of H_{0i}.

m is the number of true H_{0i} in the trial.

$\mathcal{P}(E)$ is the probability of an event E.

$\bar{E}$ is the complementary event of E.

FWER is the familywise error rate in the strong sense. It is the probability of rejecting one (or more) true null hypotheses when a mixed set (a family) of true and false null hypotheses is tested. In other words, it is the probability of making at least one Type I error in the whole set of comparisons.

α_c is the Type I error probability required for each comparison to obtain FWER $= \alpha$.

If not otherwise specified, all P-values are taken two-sided.

III. IDENTIFICATION OF THE ENDPOINTS WHERE T IS MORE EFFECTIVE THAN C

Identifying the endpoints affected by T is a multiple comparison problem. It requires the test of one hypothesis for each endpoint. Because of the many hypotheses tested, the main difficulty is to avoid the multiple testing bias without losing too much power. To reach this aim there are different sorts of methods. Some of them compute adjusted P-values using the p_i's obtained by the univariate test (even if the tests are different for different endpoints). These methods can also be used by the reader of a scientific paper reporting p_i's. Their drawback is that they are often difficult to apply to the computation of confidence intervals or of the required sample size of a clinical trial.

Some other procedures use the distribution or at least the correlation matrix of the endpoints under H_0. They either postulate a known distribution or empirically estimate it. This type of technique has some specific developments for binary endpoints.

A. *P*-Value-Based Techniques

The simplest technique of this sort is the *Bonferroni correction*, which consists in multiplying all p_i's by k. The p_{ia}'s so obtained are then compared with the chosen Type I error probability: α (e.g., 0.05). This technique keeps FWER $\leq \alpha$. Therefore from the p_{ia}'s one can draw inferences concerning each single endpoint exactly as if it were the only one tested. The Bonferroni method is based on the Bonferroni probability inequality (1), which is true for every set of k events E_j:

$$1 - \mathcal{P}\left(\bigcap^{k} E_j\right) \leq \sum^{k} \mathcal{P}(\bar{E}_j) \tag{1}$$

Let us consider a family of k endpoints. Let E_j be the event that the test statistic S_j of the jth endpoint assumes a value more probable (or with a higher probability density) than a certain s_j having P-value p. Then, the left-hand side is the probability of obtaining, at least once, a P-value equal to or lower than p in the whole set of k comparisons. Let $\bar{E}_j$ be the complement of E_j. The P-value p of the statistic s_j, by definition, is the probability of obtaining a value of S_j equally probable to or less probable (or with an equal or lower density) than s_j if H_{0j} is true. Therefore, if H_{0j} is true, $\mathcal{P}(\bar{E}_j) = p$. Moreover, if there are $m \leq k$ true H_{0j}'s, then the probability of all the corresponding events $\bar{E}_j$ is equal to p. Therefore, applying the Bonferroni inequality only to the endpoints unaffected by the treatment one obtains

$$1 - \mathcal{P}\left(\bigcap^{m} E_j\right) \leq mp \leq kp \tag{2}$$

Now, one can take p equal to the observed raw P-

value p_i of a certain H_{0i}. Then the inequality states that the probability of obtaining once or more, for the set of m true H_{0s}'s, a test statistic S_j equally probable to or less probable (or with an equal or lower density) than a value s_j having P-value p_i is lower than kp_i. By a slight generalization of the definition of P-value, we call that quantity p_{ia}, i.e., the P-value of the ith endpoint, computed by taking into account the error probability of all the comparisons. In other words, p_{ia} is the P-value adjusted for multiplicity.

Another very simple formula for the computation of adjusted p_i's, is obtained under the condition that the endpoints are stochastically independent. Its derivation is the following: The probability of obtaining, for all P_j's of the m true H_{0j}'s, values larger than p_i is $(1 - p_i)^m$, under independence. Therefore the probability of obtaining for at least one P_j corresponding to a true H_{0j} a value equal to or smaller than p_i is

$$1 - (1 - p_i)^m \leq 1 - (1 - p_i)^k \tag{3}$$

So one can take $p_{ia} = 1 - (1 - p_i)^k$.

Many improvements to the Bonferroni correction to test single endpoints have been proposed; they obtain a greater power by taking into account all the p_i's, instead of just the one to correct (2–4). Some of these are very simple to use. The general logic of Holm's step-down technique (2) is the following: Put the p_i's in increasing order and adjust the smallest by the Bonferroni correction. If the p_{ia} so obtained is larger than the chosen Type I error probability α, retain all H_{0i}'s. If, on the contrary, it is smaller than α, then reject the corresponding H_{0i}. This H_{0i} is then to be excluded from the set of possible true null hypotheses. Therefore adjust the second smallest p_i by multiplying it by the correction factor $k - 1$. Indeed, remember from Eq. (2) that to adjust p_i, it is enough to multiply it by the number of true H_{0i}'s; now, after the rejection of one hypothesis, we know that the number of true H_{0i}'s is not larger than $k - 1$. If the second p_{ia} is larger than α, then retain the corresponding hypothesis and all those having larger p_i's. If it is equal or smaller, then reject the corresponding H_{0i}, and go on with the same logic to compute the third p_{ia} by multiplying the third smallest p_i by $(k - 2)$. Compare the third p_{ia} with α, and so on. The procedure stops when a p_{ia} turns out to be nonsignificant: then retain the H_{0i} under scrutiny and the remaining H_{0i}'s.

Hochberg's step-up procedure (4) theoretically requires the stochastic independence of the test statistics used to detect the effect of T on different endpoints. However, simulation results (5,6) confirm that it is conservative, i.e., keeps the FWER smaller than the declared α, also when the test statistics are positively correlated. Hochberg and Rom (7) study its behavior when the p_i's are obtained from Z-tests (the simplest test for Gaussian endpoints with known variance). They prove theoretically that Hochberg's procedure is conservative when the test statistics are either positively correlated or examined by two-sided tests. When the Z are negatively correlated *and* the tests are one-sided with $\alpha = 0.05$, the theoretical upper bound of FWER is 0.0525.

The derivation of Hochberg's method is rather complex and is explained in three papers (3,4,8). However its application is simple: Order the p_i's in decreasing order. If the largest p_i is equal to or smaller than the wanted α, all the H_{0i}'s are rejected. Otherwise, multiply the second largest p_i by 2: If the product is equal to or smaller than α, reject the corresponding H_{0i} and all the H_{0i}'s with smaller p_i's. Otherwise, retain the corresponding H_{0i}, multiply the third largest p_i by 3, and compare it with α. If it is equal to or smaller than α, reject the corresponding H_{0i} and all the H_{0i}'s with smaller p_i's. If it is larger, then retain the corresponding H_{0i}, and so on. The procedure continues until an H_{0i} is rejected; then all H_{0i}'s with a smaller p_i than the first rejected H_{0i} are rejected as well. The p_{ia} of each endpoint is obtained from the corresponding raw p_i multiplied by its rank (in decreasing order). If this product is larger than the p_{ia} computed from the preceding p_i, it is substituted by the preceding p_{ia}. This is done to respect the nonincreasing order of the p_{ia}'s.

Hochberg's technique is always more powerful than Holm's. For an example of its application to a real clinical trial data as compared to other adjustment techniques, see Ref. 9.

Hochberg and Benjamini (10) improved Hochberg's method using an estimate $\hat{m}$ of the number of the true H_{0i}'s in the trial. The value of $\hat{m}$ is obtained by applying the graphical technique of Schweder and Spjøtfoll (11) (described in the next section of this entry) to two-sided p_i's. At the beginning of the procedure, order the p_i's in decreasing order and retain all H_{0i}'s having a p_i larger than α. Let a be the number of accepted H_{0i}'s. Then multiply the largest p_i of the H_{0i}'s not retained by the smaller of the values $a + 1$ and $\hat{m}$. If the product is smaller than α, reject the corresponding H_{0i} and all the ones with smaller p_i's. Otherwise, retain the corresponding H_{0i} and multiply the next p_i by the smaller of the values $a + 2$ and $\hat{m}$. If the product is smaller than α, reject the corresponding H_{0i} and all the H_{0i}'s with smaller p_i's. Otherwise, retain the concerned H_{0i} and multiply the next p_i by the smaller of the values $a + 3$ and $\hat{m}$, and so on. The process stops when a hypothesis is rejected; then all the following ones are rejected too.

The reason this procedure is an improvement of Hochberg's is that one never needs to multiply a p_i by a correction factor larger than the number of true H_{0i}'s. The p_{ia}'s can be defined in the following way: Each p_{ia} is the product that would test the corresponding H_{0i}, if hypothetically there wouldn't be any rejections till the smallest p_i. Whenever such product is found larger than the one preceding it in the testing procedure, the corresponding p_{ia} is taken equal to the previous one.

Brown and Russell (5) studied both the FWER control and the power of 17 different P-value-based adjustment techniques (among them there are nine step-down or step-up procedures and eight methods based on the parametric or nonparametric estimation of the distribution function of the P_i's). They simulated two different numbers of endpoints, three patterns of correlation among endpoints, three distributions of truth and falseness among the H_{0i}'s, and two different amounts of deviation from the null hypotheses. Each statistical technique was applied 10,000 times in each of the 36 simulated situations, with $\alpha = 0.05$. The authors found that the improved Hochberg's procedure (see earlier) (10) falsely rejects any true H_{0i} at most 5.16% of the time. As for the power, that method is consistently superior in all experimental situations to seven other techniques and consistently inferior to none. No other examined procedure approaches this power achievement. So, concluding their review, the authors recommend preferring the improved Hochberg's step-up procedure to the other P-value-based techniques.

Adjusted confidence intervals of outcome corresponding to step-down and step-up procedures are generally not easily available. Indeed, the estimated number of true null hypotheses isn't immediately useful for confidence interval computation. Some particular results are worked out in Hayter and Hsu (12).

It is difficult to imagine the algorithm to compute the optimal sample size of a future trial to be analyzed by the step-up or step-down procedures. However, if one plans to use the improved Hochberg method, a reasonable guess could be made in the following way. From prior knowledge, get an estimate $\hat{m}$ of the number of true H_{0i}'s in the future trial (it should be less than $k/2$; otherwise the choice of the primary endpoints of the trial needs to be reconsidered). Identify the endpoint on which the expected standardized effect of T, if present, is the smallest. Determine the minimum clinically relevant effect of T on this endpoint (it should be less than the expected effect; otherwise the endpoint shouldn't have been chosen as a primary one). Compute the required sample size to achieve the desired power (e.g., 0.8), as if the analysis were made on just this endpoint with a Type I error probability of $\alpha/\hat{m}$ (e.g., $0.05/\hat{m}$).

B. Techniques Based on the Distribution or the Correlation Structure of the Endpoint

A method based on the estimation of the correlation between endpoints is the one proposed by Tamhane (13). The adjusted p_i of each endpoint is computed by the following equation:

$$p_{ia} = 1 - (1 - p_i)^{k^{(1-\bar{\rho})}} \tag{4}$$

where k is the number of endpoints and $\bar{\rho}$ is their average sample correlation (this can be computed from the sample correlation matrix of the endpoints, which is output by any standard statistical program). The rationale behind the equation is largely commonsense: when the endpoints are completely independent, $\bar{\rho} = 0$, and Tamhane's formula turns out to be the simple correction Eq. (3) for independent tests. However, when all the correlations are 1—that is, all the endpoints are really just one endpoint (but for additive and multiplicative constants)—then the equation rightly doesn't correct the raw p_i's at all. Equation (4) has the advantage of simplicity, but it is not supported by any theoretical evidence.

Sankoh et al. (6) empirically studied the performance of two similar correction equations, based on the estimation of the correlation between endpoints. Sankoh's simulations show that they are rather liberal. For both equations the percentage of false significant results can be as high as 12% when the declared significance level of the procedure is 0.05. Sankoh proposes a method to correct their error rate, based on simulations (see later).

Tamhane's formula can be used to construct confidence intervals for the endpoint differences adjusted for multiplicity. The per-interval confidence level, $1 - \alpha_c$, can be determined by the following equation:

$$(1 - \alpha_c)^{k^{1-\bar{\rho}}} = 1 - \alpha \tag{5}$$

If each confidence interval is computed at level $1 - \alpha_c$, then the probability of getting all intervals to include the true parameters should be $1 - \alpha$. However, this value of FWER could be optimistic, depending on the validity of Tamhane's formula. Empirical validation of the technique is needed.

From Tamhane's formula one can compute the approximate sample size required for a multiple-endpoint clinical trial: one needs to know, or estimate from prior knowledge, the average correlation between endpoints. An example of sample size computation is as follows: Imagine a trial with eight endpoints, with a foreseen mean correlation of 0.7. Use Eq. (5) to compute the per-comparison significance level α_c required to achieve the desired global α. Compute α_c by solving the following equation:

$$1 - (1 - \alpha_c)^{8^{(1-0.7)}} = 0.05$$

The solution is $\alpha_c = 0.027$. Then compute the sample size required for a level-0.027 test (with the desired power) applied to the endpoint with the smallest expected standardized treatment effect. This should also be the required sample size of the trial.

Tamhane (13) also describes the *bootstrap* (or resampling) technique, which adjusts the p_i's by empirically estimating their distribution under H_0. The rationale behind the bootstrap technique goes as follows: The probability distribution of any quantity Q meant to compare the groups of the trial can be estimated from its observed frequency distribution. If the null hypothesis is true for all endpoints, then the T and C samples of the trial differ in no systematic way; they come from the same population. So under H_0, the two treatment groups can be thought of as random subsamples extracted from the total trial sample. One can artificially extract, with replacement, thousands more couples of random subsamples from the total trial sample; under H_0, they all belong to the same population as the real data. In those artificial couples of subsamples, one can observe the empirical frequency distribution of the minimum P_i of the endpoints. So the frequency with which the minimum P_i is equal to or lower than a certain p_j obtained in the original data can be observed. It estimates the probability of obtaining such a p_j or a smaller P-value for at least one endpoint if H_0 is true. In other words, it is an estimate of p_{ia}. Pooling the samples doesn't change the correlation of the endpoints. Therefore, the bootstrap adjustment estimates p_{ia}'s based on exact correlations. However, unlike the P-value-based techniques, it fails to take into account the complete set of p_i's when adjusting each of them. In other words, the adjustment is made assuming that all H_{0i}'s are true, which is probably conservative. Step-down and step-up versions of the procedure are available (14–16). The step-down version follows the logic of Holm's method: put the p_i's in increasing order. Resample data containing all endpoints to adjust the minimum p_i. Then exclude the corresponding endpoint from the data resampled to adjust the second smallest p_i. Third, also exclude the endpoint corresponding to the second p_i from the resampling, and adjust the third smallest endpoint. Go on until all the endpoints are adjusted. This method has been implemented in SAS, to correct the P-values of the t-test and the Fisher exact test for proportions, among others. It is invoked with the PROC MULTTEST command with the options BOOTSTRAP and STEPDOWN (17).

C. Techniques to Analyze Categorical and Ordinal Endpoints, Based on Discrete Probability Distributions

Some interesting methods based on the distribution of the observations are available to analyze dichotomous multiple endpoints; they are based on the "exact" Fisher test for the equality of two proportions. To use them, this test must be applied to compare the frequencies of *good* outcomes of the two arms of the trial for each endpoint. It produces discrete P-values, with the consequence that p_i's lower than certain bounds are impossible. If for a given endpoint i one gets a p_i smaller than the minimum possible P-value for every other endpoint, then no correction is needed for that p_i. Indeed it can arise by chance only in one comparison: the one where it was obtained. So no multiple-testing bias exists. If only a subset of q comparisons can generate P-values equal to or smaller than p_i, then the Bonferroni correction should take into account only these q comparisons. Indeed, p_i or a smaller P-values can arise by chance only in these q comparisons but not in the whole set of k tests. Therefore one can take $p_{ia} = qp_i$. To establish the minimum possible (one-sided) value of P_i for each comparison, the following equation is available:

$$\frac{\binom{g}{n_T}}{\binom{n}{n_T}} \text{ if } g \geq n_T \text{ and } \frac{\binom{n-g}{n_T-g}}{\binom{n}{n_T}} \text{ if } g < n_T \tag{6}$$

where n is the total number of subjects, n_T is the number treated, and g is the number of observations with the *good* outcome.

Following this logic, Mantel (18) proposed a more precise multiple comparison procedure for the Fisher test. To adjust a p_i, one adds together the largest theoretically possible P-values of all other comparisons, which are equal to or smaller than p_i. Of course, in order to follow this procedure one has to know all the possible values P_j of each comparison. They are obtained from the hypergeometric probability distribution of g_T: the number of subjects in the treated group experiencing the *good* outcome:

$$\mathcal{P}(g_T) = \frac{\binom{g}{g_T}\binom{n-g}{n_T-g_T}}{\binom{n}{n_T}} \qquad (7)$$

Consider the following example: Imagine a trial with five dichotomous endpoints (say) A, B, C, D, E. The observed p_i of endpoint A is 0.019. To correct it, get the total number g of subjects with *good* outcome, observed for endpoint B. Compute all its possible P-values, by substituting each possible value of g_T into Eq. (7). Do the same for endpoints C, D, and E. Now, among all possible P-values computed for endpoints B, C, D, and E, look for those equal to or smaller than 0.019. Imagine that the largest of such values for endpoint B is 0.016 and that the largest for endpoint D is 0.012. Endpoints C and E have no possible P-values smaller than 0.019. Then the adjusted P-value of endpoint A is $0.019 + 0.016 + 0.012 = 0.047$.

This procedure can be used only when the theoretically possible values of P_i of the k comparisons are known. This happens when small frequencies are analyzed by the Fisher exact test, but it also happens when small data sets are analyzed by nonparametric techniques, e.g., the Wilcoxon test. Therefore Mantel's adjustment method can also be applied to ordinal data.

The discrete distribution of the endpoints can be exploited together with Hochberg's improved step-up procedure. One could proceed in the following way: Start Hochberg's improved procedure as usual, by putting the p_i's in decreasing order and by retaining all H_{0i}'s corresponding to p_i's $> \alpha$. From the second step on, multiply the p_i under scrutiny by the minimum of these three numbers: $\hat{m}$ (the graphically estimated number of true H_{0i}'s), $a_p + 1$ (the number of retained hypotheses in all preceding steps + 1), and q (the number of comparisons that can generate P-values equal to or smaller than p_i). If the obtained product is larger than α, then retain the corresponding hypothesis and examine the next p_i. Otherwise, reject the corresponding hypothesis and all the following ones. This combined procedure should allow interesting power gains, because q tends to be small when a is large, and vice versa.

D. False Discovery Rate

An interesting alternative to the usual approach is the one proposed by Benjamini and Hochberg (19). They argue that it is sometimes more interesting to control the proportion of false rejections than to keep low the probability of at least one false-positive result. This seems a reasonable point of view when some tens of endpoints are to be tested and the researcher has no strong interest in any single one but rather looks for clusters of affected endpoints. The authors define the false-discovery rate as the expected proportion of rejected true H_{0i}'s among all rejected H_{0i}'s. They describe a test procedure and prove that (under stochastic independence of the endpoints unaffected by T) it warrants an expected false-discovery rate lower than a chosen value α_R (e.g., $\alpha_R = 0.05$). The procedure goes as follows: Put the raw p_i's in decreasing order. Compare the largest p_i with α_R: If it is smaller or equal, then reject all hypotheses. Otherwise, retain the corresponding hypothesis and compare the second largest p_i with $(k - 1)\alpha_R/k$. If it is smaller or equal, reject the corresponding H_{0i} and all the following H_{0i}'s. Otherwise, retain the corresponding hypothesis and compare the third largest p_i with $(k - 2)\alpha_R/k$ etc. Go on in this way until either a rejection occurs or the last p_i has been compared to α_R/k. Not surprisingly, the simulations show that, mostly when $k > 10$, this method applied with $\alpha_R = 0.05$ has a much higher power than the techniques that keep the FWER below 0.05.

IV. INFERRING THE GLOBAL SUPERIORITY OF THE TREATMENT FROM THE COMBINED EVIDENCE OF ALL ENDPOINTS

Establishing the global superiority of T can be thought of as a single-hypothesis-testing problem. The multivariate null hypothesis to be rejected H_0 declares that the true differences between the two arms are zero for *all* endpoints. Many alternative hypotheses are conceivable; it is important to choose the one that is relevant to prove in the trial at hand. The different multivariate alternative hypotheses we consider are the following.

1. $\sim H_0$ is an alternative to H_0, which contains and mixes up every possible difference of effect of T and C. Inequalities of opposite sign for different endpoints are included in $\sim H_0$. It is the complement of H_0.
2. $H_{T \geq C}$ is an alternative to H_0, which states that T has a better effect than C on at least one endpoint and has a worse effect on no endpoints.
3. H_A is an alternative hypothesis to H_0, which states that the average (defined somehow) of the

effects of T on all endpoints is better than the average effect of C.

All these hypotheses state that the effects of T and C are different, but they do not identify the affected endpoints. So they are less informative than the alternative hypotheses concerning single endpoints. However, they are proved by just one test, whereas the $\sim H_{0i}$'s are many and need many tests. For that reason, the procedures designed to prove $\sim H_0$, $H_{T \geq C}$, or H_A are often much more powerful than those used to test the H_{0i}'s. They are to be considered especially in the final stage of the research on a treatment, when one is often more interested in assessing its multidimensional efficacy, than in describing accurately the pattern of its clinical effects. For each of the three alternative hypotheses just mentioned, different statistical procedures are recommended.

To use correctly the procedures described next, be sure to measure all endpoints in such a way that positive differences always correspond to a more favorable outcome.

A. Techniques Powerful in Detecting $\sim H_0$

The alternative hypothesis $\sim H_0$ is rarely an interesting one; the discovery that some endpoints are favorably, and others unfavorably, affected by T, without identifying the concerned endpoints, can rarely have practical consequences. Anyway, two procedures can be recommended to prove $\sim H_0$. Hotelling's T^2 is the uniformly most powerful test over $\sim H_0$ (subject to some invariance conditions) when the endpoints are multivariate gaussian (20,21). This means that it detects violations of the null hypothesis, when the truth lies wherever in the specified alternative hypothesis, with higher probability than every other test. This test is routinely performed by any program of multivariate analysis of variance (like MANOVA of SPSS or GLM of SAS). To apply it, one just has to compare the two arms by such programs, inputting all the endpoints as response variables.

If the endpoints are far from Gaussian, a "*discriminant*" *logistic regression* might work better, as suggested by Cupples et al. (22). The procedure can also be used with categorical endpoints. It goes as follows: First fit a logistic model with the allocated treatment as the *response* variable and no explanatory variables. Then fit another logistic model with the same response variable, with all the endpoints input as *explanatory* variables. The P-value of the multivariate comparison between treatments is equal to the P-value that compares the fit of the two logistic models. A worked example that applies this technique can be found in Ref. 9. However, notice that with this method one cannot include prognostic covariates and stratification factors in the analysis.

B. Techniques Powerful in Detecting $H_{T \geq C}$

$H_{T \geq C}$ is certainly an interesting hypothesis to prove; however, the significance of the tests powerful in detecting it, is not a proof of it. Indeed, they can reject H_0 with a probability much larger than α, also when $H_{T \geq C}$ is far from true (this happens even if the Type I error rate is correctly kept to α). A typical situation where the tests described in this section reject H_0 when $H_{T \geq C}$ is false is when T has a strong favorable effect on some endpoints and a weaker unfavorable one on others. Therefore to use the tests described in this paragraph as a proof of $H_{T \geq C}$, one must be sure that the difference of effect between T and C is zero or is favorable to T for all endpoints.

Schweder and Spjøtfoll (11) propose a *graphical technique*. This method is not a proper statistical test but allows one to estimate the number of endpoints that are favorably influenced by T. It is based on the raw one-sided p_i's computed on the single endpoints. [In case one-sided p_i's weren't directly available, as happens with a X^2 test, then they can be computed in the following way: See in the data if the effect of T on the concerned endpoint appears to be favorable (e.g., the frequency of patients free of some discomfort appears to be increased by T). If it looks so, then: $p_{i\,\text{one-sided}} = p_{i\,\text{two-sided}}/2$; otherwise, $p_{i\,\text{one-sided}} = 1 - (p_{i\,\text{two-sided}}/2)$]. The authors propose to plot the $(1 - p_i)$'s vs. their rank for all endpoints. If H_0 is true, then the plot should be a line through the origin with slope equal to $1/(k + 1)$. The reasons for this fact are as follows: From the definition of P-value it derives that P_i is equal to or smaller than a determined value p_i, with probability equal to p_i. This means that P_i and $1 - P_i$ have a uniform probability distribution between 0 and 1. Now, when k values are obtained from a random variable uniformly distributed in the interval 0–1, they are expected to divide the unit interval into $k + 1$ equal subintervals. So if one sequentially plots k points having as ordinates the $(1 - p_i)$'s in increasing order and as abscissas the k increasing numbers, which divide the 0–1 interval into $k + 1$ equal subintervals, then one expects to find each ordinate equal to the corresponding abscissa. So the interpolating curve is expected to be a line with slope 1. If one changes the scale of the x-axis and extends each subinterval to measure 1, the abscissas become equal to the natural numbers from 1 to k. Then they are the ranks of the corresponding ordinates, i.e., of the $(1 - p_i)$'s. Moreover the former 0–1 interval extends to measure $k + 1$, so the slope of the line becomes $1/(k + 1)$ (because an increase of $k + 1$ on the x-axis corresponds to an increase of 1 on the y-axis).

If the number of true H_{0i}'s is $m < k$, then the plot should somewhere deviate from a straight line. Indeed, some $(1 - p_i)$'s should be larger than expected, corresponding to the small p_i's of the false H_{0i}'s. However, all the leftmost points of the plot probably correspond to the large p_i's of the true null hypotheses. So they should plot a line. The expected slope of this line is $1/(m + 1)$. To understand why, remember that if two points of the plot correspond to true H_{0i}'s, the expected difference of their ordinate is $1/(m + 1)$. Indeed, the $(1 - p_i)$'s that correspond to true H_{0i}'s are expected to generate $m + 1$ equal intervals between 0 and 1. So if one takes two points that correspond to true H_{0i}'s and whose abscissas differ by 1 (two consecutive points of the leftmost part of the plot), then they are on a line with expected slope $1/(m + 1)$. This fact gives a clue to as to how to estimate the number of true H_{0i}'s: One can fit a straight line to the leftmost points of the plot that do not consistently deviate from a linear trend and compute its slope b. An estimate $\hat{m}$ of the number of the true H_{0i}'s is then given by $1/b - 1$. This estimate is slightly biased toward larger-than-true values (10).

Perlman (23) found the likelihood ratio test for $H_{T \geq C}$ and derived its distribution for multivariate Gaussian endpoints with unknown covariance matrix. This test should have optimal properties; Lehmann (21) writes that the likelihood ratio tests are often asymptotically uniformly most powerful over the relevant alternative hypothesis, even in nonclassical situations. The power of Perlman's test has been empirically investigated by Follmann (24). He found that the test is actually preferable to both O'Brien's and Follmann's tests (see later) only when some endpoints are truly favorably affected by T and fewer are unfavorably affected, a situation outside the range of $H_{T \geq C}$. Perlman's test is difficult to compute and hasn't yet found its way into most known statistical programs. The following are simpler tests for multivariate Gaussian endpoints.

O'Brien's test (25) is particularly powerful if the true effects are equal on all endpoints (24). To use it, one has to compute a numerator and a denominator of a t-test statistic in the following way: Get from a computer program the k-dimensional vector $\bar{\mathbf{y}}_T$ of the outcomes of the endpoints averaged on the T sample and the analog vector $\bar{\mathbf{y}}_C$ computed on the C sample. Get the estimated covariance matrix $\hat{\mathbf{V}}$ of the endpoints, as well. Then, to obtain the numerator, multiply the inverse of the estimated covariance matrix ($\hat{\mathbf{V}}^{-1}$) by the vector of the mean differences of outcome: $\bar{\mathbf{y}}_T - \bar{\mathbf{y}}_C$, and sum the elements of the resulting vector [formally the numerator is $\mathbf{1}'\hat{\mathbf{V}}^{-1}(\bar{\mathbf{y}}_T - \bar{\mathbf{y}}_C)$, where $\mathbf{1}$ is a vector of 1's]. The denominator is given by the usual correcting factor of t-tests: $\sqrt{1/n_T + 1/n_C}$ multiplied by the square root of the sum of the elements of $\hat{\mathbf{V}}^{-1}$. [Formally, the denominator is $\sqrt{(1/n_T + 1/n_C)(\mathbf{1}'\hat{\mathbf{V}}^{-1}\mathbf{1})}$]. For two endpoints these computations can easily be performed by hand. If the endpoints are more than three, some matrix algebra program must be used. The t obtained is compared with the usual tables of critical values of t with $n_T + n_C - 2k$ degrees of freedom.

O'Brien's test can be one-sided or two-sided with the same modalities of a usual t-test. If the one-sided version turns out to be significant, the conclusion isn't necessarily that $H_{T \geq C}$ is true (though this is very likely). One has just proved that the expected value of the numerator of t, i.e., $\mathbf{1}'\hat{\mathbf{V}}^{-1}E(\mathbf{y}_T - \mathbf{y}_C)$ is positive. To see what this means, consider the case when the endpoints are independent. In that case, the numerator of O'Brien's test becomes the sum of the endpoint mean differences, each divided by the respective endpoint variance. So when the endpoints are independent, the hypothesis proved by the test states that the sum of the expected effects of T on all the endpoints, divided by their variances, is positive.

A very interesting feature of O'Brien's test is that, in its two-sided version, it can be applied to the comparison of more than two groups in stratified trials with prognostic covariates. (This can be done under the condition that the covariance matrix of the endpoints is the same within all strata, and conditional on all values of the prognostic covariates). To apply O'Brien's test in this general form, compute for each patient the quantity $u = \mathbf{1}'\hat{\mathbf{V}}^{-1}\mathbf{y}$ (where $\mathbf{y}$ is the vector of outcomes of the patient for all endpoints). Than use u as the response variable in a usual program of analysis of variance.

Simulations show that this procedure keeps the FWER fairly under control, even when the data have exponential or Cauchy distributions (25). Follmann (24) compared its power with that of Perlman's and his own test. He found that O'Brien's test is by far the most powerful to detect $H_{T \geq C}$ when the endpoints are independent. So its use is recommended for nearly independent endpoints when one is certain that, if T is truly effective, it is so for all endpoints. O'Brien's test is preferable also with moderately positively correlated endpoints, if the true effects on all endpoints are very similar.

Tang et al. (26) give a simplified approximate version of the likelihood ratio test of Perlman. The computation of Tang's test involves the Cholesky decomposition of the covariance matrix of the endpoints and other moderately complex matrix manipulations. The details are given in the original paper, together with the table of the critical values of the test statistic for up to ten endpoints. If that table isn't sufficient to analyze someone's data, the authors provide a simple formula (4.1 of their paper) for the computation of the P-value. It shouldn't be too difficult to implement that formula if the number of endpoints isn't too high. The simulations performed by Tang and colleagues compare the power of their test to that of O'Brien's. They show that O'Brien's test is superior when the true effects of T on all endpoints are similar and that their test is superior in the opposite case.

Tang's test is intrinsically one-sided, but its conception is such that it can be significant in situations in which T is not superior to C. For instance, imagine that T and C are compared on uncorrelated endpoints. In this case the test statistic is a weighted mean of the squared positive endpoint differences. If T has a strong favorable effect on one endpoint, but also an equally strong unfavorable one on the other endpoint, the test may be significant if we use it to demonstrate the superiority either of T or that of C. Therefore to apply Tang's one-sided test, it isn't sufficient to adopt the pragmatic "approve or disprove" point of view on a treatment evaluation, i.e., to merge countereffects and no effects in one null hypothesis (the logic that justifies one-sided tests in a single-endpoint trial). Indeed overwhelming countereffects do not avoid Tang's tests stating significant favorable effects.

Follmann (27) has proved that when Tang's method is applied to two positively correlated endpoints, it can become significant, even when the sample means of the two endpoints are both negatively affected by the treatment. This is not a crippling flaw of the test, however, because one could apply Tang's test at the condition that at least one of the endpoints is improved in the T sample. If this condition is not fulfilled, the null hypothesis could be accepted without further computations.

Follmann (24) conceived a *one-sided modified T^2 test* for multivariate Gaussian endpoints, which joins ease of use, theoretically exact control of the α level, and good power to detect a large spectrum of alternative hypotheses within $H_{T \geq C}$. Its procedure with significance level α is the following: Compute the mean difference between T and C of each endpoint, $\bar{y}_T - \bar{y}_C$, and sum them up over all endpoints. If the sum is negative, then retain H_0. Otherwise, compare the two samples by a usual Hotelling's T^2 test: Reject H_0 if the test statistic is larger than the 2α critical value. In other words, the P-value of Follmann's test is *half* the P-value written on the output of the T^2 computation. Follmann proves that the Type I error rate of his test is always below α. With extensive simulations, he shows that his test is more powerful in detecting $H_{T \geq C}$ than the theoretically optimal Perlman's likelihood ratio test when the endpoints are moderately positively correlated ($\rho = 0.75$). In this situation, if the true effect of T, although favorable everywhere, is rather heterogeneous among the endpoints, Follmann's test is also superior to O'Brien's and is to be recommended.

Rom (28) proposes a global one-sided test for multiple dichotomous endpoints based on their joint distribution. The author gives an example of an application of his procedure to the case of two endpoints with hypergeometric distribution. However, for three or more endpoints, Rom's method probably needs the application of bootstrap techniques to estimate the joint probability distribution of the endpoints.

C. Techniques Powerful in Detecting H_A

When one expects T to be superior to C on *average* but also to have some countereffects, the tests described in this section become relevant. Of course, according to the meaning given to the term *average superiority*, the chosen test will be different.

Follmann (24) quotes the likelihood ratio test for a H_A, defined as follows: $E(y_T - y_C) > 0$ (i.e., the arithmetic mean of the effect of T on all endpoints is positive). However, he writes that this test isn't yet available if the covariance matrix of the endpoints is unknown.

Follmann (27) proposes a very simple and appealing test for multivariate normal endpoints. A slight generalization of his method is the following: For each subject, multiply each endpoint measure by a predetermined fixed weight, and sum all the products to get a unique score. Then perform a usual analysis of variance on this score. This procedure is correct because, if the endpoints are Gaussian, every linear combination of them is Gaussian too. The test has a one-sided version, easily obtained from the two-sided one, by considering the observed score mean difference between T and C. If this difference is favorable to T, then $P_{\text{one-sided}} = P_{\text{two-sided}}/2$; otherwise, $P_{\text{one-sided}} = 1 - (P_{\text{two-sided}}/2)$. The alternative hypothesis proved by this test, if it is significant, is that the predetermined linear combination of the endpoint outcomes has a more favorable expected value for T than for C.

Follmann affirms that his method may be used when the multiple endpoints are surrogate of one "hard" endpoint difficult to measure directly. The linear combination of endpoint outcomes should be a linear model previously fitted to independent data, which turned out to be reasonably predictive of the "hard" endpoint. However, his method should also be considered when the "hard" endpoint aimed at is *impossible* to measure directly, e.g., cognitive impairment or quality of life. In this situation the weights given to each endpoint should be attributed by experts of the field. The subjective choice of the weights shouldn't be a problem, if they are clearly stated in the protocol of the analysis.

In Follmann's simulations of a two-endpoint trial, the power of his test is always superior to the mean power of the univariate tests applied to the single endpoints when both endpoints are truly favorably affected by the treatment. Moreover, Follmann finds that the form of the rejection region of his test makes sense from a substantive point of view. Indeed, whatever is the number and the correlation structure of the endpoints, all the observed outcomes leading to the rejection of the null hypothesis correspond to a better outcome score than all the outcomes leading to retain H_0.

If applied to an independently measurable "hard" endpoint, like mortality, Follmann's procedure allows one to compute confidence intervals for it. Moreover, his score can be used as the response variable of a general linear model, allowing to be taken into account, in the statistical analysis, all the trial design features: more than two arms, stratification, random factors, prognostic covariates, repeated measures, etc.

From Follmann's method, a formula can be deduced for the computation of the required sample size of a trial. Proceed as follows: Ask the field experts the minimum clinically relevant effect difference of each endpoint, supposing the other endpoints stay unaffected. Than compute the weighted mean of the minimum clinically relevant differences, using the same weights chosen for the analysis. This mean should be a meaningful minimum clinically relevant score difference. Compute the variance of the score from the variances of the endpoints and their correlation (for the computation formula, see Ref. 29). Knowing the minimum clinically relevant difference and the variance of the score, obtain the required sample size from the usual tables published for univariate trials.

O'Brien (25) proposed a *rank-based* test: the measures of each endpoint are ranked among all subjects. (A better outcome has higher rank than a worse one.) Then, for each subject, the sum of the ranks of all endpoint outcomes is computed. This sum is asymptotically Gaussian and can be used as response variable of a general linear model. The test can be one- or two-sided. The alternative hypothesis proved by the one-sided version is the following: "The probability that a subject of the T population has a better outcome on any endpoint than a subject of the C population is on average larger than 0.5." O'Brien's simulations show that the rank-sum test has a reasonable control of the Type I error rate; it rejects the true null hypothesis a maximum of 6.3% of the times for a declared $\alpha = 0.05$ in the set of situations considered by the author. The power of the rank-sum test is inferior to the power of the parametric test proposed by same author (see earlier) in situations where the last should perform at its best, i.e., when the endpoints are multivariate Gaussian and all are truly improved by similar amounts by T. When the endpoints aren't Gaussian, the rank-sum test turns out to be superior. So the rank-sum test is recommended when the endpoints are certainly far from the Gaussian distribution.

In the particular case where just one endpoint is really affected ty T, Follmann (27) empirically proved that the Bonferroni correction is optimal. However, this situation shouldn't arise if the endpoints of the trial are well chosen. If the Bonferroni correction, or other techniques devised to test the single H_{0i}'s, are used to test the global H_0, the multivariate P is equal to the smallest of the p_{ia}'s computed for the H_{0i}'s.

V. REGULATORY AGENCIES AND THE MULTIPLE-ENDPOINT ISSUE

Regulatory agencies don't give very detailed guidelines for the analysis of multiple endpoints in clinical trials. They require a fair control of the Type I error rate without giving any indication of a preferred technique. For instance, in the guidance for rheumatoid arthritis treatment evaluation (30) the FDA considers the following alternatives: (a) Three out of the four primary endpoints relevant to the pathology are significantly improved by T. (b) A multivariate statistical method is applied. (c) The efficacy variables are combined within patients (composite endpoint), and the weights used are clearly defined in the study protocol.

An example of the concern of the statisticians in the regulatory agencies for the analysis of multiple-endpoint clinical trials is expressed, for instance, by Sankoh et al. (6). They give a fair and clear account of various adjustment procedures. The authors propose as well a simulation-based method to correct the FWER of some P-value-based and ad hoc techniques. Their method is the following: Estimate the sample correlation matrix among endpoints without breaking the blind code of the trial. Simulate thousands of trials with the same correlation structure and no effect of T. Apply the adjustment defined in the protocol to each virtual trial sample, and count the frequency of rejections. If the proportion of rejections is larger than the chosen one (e.g., 0.05), apply the same adjustment again but using a smaller nominal α level (e.g., 0.04) in the formulae. Go on trying different nominal values of α till the proportion of rejections is down to the desired value (in our example, 0.05). Use the so-identified nominal α, instead of 0.05, in the adjustment procedures for the analysis of the real trial.

The authors also discuss the use of global test procedures and underscore the necessity that a global alternative hypothesis be meaningful. As Sankoh points out, global tests are certainly not appropriate when each endpoint has a completely different pathologic meaning.

REFERENCES

1. PJ Bickel, KA Doksum. Mathematical Statistics. San Francisco: Holden-Day, 1976, p 439.
2. S Holm. Scandinavian J Statist 6:65–70, 1979.
3. G Hommel. Biometrika 75:383–386, 1988.
4. Y Hochberg. Biometrika 75:800–802, 1988.
5. BW Brown, K Russell. Statist Med 16:2511–2528, 1997.
6. AJ Sankoh, MF Huque, SD Dubey. Statist Med 16: 2529–2542, 1997.
7. Y Hochberg, D Rom. J Statist Plann Inference 48:141–152, 1995.
8. RJ Simes. Biometrika 73:751–754, 1986.
9. M Comelli, C Klersy. JBS 6:115–125, 1996.
10. Y Hochberg, Y Benjamini. Statist Med 9:811–818, 1990.
11. T Schweder, E Spjøtvoll. Biometrika 69:493–502, 1982.
12. AJ Hayter, JC Hsu. JASA 89:128–136, 1994.
13. AC Tamhane. In: S Ghosh, CR Rao, eds. Handbook of Statistics: Design and Analysis of Experiments (vol 13). Amsterdam: North-Holland, 1996, pp 587–630.
14. PH Westfall, SS Young. Resampling-Based Multiple Testing: Examples and Methods for P-Value Adjustment. Vol. 1. New York: Wiley, 1993.
15. JF Troendle. JASA 90:370–378, 1995.
16. JF Troendle. Biometrics 52:846–859, 1996.
17. SAS Institute Inc., SAS/STAT Software: Changes and Enhancements Through Release 6.11. Cary, NC: SAS Institute, 1996, pp 733–739.
18. N Mantel. Biometrics 36:381–399, 1980.
19. Y Benjamini, Y Hochberg. J Royal Statist Soc 57:289–300, 1995.
20. TW Anderson. An Introduction to Multivariate Statistical Analysis. 1st ed. New York: Wiley, 1958, p 115.
21. EL Lehmann. Testing Statistical Hypotheses. 2nd ed. New York: Wiley, 1986, pp 462, 477.
22. LA Cupples, T Heeren, A Schatzkin, T Colton. Ann Intern Med 100:122–129, 1984.
23. MD Perlman. Ann Math Statist 40:549–567, 1969.
24. D Follmann. JASA 91:854–861, 1996.
25. PC O'Brien. Biometrics 40:1079–1087, 1984.
26. DI Tang, C Gnecco, NL Geller. Biometrika 76:577–583, 1989.
27. D Follmann. Statist Med 14:1163–1175, 1995.
28. DM Rom. Statist Med 11:511–514, 1992.
29. RV Hogg, AT Craig. Introduction to Mathematical Statistics. New York: Macmillan, 1978, p 177.
30. U.S. Department of Health and Human Services. Food and Drug Administration Guidance for Industry (Draft Guidance):28–29, 1998.

Mario Comelli

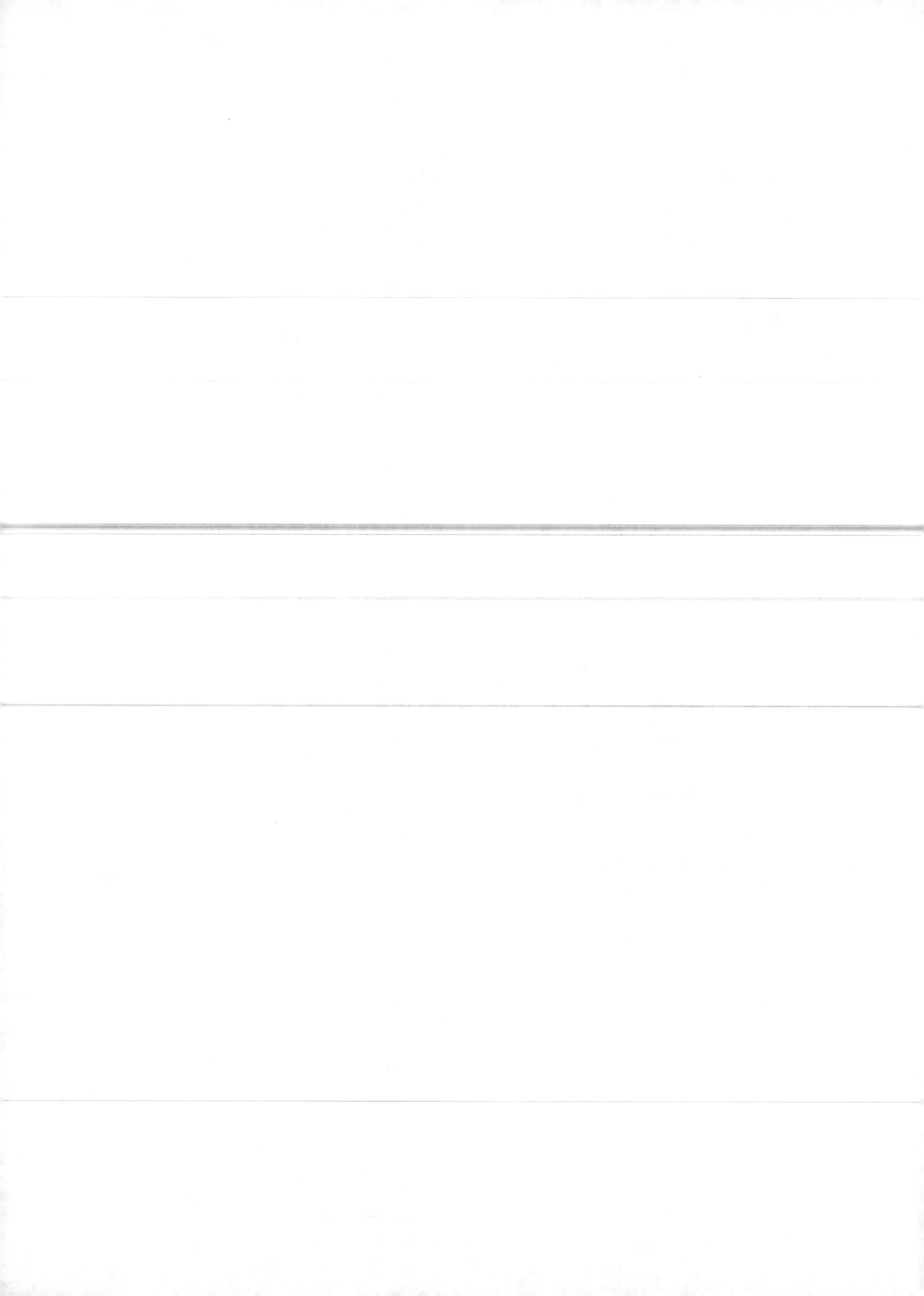

P

Parallel Design

See also *Clinical Trials*

I. INTRODUCTION

Statisticians involved in clinical research in the pharmaceutical industry encounter design of experiments almost on a daily basis while writing protocols for their clinical studies. At the outset, the primary objective(s) described in the protocols involve comparing different dose groups; to accomplish this, patients are assigned randomly to the dose groups. This general idea of assigning patients to different groups is similar to those agricultural experiments conducted by Sir Ronald Fisher, who is considered to be the father of many fundamental statistical concepts, including design of experiments, at the Rothamsted Agricultural Experiment Station near London. Sir Austin Bradford Hill was primarily instrumental in the development of controlled clinical trials in Britain. His publications in the early 1950s (3) included fundamental concepts such as random allocation, concurrent controls, eligibility of patients, treatment schedule, and statistical analysis of the clinical data.

The term *experimental design* comes from these agricultural experiments, where the experimenters sow varieties in or apply treatments to plots, and often these plots were arranged in blocks. At this time, readers are asked to take note of terms like *randomly* and *blocks*. Variations of these experiments are carried out in clinical trials when patients are randomly allocated across various treatments within each investigational site, also known as a *center*. These basic ideas will be discussed further in this entry.

As an example, imagine a clinical trial that is planned to compare the effectiveness of an experimental drug against placebo in the treatment of patients with schizophrenia, or schizoaffective disorder. In order to accomplish the primary objective, it is decided to include more than one dose of the experimental drug. As we think about this clinical trial, a number of important questions arise:

1. Are these dose groups different in efficacy?
2. Are there other characteristics of the patients that might affect the efficacy that should be controlled or investigated?
3. How many patients should be assigned to each dose group?
4. How should the patients be assigned to each dose group, and how the data should be collected?
5. What analysis method should be used on the collected data?
6. If there are differences between the dose groups, which ones are different?

It is important to answer all of these questions before concluding that a given trial is positive or not. Some of these issues will have to be specified a priori in the protocol, as required by the International Conference on Harmonization (ICH) Guidelines (ICH E-9); readers are referred to the entry on Protocol Development for more details on the protocol specification.

Though the final conclusions in any clinical trial may depend largely on the conduct of the trial itself, it is absolutely crucial that the protocol team (including clinicians and statisticians) spend enough time on the design part of the trial. This entry is intended to discuss one of the commonly used designs in clinical trials.

A. Definition

The simplest and the most commonly used designs employed in clinical trials are *parallel designs*. In a trial with parallel designs, a patient is assigned randomly to one and only one treatment, as opposed to crossover designs, in which the same patient is given two or more treatments within a trial. Differences between these two types of designs will not be discussed further in this entry.

In parallel designs, there are usually at least two treatment groups: experimental and control. In our earlier example of a study involving schizophrenia, there are several dose groups of the experimental drug and a placebo as the concurrent control. Controls in clinical trials are used as standard treatments to compare and to evaluate the effectiveness of an experimental drug. When there are no standard marketed treatments available, and if it is ethical, then placebo controls are used as standard comparators. If a standard therapy is available (e.g., anti-infectives or antihypertensives), then it may be required to use it as a control to evaluate quantitatively the efficacy and safety of an experimental drug; thus a relative risk-benefit can be specified. The importance of proper controls as a part of well-designed clinical trials is stressed in ICH Guidelines (ICH E-9), especially in their requirements for sponsors to provide at least two well-controlled trials as pivotal for the new drug approvals (NDAs). There are other types of control, such as no control and historical control, that are used less frequently in clinical trials.

It is noted in this simplest definition of parallel designs, where we simply assign patients to one and only one treatment, that no specific patient characteristics that may have a bearing on the response are taken into consideration. These patient characteristics or other trial related characteristics are typically known as *prognostic variables*. Designs that incorporate these prognostic variables (e.g., randomized block designs, analysis of covariance) will be discussed later in this entry.

B. Three Basic Concepts in Experimental Designs

The three basic principles that serve as keystones to the development of experimental designs are: *randomization, replication*, and *blocking*.

There is enough literature discussing the need for using randomization in clinical trials before treatments can be compared, but there are also references in the literature that argue against the use of randomization. It was the work of pioneers like Sir R. A. Fisher that influenced the use of randomization as the cornerstone application of statistical theory in experimental design. A classic paper by Peto et al. (8) presents a convincing argument for using randomization in clinical trials.

In a trial with multiple doses of the experimental drug and a placebo (such as the trial described in the previous section on patients with schizophrenia), patients are assigned to each group at random using a predetermined scheme. These randomization schemes can always be generated using computer algorithms. In general, there are several benefits associated with using random assignment of patients to treatments: avoidance of any bias in the results, balancing of prognostic factors, ensurance of proper conduct of trials, and validation of assumptions needed to apply certain statistical techniques.

The use of *replication*, as emphasized by Fisher, is critical to reduce variability or to get a proper estimate of the error variance. In general, *replication* means repetition of an entire experiment, or portions thereof, under similar experimental conditions. Though similar experimental conditions are desirable, it is not useful to obtain identical analytical results. To accomplish this, experiments can be repeated under two or more sets of conditions, for example, within a time window that is considered in a trial. For example, analysis of clinical trials data where measurements are collected across time points (e.g., blood pressure measurements in trials involving patients with hypertension) involves "change from baseline to final time point" as the response variable. In this case, a time window can be created for both baseline and final time point so that measurements can be repeated during the time window (e.g., three blood pressure measurements can be taken both at baseline and at final). Replicated measurements allow the experimenter to estimate accurately the experimental error or to estimate precisely the error variance if a sample mean is used in the estimation of a factor effect.

Blocking is used as a technique for increasing the precision of an experiment. A *block* is a subset of experimental units that are more homogeneous than the entire experiment itself. For example, in many clinical trials, responses to treatments are different for various groups of patients based on their age, sex, race, or initial severity of disease, which are some commonly cited prognostic variables that may have an impact on the response. Blocking (also known as *matching* in some literature) is a technique used to control the variability of the response caused by these prognostic variables. Randomized block designs are used to take blocks into account in the design. An example of a randomized block design is a paired comparison design, in which two patients who have similar characteristics are paired and the two treatments (drug and placebo in a two-group trial) are assigned to each member of the pair.

A sensible alternative to blocking as a method to control for the prognostic variables is *stratification*. A common stratification variable is center. Other prognostic variables, such as age, sex, race, and initial severity, can also be considered as stratification variables. For example, age could be grouped into, say, age <18, $18 \le \text{age} \le 64$, and age ≥ 65 as different strata. With this definition of strata for age, it is noted that patients who are homogeneous based on their age category are grouped. With regard to randomization, patients would then be randomized within each stratum separately. In clinical trials, it is very common to produce an independent set of randomization schedules for each investigational center.

Blocking and stratification are quite often used within the same data. For example, within each stratum, randomizations could be carried out using simple random samples, or blocking could be used in the randomizations. Thus, blocked randomization is a special kind of stratification. The smallest block size is the sum of the integers defined by the allocation ratio (for a 1:1 allocation, block size is 2). Use of a large number of strata may increase the chance of departure from the specified allocation ratio, unless small block sizes are used within each stratum.

II. EXAMPLES OF PARALLEL DESIGNS

A. Completely Randomized Design

The simplest version of a completely randomized design is a one-way classification model, which is a single-factor experiment with two or more levels of a design factor, such as clinical trial designed to compare several doses of an experimental drug and a placebo where all the patient data are obtained from one clinic or hospital. Mathematically, this model can be written as

$$y_{ij} = \mu_i + \varepsilon_{ij}, \quad \begin{cases} i = 1, 2, \ldots, a \\ j = 1, 2, \ldots, n_i \end{cases} \tag{1}$$

where y_{ij} is the (ij)th observation, a is the number of treatments (number of doses and placebo), and n_i is the number of observations in the ith treatment.

For the underlying model presented in Eq. (1), our interest is to test for the equality of the treatment means; that is,

$$H_0\colon \mu_1 = \mu_2 = \cdots = \mu_a$$

$$H_1\colon \mu_i \neq \mu_j \quad \text{for at least one } j, i$$

If the null hypothesis is true, then it is well known that the ratio $F = \text{MSA/MSE}$ is distributed as F with $(a - 1)$ and $(N - a)$ degrees of freedom. Definitions of the terms used are as follows:

$$\begin{aligned} \text{MSA} &= \frac{\text{SSA}}{a - 1} \qquad \text{MSE} = \frac{\text{SSE}}{N - a} \\ \text{SSA} &= \sum_i n_i(\bar{y}_{i.} - \bar{y}_{..})^2 \qquad \text{SSE} = \sum_i \sum_j (y_{ij} \cdot \bar{y}_{i.})^2 \\ N &= \sum_i n_i \qquad \bar{y}_{i.} = \sum_j \frac{y_{ij}}{n_i} \quad \bar{y}_{..} = \sum_i \sum_j \frac{y_{ij}}{N} \end{aligned} \tag{2}$$

Though these mathematical formulations would guide us in determining whether there is any signal for a difference between the treatment means, a similar analysis in clinical trials would be considered incomplete until the next step is taken, viz., responding to the primary objective specified in the protocol of showing whether or not the experimental therapy is efficacious. In other words, it is important for us to know which treatments differ from what and the magnitude of those differences. In order to answer this primary objective in a protocol, the statistical analysis will have to involve the method of *multiple comparisons*, which will be briefly described at the end of this entry and will be covered in detail in another entry.

B. Designs with Prognostic Factors

One of the most widely used designs for controlling known sources of variability due to the prognostic factors is a *randomized block design*. In the case of a large phase II/III clinical trial, the study team may be required to carry out this trial across several investigational centers. Studies like this are known as *multicenter trials*. Some of the important reasons and rationale for conducting a multicenter trial include the ability to generalize the results across different set of patients and treatment settings, and expediting the enrollment of patients in a large trial. If center is considered as a stratification variable, then individual randomizations are carried out within each center. A balanced treatment assignment within each center is generally desired.

With treatment and center (as a blocking variable) in a randomized block design with complete blocks (that is, one observation per treatment in each center), the statistical model can be written as

$$y_{ij} = \mu_i + \beta_j + \varepsilon_{ij} \quad \begin{cases} i = 1, 2, \ldots, a \\ j = 1, 2, \ldots, b \end{cases} \tag{3}$$

The model presented in Eq. (3) has no term included for a possible interaction between the treatments and centers. An interaction between treatment and center may exist if the directions of the treatment difference are different across centers. Because most of the analysis in a multicenter trial is based on pooled data from various centers, guidelines from most regulatory agencies suggest that a preliminary test for the treatment by center be carried out (ICH E-9 Guidelines). Thus, it is almost standard practice to consider the following revised model to Eq. (2):

$$y_{ij} = \mu_i + \beta_j + \mu\beta_{ij} + \varepsilon_{ij} \quad \begin{cases} i = 1, 2, \ldots, a \\ j = 1, 2, \ldots, b \end{cases} \tag{4}$$

The interaction model in Eq. (4) is also popularly known as *cell means model*.

Hypotheses involving the equality of treatment means (μ_i), center means (β_j), and existence of interaction ($\mu\beta_{ij}$) can be tested using various sums of squares terms, similar to the ones shown in Eq. (2), once the analysis of variance table is set up (2). It is common to test for the interaction first. If this test is nonsignificant, it is often taken to be a sufficient (but not necessary) justification for pooling data across centers and for concluding that the treatment differences are sufficiently consistent. If a significant interaction is identified, it is expected that the sources of the interaction, such as dropout rates and eligibility criteria, are explored and explained in the final analysis. Though it is partly based on one's judgment and what is prespecified in the statistical analysis plan in the protocol, the primary hypothesis is usually based on the two main-effects terms.

The *analysis of covariance* is another design that can be used to control the effects of prognostic variables. Imagine an experiment with a response variable y and with another variable x, where y is linearly related to x. Further, x cannot be controlled but can be observed along with y. The variable x is called a *covariate* or a *concomitant* variable. As indicated in Sec. I, some examples of covariates are demographic variables such as age, sex, race, and baseline severity of the disease. The use of covariates is restricted to quantitative variables, and it is purely a statistical method for controlling the effects of prognostic variables; no special effort is needed for any prestudy arrangements such as randomization.

For a completely randomized design with a single factor (e.g., treatments) and a covariate (e.g., baseline severity), a statistical model can be written as

$$y_{ij} = \mu + \tau_i + \beta(x_{ij} - \bar{x}_{..}) + \varepsilon_{ij} \quad \begin{cases} i = 1, 2, \ldots, a \\ j = 1, 2, \ldots, n_i \end{cases} \tag{5}$$

Note that the model in Eq. (5) assumes that the regression lines (number = a) have the same slope; that is, the regression lines are parallel. If the regression lines are not parallel, it is reasonable to expect an interaction between the treatments and the covariate. In order to accommodate this, the model in Eq. (5) can be generalized as

$$y_{ij} = \mu + \tau_i + \beta_i(x_{ij} - \bar{x}_{..}) + \varepsilon_{ij} \quad \begin{cases} i = 1, 2, \ldots, a \\ j = 1, 2, \ldots, n_i \end{cases} \tag{6}$$

For the models specified in Eqs. (5) and (6), analyses for testing various hypotheses can be carried out using the classical ANOVA procedure. A general regression approach can also be developed for this purpose (2,5).

Note that in the models referred to so far, an intrinsic assumption is made that the response is measured at one time point, which is not a reality in most clinical trials. Most clinical trials involve measuring responses from a patient across several time periods (e.g., blood pressure measurements taken at several time points in a hypertension trial). Though there are repeated measurements made on subjects, one of the models already explained is used at each time point as a standard analysis. A most widely used time point is the last observation carried forward (LOCF), with the response considered as the change from baseline to this time point.

In repeated-measurements designs, there are two sizes of experimental units, subject and subject's time. Thus, there are two parts in the model and the analysis, comparison of treatments based on subject's response and comparison of time and treatment by time interaction. Though the structure and the model considered when there are repeated measurements are similar to those of a split-plot design, the two designs operate under different assumptions. One of the main differences is that the subjects are randomly assigned to the levels of treatment, but the time intervals are not. The experimental design used for this type of data is known as *repeated-measures design*. For the analysis of such a design, one can use either a classical ANOVA approach or a multivariate approach (2).

III. ANALYSIS ISSUES

A. Multiple Comparison

In Sec. II.A, while discussing the test of significance for the equality of treatment means, it was indicated that a significant overall F-test would signal a difference in treatment means. In most clinical trials, showing the existence of an overall experimental effect is of little interest to the primary objective. In a trial involving several doses of an experimental drug and a placebo, one's primary interest is to see how many doses are different from placebo, that is, pairwise comparisons between placebo and all doses. Methods that are used to find these pairwise differences are known as *multiple comparisons*. A detailed discussion of these methods and their rationale will be covered in another entry.

B. Multiple Time Points

As discussed in Sec. II.B, if we repeatedly measure patient's responses across various time points and perform an analysis at each time point, then it can create a multiplicity issue. A repeated-measures design with follow-up multiple comparisons (2) can be used to handle this issue. As explained in the previous section, a common and effective method of handling multiple time points is to express the repeated data in terms of a single time point, for example, LOCF time point. This time point is often considered the primary time point for analyses of the primary endpoints. Other useful approaches include using the slope for each patient across time or the area under the curve (AUC) as primary endpoints. Diggle et al. (1) discuss other approaches that are useful for analyzing data with repeated measures.

C. Multiple Endpoints

It is very common that more than one endpoint (measurement) is specified as primary in phase II/III protocols. For example, in trials designed to study the efficacy and safety of an antipsychotic drug on patients with schizophrenia, one or more derived measurements from the Positive and Negative Symptoms Scale (PANSS) and the Clinical Global Assessment scales are considered primary. For these different endpoints, a separate analysis is carried out and *p*-values are reported separately. A problem with this type of approach is when to declare the drug efficacious. For instance, if there are five primary endpoints and only three of them show statistical significance, can we still claim efficacy? To avoid this, some protocols tend to specify only one endpoint as primary and the rest as secondary. In the analysis section, an attempt will then be made to explore the consistency of various results.

One approach is to consider a composite measure than can be constructed based on the primary endpoints (6). Though *p*-values are derivable for these composite measures, they are not as easy to interpret as changes or percentage changes. Multivariate approaches provide alternatives for handling multiple endpoints.

D. Sample Size

If there are two groups in the design, then sample size determination is straightforward. For normally distributed variables with a common variance σ^2, the sample size formula for each treatment group is given as

$$n = \frac{2\sigma^2(z_{\alpha/2} + z_\beta)^2}{(\mu_1 - \mu_2)^2}$$

where μ_1 and μ_2 are the true means, probability of type I error is α, and the power of the test is $1 - \beta$.

The determination of sample sizes for clinical studies with several groups (more than two) involves critical values from a *noncentral F-distribution* (7). Statisticians faced with this problem in clinical trials tend to use a practical solution; viz., they convert this into a two-group problem, where the groups chosen have the smallest difference. For this smallest difference ($\mu_1 - \mu_2$), the sample size is chosen and is considered for the entire protocol. Other issues, such as adjusting the significance level for multiple comparisons and dose–response designs, are also considered for determining sample size.

REFERENCES

1. Diggle, P. J., Liang, K-Y., Zeger, S. L. (1994). Analysis of Longitudinal Data. Oxford: Oxford University Press.
2. Fleiss, J. L. (1986). The Design and Analysis of Clinical Experiments. New York: Wiley.
3. Hill, Sir A. B. (1962). Statistical Methods in Clinical and Preventive Medicine. Edinburgh: Livingstone.
4. International Conference on Harmonization (ICH) Guidelines. (1997). E-9: Statistical Principles for Clinical Trials.
5. Mason, R. L., Gunst, R. F., and Hess, J. L. (1989). Statistical Design and Analysis of Experiments with Applications to Engineering and Science. New York: Wiley.
6. O'Brien, P. C. (1983). The appropriateness of analysis of variance and multiple comparison procedures. Biometrics 39:787–788.
7. Winer, B. J. (1962). Statistical Principles in Experimental Design. New York: McGraw-Hill.
8. Peto, R., et al. (1977). Design and analysis of randomized clinical trials requiring prolonged observation of each patient. I. Introduction and design. Br J Cancer 34: 585–612.

Mani Y. Lakshminarayanan

Patient Compliance

See also *Dropout*

I. INTRODUCTION AND DEFINITIONS

Poor compliance to prescribed medication is a common problem that can have a major impact on the success of routine patient care and on the conduct and conclusions of clinical trials. Reported clinical correlates of poor compliance include increased hospitalization in patients with hypertension (1), diminished improvements in LDL cholesterol levels in hypercholesteremic patients (2), and increased mortality in patients with acute myocardial infarction (3). The potential effects of inadequate compliance in clinical trials include underestimates of true therapeutic efficacy because medication is not taken as intended and large increases in sample size requirements because between-group differences are reduced (4–7). If the therapeutic impact of poor compliance is of sufficient magnitude, the associated reduction in statistical power can produce negative conclusions in a study that might otherwise have been positive.

Broadly speaking, a patient can be thought of as being compliant if his behavior is consistent with the recommendations of his physician or other health care provider. Poor compliance can take many forms, such as inattention to dietary or exercise recommendations; not taking the prescribed number of pills or taking them at irregular or otherwise nontherapeutic intervals; not refilling prescriptions; and not showing up at follow-up clinic visits. The focus of this review is on compliance to medication regimens.

Patients may be noncompliant for many reasons. The most obvious cause is simple forgetfulness, a circumstance that is more likely if a medication is being used for an asymptomatic condition such as hypertension. Alternatively, poor compliance may be a conscious decision based on undesired treatment side effects, on the patient's perception that the illness being treated is not serious, or on patient-specific beliefs about the potential impact of individual behaviors on health status. For example, one study of elderly diabetics (8) evaluated compliance to measures aimed at controlling diabetes using fasting blood sugar levels and diabetes-associated hospitalization rates as surrogates. Results indicated that improved control of diabetes was significantly associated with the belief that specific behaviors were likely to improve the status of one's disease and with the general perception that an individual's behavior can have a positive impact on health status. Additional reasons for poor compliance include the cost of the medication, the cognitive status of the patient, and the complexity of the drug-taking regimen. The latter two problems are particularly relevant in elderly patients, who may be unable to follow the details of a series of medication schedules for multiple chronic conditions.

II. MEASURING COMPLIANCE

The traditional approaches to measuring compliance include biological markers such as plasma concentrations and urine assays, requesting the information directly from the patient, and using pill counts that reflect the content of returned pill bottles. Each of these approaches can provide useful information. But at the same time, each has been shown to be inadequate in key respects (a) by studies that yield different compliance rate estimates when assessments are done using two or more techniques and (b) by comparisons using the newest approach to quantifying compliance: electronic monitors.

The limitations of pill counts have been emphasized by several authors (9,10). For example, Pullar and colleagues (9) assessed compliance using both returned-pill counts and plasma concentrations of phenobarbitol in 225 subjects. Among the 204 patients who returned for a scheduled follow-up visit with their pill bottles, 161 (78.9%) were classified as compliant because pill counts indicated that between 90% and 109% of prescribed medication had been taken. But 51 of these 161 patients (31.7%) had doses and body-weight-corrected plasma phenobarbitol concentrations that, according to predefined criteria, were indicative of poor compliance. Based on this large discrepancy between the two methods, the authors conclude that "return tablet count grossly overestimates compliance" and that the continued use of pill counts to measure compliance in clinical trials "cannot be justified." The inadequacy of patient interviews in measuring compliance is highlighted by reports such as that of Norell (11), who summarizes data from four studies (12–15) that measured compliance using both patient interviews and other, more objective assessment techniques. The result was that interview noncompliance rates ranged from 12% to 31%, as compared to a range of 32–82% using the more objective approach. The interview produced the lower estimate of noncompliance by amounts that ranged from 20% to 65%. In another study, Caron (16) reported that the average patient took 47% of the prescribed dose but claimed to have taken 89%.

In recent years, the development of microcircuitry has facilitated the use of miniaturized computer chips as electronic monitors that measure compliance by recording the date and time that medication is dispensed. These devices are now broadly accepted as providing the most accurate available measures of compliance. They have been attached to pill bottles (17,18), eye droppers (19), blister packs (20), and aerosol containers (21) and have confirmed the observation that traditional assessment techniques often yield substantial overestimates of compliance. And, as we discuss next, they have provided important new insights into patterns of noncompliance.

III. MAGNITUDE AND PATTERN OF POOR COMPLIANCE

While some studies have reported excellent compliance rates as high as 90% (20,22), it is well documented that compliance to medication can often be quite poor (5,17–19). Good compliance tends to be associated with medications that have few side effects, with acute and symptomatic diseases, and with recent diagnoses. In the Physicians' Health Study (22), good compliance was likely a reflection of the motivation of an interested and highly educated sample. However, when compliance is poor, it can sometimes be in the range of 50% or less (5,19). Given that the standard methods for measuring compliance tend to yield overestimates, it is reasonable to expect that many reported noncompliance percentages are a biased underestimate of the true magnitude of the problem.

While the number of doses taken may provide the best single measure of compliance to a prescribed medication regimen, it is not the only such measure. Indeed, it has become increasingly apparent that an accurate pill count may provide an inaccurate measure of patient compliance, because drugs taken with the correct total dose may simultaneously be taken at irregular and nontherapeutic intervals. Similarly, pill counts that are indicative of less than ideal compliance may understate the problem because of the same kind of irregular drug-taking patterns.

The dimensions and characteristics of irregular dosing can best be evaluated using electronic monitoring devices, which have demonstrated, for example, that compliance tends to improve in the days immediately preceding an office visit to the physician. Thus, Cramer et al. (18) found that compliance percentages were 88% during the five days immediately before a clinic visit but only 73% when no clinic visit was either imminent or had just taken place. Kass et al. (19) found similarly that compliance rates were much higher during the 24-hour period preceding an appointment than during the rest of the observation period. One implication of these observations is that even if a biological marker provides an accurate estimate of compliance immediately prior to the assay, the marker may substantially overstate the general pattern of compliance because of the tendency of patients to be more compliant when they are about to see the doctor.

A further contribution of electronic monitors to an understanding of compliance patterns lies in the observation that "drug holidays," sustained periods during which no medication is taken, are common (17–19). Thus, De Klerk et al. (17) studied the drug-taking patterns of 65 patients with ankylosing spondylitis who were prescribed once-a-day medication. The mean period of study was 225 ± 90 (SD) days, for a total of 14,607 monitored days. While 80% of prescribed doses were taken, only 61% were taken during the required 24 ± 6 hours. Extra doses were taken on 3.4% of days, while no dose was taken on 19% of monitored days. Among the nine subjects who had a high pill-taking rate of between 81% and 90%, four had a drug holiday of 13 or more consecutive days. In the Kass et al. study (19) of 184 glaucoma patients being treated with four-per-day pilocarpine eye drops, the electronic monitor found that 24.5% of patients had at least one day per month with no administration of drug, while 52.7% of patients took five or more doses during at least one 24-hour period. The fact that the nighttime dose was omitted 50% of the time in a study whose overall compliance rate was 76.0% ± 24.3% highlights the importance of patient-friendly drug-taking schedules as a compliance-enhancing strategy. The potential clinical impact of irregular drug-taking behaviors was demonstrated by Cramer et al. (23) in an electronically monitored study of epilepsy patients that found a strong tendency for seizures to occur during the period after doses were missed.

One further aspect of drug-taking patterns is that there is an apparent tendency for compliance to decrease over time (5,22,24,25). Thus, in the Physicians Health Study (22,24), compliance rates of 95.3% during a prerandomization run-in period were reduced to 83.8% during the study itself. The investigators in the

Studies of Left Ventricular Dysfunction (SOLVD, 25) reported that compliance rates were 80% at year one, 74% after two years, and 69% after three years. Even more pronounced are the results of the 10-week follow-up period of the Vaginal Infections and Prematurity (VIP) Study (5). In VIP, pill counts produced an estimated compliance rate of 85% during the first week of follow-up. But these rates decreased steadily to 47% during week nine and only 43% during the final study week.

IV. EFFECT OF POOR COMPLIANCE ON SAMPLE SIZE REQUIREMENTS IN CLINICAL TRIALS

When patients do not comply with a prescribed therapeutic regimen, a likely consequence is reduced treatment efficacy. But in the context of a clinical trial, reduced efficacy will yield a corresponding reduction in between group differences and an associated increase in the required sample size. What is surprising in all of this is the magnitude of the potential sample size increase (4–7).

To quantify the impact partial compliance may have on sample size requirements in clinical trials, Schechtman and Gordon (4) employ both a hypothetical example and the detailed compliance data provided by the Lipid Research Clinics (LRC) Coronary Primary Prevention Trial (2,26). In the hypothetical example, the goal is to compare success rates in a clinical trial where the placebo group success rate is 40% and the treatment group success rate is 60% in patients who take all of their medication as prescribed. A patient is defined as compliant if he takes at least 80% of his medication, and it is assumed that the success rate in the average compliant patients is 90% of the success rate in patients who take all of their pills. Using data from the LRC trial that translate into noncompliant patients having a cholesterol reduction of only 35.2% of the reduction in compliant patients, the hypothetical example assumes that noncompliers will receive 40% of the therapeutic benefit that is realized in compliers. Based on these assumptions, Table 1 provides a summary of Schechtman and Gordon's conclusions about the association between compliance rates and sample size. The results indicate, for example, that if the compliance rate is 70%, the sample size requirement is 1.84 times as great as it would be if all patients took all of their pills. The corresponding sample size ratio is 2.14 for a compliance rate of 60% and 2.52 if the compliance rate is 50%.

Table 1 Effect of Poor Compliance on Sample Size Requirements

Percentage of patients who comply	Sample size ratio
90%	1.40
80%	1.59
70%	1.84
60%	2.14
50%	2.52
40%	3.01

From a clinical trial where the success rate is 40% among placebo group patients and 60% in treatment group patients who take all of their pills. It is assumed that the average complier receives 90% of the benefit of subjects who take all of their pills and that the average noncomplier receives 40% of the benefit of subjects who take all prescribed medication. The sample size ratio is the ratio of the required sample size for the tabulated percentage of patients who are compliant to the corresponding sample size if all patients take all of their pills.

In the LRC trial (2,26), a treatment group that was prescribed six daily packets of cholestyramine resin, a cholesterol-lowering agent, was compared with a placebo group in a randomized double-blind trial involving 3806 hypercholesteremic men. The primary outcome measure was coronary heart disease (CHD) mortality. If the definition of compliance requires that the subject takes at least five of the six daily packets, then the reported compliance rate in the placebo group was 67.3%, as compared to 50.8% in the treatment group (2,4). Based on these compliance data, on the association between compliance rates and cholesterol reduction in the LRC trial, and on the association between cholesterol reduction and CHD mortality, it was demonstrated (4) that if the compliance rate in the treatment group (50.8%) had been as high as it was in the placebo group (67.3%), the sample size requirement in the LRC trial would have been reduced by about 41%. Moreover, the required sample size with the observed compliance rate was 2.97 times what it would have been if all patients had been compliers.

V. EFFECT OF POOR COMPLIANCE ON THE CLINICAL STATUS OF PATIENTS

Poor compliance is a problem only if there is a causative relationship between not taking pills and the clinical status of patients. On the one hand, it seems intuitively likely that partial compliance to an effective medication will increase the probability of negative clinical consequences. But at the same time, the magnitude of this causative relationship is likely to depend heavily on case-specific considerations, such as the disease in question, the half-life of the drug, and the frequency with which medication is prescribed. Moreover, because compliance may be strongly associated with other behavioral factors that influence clinical outcomes, it may be difficult to determine the degree to which poor compliance, as opposed to the behavioral correlates of poor compliance, are the cause of negative patient outcomes. In the Coronary Drug Project (27), for example, compliers to the lipid-lowering agent clofibrate had a much lower mortality rate than noncompliers to clofibrate ($P = 0.0001$). The initial suggestion is that there is a strong causal relationship between clofibrate compliance and reduced mortality. But a further look at the data suggests caution, because the same association between compliance and survival exists when you evaluate placebo-group data. Apparently, the reason for the reduced mortality among compliers was not the drug. Instead, the real explanation may be that medication compliers in the Coronary Drug Project tended also to engage in routine daily activities that improve cardiovascular health.

The message in the preceding example is that when disease status is heavily influenced by patient behaviors other than compliance, it may be difficult to determine whether poor compliance or correlated behavioral factors are the cause of negative outcomes. With this caveat in mind, we illustrate the degree to which poor compliance can indeed have serious negative clinical consequences.

LRC trial data on the association between compliance to cholestyramine and cholesterol reduction (2) are summarized in Table 2. While the table exhibits a small dose–response relationship in the placebo group, the association between cholesterol reduction and the number of packets taken is far more pronounced in the cholestyramine group. This is highlighted by the treatment-effect column of Table 2, which measures treatment benefit after subtracting off the cholesterol reduction in the placebo group. Thus, in contrast to the Coronary Drug Project, the association between poor compliance and poor outcome can be attributed overwhelmingly to the direct causative effects of not taking the drug as prescribed. In combination with LRC data on the association between CHD events and cholesterol reduction, the treatment-effect column of Table 2 has been used to estimate the expected number of CHD events that would have been prevented if LRC compliance rates had been greater than they actually were (4). The observed number of CHD events in the LRC trial was 155 in the cholestyramine group and 187 in the placebo group, suggesting that the study medication prevented about 32 events. But because of the side effects of the drug, compliance to cholestyramine (50.8%) was much lower than compliance to placebo (67.3%), where compliance is defined as taking at least five of six prescribed packets. If compliance in the treatment group had been as high as it was in the placebo group, the expected number of treatment group events would have been 145, so the number of events prevented would have increased by an expected total of 10, from 32 to 42. And if one assumes that all cholestyramine patients are in the highest compliance category of Table 2, the expected number of events prevented is increased by 24, from 32 to 56. That is, full

Table 2 Data from LRC Trial on Association Between Compliance to Medication and Cholesterol Reduction in the Cholestyramine Group and in the Placebo Group

	Placebo group		Cholestyramine group		
Mean daily packet count	*N*	% change in cholesterol	*N*	% change in cholesterol	Treatment effect
0–1	133	−3.2	294	−3.9	−0.7
1–2	79	−1.7	145	−5.4	−3.7
2–3	88	−4.0	135	−8.2	−4.2
3–4	105	−3.5	156	−11.1	−7.6
4–5	214	−4.1	205	−14.0	−9.9
5–6	1274	−5.4	965	−19.0	−13.6

The treatment effect column is the difference between the percentage change in cholesterol in the two groups.

compliance to cholestyramine is associated with an expected 75% (24/32) increase in the number of CHD events prevented by the drug, in comparison to what actually happened in the LRC trial.

The literature contains numerous other studies that associate poor compliance with negative clinical consequences, such as increased hospitalization (1) and reduced control (20) in patients with hypertension, seizures in patients with epilepsy (23), increased mortality and organ rejection in transplant recipients (28), and increased coronary heart disease event rates when beta blockers are not taken (29). The magnitude of these negative consequences is highlighted by the Joint National Committee on Evaluation, Detection, and Treatment of High Blood Pressure, which asserts that noncompliance is the most important reason for uncontrolled high blood pressure (30); by a 1984 symposium on the therapeutic consequences of noncompliance, which suggested that 125,000 annual deaths result from noncompliance in patients with cardiovascular diseases (31); and by U.S. Chamber of Commerce estimates that treating noncompliance-induced medical problems is associated with an annual cost of $13–15 billion (32).

IV. IMPROVING COMPLIANCE IN CLINICAL TRIALS

Since poor compliance to medication has the potential to reduce treatment efficacy substantially, it is not surprising that compliance-enhancing strategies are often a major component of clinical research. Broadly speaking, such strategies can be divided up into two general categories: those that precede the initiation of study therapy (2,5,22,24–26,33) and those that are implemented after randomization (2,26,34).

Compliance-enhancing strategies that precede randomization are usually focused on maximizing the likelihood that patients who enter the study will be good compliers. This is ordinarily accomplished using either a formal or an informal run-in strategy. A run-in strategy is carried out during a prerandomization period during which the likely future compliance of the patient is assessed. This may be accomplished *informally*, as it was in the LRC trial (2,26), using a screening period during which patients are expected to attend prerandomization visits aimed primarily at determining eligibility and doing preliminary assessments. If the patient does not attend these sessions or is otherwise uncooperative, the conclusion may be that the patient should not be randomized, because early behavior suggests future noncompliance. Alternatively, a formal run-in strategy may be carried out (5,22,24,25). A *formal* run-in involves a prerandomization period during which potential study subjects are asked to participate as if they had already been randomized. This specifically includes the taking of placebo medication and attendance at clinic visits that would be standard if the subject were to be randomized. Patients who do not comply up to some prespecified level are then excluded from the trial prior to randomization. It has been demonstrated that run-in strategies have increased the power of several major clinical trials (5,24). In general, run-in strategies are likely to be of benefit when poor compliance is associated with a substantial reduction in response, when poor compliance is common early in the study, when most early poor compliers could be expected to continue that pattern if randomized, when the run-in itself is inexpensive, or when the pool of eligible patients is sufficient to ensure that run-in exclusions will not compromise recruitment goals (35).

Compliance-enhancing strategies that follow randomization cover a broad range of potential activities and have been discussed by many investigators (36–40). These strategies include: educating patients about their disease and about the importance of compliance; financial and other incentives (40); simplified and flexible dosing regimens that are adjusted to fit individual patient schedules; establishing positive clinician–patient relationships; self-administered approaches, such as providing calendars to be checked off when doses are taken; specialized drug packaging, such as blister packs (20) and caps for prescription containers that sound an alarm when it is time to take the next dose (38); and encouraging family members to provide support. If there is any unifying theme in the investigation of these procedures, it is that no single approach can be successfully applied in all settings. Compliance-enhancing strategies must be tailored to the patient and to the particular circumstance. Their success depends heavily upon the ingenuity and the commitment of the health care provider.

REFERENCES

1. Maronde RF, Chan LS, Larsen F, Strandberg LR, Laventurier MF, Sullivan SR. Underutilization of antihypertensive drugs and associated hospitalization. Med Care 27:1159–1166, 1989.
2. Lipid Research Clinic Program. The Lipid Research Clinics Coronary Primary Prevention Trial Results II. Relationship of reduction in incidence of coronary heart

disease to cholesterol lowering. JAMA 251:365–374, 1984.

3. Beta-Blocker Heart Attack Trial Research Group. A randomized trial of propranolol in patients with acute myocardial infarction, 1: Mortality results. JAMA 247: 1701–1714, 1982.
4. Schechtman KB, Gordon ME. The effect of poor compliance and treatment side effects on sample size requirements in randomized clinical trials. J Biopharmaceut Stat 4:223–232, 1994.
5. Blackwelder WC, Hastings BK, Lee MLF, Deloria MA. Value of a run-in period in a drug trial during pregnancy. Controlled Clinical Trials 11:187–198, 1990.
6. Freedman LS. The effect of partial noncompliance on the power of a clinical trial. Controlled Clinical Trials 11:157–168, 1990.
7. Lachin JM, Foulkes MA. Evaluation of sample size and power for analyses of survival with allowance for nonuniform patient entry, losses to follow-up, noncompliance, and stratification. Biometrics 42:507–519, 1986.
8. Hopper SV, Schechtman KB. Factors associated with diabetic control and utilization patterns in a low-income, older adult population. Patient Educ Counseling 7:275–288, 1985.
9. Pullar T, Kumar S, Tindall H, Feely M. Time to stop counting tablets? Clin Pharmacol Ther 46:163–168, 1989.
10. Rudd P, Byyny RL, Zachary V, LoVerde ME, Titus C, Mitchell WD, Marshall G. The natural history of medication compliance in a drug trial: Limitations of pill counts. Clin Pharmacol Ther 46:169–176, 1989.
11. Norell SE. Methods in assessing drug compliance. Acta Med Scand 683(suppl):35–40, 1984.
12. Feinstein AR, Wood HF, Epstein JA, Taranta A, Simpson R, Tursky E. A controlled study of three methods of prophylaxis against streptococcal infection in a population of rheumatic children. N Engl J Med 260:697–702, 1959.
13. Bergman AB, Werner RJ. Failure of children to receive penicillin by mouth. N Engl J Med 268:1334–1338, 1963.
14. Gordis L, Markowitz M, Lilienfeld AM. The inaccuracy in using interviews to estimate patient reliability in taking medications at home. Med Care 7:49–54, 1969.
15. Hecht AB. Improving medication compliance by teaching outpatients. Nurs Forum 13:112–129, 1974.
16. Caron HS. Compliance and the case of objective measurement. J Hypertension 3(suppl I):11–17, 1985.
17. De Klerk E, Van der linden S, Van der Heijde D, Urquart J. Facilitated analysis of data on drug regimen compliance. Stat Med 16:1643–1664, 1997.
18. Cramer JA, Scheyer RD, Mattson RH. Compliance declines between clinic visits. Arch Intern Med 150: 1509–1510, 1990.
19. Kass MA, Meltzer DW, Gordon M, Cooper D, Goldberg J. Compliance with topical pilocarpine treatment. Am J Ophthalmol 101:515–523, 1986.
20. Eisen SA, Woodward RS, Miller D, Spitznagel E, Windham CA. The effect of medication compliance on the control of hypertension. J Gen Int Med 2:298–305, 1987.
21. Specter SL, Kinsman R, Mawhinney H, Siegel SC, Rachelefsky GS, Katz RM, Rohr AS. Compliance of patients with asthma with an experimental aerosolized medication: implications for controlled clinical trials. J Allergy Clin Immunol 77:65–70, 1986.
22. Steering committee of the physicians' health study research group. Final report on the aspirin component of the ongoing Physicians' Health Study. N Engl J Med 321:129–135, 1989.
23. Cramer JA, Mattson RH, Prevey ML, Ouellette V. How often is medication taken as prescribed? A novel assessment technique. JAMA 161:3273–3277, 1989.
24. Lang JM, Buring JE, Rosner B, Cook N, Hennekens CH. Estimating the effect of the run-in on the power of the Physicians Health Study. Stat Med 10:1585–1593, 1991.
25. The SOLVD Investigators. Effect of enalapril on survival in patients with reduced left ventricular ejection fractions and congestive heart failure. N Engl J Med 325:293–302, 1991.
26. Lipid Research Clinics Program. The Lipid research Clinics coronary primary prevention trial results I. Reduction in incidence of coronary heart disease. JAMA 251:351–364, 1984.
27. Coronary Drug Project Research Group. Influence of adherence to treatment and response of cholesterol on mortality in the Coronary Drug Project. N Engl J Med 303:1038–1041, 1980.
28. Schweizer RT, Rovelli M, Palmeri D, Vossler E, Hull D, Bartus S. Noncompliance in organ transplant recipients. Transplantation. 49:374–377, 1990.
29. Psaty BM, Koespell TD, Wagner EH, LoGerfo JP, Inui TS. The relative risk of incident coronary heart disease associated with recently stopping the use of beta-blockers. JAMA 263:1653–1657, 1990.
30. Joint National Committee on Detection, Evaluation, and Treatment of High Blood Pressure. The 1984 report of the Joint National Committee on Detection, Evaluation, and Treatment of High Blood Pressure. Arch Int Med 144:1045–1057, 1984.
31. Burrell CD, Levy RA. Therapeutic consequences of noncompliance. In: Improving Medication Compliance: Proceedings of a Symposium. Washington, DC: National Pharmaceutical Council, 1984, pp. 7–16.
32. Yenney SL, Behrens RA. Health Promotion and Business Coalitions. U.S. Chamber of Commerce, Washington DC, 1984.
33. The CAST Investigators. Preliminary Report: Effect of encainide and flecainide on mortality in a randomized trial of arrhythmia suppression after myocardial infarction. N Engl J Med 321:406–412, 1989.
34. Black DM, Brand RJ, Greenlick M, Hughes G, Smith J. Compliance to treatment for hypertension in elderly

patients: The SHEP pilot study. Systolic hypertension in the elderly program. J Gerontol 42:552–557, 1987.
35. Schectman KB and Gordon ME. A comprehensive algorithm for determining whether a run-in strategy is a cost-effective design modification in a randomized clinical trial. Stat Med 12:111–128, 1993.
36. Peck CL, King JN. Increasing patient compliance with prescriptions. JAMA 248:2874–2877, 1982.
37. Eraker SA, Kirscht JP, Becker MH. Understanding and improving patient compliance. Ann Int Med 100:258–268, 1984.
38. Bond WS, Hussar DA. Detection methods and strategies for improving medication compliance. Am J Hosp Pharmacol 48:1978–1988, 1991.
39. Melnikow J, Kiefe C. Patient compliance and medical research: Issues in methodology. J Gen Int Med 9:96–105, 1994.
40. Giuffrida A, Torgerson DJ. Should we pay the patient? Review of financial incentives to enhance patient compliance. Br Medical J 315:703–707, 1997.

Kenneth B. Schechtman

Pharmacodynamic Issues

See also *Clinical Pharmacology; Pharmacodynamic with Covariates; Pharmacodynamic without Covariates*

I. INTRODUCTION

There are several ways of defining *pharmacodynamics*. Here are three examples.

1. The branch of pharmacology dealing with the reactions between drugs and living systems (1)
2. The study of the biochemical and physiological effects of drugs and their mechanisms of action on the body (2)
3. The study of how drugs can affect the body (3)

After being swallowed, injected, inhaled, nasally administered, or absorbed through the skin, most drugs enter the bloodstream, circulate throughout the body, and interact with a number of target sites. Depending on its property or route of administration, a drug may act in only a specific area of the body (for example, the action of antacids is confined largely to the stomach). When a drug interacts with the target tissues (e.g., cells or organs), a desirable therapeutic effect usually takes place, which is referred to as *efficacy* or *effectiveness* in clinical trials. On the other hand, if a drug interacts with cells, tissues, or organs that are not the desired targets, side effects may occur. In clinical trials, side effects are also known as *adverse events* (AEs) or *adverse reactions* (ARs). Hence, in a phase III clinical trial pharmacodynamic (PD) endpoints are primarily safety or efficacy (or both) parameters. Efficacy endpoints are generally more heterogeneous than safety endpoints from one therapeutic area to another, because the majority of the safety endpoints are usually restricted to a limited number of specific evaluations, including physical examinations, electrocardiogram (ECG), laboratory variables, vital signs, and adverse-event reporting; drug-level monitoring may also be used to assess safety. For this reason, the focus of the statistical methodologies in this entry is on efficacy parameters and not on safety parameters.

The major organ systems in the body include cardiovascular, respiratory, nervous, skin, musculoskeletal, blood, digestive, endocrine, urinary, and reproductive. An investigational drug may be evaluated for efficacy and/or safety for one or more of these systems. An investigational new drug (IND) application containing chemistry, toxicity, and other preclinical data is submitted to the Food and Drug Administration (FDA) in the United States. Other countries utilize a similar process, allowing clinical studies to begin after a review of clinical data. After the investigational drug has been evaluated in clinical trials, a new drug application (NDA) is filed by its sponsoring drug company; this is then reviewed and, hopefully, approved by the Center for Drug Evaluation and Research (CDER) of the FDA. As of 1998, CDER had organized the Offices of Drug Evaluation into five divisions. Each division is responsible for evaluating new drugs in some specific therapeutic areas. Division I is responsible for neuropharmacological drug products, oncology drug products, and cardiorenal drug products. Division II is responsible for metabolic and endocrine drug products, pulmonary drug products, and reproductive and urologic drug products; Division III is responsible for gastrointestinal and coagulation drug products; anesthetic, critical care, and addiction drug products; and medical imaging and radiopharmaceutical drug products. Division IV is responsible for antivirus, anti-infective, and immunologic drug products. Division V is responsible for anti-inflammatory, analgesic, ophthalmologic, dermatologic, and dental drug products. It is then helpful to understand the therapeutic drug involved and which division of the CDER will be handling the NDA submission; meetings can then be coordinated between the division involved and the drug sponsor before IND application and NDA submission in order to reach mutual understanding and agreement about how the investigational drug will be conducted in clinical trials.

This entry is organized as follows: Sec. I first defines pharmacodynamics, and then, in Sec. I.A. through I.H., provides a general overview of some of the primary endpoints for some of the most common diseases within different organ systems when evaluating pharmacodynamic effects of investigational drugs used in phase III clinical trials. Section II describes a number of factors that affect the selection of statistical methods for pharmacodynamics evaluations. Section III addresses some additional issues related to Phase II/III pharmacodynamic evaluations.

These pharmacodynamic endpoints discussed in the remainder of this section are often used in clinical trials and agreed upon with regulatory health authorities for the specific indications.

A. Cardiovascular System

1. Coronary Artery Disease/Angina

Patient mortality is the most important clinical endpoint in trials assessing the beneficial effects of drugs on coronary artery disease. Such studies often are long-term (e.g., 5 or more years in duration) and may recruit thousands of patients. Patient morbidity (e.g., revascularization, stroke, nonfatal myocardial infarction) are also assessed in such trials. Subjective chest pain and exercise tolerance (stress test) are common endpoints in angina trials.

2. Hypertension

The mean change from baseline in systolic and diastolic blood pressure and cardiovascular mortality and morbidity are endpoints used in hypertension trials.

3. Congestive Heart Failure

Patient mortality, exercise tolerance, the number of hospitalizations, and cardiovascular morbidity are common endpoints in trials assessing the effects of drugs in congestive heart failure.

4. Hyperlipidemia

Low-density lipoprotein, cholesterol levels, and cardiovascular morbidity are endpoints in trials assessing the effect of lipid-lowering agents.

B. Respiratory System

1. Asthma

Asthma is a very common reaction airway disease. A clinically relevant primary endpoint used in asthma trials is a change in forced expiratory volume in 1 second (FEV1). Such FEV1 measurements are determined by spirometry evaluations.

C. Nervous System

1. Alzheimer's Disease

Cognitive and functional scales specifically designed to assess Alzheimer's disease are used as standard endpoints in Alzheimer's trials. Such trials often require that hundreds to thousands of patients be recruited.

2. Parkinson's Disease

Functional scales specifically designed to assess Parkinson's disease are typically used in evaluating drugs in this disease.

D. Skin

1. Acne

Lesion count and the nature of the lesions (e.g., papular, pustular counts) are commonly used endpoints in acne trials.

2. Psoriasis

Plaque size, induration, scaling, and redness are generally accepted endpoints in psoriasis trials. Other parameters to be considered include the extent and the severity of psoriatic lesions.

E. Musculoskeletal System

1. Osteoarthritis

Tender joints and pain-functional endpoints (e.g., Western Ontario and McMaster Universities Osteoarthritis Index, or WOMAC) are assessed in osteoarthritis trials. The WOMAC is a self-administered questionnaire that is highly reliable and sensitive to changes in health status in patients with osteoarthritis (4).

2. *Rheumatoid Arthritis*

Endpoints used in trials assessing the safety and efficacy of drugs for rheumatoid arthritis include the number of swollen and painful joints and patient and physician's global assessment of well-being. Radiographic progression of joint destruction is being assessed in current trials with rheumatoid arthritis.

F. Gastrointestinal System

1. *Peptic Ulcer Disease*

Endoscopic evidence of improvement of resolution of an ulcer is used as an endpoint in ulcer trials. Symptomatic improvement alone following administration of the investigational drug is *not* sufficient proof of benefit in phase III ulcer trials.

G. Endocrine System

1. *Diabetes*

Diabetes is a systemic disease that may affect multiple organs, including the skin (dermopathy), blood vessels (peripheral vasculopathy), kidney (nephropathy), eyes (retinopathy), and nervous system (neuropathy). Clinical trials in diabetes could assess these clinical endpoints, but they also require laboratory endpoints such as blood glucose levels and/or hemoglobin A1C.

2. *Osteoporosis*

The incidence of bone fracture is a primary efficacy endpoint in osteoporosis trials; since these trials may require a very long duration of follow-up for evaluating bone fractures (e.g., up to 5 years), bone density has been used as a surrogate marker in osteoporosis trials.

H. Other Areas

1. *Infectious Diseases*

Infections can involve any body system and therefore have many primary endpoints that can be used in clinical trials. The absence of an infectious agent by culture and mortality (e.g., sepsis trials) may be used. In addition, surrogate markers (e.g., viral load in acquired immunodeficiency syndrome [AIDS] studies) have been used.

2. *Oncology*

The incidence of patient mortality is an endpoint in many oncology trials. In those instances where there is an extremely high mortality rate, alternative endpoints such as the time of death and tumor burden can be used.

3. *Transplantation*

In recent years there has been an increase in the number of drugs approved for use in transplantation as primary or adjunctive immunosuppression for kidney, liver, and heart transplantation. Phase III clinical trials currently use the incidence of acute rejection episodes and patient and graft survival as necessary endpoints for approval by health authorities.

II. FACTORS AFFECTING THE SELECTION OF STATISTICAL METHODS FOR PHARMACODYNAMIC EVALUATIONS

The choice of statistical method employed to analyze the pharmacodynamic parameters of interest depends on a number of factors:

A. Study objective and design
B. Type and number of explanatory variables and response variables
C. Distribution of the PD parameter
D. Model assumptions
E. Therapeutic area of the study drug
F. Frequency of evaluation of the PD parameter
G. Measurements of the PD parameter
H. Regulatory concerns of the conduct of a clinical trial

A. Study Objective and Design

The objectives for phase III clinical studies may include one or more of the following four objectives (5):

1. Demonstrate/confirm efficacy.
2. Establish a safety profile.
3. Provide an adequate basis for assessing the benefit/risk relationship to support labeling.
4. Establish the dose–response relationship.

A study objective will define what PD parameter is to be considered as the primary clinical endpoint and what comparisons or investigations are deemed most clinically relevant. In phase III clinical trials, the primary PD parameter is usually an efficacy variable, because the primary objective of most phase III trials is to provide strong scientific evidence on the efficacy of the study drug. The design of the study to evaluate the investigational drug is driven by the study objective. Study design determines how the data from the clinical study are to be collected while minimizing bias and variability from possible extraneous sources in a clinical study. Additionally, a study design that appears to be the most effective one from a statistical perspective may not be suitable or feasible for the study from a clinical or pharmacological perspective.

The most popular study design in phase III is a parallel group design (known as *completely randomized* design). Other designs, such as crossover design and factorial design, are also frequently used.

Parallel group design is the simplest design of all. A parallel group design may be extended to include one additional explanatory variable that is used as a blocking factor. Such a parallel group design with a blocking factor is known as a *randomized block* design. A parallel group design may include repeated evaluation of the same PD variable over time, which is known as a *repeated measurements* design.

Usually, there are two or more treatment arms involved in clinical trials, often with one arm assigned to placebo (a *placebo-controlled* study) or to an active comparator agent (an *active-controlled* study). Placebo may not be used in any arm due to ethical or other reasons.

Each patient typically receives only one treatment in a parallel group design. In clinical trials where all the treatments are administered to each patient so that he/she can serve as his/her control, a crossover design is the study design of choice. A factorial design is highly desirable when two or more treatments are evaluated simultaneously in the same set of patients through the use of varying combinations of the treatment or when the combinations of two or more explanatory variables (including the treatment) are of interest.

B. Type and Number of Explanatory Variables and Response Variables

Explanatory variables are also known as *independent* variables or simply as *input* variables. *Response* variables are also called *dependent* variables or *output* variables. The response variables throughout this entry are referred to as clinical responses, i.e., the PD parameter.

Generally, any data can be classified as either categorical or numerical. A *categorical* variable is either *nominal*, when it does not have a natural ordering (e.g., gender), or *ordinal*, when it is associated with a certain order (e.g., condition of disease, such as mild, moderate, or severe). A numerical variable can be either continuous or discrete. Theoretically, a *continuous* variable is not countable and can take on any possible value in a certain interval. On the other hand, a *discrete* variable is countable and in many cases assumes only a finite number of values. A categorical variable, in fact, can be treated as a numerical variable when we can artificially assign numerical values to the categories of a categorical variable. Of course, it is understood that the numerical values of a categorical variable with a nominal scale do not have any meaning in terms of ordering. Based on this argument, we can simply say that a PD variable can be classified as either continuous or discrete (with a nominal scale or an ordinal scale) (6,7).

In reality, all variables collected from clinical trials should be classified as discrete variables, because (a) they are collected from a finite number of patients and (b) they are countable. Therefore, the fundamental idea of determining whether a variable is continuous or discrete is dependent upon the number of different values obtained from that variable of interest.

If a variable takes on a small number of values (there is no specific rule as to how small is *small*—it should be relative to the sample size), we can conclude that the variable is discrete. In contrast, if data values encompass a relatively wide range of different values, the variable can be considered continuous.

A commonly used statistical method for evaluating a continuous variable if the data are normally distributed is analysis of variance. In contrast, Fisher's exact test, or chi square test, may be applied for categorical-data analysis (i.e., a discrete variable with a nominal or ordinal scale). A continuous variable may be classified as a discrete variable with an ordinal scale. For example, we can categorize a response to antihypertensive therapy in terms of the systolic blood pressure's being either no less than 20% reduction from baseline or less than 20% reduction from baseline.

Additionally, the number of variables (either explanatory or response) or the number of levels for each variable in a clinical trial also affects the selection of certain statistical methods. Here are some examples.

1. A two-sample t-test may be employed if there are only two levels for the only explanatory variable with a continuous scale (i.e., two doses for the treatment), but the same t-test cannot be applied if there are three or more treatment levels. In the case in which there are three or more treatment levels, an analysis of variance may be applied.

2. Two-way analysis of variance or analysis of covariance may be employed if two explanatory variables are involved.
3. A multivariate approach may be considered if there is more than one primary PD parameter to be considered simultaneously.

C. Distribution of the Pharmacodynamic Parameter

A parametric approach is generally preferred when the PD variable can be justified to be fairly normally distributed. When the continuous PD variable is skewed or not normally distributed, a nonparametric method may be used. In general, a parametric method is more powerful than a nonparametric method. Note that the power of a test is the probability that the null hypothesis is rejected when it is not true.

Sometimes the skewness of the data is attributed to the existence of outliers. The discussion on handling outliers will be detailed in Sec. II.H, because the consideration of removing outlier(s) from a statistical analysis is based not only on statistical concerns but also on regulatory concerns.

As for discrete variables, the most common distributions are binomial, multinomial, and Poisson. Logit (log of odds) models are used with binomial and multinomial distributions. The log-linear model is used with the Poisson distribution.

D. Model Assumptions

A parametric approach is used in a conventional statistical model along with a number of assumptions. The assumption of normality is only the first step to determine the suitability of a parametric method. There are other conditions required for a specific model, including the following.

1. *Independence*: Most of the models require independence of the data. This means that the observations from different treatments are obtained from different patients. A different model needs to be considered when observations are dependent, such as the data from a study with a crossover design or from a study with repeated measures.
2. *Homogeneous Variance*: Equal variability is required in the approach with conventional analysis of variance. When this requirement is not met, there are two possible solutions: (a) use an algebraic transformation such as log to reduce the observations in a much smaller scale, in hopes of achieving the equality of variance; (b) employ a more advanced method, such as a generalized F-test (8), that does not require equal variance in the model. However, the software program to accommodate such methodologies for exact tests does not exist in SAS.

E. Therapeutic Area of the Study Drug

In almost all cases, time should be considered a continuous variable. Intuitively, we may think about applying analysis of variance to analyze "time" when it is the primary PD parameter, but this is not always the case. For example, in a clinical trial to assess an analgesic drug, the onset time (the time that the degree of pain starts to diminish) can be a primary PD parameter. Analysis of variance can be applied to compare the mean onset time among different treatment groups. However, in cancer trials, survival analysis may be more appropriate (e.g., in a cancer patient, the time to survival after a treatment). In summary, the mean time can be generally compared with a t-test or analysis of variance, but the time to the first predefined event is usually analyzed with a survival analysis.

F. Frequency of Evaluation of the Pharmacodynamic Parameter

When a PD parameter is evaluated more than once throughout a clinical trial, it is more appropriate to employ a method that takes into account the repeated measures. A method such as one-way analysis of variance with repeated measures may be appropriate for such a trial.

G. Measurements of Pharmacodynamic Parameters

The measurement of a primary PD parameter in a clinical trial protocol must be clearly defined. For example, it is not adequate to indicate that mortality is the primary parameter in a cancer trial. Mortality can be assessed in many ways: the proportion of patients alive at a predetermined point in time, the overall survival curve, the mean survival time during a fixed time interval, or a combination of assessments.

When a continuous PD variable has been determined, the change from baseline or the percentage change from baseline may be of most interest. One major concern with such a measure is the direction of the change, which may be above or below baseline, so the average of such a measure from all patients could be misleading; for example, the elevation or decrease of blood glucose from baseline is an important indication for a diabetes trial assessing a hypoglycemic agent. On the other hand, if only the magnitude of the change (or percentage change) is considered, the probability distribution of this absolute value of this derived PD variable may not be easily identified and would greatly impact the conduct of the hypothesis testing for the trial.

As discussed earlier, a continuous variable can be converted into a discrete variable with an ordinal scale. When such a conversion is deemed clinically relevant, the employment of statistical methods should be adjusted accordingly.

H. Regulatory Concerns about the Conduct of a Clinical Trial

In January 1997 the FDA published a draft guideline on statistical principles for clinical trials (9). It addresses statistical considerations about handling a number of potential issues for a clinical trial. They include the following.

1. Type of Comparison

This is closely related to the study objective and will determine how the hypothesis testing should be set up: a two-sided test, a one-sided test (a right-tailed test or a left-tailed test). The following three possible comparisons are usually demonstrated in phase III trials.

a. Superiority Trials

These are meant to demonstrate that the study drug is superior to placebo in a placebo-controlled study or is superior to an active control treatment. Generally, superiority trials are assumed, unless it is explicitly stated otherwise.

b. Equivalence Trials

Some examples:

1. Pharmacokinetic bioequivalence trials to show two formulations are bioequivalent or a generic drug is bioequivalent to an innovator drug
2. Show equivalence between a comparator agent at one dose level and the study drug at a lower dose level
3. Show equivalence between two different dosing regimens of the study drug with an identical total dose, such as 150 mg twice a day (b.i.d.) vs. 100 mg (t.i.d.) three times a day.

c. Noninferiority Trials

These can be seen in many active-controlled studies to show that the efficacy of an investigational drug is no worse than that of the comparator drug. In some situations, a trial can be designed to add a placebo arm to an active-controlled study so that the objectives of the study are (a) to establish superiority to the placebo and (b) to show a similar efficacy (and/or safety) profile to the comparator agent.

2. Patient Population Included in the Statistical Analysis

This does not affect the decision making in terms of statistical approaches, but it would certainly influence the results and interpretation of the results. The following two types of patient populations are generally allowed in the statistical analysis.

a. All Randomized Patients

An analysis based on this population is also well known as an "intent-to-treat" analysis. In theory, the analysis should include all eligible patients who were randomized to the study treatments. In practice, it often excludes patients who did not receive any dose of study treatments (including placebo) or did not have any data postrandomization. Two major problems arise in this case: First, should we include patients who may fall into one of the following problems: protocol violation; error in treatment assignment; use of prohibited medications during the study? Second, in which treatment group should a patient be placed in the statistical analysis if he is not treated with the assigned study drug? For example, should the patient be in treatment group A or B if he actually received treatment B but was randomized to treatment A? There is no clear direction in the guidelines as to how to resolve these problems; they should be handled case by case.

b. Per-Protocol Patients

An analysis using this patient population is referred to as *evaluable patients* analysis. *Evaluable patients* should be defined as concisely as possible in the protocol or at latest before the unblinding of the treatment. Any change to the definition after unblinding has to be justified.

3. *Handling Missing Values and Outliers*

Missing values do occur in almost every clinical trial. Unfortunately, there is no universally applicable method for handling missing values. The best way is to predefine the methods to be used for missing values in the analysis plan, especially for the study that may involve a substantial number of missing values. As with the handling of missing values, the statistical method for detecting outliers should be well described before the treatment codes are unblinded. The removal of outliers from statistical analysis will be more convincing when the outliers can be justified clinically as well as statistically.

I. Summary

A selection of a specific statistical method for evaluating PD parameters in a clinical trial should be carefully based on all of the factors just discussed. This will help a sponsoring pharmaceutical company to get the study drug approved by the FDA by employing an appropriate statistical method (it may be a very simple one) that is acceptable from statistical, medical, and regulatory standpoints.

III. OTHER ISSUES

There are a number of additional issues that have received specific attention by the FDA for evaluations of new drug. Since there is no absolute statistical method to handle these specific topics, these will be discussed from clinical and regulatory perspectives.

A. Surrogate Markers

Surrogate markers are now being considered as appropriate target endpoints for various diseases in clinical trials, especially in those diseases where the surrogate marker has been validated (e.g., acute rejection episodes as a predictor of chronic organ rejection), where medical need may justify a rapid assessment (e.g., T-cell count due to the high mortality from AIDS), or where the duration of a trial (bone fracture endpoint in osteoporosis trials) or the number of patients required for enrollment (e.g., rare medical conditions) may be problematic.

A *surrogate* endpoint is an endpoint that is intended to relate to a clinically important outcome but does not in itself measure a clinical benefit. Surrogate endpoints may be used as primary endpoints when appropriate (when the surrogate is reasonably likely or well known to predict a clinical outcome (10)). The clinical conclusions may be misleading if the selected surrogate endpoints for the studies are not vigorously established. Fleming (11) discusses details of using surrogate markers in clinical trials, and Fleming and DeMets (12) present examples of surrogate endpoints that failed to be validated in clinical trials.

B. Subgroup Considerations

Particular subgroups deserve special PD consideration because of unique characteristics, thereby requiring the need to evaluate such subgroups separately in clinical trials. These subgroups include the elderly and pediatric populations, different races, gender differences, and medical conditions that could affect PD interpretation. It would therefore be appropriate in these subgroups to analyze the PD effects of a drug separate from the rest of the trial population.

1. *Elderly Population*

It is well known that the elderly population is different from the younger population with respect to sensitivity to many drugs (13). For example, patients treated with nonsteroidal anti-inflammatory agents have a higher incidence of gastrointestinal bleeding than do younger patients. The FDA also provides guidance on the need for collecting information on geriatric patients (14). There is no clear definition of a geriatric population; normally this includes patients 65 years and older. However, the FDA encourages drug companies to include patients in clinical studies, if possible, who are 75 years and older.

2. *Pediatric Population*

Body surface area is different in very young children, compared to adolescents or adults; this often requires dosing of a drug on a mg/kg basis compared to adults, among whom body surface area is not dramatically different. Bowel length is much shorter in very young children, which is well known to affect the pharmacokinetics of a drug and therefore the PD endpoints in a clinical trial.

When pediatric patients are to be considered for participation in clinical trails, safety data from adult patients should be available (15). In addition, for a drug expected to be used in children, evaluation should be made in the appropriate group. When a drug is to be evaluated in children, it is usually appropriate to begin with older children before extending the trial first to younger children and then to infants (16).

3. *Racial Differences*

When a new drug is to be developed globally, the drug may be tested among different ethnic groups. It is highly suggested by the FDA that such a drug be characterized as ethnically sensitive or insensitive. A new chemical entity (NCE) is said to be sensitive to ethnic factors when its pharmacokinetic, PD, or other characteristics suggest the potential for clinically significant impact by intrinsic and/or extrinsic ethnic factors on safety, efficacy, or dose response. *Extrinsic* ethnic factors are factors associated with the patient's environment and/or culture. Extrinsic factors tend to be less genetically and more culturally and behaviorally determined. Examples are medical practice, diet, socioeconomic status, and exposure to pollution and sunshine. *Intrinsic* ethnic factors are characteristics associated with the drug recipient. These are factors that help to define and identify a subpopulation and that may influence the ability to extrapolate clinical data between ethnic groups (or between regions). Examples of intrinsic ethnic factors include polymorphism, age, gender, height, weight, lean body mass, body composition, and organ dysfunction (17).

The following is an example of a drug showing its sensitivity to ethnic factors: African Americans have been reported to have less drug bioavailability from immunosuppressive drugs administered after transplantation, which has been attributed to lower absorption compared to the Caucasian population. This pharmacokinetic difference has a significant impact on the PD endpoint of acute rejection, because African Americans who receive kidney transplants are known to have poorer graft survival (18).

4. *Gender Differences*

Due to unique anatomy and physiology, women are known to have complex pharmacokinetics different from men (19). Although pharmacokinetic differences between men and women have been extensively reported, less is known about PD differences between men and women.

Issues related to pregnancy and oral contraceptive use appear to be important (20). Pregnancy may increase the metabolism of anti-epileptic drugs, and oral contraceptives are well known to interfere with the metabolism of many drugs; on the other hand, certain drugs may impair contraceptive efficacy. Women have been reported to be more prone to developing torsades de pointes from drugs such as quinidine and procainamide than men, and women have been reported to have less sensitivity to insulin than men. Ongoing research in this area will delineate additional PD differences between men and women.

5. *Patients with Impaired Renal Function*

After entering the body, a drug is eliminated either by excretion or by metabolism. While elimination can occur via any of several routes, most drugs are cleared either by elimination of unchanged drug by the kidney or by metabolism in the liver. For a drug eliminated primarily via renal excretory mechanisms, impaired renal function may alter its PK and PD to an extent that the dosage regimen needs to be changed from that used in patients with normal renal function. While the most obvious type of change arising from renal impairment is a decrease in renal excretion (or possibly renal metabolism) of a drug or its metabolites, renal impairment has also been associated with other changes, such as changes in hepatic metabolism, plasma protein binding, and drug distribution. These changes may be particularly prominent in patients with severely impaired renal function and have been observed even when the renal route is not the primary route of elimination of a drug (21).

C. Meta-Analysis

The results from more than one clinical trial may be combined in order to draw a more robust clinical conclusion. A meta-analysis of clinical trials may also be necessary when data exist from multiple studies that are each insufficiently powered to reach statistically significant conclusions or if each study has a limited number of enrolled patients (due to logistical or other reasons). There is a great amount of literature published in this area [e.g., Hedges and Olkin (22), Hunter and Schmidt (23), and Geller and Proschan (24)].

D. Pharmacokinetic/Pharmacodynamic (PK/PD) Relationships

During recent drug development and evaluation efforts, PK/PD relationships have been a focus, because the combination of these two areas, through an appropriate modeling, will lead to the therapeutically most relevant relationship between pharmacological effect and time and will produce an optimal dose recommendation. Pharmacokinetic data and PD data can be linked directly or indirectly. The direct or indirect link will play an important role in determining an appropriate mathematical PK/PD model (25).

REFERENCES

1. Merriam Webster's Collegiate Dictionary. 10th ed. Springfield, MA: Merriam-Webster, 1994, p 871.
2. AG Goodman, LS Goodman, TW Rall, F Murad F. The pharmacological basis of therapeutics. 7th ed. New York: MacMillan, 1985.
3. Merck Manual of Medical Information. Home ed. PA: Merck, 1997, pp 31–34.
4. N Bellamy, WW Buchanan, CH Goldsmith, J Campbell, LW Stitt. Validation study of WOMAC: a health status instrument for measuring clinically important patient relevant outcomes to antirheumatic drug therapy in patients with osteoarthritis of the hip or knee. J Rheumatol 15:1833–1840, 1988.
5. International Conference on Harmonization E8: Guidance on General Considerations for Clinical Trials, 1997.
6. MR Selwyn. Principles of Experimental Design for the Life Science. Boca Raton, FL: CRC Press, 1996.
7. A Agresti. Categorical Data Analysis. New York: Wiley, 1990, pp 2–7.
8. S Weerahandi. Exact Statistical Methods for Data Analysis. New York: Springer Verlag, 1994.
9. International Conference on Harmonization E9: Note for Guidance on Statistical Principles for Clinical Trials, 1997.
10. DJ Sheskin. Handbook of Parametric and Nonparametric Statistical Procedures. Boca Raton, FL: CRC Press, 1996.
11. TR Fleming. Surrogate Endpoints in Clinical Trials. Drug Inf J 30:545–551, 1996.
12. TR Fleming, DL DeMets. Endpoints in clinical trials: are we being misled? Ann Intern Med 125:605–613, 1996.
13. CG Swift. Pharmacodynamics: changes in homeostatic mechanisms, receptor and target organ sensitivity in the elderly. Br Med Bull 46:36–52, 1990.
14. International Conference on Harmonization E7: Studies in Support of Special Populations: Geriatrics, 1994.
15. International Conference on Harmonization M3: Nonclinical Safety Studies for the Conduct of Human Clinical Trials for Pharmaceuticals, 1997.
16. International Conference on Harmonization E8: Guidance on General Considerations for Clinical Trials, 1997.
17. International Conference on Harmonization E5: Ethnic Factors in the Acceptability of Foreign Clinical Data, 1997.
18. S Katznelson, DW Gjertson JM Cecka. The effect of race and ethnicity on kidney allograft outcome. In JM Cecka, PI Terasaki, eds. Clinical Transplants 1995. Los Angeles: UCLA tissue typing Laboratory, 1995, p 379.
19. CV Fletcher, EP Acosta, JM Strykowski. Gender differences in human pharmacokinetics and pharmacodynamics. J Adolescent Health 15:619–629, 1994.
20. RZ Harris, LZ Benet, JB Schwartz. Gender effects in pharmacokinetics and pharmacodynamics. Drug 50: 222–239, 1995.
21. International Conference on Harmonization: Pharmacokinetics and Pharmacodynamics in Patients with Impaired Renal Function: Study Design, Data Analysis, and Impact on Dosing and Labeling, 1997.
22. LV Hedges, I Olkin. Statistical Methods for Meta-Analysis. New York: Academic Press, 1985.
23. JE Hunter, FL Schmidt. Methods of Meta-Analysis. Beverly Hills, CA: SAGE Publications, 1990.
24. NL Geller, M Proschan. Meta-analysis of clinical trials: a consumer's guide. J Biopharm Statist 6:377–3395, 1996.
25. H Derendorf, G. Hochhaus. Handbook of Pharmacokinetic/Pharmacodynamic Correlation. Boca Raton, FL: CRC Press. pp 79–120, 1995.

Cheng-Tao Chang
Robert L. Wong

Pharmacodynamics with Covariates

See also *Pharmacodynamic Issues*

I. INTRODUCTION

In this entry, we will discuss several univariate statistical methods that are used mostly for phase III clinical studies for considering explanatory factors other than treatment variable and pharmacodynamic (PD) response. We specify the underlying conditions (including study design, model assumptions, and number of data points per patient) that apply to each method. Note that the *explanatory variable* refers to treatment (or dosage level) and any other variable used as the stratified variable or covariate. In this entry, we will classify the explanatory variable as a "treatment" variable or as an "other explanatory" variable. *Response variable* refers to the PD parameter, and we will consider only the clinical trials with a single primary PD parameter.

We will divide up clinical trials into five groups according to the type of PD parameter and the number and type of explanatory variables. Table 1 displays the summary of these five groups.

The discussion of each statistical method will cover purpose, statistical model, hypothesis testing, test statistic, rejection region, confidence interval, two-sided p-value, description of and notes on the method, and any related SAS procedure. (SAS, the most popular statistical software package, is also accepted by the Food and Drug Administration as one of the standard packages.

The statistical methods discussed for each group are listed in Table 2.

II. STATISTICAL METHODS

A. Group 1

1. Two-Way Analysis of Variance

Study Design: Parallel group design with a blocking factor

Model Assumptions: Normal distribution, independence, equal variance for all treatments

Number of Observations per Patient for PD Parameter: 1

Purpose:

a. To compare the equality of treatment means
b. To compare the equal block effect
c. To determine the interaction effect

Statistical Model:

$$y_{ijm} = \mu + \tau_i + \eta_j + \tau\eta_{ij} + \varepsilon_{ijm}, \quad \text{for} \quad i = 1, 2, \ldots, k; j = 1, 2, \ldots, s; \quad m = 1, 2, \ldots, n_{ij}$$

where

y_{ijm} = PD value for patient m in treatment group i and block j

μ = unknown overall population mean

τ_i = unknown treatment effect for treatment group i

η_j = unknown jth block effect

$\tau\eta_{ij}$ = interaction effect of ith treatment with jth block

ε_{ijm} = random error associated with patient m in treatment group i and block j

Analysis of Variance Summary: See Table 3.

Computation for Sum of Squares:

$$SS(T) = \sum_{i=1}^{k} \sum_{j=1}^{s} \sum_{k=1}^{n_{ij}} (\overline{y_{i..}} - \overline{y_{...}})^2$$

$$SS(B) = \sum_{i=1}^{k} \sum_{j=1}^{s} \sum_{k=1}^{n_{ij}} (\overline{y_{.j.}} - \overline{y_{...}})^2$$

$$SS(TB) = \sum_{i=1}^{k} \sum_{j=1}^{s} \sum_{k=1}^{n_{ij}} (\overline{y_{ij.}} - \overline{y_{i..}} - \overline{y_{.j.}} + \overline{y_{...}})^2$$

$$SS(E) = \sum_{i=1}^{k} \sum_{j=1}^{s} \sum_{k=1}^{n_{ij}} (y_{ijk} - \overline{y_{ij.}})^2$$

Table 1 Classification for Model Grouping

	PD variable		Explanatory variables: Treatment variable	Explanatory variables: Other explanatory variables		
	Continuous (C) vs. discrete (D)	No. of values[a]	No. of values	No. of variables	Continuous (C) vs. discrete (D)	No. of values
1	C	—	2 or more	1	D	2 or more
2	C	—	2 or more	1	C	—
3	D	2	2	1	D	2 or more
4	D	2	2 or more	1 or more	C or D	2 or more
5	C	—	2 or more	1 or more	C or D	2 or more

[a]When "C" is specified, this column is not needed.

Table 2 Statistical Methods for Each Group

Group	Statistical Methods
1	Two-way analysis of variance Crossover design One-way analysis of variance with repeated measures
2	Linear regression One-way analysis of covariance
3	Cochran–Mantel–Haenszel test
4	Logistic regression
5	Cox proportional hazards analysis

$$SS(TOT) = \sum_{i=1}^{k}\sum_{j=1}^{s}\sum_{k=1}^{n_{ij}} (y_{ijk} - \overline{y_{...}})^2$$

$$\overline{y_{i..}} = \sum_{j=1}^{s}\sum_{k=1}^{n_{ij}} y_{ijk}/n_{i.}, \quad n_{i.} = \sum_{j=1}^{s} n_{ij}$$

$$\overline{y_{.j.}} = \sum_{i=1}^{k}\sum_{k=1}^{n_{ij}} y_{ijk}/n_{.j}, \quad n_{.j} = \sum_{i=1}^{k} n_{ij}$$

$$\overline{y_{ij.}} = \sum_{k=1}^{n_{ij}} y_{ijk}/n_{ij}$$

$$\overline{y_{...}} = \sum_{i=1}^{k}\sum_{j=1}^{s}\sum_{k=1}^{n_{ij}} y_{ijk}/n_{..}, \quad n = \sum_{i=1}^{k}\sum_{j=1}^{s} n_{ij}$$

Hypothesis Testing for Treatment Effect:

a. $H_0: \tau_i = 0$ for all i vs. $H_a: \tau_i \neq 0$ for some i if treatment effect fixed
b. $H_0: \sigma_\tau^2 = 0$ vs. $H_a: \sigma_\tau^2 \neq 0$ if treatment effect random.

Test Statistic for Treatment Effect:

$$F_0 = \frac{MS(T)}{MS(E)} \quad \text{for fixed-treatment effect}$$

$$F_0 = \frac{MS(T)}{MS(TB)} \quad \text{for random-treatment effect.}$$

Rejection Region at α Level:

$F_0 > F_{(k-1,N-k),1-\alpha}$ for fixed-treatment effect

$F_0 > F_{(k-1,(k-1)(s-1)),1-\alpha}$

for mixed- or random-treatment effect

Two-Sided p-Value:

$1 - \text{PROBF}(F_0, k-1, N-sk)$

for fixed-treatment effect

$1 - \text{PROBF}(F_0, k-1, (k-1)(s-1))$

for random-treatment effect

Table 3 Analysis of Variance Summary for Parallel Group Design with a Blocking Factor

Source of variation	Degrees of freedom	Sum of squares	Mean squares
Treatment	$k - 1$	SS(T)	MS(T)
Block	$s - 1$	SS(B)	MS(B)
Treatment × block	$(k-1)(s-1)$	SS(TB)	MS(TB)
Error	$N - sk$	SS(E)	MS(E)
Corrected total	$N - 1$	SS(TOT)	—

Hypothesis Testing for Block Effect:

a. $H_0: \eta_i = 0$ for all i vs. $H_a: \eta_i \neq 0$ for some i if treatment effect fixed
b. $H_0: \sigma_\eta^2 = 0$ vs. $H_a: \sigma_\eta^2 \neq 0$ if treatment effect random

Test Statistic for Block Effect:

$$F_0 = \frac{MS(B)}{MS(E)} \quad \text{for fixed-block effect}$$

$$F_0 = \frac{MS(B)}{MS(TB)} \quad \text{for random-block effect}$$

Rejection Region at α Level:

$F_0 > F_{(s-1,N-k),1-\alpha}$ for fixed-block effect

$F_0 > F_{(s-1,(k-1)(s-1)),1-\alpha}$ for random-block effect

Two-Sided p-Value:

$1 - \text{PROBF}(F_0, s-1, N-sk)$

for fixed-block effect

$1 - \text{PROBF}(F_0, s-1, (k-1)(s-1))$

for random-block effect

Hypothesis Testing for Treatment-by-Block Interaction Effect:

a. $H_0: \tau\eta_{ij} = 0$ for all i, j vs. $H_a: \tau\eta_{ij} \neq 0$ for some i, j for fixed model
b. $H_0: \sigma_{\tau\eta}^2 = 0$ vs. $H_a: \sigma_{\tau\eta}^2 \neq 0$ for mixed or random model

Test Statistic for Interaction Effect: $F_0 = MS(TB)/MS(E)$ for any model
Rejection Region: $F_0 > F_{((k-1)(s-1),N-sk),1-\alpha}$
Two-Sided p-Value: $1 - \text{PROBF}(F_0, (k-1)(s-1), N-sk)$
Related SAS Procedure: PROC GLM (or MIXED or ANOVA)
Description of and Notes on the Method:

a. This model requires one other explanatory variable (a discrete variable) that serves as the blocking factor in the model.
b. The model may apply to the situation in which two drugs are the treatment and blocking factors, and the combination of the two drugs is the interaction effect. In this case, the treatment randomization schedule is very likely to be generated with a full consideration of each factor and the combination, and the sample size for the study is estimated appropriately to show a sufficiently high power (≥0.80). For those situations where the blocking factor is a stratified variable, such as gender, race, age group, or underlying disease condition, the treatment randomization schedule in many cases does not take into account the stratification, so the near-equal number of sample size for each stratified variable may not occur. In addition, the sample size calculation, again in most cases, is based on the treatment difference only. Hence the result from the study may not provide strong power for the statistical significance for the blocking effect as well as for the interaction effect.
c. The treatment randomization schedule is normally based on the number of *k* groups if the blocking factor is a stratified factor, and it will be based on the number of *ks* groups for a drug combination study.
d. If the blocking factor represents the center effect for a multicenter study, the sample size for each center may vary to a great degree. In some cases, several small centers may be collapsed into a single "dummy" center in order to reduce the degrees of freedom for center effect and to increase the sample size for each center. In reality, the treatment effect will be difficult to conclude in a trial where the treatment-by-center interaction is statistically significant for a trial with a large number of centers and the treatment effect goes in different directions among centers (e.g., in a trial with 50 centers and two treatment groups, 50% of the 50 centers showed that treatment A is superior to treatment B, and the remaining 50% centers showed that treatment A is inferior to treatment B).
e. The model is *fixed* if both treatment and block effects are fixed; the model is *random* when both treatment and block effects are random. It is a *mixed* model when one of the two effects (treatment or block) is random and the other is fixed.
f. The model is *balanced* when all the cells have the same number of observations; otherwise, it is *unbalanced*.
g. There are four ways of computating sum of squares in SAS for unbalanced data, known as Type I through Type IV. In general, Type III sum of squares is selected (1,2).
h. The interaction term can be ignored if every cell contains only one observation or the interaction effect is insignificant. In this case, the model is reduced to a model with a randomized block design.
i. If the treatment effect represents different dose levels (including the placebo) of a study drug, it is desirable to investigate the trend effect, such as linear or quadratic. This can also apply to the blocking factor if it is measured on an ordinal scale. In such a case, the interaction term can be decomposed into the combination of those trends, such as $T_{linear} \times B_{linear}$ or $T_{linear} \times B_{quadratic}$.
j. PROC ANOVA in SAS can be applied to a balanced case only. PROC GLM is developed mainly for the use of the fixed-effect model. PROC MIXED is a new SAS procedure designed to handle the more complicated, so-called mixed model.
k. Confidence intervals for each level of treatment factor (and/or blocking factor) as well as for the mean difference between two levels for the same factor can be constructed via a *T*-distribution. But note that the standard error used in the confidence interval depends on the type of factor (fixed or random).
l. Multiple comparisons may be desirable to determine pairwise differences between various levels of the treatment effect if the treatment effect is deemed significant.
m. There are a number of other methodologies proposed for combination drug trials, such as response surface method (3–5).
n. This two-way analysis of variance can be extended to any higher-order (≥3) of analysis of variance (i.e., a study with two or more "other explanatory variables"). With a higher-order analysis of variance, many different types of interaction exist.

2. *Crossover Design (Using a 2 × 2 Crossover Design for Illustration)*

Study Design: Crossover design

Model Assumptions: Normal distribution, dependence, equal variance for all treatments, compound symmetry

Number of Observations per Patient for PD Parameter: 2

Purpose: To compare the equality of treatment means

Statistical Model:

$$y_{ijm} = \mu + Q_i + S(Q)_{m(i)} + P_j + \tau_{(i,j)} + \varepsilon_{ijm}, \quad \text{for } i,j = 1, 2;\ m = 1, 2, \ldots, n_i$$

where

y_{ijm} = PD value for patient m in period j in treatment sequence i

μ = overall population mean

Q_i = effect of ith sequence

$S(Q)_{m(i)}$ = error term associated with mth subject in sequence i

$\tau_{(i,j)}$ = treatment effect in period j for treatment sequence i

P_j = effect for jth period

ε_{ijm} = random error associated with patient m in treatment sequence i and period j

Analysis of Variance Summary: See Table 4.

Computation for Sum of Squares:

$$\text{MS(Q)} = \sum_{i=1}^{2}\sum_{j=1}^{2}\sum_{m=1}^{n_i}(\overline{y_{i..}} - \overline{y_{...}})^2$$

$$\text{MS(S(Q))} = \frac{\sum_{i=1}^{2}\sum_{j=1}^{2}\sum_{m=1}^{n_i}(\overline{y_{i.m}} - \overline{y_{i..}})^2}{n_1 + n_2 - 2}$$

$$\text{MS(T)} = \sum_{i=1}^{2}\sum_{j=1}^{2}\sum_{m=1}^{n_i}(\overline{y_{ij.}} - \overline{y_{...}})^2$$

$$\text{MS(P)} = \sum_{i=1}^{2}\sum_{j=1}^{2}\sum_{m=1}^{n_i}(\overline{y_{.j.}} - \overline{y_{...}})^2$$

$$\text{SS(TOT)} = \sum_{i=1}^{2}\sum_{j=1}^{2}\sum_{m=1}^{n_i}(y_{ijk} - \overline{y_{...}})^2$$

Hypothesis Testing for Treatment Effect:

$$H_0\text{: } \tau_1 = \tau_2 \quad \text{vs.} \quad H_a\text{: } \tau_1 \neq \tau_2$$

Test Statistic for Treatment Effect: F_0 = MS(T)/MS(E)

Rejection Region at α *Level*: $F_0 > F_{(1, n_1+n_2-2, 1-\alpha)}$

Two-Sided P-Value: 1 − PROBF(F_0, 1, $n_1 + n_2 - 2$, $1 - \alpha$)

Table 4 Analysis of Variance Summary for 2 × 2 Crossover Design

Source of variation	Degrees of freedom	Mean squares
Sequence	1	MS(Q)
Subject (Sequence)	$n_1 + n_2 - 2$	MS(S(Q))
Treatment	1	MS(T)
Period	1	MS(P)
Error	$n_1 + n_2 - 2$	MS(E)
Corrected total	$2(n_1 + n_2) - 1$	MS(TOT)

Hypothesis Testing for Period Effect:

$$H_0\text{: } P_1 = P_2 \quad \text{vs.} \quad H_a\text{: } P_1 \neq P_2$$

Test Statistic: F_0 = MS(P)/MS(E)

Rejection Region at α *Level*: $F_0 > F_{(1, n_1+n_2-2, 1-\alpha)}$.

Two-Sided P-Value: 1 − PROBF(F_0, 1, $n_1 + n_2 - 2$)

Related SAS Procedure: PROC GLM (or MIXED)

Description of and Notes on the Method:

a. We use a 2 × 2 crossover design as an example for illustration.
b. One discrete "other explanatory variable" is required. The "period" factor is the "other explanatory variable" in this model.
c. This is used to compare two treatments with the removal of the time effect, since each treatment is administered in both dosing periods to all patients. Each patient serves as his/her control.
d. There are two major sources of variation: intersubject variation and intrasubject variation. In this model, the intersubject variation is composed of two components: sequence and patient within sequence; the intrasubject variation contains three components: treatment, period, and error.
e. There are two types of variability associated with crossover design: between-patient variability and within-patient variability. MS(S(Q)) in the analysis of variance refers to the between-patient variance, and MS(E)) refers to the within-patient variance.
f. This model is commonly used in a phase I PK bioequivalance study.
g. It is important that the study allow a suffi-

ciently long washout period between the two treatment administrations to avoid the carryover effect.

h. This model assumes no carryover effect.
i. It is important that a baseline value be obtained in each period.
j. The model can be extended to a higher-order number of treatments and time periods. Note that using a higher-order crossover study may potentially increase the dropout rate, because the duration of study is prolonged; and when an odd number of treatments is used for a clinical study, it would be necessary to employ William's design in order to maintain the equal number of appearances for each ordered pair of treatments.
k. The crossover design is not suitable for study drugs with long terminal half-lives or for study drugs with irreversible PD responses.
l. In theory, *repeated measures* is defined as a series of the same information (data) taken under the same experimental condition from the same patient over time. Therefore, strictly speaking, this crossover design should not be classified as a case of a repeated-measures design.
m. *Compound symmetry* means that the correlations between each pair of two time points are the same. Mauchly's criterion can be used for this purpose, which is also available in SAS GLM REPEATED PROCEDURE (6).
n. When there are repeated measures for each period, the analysis proposed by Wallenstein and Fisher can be used (7).
o. This analysis assumes a fixed-effect model.
p. A confidence interval for treatment mean difference can be obtained.
q. In general, patients with missing values are removed from the statistical analysis.

3. *One-Way Analysis of Variance with Repeated Measures*

Study Design: Parallel group design with a time factor

Model Assumptions: Normal distribution, independence, equal variance, compound symmetry

Number of Observations per Patient for PD Parameter: Multiple

Purpose:

a. To determine each treatment profile over a period of time
b. To compare the treatment effects

Statistical Model:

$$y_{ijm} = \mu + \tau_i + S_{j(i)} + T_m + \tau T_{im} + \varepsilon_{ijm} \quad \text{for}$$
$$i = 1, 2, \ldots, k; j = 1, 2, \ldots, n_i; m = 1, 2, \ldots, p$$

where

y_{ijm} = PD value for patient j at time m in treatment group i
μ = overall population mean
τ_i = treatment effect for treatment group i
$S_{j(i)}$ = effect for jth patient in treatment group i
P_m = effect at mth time point
τP_{im} = interaction effect of ith treatment with mth time point
ε_{ijm} = random error associated with patient j in treatment group i at time j

Analysis of Variance Summary: See Table 5.
Computation for Sum of Squares:

$$SS(T) = \sum_{i=1}^{k} \sum_{j=1}^{n_i} \sum_{m=1}^{p} (\overline{y_{i..}} - \overline{y_{...}})^2$$

$$SS(BET) = \sum_{i=1}^{k} \sum_{j=1}^{n_i} \sum_{m=1}^{p} (\overline{y_{ij.}} - \overline{y_{i..}})^2$$

Table 5 Analysis of Variance Summary for Parallel Group Design with a Time Factor

Source of variation	Degrees of freedom	Sum of squares	Mean squares
Treatment	$k - 1$	SS(T)	MS(T)
Patient (treatment)	$N - k$	SS(BET)	MS(BET)
Time	$p - 1$	SS(TIME)	MS(TIME)
Treatment × Time	$(k - 1)(p - 1)$	SS(T × Time)	MS(T × Time)
Error	$(N - k)(p - 1)$	SS(W)	MS(W)
Corrected total	$Np - 1$	SS(TOT)	

$$\text{SS(TIME)} = \sum_{i=1}^{k}\sum_{j=1}^{n_i}\sum_{m=1}^{p}(\overline{y_{..m}} - \overline{y_{...}})^2$$

$$\text{SS(T} \times \text{TIME)} = \sum_{i=1}^{k}\sum_{j=1}^{n_i}\sum_{m=1}^{p}(\overline{y_{i.m}} - \overline{y_{i..}} - \overline{y_{..m}} + \overline{y_{...}})^2$$

$$\text{SS(TOT)} = \sum_{i=1}^{k}\sum_{j=1}^{n_i}\sum_{m=1}^{p}(\overline{y_{ijm}} - \overline{y_{...}})^2$$

$$\text{SS(W)} = \text{SS(TOT)} - \text{SS(T)} - \text{SS(BET)} - \text{SS(TIME)} - \text{SS(T} \times \text{TIME)}.$$

where

$$\overline{y_{i..}} = \sum_{j=1}^{n_i}\sum_{m=1}^{p} y_{ijm}/pn_i$$

$$\overline{y_{..m}} = \sum_{i=1}^{k}\sum_{j=1}^{n_i} y_{ijm}/N$$

$$\overline{y_{ij.}} = \sum_{m=1}^{p} y_{ijm}/p$$

$$\overline{y_{i.m}} = \sum_{j=1}^{n_i} y_{ijm}/n_i$$

$$\overline{y_{...}} = \sum_{i=1}^{k}\sum_{j=1}^{n_i}\sum_{k=1}^{p} y_{ijm}/pN \quad \text{and} \quad N = \sum_{i=1}^{k} n_i$$

Hypothesis Testing for Treatment Effect:

$$H_0\text{: } \tau_i = 0 \text{ for all } i \quad \text{vs.} \quad H_0\text{: } \tau_i \neq 0 \text{ for some } i$$

Test Statistic for Treatment Effect: F_0 = MS(T)/MS(BET)
Rejection Region for Treatment Effect: $F_0 > F_{(k-1,N-k),1-\alpha}$
Two-sided p-Value for Treatment Effect: 1 − PROBF(F_0, $k - 1$, $N - k$)
Hypothesis Testing for Time Effect:

$$H_0\text{: } T_m = 0 \text{ for all } m \quad \text{vs.} \quad H_0\text{: } T_m \neq 0 \text{ for some } m$$

Test Statistic for Time Effect: F_0 = MS(TIME)/MS(W)
Rejection Region for Time Effect: $F_0 > F_{(p-1,(N-k)(p-1)),1-\alpha}$
Two-sided p-Value for Time Effect: 1 − PROBF(F_0, $p - 1$, $(N - k)(p - 1)$)
Hypothesis Testing for Treatment-by-Time Interaction Effect:

$$H_0\text{: } \tau T_{im} = 0 \quad \text{for all } i, m \quad \text{vs.}$$
$$H_0\text{: } \tau T_{im} \neq 0 \quad \text{for some } i, m$$

Test Statistic for Interaction Effect: F_0 = MS(T × TIME)/MS(W)
Rejection Region for Interaction Effect: $F_0 > F_{((k-1)(p-1),(N-k)(p-1)),1-\alpha}$
Related SAS Procedure: PROC GLM or PROC MIXED
Description of and Notes on the Method:

a. One discrete "other explanatory variable" is required. The "time" factor is the "other explanatory variable" in this model.
b. Repeated measures can be collected in at least two ways:
   i. They are over a certain period of time within the same visit (e.g., measurement of blood pressure over 24 hours posttreatment at one clinic visit). In this case, there is only one baseline.
   ii. They are collected at a number of visits, with one observation per visit. In this case, a baseline at each visit is highly desirable.
c. Compound symmetry has to be examined to ensure the validity of the model.
d. The preceding discussion points apply to the fixed model only; i.e., both treatment and time effects are fixed. This is generally what the model assumes for phase III clinical trials, because the dosage levels and time points are selected with clinical reasons.
e. When a mixed or a random model is encountered, then the test statistic employed to test the time effect is

$$F_0 = \frac{\text{MS(TIME)}}{\text{MS(T} \times \text{TIME)}}$$

As for the test of the treatment effect, a pseudo-F-test of the form

$$\frac{\text{MS(T)}}{\text{MS(BET)} + \text{MS(T} \times \text{TIME)} - \text{MS(W)}}$$

can be obtained, which approximates an F-distribution with ($k - 1$, w) degrees of freedom, where (8,9)

$$w = \frac{[\text{MS(BET)} + \text{MS(T} \times \text{TIME)} - \text{MS(W)}]^2}{\frac{(\text{MS(BET)})^2}{N-k} + \frac{(\text{MS(T} \times \text{TIME)})^2}{(k-1)(p-1)} + \frac{(\text{MS(W)})^2}{(N-k)(p-1)}}$$

f. There is another way to deal with a repeated measure, consider "individual time point analysis." This means the data are analyzed separately at each time point, with an adjustment of α level (10,11).

g. An overlaid response-time plot is useful to visualize the trend of the PD response for between- and within-treatments comparisons.

h. One major drawback for repeated-measures design is a higher possibility of encountering missing values. Four "imputation" procedures are commonly used:
   i. Remove patients with missing values. This allows a straightforward statistical analysis. However, those incomplete patients may provide important information for the treatment.
   ii. Estimate the missing values by minimizing the sum of squares due to error. This is appropriate statistically only when very few missing values are observed in the study.
   iii. Replace the missing values with the last available observation. This is known as the "last observation carried forward" procedure, as suggested by Gillings and Koch (12).
   iv. Replace the missing values with the mean of all valid observations.

i. Two major sources of variation, as with the model in Sec. II.A.2, are observed: intersubject variation and intrasubject variation. In this model, the intersubject variation contains the variation due to treatment and due to the patient within treatment; the intrasubject variation contains three components: time, treatment-by-time, and error. The mean square due to error is the estimate of intrasubject variance, and the mean square due to patient (within treatment) is the estimate of intersubject variance.

j. The model can be generalized to a higher-order analysis of variance/covariance (i.e., one or more "other explanatory variables") with repeated measures.

B. Group 2

1. Simple Linear Regression

Study Design: Parallel group design
Model Assumptions: Normal distribution, independence
Number of Observations per Patient for PD Parameter: 1

Purpose:

a. To determine the relationship between a PD variable and a prespecified explanatory variable
b. To compare the relationship between the two treatment groups

a. Within-Treatment-Group Analysis

Statistical Model:

$$y_{ij} = \alpha_i + \beta_i x_{ij} + \varepsilon_{ij}, \qquad \text{for} \quad i = 1, 2, \ldots, k;\ j = 1, 2, \ldots, n_i$$

where

y_{ij} = PD value for patient j in treatment group i
α_i = unknown intercept for treatment group i
β_i = unknown slope for treatment group i
x_{ij} = value of covariate for patient j in treatment group i
ε_{ij} = random error associated with patient j in treatment group i

Point Estimation of Unknown Parameters for a Fixed i:

$$\hat{\beta}_i = \frac{S^i_{xy}}{S^i_{xx}}, \qquad \hat{\alpha}_i = \bar{y}_{i.} - \hat{\beta}_i \bar{x}_{i.}, \qquad \hat{y}_{ij} = \hat{\alpha}_i + \hat{\beta}_i x_{ij}$$

$$r_i = \frac{S^i_{xy}}{\sqrt{S^i_{xx} S^i_{yy}}}$$

where

$$S^i_{xy} = \sum_{j=1}^{n_i} (x_{ij} - \bar{x}_{i.})(y_{ij} - \bar{y}_{i.})$$

$$S^i_{xx} = \sum_{j=1}^{n_i} (x_{ij} - \bar{x}_{i.})^2 \quad \text{and} \quad S^i_{yy} = \sum_{j=1}^{n_i} (y_{ij} - \bar{y}_{i.})^2$$

Hypothesis Testing for α_i:

$$H_0\colon \alpha_i = 0 \qquad \text{vs.} \qquad H_a\colon \alpha_i \neq 0$$

Test Statistic:

$$T_0 = \frac{\hat{\alpha}_i}{\sqrt{\left(\dfrac{\sum_{j=1}^{n_i} (y_{ij} - \hat{y}_{ij})^2}{n_i - 2}\right)\left(\dfrac{1}{n_i} + \dfrac{\bar{x}^2_{i.}}{S^i_{xx}}\right)}}$$

Rejection Region: $|T_0| > t_{(n_i-2,1-\alpha/2)}$
Two-Sided p-Value: $2(1 - \text{PROBT}(|T_0|, n_i - 2))$

Hypothesis Testing for β_i:

$$H_0\colon \beta_i = 0 \quad \text{vs.} \quad H_a\colon \beta_i \neq 0$$

Test Statistic:

$$T_0 = \frac{\hat{\beta}_i \sqrt{S^i_{xx}}}{\sqrt{\left(\frac{1}{n_i - 2} \sum_{j=1}^{n_i} (y_{ij} - \hat{y}_{ij})^2\right)}}$$

Rejection Region: $|T_0| > t_{(n_i-2,1-\alpha/2)}$
Two-sided p-Value: $2(1 - \text{PROBT}(|T_0|, n_i - 2))$
Hypothesis Testing for ρ_i *(Correlation Coefficient)*:

$$H_0\colon \rho_i = 0 \quad \text{vs.} \quad H_a\colon \rho_i \neq 0$$

Test Statistic:

$$T_0 = \frac{r_i\sqrt{n_i - 2}}{\sqrt{1 - r_i^2}} \quad \text{for small } n_i\ (n_i < 20)$$

$$Z_0 = \frac{\frac{1}{2}\ln\frac{1 + r_i}{1 - r_i}}{\sqrt{\frac{1}{n_i - 3}}} \quad \text{for large } n_i\ (n_i \geq 20)$$

Rejection Region at α *Level*: $|T_0| > t_{(n_i-2,1-\alpha/2)}$ or $|Z_0| > Z_{(1-\alpha/2)}$
Two-Sided p-Value: $2(1 - \text{PROBT}(|T_0|, n_i - 2))$ or $2(1 - \text{PROBNORM}(|Z_0|)$
Analysis of Variance Approach (for Treatment Group i): See Table 6
Computation for Sum of Squares:

$$\text{SS(X)} = \frac{(S^i_{xy})^2}{S^i_{xx}} \quad \text{and} \quad \text{SS(E)} = \sum_{j=1}^{n_i} (y_{ij} - \hat{y}_{ij})^2$$

Hypothesis Testing for No Relationship Between PD Parameter and Explanatory Variable:

$$H_0\colon \beta_i = 0 \quad \text{vs.} \quad H_a\colon \beta_i \neq 0$$

Test Statistic for No Relationship Between PD Parameter and Explanatory Variable: F_0 = MS(X)/MS(E)

Table 6 Analysis of Variance Summary for Treatment Group *i*

Source of variation	Degrees of freedom	Sum of squares	Mean squares
Regression	1	SS(X)	MS(X)
Error	$n_i - 2$	SS(E)	MS(E)
Corrected total	$n_i - 1$	SS(TOT)	

Rejection Region at α *Level*: $F_0 > F_{(1,n_i-2),1-\alpha}$
Two-Sided p-Value: PROBF(F_0, 1, $n_i - 2$)
Related SAS Procedure: PROC REG (or GLM)

b. *Between-Treatment-Groups Analysis*

Hypothesis Testing for $\rho_i - \rho_m$:

$$H_0\colon \rho_i = \rho_m \quad \text{vs.} \quad H_a\colon \rho_i \neq \rho_m$$

Test Statistic:

$$Z_0 = \frac{\frac{1}{2}\ln\frac{1 + r_i}{1 - r_i} - \frac{1}{2}\ln\frac{1 + r_m}{1 - r_m}}{\sqrt{\frac{1}{n_i - 3} + \frac{1}{n_m - 3}}}$$

Rejection Region: $|Z_0| > z_{(1-\alpha/2)}$
Two-Sided p-Value: $2(1 - \text{PROBNORM}(|Z_0|)$
Hypothesis Testing for $\beta_i - \beta_m$:

$$H_0\colon \beta_i = \beta_m \quad \text{vs.} \quad H_a\colon \beta_i \neq \beta_m$$

Test Statistic for $\beta_i - \beta_m$:

a. For small sample size,

$$T_0 = \frac{\hat{\beta}_i - \hat{\beta}_m}{S_{\text{pool}}\sqrt{\frac{1}{S^i_{xx}} + \frac{1}{S^m_{xx}}}}$$

where

$$S^2_{\text{pool}} = \frac{\sum_{j=1}^{n_i} (y_{ij} - \hat{y}_{ij})^2 + \sum_{j=1}^{n_m} (y_{mj} - \hat{y}_{mj})^2}{n_i + n_m - 4}$$

b. For large sample size ($n_i > 25$ and $n_m > 25$),

$$Z_0 = \frac{\hat{\beta}_i - \hat{\beta}_m}{\sqrt{\frac{\frac{1}{n_i - 2}\sum_{j=1}^{n_i} (y_{ij} - \hat{y}_{ij})^2}{S^i_{xx}} + \frac{\frac{1}{n_m - 2}\sum_{j=1}^{n_m} (y_{mj} - \hat{y}_{mj})^2}{S^m_{xx}}}}$$

Rejection Region at α *Level*: $|T_0| > t_{(n_i+n_m-4,1-\alpha/2)}$ or $|Z_0| > Z_{(1-\alpha/2)}$
Two-Sided p-Value: $2(1 - \text{PROBT}(|T_0|, n_i + n_m - 4))$ or $2(1 - \text{PROBNORM}(|Z_0|)$
Description of and Notes on the Method:

a. The explanatory variable for this model is specified as continuous. In fact, regression analysis will work fine with discrete explanatory variables in an ordinal scale with a reasonable number of different values.

b. The regression analysis focuses on the relationship between the identified explanatory variable and the PD parameter within the treatment group. That is different from an analysis of covariance (discussed in the next section), which focuses more on the between-treatments comparison. In addition, this regression model does not assume that all treatment groups have equal slope.

c. It can be seen that there are two approaches to deal with within-treatment regression analysis. But both methods for testing a zero slope are the same (one using a T-test, the other using an F-test).

d. There are two measures of interest with regard to the relationship between the PD parameter and the explanatory variable: one is measured as a linear slope, the other is expressed as a correlation coefficient.

e. Confidence intervals for intercept and slope can be constructed.

f. It is often desirable to obtain a confidence interval for the mean response at a fixed value (say, at x_0) of the explanatory variable for treatment group i, which is given by:

$$\overline{y_{i.}} + \hat{\beta}_i(x_0 - \overline{x_{i.}}) \pm t_{(n_i-2,1-\alpha/2)} \cdot \sqrt{\left(\frac{\sum_{j=1}^{n}(y_{ij} - \hat{y}_{ij})^2}{n_i - 2}\right)\left(\frac{1}{n_i} + \frac{(x_0 - \overline{x_{i.}})^2}{S_{xx}^i}\right)}$$

g. A regression plot with the predicted line is always made for this analysis. In addition, it is usually of great interest to construct confidence bands for the regression line.

h. ρ_i = unknown population correlation coefficient for treatment group i; and r_i = observed correlation coefficient for treatment group i.

i. When only one explanatory variable is employed, we have a simple linear regression, which is discussed in this section.

j. If more than one observation occurs for most of the values for the explanatory variable, the error term in the ANOVA can be partitioned into two components: pure error (PE) and lack of fit (LOF). SS(PE) can be calculated by summing up the sum of squares for all PDs at the same "x" value, and d.f.(PE) = u (where u = number of different values of "x"). Thus SS(LOF) = SS(Error) − SS(PE), and d.f.(LOF) = d.f.(Error) − d.f.(PE). In such a case, a test of the adequacy of the model can be made by means of the test statistic of the form F = MS(LOF)/MS(PE). Note that the denominator of the test statistic for X remains unchanged (that is, MS(Error).

k. When more than one explanatory variable is employed, we have a multiple linear regression. The d.f. for the source of "regression" in the ANOVA table is equal to the number of explanatory variables. Associated with the multiple regression model are multiple correlation, partial correlation, overall F-test, and partial F-test. The details can be found in any statistics book that discusses the topic of multiple regression (e.g., Ref. 13).

l. In a simple regression case, a second-order or higher term may also be needed for the model, which is known as *polynomial regression* (13).

m. A nonparametric method using ranks may be applied to regression analysis (14).

n. Linearity is the simplest situation between the two variables; it is much more complicated when the relationship is nonlinear (15,16).

2. *One-Way Analysis of Covariance*

Study Design: Parallel group design.

Model Assumptions: Normal distribution, independence, equal variance, equal slope for all treatment groups

Number of Observations for PD Parameter per Patient: 1

Purpose: To compare the mean adjusted for covariate(s) of all treatment groups are equal.

Statistical Model:

$$y_{ij} = \alpha_i + \beta x_{ij} + \varepsilon_{ij} = \mu + \tau_i + \beta x_{ij} + \varepsilon_{ij},$$
$$\text{for } i = 1, 2, \ldots, k; j = 1, 2, \ldots, n_i$$

where

y_{ij} = PD value for patient j in treatment group i

α_i = unknown intercept for treatment group i

β = unknown common slope

μ = overall population mean

τ_i = treatment effect for treatment group i

x_{ij} = value of the covariate for patient j in treatment group i

ε_{ij} = random error associated with patient j in treatment group i

Analysis of Variance Summary: See Table 7.
Computation for Sum of Squares:

$$SS(TOT) = S_{yy}$$

$$SS(E) = \frac{\left(\sum_i S^i_{yy}\right)\left(\sum_i S^i_{yy}\right) - \left(\sum_i S^i_{xy}\right)^2}{\left(\sum_i S^i_{xx}\right)}$$

$$SS(T) = \frac{S_{yy}S_{xx} - S^2_{xy}}{S_{xx}} - SS(E)$$

$$SS(X) = \left(\sum_i S^i_{yy}\right) - SS(E)$$

where

$$S_{yy} = \sum_{i=1}^{k}\sum_{j=1}^{n_i}(y_{ij} - \bar{y}_{..})^2$$

$$S_{xx} = \sum_{i=1}^{k}\sum_{j=1}^{n_i}(x_{ij} - \bar{x}_{..})^2$$

$$S_{xy} = \sum_{i=1}^{k}\sum_{j=1}^{n_i}(x_{ij} - \bar{x}_{..})(y_{ij} - \bar{y}_{..})$$

$$S^i_{yy} = \sum_{j=1}^{n_i}(y_{ij} - \bar{y}_{i.})^2$$

$$S^i_{xx} = \sum_{j=1}^{n_i}(x_{ij} - \bar{x}_{i.})^2$$

$$S^i_{xy} = \sum_{j=1}^{n_i}(x_{ij} - \bar{x}_{i.})(y_{ij} - \bar{y}_{i.})$$

$$\bar{y}_{i.} = \sum_{j=1}^{n_i} y_{ij}/n_i$$

$$\bar{y}_{..} = \sum_{i=1}^{k}\sum_{j=1}^{n_i} y_{ij}/N$$

$$\bar{x}_{i.} = \sum_{j=1}^{n_i} x_{ij}/n_i, \quad \bar{x}_{..} = \sum_{i=1}^{k}\sum_{j=1}^{n_i} x_{ij}/N$$

$$N = \sum_{i=1}^{k} n_i$$

Point Estimation for the Unknown Parameters for a Fixed i:

$$\hat{\beta} = \frac{S_{xy}}{S_{xx}}, \qquad \hat{\alpha}_i = \bar{y}_{i.} - \hat{\beta}\bar{x}_{i.}, \qquad \hat{y}_{ij} = \hat{\alpha}_i + \hat{\beta}x_{ij}$$

Hypothesis Testing for Treatment Effect:

Table 7 Analysis of Variance Summary for Parallel Group Design

Source of variation	Degrees of freedom	Sum of squares	Mean squares
Treatment	$k - 1$	SS(T)	MS(T)
Covariate	1	SS(X)	MS(X)
Error	$N - k - 1$	SS(E)	MS(E)
Corrected total	$N - 1$	SS(TOT)	

a. H_0: $\tau_i = 0$ for all i, vs. H_0: $\tau_i \neq 0$ for some i if treatment effect is fixed
b. H_0: $\sigma^2_\tau = 0$ vs. H_a: $\sigma^2_\tau \neq 0$ if treatment effect is random

Test Statistic for Treatment Effect: F = MS(T)/MS(E)
Rejection Region at α Level: $F > F_{(k-1,N-k-1),1-\alpha}$
Two-Sided p-Value: 1 − PROBF(F_0, $k - 1$, $N - k - 1$)
100(1 − α)% Confidence Interval for the Adjusted Mean Response for Treatment Group i:

$$\hat{\alpha}_i + \hat{\beta}\bar{x}_{..} \pm t_{(N-k-1,1-\alpha/2)} \cdot \sqrt{MS(E)\left(\frac{1}{n_i} + \frac{(\bar{x}_{i.} - \bar{x}_{..})^2}{S_{xx}}\right)}$$

100(1 − α)% Confidence Interval for the Difference of Adjusted Mean Response Between Treatment Group i and Treatment Group j:

$$\hat{\alpha}_i - \hat{\alpha}_j \pm t_{(N-k-1,1-\alpha/2)} \cdot \sqrt{MS(E)\left(\frac{1}{n_i} + \frac{1}{n_j} + \frac{(\bar{x}_{i.} - \bar{x}_{j.})^2}{S_{xx}}\right)}$$

Hypothesis Testing for Slope:

$$H_0: \beta = 0, \quad \text{vs.} \quad H_0: \beta \neq 0$$

Test Statistic for Slope: F = MS(X)/MS(E)
Rejection Region at α Level: $F > F_{(1,N-k-1),1-\alpha}$
Two-Sided p-Value: 1 − PROBF(F_0, 1, $N - k - 1$)
Related SAS Procedure: PROC GLM (or MIXED)
Detailed Descriptions:

a. When is a *covariate* (an uncontrollable variable that is highly correlated with the PD parameter but not directly affected by the treatment) is taken into consideration in a clinical trial, analysis of covariance is a way to adjust for the covariate.

b. This model can be extended to any number of covariates.

c. Since a requirement for this model is the common slope among all treatment groups, a preliminary test on the common slope is recommended that can be done with the following model in SAS:

$$y_{ij} = \mu + \tau_i + \beta\tau_i + \varepsilon_{ij}$$

The analysis of covariance is not appropriate if the interaction term of $\beta\tau_i$ is significant.

d. Another way of handling a covariate in a situation where the covariate is the baseline is to adjust this value from the postdose PD value first and then to apply analysis of variance or other appropriate methods.

e. The model is reduced to a linear regression if the treatment effect is insignificant. On the other hand, the model is reduced to a one-way analysis of variance if the slope associated with the covariate is insignificant.

C. Group 3

1. *Cochran–Mantel–Haenszel Test*

Study Design: Parallel group design
Model Assumptions: Binomial distribution, independence
Number of Observations per Patient for PD Parameter: 1
Purpose: To compare the proportions of the two treatment groups based on stratified samples
Hypothesis Testing:

$$H_0: \pi_1 = \pi_2 \quad \text{vs.} \quad H_a: \pi_1 \neq \pi_2$$

Test Statistic:

$$M_0 = \frac{\left(\left|\sum_I \left(n_{I11} - \frac{n_{I+1}n_{I1+}}{n_{I++}}\right)\right| - 0.5\right)^2}{\sum_I \left(\frac{n_{I+1}n_{I1+}n_{I+2}n_{I2+}}{n_{I++}^2(n_{I++} - 1)}\right)}$$

Rejection Region at α *Level*: $M_0 > \chi^2_{1,1-\alpha}$
Two-Sided p-Value: $1 - \text{PROBCHI}(M_0, 1)$
Observed Frequency Table: See Table 8.
Related SAS Procedure: PROC FREQ
Description of and Notes on the Method:

a. The stratum may represent study centers, gender, or patients' underlying condition (e.g., diabetics vs. nondiabetics), etc.

Table 8 Observed Frequencies for Group 3

Stratum I	Treatment 1	Treatment 2	Row total
Responder	n_{I11}	n_{I12}	n_{I1+}
Nonresponder	n_{I21}	n_{I22}	n_{I2+}
Column total	n_{I+1}	n_{I+2}	n_{I++}

b. The quantity "0.5" in the test statistics represents a correction for continuity, which may not be used.

c. An interaction may exist if the difference of the two proportions varies from one stratum to another. Under this circumstance, each stratum shall be analyzed separately.

d. The number of strata is best kept to a minimum, for at least two reasons:
   i. To avoid small or empty cells
   ii. Because the interaction term is difficult to interpret if significant

e. This test can be generalized to cases with k treatments ($k > 2$), r responses ($r > 2$), and s strata ($s > 2$) (17). However, from practical point of view, all k, r, and s shall be kept as small as possible, for the same two reasons described in point (d).

D. Group 4

1. *Logistic Regression*

Study Design: Parallel group design
Model Assumptions: Binomial distribution, independence
Number of Observations per Patient for PD Parameter: 1
Purpose: To determine the correlation between the proportion of the PD parameter and explanatory variable(s)
Statistical Model:

$$E\left(\log\left(\frac{\pi(\mathbf{x})}{1 - \pi(\mathbf{x})}\right)\right) = \alpha_0 + \sum_i \beta_i x_i$$

Hypothesis Testing for one correlation β_i:

$$H_0: \beta_i = 0 \quad \text{vs.} \quad H_a: \beta_i \neq 0$$

Test Statistic for β_i:

$$\chi_w^2 = \left(\frac{\hat{\beta}_i}{\text{SE}(\hat{\beta}_i)}\right)^2$$

Rejection Region for β_i *at* α *level*: $\chi_w^2 > \chi_{1,1-\alpha/2}^2$
Hypothesis Testing for a subset of model parameters $\boldsymbol{\beta}_q$:

$$H_0: \boldsymbol{\beta}_q = 0 \quad \text{vs.} \quad H_a: \boldsymbol{\beta}_q \neq 0$$

Test Statistic for $\boldsymbol{\beta}_q$:

$$\chi_w^2 = \mathbf{b}_q'(\text{COV}(\mathbf{b}_q))^{-1}\mathbf{b}_q$$

Rejection Region for $\boldsymbol{\beta}_q$ *at* α level: $\chi_w^2 > \chi_{q,1-\alpha/2}^2$
Related SAS Procedure: PROC LOGISTIC
Description of and Notes on the Method:

a. In this method, there are two responses for the PD parameter, with 0 indicating the occurrence of the event of interest, and 1 indicates nonoccurrence.
b. The notations in the model are defined as follows:

x_1 = treatment effect (if there are two treatments, such as active vs. control, we define $x_1 = 1$ for active drug and $x_1 = 0$ for placebo

$x_2, \ldots, x_k = k - 1$ other explanatory variables ($k \geq 1$), which can be continuous or discrete

$\pi(\mathbf{x})$ = probability of having event associated with $\mathbf{x} = (x_1, x_2, \ldots, x_k)$

$E(Y)$ = expected value of Y

α_0 = intercept

β_i = unknown true correlation (slope) between $\pi(\mathbf{x})$ and x_i

$\boldsymbol{\beta}_q = (\beta_1, \beta_2, \ldots, \beta_q)'$ is a subset of $\boldsymbol{\beta} = (\beta_1, \beta_2, \ldots, \beta_k)'$, $(q \leq k)$

$\mathbf{b}_q = (b_1, b_2, \ldots, b_q)'$ is an estimate of $\boldsymbol{\beta}_q = (\beta_1, \beta_2, \ldots, \beta_q)'$

$\text{COV}(\mathbf{b}_q)$ = variance-covariance matrix for $\mathbf{b}_q$,

$\text{SE}(\hat{\beta}_i)$ = standard error of $\hat{\beta}_i$

c. The test statistic is called the Wald statistic, which is asymptotically chi square distributed with 1 degree of freedom for a sufficiently large sample size. The first test statistic is a special case of the second test statistic when $q = 1$.
d. There is another common test for logistic regression, which is well known as the maximum log-likelihood test, which is asymptotically chi square distributed with q d.f.
e. We should first test β_i for any $i \geq 2$ to determine which covariates are correlated with the proportion of the PD response. Then we test β_1 for the proportion between the two treatment groups.
f. Note that

$$\pi(\mathbf{x}) = \frac{1}{1 + \exp\left(-\left[\alpha_0 + \sum_i \beta_i x_i\right]\right)}$$

Hence in the case of two treatments we have

$$\pi(x_1 = 0) = \frac{1}{1 + \exp\left(-\left[\alpha_0 + \sum_{i\geq 2} \beta_i x_i\right]\right)} \quad \text{and}$$

$$\pi(x_1 = 1) = \frac{1}{1 + \exp\left(-\left[\alpha_0 + \beta_1 + \sum_{i\geq 2} \beta_i x_i\right]\right)}.$$

g. Again, in the case of two treatments, the odds for the control group is given by

$$\text{odds (control)} = \frac{\pi(x_1 = 0)}{1 - \pi(x_1 = 0)} = \exp\left(\alpha_0 + \sum_{i\geq 2} \beta_i x_i\right)$$

$$\text{odds (active)} = \frac{\pi(x_1 = 1)}{1 - \pi(x_1 = 1)} = \exp\left(\alpha_0 + \beta_1 + \sum_{i\geq 2} \beta_i x_i\right)$$

Hence the odds ratio of the active group relative to the control group is

$$\text{odds (active)/odds (control)} = \exp(\beta_1)$$

This implies that β_1 is a measure of the odds ratio. The $1 - \alpha\%$ confidence interval for exp (β_1) can be obtained in a conventional way as follows: exp $(\hat{\beta}_1 \pm z_{1-\alpha/z}\, \text{SE}(\hat{\beta}_1))$.
h. Based on the discussion in point (g), the odds ratio can be generalized for any covariate i, which is equal to exp (β_i).
i. Logistic regression is analogous to analysis of covariance. Analysis of covariance is for the comparison of mean responses adjusted for covariate(s), while logistic regression is for the comparison of proportion responses adjusted for covariate(s).

E. Group 5

1. Cox Proportional Hazards Analysis

Study Design: Parallel group design
Model Assumptions: Independence

Number of Observations per Patient for PD Parameter: 1

Purpose: To determine and compare the survival time between the PD parameter and explanatory variable(s).

Statistical Model: $h(t) = \lambda(t) \exp(\Sigma_i \beta_i x_i)$

Hypothesis Testing for one correlation β_i:

$$H_0: \beta_i = 0 \quad \text{vs.} \quad H_a: \beta_i \neq 0$$

Test Statistic for β_i:

$$\chi_w^2 = \left(\frac{\hat{\beta}_i}{\text{SE}(\hat{\beta}_i)}\right)$$

Rejection Region for β_i *at* α *level*: $\chi_w^2 > \chi^2_{1,1-\alpha/2}$

Hypothesis Testing for a Subset of Model Parameters $\boldsymbol{\beta}_q$:

$$H_0: \boldsymbol{\beta}_q = 0 \quad \text{vs.} \quad H_a: \boldsymbol{\beta}_q \neq 0$$

Test Statistic for $\boldsymbol{\beta}_q$: $\chi_w^2 = \mathbf{b}_q'(\text{COV}(\mathbf{b}_q))^{-1}\mathbf{b}_q$

Rejection Region for $\boldsymbol{\beta}_q$ *at* α *level*: $\chi_w^2 > \chi^2_{q,1-\alpha/2}$

Related SAS Procedure: PROC PHREG

Description of and Notes on the Method:

a. In this method, there are two responses for the PD parameter, with 0 indicating the occurrence of the event of interest and 1 indicating non-occurrence.

b. The notations in the model are defined as follows:

x_1 = treatment effect (in the case of two treatments, we may deine $x_1 = 1$ for active drug and $x_1 = 0$ for placebo)

$x_2, \ldots, x_k = k - 1$ other explanatory variables ($k \geq 1$), which can be continuous or discrete

$h(t)$ = hazard function at time t associated with $\mathbf{x} = (x_1, x_2, \ldots, x_k)$

$\lambda(t)$ = unspecified initial hazard function at time t

β_i = correlation (slope) between $\pi(\mathbf{x})$ and x_i

$\boldsymbol{\beta}_q = (\beta_1, \beta_2, \ldots, \beta_q)'$ is a subset of $\boldsymbol{\beta} = (\beta_1, \beta_2, \ldots, \beta_k)'$

$\mathbf{b}_q = (b_1, b_2, \ldots, b_q)'$ is an estimate of $\beta_q = (\beta_1, \beta_2, \ldots, \beta_q)'$

$\text{COV}(\mathbf{b}_q)$ = variance-covariance matrix for $\mathbf{b}_q$

$\text{SE}(\beta_i)$ = standard error of estimate of β_i

c. The test statistic, called the Wald statistic, was discussed in Sec. II.D.1.

d. The log-likelihood test is also common for this analysis, in addition to the Wald chi square test.

e. We should first test β_i for any $i \geq 2$ to determine which covariates are correlated with the hazard function. Then we test β_1 between the two treatment groups.

f. Cox proportional hazards analysis is analogous to logistic regression, where logistic regression is dealing with the proportion of responses for treatment groups, with adjustment for covariate(s), and Cox proportional hazards model is dealing with survival time responses for treatment groups, with adjustment for covariate(s). On the other hand, Cox proportional hazards analysis is also similar to the log-rank test, because both tests are used for analyzing survival time. The log-rank test, however, does not take into account any covariate, whereas the Cox proportional hazards analysis does.

g. It can be shown that $\exp(\hat{\beta}_i)$ is an estimated risk ratio for covariate i and is independent of time and other covariates.

REFERENCES

1. SR Searle. Linear Models for Unbalanced Data. New York: Wiley, 1987.
2. RC Littell, RJ Freund, PC Spector. SAS System for Linear Models, Cary, NC: SAS Institute, 1991.
3. WH Stewart. Application of response surface methodology and factorial designs to clinical trials for drug combination development. J Biopharm Stat 6:219–230, 1996.
4. HMJ Hung. On evaluation of multiple-dose combination drugs in factorial clinical trials. Amer Stat Assoc, Proceedings of the Biopharm Section, pp 25–32, 1996.
5. HMJ Hung. Global tests for combination drug studies in factorial trials. Stat Med 15:233–247, 1996.
6. SAS/STAT User's Guide. Version 6. 4th ed. Vol. 2. Cary, NC: SAS Institute, 1990.
7. SW Wallenstein, AC Fisher. The analysis of the two-period repeated measurements crossover design with application to clinical trials. Biometrics 33:261–269, 1977.
8. SAS/STAT User's Guide. Version 6. 4th ed. Vol. 2. Cary, NC: SAS Institute, 1990, pp 509–512.
9. CR Hicks. Fundamental Concepts in the Design of Experiments. 3rd ed. New York: CBS College, 1982, pp 210–224.
10. Y Hochberg. A sharper Bonferroni procedure for multiple tests of significance. Biometrika 75:800–802, 1988.

11. P Wright. Adjusted *P*-values for simultaneous inference. Biometrics 48:1005–1013, 1992.
12. D Gillings, G Koch. The application of intention-to-treat to the analysis of clinical trials. Drug Inform J 25: 411–424, 1991.
13. DG Kleinbaum, LL Kupper. Applied Regression Analysis and Other Multivariable Methods. Boston: Duxbury Press, 1978.
14. D Birkes, Y Dodge. Alternative Methods of Regression. New York: Wiley, 1993, pp 111–142.
15. AR Gallant. Nonlinear Statistical Methods. New York: Wiley, 1989.
16. DM Bates, DG Watts. Nonlinear Regression Analysis and Its Application. New York: Wiley, 1988.
17. A Agresti. Categorical Data Analysis. New York: Wiley, 1990, pp 234–235.

Cheng-Tao Chang
Robert L. Wong

Pharmacodynamics with No Covariates

See also *Pharmacodynamic Issues*

I. INTRODUCTION

In this entry, we will discuss several univariate statistical methods that are used mostly for phase III clinical studies for considering one independent and one dependent variable. This entry will serve as a review for biostatisticians and as a utility tool for nonstatisticians. In the following discussions, we will specify the underlying conditions (including study design, model assumptions, and number of data points per patient) that apply to each method. Note that *explanatory variable* refers to treatment (or dosage level). *Response variable* refers to the PD parameter, and we will consider the clinical trials with a single primary PD parameter.

We will divide up clinical trials into four groups according to the type of PD parameter and the number and type of explanatory variables. Table 1 displays the summary of these four groups. The statistical methods for the groups in Table 1 may apply to single-center studies or to multicenter studies with no consideration of a "center" effect. However, in general it is advised that an investigation of the center effect as well as the center-by-treatment effect be performed *before* data from all centers in a multicenter clinical study are combined and analyzed. Statistical methods for considering the center effect and/or other effects, such as demographic data and baseline evaluations, are discussed in the entry on *Pharmacodynamics with Covariates*.

Table 1 Classification for Model Grouping

Group	PD variable: Continuous (C) vs. discrete(D)	PD variable: No. of values[a]	Treatment variable: No. of values
1	C	—	2
2	C	—	2 or more
3	D	2	2
4	D	2 or more	2 or more

[a]When "C" is specified, this column is not needed.

A discussion of each statistical method will cover purpose, statistical model, hypothesis testing, test statistic, rejection region, confidence interval, two-sided *p*-value, description of and notes on the method, and any related SAS procedure. (SAS, the most popular statistical software package, is also accepted by the Food and Drug Administration as one of the standard packages.)

The statistical methods discussed for each group are listed in Table 2.

II. STATISTICAL METHODS

A. Group 1

1. *Two-Sample t-Test*

Study Design: Parallel group design (or completely randomized design)
Model Assumptions: Normal distribution, independence, and equal variance
Number of Observations per Patient for PD parameter: 1

Table 2 Statistical Methods for Each Group

Group	Statistical methods
1	Two-Sample t-test
	Paired t-test
	Wilcoxon's rank-sum test
	Wilcoxon's signed-rank test
2	One-way analysis of variance
	Kruskal–Wallis test
	Log-rank test
3	Fisher–Exact test
	McNeMar's Test
4	Chi square test

Purpose: To compare the equality of two treatment means

Hypothesis Testing:

$$H_0\colon \mu_1 = \mu_2 \quad \text{vs.} \quad H_a\colon \mu_1 \neq \mu_2$$

Test Statistic:

$$T_0 = \frac{(\bar{Y}_1 - \bar{Y}_2)}{S\sqrt{\dfrac{1}{n_1} + \dfrac{1}{n_2}}}$$

Rejection Region at α Level: $|T_0| > t_{(n_1+n_2-2),(1-\alpha/2)}$

100(1 − α)% Confidence Interval:

$$\bar{Y}_1 - \bar{Y}_2 \pm t_{(n_1+n_2-2),(1-\alpha/2)}\left(S\sqrt{\frac{1}{n_1} + \frac{1}{n_2}}\right)$$

Two-Sided p-Value: $2 \times (1 - \text{PROBT}(|T_0|, n_1 + n_2 - 2))$

Related SAS PROC Procedure: PROC TTEST

Description of and Notes on the Method:

a. Each patient is assigned to one of the two treatment groups and contributes only one observation to the entire data set.
b. The data collected from the two treatment groups are independent because they are from different patients.
c. μ_1, μ_2 represent the unknown population means for treatment groups 1 and 2, respectively.
d. $\bar{Y}_1$, $\bar{Y}_2$ are the sample means for treatment groups 1 and 2, respectively; n_1 and n_2 are the sample sizes for treatment groups 1 and 2, respectively.
e. S is the pooled standard deviation, calculated via

$$\sqrt{\frac{(n_1 - 1)S_1^2 + (n_2 - 1)S_2^2}{n_1 + n_2 - 2}}$$

where S_1^2 and S_2^2 are the variances of treatment groups 1 and 2, respectively.
f. PROBT (x,df) expressed in the two-sided p-value is the cumulative probability function for the t-distribution in SAS.
g. $t_{(n_1+n_2-2),(1-\alpha/2)}$ is the numerical value at which the cumulative probability is $1 - \alpha/2$ under a t-distribution with $n_1 + n_2 - 2$ degrees of freedom.
h. The hypothesis testing can be modified from a two-sided test to a one-sided test, depending on the study objective.
i. The p-value of a one-sided test can be obtained by halving the p-value of the two-sided test if the test statistic is positive and the alternative hypothesis is a right-tailed test (H_a: $\mu_1 > \mu_2$), or the test statistic is negative and the alternative hypothesis is a left-tailed test (H_a: $\mu_1 < \mu_2$).
j. The rejection region for the right-tailed test (H_a: $\mu_1 > \mu_2$) is $T_0 > t_{(n_1+n_2-2),(1-\alpha)}$; the rejection region for the left-tailed test (H_a: $\mu_1 < \mu_2$) is $T_0 < t_{(n_1+n_2-2),\alpha}$.
k. A preliminary two-sided F-test on equal variances of the two treatment groups is highly suggested. When the null hypothesis of equal variance is rejected, the test statistic listed earlier may not be applied. Instead, you may employ an approximate t-test statistic of the form

$$T_0 = \frac{(\bar{Y}_1 - \bar{Y}_2)}{\sqrt{\dfrac{S_1^2}{n_1} + \dfrac{S_2^2}{n_2}}}$$

with the approximate degrees of freedom (1)

$$\text{df} = \frac{\left(\dfrac{S_1^2}{n_1} + \dfrac{S_2^2}{n_2}\right)^2}{\left(\dfrac{\left(\dfrac{S_1^2}{n_1}\right)^2}{(n_1 - 1)}\right) + \left(\dfrac{\left(\dfrac{S_1^2}{n_2}\right)^2}{(n_2 - 1)}\right)}$$

2. *Paired t-test*

Study Design: Parallel group design

Model Assumptions: Normal distribution, dependence

Number of Observations per Patient for PD Parameter: 2

Purpose: To compare the equality of two observations within the same treatment group

Hypothesis Testing:

$$H_0: \mu_d = 0 \quad \text{vs.} \quad H_a: \mu_d \neq 0$$

Test Statistic:

$$T_0 = \frac{\bar{Y}_d}{\frac{S_d}{\sqrt{n}}}$$

Rejection Region at α Level: $|T_0| > t_{(n-1),(1-\alpha/2)}$
100(1 − α)% Confidence Interval: $\bar{Y}_d + t_{(n-1),(1-\alpha/2)}S_d/\sqrt{n}$
Two-Sided p-Value: $2 \times (1 - \text{PROBT}(|T_0|, n - 1))$
Related SAS Procedure: PROC MEANS
Description of and Notes on the Method:

a. This test, in general, applies to the comparison of two readings under two conditions for the same treatment from the same sample of patients. The two readings are dependent and highly correlated, since they are obtained from the same patient sample. Normally, this test is more useful for within-treatment comparison than for between-treatments comparison. Two examples of such an application are: baseline vs. post-treatment; two post-treatment results at two different time points.
b. One reason why this paired *t*-test is not an ideal test for comparing two treatments if the two evaluations are collected under two different treatments for the same patients is that one treatment is always administered before (or after) the other, so the treatment effect may be confounded with the order/time effect. It is more appropriate to design such a study in a crossover fashion, so every patient can receive both treatments. (Please refer to Sec. I.A.2 in the entry on *Pharmacodynamics with Covariates* for a detailed discussion of crossover design).
c μ_d = population mean of paired difference

$\bar{Y}_d$ = sample mean of μ_d

n = number of pairs

d. Patients with only one observation are removed from the analysis.
e. A one-sided paired test can also be employed in a similar way to a two-sample *t*-test.

3. *Wilcoxon's Rank-Sum Test*

Study Design: Parallel group design
Model Assumptions: Independence
Number of Observations per Patient for PD Parameter: 1
Purpose: To compare the equality of two treatment medians
Hypothesis Testing:

$$H_0: M_1 = M_2 \quad \text{vs.} \quad H_a: M_1 \neq M_2$$

Test Statistic:

$$Z_0 = \frac{\text{sign}\left(\left|R_1 - \frac{n_1(N+1)}{2}\right| - 0.5\right)}{\sqrt{\frac{n_1 n_2}{12}\left(N + 1 - \frac{\sum_i m_i(m_i^2 - 1)}{N(N-1)}\right)}}$$

Rejection Region at α Level: $|Z_0| > Z_{1-\alpha/2}$.
Two-Sided p-Value: $2 \times (1 - \text{PROBNORM}(|Z_0|)$
Related SAS Procedure: PROC NPAR1WAY.
Description of and Notes on the Method:

a. This parallels the two-sample *t*-test, with the following two major differences:
   i. Normal distribution of the PD parameter is not required.
   ii. Comparison is based on the median, not on the mean.
b. A preliminary test of normality can be performed to determine whether the data are normally distributed. A Shapiro–Wilk or Kolmogorov can be used, which is available in the PROC UNIVARIATE procedure of SAS (2).
c. In order to employ this nonparametric test, the data from both treatment groups are first ranked from lowest to highest. The average rank is assigned to tied values if applicable.
d. R_1 represents sum of ranks associated with treatment 1, n_1 and n_2 represent the sample size for treatments 1 and 2, respectively. $N = n_1 + n_2$.
e. m_i represents the number of ties for the *i*th value that has ties.
f. $\text{sign}(|y| - a) = \begin{cases} y - a & \text{if } y > 0 \\ -(|y| - a) & \text{if } y < 0 \end{cases}$
g. PROBNORM(x) is the cumulative probability function at x for the standard normal distribution in SAS, and $Z_{1-\alpha/2}$ is a numerical value at which the cumulative probability is $1 - \alpha/2$ under a standard normal curve.
h. "0.5" in the test statistic is known as "correction for continuity." It is not always necessary to include it in the test statistic. When it is not used, the numerator of the test statistic can be

simplified as $R_1 - n_1(N + 1)/2$ without the absolute-value sign.

i. The Wilcoxon rank-sum test is shown to be equivalent to another nonparametric test for two independent samples, i.e., the Mann–Whitney U test, when the continuity for correction is not used (3; 4, p. 406).

j. A one-sided Z-test may be performed in the same manner as for a t-test.

k. For $n > 30$, we may feel free to use a two-sample t-test instead of this nonparametric test if the data are symmetric, based on the central limit theorem that the sample mean tends to be normally distributed for large sample sizes (>30) regardless of the underlying distribution of the data.

l. Another approximation test procedure to employ is a two-sample t-test with the ranked data (5).

4. *Wilcoxon's Signed-Rank Test*

Study Design: Parallel group design
Model Assumptions: Dependence
Number of Observations per Patient for PD parameter: 2
Purpose: To compare the median of two observations within the same treatment group
Hypothesis Testing:

$$H_0: M_d = 0 \quad \text{vs.} \quad H_a: M_d \neq 0$$

Test Statistic:

$$T_0 = \frac{\left(\frac{R^+ - R^-}{2}\right)\sqrt{n-1}}{\sqrt{\left(\frac{n^2(n+1)(2n+1)}{24} - \frac{n\sum_i (m_i^3 - m_i)}{48}\right) - \left(\frac{R^+ - R^-}{2}\right)^2}},$$

for small sample size

$$Z_0 = \frac{\text{sign}\left(\left|\min\{R^+, R^-\} - \frac{n(n+1)}{4}\right| - 0.5\right)}{\sqrt{\frac{n(n+1)(2n+1)}{24} - \frac{\sum_i (m_i^3 - m_i)}{48}}}$$

for large sample size

Rejection Region at α Level: $|T| > T_{(n-1),(1-\alpha/2)}$ or $|Z_0| > Z_{1-\alpha/2}$
Two-Sided p-Value: $2 \times (1 - \text{PROBT}(|T_0|, n - 1))$ or $2 \times (1 - \text{PROBNORM}(|Z_0|))$
Related SAS Procedure: PROC UNIVARIATE

Description of and Notes on the Method:

a. This test deals with paired data when the normality assumption cannot be fulfilled.

b. Wilcoxon's sign-ranked test is the nonparametric version of the paired t-test.

c. M_d = population mean of paired difference
n = number of nonzero pairs

d. The differences are ranked from lowest to largest with respect to the absolute value, and zeros are ignored.

e. R^+ = sum of ranks related to positive differences, and
R^- = sum of ranks related to negative differences

f. $\min\{R^+, R^-\}$ = smaller value of R^+ and R^-.

g. m_i represents the number of ties for the ith value that has ties.

h. The value of $\min\{R^+, R^-\} - n(n + 1)/4$ is always nonpositive.

i. The "0.5" correction for continuity in the test statistic is not always needed. The absolute-value sign should be removed in the test statistic if "0.5" is not used.

j. $$\text{sign}\left(\left|\min\{R^+, R^-\} - \frac{n(n+1)}{4}\right| - 0.5\right) = \begin{cases} \left|\min\{R^+, R^-\} - \frac{n(n+1)}{4}\right| - 0.5 & \text{if } R^+ > R^- \\ -\left(\left|\min\{R^+, R^-\} - \frac{n(n+1)}{4}\right| - 0.5\right) & \text{if } R^+ < R^- \end{cases}$$

k. A one-sided test can be performed in a similar manner to that for a t-test.

B. Group 2

1. *One-Way Analysis of Variance*

Study Design: Parallel design
Model Assumptions: Normal distribution, independence, equal variance for all treatments
Number of Observations per Patient for PD Parameter: 1
Purpose: To compare the equal means of all treatment groups.
Statistical Model:

$$y_{ij} = \mu + \tau_i + \varepsilon_{ij}, \quad \text{for } i = 1, 2, \ldots, k; \quad j = 1, 2, \ldots, n_i$$

where

y_{ij} = PD value for patient j in treatment group i
μ = unknown overall population mean
τ_i = unknown treatment effect for treatment group i
ε_{ij} = random error associated with patient j in treatment group i

Analysis of Variance Summary: See Table 3.
Computation for Sum of Squares:

$$SS(T) = \sum_{i=1}^{k} n_i(\overline{y}_{i.} - \overline{y}_{..})^2, \quad SS(E) = \sum_{i=1}^{k}\sum_{j=1}^{n_i} (y_{ij} - \overline{y}_{i.})^2$$

$$SS(TOT) = \sum_{i=1}^{k}\sum_{j=1}^{n_i} (y_{ij} - \overline{y}_{..})^2, \quad MS(\text{Source } X) = \frac{SS(\text{Source } X)}{\text{d.f.}(\text{Source } X)}$$

$$\overline{y}_{i.} = \sum_{j=1}^{n_i} y_{ij}/n_i, \quad \overline{y}_{..} = \sum_{i=1}^{k}\sum_{j=1}^{n_i} y_{ij}/N, \quad N = \sum_{i=1}^{k} n_i$$

Hypothesis Testing for Treatment Effect:

a. H_0: $\tau_i = 0$ for all i, vs. H_a: $\tau_i \neq 0$ for some i if treatment effect fixed
b. H_0: $\sigma_\tau^2 = 0$ vs. H_a: $\sigma_\tau^2 \neq 0$ if treatment effect is random

Test Statistic for Treatment Effect: F_0 = MS(T)/MS(E)
Rejection Region at α *Level*: $F_0 > F_{(k-1,N-k),1-\alpha}$
Two-Sided p-Value: $1 - \text{PROBF}(F_0, k-1, N-k)$
100(1 − α*)% Confidence Interval for Treatment i*:

$$\overline{Y}_{i.} \pm t_{(N-k,1-\alpha/2)}\left(\sqrt{\frac{MS(E)}{n_i}}\right)$$

100(1 − α*)% Confidence Interval for the Difference Between Treatments i and i′*:

$$\overline{Y}_{i.} - \overline{Y}_{i'.} \pm t_{(N-k,1-\alpha/2)}\left(\sqrt{MS(E)\left(\frac{1}{n_i} + \frac{1}{n_{i'}}\right)}\right)$$

Related SAS Procedure: PROC ANOVA or PROC GLM
Description of and Notes on the Method:

Table 3 Analysis of Variance Summary for Parallel Group Design

Source of variation	Degrees of freedom (d.f.)	Sum of squares (SS)	Mean squares (MS)
Treatment	$k - 1$	SS(T)	MS(T)
Error	$N - k$	SS(E)	MS(E)
Corrected Total	$N - 1$	SS(TOT)	—

a. PROBF(x,df1,df2) is the cumulative probability function for an F-distribution in SAS, and $F_{(df1,df2),\alpha}$ represents a numerical value at which the cumulative probability function is α under an F-distribution with (df1,df2) degrees of freedom where x = a random value.
b. When the treatment effect is fixed, it implies that $\sum_i T_i = 0$. When it is random, it is normally distributed with a variance of σ_τ^2.
c. Typical examples in this group are a placebo-controlled clinical trial with two or more dosage levels of the investigational drug, and an active-controlled study with a study drug at one or more dosage levels and a comparator agent at one or more dosage levels. Several important things that this approach always brings are:
   i. Departure of homogeneity of variance: first, use Bartlett's test to determine whether the variances are homogeneous (6, pp 106–107). If not, log-transformed data or a generalized F-test, as discussed on page 360, may be used.
   ii. Multiple/pairwise comparison: In general, pairwise comparisons are needed when an overall F-test for treatment difference is significant and it is desirable to determine which pair(s) of the treatments contributes to the differences. Many procedures have been proposed, including Fisher's LSD (least significant difference) test, Bonferroni–Dunn test, Tukey's HSD (honestly significant difference) test, Newman–Keuls test, Scheffe test, and Waller–Duncan test. Another test, particularly useful for comparing several treatments with a control, is the Dunnett test (1955) (7, p 945).
   iii. Dose response: It is well known that when a dose response is desirable in a clinical trial (i.e., an investigation on how the efficacy of the drug is related to the dosage given in the trial), a common design is to allow each patient to receive all or part of doses available in the study, in order to reduce variability. However, it is not uncommon to use a parallel group design for investigating the dose response. The trend of dose response can be evaluated by setting up orthogonal linear contrasts (linear, quadratic, cubic, quartic, etc.) of the treatment means (8, pp 117–138).
d. The treatment effect can be fixed or random, depending on how the treatment levels are se-

lected in the study. In either case, the test statistic remains the same, but the implication of the statistical conclusion will vary.

e. When k (the number of treatment groups) = 2, the F-test is exactly the same as for a two-sample t-test.

f. This F-test for analysis of variance is nondirectional. More specifically, it would signal a "difference," but without showing the direction of either a "superiority" or an "inferiority." However, with the aid of multiple comparisons, a better picture of the direction may be observed.

g. PROC ANOVA is mainly for the equal-sample-size case, and Proc GLM can handle unequal sample sizes.

2. *Kruskal–Wallis Test*

Study Design: Parallel group design
Model Assumptions: independence
Number of Observations per Patient for PD Parameter: 1
Purpose: To compare the equality of the median for all treatment groups.
Hypothesis Testing:

$$H_0: M_1 = M_2 = \cdots = M_k \quad \text{vs.} \quad H_a: M_i \neq M_j \text{ for some } i, j$$

Statistical Method:

$$K_0 = \frac{\dfrac{12}{N(N+1)}\left(\sum_{i=1}^{k} \dfrac{R_i^2}{n_i}\right) - 3(N+1)}{1 - \dfrac{\sum_{j=1}^{g} m_j(m_j^2 - 1)}{N(N^2 - 1)}}$$

Rejection Region at α Level: $K_0 > \chi^2_{(k-1),1-\alpha}$
Two-Sided p-Value: PROBCHI(K_0, $k - 1$)
Related SAS Procedure: PROC NPAR1WAY WILCOXON
Description of and Notes on the Method:

a. The Kruskal–Wallis test is a nonparametric version of analysis of variance to deal with PD parameters with more than two treatments. It is an extension of the Wilkcoxon rank-sum test, just like analysis of variance is an extension of the two-sample t-test.

b. The data are ranked from the lowest to highest over the combined data set.

c. Let y_{ij} = the observation of the PD parameter for patient j in treatment group i, r_{ij} = the corresponding rank for y_{ij}, n_i = sample size for group i. Then

$$R_i = \sum_{j=1}^{n_i} r_{ij} \quad \text{and} \quad N = \sum_{i=1}^{k} n_i$$

d. Assume there are g categories in which m_j observations are tied in the jth category.

e. PROBCHI(x) is a cumulative probability function at x for a chi square distribution in SAS, and $\chi^2_{(k-1),1-\alpha}$ is a numerical value at which the cumulative probability is $1 - \alpha$ for a chi square distribution with $k - 1$ degrees of freedom.

f. For k (the number of treatment groups) = 2, the Kruskal–Wallis Test is identical to the Wilkcoxon rank-sum test.

g. Another approximation test procedure to employ is a one-way analysis of variance with the ranked data.

h. Pairwise comparisons may be employed in two ways if the overall Kruskal–Wallis test is significant (9):
   i. Using the Wilkcoxon rank-sum test for each pair of treatment groups, with a caution of adjustment for the α-level.
   ii. Using a method similar to the Bonferroni–Dunn test for analysis of variance by identifying the minimum required difference between the means of the ranks of any two groups

3. *Log-Rank Test (Mantel–Cox Test)*

Study Design: Parallel group design
Model Assumptions: Independence
Number of Observations per Patient for PD parameter: 1
Purpose:

a. To estimate the survival rate for each treatment group

b. To compare the survival curves over time for two or more treatment groups

Kaplan–Meier Estimate for $S_i(t)$ for Treatment Group i:

$$\hat{S}_i(t) = \prod_{j=1}^{t} \left(\frac{n_{ij} - d_{ij}}{n_{ij}}\right)$$

100(1 − α)% Confidence Interval for $S_i(t)$:

$$\hat{S}_i(t) \pm Z_{1-\alpha/2} \sqrt{\left(\frac{\hat{S}_i(t)(1 - \hat{S}_i(t))}{n_{i0} - c_{it}}\right)}$$

Hypothesis Testing for Survival Curves:

$$H_0: S_1(t) = S_2(t) = \cdots = S_k(t) \quad \text{for all time } t$$
$$\text{vs.} \quad H_a: S_i(t) \neq S_j(t) \quad \text{for some } i, j \text{ and } t$$

Statistical Method:

$$Q_0 = \sum_{i=1}^{k} \frac{(d_{i+} - e_{i+})^2}{e_{i+}}$$

where

$$d_{i+} = \sum_{j=1}^{b} d_{ij}, \quad e_{i+} = \sum_{j=1}^{b} e_{ij}, \quad e_{ij} = \frac{d_{+j}n_{ij}}{n_{+j}}$$

$$d_{+j} = \sum_{i=1}^{k} d_{ij}, \quad n_{+j} = \sum_{i=1}^{k} n_{ij}$$

Rejection Region at α Level: $Q_0 > \chi^2_{(k-1),1-\alpha}$
Two-Sided p-Value: $1 - \text{PROBCHI}(Q_0, k - 1)$
Related SAS Procedure: PROC LIFETEST
Description of and Notes on the Method:

a. This test applies to the PD parameter when the time to a specified event (e.g., death) is of most interest. This test allows a comparison of the Kaplan–Meier survival curve between two or more treatment groups. Patients who discontinue from the study before the first occurrence of the predefined event or patients who never experienced the predefined event during the entire course of the study are called *censored.*
b. The study is divided into b time periods, denoted by t for $t = 1, 2, \ldots, b$.
c. $S_i(t)$ = survival rate at time t for treatment group i
   d_{ij} = number of patients experiencing predefined event for treatment group i at time j
   n_{ij} = number of patients for treatment group i at beginning of time j (i.e., number of patients at risk at time j)
   n_{i0} = number of patients recruited to treatment group i
   c_{it} = number of patients censored for treatment group i before time t

Table 4 displays information on the number surviving and not surviving for each treatment group for any given time j.

d. There are several methods for obtaining the standard deviation for the Kaplan–Meier proportion. The one shown in the confidence interval was given by Peto (1984) (10).

Table 4 Frequency Table for Survival and Nonsurvival

	Survival		
At time j	Yes	No	Row subtotal
Treatment 1	d_{1j}	$n_{1j} - d_{1j}$	n_{1j}
Treatment 2	d_{2j}	$n_{2j} - d_{2j}$	n_{2j}
⋮	⋮	⋮	⋮
Treatment k	d_{kj}	$n_{kj} - d_{kj}$	n_{kj}
Column subtotal	d_{+j}	$n_{+j} - d_{+j}$	n_{+j}

e. Another common test for survival curves is known as the *log maximum-likelihood test.*

C. Group 3

1. Fisher-Exact Test

Study Design: Parallel group design
Model Assumptions: Binomial distribution, independence
Number of Observations Per Patient for PD parameter: 1
Purpose: To compare the proportion of responders between two treatment groups
Hypothesis Testing:

$$H_0: \pi_1 = \pi_2 \quad \text{vs.} \quad H_a: \pi_1 < \pi_2$$

Test Statistic:

$$p_0 = \sum_{x \le x_1} \frac{\binom{n_1}{x}\binom{n_2}{r-x}}{\binom{N}{r}}$$

Rejection Region at α Level: $p_0 < \alpha$
Observed Frequency Table: See Table 5.
Related SAS Procedure: PROC FREQ
Description of and Notes on the Method:

a. The "x" in the test statistic refers to the count in the (1,1) position of the frequency table, and it is obtained by fixing n_1, n_2, N, and r in the observed frequency table.
b. $\binom{x}{y} = \dfrac{x!}{y!(x - y)!}$
c. The requirement of this method is that all of the values for the PD parameter be classified into two responses, regardless of the original measurement of the PD parameter.

Table 5 Observed Frequency Table

	Treatment 1	Treatment 2	Row total
Responder	x_1	x_2	$r\ (= x_1 + x_2)$
Nonresponder	y_1	y_2	$y_1 + y_2 = N - r$
Column total	$n_1\ (= x_1 + y_1)$	$n_2\ (= x_2 + y_2)$	$N\ (= n_1 + n_2)$

d. This test generally applies to studies with small sample size (<20).
e. π_i = the true unknown rate for responders for treatment group i.
f. When H_a: $\pi_1 > \pi_2$ is desirable, we simply change the summation "$x \le x_1$" to "$x \ge x_1$" in the test statistic.
g. When H_a: $\pi_1 \ne \pi_2$ is desirable, the p-value is obtained by first computing the probabilities for all possible combinations having the same row and column totals as the observed frequency table. Then we add the probabilities for combinations that have a probability smaller than or equal to the observed one.
h. This test can be generalized into a study with more than two responses and more than two treatment groups.

2. *McNeMar's Test*

Study Design: Parallel group design
Model Assumptions: Binomial distribution, dependence
Number of Observations per Patient for PD Parameter: 2
Purpose: To compare the proportions of responders between two treatment groups
Hypothesis Testing:

$$H_0\colon \pi_1 = \pi_2 \quad \text{vs.} \quad H_a\colon \pi_1 \ne \pi_2$$

Test Statistic:

$$Q_0 = \frac{(|B - C| - 1)^2}{B + C}$$

Rejection Region: $Q_0 > \chi^2_{1,1-\alpha}$
Two-Sided p-Value: 1 − PROBCHI(Q_0, 1)
100(1 − α)% Confidence Interval for $\pi_1 - \pi_2$:

$$\frac{B - C}{n} \pm z_{(1-\alpha/2)} \frac{1}{n} \sqrt{B + C - \frac{(B - C)^2}{n}}$$

Observed Frequency Table: See Table 6.
Related SAS Procedure: PROC FREQ
Description of and Notes on the Method:

a. Each patient contributes two observations under two conditions, such as before and after treatment, or under two different treatments, or at two different time points.
b. This assumes patients' well-being with regard to the PD parameter are the same before treatments. This assumption, in reality, may not be true.
c. π_1 is estimated by $(A + B)/N$, and π_2 is estimated by $(A + C)/N$.
d. The quantity "1" in the test statistic is a correction for continuity, which may not be needed for the analysis.
e. The chi square test statistic without the correction for continuity is also equivalent to a Z-test statistic of the form $Z_0 = (B - C)/\sqrt{B + C}$,

Table 6 Observed Frequency Table

Trt 2 / Trt 1	Responder	Nonresponder	Row total
Responder	A	B	$A + B$
Nonresponder	C	D	$C + D$
Column Total	$A + C$	$B + D$	$N\ (= A + B + C + D)$

Table 7 Observed Frequency Table ($m \times k$ Table)

	Treatment 1	Treatment 2	...	Treatment k	Row total
Response 1	n_{11}	n_{12}	...	n_{1k}	n_{1+}
Response 2	n_{21}	n_{22}	...	n_{2k}	n_{2+}
⋮	⋮	⋮	...	⋮	⋮
Response m	n_{m1}	n_{m2}	...	n_{mk}	n_{m+}
Column total	n_{+1}	n_{+2}	...	n_{+k}	n_{++}

which follows a standard normal distribution and is useful for a one-sided test.

f. The McNemar test has been extended to a $k \times k$ ($k > 2$) case in the Stuart–Maxwell test (11) (i.e., for studies with k treatment groups and k responses).

D. Group 4

1. (Pearson's) Chi Square Test

Study Design: Parallel group design

Model Assumption: Multinominal distribution, independence

Number of Observations per Patient for PD Parameter: 1

Purpose: To determine and compare the relationship between the PD parameter and the explanatory variable

Hypothesis Testing:

$$H_0: \pi_{ij} = (\pi_{i.})(\pi_{.j}) \quad \text{for all } i,j \quad \text{vs.}$$

$$H_a: \pi_{ij} \neq (\pi_{i.})(\pi_{.j}) \quad \text{for some } i,j$$

Test Statistic:

$$Q_0 = \sum_{j=1}^{k} \sum_{i=1}^{m} \frac{\left(\left| n_{ij} - \frac{n_{+j}n_{i+}}{n_{++}} \right| - 0.5 \right)^2}{\frac{n_{+j}n_{i+}}{n_{++}}}$$

Rejection Region at α Level: $Q_0 > \chi^2_{(m-1)(k-1),1-\alpha}$

Two-Sided p-Value: 1 − PROBCHI(Q_0, (m − 1)(k − 1))

Observed Frequency Table ($m \times k$ Table): See Table 7

Related SAS Procedure: PROC FREQ.

Description of and Notes on the Method:

a. π_{ij} = unknown true proportion for having response i in treatment group j

$\pi_{i.}$ = unknown true proportion for having response i regardless of treatment group

$\pi_{.j}$ = unknown true proportion for treatment group j regardless of responses

b. The null hypothesis indicates the independence between treatment group and response.

c. There is another type of hypothesis testing for a chi square test concerning a test for homogeneity; it means all proportions for all cells are the same.

d. The quanity "0.5" in the test statistic is the "continuity correction." An absolute value for the difference in the denominator is always needed when the "continuity correction" is considered. The "continuity correction" in the test statistic may not be needed.

e. There are a number of arguments about using the chi square test (12):

   i. Do not use the chi square test when $n_{++} < 20$.

   ii. All n_{ij}'s are at least 5 if $20 \leq n_{++} < 40$.

   iii. All expected cell frequencies are at least 1 if $n_{++} \geq 40$.

   iv. For a 2×2 table with $n_{++} < 20$, Fisher-exact test is suggested.

f. Another approach to handling an $m \times k$ table is to employ a likelihood-ratio chi square test with the test statistic expressed as follows:

$$G^2 = 2 \sum_{j} \sum_{i} n_{ij} \log \left(\frac{n_{ij}}{\frac{n_{i+}n_{+j}}{n_{++}}} \right)$$

This test statistic also follows a chi square distribution with (m − 1)(k − 1) d.f.

REFERENCES

1. FW Satterthwaite. An approximate distribution of estimates of variance components. Biometrics Bull 2:110–114, 1946.

2. SAS Procedures Guide. Version 6. 3rd ed. Cary, NC: SAS Institute, 1990, pp 617–634.
3. EL Lehmann. Nonparametrics: Statistical Methods Based on Ranks. San Francisco: Holden-Day, 1975, p 12.
4. DJ Sheskin. Handbook of Parametric and Nonparametric Statistical Procedures. Boca Raton, FL: CRC Press, 1996.
5. WJ Conover, RL Iman. Rank Transformations as a bridge between parametric and nonparametric statistics. Am Statistician 35:124–129, 1981.
6. BJ Winer, DR Brown, KM Michels. Statistical Principles in Experimental Design. 3rd ed. New York: McGraw-Hill, 1991.
7. SAS/STAT User's Guide. Version 6. 4th ed. Vol 2. Cary, NC: SAS Institute, 1990.
8. DA Berry. Statistical Methodology in the Pharmaceutical Sciences. New York: Marcel Dekker, 1990.
9. DJ Sheskin. Handbook of Parametric and Nonparametric Statistical Procedures. Boca Raton, FL: CRC Press, 1996, pp 402–405.
10. MKB Parmar, D Machin. Survival Analysis: A Practical Approach. New York: Wiley, 1995, pp 36–40.
11. JL Fleiss. Statistical methods for Rates and Proportions. New York: Wiley, 1973, pp 77–79.
12. DJ Sheskin. Handbook of Parametric and Nonparametric Statistical Procedures. Boca Raton, FL: CRC Press, 1996, p 218.

Cheng-Tao Chang
Robert L. Wong

Pharmacoeconomics

See also *Clinical Trials; Post Marketing Surveillance*

I. INTRODUCTION

The evaluation of pharmaceutical and biological products has traditionally focused on considerations of safety and efficacy. The methods for product development have been enhanced, and federal regulation requires that studies of safety and efficacy be conducted before a product is approved and marketed. In recent years, the information needs of health care decision makers has expanded beyond the traditional measures of safety and efficacy to include patient outcomes, cost, and cost-effectiveness. The primary motivation for this shift is a growing concern over the cost of health care; the current availability of health-related interventions exceeds society's ability and willingness to pay. One important consequence in the changing focus of health care decision making is the need for appropriate methods and decision rules to guide choices toward interventions most likely to yield the greatest value for the level of investment. *Health economics* is a field of study that applies economic concepts to health and health care systems. An important topic in health economics is the evaluation of heath care treatments and programs on the basis of costs (inputs) and consequences (outputs). *Pharmacoeconomics* focuses on economic evaluations of the use of pharmaceutical products, programs, and services. The purpose of this entry is to provide an overview of the terminology, family of technologies, and key methodologic features of health economic evaluations. The primary focus is on the statistical considerations for planning and analyzing studies and interpreting the results.

II. IMPORTANCE OF ECONOMIC EVALUATION

The concern over the rising costs of health care has prompted several initiatives that influence the price, availability, or use of medicines. Many hospital and managed care pharmacies have a strict budgeting process for providing pharmaceutical products and services. Suboptimal decisions are sometimes made in this setting when the pharmacy decisions are not coordinated with the larger health care system. For example, it is possible that the use of a new, higher-priced antibiotic would cause an increase in the pharmacy budget yet may reduce costs overall if the use of the antibiotic were associated with shorter lengths of hospital stay or a reduced rate of reinfection. For this reason, many hospitals and managed care organizations have lists, called *formularies*, of selected pharmaceuticals and the dosages judged to be the most appropriate or cost advantageous for patient care. Physicians who practice as part of a managed care provider organization are required or encouraged to prescribe from the formulary. Many managed care organizations also recommend specific programs of care as a guideline for practicing physicians. Many countries maintain drug formularies at the national level to determine the reimbursement level for drugs. Other initiatives are specifically directed at prices; one example used in Germany and the Netherlands is a practice called *reference-based pricing*, in which product reimbursement is set at the same level for all drugs in a therapeutic class. The reference price is usually set according to the least expensive drug in the class. Reference-based pricing is becoming more common worldwide as a way to control health care costs.

Another sign of the increasing importance of economic evaluations in health care decision making has been the emergence of guidelines for the conduct of cost-effectiveness analyses. The purpose of the guidelines has been to encourage a rational diffusion and application of medical technologies. Australia (6) and Ontario, Canada (50), have proposed guidelines for the provision of cost-effectiveness data before drugs are listed in the national or provincial formulary. The Australian guidelines are now in operation, and the Ontario guidelines have been subsumed in a federal initiative. Several European countries are considering guidelines (18). The U.S. Food and Drug Administration (FDA, 24) applies strict standards to claims of cost-effectiveness by pharmaceutical companies. These standards include the use of adequate and well-controlled studies demonstrating evidence for the comparative claims about the drugs in question. Major medical journals, including *The New England Journal of Medicine* (35) and the *British Medical Journal* (15) have issued guidelines for the design and reporting of economic evaluations considered for publication. Most recently, the Panel on Cost Effectiveness in Health and Medicine, sponsored by the U.S. Public Health Service, reported recommendations on standardized methods for conducting cost-effectiveness analyses (27,61). While most guidelines have dealt specifically with the methodologic aspects of studies, Hillman et al. (30) discussed guidelines for the conduct of studies sponsored by pharmaceutical companies.

These initiatives have also resulted in an increased level of competition among pharmaceutical companies to make comparative claims on the basis of pharmacoeconomic outcome variables, including resource use, cost, and cost-effectiveness. Many clinical development programs now include pharmacoeconomic objectives, which can greatly affect the overall design and implementation of clinical trials. Nonclinical studies of disease epidemiology and the costs of treating the targeted diseases are also common in clinical development programs as a way of supplementing trial data. These changes have broadened the traditional roles and responsibilities of the clinical trial statistician.

III. TYPES OF ANALYSES

Economic evaluations of health care technologies involve a formal assessment of the costs and consequences of medical treatments or treatment programs. The major components of an economic evaluation include identifying, measuring, and valuing both the resources applied to the health care program (inputs) and the health improvements derived from the health care program (outputs). The theoretical foundation of the assessment is based on the idea of *economic efficiency*, in the sense of optimizing health outcomes for given fixed levels of resources consumed (65,25). The basic model recognizes that all health care interventions involve costs (measured in terms of the economic value of the resources used to treat or prevent illness) and health consequences (such as improvements in health status) and may include savings or reductions in future health care costs.

One type of study often conducted at the start of a health economic research program is a *burden-of-illness study*. This form of analysis typically combines epidemiologic data on the incidence and prevalence of the disease with cost data in order to estimate the total direct and indirect costs attributable to a disease incurred by the population over a defined period. Burden-of-illness studies are useful for influencing prioritization by alerting policy makers and health care professionals to the importance of a problem. Several important issues have been raised about the conduct and use of these studies. Drummond (12) noted that their use is limited because they are essentially noncomparative and therefore do not indicate whether more or less resource should be devoted to treatment. In addition, the methods for estimating the value of lost productivity and reduced productivity are thought possibly to exaggerate those costs. At the other extreme, burden-of-illness studies do not account for the intangible costs related to a diminished quality of life or for the impact of the disease on others in the form of grief. They can, however, provide a benchmark against which a new intervention can be evaluated.

Comparative economic evaluations provide the most useful information for guiding health care decisions. Researchers have developed several forms of comparative economic analysis, which can broadly be classified into the following groups (17): cost analysis, cost–consequences analysis, cost-effectiveness analysis, cost–utility analysis, and cost–benefit analysis.

A complete evaluation of therapy costs is needed for all forms of economic analysis. In a *cost analysis*, or *cost-minimization analysis*, the comparison is based on an accounting of the resources used as input to the therapy or health care program. This form of analysis is used when the effectiveness of the candidate and comparative therapies can be assumed to be equal and when the comparison is made strictly on an evaluation of costs. Although therapies are rarely equal in terms of effectiveness, cost minimization can provide a reasonable approximation in some cases. For example, de Lissovoy et al. (10) performed a cost analysis of imipenem and cilastatin compared with tobramycin plus clindamycin for the treatment of patients suspected of having acute intra-abdominal infection. The study compared hospital treatment costs during an episode of infection for patients enrolled in a clinical trial comparing the effectiveness of the two therapy groups. Costs of hospital stays in regular and intensive care units, medical procedures, antibiotics, other drugs, and laboratory evaluation were included in the analysis. The mean total costs did not differ significantly between the two treatment groups ($12,040 for imipenem and cilastatin and $11,220 for tobramycin plus clindamycin); however, the study showed that costs were significantly affected primarily by the severity of illness, as measured by the Acute Physiology and Chronic Health Evaluation acute physiology score. Use of a cost-minimization analysis for this study was conservative for imipenem and cilastatin because the principal finding in the main clinical study was a lower incidence of treatment failure in patients in the imipenem and cilastatin group (P = .043). This is not only an important health outcome but an additional source of costs that was not followed in the main study.

A *cost–consequences analysis* extends the cost-minimization analysis by also considering health outcomes. This study displays complete but separate enumerations of the costs and health outcomes (measured in natural units) of the candidate therapies. Because the results are presented in a disaggregated form, the overall analysis of the study is left to the user; no specific value is placed on the outcomes. Cost–consequences analyses are most useful to analysts who have the technical ability to use the reported results as a basis for applying their own methods of evaluation.

A *cost-effectiveness analysis* measures the health outcomes of the candidate treatment programs in natural units. Examples include variables such as life-years gained, fractures avoided, and coronary events avoided. In the classic form of cost-effectiveness analysis, the results are expressed by the ratio cost per unit of health benefit gained as a measure of efficiency. For example, Oster and Epstein (51) used an epidemiologic model based on the Framingham Heart Study to evaluate the cost-effectiveness of pharmacologic therapy for reducing cholesterol levels. They estimated that for a 40- to 44-year-old man with elevated serum cholesterol levels, therapy with the pharmaceutical agent cholestyramine costs $76,500 per additional life-year gained.

A *cost–utility analysis* is a form of cost-effectiveness analysis that values the consequences of treatment programs by using health state preference scores or utility weights. In general terms, this method permits assessments on the quality of life-years gained, not just the crude number of years. This approach is particularly useful for evaluating treatments that may extend life expectancy (possibly by preventing fatal disease in the future) but may be associated with adverse experiences in the short term. Cost–utility analyses also provide a common metric for overall reductions in morbidity and mortality. For the broader policy perspective, this allows comparisons across treatments for different diseases or even for comparisons of treatment versus prevention strategies. In the U.S. literature, the term *cost–utility analysis* is rarely used.

A *cost–benefit analysis* evaluates health outcomes in monetary terms rather than in terms of patient outcome. This approach makes the measure of health benefit commensurate with resource costs. This may be the broadest form of analysis, because it allows direct assessment of the benefits of the treatment program with respect to the costs and has the potential to capture significant nonhealth benefits. Two methods for assigning monetary value to health states are used most often (31,17). The *human capital method* evaluates health benefits in terms of the economic productivity gained from the increased survival or decreased morbidity. Wage rates are typically used for the analysis. The *willingness-to-pay method* defines the maximum amount of money an individual is prepared to spend to secure a defined health state. Both methods present significant analytic challenges (17,27), but the willingness-to-pay method is considered by most economists to be the more theoretically correct.

IV. DESIGN CONSIDERATIONS

A. Study Perspective

The perspective of the economic evaluation determines which costs and health outcomes are relevant and plays an important role in how they should be valued. Possible study perspectives include those of the patient, the employer, the health care provider, the hospital, the managed care organization, the insurance company, and society at large. Gold et al. (27) recommend using the *societal perspective*, because it is the most comprehensive and is relevant to the broad allocation of resources. This perspective includes a complete evaluation of the health effects and costs experienced by anyone significantly affected by the intervention. Economic evaluations of pharmaceutical products often use a narrower perspective that would be relevant to specific customer segments. The study perspective is an important design feature and should be defined early in the study process. Drummond et al. (17) recommend performing the evaluation for various sectors and combining results to allow the study potentially to address questions from several perspectives.

B. Costing Methods

Three general categories of costs may be examined within an economic evaluation of a health care technology: direct costs, indirect costs, and intangible costs.

Direct costs include the value of medical resources used in the provision of the treatment and the consequences of the treatment, in terms of both beneficial effects and unwanted side effects. For pharmaceutical programs, direct costs would include the costs of drugs, medical staff, medical facilities, medical procedures, diagnostic and laboratory tests, and supplies. When appropriate, direct nonmedical costs, which are borne by the patient seeking care, may be included in the analysis. These would include the costs of transportation, child care, housekeeping, and so forth. Several authors have discussed methods for assigning direct costs to units of health care resources (e.g., Refs. 17 and 40); such discussion is beyond the scope of this entry. Most studies capture patient-specific data on the health care resources used for treatment and then conduct separate studies to estimate costs for these resources. Patient-level cost data can then be computed by combining the cost and resource data from the trial.

Indirect costs include the value of productivity lost because of the underlying illness or its treatment. Recently, these costs have been referred to as *productivity costs* (17) to avoid confusion with the term as it is used by accountants for overhead costs. They include the costs associated with lost or impaired ability to work or to engage in leisure activity because of illness and lost economic productivity caused by premature death. Historically, the *human capital approach* has been used to estimate productivity costs, but this method has been criticized for its use of gross earnings that include employee benefits. It is argued that using earnings raises both ethical and equity concerns, because this method does not place a value on unpaid employment (e.g., homemaking) and may undervalue the impact of the disease on women and the elderly. In addition, the method may overestimate the true costs, because short-term losses in productivity may be lower than average wage rates. For longer-term losses, companies can replace laborers to minimize losses in production. In this context, Koopmanschap and Rutten (36) proposed a *friction cost* approach based on the notion that the amount of production lost depends on the time span that organizations need to restore the initial production level. Their estimates for the Netherlands in 1988 show that the human capital method can provide estimates that exceed the friction cost by a factor of 10. Luce et al. (40) recommend that productivity costs be estimated and reported separately in cost-effectiveness analysis to avoid double-counting the beneficial effects of treatment.

Outcomes such as pain, emotional well-being, social functioning, activities of daily living, mobility, and cognitive functioning are all important considerations in the economic evaluation of a pharmaceutical treatment program. Failure to include these factors reduces the credibility of the assessment and may threaten its adoption by decision makers. The value associated with these types of outcomes has been called *intangible costs* (17). These measures are often captured in clinical trials by using validated quality-of-life instruments. For purposes of economic evaluation, their impact is included as part of the effectiveness measure in a cost–utility analysis. Monetary values of intangible outcomes would be (implicitly) included in the willingness-to-pay method.

C. Outcome Measures

Because economic evaluation focuses on efficiency, measuring the benefit of the intervention in terms of health outcomes is as important as measuring benefit in terms of cost. Several approaches have been used, and there is no clear consensus on the endpoints to use for the evaluations (13). The choice is often driven by complexities in the measurement instruments and by possible limitations in the design of the clinical programs. Short-term *surrogate*, or *intermediate*, *measures* of response to treatment are attractive in these situations when they provide reliable indicators of long-term patient outcome. For example, in the area of cardiovascular disease, meta-analysis (29) and long-term clinical trials (56,60) have shown that lowering serum cholesterol levels confers the benefits of a reduced risk for coronary heart disease and increased survival and thus supports the cholesterol hypothesis. These findings established the rationale for a cost-effectiveness analysis based on intermediate endpoints by Schulman et al. (57). These researchers reported a comparative analysis of various lipid-lowering drugs by plotting percentage reduction in cholesterol levels by the 5-year cost of patient management associated with the use of each drug. Economic evaluations based on intermediate endpoints assume that the effect of therapy on the ultimate clinical endpoint is fully captured through an established relationship with the surrogate marker. This assumption can be difficult to validate without the appropriate long-term clinical trials.

Cost-effectiveness analyses can also be based on different types of *clinical endpoints*, such as coronary events avoided or fractures avoided. An advantage of using endpoints prevented as the effectiveness measure is that they are closely related to the clinical objectives of therapy. This makes ratios such as cost per event avoided a meaningful measure for decision makers. When the medical condition is potentially life-threatening, *mortality endpoints* such as deaths avoided are applicable. One common measure of cost-effectiveness that has been used for medical evaluations is the *cost per life-year gained*. The advantage of using life-years gained is that this measure has a very clear definition as the area under the cumulative probability density function for survival. As such, it takes into account the number of life-years during which the risk for death is reduced. The obvious disadvantage of mortality endpoints is that they do not apply to the evaluation of treatments for non-life-threatening illnesses and do not consider the patient's *quality of life*.

The *quality-adjusted life-year* (*QALY*) method (62) is an attempt to combine the concepts of quality and length of survival. The QALY method uses measures of health outcome that assign, for each period of time, an overall measure of a patient's quality of life by using a weight defined from 0 (death) to 1 (perfect health). Health states with diminished quality are scored less than 1. The number of QALYs is simply the sum of the QALY weights over the time horizon for the evaluation. The number of QALYs gained is a popular measure of effectiveness used in cost–utility analyses.

The QALY concept is not without controversy. This measure can be interpreted as the number of years of perfect health a patient values equivalently to their years with diminished quality. For example, a QALY of 1 could be derived from a single year with perfect life or from several years with diminished quality. Mehrez and Gafni (44) argue that patients would value the same number of QALYs differently depending on the path by which the QALYs are accumulated. They proposed a *healthy years equivalent* (HYE) measure that is derived by using a complex two-step procedure to determine patient preferences for different paths for accumulating QALYs. At the other extreme, Cox et al. (9) argue that the QALY approach is overly complex and that simpler measures addressing the dimensions of quality of life and survival separately should be used.

The endpoint chosen is an important factor in economic evaluations. Drummond et al. (14) showed that the choice of endpoint can have an important impact on the ranking of alternative therapies for lipid lowering. The ranking depended on the value placed on life-years after coronary heart disease developed, the effect that drug therapy had on quality of life, and even the discount rate at which future QALYs were valued. Ultimately, the selection of endpoints depends on the nature of the disease and treatment program being studied, and, if this can be determined, what is most relevant to the decision makers.

D. Discounting

Economic evaluations should include the value of all costs, adverse experiences, and benefits that are the direct consequence of the disease or treatment program under study. This may include a stream of costs and health effects over a long period. It is important to recognize that the timing of costs and benefits may differ over time and may differ according to the programs under study. For example, the cost of drug therapy to reduce serum cholesterol levels is incurred with the initiation of therapy, but the benefits are realized through a reduced risk for coronary heart disease in later years. In addition, comparisons of drug therapy with surgical procedures need to account for the different timing of costs and benefits.

The method of *discounting to present value* is used to account for the timing of the stream of costs and health benefits. Computationally, $PV(X) = \Sigma X_i/(1 + r)^i$ gives the present value for the stream of costs or health effects, denoted by X_i for year i, using a discount rate r. The value assigned to r is an important and controversial design consideration, as is the notion of discounting both future costs and future health effects at the same rate. The Panel on Cost-Effectiveness in Health and Medicine (61) recommends discounting costs and health outcomes at the same 3% discount rate. A riskless 3% discount rate was chosen to be consistent with the real rate of return on long-term U.S. government bonds. To maintain comparability with previously published analyses, they recommend also using a 5% rate and conducting a sensitivity analysis for rates ranging from 0% to 7%. Undiscounted gains in life expectancy and quality-adjusted life expectancy can be reported separately, since they are often of interest to the reader.

V. RESEARCH METHODS

A variety of research methods have been used for economic evaluations, including modeling designs and studies using clinical-economic trials that involve prospective or retrospective data collection.

A. Modeling Studies

Modeling studies have employed two main types of methods to simulate patient outcomes and cost by using assumptions about the effects of therapy. Clinical *decision-analytic models* systemically lay out all probable pathways and consequences of the candidate treatment programs in a decision tree. The likelihood and expected cost of each pathway are estimated by use of data from clinical trials or epidemiologic investigations. These models have been used mostly for treatments for acute conditions, which may resolve in a short period. *Markov models* are used for diseases and treatments that are characterized by recurrent disease states or treatment programs.

Epidemiologic models have been used for disease areas in which the outcomes are related to sets of risk factors and the clinical outcomes may take years to develop. Several models for the treatment of elevated cholesterol levels have been developed on the basis of data from the Framingham Heart Study (51,28,26). These models estimate the risk for coronary heart disease according to pretreatment and posttreatment cholesterol levels by using risk equations derived from the Framingham study and by using estimates of treatment efficacy derived from short-term clinical trials. Similar types of models have been developed for osteoporosis and HIV disease on the basis of epidemiologic links between clinical outcomes and risk factors that can be altered by treatment.

B. Clinical-Economic Trials

The collection and analysis of economic data in clinical trial settings is becoming more common. These types of studies have been called *piggyback studies*, a term denoting that the scope of the trial was broadened to address economic questions on top of the clinical evaluation. Typically, they are undertaken during phase III of the drug development process. Many of these studies have the advantages of being multicenter, randomized, double-blinded, and placebo-controlled, features that offer a high degree of internal validity to allow researchers to address specific questions about the safety and efficacy of the candidate treatment.

Studies of this type have important drawbacks for economic evaluations (13,55). They typically have insufficient sample size for addressing questions about cost and, more important, lack generalizability and external validity because of the use of a comparator (placebo) that is typically inappropriate. Moreover, the studies are usually conducted in specialized medical centers with restrictive patient populations, and they define patient care by using rigorous clinical protocols in which the physician is blinded to knowledge of treatment. All of these factors limit the usefulness of these studies in providing data to answer questions about the cost and consequence of therapies as they will be used in real-world settings.

Schulman et al. (59) and Mauskopf et al. (43) have discussed strategies for undertaking economic assessment within the phase I through phase IV clinical development program. Drummond and O'Brien (16) and O'Brien and Drummond (48) addressed issues of clinical and statistical significance in the context of socioeconomic evaluations of medicines. They recommend using confidence interval methods instead of strict accept/reject hypotheses as a way to permit the readers of the published findings to assess the magnitude and significance of the observed differences. Schulman et al. (58) discussed their attempts to overcome these limitations in the economic evaluation of the Flolan® International Randomized Survival Trial (FIRST), a multinational phase III clinical trial of Flolan® in the treatment of congestive heart failure. Freemantle and Drummond (23) addressed the issue of blinding in clinical-economic trials, noting that blinding may lead to artificialities in treatment protocols that may affect costs and potential benefits.

An alternative to the piggyback design is the *randomized clinical-economic trial* done expressly for cost-effectiveness analysis. Oster et al. (52) discussed the design of a randomized trial intended to assess effectiveness and cost of two cholesterol treatment programs that involved stepped-care regimens with several drug choices. One treatment program was modeled after the 1988 guidelines of the National Cholesterol Education Program. This stepped-care approach recommended initiating niacin therapy and then introducing a series of other pharmacologic agents until the low-density lipoprotein cholesterol levels attained target levels. The other program was a regimen that started with the HMG-CoA reductase inhibitor lovastatin and allowed changes in therapy if the target level was not attained or if the patient preferred to switch therapy. Lovastatin was also allowed in the final step of the stepped-care regimen. The study was conducted in a managed care setting; after randomization, patient management approximated usual clinical practice. Provider compliance with the treatment plans was encouraged but not enforced, and patients paid for medication as they customarily would. Another feature of the study is that it used the intention-to-treat principle for both design and analysis, because patients in the different program groups may actually have been prescribed the same drug at some points in the trial.

Some of the limitations of clinical-economic trials can also be overcome by applying prediction models to data collected during the main trial. This allows the analysis to consider, for example, questions about longer time horizons or specific patient subgroups. Mark et al. (42) supplemented data from the 1-year GUSTO (Global Utilization of Streptokinase and TPA for Occluded Arteries) study with data on projected life expectancy and hospital cost data to evaluate the cost-effectiveness of tissue plasminogen activator compared with streptokinase for acute myocardial infarction. By using cost-effectiveness analysis, the authors concluded that the added costs of tissue plasminogen activator therapy were consistent with those of other well-accepted medical technologies when evaluated in comparison to the benefits of therapy. Similarly, the Diabetes Control and Complications Trial (DCCT) group (11) examined the cost-effectiveness of aggressive approaches to the management of insulin-dependent diabetes mellitus. They used a Monte Carlo simulation model that was based on clinical endpoint results of the DCCT and was supplemented with other clinical, epidemiologic, and cost studies. Both of these studies used novel modeling techniques and secondary sources of data to overcome limitations in the designs of the main studies to answer the relevant economic questions.

VI. METHODS OF ANALYSIS

Much attention has recently focused on statistical methods for assessing uncertainty in health economic evaluations (17,41,47). Uncertainty in the estimate of program costs and effects arises from two primary sources. The first is uncertainty in assumptions about the form of the underlying economic model, including values for the key model parameters, such as the time horizon and the discount rate. In other situations, economic models use point estimates for parameters relating to patient outcomes or cost. O'Brien et al. (49) call this first type of evaluation a *deterministic analysis* and recommended that uncertainty be assessed by using a comprehensive *sensitivity analysis*. Briggs et al. (2) discuss the different forms of sensitivity analysis that can be applied. These may include simple reporting of the results over a range of parameters, analysis of extremes, or a threshold analysis. Probabilistic sensitivity analyses are also used in some studies.

The second source of uncertainty is the result of sampling variability in model variable estimates, which are derived from studies of individual patients. This type of evaluation is called a *wholly stochastic analysis*, and formal statistical analyses can be used to estimate variability and the 95% confidence limits of the key economic parameters. Evaluations that involve uncertainty from both model parameters and sampling variability are called *partially stochastic analyses*. The rest of this section briefly discusses the statistical considerations involved with analyzing cost data and cost-effectiveness evaluations.

A. Analysis of Cost Data

Analysis of cost data should first determine whether treatment programs differ in the actual resources utilized (e.g., hospitalization, procedures, and drug use). A candidate treatment may reduce one type of resource but increase another type. Computing patient cost involves combining resource data with prices in a way that is consistent with the study perspective. The statistical analysis of cost data can present several potential challenges. The distribution of costs is typically right-skewed, with a few patients incurring very large costs. Use of summary measures that are insensitive to extreme values, such as the median, are usually not desirable to organizations attempting to manage total aggregate patient costs. Patients with high resource use are substantial contributors to aggregated treatment costs but have a small influence on the median.

Cost distributions also tend to exhibit large variability that can pose problems for the provision of accurate and precise estimates of cost differences with the appropriate level of statistical power. The variability in cost data often increases with mean cost to such a degree that the usual normalizing and variance-stabilizing transformations are not helpful. For these reasons, nonparametric methods, resampling methods (e.g., bootstrap methods), and randomization tests may be preferred to parametric tests done by using analysis of variance.

Another important complication relates to patient cost data that are subject to right censoring. Censoring occurs whenever study data on patient outcomes collected over time are incomplete. Difficulties arise in the analysis because only partial data on the patient are available. This situation can arise because of study design considerations (e.g., a clinical trial uses a rolling enrollment period and a fixed study termination) or because a patient is lost to follow-up. The latter can occur for many reasons: for example, patients may transfer to other health care providers or drop out of the study because of the treatment's poor efficacy or adverse effects. Death is an interesting type of patient censoring. Some authors consider patient death a censored observation for cost analysis. For example, Dudley et al. (20) suggested that for cost studies of surgical procedures, deaths should be treated as censored observations so that hospitals with high mortality rates and therefore lower costs are not rewarded. The alternative argument is that patient death presents a well-defined endpoint for the accumulation of costs and therefore is uncensored.

Recently, health economists have shown interest in statistical methods to estimate mean costs when data for some patients are right censored. Dudley et al. (20) compared several analytic models, including both parametric and nonparametric survival models for estimating the effects of clinical factors on the hospital costs of coronary bypass graft surgery. Fenn et al. (20) pointed out the biases that result when the censored observations are either excluded from the calculations (called the *uncensored* method) or included as complete observations (the *full sample* method). They suggest using the Kaplan–Meier estimator to incorporate the censoring information. However, this approach can also result in a substantial bias (3).

Lin et al. (39) proposed a method that uses the sample mean costs for patients who die within distinct time intervals defined for the study (e.g., monthly or yearly intervals). These interval means are combined with the Kaplan–Meier estimate of dying within the interval to obtain an estimate of mean cost. Carides (4) improved on the method of Lin et al. by using a two-stage estimator that first fits the uncensored patient cost data to an appropriate statistical model using linear or nonlinear regression and then applies Kaplan–Meier survival estimation to the fitted values. Both methods are unbiased, but the two-stage method gains efficiency from the smoothing that occurs during the regression stage.

B. Cost-Effectiveness Ratios

The cost-effectiveness of a candidate treatment or treatment program is typically evaluated by using the *incremental cost-effectiveness ratio*, defined by the equation

$$R = \frac{\Delta C}{\Delta E} = \frac{C_B - C_A}{E_B - E_A}$$

where C_A and C_B are the mean costs associated with treatments A and B and E_A and E_B are the mean health effects associated with these treatments. The *incremental cost-effectiveness ratio* R measures the additional cost per additional unit of health outcome for therapy B relative to therapy A.

Point estimates of R, denoted by $\hat{R} = \Delta\hat{C}/\Delta\hat{E}$, have typically been reported for clinical trials by using average between-treatment costs and effects. Recent reports in the literature have emphasized the need to provide 95% confidence interval estimates of R to account for the inherent variability of the estimates. Computing interval estimates of R is also important because the limited sample sizes used in clinical trials are usually not sufficient either to detect differences in mean cost between therapies or to get precise estimates of cost-effectiveness ratios.

Several methods have been proposed for estimating cost-effectiveness ratios using 95% confidence intervals. These include the use of Taylor series approximation for the variance of $\hat{R}$ (49) and application of Fieller's theorem (5,66,37) for estimating 95% confidence limits for the ratio of two random variables. A box method was proposed by Wakker and Klaassen (64) and extended by Laska et al. (37). This method provides a conservative upper-bound estimate of R by using the separate 95% confidence interval estimates of ΔC and ΔE. In addition, a bootstrap estimate of R was suggested by O'Brien et al. (49) and further developed by Chaudhary and Stearns (5). The latter authors provide a nice overview of the Taylor series, Fieller's theorem, and the bootstrap method. They also give specific formulations to apply to each of those methods. Polsky et al. (54) used simulation methods to compare the box method, Taylor series, Fieller's theorem, and bootstrap for this application. They conclude that the bootstrap and Fieller's theorem were more dependably accurate than the others because they allow for the asymmetric shape of the sampling of distribution of $\hat{R}$. Moreover, the box method does not account for the correlation between costs and effects.

Most of the difficulties encountered with the available methods for confidence interval estimates of cost-effectiveness ratios are related to situations in which ΔE is numerically small. The problem is heightened when the estimate of R is based on estimates of ΔE from clinical trials. Even if the null hypothesis that $\Delta E = 0$ is rejected, the treatment effects on the health outcome variable may be numerically small or modest. All of the available methods deal with the problem of small denominators in different ways. The accuracy of the Taylor series approximation depends on having small coefficients of variation for the separate estimates of between treatment differences in both costs and effects. The accuracy of the approximation begins to fail as the between-treatment differences approach 0. Laska et al. (37) noted the complexity in interpreting interval estimates obtained by use of Fieller's theorem because this method depends on whether the individual confidence interval estimate of ΔE excludes 0. It is possible to have inclusion limits for R (R_L,R_U) if the confidence interval for ΔE excludes 0, and exclusion limits of the form "not (R_L,R_U)" if the confidence interval for ΔE includes 0. The condition that the entire 95% confidence interval estimate of the denominator ΔE not including 0 is equivalent to rejection of the null hypothesis that the treatments do not differ in mean effectiveness at the $\alpha = .05$ level. It is also possible to obtain confidence intervals for R that consist of the whole line $(-\infty,\infty)$ if the coefficients of variation for both ΔE and ΔC are large. Wakker and Klaassen (64) suggest that an estimate of R is useful only when the data are sufficient to allow the conclusion that ΔE is positive.

Cook and Heyse (8) propose a method for estimating R and its 95% confidence interval through an angular transformation. Figure 1 shows the angular transformation done by using the cost-effectiveness plane (1). The slopes of the rays connecting points A and B to the origin are sometimes called the *average cost-effectiveness ratios*. The incremental cost-effectiveness ratio R is simply the slope of the line segment connecting the mean costs (C) and health outcomes (E) for treatments A and B. Redefining the origin of the cost-effectiveness plane to the point (E_A,C_A), θ is simply the angle connecting the $\Delta E > 0$ axis to the $A{-}B$ line segment. Table 1 shows how θ can be used to facilitate the interpretation of cost-effectiveness ratios.

Values of θ in quadrants II or IV indicate that one of the treatments dominates the other, in the sense that one treatment has a greater effect and lower cost. Values of θ in quadrants I or III indicate that the two treatments trade off costs and effects. Use of θ allows a specific interpretation of the relative direction of treatments A and B. The ratio R does not allow a spe-

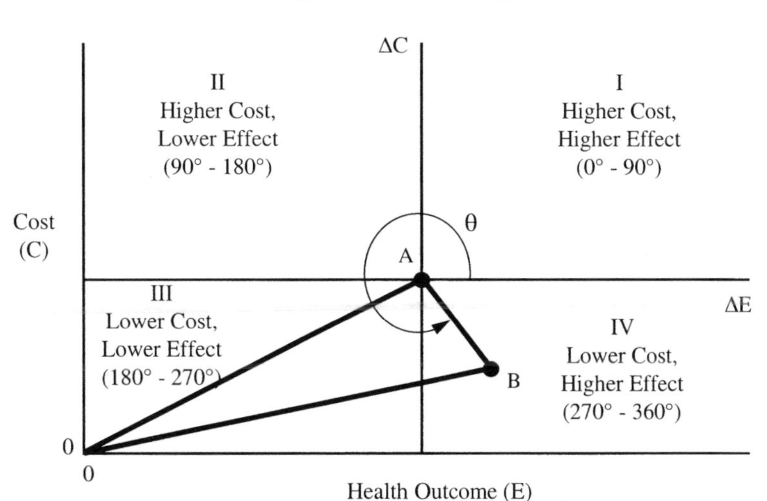

Fig. 1 Cost-effectiveness angle (θ): program B relative to program A.

Table 1 Use of θ to Help Interpret Cost-Effectiveness Ratios

Direction of ΔC and ΔE	Range for angle	Quadrant
$\Delta C \geq 0$; $\Delta E \geq 0$	0–90°	I
$\Delta C \geq 0$; $\Delta E \leq 0$	90–180°	II
$\Delta C \leq 0$; $\Delta E \leq 0$	180–270°	III
$\Delta C \leq 0$; $\Delta E \geq 0$	270–360°	IV

cific interpretation. For example, the ratio cannot distinguish between quadrants II and IV when R is negative, and it cannot distinguish between quadrants I and III when R is positive. However, the interpretation of the cost-effectiveness of therapy B relative to A depends on the quadrant.

For purposes of estimation, the method uses standardized differences in average cost and effect for the angular transformation. This is given by the equation

$$\theta = \begin{cases} \tan^{-1}\,[\Delta\hat{C}/s_{\Delta\hat{C}},\ \Delta\hat{E}/s_{\Delta\hat{E}}] \text{ if } \Delta\hat{C} \geq 0 \\ 180° + \tan^{-1}\,[\Delta\hat{C}/s_{\Delta\hat{C}},\ \Delta\hat{E}/s_{\Delta\hat{E}}] \text{ if } \Delta\hat{C} < 0 \end{cases} \tag{1}$$

where $s_{\Delta\hat{C}}$ and $s_{\Delta\hat{E}}$ are the standard errors of the between-group differences in mean costs and effects, respectively. The transformation θ stabilizes the variability of the ratio, especially when the effects are numerically small. The angle θ and its standard error can be estimated by using the jackknife (45) or the bootstrap (21). The grouped jackknife method (46) may offer some advantages over the bootstrap because it can easily adjust for multiple centers by doing the calculations leaving out one center at a time. Estimates of ratios can be computed by back-transforming from θ to R through Eq. (1).

C. Illustration: Scandinavian Simvastatin Survival Study

The Scandinavian Simvastatin Survival study (4S) was a randomized, multinational, double-blind, placebo-controlled trial of simvastatin for hypercholesterolemic patients with existing heart disease (56). The study consisted of 4444 patients from Denmark (n = 713), Finland (n = 868), Iceland (n = 157), Norway (n = 1025), and Sweden (n = 1681) who previously had myocardial infarctions or stable angina pectoris. The median duration of patient follow-up was 5.4 years. The study showed that simvastatin was associated with a 30% reduction in the risk for death from any cause (P = .0003). Data on hospital admissions were prospectively collected to evaluate the effect of simvastatin on health care resource use. Pedersen et al. (53) reported that patients randomly assigned to simvastatin had 34% fewer hospital days than patients assigned to placebo ($P < .0001$).

Other investigators have based formal cost-effectiveness analyses of cholesterol lowering with simvastatin on 4S (53,32). It is not our intention to reconsider these evaluations but rather to illustrate the statistical methods. We focus on the cost-effectiveness measure cost per death averted. Costs were assigned to simvastatin use and hospitalization by using Swedish prices converted to U.S. dollars. Simple summaries of the 4S results are given in Table 2. Simvastatin was associated with fewer patient deaths during the trial, but this positive health effect came at additional patient costs. The overall full-sample cost-effectiveness ratio was $59,201 per additional death averted. Table 3 gives a comparison of several methods for estimating R with 95% confidence intervals. Most of the methods yield similar results, because the health effect in this study was substantial. The Taylor series method appears to differ from the others, because it provides symmetric limits, which are probably not appropriate for ratios of this type.

Recall that 4S was a multinational study. Because of the potential for large variability among countries in the utilization and cost of health care resources, the appropriateness of combining economic data across the countries should be assessed. Cook et al. (7) provided a test for homogeneity of the individual country ratios that can be used in this case. Figure 2 shows the individual cost-effectiveness ratio estimates per country from 4S and the 95% confidence intervals. These ratios appear to be homogeneous (P = 0.73), justifying the appropriateness of the overall pooled ratio estimate given in Fig. 2 and Table 3.

VII. INTERPRETING THE RESULTS

Drawing conclusions from the results of cost-effectiveness analyses can also be complex, because doing so usually involves some evaluation of the trade-offs between costs and health outcomes. The exception would be therapies that are associated with both lower costs and greater health outcomes. These therapies are said to dominate the comparator therapy, and interpreting the results in these cases is straightforward when a cost–consequences analysis is used. However, most

Table 2 Summary of Patient Costs and Survival in 4S

	Placebo	Simvastatin	Difference $(S - P)$
Sample Size	2223	2221	
Proportion surviving to end of trial	.885	.918	.033
Average cost/patient	$3131	$5098	$1966

Table 3 Comparison of Methods to Estimate R and 95% CI

Method	Estimate of R	95% confidence interval
Full sample		
Taylor Series	$59,201	(26,579, 91,824)
Fieller's	$59,201	(37,369, 126,690)
Bootstrap		
Untransformed	$63,992	(35,612, 126,819)
Using θ	$58,146	(35,612, 126,819)
Jackknife using θ		
Grouping by center	$59,364	(38,968, 116,596)
Stratifying by country	$62,653	(36,723, 180,691)

R = incremental cost per death averted.

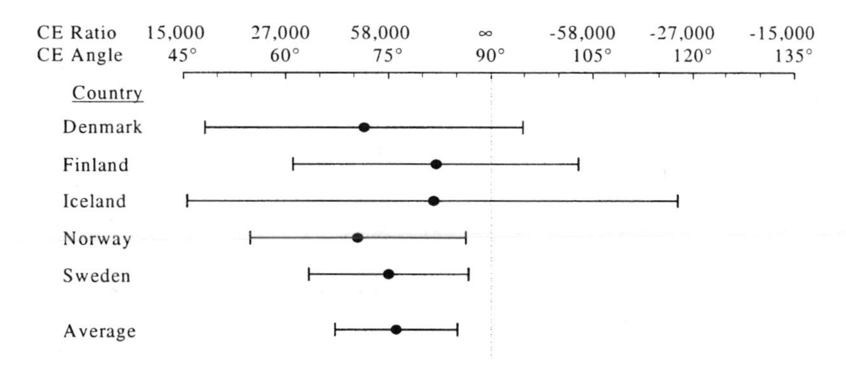

Fig. 2 Ninety-five percent confidence intervals for the incremental cost per additional survivor in the Scandinavian Simvastatin Survival Study, by country and for the overall country average. CE = cost-effectiveness.

evaluations will need to compare programs that favor one on cost and the other on effectiveness. In these cases, the analyst needs to base a conclusion on the magnitude of the incremental cost-effectiveness ratio. One method proposed by Laupacis et al. (38) for Canada was to classify technologies into one of five grades for recommendation on the basis of their incremental cost per QALY gained. The authors did acknowledge, however, that cost-effectiveness should not be the sole basis on which to recommend the acceptance of a new therapy.

Another method often used in published studies compares the cost-effectiveness ratio for the therapy under study to ratios for a broad range of well-accepted medical therapies. For example, Johannesson et al. (32) reported that the cost per life-year gained in patients with coronary heart disease who received simvastatin therapy ranged from $3,800 for 70-year-old men with total cholesterol levels of 309 mg/dL to $27,400 for 35-year-old women with total cholesterol levels of 213 mg/dL. The authors concluded that the radios were well within the range that was considered cost-effective in other studies. Rankings of the cost per life-year or cost per QALY gained for a broad range of therapies, referred to as *league tables*, have been published (19). League tables should be interpreted with caution, because the methods used for the individual evaluations may be very different.

A. Bayesian Statistical Inference

It is widely recognized by economists and statisticians that simple point estimates of the mean costs and benefits are not adequate to describe the results of economic evaluations. Most of the available methods provide confidence interval estimates of the means, or ratios of the means, by incorporating estimates of variability into the evaluation. Other classic methods of statistical inference using formal hypothesis testing may have limitations when applied to studies with inadequate power to detect meaningful differences. Some authors (34) have argued for the use of formal *Bayesian methods* that elicit prior beliefs by using expert opinion.

Some evaluations are beginning to apply *empirical Bayes methods* to cost-effectiveness analysis. Jones (33) recommended reporting the empirical probability density function for the economic variable of interest to enable the user to make a more informed interpretation about the uncertainty in the estimate. Other forms of analysis may simply estimate the probability that the cost-effectiveness ratio falls within each of the four quadrants of the cost-effectiveness plane. One useful graphical display for summarizing the results of a cost-effectiveness analysis is the *cost-effectiveness acceptability curve* proposed by van Hout et al. (63). The curve provides an estimate of the probability (plotted on the Y-axis) that the cost-effectiveness ratio for a candidate therapy is favorable with respect to a target ratio R (plotted on the X-axis). Their method assumes that costs and benefits are bivariate normal and uses point estimates of the variables to integrate over the region of the cost-effectiveness plane that gives ratios as favorable as R. An alternative that does not make any distributional assumptions is a simple cumulative probability plot of a rank ordering of the bootstrap resampled cost-effectiveness ratios from the most to the least favorable. This nonparametric method was applied to the 4S data, and the resulting cost-effectiveness acceptability curve is shown in Fig. 3. For example, Fig. 3 could be used to show that there is greater than a 90% probability that simvastatin therapy is associated with a cost-effectiveness ratio at least as favorable as $100,000 per death avoided. Results of this type are of interest to decision makers, who may establish policy on the basis of the magnitude of estimated cost-effectiveness ratios.

VIII. FUTURE DIRECTIONS

The concern over the costs of health care will continue to grow as more novel health care technologies become available to a growing elderly population. In this environment, health care providers will attempt to identify and apply treatment programs that offer the greatest benefit for the fixed level of investment in health care. Formal economic evaluations will be looked on to help guide these decisions. Pharmaceutical companies will need to enhance their clinical development programs significantly to address the needs of the changing marketplace. More emphasis will be placed on studies of patient outcomes, including quality of life and cost-effectiveness, much earlier in the product development program. Moreover, as guidelines and formal requirements for cost-effectiveness analysis are put in place, there will be a growing need to assemble specific documentation of the socioeconomic impact of the candidate product. This information would also be helpful in marketing new products to sophisticated managed care organizations.

Statisticians working in the pharmaceutical industry will need to address the complexities that the changes will bring to the development program. There will be

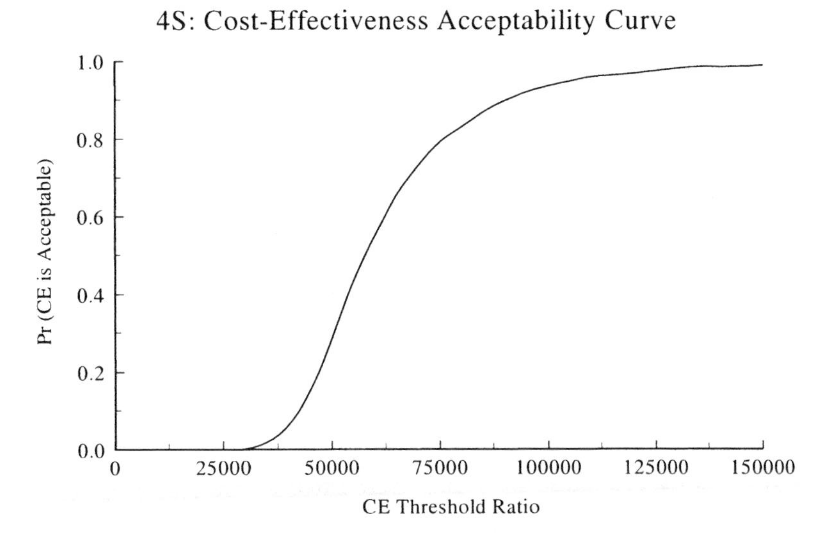

Fig. 3 Cost-effectiveness (CE) acceptability curve in the Scandinavian Simvastatin Survival Study. The graph plots the probability that for Simvastatin, the cost per additional survival would be deemed acceptable for a range of cost-effectiveness ratio threshold.

a shift away from formal hypothesis testing to methods of estimation, including the use of statistical models to help identify the best uses of the candidate therapies. There will also be opportunities to work on data sources outside of the clinical program to provide the most comprehensive evaluation for the product.

REFERENCES

1. Black W. C. The CE plane: A graphic representation of cost-effectiveness. Medical Decision Making 10:212–214, 1990.
2. Briggs A., Schulpher M., Buxton M. Uncertainty in the economic evaluation of health care technologies: the role of sensitivity analysis. Health Econ 3:95–104, 1994.
3. Carides G. W., Heyse J. F. Nonparametric estimation of the parameters of cost distributions in the presence of right-censoring. Proceedings of the Biopharmaceutical Section, American Statistical Association Annual Meetings, 1996, pp 186–191.
4. Carides G. W. Estimation of mean treatment costs in the presence of right-censoring. Ph.D. dissertation, Temple University, May 1998.
5. Chaudhary M. A., Stearns S. C. Estimating confidence intervals for cost effectiveness ratios: an example from a randomized trial. Stat Med 15:1447–1458, 1996.
6. Commonwealth of Australia. Guidelines for the pharmaceutical industry on preparation of submissions to the Pharmaceutical Benefits Advisory Committee: Including economic analyses. Canberra, Australia: Department of Health and Community Services, 1995.
7. Cook J. R., Drummond M., Glick H., Heyse J. F. Analyzing economic data from multinational clinical trials: issues and recommendations. Technical Report, Clinical Biostatistics and Research Data Systems, Merck Research Laboratories, 1998.
8. Cook J. R., Heyse J. F. Use of an angular transformation for ratio estimation in cost-effectiveness analysis. Technical Report. Clinical Biostatistics and Research Data Systems, Merck Research Laboratories, 1998.
9. Cox D. R., Fitzpatrick R., Fletcher A. E., Gore S. M., Spiegelhalter D. J., Jones D. R. Quality of life assessment: can we keep it simple? J Roy Statist Soc A 155: 353–393, 1992.
10. de Lissovoy G., Elixhauser A., Luce B. R., Weschler J., Mowery P., Reblando J., Solomkin J. Cost analysis of imipenem-cilastatin versus clindamycin with tobramycin in the treatment of acute intra-abdominal infection. PharmacoEcon 4:203–214, 1993.
11. Diabetes Control and Complications Trial Research Group. Lifetime benefits and costs of intensive therapy as practiced in the Diabetes Control and Complications Trial. JAMA 276:1409–1415, 1996.
12. Drummond M. R. Cost-of-illness studies, a major headache? PharmacoEcon 2:1–4, 1992.
13. Drummond M. F., Davies L. M. Economic analysis alongside clinical trials: revising the methodological principles. Int J Technol Assessment Health Care 7: 571–573, 1991.
14. Drummond M. F., Heyse J. F., Cook J. R., McGuire A. Selection of endpoints in economic evaluations of coronary heart disease interventions. Med Decision Making 13:184–190, 1993.

15. Drummond M. F., Jefferson T. O. On behalf of the BMJ Economic Evaluation Working Party. Guidelines for authors and peer reviewers of economic submission to the BMJ. Br Med J 313:275–283, 1996.
16. Drummond M. F., O'Brien B. J. Clinical importance, statistical significance and the assessment of economic and quality of life outcomes. Health Econom 2:205–212, 1993.
17. Drummond M. F., O'Brien B. J., Stoddart G. L., Torrance G. W. Methods for Economic Evaluation of Heath Care Program. 2nd ed. New York: Oxford University Press, 1997.
18. Drummond M. F., Rutten F., Brenna A., Pinto C. G., Horisberger B., Jonsson B., Le Pen C., Rovira J., Graf von der Schulenburg M., Sintonen H., Torfs K. Economic evaluation of pharmaceuticals, a European perspective. PharmacoEconom 4:173–186, 1993.
19. Drummond M. F., Torrance G. W., Mason J. Cost-effectiveness league tables: more harm than good? Soc Sci Med 37:33–40, 1993.
20. Dudley R. A., Harrell F. E., Smith L. R., Mark D. B., Califf R. M., Pryor D. B., Glower D., Lipscomb J., Hlatky M. Comparison of analytic models for estimating the effect of clinical factors on the cost of coronary artery bypass graft surgery. J Clin Epidemiol 46:261–271, 1993.
21. Efron B., Gong G. A leisurely look at the bootstrap, jackknife and cross-validation. Am Statistician 37:36–48, 1983.
22. Fenn P., McGuire A., Phillips V., Backhouse M., Jones D. The analysis of censored treatment cost data in economic evaluation. Medical Care 33:851–863, 1995.
23. Freemantle N., Drummond M. F. Should clinical trials with concurrent economic analyses be blinded? JAMA 227I:63–64, 1997.
24. Food and Drug Administration. Comparing treatments: safety, effectiveness and cost-effectiveness. Bethesda, MD: National Institutes of Health, 1995
25. Garber A. M., Phelps C. E. Economic foundations of cost-effectiveness analysis. J Health Econ 16:1–31, 1997.
26. Glick H., Heyse J. F., Thompson D., Epstein R. S., Smith M. E., Oster G. A model for evaluating the cost-effectiveness of cholesterol-lowering treatment. Int J Technol Assessment Health Care 8:719–734, 1992.
27. Gold M. R., Siegel J. E., Russell L. B., Weinstein M. C., eds. Cost-Effectiveness in Health and Medicine. New York: Oxford University Press, 1996.
28. Goldman L., Weinstein M. C., Goldman P. A., Williams L. W. Cost-effectiveness of HMG-CoA reductase inhibition for primary and secondary prevention of coronary heart disease. JAMA 265:1145–1151, 1991.
29. Gould A. L., Rossouw J. E., Santanello N. C., Heyse J. F., Furberg C. D. Cholesterol reduction yields clinical benefit: a new look at old data. Circulation 91:2274–2282, 1995.
30. Hillman A. L. Eisenberg J. M., Pauly M. V., Bloom B. S., Glick H., et al. Avoiding bias in the conduct and reporting of cost-effectiveness research sponsored by pharmaceutical companies. N Engl J Med 324:1362–1365, 1991.
31. Johannesson M., Jonsson B., Karlsson G. Outcome measurement in economic evaluation. Health Econ 5: 279–296, 1996.
32. Johannesson M., Jonsson B., Kjekshus J., Olsson A. G., Pedersen T. R., Wedel H., for the Scandinavian Simvastatin Survival Study Group. Cost-effectiveness of simvastatin treatment to lower cholesterol levels in patients with coronary heart disease. N Engl J Med 336: 332–336, 1997.
33. Jones D. A. The role of probability distributions in economic evaluations. Br J Medical Econ 8:137–146, 1995.
34. Jones D. A. A Bayesian approach to the economic evaluation of health care technologies. In: B. Spilker, ed. Quality of Life and Pharmacoeconomics in Clinical Trials. 2nd ed. Philadelphia: Lippincott-Raven, 1996, pp 1189–1196.
35. Kassirer J. P., Angell M. The Journal's policy on cost-effectiveness analysis (editorial). N Engl J Med 331: 669–670, 1994.
36. Koopmanschap M. A., Rutten F. F. H. Indirect costs in economic studies, confronting the confusion. PharmacoEcon 4:446–454, 1993.
37. Laska E. M., Meisner M., Siegel C. Statistical inference for cost-effectiveness ratios. Health Econ 6:229–242, 1997.
38. Laupacis A., Feeny D., Detsky A. S., et al. How attractive does a new technology have to be to warrant adoption and utilization? Tentative guidelines for using clinical and economic evaluations. Can Med Assoc J 146:473–481, 1992.
39. Lin D. Y., Fewer E. J., Etzioni R., Wax Y. Estimating medical costs from incomplete follow-up data. Biometrics 53:113–128, 1997.
40. Luce B. R., Manning W. G., Siegel J. E., Lipscomb J. Estimating costs in cost-effectiveness analysis. In: Gold M. R., Siegel J. E., Russell L. B., and Weinstein M. C., eds. New York: Oxford University Press, 1996, pp 176–213.
41. Manning W. G., Fryback D. G., Weinstein M. C. Reflecting uncertainty in cost-effectiveness analysis. In: Gold M. R., Siegel J. E., Russell L. B., and Weinstein M. C., eds. Cost-Effectiveness in Health and Medicine. New York: Oxford University Press, 1996, pp 246–275.
42. Mark D. B., Hlatky M. A., Califf R. M., Naylor C. D., Lee K. L., Armstrong P. W., Barbash G., White H., Simoons M. J., Nelson C. L., Clapp-Channing N., Knight D., Harrell F. E., Simes J., Topol E. J. Cost effectiveness of thrombolytic therapy with tissue plasminogen activator as compared with streptokinase for acute myocardial infarction. N Engl J Med 332:1418–1424, 1995.

43. Mauskopf J., Schulman K., Bell L., Glick H. (1996). A strategy for collecting pharmacoeconomic data during phase II/III clinical trials. PharmacoEcon 3:264–277, 1996.
44. Mehrez A., Gafni A. Quality-adjusted life years, utility theory, and healthy-years equivalents. Medical Decision Making 9:142–149, 1989.
45. Miller R. G. The jackknife–a review. Biometrika 61:1–15, 1974.
46. Mosteller F., Tukey J. W. Data Analysis and Regression: A Second Course in Statistics. Reading, MA, Addison-Wesley, 1977.
47. Mullahy J, Manning W. Statistical issues in cost-effectiveness analyses. In: Sloan F., ed. Valuing Healthcare Costs, Benefits, and Effectiveness of Pharmaceuticals and Other Medical Technologies. New York: Cambridge University Press, 1994, pp 149–241.
48. O'Brien B. J., Drummond M. F. Statistical versus quantitative significance in the socioeconomic evaluation of medicines. PharmacoEcon 5:389–398, 1994.
49. O'Brien B. J., Drummond M. F., LaBelle R. J., Willan A. R. In search of power and significance: issues in the design and analysis of stochastic cost-effectiveness studies in health care. Medical Care 32:150–163, 1994.
50. Ontario Ministry of Health. Ontario guidelines for economic analysis of pharmaceutical products. Toronto, Canada: Ministry of Health, 1994.
51. Oster G., Epstein A. M. Cost-effectiveness of antihyperlipidemic therapy in the prevention of coronary heart disease. JAMA 258:2381–2387, 1987.
52. Oster G., Borok G. M., Menzin J., Heyse J. F., Epstein R. S., Quinn V., Benson V., Dudl R. J., Epstein A. A randomized trial to assess effectiveness and cost in clinical practice: rationale and design of the Cholesterol Reduction Intervention Study (CRIS). Controlled Clin Trials 16:3–16, 1995.
53. Pedersen T. R., Kjekshus J., Berg K., Olsson A. G., Wilhelmsen L., Wedel H., Pyorala K., Miettinen T., Haghfelt T., Faergeman O., Thorgeirsson G., Jonsson B., Schwartz J. S., for the Scandinavian Simvastatin Survival Study Group. Cholesterol lowering and the use of healthcare resources, results of the Scandinavian Simvastatin Survival Study. Circulation 93:1796–1802, 1996.
54. Polsky D., Glick H. A., Willke R., Schulman R. Confidence intervals for cost-effectiveness ratios: a comparison of four methods. Health Econ 6:243–252, 1997.
55. Powe N. R., Griffiths R. I. The clinical-economic trial: promise, problems, and challenges. Controlled Clin Trials 16:377–394, 1995.
56. Scandinavian Simvastatin Survival Study Group. Randomized trial of cholesterol lowering in 4444 patients with coronary heart disease: the Scandinavian Simvastatin Survival Study (4S). Lancet 344:1383–1389, 1994.
57. Schulman K. A., Kinosian B., Jacobson T. A., Glick H., Willian M. K., Koffer H., Eisenberg J. M. Reducing high blood cholesterol level with drugs. JAMA 264: 3025–3033, 1990.
58. Schulman K. A., Buxton M., Glick H., Sculpher M., Guzman G., Kong J., Backhouse M., Mauskopf J., Bell L., Eisenberg J. M., and the FIRST Investigators. Results of the economic evaluation of the FIRST study, a multinational prospective economic evaluation. Int J Technol Assessment Health Care 4:698–713, 1996.
59. Schulman K. A., Llana T., Yabroff K. R. Economic assessment within the clinical development program. Medical Care 34:DS89–DS95, 1996.
60. Shepherd J., Cobbe S. M., Ford I., Isles C. G., Lorimer A. R., MacFarlane P. W., McKillop J. H., Packard C. J. Prevention of coronary heart disease with pravastatin in men with hypercholesterolemia. N Engl J Med 20: 1301–1307, 1995.
61. Siegel J. E., Torrance G. W., Russell L. B., Luce B. R., Weinstein M. C., Gold M. R., and the members of the Panel on Cost-Effectiveness in Health and Medicine. Guidelines for Pharmacoeconomic Studies, recommendations from the panel on cost effectiveness in health and medicine. PharmacoEcon 11:159–168, 1997.
62. Torrance G. W., Feeny D. Utilities and quality adjusted life years. Int J Technol Assessment Health Care 5:559–575, 1989.
63. Van Hout B. A., Al M. J., Gordon G. S., Rutten F. F. H. Costs, effects and C/E-ratios alongside a clinical trial. Economic Evaluations 3:309–319, 1994.
64. Wakker P., Klaassen M. P. Confidence intervals for cost/effectiveness ratios. Health Econ 4:373–381, 1995.
65. Weinstein M. C., Stason W. B. Foundations of cost-effectiveness analysis for health and medical practices. N Engl Med 296:716–721, 1977.
66. Willan A. R., O'Brien B. J. Confidence intervals for cost-effectiveness ratios: an application of Fieller's theorem. Health Econ 5:297–305, 1996.

Joseph F. Heyse
John R. Cook
Michael F. Drummond

Placebo Effect

See also *Adjustment for Covariates; Confounding and Interaction*

I. INTRODUCTION AND BACKGROUND

Prior to the advent of modern medicine, many medications prescribed as a part of medical practice were subsequently found to be pharmacologically inert. Despite this fact, such inactive medication, often referred to as a *placebo*—which literally means "I shall please"—appeared to be efficacious for a large number of medical problems. Shapiro (1) has suggested that most medical practice prior to the 17th century was just an exploitation of this "placebo effect." This is reflected in Mothby's *New Medical Dictionary* written in 1785, which defined *placebo* as "a commonplace method or medicine" (2). Toward the end of the 18th century, when the medical community was starting to doubt the effectiveness of some commonly prescribed medications, the term *placebo* was viewed more derisively. For example, the 1811 edition of Harper's *Medical Dictionary* defined *placebo* as "an epithet given to any medium adopted more to please than to benefit the patient" (3).

In the last 50 years or so, much has been learned and written about the construct known as the *placebo effect*. Much of this large body of knowledge about the effects of placebo has been gleaned from the use of placebo as a control in clinical trials. A response to placebo has been reported for diseases involving nearly all organ systems within the body. Satisfactory response to placebo has been reported to be at least 30% or more for a side variety of disorders and often as high as 70% (4). This had led some in the medical community to recommend placebo as a legitimate form of treatment (5). Despite the amount of research on this phenomenon and the extensive reports of responses to placebo, there still is not a good understanding about the mechanism and causes of the placebo effect. In fact, these is still some confusion and a lack of consensus on what constitutes the placebo and a placebo response.

Beecher (6) defined the placebo as a pharmacologically inert substance. This is consistent with the definition given in Stedman's *Medical Dictionary*, 26th edition (7), which appears as (a) "an inert substance given as a medicine for its suggestive effect," (b) "an inert compound identical in appearance to material being tested in experimental research, which may or may not be known to the physician and/or patient, administered to distinguish between drug action and suggestive effect of the material under study." In a pharmacological setting it is common to expand the first definition by equating an inert substance to all factors other than the active pharmacological agent. However, this definition implies that any form of specific nonpharmacological intervention, such as psychotherapy, is a form of placebo.

In order to avoid such erroneous classification, more expanded definitions of placebo attempt to differentiate between specific actions of an agent and its nonspecific effects—the latter being defined as placebos. Shapiro (8) proposed a broad definition of placebo that includes nonpharmacological cases as well. He defines the placebo as any therapy (or component of therapy) that is deliberately or knowingly used for its nonspecific, psychological, or psychophysiological effect, or that is used unknowingly for its presumed or believed effect on a patient, symptom, or illness, but which, unknown to a patient and therapist, is without specific activity for the condition being treated. This definition encompasses the use of placebo as a control in clinical trials as well as a treatment in medical practice. However, classifying a therapy's effect as nonspecific or specific is often difficult in practice, particularly when the therapy is nonpharmacological. This has lead some authors to conclude that one can not define placebo in a logically consistent way (9). Still others have proposed more elaborate and lengthy definitions that have attempted to avoid the alleged contradictions inherent in the specific and nonspecific distinction (10).

As a result of various definitions of placebo, there is also much lack of agreement on what a placebo effect is and what it isn't. Some examples of definitions of the placebo effect are: (a) the improvement that occurs in a patient who is being treated with placebo (11), (b) a significant therapeutic response to "inert" or "placebo" substances (12), (c) an effect that is produced by a medical procedure's therapeutic intent and not its specific nature (13). What some authors consider factors or causes of the placebo effect, others claim to be alternative effects that are not part of the placebo response. Since the focus of this entry is on pharmacological agents in clinical trials, *placebo* will be defined as simply an inert pill or injection masking as an active agent; correspondingly, the *placebo effect* will be defined as a therapeutic response to the treatment with placebo. While items such as spontaneous remission (i.e., natural course of healing) and regression toward the mean would generally not be considered part of the placebo effect, we will consider such phenomena, because they can play a key role in the observed improvement of placebo-treated subjects.

II. REGULATORY REQUIREMENTS

Before a new drug can be approved for use in nearly any country throughout the world, it must be demonstrated that it is safe and effective. For nearly 30 years, the Food and Drug Administration (FDA), through the Federal Register, has mandated that new drugs must be evaluated via randomized, blinded, controlled clinical trials. Similar requirements have been adopted by many other Western countries. To gain approval in the United States, a drug must be shown to be efficacious, usually in at least two adequate and well-controlled trials. While the guidelines do not specifically mandate the use of placebo as a control, placebo usually is the standard for comparison. To demonstrate efficacy in a placebo-controlled trial, one has to show superiority to placebo; i.e., the drug effect can add more to the underlying placebo effect. In many studies in which the placebo group has improved substantially, even drugs that are known to be efficacious do not show statistical superiority over placebo. For example, sertraline, a widely used efficacious drug for major depressive disorder (MDD) was found to be statistically superior to placebo in just two of the five placebo-controlled trials that were conducted in MDD as part of the approval process (14). In the last 17 years, there has been only two anxiolytics approved by the FDA. It is speculated that it is just too difficult to demonstrate superiority to placebo in generalized anxiety disorder, where the placebo response rate is notoriously high. Hence, it is assumed that if one could reduce the placebo response, it would be easier to demonstrate that an effective agent is superior to placebo. To that end, one first needs to consider the mechanisms that are associated with the placebo response.

III. FACTORS INFLUENCING THE PLACEBO RESPONSE

Factors and mechanisms associated with the placebo response can be crudely classified into two types (12): environmental factors and patient factors. The former include the clinical setting, the attitude of the medical care personnel, and the expectation of the principle investigator. The latter refer to patient history, patient personality traits, and the patient's experience and expectations. Environmental and patient factors are related, and their interaction in the form of a doctor–patient relationship can create a very important social psychological phenomenon that is thought to be highly influential on the placebo effect.

Clinical settings and medical practitioners that convey empathy, warmth, interest, and understanding are generally thought to contribute to nonspecific positive therapeutic response, especially in disorders in which the measures of response to treatment are more subjective. If the clinician has an enthusiastic attitude and a positive expectation, this is likely to increase the response to placebo. On the other hand, a practitioner or experimenter that shows aloofness, curtness, and pessimism will create an environment that will probably diminish the placebo effect.

Patient factors include personality, history, and demographic variables. Unlike the environmental factors, these items are generally not controllable. It has been stated by several researchers (15)—with varying degrees of evidence—that patients that are more acquiescent and suggestible are more likely to respond to placebo. The patient's expectations and attitude about the treatment, which are shaped to a large degree by a confluence of his or her personality and history, also are believed to be associated with the degree of placebo response. For example, patients that want to be in control are felt to be more likely to respond to placebo, since the act of taking a pill is an action that will make them feel in charge. Intuitively, patients that have had favorable experiences with medication are thought to be more responsive to placebo. In contrast, patients that are highly skeptical of much of modern medicine or generally shun conventional medical treatment have been found to be placebo nonresponders (16). Conditioning—which can be influenced by experience and expectation—has also been implicated as a cause of the placebo effect. For example, airway resistance (17) and gastric contractibility (18) have been reported to be influenced by placebo. Other physiological variables that may be influenced by classical conditioning, and thus potentially be affected by placebo, include heart rate, blood pressure, EKG patterns, vasomotor activity, and the immune response (19,20). It has been reported that placebo has been found to produce endorphins in pain studies (22,23), although the validity of this finding has been challenged (4,24,25).

Bush (16) has categorized factors that influence the placebo effect into two classes: bound and manipulable. *Bound* factors depend on the history and personality of the patient and clinicians and are viewed as generally unalterable. *Manipulable* factors, which include most of the environmental factors discussed in the previous paragraph, are viewed as alterable to a large degree. Further examples of these factors include the shape, color, and characteristics of the active and placebo agents. Schapira et al. (25) studied the use of oxazepam in either a green, yellow, or red tablet in 48 patients; they reported that green tablets tended to be more effective for anxiety, while yellow pills seemed to more effective for depressive symptoms. Blackwell et al. (26) suggested that multiple pills and pills of larger size produced a larger placebo effect. Thomson (27) insinuated that the "active placebo" (one that has similar side effects to the active agent it is to mimic) atropine amplified the placebo response in tricyclic-antidepressants trials.

IV. STATISTICAL ISSUES

A. Factors Associated with the Placebo Response

Improvement in response to placebo does not imply that placebo or associated nonspecific effects have generated the therapeutic benefit. For example, improvement in some cases is the result of spontaneous remission. One very important factor that probably accounts for a large degree of the improvement in the placebo group in clinical trials is the phenomenon of regression toward the mean, which was first described by Galton (28). *Regression toward the mean* is the tendency for an extreme variable value to move closer to the mean when the variable is remeasured. McDonald and Mazzuca (29) argued that much of the improvement observed in the placebo group in clinical trials is attributed to regression toward the mean. To explain regression toward the mean mathematically, let the random variables Y_1 and Y_2 represent some response variable for the same individual at two separate time points. Further, for the sake of simplicity, assume that the joint distribution of Y_1 and Y_2 is bivariate normal with a common mean μ, common variance σ^2, and correlation coefficient ρ. Then

$$E(Y_2 - Y_1|_{Y_1=y_1}) = (\mu - y_1)(1 - \rho)$$

Note that if $y_1 > \mu$, then $E(Y_2 - Y_1) < 0$; moreover, the expected decrease from the first measure to the second is proportional to $|\mu - y_1|$, i.e., to how extreme the first value is. This expected decrease (when $y_1 > \mu$) is also proportional to $1 - \rho$; hence one would expect a larger decrease as the correlation becomes less positive (repeated measures on the same individual are almost always positively correlated). As a result, outcome variables that are less reliable—or reproducible—will be more prone to the effects of regression toward the mean.

Since detailed information concerning μ, σ^2, and ρ for many laboratory variables is available in the medical literature, McDonald and Mazzuca (29) calculated the expected change in a repeat measure when the first measure was greater than $\mu + 3\sigma$. For 15 laboratory variables, they found the expected decrease ranged from 2.5% to 26%, which agreed well with empirical data from 12,000 patients obtained via hospital medical records.

Since entrance criteria for clinical trials are usually based on the subject's presenting either a high or low value at baseline, regression toward the mean is likely to explain a large degree of the improvement observed in clinical trials, particularly when the response variable is fairly subjective and not highly reliable.

When the outcome variable is rather subjective, *observer bias* is another factor that could potentially contribute to the improvement attributed to placebo. This factor has been largely ignored in the plethora of literature about the placebo response. Indications of observer bias appear in the data from clinical trials that incorporate a single-blind placebo run-in phase (this design feature will be discussed further in the next section) in order to screen for "placebo responders" prior to randomization into the double-blind phase of the study (30). It has been the authors' experience, based on many placebo-controlled trials using placebo run-ins, that the degree of improvement on the average (or the percentage of "responders") during this initial phase is generally far less than that experienced by the patients randomized to placebo during a comparable period of time within the double-blind portion of the study. While this observation could in part be due to regression toward the mean, there may be an upward bias in the measurements (i.e., more abnormal) at the end of the single-blind placebo phase of the trial for two reasons: first, the observer or investigator knows the subjects are receiving placebo; second, there may be incentive for the investigative site to reach their recruitment goals. On the other hand, during the double-blind portion, there may be a bias toward more improved measures or ratings, since the investigator may speculate that the subject is on an effective medication and hence is expecting an improvement.

B. Design Considerations

One design feature frequently incorporated into clinical trials (e.g., hypertension and psychopharmacology trials), as already mentioned, is the use of a single-blind placebo run-in. One of the main purposes of this run-in phase is to screen out subjects that show a substantial improvement with placebo. Despite the lack of critical review of such a design, a placebo run-in is almost always used in depression studies. The few investigations that have attempted to assess this design have generally not found the placebo run-in to be effective at reducing the placebo response rate (30). Trivedi and Rush (31) conducted a meta-analysis on 101 blinded placebo-controlled depression studies that were conducted from 1959 to 1990. After cross-classifying the trials by the type of active drug (i.e., trycyclics, monoamine oxidase inhibitors, and serotonin reuptake inhibitors) and whether they were outpatient or inpatient, they found few statistically significant differences in the placebo response rates between trials that had a run-in and those that did not.

Regression toward the mean is one reason that the placebo run-in appears not to be effective at reducing the response to placebo. Subjects are still entered into the study on the basis of having an extreme baseline value for the response of interest. In the absence of other reasons for extreme baseline values—such as observational bias—one approach to minimize the effects of regression toward the mean would be to take multiple measurements during a screening period. These measures could then be averaged to obtain a baseline value. Another approach would be to take two "baseline" measurements (that would perhaps be at least within a week of each other): the first would be used for determining eligibility (i.e., the subjects would have to meet some criterion for disease severity) and the other used as the baseline measurement for the purpose of calculating change.

As previously implied, observational bias may be another reason that single-blind placebo run-ins may not be useful at reducing the placebo effect. In studies in which the primary outcome variable is subjective, one possible design that would address observational bias would be one that had a random-length double-blind placebo-run. One could perceive of many variations of such a design. One possibility for comparing one active agent to placebo is as follows: After a 1-week single-blind placebo period, subjects still meeting the eligibility criterion (i.e., an extreme value for the outcome of interest) would be randomized in double-blind fashion to placebo (95% of the subjects) or active drug (5% of the subjects). After a 1-week period (and a clinical assessment), half of the placebo subjects would be randomized to placebo while the other half continues to receive placebo. The primary analysis for efficacy would include only those subjects randomized at week 2 that still continued to meet the entrance criterion. Week 2 would be considered baseline. One could also conduct a secondary analysis that would include all randomized subjects (week 1 would be considered baseline for the 5% of the subjects randomized to active drug at week 1; week 1 might also be considered to be the baseline for all other subjects in such a secondary analysis). This design, as well as variations of a random-length placebo period, would probably be even more effective at reducing observational bias if the precise nature of the run-in and the data analysis were not completely spelled out in the protocol. Although this may cause some ethical and logistic concerns, it might be sufficient just to state in the protocol that a randomization to the two treatments will occur over several weeks. The details of the planned analysis could be contained in another document, separate from the procedures of the protocol.

There is yet another way to reduce observational bias in outcomes that are measured on some type of calibrated instrument (e.g., sphygmomanometer, spectrophotometer, scales for weight): recalibrate in such a way that one randomly sets the "zero mark." For instance, a diastolic blood pressure of 80 mm Hg would appear as 100 on an sphygmomanometer in which zero was set at 20. Such tactics would be most useful for reducing observational bias in single-blind studies.

V. ATTEMPTS AT PREDICTING PLACEBO RESPONSE IN DEPRESSION

In much of clinical research, particularly in psychopharmacology, there is a perception that the placebo response is becoming more pronounced (32), which has led many researchers to believe that it is becoming increasingly more difficult to demonstrate the efficacy of new medications. By the nature of clinical trials, usually it is implicitly assumed that the placebo effect and the therapeutic effect of an active medication are additive, but this may not be the case in many instances, particularly when the placebo effect is large and there is a ceiling effect on the degree of improvement. While there has been a large amount published about qualitatively describing "placebo responders," only in the last decade or so has there been much effort at attempting to predict placebo response on an individual basis based on screening information (33–39). If one could predict placebo response based on demographic or medical history information, then one could exclude likely "placebo responders" prior to entry into the study. Most of these efforts at finding predictors of placebo response in psychiatric disorders have entailed retrospective, exploratory analyses of small data sets comprised of one to a few studies. As a result, there has been a smattering of reports of "statistically significant" differences between "placebo responders" and nonresponders from one investigation to another but limited agreement among the reports. Following are some of the variables that have been cited from two or more published studies as being related to placebo response in depression. The likelihood of being a placebo responder increases as (a) the duration of the current episode decreases, (b) the degree of precipitating stress increases, and (c) the number of previous treatments for depression decreases.

In unpublished research conducted by these authors and colleagues, an artificial neural network (ANN) was trained (using a data set containing patients from four depression studies) to predict which subjects—based on some of the variables listed previously as well as others—would respond to placebo (experience a 50% or more decrease in the Hamilton Depression Rating Scale). While the ANN seemed to have some success for predicting placebo response, it did not appear to enhance the difference between the active drug and placebo. In fact, removing the predicted "placebo responders" from the analysis of a validation data set decreased the average improvement of the active drug more than that of the placebo group. This effect has also been reported by Reimherr et al. (30). In depression, this suggests that "placebo responders" also show a large response to active drug and that removal of such subjects leaves a population of subjects that are in general more treatment resistant. For a binary outcome, it has been speculated by Day (40) that in some instances there may be an optimal level of placebo response for the purpose of detecting a difference between placebo and active drug. For example, the placebo response and the therapeutic effect due to active drug might be additive on a logistic scale. If the true odds ratio between active and placebo is 3.86, then increasing the placebo response from 0.1 to 0.2 results in an increase in the active drug response from 0.3 to 0.49. It is also possible in certain cases that the "placebo response" and the "therapeutic response" might be more than just additive on measurement scales that are commonly used for continuous outcomes.

REFERENCES

1. AK Shapiro. Behav Sci 5:398–430, 1960.
2. JG Howells, ed. The Placebo Response in Modern Perspectives in World Psychiatry. Vol 2. Edinburg, Scotland: Oliver and Boyd, 1968.
3. RL Stanford, TH Schmitz. In: Stat Pharmaceutical Industry. CR Buncher, JY Tsay, eds. New York: Marcel Dekker, 1981, pp 231–250.
4. GS Kienle, H Keine. Alternative Therapies 2:39–54, 1996.
5. WA Brown. Neuropsychopharmacology 10:265–288, 1994.
6. HK Beecher. JAMA 159:1602–1606, 1955.
7. M Spaycar, ed. Stedman's Medical Dictionary. 26th ed. Baltimore: Williams and Wilkins, 1995.
8. AK Shapiro. J Clin Pharm, March–April:73–78, 1970.
9. PC Gøtzsche. Lancet 344:925–926, 1994.
10. A Grünbaum. Psychol Med 16:19–38, 1986.
11. E Schweizer, K Rickels. J Clin Psych 58(suppl 11):30–38, 1997.
12. MA Piercy, JJ Sramek, NM Kurtz, NR Cutler. Ann Pharmacotherapy 30:1013–1019, 1996.
13. R Liberman. J Chronic Disease 15:761, 1962.
14. Summary Basis of Approval—Zoloft Tablets. FDA, Dec 1991.
15. TJM Cleophas. JAMA 273:283, 1995.
16. PJ Bush. J Am Pharmaceutical Asso 14:671–674, 1974.
17. T Luparello, N Leist, CH Lourie. Psychosom Med 32: 509–513, 1970.
18. RA Sternbach. Psychophysiology 1:67–72, 1964.
19. D McDonald, J Stern, W Hahn. J Psychosom Res 7: 97–106, 1963.
20. JL Straus, SVA Cavanaugh. Psychosomatics 37:315–326, 1996.
21. M Heylin. Chem Eng News 56:24, 1978.
22. JD Levine, NC Gordon, HL Fields. Lancet 2:54–657, 1978.
23. A Goldstein, P Grevert. Lancet, Dec 23, 30:1385, 1978.
24. PD Wall. Ciba Found Symp 174:817–211, 1993.
25. K Shapira, HA McClelland, NR Griffiths. BMJ 2:446–449, 1970.
26. B Blackwell, SS Bloomfield, CR Buncher. Lancet 1: 1279–1282, 1972.
27. R Thomson. Br J Psychiatry 140:64–68, 1982.
28. F Galton. J Anthropological Inst Great Britain Ireland 15:246–263, 1885–1886.
29. CJ McDonald, SA Mazzuca. Stat Medicine 2:417–427, 1983.
30. FW Reimherr, MF Ward, WF Byerley. Psych Res 30: 191–199, 1989.
31. MH Trivedi, J Rush. Neuropsychopharmacology 11: 33–43, 1994.
32. EH Uhlenhuth, W Matuzas, TD Warner, PM Thompson. Psychopharmaceutical Bull 33:31–39, 1997.
33. WA Brown, BE Dornseif, JF Wernicke. Psych Res 26: 259–264, 1988.
34. WA Brown. Psychopharmaceutical Bull 24:14–17, 1988.
35. KE Woodin. Positive placebo response in clinical trials

of depressed outpatients. PhD dissertation, Univ of Massachusetts, Amherst, 1990.
36. A Khan, WA Brown. Psychopharmaceutical Bull 27: 271–274, 1991.
37. WA Brown, MF Johnson, M Chen. Psych Res 41:203–214, 1992.
38. MK Shear, AC Leon, MH Pollack, JF Rosenbaum, MB Keller. Psychopharmaceutical Bull 31:273–278, 1995.
39. T Taiminen, E Syvälahti, S Saarijärvi, H Niemi, H Lehto, V Ahola, R Salokangas. J Nerv Ment Dis 184: 109–113, 1996.
40. SJ Day. Stat Med 7:1187–1194, 1988.

Thomas R. Stiger
Edward F. C. Pun

Postmarketing Surveillance

See also *Pharmacoeconomics*

I. INTRODUCTION

Drugs are not approved for general use until they go through an elaborate testing period. The "gold standard" of randomized, controlled trials must be utilized for drugs to be granted approval. However, once a drug is released on the market, it can be used for a variety of illnesses that were not in the original randomized testing process. It can be prescribed to a general population that was previously excluded from the study subjects. Therefore, there are several questions that must be answered through postmarketing surveillance.

The first such problem is to determine the effects of the drug on populations previously excluded from the randomized trials. Two populations that are virtually routine to exclude are children and pregnant women. Studies generally recruit few in minority populations (1). This can usually be extended to the exclusion of women who might become pregnant, that is, to exclude women of child-bearing age. However, once a drug is approved for use, exclusion criteria no longer exist. A drug can be prescribed for anyone for any problem. A physician may have to assume liability when prescribing a drug for a nonapproved use, unless sufficient support exists in the medical literature for treatment. There is no liability for extending use to a segment of the population not previously studied.

Clinical trials by their nature must have an ending point. Usually, even phase III clinical trials last a few weeks, possible a few months, rarely several years. However, particularly if the medication is prescribed for chronic problems, patients can take the medication for 30, 40, or 50 years. There might be possible long-term effects that are not seen in the short-term trials. Also, adverse reactions are rare; they will not always become apparent in clinical trials. The power analysis that determines the sample size for the clinical trial is defined in terms of the main effect; a much larger sample would be required to examine rare occurrences. This must be carefully monitored in postmarketing surveillance.

Another problem to be examined is that of cost-effectiveness. A new drug does not ordinarily treat a new illness; it usually has a claim to be more effective or safer than drugs currently in use. However, sometimes the effects are relatively similar. In this case, it is necessary to examine the cost of providing the benefit to the patient. This becomes particularly crucial as health management plans are devising formularies and limiting the drugs that can be used for treatment.

Each of these problems must be addressed through methodology and through data collection concerning drugs as they are used on the general population in postmarketing surveillance. For the researcher, statistical methods and data-mining techniques that can be used on observational data must be used. For the practicing clinician, there must be a way to record data from patients. For both, there is need to understand information in the medical literature that reports on postmarketing surveillance results.

II. DATA COLLECTION

The quality of the data generally depends upon the examination of the four parameters of hypothesis testing: significance level, power, effect size, and sample size. There are a number of ways that data can be collected

Table 1 Quality of Evidence Collected in Postmarketing Surveillance

Level of evidence	Description	Treatment recommendation
I	Large, phase IV randomized, controlled trials	A. Should prescribe treatment to patients
II	Smaller randomized, controlled trials with low power so that the results are not necessarily conclusive	B. Should prescribe treatment to patients
III	Nonrandomized trial; contemporaneous controls	C. In the absence of severe adverse effects, should prescribe treatment to patients
IV	Nonrandomized trial; historical controls	C
V	Uncontrolled studies, case series	C

Source: Ref. 2.

in postmarketing surveillance. Table 1 provides a chart of the quality of evidence (2). When there is any evidence to indicate that a patient might benefit without undue risk, the treatment should be tried, particularly if that treatment is not only very low risk but also of low cost.

It is important to examine any study for methodological flaws or for insufficient sample size, particularly if the results are negative. The power of a statistical test is the probability that it will lead to the rejection of the null hypothesis. It is desirable, although impossible, to have a small level of significance and a large power simultaneously without specifying a level of indifference (where two drug treatments are so close in effect that the investigator is indifferent as to outcome). Then the sample size is chosen. However, when small numbers of patients are used, the power will be low.

Observational studies (which include all studies without randomized controls) are particularly subject to confounding. This occurs when two variables examined are directly related to a third variable unexamined. The possible existence of a confounding factor can completely void the conclusions of the study. Therefore great care must be taken to collect data on any known or suspected confounders.

III. POSTMARKETING QUESTIONS

Traditional methods of statistical analysis have focused on a very small number of variables (3) utilizing a technique such as Bayesian analysis or logistic regression. More recently, data-mining techniques such as neural networks, fuzzy logic, and decision trees have been used:

> Traditional model building is based on the use of rigorous statistical techniques such as multiple linear regressions. In contrast, neural networks are "adaptive computing," in that they learn from data to build a model to explain an application or biological system. (There are no a priori assumptions of a linear relationship among variables.) A neural network is designed from the application, not the other way around. Applications may be simple or complex and linear or nonlinear such as in finance or other noisy systems (i.e., highly variable) where the relationship between variables is not in a straight line (4).

And stated in Ref. 5:

> The experienced statistician, perhaps the most capable of guiding the development of automated tools for data analysis, may also be the most acutely aware of all the difficulties that can arise when dealing with real data. This hesitation has bred skepticism of what automated procedures can offer and has contributed to the strong focus by the statistical community on model estimation to the neglect of the logical predecessor to this step, namely model identification.
>
> The statistician's tendency to avoid complete automation out of respect for the challenges of the data, and the historical emphasis on models

with interpretable structure, has led that community to focus on problems with a more manageable number of variables (a dozen, say) and cases (several hundred typically) than may be encountered in statistical problems, which may be orders of magnitude larger at the outset. With increasingly huge and amorphous databases, it is clear that methods for automatically hunting down possible patterns worthy of fuller, interactive attention are required. The existence of such tools can free one up to, for instance, posit a wider range of candidate data features and basis functions (building blocks) than one would wish to deal with, if one were specifying a model structure "by hand."

It is becoming increasing clear that the more traditional techniques are not adequate in postmarketing surveillance. The algorithms developed through data mining examine databases with a set of built-in processes that automatically test hypotheses to generate interesting and unexpected rules and graphs that characterize the database. The automatic hypotheses-formation and -testing cycle continues until important rules and patterns emerge. It is up to the investigator to verify the significance of each rule and pattern.

Once the model has been identified, the verification process needs to begin. An outline of the process is given in Table 2 (6).

A. Drug Interactions

There are a number of statistical models that can be used to determine risk factors associated with medication use in the general population. The primary difficulty is to determine the approximate size of the population taking the medication. Unlike other developed

Table 2 Process for Verifying an Identified Model

Procedure	Rationale
Developing an understanding of the application domain, the relevant prior knowledge, and the goals of the end user	What are the goals? What performance criteria are important? Will the final produce of the process be classification, visualization, summarization, prediction? Is understandability an issue? What is the trade-off between simplicity and accuracy?
Creating a target data set, selecting a data set, or focusing on a subset of variables or data samples on which discovery is to be performed.	This involves consideration of homogeneity of the data, change over time, sampling strategy, etc.
Data cleaning and preprocessing	This involves the removal of noise or outliers, deciding on strategies for handling missing data fields, etc.
Data reduction and transformation	This involves finding useful features to represent the data, depending on the goal of the task; using dimensionality reduction or transformation methods to reduce the effective number of variables under consideration or to find invariant representations for the data; and projecting the data onto spaces in which a solution is likely to be easier to find
Choosing the data-mining task	Deciding whether the goal is classification, regression, clustering, summarization, modeling, etc.
Data mining	The actual analysis, automated and most often done with software; includes neural network analysis
Evaluating output	Determining what is actually knowledge and what is "fool's gold." Knowledge must be filtered from other outputs. This can be done by using statistical checks, visualization, or expertise.
Incorporating knowledge into the performance system	Checking and resolving potential conflicts with previously extracted knowledge

Source: Ref. 6.

nations (Price, et al., 1996), the United States does not have a national prescription registry to make it possible to track such information.

Once a population estimate has been made, the following variables need to be collected (7):

D = dosage and schedule taken over the time interval [0,T]

A = set of patient attributes

E = adverse event type

Then it is possible to estimate the conditional probability of E given the values of D, A, and T. There are a number of methods suggested for making an estimate. One that takes into consideration the fact that reported data are often dependent in various ways is that given in Cerrito and Cerrito (8). Define the vector (X,Y,T), where

X = drug benefit

Y = drug risk (Table 3)

T = duration of medication use

The data collected are assumed to be from a weakly dependent set. In this way, the patient attributes needed in the model of Ref. 7 can be omitted. Then the probability density function of the vector (X,Y,T) can be estimated by

$$\hat{f}(x,y,t) = \frac{1}{na_n^3} \sum_{j=1}^{n} k\left(\frac{x - X_j}{a_n}, \frac{y - Y_j}{a_n}, \frac{t - T_j}{a_n}\right)$$

where

$k(\cdot\cdot\cdot)$ = known multivariate function, called the kernel function;

$(X_1,Y_1,T_1), \ldots, (X_n,Y_n,T_n)$ = collected observations

a_n = constant smoothing parameter depending on n

Suppose it is only possible to define risks in terms of discrete categories $\{1, 2, \ldots, m\}$. Suppose category 1 represents the most serious risk and category m represents the least serious so that the risks can be ordered. The standard scale is as shown in Table 3 (9). Then a discrete estimator developed by Ahmad and Cerrito (10) can estimate the discrete risks at each interaction level:

$$\hat{f}(x,y,t) = \frac{1}{na_n^2} \sum_{j=-\infty}^{\infty} W(\hat{s}_n,i,j) \sum_{1 \in C_n(l)} k\left(\frac{x - X_l}{a_n}, \frac{t - T_l}{a_n}\right)$$

where $C_n(j)$ = {indices l such that (X_l,Y_l,T_l) is such that Y_l is in category j). The function $W(s,I,j)$ is a discrete window weight function defined on S^xJ^xJ, where S is an interval on the real line and $J = \{\ldots, -1, 0, 1, \ldots\}$ satisfying (11)

$$W(0,i,j) = \begin{cases} 0 & i \neq j \\ 1 & i = j \end{cases}$$

It has been shown (12) that the choice of the kernel function will not significantly alter the outcome. The choice of the smoothing parameter can have significant impact. Techniques have been developed to optimize the choice of this parameter. However, since the outcomes are discrete, various smoothing parameters can be attempted and investigated using ROC curves (for exactly two outcomes) (13) or by using the standard training/testing/validation process routinely used in neural network analysis (14). The misclassification rate (m1) of the training set is compared to the misclassification rate of the testing set (m2). If m1 < m2 then the value of a_n is adjusted upward. The goal is to have m1 and m2 within a small interval of each other.

The commercially available NeuralWare II from AspenWorks, Inc., does perform kernel density estimation. In addition, PROC DISCRIM in SAS from SAS Institute, Inc., will perform kernel density estimation. For an adaptation of the discrete-continuous parameter, the symbolic computational software Mathematica from Wolfram, Inc., can easily be used to develop the necessary programming code to perform the required algorithm. Suggestions of such Mathematica code are given in the appendix to this entry.

B. Excluded Populations

Although some populations (pregnant women and children) are routinely excluded from clinical trials because of risk and liability issues, still other groups are indirectly excluded because of small numbers of volunteers. Yet it is clear that race and ethnicity do have an impact on the pharmacokinetics of drugs (1). Bioavailability (F) has three determinants, $F = f_a f_g f_h$, where

f_a = fraction of drug absorbed

f_g = fraction escaping gut metabolism

f_h = fraction escaping hepatic first-pass metabolism

Studies that have compared drug absorption for different races have demonstrated that there is no difference when drug absorption is passive. However, when the drug undergoes active absorption, there are statistically significant differences. Studies comparing racial groups have had different outcomes for different drugs, making generalizations impossible (15). Differences in absorption can result in differences in effect for different populations. In this case, the model defined in Sec. 3.A need to be expanded to include a vector Z of patient characteristics:

$$\hat{f}(x,y,z,t) = \frac{1}{na_n^3} \sum_{j=-\infty}^{\infty} W(\hat{s}_n,i,j) \sum_{l \in C_n(j)} k\left(\frac{x - X_l}{a_n}, \frac{z - Z_l}{a_n}, \frac{t - T_l}{a_n}\right)$$

Another population generally excluded from clinical trials is the elderly. For this population, males average 13.26 different prescriptions; females average 16.25 (16). Interaction studies typically examine 2–3 drugs in combination. It is necessary to develop tools that will investigate medications prescribed in such large numbers.

C. Cost-Effectiveness Analysis

Cost-effectiveness is a broad term that encompasses three major types of models (17):

- *Cost–benefit studies.* Both costs and outcomes are measured in dollars.
- *Cost-effectiveness studies.* Outcomes are quantified in terms of health status measures, such as the number of symptom-free days.
- *Cost–utility studies.* Effectiveness is measured in terms of utility, such as quality-adjusted life-years.

There are also different perspectives taken with regard to cost analysis. The broadest perspective is societal, where the interest is to define the greatest good for the greatest number of people (while possibly increasing risks for some individuals). An intermediate perspective belongs to an insurer or a health maintenance organization. The most narrow perspective belongs to an individual patient or an individual provider. These perspectives can sometimes be at odds.

The different perspectives will clash, for the broader viewpoints will sometimes be at the expense of individuals. Health maintenance organizations tend to deny any treatment without Level I (or possible Level II) evidence. Oregon attempted to prioritize its Medicaid funding based on optimizing results from a societal perspective. The program failed because the societal perspective often differed considerably from the individual perspective (18).

In each case, an attempt is made to quantify health outcomes, usually assuming a linear relationship. In addition, most cost-effectiveness analysis depends upon rough estimates of probabilities, usually provided by the "Delphi interview method." This involves asking a group of experts for their perceptions. Real data are rarely used to develop the model. Instead, a sensitivity analysis is performed. The estimates given by the experts are altered to determine just how much deviation will still give the same outcomes (a measure of robustness). However, as in the case of peptic ulcers, if the experts were unaware of the actual cause of an illness, the sensitivity analysis would not provide an accurate response. Therefore, cost-effectiveness analysis will always support conventional belief.

The model for examining cost takes a form similar to the one described in Sec. 3.A. There is a vector (E,C,Z), where E = effects (both benefits and risks), C = costs, and Z = patient characteristics. Then if a linear relationship is assumed, it is defined by the equation

$$e_{ij} = G_j(c_{ij}) + \varepsilon_{ij}$$

However, costs from a societal perspective are assumed to stay within 2 standard deviations of the mean. Unfortunately, health care has much to do with outliers, i.e., individuals at the extreme end of need, who use

Table 3 Standard Scale for Risk

Significance rating	Severity	Documentation
1	Major: The effects are potentially life-threatening or capable of causing permanent damage	Suspected or established
2	Moderate: The effects may cause a deterioration in a patient's clinical status; additional treatment may be necessary	Suspected or established
3	Minor: The effects are usually mild; consequences may be bothersome or unnoticeable	Suspected or established
4	Major/moderate	Possible
5	Minor	Possible
	Any	Unlikely

Source: Ref. 9.

health care resources way beyond this preset norm. Therefore it is preferable to use a nonlinear model with

$$\hat{f}(e,c,z) = \frac{1}{na_n^2} \sum_{j=-\infty}^{\infty} W(\hat{s}_n,i,j) \sum_{l \in D_n(l)} k\left(\frac{c - C_l}{a_n}, \frac{z - Z_l}{z_n}\right)$$

assuming that E takes discrete values.

IV. EXAMPLES FROM THE MEDICAL LITERATURE

A. Surveillance of *Helicobacter pylori* Treatments

The problem of peptic ulcers is one of extreme interest. For years the standard treatment was to use an acid inhibitor such as omeprazole or ranitidine. It was assumed that the underlying cause of ulcers was stress. Therefore, it was also assumed that a maintenance dose of medication was required to prevent recurrence or that stress-reduction therapy would be needed to prevent additional problems. It was also assumed that bacteria could not survive in the stomach, so until recently no studies were conducted to investigate the possibility of infection (19). The discovery of *Helicobacter pylori* in the stomach completely changed the entire course of investigation. Therefore, it is possible to examine all aspects of postmarketing surveillance in the investigation of the treatment and outcome of peptic ulcer disease, particularly since all medications used to treat the problem were fairly common and already approved for use for other illnesses.

1. *Drug Interactions*

Once it was discovered that *Helicobacter pylori* could survive in the acidic environment of the stomach, studies were conducted to determine a treatment that could successfully eradicate it. All studies involved postmarketing surveillance, for all drugs tested were already approved for use. Decisions were made early on to use drugs in combination: various antibiotics and acid suppressors. Penston (20) focused on the combination of omeprazole, amoxicillin, and metronidazole. Others (21–24) examined other combinations. No study has shown that these combinations put patients at risk for serious adverse reactions. Table 4 (2) summarizes the level of evidence for eradicating *H. pylori* for a variety of gastrointestinal complaints. Treatment is recommended for levels III–V, since it has low risk and low cost and there is little indication that another treatment is more effective (Table 4).

Another type of interaction to examine occurs when different problems occur simultaneously. Stress ulcers still occur, particularly in an intensive care unit. Rune (25) explored the possibility that *H. pylori* contributed to the incidence of stress ulcers. Examining 874 patients (sufficient numbers for 90% power), the presence of *H. pylori* was found to be statistically significant.

2. *Long-Term Follow-Up*

Since acid suppressors were developed as the standard treatment for peptic ulcers, a study (26) was conducted to determine the long-term treatment results for peptic ulcers. The results demonstrated clearly that the traditional measures of acid suppression and surgery were somewhat temporary; the majority of patients still had gastric problems. This study was not completed until years after the discovery of *H. pylori*. Yet none of the subjects followed was given a test for the presence of the bacteria; none were treated for the infection. It is very speculative that this occurred, because physicians are still generally reluctant to prescribe triple therapy (27) or because of a reluctance to stop the study early. However, if this study had been initiated at an earlier time, there should not have been so much complacency concerning the effectiveness of acid suppressors in treatment. Patients with recurring problems should have been so casually dismissed as having stress disorders. No matter how much confidence exists in a par-

Table 4 Level of Evidence for Eradicating *H. pylori*

Complaint	Level of evidence	Recommendation
Peptic ulcer	I	Treat
Bleeding duodenal ulcer	I	Treat
Nonulcer dyspepsia	II	Treat
Early gastric cancer	III	Treat
Gastric mucosa–associated lymphoid tissue	V	Treat
Gastroesophageal reflux disease	Below level V	Do not treat

Source: Ref. 2.

ticular treatment, there will always be unknowns that can drastically change the perspectives.

3. *Excluded Populations*

Caraco et al. (15) conducted a study to compare absorption rates of omeprazole between Chinese and Caucasian subjects. The study concluded that the Chinese had greater bioavailability of the drug. It is suggested that Chinese patients should routinely receive a lower dose of the drug to compensate for the greater bioavailability. It is not suggested what that dose should be. It also remains unknown just how this absorption rate of omeprazole affects the triple therapy routinely used to eradicate *H. pylori*.

Another study (28) was conducted to investigate the problem of ulcers in the elderly population. A sample of 121 patients ranging in age from 61–89 years was treated with omeprazole and with one or two antibiotics for 3 or 7 days. There was a total of six different combinations. This study has sufficient power (80%) to detect a 15% difference in outcome with a 5% level of significance. The study also resulted in 5% of the subjects with adverse reactions to the medications. The optimal treatment was omeprazole, clarithromycin, and metronidazole for 7 days.

4. *Cost-Effectiveness Analysis*

Since acid suppressors were developed to treat ulcers, a cost-effectiveness analysis was conducted to compare this treatment with the triple therapy used to eradicate *H. pylori* (29). This study obviously had to be done from a societal perspective, since the individual perspective would be a preference for complete eradication of the cause of the ulcers rather than to use lifelong maintenance therapy to relieve symptoms. The study did conclude that the societal perspective was identical to the individual perspective. Therefore, from all perspectives, eradication is cost-effective. Yet it is still little prescribed (27).

Greenberg et al. (30) examined the need for routine endoscopy for patients with gastrointestinal problems. It is suggested that a trial of omeprazole should be made prior to routine endoscopy to reduce costs and to increase the effectiveness of endoscopy. For patients for whom omeprazole relieves symptoms but does not heal, a seriology test for the presence of *H. pylori* should be done in place of endoscopy. O'Malley et al. (31) investigated cost-effectiveness in a different way. The claim is made that many patients presenting with gastrointestinal complaints actually have psychiatric disorders. The claim is then made that screening for psychiatric disorders prior to upper endoscopy is cost-effective. However, of the patients diagnosed with a psychiatric disorder, 26% also had a medical problem. Of the patients with a disorder, 44% were diagnosed with somatoform that should not be diagnosed without first screening for a medical disorder. Out of the 116 patients in the study, none were tested for the presence of *Helicobacter pylori*.

B. Surveillance of Fluoxetine

There are a number of papers in the medical literature discussing postmarketing surveillance of this drug. Some (32–34) discuss anecdotal case studies involving the use of fluoxetine. Without statistical evidence, it is not known whether the adverse effects reported in these editorials are actually caused by fluoxetine or are just coincidental. Even if real, there is no way of knowing the actual risk, nor is there any way of knowing whether the risk is for a particular segment of the population. The most such letters can do is to make a practicing physician aware of the possibility of an adverse reaction.

Many of the postmarketing studies have examined segments of the population that were excluded from trials prior to FDA approval (35–36). Chambers et al. (36) provide level III evidence. Consider the method of recruiting subjects:

> From 1989 through 1995, the California Teratogen Information Service and Clinical Research Program received approximately 1500 calls requesting information on the potential teratogenic effects of fluoxetine. An estimated one-third of these inquiries were made by pregnant women currently taking the drug. We selected 288 of these women for inclusion in the study on the basis of accessibility by telephone and willingness to participate. During this same period, pregnant women who called the program with questions about drugs and procedures not considered teratogenic—including acetaminophen use, dental radiography, and limited alcohol ingestion (<1 oz [30 ml] of 100 percent alcohol per week before pregnancy was recognized)—were asked to enroll in the study as a control group. From this group, 254 women were selected as controls because their inquiries were closest in time to those of the women taking fluoxetine. The majority of women in each group were enrolled in the study during the first trimester of their pregnancies, and all were enrolled before any outcomes of the

pregnancy were known, including knowledge of conditions that were diagnosed prenatally.

Each woman who enrolled in the study completed a questionnaire that included her history of previous pregnancies and family medical history, socioeconomic and demographic information for each woman and her partner, and exposures during the current pregnancy. The exposure history included dosages, dates, and indications for all medications; use of caffeine; use of supplemental vitamins; occupational exposures; infectious or chronic diseases; prenatal testing or other medical procedures; and use of recreational drugs, tobacco, and alcohol. Each woman was provided with a diary in which she was asked to keep a record of any additional exposures that might occur before delivery. We supplemented this record by calling the women throughout their pregnancies.

Only women who called for information were asked to participate. It was not determined if there were differences in the two groups of women, specifically in any socioeconomic variables; the control group was matched by length of gestation only. A power analysis is needed to determine if there is a sufficient sample size to detect differences in rare events. A severe adverse effect can be expected to occur in the population only 1–2% of the time. To detect even a 4% difference (from 10 per 1000 women to 14 per 1000 women) requires a total sample size of 1000 patients. There were only 482 women enrolled in the study. Moreover, fewer infants were evaluated for severe reactions, reducing the sample size even more. This study was totally unable to determine if real differences in safety existed. In addition, a subgroup analysis was performed based upon trimester of exposure. Subgroup analysis should always be used to generate hypotheses, never to make inference (Bailor, 1994). Yet this paper makes conclusions based upon this subgroup analysis.

V. DISCUSSION

There are a number of studies that must be conducted once a drug has been approved and marketed: long-term follow-up, additional diagnoses, previously excluded populations, and cost analysis. Most of these studies will involve the collection of observational data instead of randomized clinical trials. Statistical methods that can safely utilize such observational data must be used to generate a hypothesis. These methods include data mining and neural network analysis. Once the hypothesis has been generated, it must be validated by additional data.

APPENDIX

A. Mathematica Program for Univariate Kernel Density Estimation

```
s[x_]: = (0.75*(1 − 0.2*x^2))/Sqrt[5]/;x^2 < 5
s[x_]: = 0/;x^2 > = 5
r[x_,y_,a_]: = s[(x−y)/a]
f[x_,a_]: = Sum[r[x,arr[[i]],a], {i. 1, Length[arr]}]/
(a*Length[arr])
u[x_,a_]: = f[x − Mean[tab])/StandardDevi-
ation[tab],a]/standardDeviation[tab]
```

B. Mathematica Program for Bivariate Kernel Density Estimation

```
t[x_,y_]: = (3*(1 − x^2 − y^2))/Pi/; x^2 + y^2
< 1
t[x_,y_]: = 0/; x^2 + y^2 > = 1
h[x_,y_,u_,v_,a_: = t[(x − u)/a, (y − v)/a]
g[x_,y_,a_]: = Sum[h[x,y,arr1[[i]], arr2[[i]], a], {i,
1, Length[arr1]}]/(a^2(Length[arr1])
b[x_,y_,a_: = N[g[(x − Mean[tab 1])/Standard-
Deviation[tab1],(y − Mean[tab2])/Standard-
Deviation[tab2],a]]/(StandardDevia-
tion[tab1]*StandardDeviation[tab2])
```

C. Mathematica Program for Discrete-Continuous Estimation

```
w[s_,i_,j_]: = .5(1 − s)s^Abs[i − j]/;Abs[i−j]>
= .5
w[s_,i_,j_]: = 1 − s/;Abs[i − j] < .5
n: = Sum[arr[[j]],{j,l,Length[arr]}]
p[s_,i_]: = Sum [w[s,i,j]*arr[[j]],{j,1,Length[arr]}]/n
```

REFERENCES

1. Johnson JA. Influence of race or ethnicity on pharmacokinetics of drugs. J Pharm Sciences 86:1328–1333, 1997.
2. Howden CW. For what conditions is there evidence-based justification for treatment of *Helicobacter pylori* infection? Gastroenterol 113:S107–S112, 1997.
3. Matheus CJ, Piatetsky-Shapiro G, McNeill D. Selecting and reporting what is interesting. In: WM Fayyad, G Piatetsky-Shapiro, P Smyth, R Uthurusamy, eds. Ad-

vances in Knowledge Discovery and Data Mining. Cambridge, MA: MIT Press, 1996.
4. Weiss SI, Kulikowski C. Computer Systems That Learn: Classification and Prediction Methods from Statistics, Neural networks, Machine Learning, and Expert Systems. San Francisco: Morgan Kaufman, 1991.
5. Elder JF, Pregibon D. A statistical perspective on knowledge discovery in databases. In: WM Fayyad, G Piatetsky-Shapiro, P Smyth, R Uthurusamy, eds. Advances in Knowledge Discovery and Data Mining. Cambridge, MA: MIT Press, 1996.
6. Fayyad U, Piatetsky-Shapiro G, Smyth P. The KDD process for extracting useful knowledge from volumes of data. Communications ACM 39:27–34, 1996.
7. Lane D, Rawson N. Inferential problems in postmarketing surveillance. In: DB Owen, ed. Statistical Methodology in the Pharmaceutical Sciences. New York: Marcel Dekker, 1990.
8. Cerrito PB, Cerrito JC. Assessing risk in postmarketing surveillance. J Biopharm Stat 1:221–235, 1991.
9. Tatro DS. Drug Interaction Facts. 3rd ed. St. Louis, MO: Facts and Comparisons, 1992.
10. Ahmad IA, Cerrito PB. Nonparametric estimation of joint discrete-continuous probability densities with econometric applications. J Statistical Planning 41: 349–363, 1993.
11. Wang M, van Ryzin J. A class of smooth estimators for discrete distributions. Biometrika 68:301–309, 1981.
12. Silverman BW. Density Estimation for Statistics and Data Analysis. London: Chapman and Hall, 1986.
13. Knuiman MW, Vu HT, Segal MR. An empirical comparison of multivariable methods for estimating risk of death from coronary heart disease. J Cardiovasc Risk 4:127–134, 1997.
14. Kasabov NK. Foundations of Neural Networks, Fuzzy Systems, and Knowledge Engineering. Cambridge, MA: MIT Press, 1996.
15. Caraco Y, Lagerstrom PO, Wood AJJ. Ethnic and genetic determinants of omeprazole disposition and effect. Clin Pharm Therap 60:157–167, 1996.
16. Moeller J, Mathiowetz N. Prescribed medicines: A summary of use and expenditures by Medicare beneficiaries. National Medical Expenditure Survey Findings 3. DHHS Publication No. (PHS) 89-3448. Rockville, MD: National Center for Health Services Research, 1989.
17. Laska EM, Meisner M, Siegel C. Statistical inference for cost-effectiveness ratios. Health Econ 6:229–242, 1997.
18. Ubel PA, Loewenstein G, Scanlon D, Kamlet M. Individual utilities are inconsistent with rationing choices: a partial explanation of why Oregon's cost-effectiveness list failed. Medical Decision Making 16:108–116, 1996.
19. Calam J. The somatostatin-gastrin link of *Helicobacter pylori* infection. Ann Med 27:569–573, 1995.
20. Penston JG. Review article: clinical aspects of *Helicobacter pylori* eradication therapy in peptic ulcer disease. Alimen Pharm Therap 10:469–486, 1996.
21. Tytgat GN. No *Helicobacter pylori*, no *Helicobacter-pylori*-associated peptic ulcer disease. Alimen Pharm Therap 9(Suppl 1):39–42, 1995.
22. Fendrick AM. Outcomes research in *Helicobacter pylori* infection. Alimen Pharm Therap 11(Suppl 1):95–101, 1997.
23. Lim AG, Walker C, Chambers S, Gould SR. *Helicobacter pylori* eradication using a 7-day regimen of low-dose clarithromycin, lansoprazole and amoxycillin. Alimen Pharm Therap 11:537–540, 1997.
24. Munnangi S, Sonnenberg A. Time trends of physician visits and treatment patterns of peptic ulcer disease in the United States. Arch Intern Med 157:1489–1494, 1997.
25. Rune SJ. Treatment strategies for symptom resolution, hearling, and *Helicobacter pylori* eradication. Scand J Gastroenterol 205(Suppl):45–47, 1994.
26. Riester KA, Peduzzi P, et al. Statistical evaluation of the role of *Helicobacter pylori* in stress gastritis: applications of splines and bootstrapping to the logistic model. J Clin Epidemiol 50:1273–1279, 1997.
27. Malliwah JA, Tabaqchali M, Watson J, Venables CW. Audit of the outcome of peptic ulcer disease diagnosed 10 to 20 years previously. Gut 38:812–815, 1996.
28. Pilotto A, et al. Cure of *Helicobacter pylori* infection in the elderly: effects of eradication on gastritis and serological markers. Alimen Pharm Therap 10:1021–1027, 1996.
29. Jonsson B, Karlsson G. Economic evaluation in gastrointestinal disease. Scand J Gastroenterol 220:44–51, 1996.
30. Greenberg PD, Koch J, Cello JP. Clinical utility and cost effectiveness of *Helicobacter pylori* testing for patients with duodenal and gastric ulcers. Am J Gastroenterol 91:228–232, 1996.
31. O'Malley PG, et al. The value of screening for psychiatric disorders prior to upper endoscopy. J Psychosomatic Res 44:279–287, 1998.
32. Benazzi F. Dangerous interaction with nefazodone added to fluoxetine, desipramine, vanlafaxine, valproate and clonazepam combination therapy. J Psychopharmacol 11:190–191, 1997.
33. Benazzi F. Severe anticholinergic side effects with venlafaxine-fluoxetine combination. Canadian J Psychiatry 42:980–981, 1997.
34. Tiller JW, Johnson GF, Burrows GD. Moclobemide for depression: an Australian psychiatric practice study. J Clinical Psychopharm 15(4 Suppl 2):31S–34S, 1995.
35. Goldstein DJ, Corbin LA, Sundell KL. Effects of first-trimester fluoxetine exposure on the newborn. Obstet Gyn 89(5 pt I):713–718, 1997.
36. Chambers CD, et al. Birth outcomes in pregnant women taking fluoxetine. N Engl J Med 335:1010–1015, 1996.

Patricia B. Cerrito

Power

See also *Sample Size Determination; Statistical Significance*

I. SOME DEFINITIONS

The *power* of a statistical test is the probability of rejecting the null hypothesis when it is false. Closely related to statistical power is the false-negative rate, which is often referred to as the Type II error or beta error. The false-negative rate equals 1 minus the power. It is defined as the probability of not claiming statistical significance when the null hypothesis is false. A fundamental goal of every research protocol should be to define a study that has adequate statistical power and, thereby, a sufficiently small Type II error rate. While there are no universally accepted guidelines, it is broadly accepted that a prospective study should not be initiated and is unlikely to be funded unless the statistical power that is associated with primary hypothesis tests is projected to be in the 0.8–0.9 range.

II. CORRELATES OF STATISTICAL POWER

Statistical power depends in obvious ways on several characteristics of the statistical test being performed. In addition to the sample size association, which is discussed next, the power of a statistical test depends also on the size of the between-group difference and on the amount of noise in the data. For example, the greater the difference between groups that are to be compared, the easier it is to establish statistical significance. Thus, statistical power is directly related to the magnitude of the failure of the null hypothesis. Similarly, the greater the variability in a quantitative measure, the greater is the noise in the data and the harder it is to establish significance. That is, an increase in the variance of a parameter implies a decrease in the statistical power of the associated hypothesis test.

III. STATISTICAL POWER AND SAMPLE SIZE

The most important correlate of statistical power is the sample size. An evaluation of the close relationship between statistical power and sample size should be a key component of the design of every research protocol and of the interpretation of the resulting publications. Unfortunately, repeated reviews of the literature suggest that low power and inadequate sample size are extremely common, if not routine, in the medical literature (1–4). For example, Freiman and colleagues (1) reviewed 71 randomized controlled trials whose primary conclusion was negative. They found that 67 of the 71 trials (94.4%) had greater than a 10% probability of missing a true 25% therapeutic improvement. That is, the power to detect a 25% treatment benefit was nearly always less than 0.9. The power to detect a very large treatment benefit of 50% was less than 0.9 in 70.4% of the reviewed papers. And while these authors do not present the precise results, it is clear from their tabulations that the power of many if not most of the reviewed hypothesis tests was such that clinically important differences in primary endpoints were more likely to be missed than to be reported as statistically significant. This suggestion that statistical power in the published literature is often extremely low is highlighted by Schechtman et al. (4), whose review of 37 papers on the surgical treatment of sleep apnea found that the median sample size would yield a power of under 0.2 to test key hypotheses of interest. When the power is this low, the claim that "there is no significant differences between groups" translates roughly into the more accurate assertion that "we have no idea if the two treatments are different in clinically important ways because the sample size was too small to address the question."

In addition to confirming that many clinical trials are too small to detect large and clinically important differences, Moher and colleagues (2) have noted that even in the most prestigious and carefully reviewed journals, there is a strong tendency to omit all discussion of sample size and statistical power. These authors reviewed the 383 randomized trials published in the *New England Journal of Medicine*, the *Journal of the American Medical Association*, and the *Lancet* in the years 1975, 1980, 1985, and 1990. They found that only 32% of the trials with negative results discussed statistical power and sample size calculations. This means that unless the reader is energetic enough to do these computations independently, it may be difficult to determine whether negative conclusions mean that there are unlikely to be clinically important differences between groups or, alternatively, that the study was too small to address the question. This uncertainty is compounded by the fact that only 20 of the 102 negative studies that were reviewed in the Moher paper made any statement as to the clinical significance of observed differences. The key bright spot in this report is that while the 0% rate at which sample size considerations were discussed in the reviewed 1975 publications mirrors the poor rates reported elsewhere at about that time (5,6), the more recent rate is a substantially improved 43% in 1990.

IV. DETERMINING WHETHER A STUDY IS INADEQUATELY POWERED

Because they are discussed elsewhere in this volume, we do not focus on the many standard formulae that are commonly used to compute power and sample size. However, we demonstrate how an astute reader can quickly decide whether a study is too small using a standard unpaired *t*-test as an illustration. Considerations similar to those described next can be applied in other settings.

The required per-group sample size N for an unpaired *t*-test with the same number of patients in each group is computed using the equation

$$N = \frac{2(Z_\alpha + Z_\beta)^2}{\Delta^2}$$

where Z_α is the standard normal deviate for a test with significance level α (e.g., $Z_\alpha = 1.96$ for a two-sided test at the 0.05 level of significance), Z_β is the normal deviate associated with a Type II error rate of β or a power of $1 - \beta$ (e.g., $Z_\beta = 1.28$ for a power of 0.9), and $\Delta = (\mu_2 - \mu_1)/\sigma$, where μ_1 and μ_2 are the means in the two groups and σ is the common standard deviation. The foregoing expression can be adjusted to compute the sample size in one of two groups when patient allocation is unequal by multiplying by $(r + 1)/r$, where r is the ratio of the total sample size to the sample size in the group of interest.

It follows from the foregoing equation that for a given significance level and a given statistical power, the required sample size depends totally on Δ, the ratio of the difference in means divided by the common standard deviation. Using the equation or the sample size tables that are presented in many statistics texts, it is easy to determine whether the sample size in a given study is reasonable. Thus, assuming a two-sided test at the 0.05 level of significance, a Δ of 1 means that 23 patients per group will yield a power of 0.9. If you are reading a paper where $\Delta = 1$ and the sample size is 50 per group, it's overkill insofar as the relevant hypothesis test is concerned. If there are 10 patients in each group, the study is far too small unless Δ is in the 1.5 range. For smaller Δ's in the 0.5 range, 70 or so subjects must be studied in each group to yield adequate power. If a study with a Δ of 0.5 enrolls only 20 patients per group, a common state of affairs in much of the literature, you know immediately that it is underpowered and that nonsignificant results have very little meaning. An astute reader should apply this type of reasoning to every manuscript. A well-written paper should discuss power and sample size so the reader does not have to fill in the blanks.

REFERENCES

1. Freiman JA, Chalmers TC, Smith H, Jr, Kuebler RR. The importance of bets, the type II error and sample size in the design and interpretation of the randomized control trial: survey of 71 "negative" trials. N Eng J Med 299: 690–694, 1978.
2. Moher D, Dulberg CS, Wells GA. Statistical power, sample size, and their reporting in randomized controlled trials. JAMA 272;122–124, 1994.
3. Williams HC, Seed P. Inadequate size of "negative" trials in dermatology. Br J Dermatol 128:317–326, 1993.
4. Schechtman KB, Sher AE, Piccirillo JF. Methodological and statistical problems in sleep apnea research: the literature on uvulopalatopharyngoplasty. Sleep 18:659–666, 1995.
5. Chalmers TC, Silverman B, Shareck EP, Ambroz A, Schroeder B, Smith H Jr: Randomized control trials in gastroenterology with particular attention to duodenal ulcer (in Report to the Congress of the United States of the National Commission on Digestive Diseases. Vol IV: Reports of the Workgroups: Part 2B: Subcommittee on Research-Targeted and nondirected, pp 223–255). Publ no. NIH 79-1885. Bethesda, MD: National Institute of Arthritis, Metabolism, and Digestive Diseases, 1978.
6. Mosteller F, Gilbert JP, McPeek B. Reporting standards and research strategies for controlled trials: agenda for the editor. Controlled Clin Trials 1:37–58, 1980.

Kenneth B. Schechtman

Protocol Development

See also *Clinical Trial Process; Clinical Trials*

I. INTRODUCTION

Readers who wish to review in depth the advice of the International Conference on Harmonization will peruse for themselves the guidance documents promulgated by that body (1–3). It is the purpose of this entry to abstract that guidance as a convenience, and also to focus that guidance toward the point of view of the statistical practitioner in biopharmaceutical clinical trials.

II. NATURE AND PURPOSE OF THE PROTOCOL FOR A CLINICAL TRIAL

The protocol for a clinical trial is "[a] document that describes the objective(s), design, methodology, statistical considerations, and organization of a trial" (1). It describes the medical conjecture, the subject population, the treatment regimens, the data gathering, and the analysis and evaluation of the data.

The primary purpose of this entry is to describe the function of the protocol as the portrayal of the statistical design and data analysis for a medical experiment. However, a clinical trial is a complex enterprise. The clinical trials statistician should be aware that the protocol also serves a number of administrative and organizational purposes. As a document, the protocol is a contract between the sponsor and the investigator (1). The protocol identifies the agencies and individuals involved in the project and describes their responsibilities and obligations (e.g., for serious adverse event reporting, availability of records for inspection). It serves as the primary means for the Institutional Review Board to ensure the ethical purpose and conduct of the trial and should contain a statement that the trial will be conducted in compliance with good clinical practice and the applicable regulatory requirements. The protocol summarizes the nature of the known and potential risks and benefits to the subjects. It requires informed consent and adherence to the principles in the Declaration of Helsinki. Certain sections of a protocol will typically contain instructions concerning the handling of specimens for laboratory analysis or the administration of special (or even routine) medical procedures.

Most institutions, and nearly all pharmaceutical companies, have standard outlines and naming/numbering conventions for protocols. Sections generally include, but are by no means limited to, those in the following abbreviated list. There are a number of reasonable ways to organize these topics within the document, and different sponsors will choose different conventions. However, nearly everyone covers at least the following topics somewhere in their protocols.

Introduction: Describes the test article, synopsizes what is known about its effects and previous research, and motivates the current project.

Objective: Sharpens the medical conjecture.

Subject selection criteria: Defines the subject eligibility criteria in such a way as to link to a target population of prospective users of the treatment and protect the safety of study subjects.

Study medication: Thoroughly describes the test and control treatments, their packaging, and the mechanism for treatment masking.

Plan and procedures: Lays out the schedule of visits, doses, tests and evaluations to be performed on the subjects by the investigative staff, rules around discontinuing subjects, reporting of serious adverse events, and administration of concomitant or rescue medication.

Efficacy and safety endpoints and criteria: Describes the primary and secondary efficacy and safety endpoints and how those data are going to be collected and scored. Study monitoring and data-collection quality controls might be described in this section.

Data analysis and statistical considerations: Defines primary and secondary variables; describes the primary and secondary analyses and the statistical design considerations. This is the topic of upcoming Sec. III.

Committees: Gives the composition, charter, and operating rules for data monitoring committees, endpoint classification committees, steering committees, executive committees, and the like.

Pharmacoeconomic investigations: Contains many of the same elements as the main protocol; often inserted as a subprotocol.

Sponsors should have established procedures for generating, approving, and amending protocols. These procedures should state who is to be involved in the process and what role they are to play. The ICH section E6 on Good Clinical Practice (1) asserts that

> [t]he sponsor should utilize qualified individuals (e.g., biostatisticians, clinical pharmacologists, and physicians) as appropriate, throughout all stages of the trial process, from designing the protocol and CRF's and planning the analyses to analyzing and preparing interim and final clinical trial/study reports.

It is a sound practice to require that the statistical section be contributed by a qualified professional statistician.

III. CONTENT OF THE STATISTICAL SECTION OF A PROTOCOL FOR A CLINICAL TRIAL

The advice of the ICH is to prespecify the analysis: "The extent to which the procedures in the protocol are followed and the primary analysis is planned a priori will contribute to the degree of confidence in the final results and conclusions of the trial" (2). Hence, the a priori specification of the analysis is an important function of the statistical analysis section of a protocol.

A. Primary Variable(s) and Analyses

"[T]he protocol should identify the primary [efficacy variables] with an explanation of why they were chosen" (3). "Only results from analyses envisaged in the protocol (including amendments) can be regarded as confirmatory" (2).

Because most clinical trials collect a number of different types of variables, there will typically be a number of different analyses conducted on the data from a given trial. "The protocol should make clear a distinction between aspects of a trial that will be used for confirmatory proof and the aspects that will provide data for exploratory analysis" (2). One means of drawing this distinction, and of addressing the multiplicity issue besides, is via the definition, in the statistical section of the protocol, of a primary variable and a primary analysis. The terms *primary*, *secondary*, *exploratory*, and the like are probably not susceptible to precise mathematical definition. However, one sense in which those in the clinical trials field tend to understand them is in terms of this confirmatory/exploratory distinction. The *primary* variable (or endpoint) is the variable in terms of which the investigators intend either to confirm or to refute the conjecture upon which the trial is based. Because it is often possible to analyze a given variable in a number of different ways, it is also recommended to specify a (stochastic) model and an analysis strategy for the primary variable.

What, then, is entailed in the specification of the primary variable and its analysis in the statistical analysis section of a protocol? The subject of clinical trials is so vast that no list could lay credible claim to being absolutely comprehensive; however, pharmaceutical trial statisticians have generally found the following list of considerations to be useful.

1. Primary Variable Definition

This is the precise mathematical definition of the primary outcome variable. In many instances this variable is not raw data but is derived according to some algorithm from the raw data. It it important to state that algorithm; e.g., say, "the return-visit-3 measurement minus the baseline-visit measurement as a percentage of the baseline" as opposed to, "the percentage change from baseline." Mathematical notation is appropriate and useful; but it should be supplemented by a verbal description. Note that, in this example, there is an implicit understanding of the definition of *baseline*. It is usually helpful to state explicitly the definition of *baseline* and other time-point dependencies; they may not be as obvious as one might suppose.

2. Primary Statistical Objective

This involves a translation of the primary objective of the investigation into statistical terms: a statistical model and an accompanying hypothesis testing or estimation objective. If it is a hypothesis test, as is frequently the case, the null and alternative hypotheses should be stated in terms of the aforementioned primary statistical model.

"It is vital that the protocol of a trial designed to demonstrate equivalence or noninferiority contain a clear statement that this is its explicit intention" (2).

3. Primary Analyses

These are the analyses, or statistics, that will produce the answers to the primary investigational conjecture. The ICH E9 document recommends that "all the principal features of the proposed confirmatory analysis of the primary variables" (2) be laid out in the protocol. It is not unreasonable that different, equally competent analysts might make differing judgments about what is an appropriate level of analytical detail to include in a protocol, i.e., what constitutes a "principal feature" of the analysis. Further detail can be given in an accompanying analysis plan document. Most practitioners would state in the protocol the main statistic upon which the primary analysis will be based, e.g., least-squares means, log-likelihood ratio, some ranks-based test. If not prespecified, covariate adjustments, variable transformations, and methods to handle dropouts or missing data can easily lead to controversy later. It is appropriate and useful to employ mathematical notation.

4. *Decision Rules*

A description of the statistical decision space and decision rules is essential. "The protocol should . . . designate the pattern of significant findings or other method of combining information that would be interpreted as supporting efficacy" (3).

Given that one has calculated a set of statistics, one must state a priori how this set of statistics will be used to evaluate the primary conjecture. Whether hypothesis tests are one- or two-sided and their significance levels, or outcomes that will lead the experimenters to conclude that treatments are equivalent, would be included with this topic. If a number of statistical hypotheses are being evaluated, give the map from the subsets of possible outcomes to the overall substantive conclusions.

The separation of the analytic enterprise into "primary analyses" and "decision rules" may seem forced; and the exposition in a given protocol may or may not be best served by observing the distinction. Here, the rubric of analysis emphasizes the arithmetic, whereas the rubric of decision rule emphasizes the inference to be made from that arithmetic. It is useful, at least, to do the thinking around the analytic strategy in these terms.

5. *Error Control*

The method for controlling the error rates must be set up. Particularly when a number of hypotheses are being tested, the method for controlling the Type I error rates, and at what level they will be controlled, should be stated. The method of management of the joint coverage probabilities for sets of interval estimates would be given. Type II error rates might also be addressed.

6. *Analysis Subsets*

A description of the subset of subjects whose data will contribute to the primary analysis should be included. This is where such terms as "Intention-to-treat," "per-protocol," and the like come in. It is advisable not to assume that all readers have the same understanding of these terms, but rather to explicitly spell out the membership requirements for (at least) the primary analysis subset.

B. Further Topics

1. *Design*

The statistical aspects of the experimental design should be set out. This section gives the reasoning that justifies the planned number of subjects. Working estimates of anticipated treatment effects and variability would be given, along with their sources if available. Clinically meaningful treatment effects that the study is targeted to detect or, in the case of equivalence trials, exclude, would be stated. If the design is something besides one-period, double-blind, parallel-group (e.g., a crossover or dose escalation) or if the design allows treatment transitions, then the statistical implications of these characteristics should be accounted for.

Sometimes trials are designed to be stopped when a certain amount information has been accumulated. This is a statistical design feature that may not be obvious and ideally should be explicitly stated.

In a multicenter study, it may be useful to state the number of subjects planned per site.

2. *Interim Analyses*

The timing, purpose, and scope of planned interim analyses needs to be stated. If decisions about stopping the trial, extending the trial, or otherwise modifying the design are to be based on interim analyses, then the decision criteria should be spelled out. Spelling out comprehensively and in detail the process by which accumulating data will be examined will obviate concerns about bias and scientific validity. This material could be rather extensive. See the entry on *Interim Analysis* for further advice on what to include.

3. *Randomization and Blinding*

A short description of the randomization scheme should be included. It is not recommended that the block size be revealed in the protocol. A small proportion of trials has relatively more complex randomization schemes, and in these cases, the description would be appropriately more detailed. If the trial is not of the usual double-blind variety, then it may be helpful to argue for the type of treatment masking that is planned. The point is to address concerns that critics of the trial may raise regarding bias.

4. Problem Data

Conditions under which data are invalid or would otherwise create analytical anomalies should be described. If it is possible to anticipate such conditions, then it may be worth describing them in the protocol and prespecifying methods to account for these data analytically. Outliers, missing observations, premature terminations, losses to follow-up, and protocol violations are some of the general categories of difficulties that can be anticipated. Another common issue is created by low-enrolling centers in a multicenter study where the analysis is blocked by center. Some sponsors may prefer to base their analytical approach to problem data on a treatment-masked review of the data conducted prior to any actual analysis. If plans for handling the problem data stem from such a blinded preanalysis review, and the plans are documented, then concerns around potential bias do not arise.

The ICH E9 document advises that "[t]he protocol should also specify procedures aimed at minimizing any anticipated irregularities in study conduct that might impair a satisfactory analysis" (2). This material may fit more naturally in the plan and procedures section of the protocol than in the statistical analysis section, however.

5. Safety Data

The planned analyses of safety data should be set out. According to conventions used by many sponsors, this description is terse, emphasizing summarization and display. However, if there are known safety issues, it is clearly in a sponsor's interest to prespecify in some detail an inferential strategy to address them. General analytical issues around safety data include the definition of incidence (per subject, per dose, per subject-year, etc.) and how to treat data from subjects who have no or minimal exposure to the study treatments.

6. Secondary Endpoints and Analyses

"Secondary variables are either supportive measurements related to the primary objective or measurements of effects related to the secondary objective" (2). The testing of the primary efficacy conjecture in some subpopulation is a common secondary objective.

Secondary endpoints can be defined, and their analyses planned, in an exposition that parallels that for the primary variables: variable definition, statistical model, statistical objective, analytic strategy, and decision rules. The level of detail for secondary analyses is frequently reduced, as compared to the primary analyses, and they can be discussed as a group in order to make the exposition efficient and readable. However, just as for primary analyses, it is obvious that the statement of secondary conjectures and analyses a priori is an important factor in lending credibility to the conclusions.

7. Some Miscellaneous Topics

Computer packages: The programs used to run the analyses, if not standard packages

Model validation statistics: Goodness-of-fit tests, deviance measures, and the like

REFERENCES

1. International Conference on Harmonization. E6 Good Clinical Practice: Consolidated Guideline. May 1997, www.fda.gov/cder/guidance.
2. International Conference on Harmonization. Statistical Principles for Clinical Trials (Draft). May 1997, www.fda.gov/cder/guidance.
3. International Conference on Harmonization. E3 Structure and Content of Clinical Study Reports. July 1996, www.fda.gov/cder/guidance.

Robert D. Chew

Process Validation

See also *USP Tests*

I. INTRODUCTION

A. Scope

The Food and Drug Administration (FDA) defines validation in their 1987 publication *Guideline on General Principles of Process Validation* (1) as "establishing documented evidence which provides a high degree of assurance that a specific process will consistently produce a product meeting its predetermined specifications and quality characteristics." Since this definition contains the phrases "high degree of assurance" and "consistently produce," which have statistical implications, statisticians have been involved in developing acceptance limits and sampling plans as well as analyzing validation data for process validation programs.

Although process validation is applied to all product forms, the focus of this entry will be on the statistical techniques that aid in the validation of solid oral dosage forms, i.e., tablets and capsules. A general discussion of process validation will also be provided. This entry should be of most interest to statisticians who want an overview of the concepts of process validation and a discussion of those statistical techniques that are most relevant. By including a number of examples, the chapter is written so that those who conduct validation studies can obtain appropriate acceptance criteria for practical use.

B. History

Validation started in the early 1970s with assay verification. During the mid-1970s, most of the attention given to validation dealt with sterile processes, because of the critical nature of injectable products. In the early 1980s, the FDA turned its attention to such nonsterile processes as solid dosage, semisolid dosage, liquids, suspensions, and aerosols. The draft process validation guideline was made available for comment in March 1983 and issued in May 1987. The guideline was written to help industry comply with the 1978 revision of the "Current Good Manufacturing Practices (cGMP) Regulations for Finished Pharmaceuticals" and the "Good Manufacturing Practice Regulations for Medical Devices." This was the first time that the word *validation* had appeared in the regulations. Since following the cGMP is a legal requirement, failure to comply would constitute a violation of the Food, Drug and Cosmetic Act.

C. Overview

Many authors have written about process validation (2–16). This introduction is a compilation from these papers. Process validation is performed on new drug products (both prescription and over-the-counter), active pharmaceutical ingredients, products involved in technology transfer (e.g., to a new manufacturing site), marketed products that have not been validated, and previously validated products that incurred a major change (e.g., in a processing step, a new raw material source for the active, manufacturing equipment, or process parameters).

Elements of the validation concept should be incorporated at each stage of the product and process development continuum. Identification of the critical process variables and studies aimed at establishing optimal ranges for these variables should be completed during the prevalidation phases. Statistics can play a major role in the prevalidation phases of assay validation, in the setting of specifications, and in formula screening/optimization studies. Some of the critical variables that need to be studied are: particle size of the drug substance, bulk density of the active drug/excipients, blender load, binder concentration, amount and spray rate of granulating fluid, blender speed and blending times, moisture content of the dried granulation, particle size of the lubricated granulation, and lubricant blending times. In fact, a number of major steps are involved in the development of stable, bioavailable, manufacturable and cost-effective solid oral dosage forms. These include the physicochemical characterization of the drug substance, formulation development of commercial dosage forms, development and validation of analytical test methods, specification setting, scale-up and process optimization, development of stability data on commercial formulations packaged in the container/closure systems to be marketed, and process validation. During the validation phase, sampling plans are needed as well as acceptance criteria. After the data are collected, a statistical review may be needed to evaluate the data against the preset acceptance criteria.

D. Types of Validation

There are three different types of process validation: retrospective, concurrent, and prospective. *Retrospective* validation covers those situations where a product is already marketed without a documented process validation program. Retrospective validation was not intended to be a method of validating new processes, but rather a safety net that could be used to validate products already on the market at the time the validation requirements were put into the regulations. Retrospective validation has been used when there have been no changes in formulation, procedure, or equipment and when the efficiency of the process is able to be judged from in-process and finished product results. In retrospective validation, it is not sufficient to examine only pass/fail records on previously manufactured lots; rather, a statistical analysis of quantitative data of product attributes is necessary to predict process capability. Retrospective process validation involves using historical data to provide documented evidence that a system does what it purports to do. Since most companies have completed their application of retrospective validation to existing products, it will not be discussed in this entry. However, a discussion of some analysis techniques that might be used when performing a retrospective validation can be found in Kohberger (17).

Prospective process validation involves proving that the process does what it purports to do based on a preplanned protocol and before the process is put to commercial use. It involves the collection and evaluation of data on at least three consecutive successfully manufactured batches matching routine production in all aspects of production and control. The statistical aspects of prospective validation will be discussed in this entry. The protocol should state the process steps, identify the controls to be used, specify the variables to be monitored, state what samples are to be taken both for testing and as contingency, specify both the product performance characteristics/attributes to be evaluated and their acceptance criteria, and refer to the test methods to be used. Historical data from formulation development and scale-up studies may help to set these criteria.

Concurrent validation is similar to prospective validation except that batches are validated and released one batch at a time. Concurrent validation is accepted only in limited situations, e.g., for orphan drugs where it is deemed an economic hardship to produce three batches at one time when they cannot all be sold. The validation procedures and acceptance criteria are similar to what would be done for a prospective validation.

II. SAMPLING

A manufacturing process may involve drying and granulation steps as well as intermediate and final mixing steps. Once a blend has been mixed, it may be transported to another location for screening or tableting (or encapsulation). Sampling can be performed at any of these steps in the manufacturing process. Samples can be taken when the blend is in a mixer, while being discharged from the mixer, when it is in a transport container (e.g., drum), throughout tablet compression or encapsulation, and after film coating (if appropriate). Sampling plans need to be developed at each of these stages.

There are two sampling plans that are generally used when testing blends or final product. In the first plan (Sampling Plan 1), a single test result is obtained from each location sampled. For example, in a blending step, a single test result would be obtained from each of a number of different locations within the blender. In a drum, a single test result might be obtained from the different locations within the drum or from each of a number of different drums. For final tablets, a single tablet may be tested from various time points throughout the tableting run. In the second plan (Sampling Plan 2), more than one test result is obtained from each of the sampled locations. For example, during the tableting operation, if a cup is placed under the tablet press at specific time points during the tableting run, several of the tablets from each cup sample would be tested for content uniformity. Sampling Plan 2 allows for estimation of between-location and within-location variability.

It is assumed for the remainder of this entry that the same number of units is tested from each of the sampled locations (i.e., it is a balanced sampling plan). Regardless of what sampling plan is used to determine testing, multiple units are normally collected at each of the sample locations during validation to serve as contingency samples for possible later testing.

A. Powder Blend Sampling

Based on the interpretation of the Wolin Court decision (*U.S. vs Barr Laboratories*), the allowable size of sample taken from powder blends has been set at no more than three times the dosage unit weight. A perplexing problem facing oral solid dosage form manufacturers today is the difficulty in applying this unit dose sampling to blend uniformity validation because of the current limitations in sampling technologies. An excellent discussion of blend sampling is given in the Parenteral Drug Association (PDA) Technical Report on Blend Uniformity (18). Much of the following discussion is taken from that paper.

There is a great deal of frustration amongst oral solid dosage form manufacturers caused by unit dose sampling of blends. Companies have obtained very uniform results when testing the finished dosage form (i.e., tablets or capsules) while obtaining highly variable results when attempting to comply with the current FDA position on blend uniformity sampling.

It is generally recognized that a thief is far from an ideal sampling device, due to a propensity to provide nonrepresentative samples, i.e., the sample has significantly different physical and chemical properties from the powder blend from which it was withdrawn. Although simple in concept, demonstrating blend uniformity is complicated by this potential for sampling error. The current technology does not yet provide a method for consistently obtaining small representative samples from large static powder beds. Using X-ray fluorescence and near-infrared spectroscopy methods to measure blend uniformity directly, it is hoped that these problems may soon be overcome.

As stated in the PDA Technical Report (18), sampling error can be influenced by: (a) the design of the thief, (b) the sampling technique, and/or (c) the physical and chemical properties of the formulation. The physical design of the thief can affect sampling error, since the overall geometry of the thief can influence the sample that is collected. Surface material can fall down the side slit of a longitudinal thief as it is inserted into a powder bed. Sampling technique can also have an impact on sampling error. As the thief is inserted into a static powder, it will distort the bed by carrying material from the upper layers of the mixture downward toward the lower layers. The angle at which the thief is inserted into the powder bed can also influence sampling error. Another factor that can affect sampling error is the physical and chemical properties of the formulation. The force necessary to insert a long thief into a deep powder bed can be appreciable. This force, depending on the physical properties of the formulation, can lead to compaction, particle attrition, and further distortion of the bed. Ideally, the thief should be constructed from materials that do not preferentially attract the individual components of the formulation. In general, the potential for sampling error increases as the size of the sample and/or the concentration of drug in the formulation decreases. Samples obtained using thief probes can be subject to significant errors.

B. Finished Product Sampling

Two of the most common tests for finished product that have acceptance criteria are content uniformity and dissolution. The United States Pharmacopeia (USP) requirements for content uniformity for both tablets and capsules as well as for dissolution are summarized in Tables 1–3.

When collecting samples to evaluate these tests, it is important to maintain the location identity of all samples taken and to maintain this identity throughout the testing regimen. Validation is the one time when the exact location in the batch is known for each of the individual dosage units tested. By showing that each of the sample locations tested provides acceptable results, a justification is developed for the later combining of tablets or capsules into a QC composite sample for the release testing of future batches. If a two-sided press is used for tableting, the identity of the side of the press from which the samples were taken should also be maintained.

It is recommended that individual dosage units be tested from as many different sample locations as possible. The number of units tested could even be tied to run length, with more units tested when the run length goes across multiple shifts. A discussion of the effect of sample size on one of the methods discussed, the Bergum approach, is provided in Sec. III.F.2. Because Sampling Plan 2 allows for the estimation of both between-location and within-location variability, this plan

Table 1 USP XXIII Content Uniformity Test Requirements for Tablets[a]

Stage	Number tested	Pass stage if:
S_1	10	Each of the 10 units lies within the range of 85.0%–115.0% of label claim, and the relative standard deviation (or RSD) is less than or equal to 6.0%
S_2	20	No more than 1 unit of the 30 units (S_1 + S_2) is outside the range of 85.0%–115.0% of label claim, no unit is outside the range of 75.0%–125.0% of label claim, and the relative standard deviation (or RSD) of the 30 units (S_1 + S_2) does not exceed 7.8%

[a]Where average of potency limits is 100.0% or less.

Table 2 USP XXIII Content Uniformity Test Requirements for Capsules[a]

Stage	Number tested	Pass stage if:
S_1	10	Not more than 1 of the 10 units lies outside the range of 85.0%–115.0% of label claim, no unit is outside the range of 75.0%–125.0% of label claim, and the relative standard deviation (or RSD) is less than or equal to 6.0%
S_2	20	No more than 3 of the 30 units ($S_1 + S_2$) are outside the range of 85.0%–115.0% of label claim, no unit is outside the range of 75.0%–125.0% of label claim, and the relative standard deviation (or RSD) of the 30 units ($S_1 + S_2$) does not exceed 7.8%

[a]Where average of potency limits is 100.0% or less.

is generally recommended when testing individual dosage units for content uniformity. For dissolution, one might choose either Sampling Plan 1 or Sampling Plan 2, depending upon how many total units are tested.

III. STATISTICAL TECHNIQUES/APPROACHES

Since the start of validation in the late 1970s, there has been little published on the statistical aspects of conducting a successful process validation. What follows are some of the statistical techniques that have been either suggested in the literature or used in practice when conducting validation studies. Their use in the development of validation criteria will be discussed in Sec. IV. The Bergum approach is discussed in a paper by Bergum (19). A discussion of the other techniques can be found in Hahn and Meeker (20). Other techniques that have been applied to validation data but are not discussed in detail in this entry are analysis of variance (ANOVA) and process capability analysis. In the following subsections, let $\bar{X}$ and s denote, respectively, the mean and standard deviation of a sample of size n and let t and F be the critical values for the t- and F-distributions with their associated degrees of freedom and confidence levels. Let MSB be the between-location mean square from the one-way ANOVA.

Table 3 USP XXIII Dissolution Test Requirements

Stage	Number tested	Pass stage if:
S_1	6	Each unit is not less than $Q + 5\%$
S_2	6	Average of 12 units ($S_1 + S_2$) is equal to or greater than Q, and no unit is less than $Q - 15\%$
S_3	12	Average of 24 units ($S_1 + S_2 + S_3$) is equal to or greater than Q, not more than 2 units are less than $Q - 15\%$, and no unit is less than $Q - 25\%$

A. Tolerance Interval

A *tolerance interval* is an interval that contains at least a specified proportion P of the population with a specified degree of confidence, $100(1 - \alpha)\%$. This allows a manufacturer to specify that at a certain confidence level at least a fraction of size P of the total items manufactured will lie within a given interval. The form of the equation is:

$$\bar{X} \pm ks$$

where k = tabled tolerance factor and is a function of $1 - \alpha$, P, n, and whether it is a one- or two-sided interval.

B. Prediction Interval

A number of prediction intervals can also be generated. A *two-sided prediction interval for a single future observation* may be of interest. This is an interval that will contain a future observation from a population with a specified degree of confidence, $100(1 - \alpha)\%$. The form of this equation is:

$$\bar{X} \pm ks$$

where $k = t_{1-\alpha/2,n-1}\sqrt{1 + 1/n}$

Another type of prediction interval that might be of interest is a one-sided upper prediction interval, to contain the standard deviation of a future sample of m observations, again with a specified degree of confidence, $100(1 - \alpha)\%$. This is called the *standard deviation prediction interval* (SDPI). The form of this equation is:

$$S\sqrt{F_{1-\alpha,m-1,n-1}}$$

C. Confidence Interval

Confidence intervals can be generated for any population parameter. Specifically, a two-sided confidence interval about the mean is an interval that contains the true unknown mean with a specified degree of confidence, $100(1 - \alpha)\%$. The form of this equation, which depends on the sampling plan, is as follows:

Sampling Plan 1:

$$\bar{X} \pm ks$$

where $k = t_{1-\alpha/2,n-1}/\sqrt{n}$

Sampling Plan 2:

$$\bar{X} \pm k\sqrt{MSB}$$

where

$$k = t_{1-\alpha/2,\#\text{locations}-1}/\sqrt{(\#\ locations)(\#\ per\ location)}$$

Note: for any stated confidence level, the confidence interval about the mean is the narrowest interval, the prediction interval for a single future observation is wider, and the tolerance interval (to contain 95% of the population) is the widest.

D. Variance Components

Variance components analysis has been used in a number of applications within the pharmaceutical industry. The power of this statistical tool is the separation or partitioning of variability into nested components. The approach requires using Sampling Plan 2 so that the between-location and within-location variance components can be estimated. These estimates can be calculated using one-way analysis of variance (ANOVA). The within-location variance is estimated by the mean square error, whereas the between-location variance is estimated by subtracting the mean square error from the mean square between locations and then dividing by the number of observations within each location. When applied to the blending operation, the method allows us to determine the between-location variance, which quantifies the distribution of active throughout the blend, and the within-location variance, which in turn is composed of sampling error, assay variance, and a component related to the degree of mixing on the "micro" scale. The total variance in the container or mixer is the sum of the two variance components. Similarly, one may also determine these components in the product. Here, the within-location variance will again consist of the assay variance, the sampling error, and the "micro" mixing component, in addition to the weight variation. The between-location component is that variance associated with macro changes in the blend environment. It is this component that reflects the overall uniformity of the blend and is minimized when optimum blender operation is achieved.

E. Simulation

Monte Carlo simulation can be used to estimate the probability of passing multiple-stage tests such as content uniformity and dissolution. This technique is performed by generating computer-simulated data from a specific probability distribution (e.g., normal) and then using these generated sample data as if they were actual observations. The multiple-stage test is then applied to the data. This process can be repeated many times to evaluate various test properties (e.g., determining the probability of passing the multiple-stage test for specific values of the population mean and standard deviation of a normal distribution).

F. Bergum Approach

Bergum (19) published a method for constructing acceptance limits that relates the acceptance criteria directly to multiple-stage tests such as the USP content uniformity and dissolution tests. These acceptance limits are defined to provide, with a stated confidence level $(1 - \alpha)100\%$, a stated probability (P) of passing the test. For example, one can make the statement that, with 95% confidence, there is at least a 95% probability of passing the USP test. Both the USP content uniformity and the USP dissolution tests have been evaluated. In each case, the required limits are provided in "acceptance tables," which are computer generated. These tables change with the confidence level $(1 - \alpha)$, the probability bound (P), the sample size (n), and whether tablets or capsules are being evaluated (for content uniformity) or the Q value (for dissolution). Confidence levels as well as values for P are typically 50%, 90%, or 95%. The PDA Technical Report (18) suggests the use of a 90% confidence level to provide 95% coverage. The FDA prefers a 95% confidence level. A 50% confidence level can be considered a "best estimate" of the coverage. A SAS program has been written to construct these acceptance limit tables for the USP content uniformity and dissolution tests for both Sampling Plans 1 and 2. The computer program can be obtained by contacting James Bergum at Bristol-Myers Squibb.

1. *Statistical Basis*

The Bergum approach uses the fact that these are multiple-stage tests. A multiple-stage test is a test with several stages, where each stage has requirements for passing the test. As can be seen in Tables 1–3, the USP content uniformity and dissolution tests are multiple-stage tests with multiple criteria at each stage. The lower bound, LBOUND (also called P), for the probability of passing the USP content uniformity and dissolution tests uses the following relationship:

$$\text{Prob(passing USP test)} \geq \max\{\text{Prob(passing } ith \text{ stage)}\}$$

where $i = 1$ to S (S = number of stages in USP test). One requirement for this inequality to hold for a multiple-stage test is that failure of the overall test at any stage also results in failure of the overall test at any subsequent stage.

Assume that the test results follow a normal distribution with mean μ and standard deviation σ. Sigma (σ) is the standard deviation of a single observation. For a given value of μ and a given value of σ, LBOUND can be determined by calculating the probability of passing all of the requirements at each stage. Figures 1 and 2 compare the 95% contours for the calculated bound LBOUND and for the true probability of passing the USP test, calculated by simulation. If μ and σ are on the 95% LBOUND contour, then at least 95% of the samples tested using the USP test would pass the test. These figures show how close the calculated bounds are to the simulated results for both the USP content uniformity and dissolution tests.

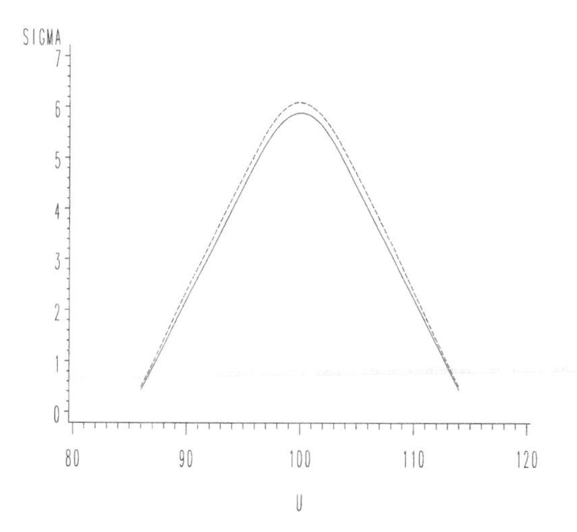

Fig. 1 95% contour plots for probability of passing USP content uniformity test for tablets. Solid line indicates computed LBOUND; dashed line indicates simulation result.

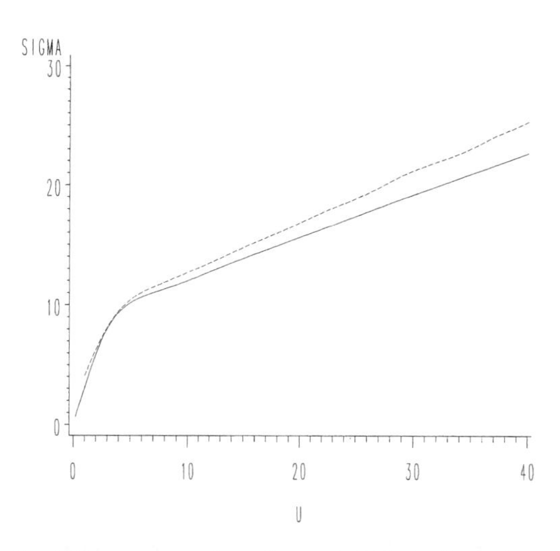

Fig. 2 95% contour plots for probability of passing USP dissolution test for tablets. Solid line indicates computed LBOUND; dashed line indicates simulation result.

The LBOUND can be used to develop acceptance criteria by constructing a simultaneous confidence interval for μ and σ from the data. If a 90% confidence interval were constructed for μ and σ and the entire interval were below the 95% LBOUND, then, with 90% confidence, at least 95% of the samples tested would pass the USP test. For Sampling Plan 1, the sample mean and sample standard deviation estimate the population parameters μ and σ. A simultaneous confidence interval for μ and σ is given in Lindgren (21). Since the variance of a single observation using Sampling Plan 2 is the sum of the between-location and within-location variances, σ (i.e., the standard deviation of a single observation) is estimated by calculating the square root of the sum of the between- and within-location variance components. A confidence interval for σ is given by Graybill and Wang (22). The simultaneous confidence interval for μ and σ is constructed by using a Bonferroni adjustment on the two individual confidence intervals for μ and σ. Once the confidence interval is constructed, it must fall completely below the LBOUND specified. An acceptance limit table can be generated by finding the largest sample standard deviation for a fixed sample mean such that the resulting confidence interval remains below the prespecified LBOUND.

2. *Effect of Sample Size*

Through the use of operating characteristic (OC) curves, the effect of sample size on the ability to pass the Bergum approach can be evaluated. The OC curves provide estimates of the probability of passing the Ber-

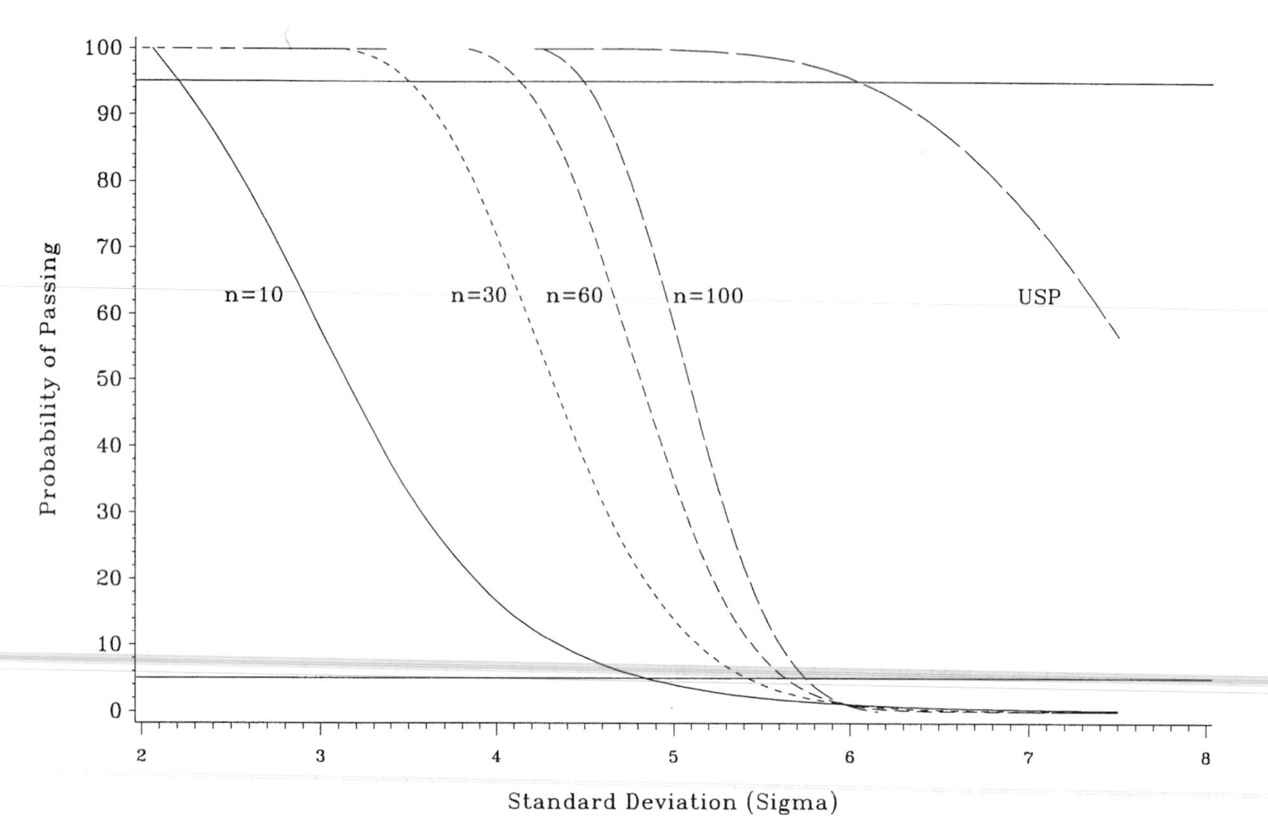

Fig. 3 Probability of meeting Bergum acceptance table for Sampling Plan 1, mean = 100.0%, 90% assurance/95% coverage, tablets.

gum approach over a number of different population mean and standard deviation values. Figures 3 and 4 provide OC curves for specific sample sizes using Sampling Plans 1 and 2, respectively. For these plots a mean of 100% was assumed with the tablet dosage form. A confidence level of 90% to obtain 95% coverage was also used. The estimated probability of passing the USP content uniformity test was included for comparison.

Figure 3 provides the OC curves using Sampling Plan 1 for sample sizes of 10, 30, 60, and 100. As expected, the probability of passing the acceptance limit table increases as the sample size increases. For example, if n = 30, the probability of passing the acceptance limit table for tablets when sigma is 4.0% is approximately 75%. To increase the probability of passing the Bergum approach with this type of true quality, a larger sample size would be needed.

Figure 4 provides the OC curves using Sampling Plan 2 for a sample size of 60 but with different numbers of locations sampled; the results are compared to the use of Sampling Plan 1 without any replication. It is assumed for this plot that half of the total variation is due to between-location variance and half is due to within-location variance (i.e., factor = 0.5). Note that the number of locations has a significant effect on the probability of passing the acceptance limit table. This effect would have been even larger if the percent of variation due to locations was assumed to have been something greater than one-half. It is recommended that when using Sampling Plan 2, the number of locations used be as large as possible. For example, if a total of 60 tablets are sampled across the batch, it is better to sample 3 from each of 20 locations than 20 from each of 3 locations.

IV. COMPARISON OF ACCEPTANCE CRITERIA

There are a number of tests that are performed during validation. In the blends, the primary interest is in

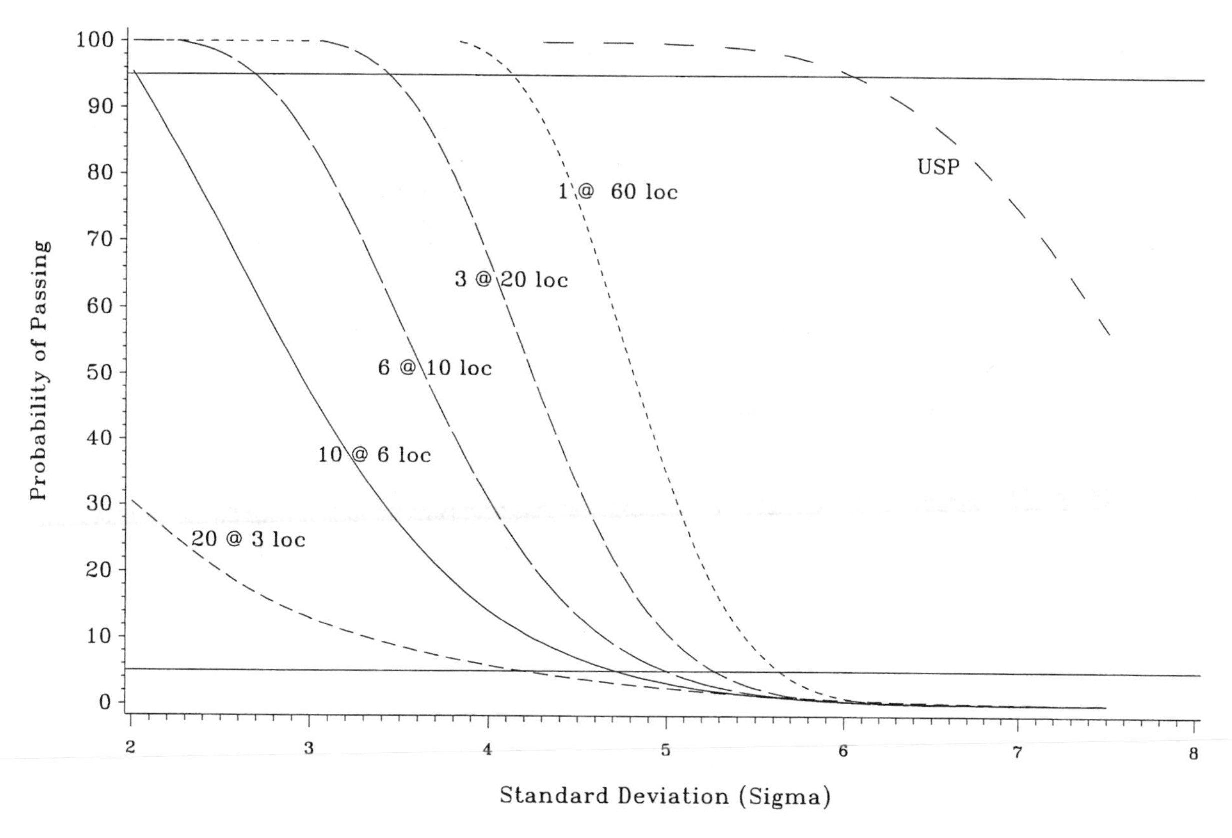

Fig. 4 Probability of meeting Bergum acceptance table for Sampling Plan 2, mean = 100.0%, 90% assurance/95% coverage, factor = 0.5, tablets.

showing that the blend is uniform in drug content. Uniformity can also be evaluated in the drums to ensure that segregation or demixing did not occur during transfer. The overall potency is generally not considered a critical variable in the blends, since it is neither enhanced nor diminished by additional mixing. However, there may be a concern with potency loss during processing or storage between processing steps, e.g., after emptying the blended powder into transports or as a result of tablet compression. In the final product, content uniformity and dissolution are of primary interest, and to a lesser extent, potency. Acceptance limits are generally developed for these tests. Other tests that are performed, such as particle size, bulk/tap density and flow, hardness, friability and weight variation, may or may not have formal, statistically derived acceptance limits.

There are almost as many approaches to validation as there are companies performing validation. What follows is a discussion of some of the methods of statistical analysis along with their advantages and disadvantages. Two proposals, one from the FDA for blends and another by the Parenteral Drug Association (PDA), called a "holistic" approach to validation, are also discussed. The advantages and disadvantages of these methods are listed in Tables 4A and 4B for powder blends and finished product (tablets/capsules), respectively. Note that what might be an advantage to one person can be a disadvantage to the next.

A. Powder Blends

1. Blend Uniformity

a. Approach of the Food and Drug Administration

The FDA has proposed the following acceptance criteria for blend uniformity testing (23):

Each individual sample should meet compendial assay limits (e.g., 90.0%–110.0%).
The RSD should be no greater than 5%.
A minimum of 10 samples should be tested.

Table 4A Advantages and Disadvantages of Various Statistical Techniques for Powder Blends

Test	Advantages	Disadvantages
Blend Uniformity:		
1. FDA approach	Accepted by the FDA	Not statistically based Penalized for large n Adversely affected by constant loss of potency
2. SDPI	Rewarded for larger n Not affected by constant loss of potency Tied to part of Stage 1 USP CU test	Difficult to apply with Sampling Plan 2 Not tied directly to full USP CU test
3. Tolerance interval	Easy to calculate Rewarded for larger n	Not tied directly to full USP CU test Difficult to apply for Sampling Plan 2 Factors can be hard to find for non-standard coverage probabilities
4. Bergum approach	Rewarded for larger n Tied directly to USP overall test Easy table look-up Provides high assurance of passing USP test	Computer program required (but can be provided) Difficult to pass using Sampling Plan 2 with few locations
5. Holistic approach	Provides chance to recover from variable blender results	Substitution of variance components concept may be a hard sell Degrees of freedom can be significantly reduced

The samples should include potential "dead spots." The tighter RSD requirement is to allow for additional variation from possible demixing and weight or fill variation. It is stated (23) that just meeting the USP content uniformity criteria is not appropriate for blends. The blend criteria also should not be relaxed because of sampling difficulties. It is more appropriate to change the sampling procedure to ensure accurate results.

The advantages of these criteria are that they are easily understood and implemented and any firm that meets them would be highly confident of satisfactorily passing a GMP or preapproval inspection. However, these criteria have a number of disadvantages. A firm is penalized for taking more samples, since the probability of finding an out-of-range sample increases accordingly. These criteria also assume that current sampling practices can always provide a consistent collection of unit dose samples representative of the powder blend.

b. Standard Deviation Prediction Interval (SDPI)

Since uniformity is of primary interest in powder blend validation and because of a concern that a constant sampling error can occur, one approach is to base the criteria only on variability. The standard deviation prediction interval (SDPI) allows one to predict, from a sample of size n, and with a specified level of assurance, an upper bound on the standard deviation of a future sample of size m from the same population. This approach is recommended in the PDA paper on blend uniformity (18).

By setting the future sample size m to 10, which is the stage-1 sample size for the USP content uniformity test, and by requiring that the upper bound on the standard deviation of a future sample of size 10 be less than 6.0%, which is the USP stage-1 RSD requirement, the SDPI approach can be tied to the USP content uniformity test. The SDPI equation in the Statistical Techniques/Approaches section can be rearranged to obtain the following equation:

Table 4B Advantages and Disadvantages of Various Statistical Techniques for Finished Product (Tablets/Capsules)

Test	Advantages	Disadvantages
CU/dissolution:		
1. Simulation	Tied directly to USP Can be tied to either Stage 1 or full test Can handle nonsymmetric potency limits	Not a function of n Does not provide high assurance of passing USP test (only point estimate)
2. Tolerance interval	Easy to calculate Rewarded for larger n	Not tied directly to USP overall test Difficult to apply for Sampling Plan 2 Factors can be hard to find for non-standard coverage probabilities
3. Bergum approach	Rewarded for larger n Tied directly to USP overall test Easy table look-up Provides high assurance of passing USP test	Does not directly address nonsymmetric potency limits Computer program required (but can be provided) Difficult to pass using Sampling Plan 2 with few locations
Potency (composite assay):		
1. Confidence interval	Provides strong statement that overall batch average potency is acceptable	Does not provide assurance that a given assay result will meet requirements
2. Tolerance/prediction interval	Provides strong statement that an individual assay result will meet assay requirements	Requires a large number of composite assays

$$S_{cr} = \frac{S_m}{[F_{1-\alpha,m-1,n-1}]^{1/2}}$$

where

n = size of current sample
s_{cr} = critical standard deviation
s_m = upper bound of a future sample of size m
$1 - \alpha$ = confidence level (e.g., 0.90)
F = critical F-value

S_{cr} becomes the maximum acceptable sample standard deviation to meet the acceptance criteria. If the sample standard deviation, s_n, is less than s_{cr}, then we are guaranteed, with a minimum assurance of $100(1 - \alpha)\%$, that the upper prediction bound for a future sample of size 10 will not be greater than 6.0% of the target concentration.

c. Tolerance Interval Approach

To use the tolerance interval as an acceptance criterion, the confidence level $100(1 - \alpha)\%$ and coverage level P need to be chosen. One approach is to assume that the blend samples are the same as the resulting final product from that blend. To tie the tolerance interval to the USP content uniformity test, one choice for capsules might be to use a coverage of 90%, since the USP allows three capsules out of 30 to be outside 85%–115% of label claim. If the tolerance interval is completely contained within the 85%–115% interval, this acceptance criterion would be met. For tablets, the coverage level would be approximately 96.7% (29/30), since only one tablet out of 30 is allowed out of 85%–115% of claim. This approach is not as appealing for application to blends, since, without the weight variation of the finished dosage form, there is no reason that blends that go into the capsules should be any looser than the blends that go into the tablets. Use of the interval associated with tablets may be preferred. Another choice of how to define the coverage P is discussed in Ref. 18. Although difficult to find tolerance factors for nonstandard coverage levels in published tables, they can be generated using the interval statement in the SAS/QC® procedure CAPABILITY (24).

Tolerance intervals assume that sampling is done using Sampling Plan 1. There is only one variance component used to estimate the variance of a single observation (i.e., the sample variance). The degrees of freedom used to determine the tolerance factor k is the degrees of freedom associated with the sample variance. However, if Sampling Plan 2 is used, there are two variance components used to estimate the variance of a single observation. Therefore, the degrees of freedom must be approximated. This can be done using Satterthwaite's approximation (25).

d. Bergum Approach

Since the USP content uniformity test is applied only to the finished product, application of the Bergum approach requires that the acceptance limits for the blend be tied to either the capsule or the tablet USP test. The same points mentioned earlier in Sec. IV.A.1.c, "Tolerance Interval Approach," are appropriate when deciding whether to apply the tablet or capsule test criteria to the blend data. In addition, for any of the approaches, each result is generally expressed in percentage of label claim as a percentage of active in a theoretical tablet weight. An alternative to the SDPI approach (which is not dependent upon the mean) is to express each result as a percentage of the sample mean and to then apply the Bergum approach. This has the effect of removing the mean effect and just evaluating the variability. If this were done, then the acceptance limit would be the RSD associated with a sample mean of 100%.

e. Holistic Approach

The PDA report (18) proposes a "holistic" approach to the validation, in which means and variances of the blend are compared to the means and variances of the final product. The validation is considered successful if all criteria are met for both the blend and the final dosage form. If the final product fails, the validation is unsuccessful, regardless of the blend results. However, there are situations where the final product is acceptable and the blend is unsuccessful but the true blend uniformity can be deemed acceptable. This is because the inconsistent results might be due to sampling error when sampling the blend. The blend may have good location-to-location variability, but, because of sampling errors, the within-location error causes the blend results to fail. One approach given by the PDA paper (18) is to use "analysis by synthesis." To employ this technique, Sampling Plan 2 must be used for both the blend and the final dosage form so that the between- and within-location variance components can be estimated. These variance components, as well as the total variance, can be tested statistically using an F-test to determine if there is a significant difference between the variances at the two stages. If the within-location variance component in the blend is significantly higher than that in the final product, then the within-location variance component for the final product is substituted for the within-location variance component of the blend, in an attempt to remove the effect of sampling error in the blend sample results. This reduced overall variance for the blend is compared to the acceptance criteria.

2. Average Potency

There may be a desire to assess possible potency loss between the different sample stages. There also may be an interest in assessing whether the average potency results are at the target potency. At the blender stage, the average potency can be determined either from taking the average of several potency assays or by using the average of the uniformity values if it is felt that there are no sampling issues associated with the smaller sample quantity and if the assay and content uniformity methods are the same. If it is not clear whether there will be sampling issues during validation, it is suggested, when possible, that a formulation study be conducted prior to the validation to determine whether the smaller sample quantity will provide consistent uniformity results and, if not, what sample quantity will produce consistent results. With this support in hand, the smallest sample quantity that will provide consistent results should be used for the validation. It is understood, however, that it may not always be possible to conduct such prevalidation studies.

If a comparison across stages is to be performed, it is recommended that all powder results be reported as percentage of label claim, and not as a percentage of theoretical, so as to provide a direct comparison of the average results to finished product. One must remember that the potency results obtained prior to any adding of lubricant must be adjusted down to account for the fact that the lubricant was not included at the time of sampling. To compare the average uniformity or potency results across stages, one can require that the averages at each stage be within some stated amount of each other or of target. Statistically based techniques such as ANOVA or confidence intervals using the variation of the data and a stated assurance level can also be used.

B. Finished Product

1. *Content Uniformity/Dissolution*

a. *Bergum Approach*

The Bergum approach is written specifically for tablets or capsules. This approach is recommended in the PDA paper (18) for final product testing. For content uniformity, when the potency limits are not symmetrical about 100% of label claim, the USP content uniformity test allows the individual results to be expressed as either a percentage of the label claim, the found mean, or the average of the upper and lower potency specifications, depending on the value of the sample mean. Acceptance limits have not been constructed for this more complicated situation. One approach to this problem is to evaluate the content uniformity results twice. First express the sample mean as a percentage of label claim, and then express the mean as a percentage of the average of the potency specifications. To pass the acceptance limits, both means must meet the acceptance criteria. To use the dissolution acceptance limit tables, the value of Q is required.

b. *Simulation*

One approach is to assume the sample mean and standard deviation are the true population mean and standard deviation, to provide a ''best estimate'' of the true probability of passing. This has the advantage that it can provide estimates of the probability of passing at any stage and can handle the nonsymmetric potency shelf life limits in the content uniformity test. The disadvantage is that is does not provide a bound on the probability with high assurance and is not a function of sample size. It can provide a good summary statistic of the content uniformity data, however.

c. *Tolerance Interval*

See Sec. IV.A.1.c for comments regarding the use of tolerance intervals as acceptance limits for content uniformity data. Tolerance intervals can also be used as acceptance limits for dissolution. Since the USP dissolution test for stage 1 is that all six capsules be greater than $Q + 5$, the tolerance interval could be tied to the USP test by requiring that the lower bound on the tolerance interval be greater than $Q + 5$. To obtain a 95% probability of passing at stage 1, the coverage P of the tolerance interval would need to be $(0.95)^{1/6}$, or 0.991. Using a tolerance interval based on stage 1 of the USP test can be very restrictive.

d. *Confidence Interval*

Confidence intervals are not recommended for evaluating content uniformity data. However, an approach that is less restrictive than tolerance intervals for evaluating dissolution data is to base the acceptance limits on meeting the second and third stages of the USP dissolution test. Both the second and third stages require that the sample mean be less than Q. Therefore, a lower one-sided confidence interval for the population mean could be used as an acceptance limit. The criterion is that the lower bound on the confidence interval be greater than Q.

2. *Potency*

Potency can also be evaluated during validation. It is assumed that some number of composite assays are tested during validation. One criterion might be to generate a $100(1 - \alpha)\%$ confidence interval about the mean using all the potencies collected. This interval will contain the true batch potency, with $100(1 - \alpha)\%$ confidence. This interval should be contained within the potency ''in-house'' or release limits. Enough potencies should be looked at to have sufficient power that this interval will be contained within the desired limits.

Meeting the foregoing criterion should not be interpreted to mean that an individual composite potency assay will meet the ''in-house'' limits with high assurance. If this is desired, a prediction interval for a single future observation or, better yet, a tolerance interval should be used. The validation specialist should be cautioned that additional composite assays might need to be tested to meet either one of these criteria with high confidence.

At a minimum, each of the composite assay results obtained should fall within the desired limits, either the potency shelf specifications or the potency ''in-house'' (or release) limits. The ''in-house'' limits are felt to be the more appropriate, since these are the limits that ensure that the product will meet the shelf limits throughout expiry.

3. *Other Validation Issues*

Validation data should be plotted whenever possible. For example, content uniformity and dissolution can be plotted against the sample locations. This allows for a visual check for trends. A criterion requiring "no trends of note" or that some specific trend rule be met [such as Nelson's mean square successive difference trend test (26)] might be included as part of the acceptance criteria. Some companies use more of a process-capability approach to determine consistency of test results across sample locations.

It is desirable that samples be sent to the laboratory for testing in a designed and ordered way, to be able to separate laboratory effects from process effects if it becomes necessary. For example, if four units were to be tested from each sample location, send one-half of the units from each location to the laboratory on each of two days. In practice, the laboratory may resist doing this.

Weights of individual dosage units should be obtained for every unit tested for both content uniformity and dissolution at the time the units are tested. This may be useful information for later investigation if unacceptable test results are obtained.

For coated products, since this is the finished form, sampling and testing should also be conducted. However, the emphasis is usually on the cores, where the sample identity across the batch is known and can be evaluated. At the coated stage, the effect of the coating solution on dissolution is probably of most interest. Individual coating pans, either all of them or some portion of them, should be sampled and tested, with pan number identity maintained.

V. EXAMPLES

The two examples given in this section demonstrate the application of some of the statistical techniques described in previous sections using both Sampling Plans 1 and 2. Example 1 uses Sampling Plan 1 and Example 2 uses Sampling Plan 2. In each example, samples are taken from the blend and from the final product (capsules were chosen). Samples from both the blend and final capsules are tested for content uniformity. The final capsules are also tested for dissolution. We assume that the USP dissolution specification for this immediate release product has a Q of 85% at 30 minutes. Suppose the blend samples are taken from a V-blender. This type of blender looks like a V with a left and a right side of the V. Samples are taken from the front and back of each side of the blender from the top, middle, and bottom of the granulation, for a total of 12 locations. Assume that the data are in percentage of label claim units. Although a 90% confidence level is used throughout the example, 95% is also a typical confidence level. For the Bergum approach, a 95% probability of passing is used throughout. All tolerance factors were calculated using the interval statement in the SAS/QC® procedure CAPABILITY (24).

A. Example 1 (Sampling Plan 1)

Blend. Using Sampling Plan 1, a single content uniformity result is obtained from each location in the V-blender, with the following results:

Blend Data Display

	Side	
Location	Left	Right
Front		
Top	100.76	92.53
Middle	97.17	98.22
Bottom	95.64	101.91
Back		
Top	100.88	98.97
Middle	97.93	96.30
Bottom	95.63	97.13
Mean = 97.76		
Std Dev = 2.64		
RSD (%) = 2.70		

For the tolerance interval approach, a 90% coverage is used, since capsules are being evaluated (see Sec. IV.A.1.c). The 90% two-sided tolerance interval to capture 90% of the individual content uniformity results is $97.76 \pm 2.406(2.64) = (91.41, 104.11)$. Since the interval is completely contained within the 85%–115% range, the criterion is met.

Note: as mentioned in Sec. IV.A.1.c, if the coverage level associated with tablets (96.7%) were used instead of the coverage level associated with capsules (90.0%), the tolerance factor would be 3.112 and the tolerance interval would be (89.54, 105.98). This, too, would meet the criterion.

The s_{cr} based on the SDPI is $6.0/\sqrt{2.27} = 3.98\%$. Since the standard deviation for the example is 2.64%, which is less than s_{cr}, this sample meets the acceptance criterion.

To use the acceptance limits proposed by Bergum, an acceptance limit table is generated to give the upper bound on the sample RSD for various values of the sample mean. For this example, the table was constructed for capsule content uniformity using a 90% confidence level with a lower bound (LBOUND) of 95%. A portion of the acceptance limit table is as follows:

Mean (% claim)	RSD (%)
97.5	3.64
97.6	3.66
97.7	3.68*
97.8	3.70

*denotes table entry of interest

The sample mean for this example is 97.76%, so the upper limit for the sample RSD is 3.68% (marked with *). It is recommended that the means always be rounded to the more restrictive RSD limit so that the assurance level and lower bound specifications are still met. So in this case, 97.76% is rounded to 97.7%. Therefore, since the sample RSD of 2.70% is less than the critical RSD of 3.68, the acceptance criterion is met. This means that, with 90% assurance, at least 95% of samples taken from the blender would pass the USP content uniformity test for capsules. As mentioned in Sec. IV.A.1.c, if the USP tablet criterion were evaluated instead of the capsule criterion, the upper limit for the sample RSD would be 2.98% and the criterion would also pass.

Capsules. Assume that during encapsulation, a sample was taken at each of 30 locations throughout the batch. One capsule from each location was tested for content uniformity and one for dissolution, with the following results:

Data Display

CU:

99.19	96.38	98.82	98.53	94.37
97.33	95.97	101.32	97.78	97.03
97.05	94.39	100.85	97.77	95.42
95.42	96.73	101.29	96.80	103.03
99.23	97.28	97.52	100.26	95.27
97.36	91.77	98.23	98.07	98.35

Mean = 97.63
Std Dev = 2.34
RSD (%) = 2.40

Data Display

Dissolution:

93.78	94.65	87.83	96.81	92.57	87.68
92.17	88.01	96.59	101.46	93.75	99.44
95.27	92.47	98.46	96.34	93.52	90.73
92.75	94.53	88.72	89.58	97.37	96.41
90.93	96.11	93.41	96.60	94.45	92.82

Mean = 93.84
Std Dev = 3.47
RSD (%) = 3.69

A 90% tolerance interval to capture 90% of the individual content uniformity test results is 97.63 ± 2.025(2.34) = (92.89, 102.37). Since this interval is contained within the 85%–115% interval, the criterion is met.

Using a criterion based on passing stage 1 of the USP dissolution test, a lower one-sided 90% tolerance interval to capture 99.1% of the individual dissolution values is 93.84 − 2.930(3.47) = 83.67. Using this criterion, dissolution would fail, since the lower bound is less than $Q + 5$, which is 90.

Using a criterion based on stages 2 and 3 of the USP dissolution test, a lower one-sided 90% confidence interval for the population mean is $93.84 - 1.311(3.47)/\sqrt{30} = 93.01$. Since the lower bound on the confidence interval for the mean is greater than Q, these results would pass the criterion.

The Bergum acceptance limit table for capsule content uniformity and dissolution are as follows:

Content Uniformity ($n = 30$)

Mean (% claim)	RSD (%)
97.5	4.69
97.6	4.70*
97.7	4.72
97.8	4.73

Dissolution ($n = 30$)

Mean (% claim)	RSD (%)
93.6	8.55
93.8	8.60*
94.0	8.65
94.2	8.70

Since the sample RSD values of 2.40% for content uniformity and 3.69% for dissolution are less than the corresponding acceptance limits from the tables of 4.70% and 8.60%, both tests pass the acceptance criterion.

B. Example 2 (Sampling Plan 2)

Blend. Two samples are taken from each location in the V-blender, with the following results:

Location	Data (% label) 1	2	Summary Statistics Mean	Variance	Std Dev
1	90.45	99.19	94.82	38.12	6.17
2	95.90	99.33	97.62	5.88	2.42
3	89.86	99.18	94.52	43.43	6.59
4	96.88	92.55	94.71	9.37	3.06
5	98.40	94.23	96.32	8.69	2.95
6	100.03	106.50	103.27	20.93	4.57
7	93.74	96.36	95.05	3.43	1.85
8	106.43	100.24	103.34	19.16	4.38
9	101.72	97.18	99.45	10.31	3.21
10	97.32	99.64	98.48	2.69	1.64
11	100.58	98.39	99.48	2.40	1.55
12	90.49	95.48	92.99	12.45	3.53

To apply the tolerance interval, SDPI, and Bergum approaches, it is necessary to compute the following variance components:

Variance Components		
Source	Mean Square	Estimate (Std Dev)
Between	23.20	2.056
Within	14.74	3.840
Total		4.356

The estimated standard deviation of a single observation is 4.356.

To use the tolerance interval approach, the Satterthwaite approximate degrees of freedom (d.f.) is 21.48. The 90% tolerance interval to capture 90% of the individual capsule content uniformity results is 97.50 ± 2.112(4.356) = (88.30, 106.70). The tolerance factor was determined using linear interpolation. This would meet the criterion, since the interval is completely contained within the interval 85%–115%.

As mentioned in Sec. IV.A.1.c, if the coverage level associated with tablets (96.7%) were used instead of the coverage level associated with capsules (90.0%), the tolerance factor would be 2.731 and the tolerance interval would be (85.60, 109.40). This would just barely meet the acceptance criterion.

The s_{cr} using the SDPI is $6.0/\sqrt{1.94} = 4.31$ using the degrees of freedom from the Satterthwaite approximation. The sample standard deviation (4.356) does not pass this criterion.

The Bergum approach requires calculating the standard deviation of the location means, the within-location standard deviation, and the overall mean:

Mean = 97.50

SE (within location std dev) = 3.84

Std dev of location means = 3.41

The standard deviation of location means is computed by taking the standard deviation of the location means. It is *not* the between-location variance component.

A portion of the acceptance limit table generated to meet the capsule criterion is as follows:

	Standard Deviation of Location Means					
	3.3		3.4		3.5	
SE	LL	UL	LL	UL	LL	UL
3.7	98.7	101.9	99.5	101.5	100.5	101.0
3.8	99.0	101.8	99.8	101.3	100.9	100.9
3.9	99.3	101.6	100.2	101.2	.	.*

The lower (LL) and upper (UL) acceptance limits for the sample mean are given for various values of the standard deviation of location means and the within-location standard deviation (SE). For our example, after rounding the standard deviation estimates up to the more restrictive values, the combination of 3.5 for the standard deviation of location means and SE of 3.9 is off the table. So this combination has too large a combination of standard deviations to pass the criterion. Therefore, the criterion fails. If the USP tablet criterion were evaluted instead of the capsule criterion, this would be even more restrictive and would also fail the criterion.

Capsules. Suppose that four capsules are tested at each of 15 locations throughout the batch for content uniformity and dissolution, with the following results:

CU

Location	Data (% label) 1	2	3	4	Summary Statistics Mean	Variance	Std Dev
1	97.08	99.72	98.37	93.50	97.17	7.13	2.67
2	99.72	100.32	101.01	100.29	100.33	0.28	0.53
3	99.90	98.27	98.88	97.96	98.75	0.73	0.85
4	92.78	92.17	93.44	91.22	92.40	0.89	0.94
5	96.32	96.61	95.66	97.20	96.45	0.42	0.64
6	100.97	102.17	99.06	98.80	100.25	2.57	1.60
7	97.02	95.35	98.65	95.98	96.75	2.08	1.44
8	99.39	98.81	98.63	98.06	98.72	0.30	0.55
9	99.59	97.80	97.67	95.95	97.75	2.21	1.49
10	97.97	98.54	100.26	98.74	98.88	0.96	0.98
11	96.09	97.61	95.49	97.50	96.67	1.10	1.05
12	98.87	97.81	97.28	98.80	98.19	0.60	0.78
13	101.10	102.60	100.48	98.62	100.70	2.71	1.65
14	100.80	100.34	98.49	100.93	100.14	1.27	1.13
15	99.70	100.09	100.14	99.20	99.78	0.19	0.43

Variance Components

Source	Mean Square	Estimate (Std Dev)
Between	18.486	2.057
Within	1.563	1.250
Total		2.407

A 90% tolerance interval to capture 90% of the individual content uniformity results using the Satterthwaite approximation of 21.56 degrees of freedom is 98.20 ± 2.111(2.407) = (93.12, 103.28). The tolerance interval indicates that the capsules have good content uniformity.

The descriptive statistics to use the Bergum approach are:

Mean = 98.20

SE (within-location std dev) = 1.25

Std dev of location means = 2.15

The portion of the table for this combination of results is:

Standard Deviation of Location Means

SE	2.0 LL	2.0 UL	2.1 LL	2.1 UL	2.2 LL	2.2 UL
1.2	91.7	108.3	92.0	108.0	92.3	107.7
1.3	91.8	108.2	92.1	107.9	92.4	107.6*
1.4	91.8	108.2	92.1	107.9	92.4	107.6

The lower and upper acceptance limits for the mean are 92.4 and 107.6. Since 98.2 falls within the interval, the capsules pass the acceptance criterion.

Dissolution

Location	Data (% released) 1	2	3	4	Summary Statistics Mean	Variance	Std Dev
1	101.4	99.5	92.9	94.9	97.16	15.55	3.94
2	106.6	101.4	98.0	100.0	101.51	13.53	3.68
3	103.9	100.6	95.3	100.5	100.07	12.64	3.56
4	96.6	93.5	92.6	94.5	94.28	2.89	1.70
5	89.4	93.1	84.6	92.4	89.89	14.97	3.87
6	90.9	90.7	93.2	91.9	91.67	1.39	1.18
7	93.8	92.6	94.8	99.8	95.27	10.08	3.17
8	99.8	98.6	98.1	92.4	97.23	11.03	3.32
9	92.4	96.0	98.4	88.8	93.90	17.86	4.22
10	100.8	99.5	90.6	99.0	97.50	21.50	4.64
11	95.9	98.2	95.9	95.9	96.47	1.39	1.18
12	103.8	103.4	100.8	104.0	102.99	2.28	1.51
13	95.2	92.2	96.1	94.2	94.43	2.88	1.70
14	96.4	98.7	95.4	101.7	98.03	7.69	2.77
15	95.7	96.7	96.2	95.9	96.13	0.17	0.41

Variance Components

Source	Mean Square	Estimate (Std Dev)
Between	48.253	3.130
Within	9.056	3.009
Total		4.342

A 90% one-sided tolerance interval to capture 99.1% of the individual dissolution values using the Satterthwaite approximation of 31.13 degrees of freedom is 96.44 − 2.907(4.342) = 83.82. The tolerance interval indicates that the capsules are not assured of passing stage 1 of the USP dissolution test. The confidence interval approach based on stages 2 and 3 of the USP dissolution test has a lower bound for the population mean of 96.44 − $1.345\sqrt{48.25}/\sqrt{60} = 95.23$. Since the lower bound of 95.23 is greater than Q, the criterion is met.

Using the Bergum approach, the descriptive statistics are:

Mean = 96.44
SE (within-location std dev) = 3.01
Std dev of location means = 3.47

The portion of the acceptance limit table for this combination of results is:

	Standard Deviation of Location Means		
SE	3.25	3.50	3.75
2.75	88.80	89.10	89.40
3.00	88.90	89.10	89.40
3.25	88.90	89.20*	89.40

The lower acceptance limit for the mean is 89.20%. Since 96.44 is greater than 89.20, the capsules pass the acceptance criterion for dissolution.

C. Analysis by Synthesis

Notice that in Example 2, the blend failed content uniformity but the capsules passed. The approach given in the PDA paper (18) applies an analysis by synthesis as follows:

1. Calculate the variance components for the blend and final capsules:

	Std Dev	
Variance components	Blend	Capsules
Between location	2.056	2.057
Within location	3.840	1.250
Total	4.356	2.407

2. Compare variance components:

Within-location standard deviations: Compare 3.84 in the blend to 1.25 in the capsules. The F-test two-sided p-value is less than 0.001, indicating a significant reduction in within-location standard deviation.

Total variance: Compare 4.356 in the blend to 2.407 in the capsules. The F-test two-sided p-value is less than 0.01, indicating a reduced overall variation in the capsules.

3. Substitute the capsule within location for blend within location:

Variance components:

Blend between-location std dev = 2.056
Capsule within-location std dev = 1.250
Total = 2.406

In this example, this reduces the total standard deviation for the blend from 4.356 to 2.406. The Satterthwaite degrees of freedom is 2.00. It is noted in the PDA technical report (18) that "sometimes this [i.e., using Satterthwaite approximation] will result in a number less than any of the d.f. associated with the individual mean-square terms used in the computation. It is suggested that in such cases the d.f. be selected to be no less than the lesser of these mean-square d.f.'s." This occurred in the preceding example, and so 11 was selected as the appropriate degrees of freedom for the total synthesized variance. With this, an s_{cr} of 3.98 is obtained and the blend passes.

VI. CONCLUSIONS

A number of statistical techniques are described for possible use in the analysis of prospective process validation data of tablets and capsules and some of their advantages and disadvantages discussed. Detailed examples are provided to aid in the understanding of many of the techniques discussed. The authors hope that this entry will stimulate the use of the outlined statistical approaches for the analysis of validation data by industry and their acceptance by the FDA. For powder blends, industry feels compelled to use the FDA approach, for it is most likely to be accepted by the FDA. However, a number of other approaches, such as the SDPI and the Bergum approaches, are also more constraining than the USP test while providing a sound statistical basis for the development of acceptance criteria. For finished product testing of content uniformity and dissolution, the Bergum approach offers a number of advantages that the authors believe should be considered. It is hoped that this entry will not only encourage an increase in the use of statistical techniques for the analysis of validation data but also spur discussion of the relative merits of the various techniques.

REFERENCES

1. FDA. Guideline on General Principles of Process Validation. Rockville, MD: May 1987.
2. Nash, RA. Process validation for solid dosage forms. Pharmaceutical Technol 3:105–107, 1979.
3. Nally, JD. Validation guidelines—industry's perspective. Pharmaceutical Eng 4:21–24, 26, 27, 30, 32, 1984.
4. Chapman KG. PAR approach to process validation. Pharmaceutical Technol 8:22, 24, 26, 28–29, 32–34, 36, 1984.
5. Fry, EM. General principles of process validation. Pharmaceutical Eng 4:33–36, 1984.
6. Fry, EM. Process validation policy. Pharmazeutische Industrie (Germany) 46:601–605, 1984.
7. Fry, EM. Process validation: the FDA's viewpoint. Drug Cosmet Ind 137:46–51, 1985.
8. PMA Validation Advisory Committee. Process validation concepts for drug products. Pharmaceutical Technol 9:78, 80, 82, 1985.
9. Edwards, CM. Validation of solid dosage forms: FDA view. Drug Devel Industrial Pharmacy 15:1119–1133, 1989.
10. Nash, RA. Response to Recent GMP-Validation Interpretation. Clin Res Regulatory Affairs 10:253–264, 1993.
11. FDA. Guide to Inspections of Validation of Cleaning Processes. Rockville, MD: July 1993.
12. Berry, IR, Nash, RA, eds. Pharmaceutical Process Validation. 2nd ed. New York: Marcel Dekker, 1993.
13. O'Shea, E. Aspects of process validation in the manufacturing industry. Irish Pharmacy J 73:107–110, 112, 1995.
14. Berman, J, Planchard, JA. Blend uniformity and unit dose sampling. Drug Develop Industrial Pharmacy 21: 1257–1283, 1995.
15. Murthy, KS, Bozzone, S, Maximos, AS. Process validation of solid oral dosage forms. Pharmaceut Eng 16: 42–44, 46, 50–52, 54–58, 1996.
16. Chapman, KG. A suggested validation lexicon. Pharmaceut Technol 7:51–57, 1983.
17. Kohberger, RC. In: KE Peace, ed. Biopharmaceutical Statistics for Drug Development. New York: Marcel Dekker, 1988, pp 605–629.
18. Blend uniformity analysis: validation and in-process testing. Technical Report No. 25. J Pharmaceut Sci Technol (suppl) 51:1997.
19. Bergum, JS. Constructing acceptance limits for multiple stage tests. Drug Develop Industrial Pharmacy 16: 2153–2166, 1990.
20. Hahn, GJ, Meeker, WQ. Statistical Intervals: A Guide for Practitioners. New York: Wiley, 1991.
21. Lindgren, BW. Statistical Theory. New York: Macmillan, 1968.
22. Graybill, FA, Wang, CM. Confidence intervals on nonnegative linear combinations of variances. J Am Statistical Assoc 75:869–873, 1980.
23. Dietrick, J. Special Forum on Blend Analysis. Presentation sponsored by PDA, Rockville, MD, January 1996.
24. SAS/QC Software®: Usage and Reference, Version 6. 1st ed. Vol. 1. Cary, NC: SAS Institute, 1995, pp 175–186.
25. Satterthwaite, FE. An approximate distribution of estimates of variance components. Biometrics 6:110–114, 1946.
26. Nelson, LS. The mean successive difference test. J Qual Technol 12:174–175, 1980.

James Bergum
Merlin L. Utter

R

Release Targets

See also *USP Tests*

I. INTRODUCTION

Manufacturers of drug products must demonstrate that their products meet with United States Pharmacopeia specifications before the products can be released for sale. One of the important specifications is the lower registration limit (LRL) on a drug's potency. After a drug product is manufactured, the natural degradation process gradually reduces the drug's potency. A drug's potency must remain above the LRL at the end of its registered shelf life. An important factor in determining the potency at the end of its shelf life is the drug's potency at time zero, or call it *initial potency*. Manufacturers need to set an in-house specification on initial potency that is higher enough that a batch that satisfies this specification at time zero will meet the LRL with high certainty. This in-house specification is called the *release limits* (RL) or the *release targets*. A typical scenario is illustrated in Fig. 1.

Some statistical approaches have being developed for determining the release limits of initial drug potency. Allen et al. (1) proposed a simple methodology for calculating the release limits. Greg Wei (2) suggested a similar but more statistically justifiable approach. Jun Shao and Shein-Chung Chow (3) adopted a Bayesian decision theory approach to construct the release limits. This entry will describe only the methods of Allen et al. and of Wei. The Bayesian method of Shao and Chow will not be covered here because of its technical sophistication to most practitioners.

II. ALLEN, DUKES, AND GERGER METHOD

The idea of their approach is very intuitive. In order to ensure that potency over the shelf life is above the registration limit, the release limit is simply set to the registration limit plus a "buffer." The buffer consists of the potency decrease over the shelf life, the variability associated with the rate of decrease, and the variability of assay. These quantities are estimated from stability studies using linear regression analysis.

A. Case I: Product with No Degradation

The expression for the release limit is shown in the following equation:

$$\mathrm{RL} = \mathrm{LRL} + t\frac{S}{\sqrt{n}}$$

where

RL = release limit
LRL = lower registration limit
S = assay standard deviation
DF = degrees of freedom
t = 95% confidence (one-sided) t-value with DF
n = number of replicate assays used for batch release

B. Case II: Product with Degradation

For products whose potency decreases over the shelf life due to a degradation process, RL can be calculated

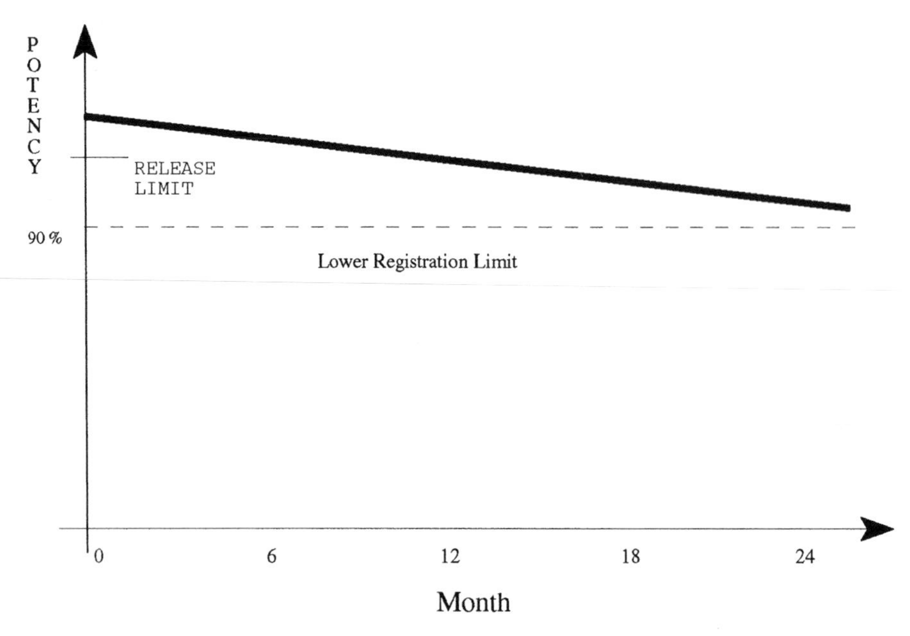

Fig. 1 Release limit of drug potency.

from the following expression:

$$RL = LRL + bT + t\sqrt{S_T^2 + \frac{S^2}{n}}$$

where

b = average degradation rate
T = shelf life
S_T = standard error of (bT)

The degrees of freedom used to determine t can be calculated by means of the Satterthwaite approximation.

C. Case III: Product Requiring Reconstitution Prior to Use

Some products must be reconstituted before use, in which case the potency decrease over the shelf life and its variability must be accounted for in the RL calculation. Release limits can be calculated via the following expression:

$$RL = LRL + bT + rT + t\sqrt{S_T^2 + S_R^2 + \frac{S^2}{n}}$$

where

r = average degradation rate of reconstituted solution
S_R = standard error of average degradation rate of reconstituted solution

The degrees of freedom used to determine t can be calculated by means of the Satterthwaite approximation.

III. CONDITIONAL AND UNCONDITIONAL METHODS

The Allen Dukes, and Gerger method has strong appeal because of its simplicity and intuitiveness. However, their approach lacks a rigorous statistical justification. Wei (2) proposed a more statistical means for RL determination. His method is briefly described next.

A. Decision Rules

First, two types of release limits need to be defined via the following decision rules, with each expressed in a probability statement. The basic idea is that RL should be set at a value such that the probability of failing LRL at the end of shelf life is controlled.

Conditional Rule. Letting $y_{i0.}$ be an observed sample average of time zero potency, find a release limit such that if $y_{i0.} \geq$ RL, then

$$P(\text{lower 95\% CI of true potency at time } T < \text{LRL} \mid y_{i0.}) \leq \alpha$$

This rule defines the release limit as a value such that if the potency measurement at time zero is above this value, then the probability of failure for this batch at time T will be no more than α.

Unconditional Rule. Find RL such that

$$P(\text{lower 95\% CI of true potency at time } T < \text{LRL} \mid \text{RL} \le y_{i0.} \le M) \le \alpha$$

This rule defines the release limit as a value such that as long as the potency measurement at time zero remains above this value and below the upper limit M, the probability of failure will be no more than α. The release limits defined under both the conditional rule and the unconditional rule control the chance of failure under a given α. However, the two rules have different meanings. The release limits based on the conditional rule, denoted by RL(C), keeps the chance of failure for *each* released batch under α, whereas the release limits based on the unconditional rule, denoted by RL(UC), keeps the chance of failure for *all* future released batches under α.

B. Release Limits Based on the Conditional Rule, RL(C)

Assume the observed potency follows the following model with parameters b, σ_α^2, σ_β^2, and σ_e^2. Then for the kth replicate from the ith batch at time t,

$$y_{itk} = \mu + \alpha_i + (b + \beta_i)t + e_{itk},$$
$$t = 0, t_1, t_2, \ldots, T, \qquad k = 1, 2, \ldots, n_t$$

where

y_{itk} = assayed potency of kth replicate, ith batch, at time t
μ = mean batch potency at time zero
α_i = random batch effect on potency at time zero, $\alpha_i \sim N(0, \sigma_\alpha^2)$
b = average potency change rate per time unit
β_i = random batch effect on slope, $\beta_i \sim N(0, \sigma_\beta^2)$
t = sampling time
e_{itk} = total random assay error, $e_{itk} \sim N(0, \sigma_e^2)$

and where α_i and β_i are independent and (α_i, β_i) and e_{itk} are independent.

The same model can also be expressed as

$$y_{itk} = \mu_i + b_i t + e_{itk}, \qquad t = 0, t_1, t_2, \ldots, T,$$
$$k = 1, 2, \ldots, n_t$$

where $\mu_i = \mu + \alpha_i$ and $b_i = b + \beta_i$.

Having observed the sample average of assay values at time zero, the future observed assay value at time t follows a new model:

$$y_{itk}^* = y_{i0} + \eta_i + (b + \beta_i)t + e_{itk}, \tag{1}$$
$$t = t_1, t_2, \ldots, T, \qquad k = 1, \ldots, n_t$$

where $\eta_i \sim N(0, \sigma_e^2/n_0)$.

Let

$$Y_{it} = \begin{pmatrix} y_{it1} \\ y_{it2} \\ \vdots \\ y_{itn_t} \end{pmatrix}, \quad X_{1i} = \begin{pmatrix} 1 \\ 1 \\ \vdots \\ 1 \end{pmatrix}, \quad X_{2i} = \begin{pmatrix} t \\ t \\ \vdots \\ t \end{pmatrix}, \quad E_{it} = \begin{pmatrix} e_{it1} \\ e_{it2} \\ \vdots \\ e_{itn_t} \end{pmatrix}$$

$$Y_i = \begin{pmatrix} Y_{i0} \\ Y_{it_1} \\ \vdots \\ Y_{iT} \end{pmatrix}, \quad X_1 = \begin{pmatrix} X_{10} \\ X_{1t_1} \\ \vdots \\ X_{1T} \end{pmatrix}, \quad X_2 = \begin{pmatrix} X_{20} \\ X_{2t_1} \\ \vdots \\ X_{2T} \end{pmatrix}, \quad E_i = \begin{pmatrix} E_{i0} \\ E_{it_1} \\ \vdots \\ E_{iT} \end{pmatrix}$$

$$Y_i^* = \begin{pmatrix} Y_{it_1} \\ Y_{it_2} \\ \vdots \\ Y_{iT} \end{pmatrix}, \quad X_1^* = \begin{pmatrix} X_{1t_1} \\ X_{1t_2} \\ \vdots \\ X_{1T} \end{pmatrix}, \quad X_2^* = \begin{pmatrix} X_{2t_1} \\ X_{2t_2} \\ \vdots \\ X_{2T} \end{pmatrix}, \quad E_i^* = \begin{pmatrix} E_{it_1} \\ E_{it_2} \\ \vdots \\ E_{iT} \end{pmatrix}$$

In vector form Eq. (1) becomes

$$Y_i^* = (X_1^* X_2^*) \begin{pmatrix} y_{i0} \\ b \end{pmatrix} + X_1^* \eta_i + X_2^* \beta_i + E_i^*$$

The point estimate of potency and its variance at time T under the new model can now be calculated as

$$\hat{Y}_{iT}^* = y_{i0} + \hat{b}_i T$$

$$\text{var}(\hat{Y}_{iT}^*) = T^2 \text{ var}(\hat{b}_i) = T^2 (X_2^{*\prime} X_2^*)^{-2} X_2^* \Sigma^* X_2^{*\prime}$$
$$= T^2 \left[\sigma_\beta^2 + (X_2^{*\prime} X_2^*)^{-1} \sigma_e^2 \left(1 + \frac{N}{n_0}\right) \right]$$

where $N = n_{t1} + n_{t2} + \cdots + n_T$.

To control the probability of failure for this batch, it is sufficient to keep the probability of the 95% lower confidence interval for mean potency at time t below LRL under a certain level (say, α). It is clear this probability is a decreasing function in y_{i0}:

$$P(\hat{Y}_{iT} - z_{0.05}\sqrt{\text{var}(\hat{Y}_{iT})} < \text{LRL} \mid y_{i0})$$

RL is the minimum value of y_{i0}, such as

$$P(\hat{Y}_{iT} - z_{0.05}\sqrt{\text{var}(\hat{Y}_{iT})} < \text{LRL} \mid y_{i0}) \le \alpha$$

In fact, the solution for RL can be expressed as

$$\text{RL(C)} = \text{LRL} + bT + (z_\alpha + z_{0.05})T \cdot \sqrt{\sigma_\beta^2 + (X_2^{*\prime} X_2^*)^{-1} \sigma_e^2 \left(1 + \frac{N}{n_0}\right)}$$

When samples are drawn only at time zero and at the end of study, the preceding expression is reduced to

$$\mathrm{RL(C)} = \mathrm{LRL} + bT + (z_\alpha + z_{0.05}) \cdot \sqrt{\sigma_\beta^2 T^2 + \sigma_e^2 \left(\frac{1}{n_T} + \frac{1}{n_0}\right)}$$

Notice the similarity between this expression and the expression in Case II of the Allen, Dukes, and Gerger method. The release limits based on the conditional rule [RL(C)] is batch specific. For each batch, y_{i0} will be obtained and compared to RL(C) to decide whether the batch can be released. For every batch released in such a way, the probability of failure at time T is controlled. This method does not utilize any prior information about distribution of time-zero potency among batches.

C. Release Limits Based on the Unconditional Rule, RL(UC)

Now, suppose the distribution of time-zero potency among batches is known. Recall the following model:

$$y_{itk} = \mu_i + b_i t + e_{itk}, \qquad t = 0, t_1, t_2, \ldots, T, \quad k = 1, 2, \ldots, n_t$$

Use the same vector notation as before and let $X_i = (X_{1i} X_{1i})$. Then the lower 95% CI for the true potency at time T is

$$\text{Lower 95\% CI for true potency at time } T = \hat{Y}_{iT} - z_{0.05}\sqrt{\mathrm{var}(\hat{Y}_{iT})}$$

Since the lower 95% CI for the true potency at time T follows a normal distribution, as can be seen from the foregoing expression, the probability of failure at time T can be calculated as follows:

$$P(\text{95\% lower CI of true potency at time } T < \mathrm{LRL})$$

Since the estimated time-zero potency is negatively correlated with the predicted potency at time T, the probability of failure at time T is negatively associated with the release limits at time zero. The basic idea is that for those batches that pass the release limits, the probability of failure at time T will be under an allowable value. The release limits RL(UC) can then be solved from the following inequality, in which Y_{i0} is the sample mean assay value:

$$P(\text{95\% lower CI of true potency at time } T < \mathrm{LRL} \mid \mathrm{RL(UC)} \le Y_{i0} \le M) \le \alpha \qquad (2)$$

or

$$P(\hat{Y}_{iT} - z_{0.05}\sqrt{\mathrm{var}(\hat{Y}_{iT})} < \mathrm{LRL} \mid \mathrm{RL(UC)} \le Y_{i0} \le M) = \frac{P(\hat{Y}_{iT} - z_{0.05}\sqrt{\mathrm{var}(\hat{Y}_{iT})} < \mathrm{LRL} \cap \mathrm{RL(UC)} \le Y_{i0.} \le M)}{P(\mathrm{RL(UC)} \le Y_{i0} \le M)}$$

Under normality, RL(UC) can then be solved using the joint distribution of Y_{i0} and the predicted Y_{iT}, as given here:

$$\begin{pmatrix} Y_{i0} \\ \hat{Y}_{iT} - z_{0.05}\sqrt{\mathrm{var}(\hat{Y}_{iT})} \end{pmatrix} \sim N\left(\begin{pmatrix} \mu \\ \mu + bT - z_{0.05}\sqrt{\mathrm{var}(\hat{Y}_{iT})} \end{pmatrix}, \ \mathrm{cov}\begin{pmatrix} Y_{i0} \\ \hat{Y}_{iT} \end{pmatrix}\right)$$

and

$$\mathrm{var}(Y_{i0}) = \sigma_\alpha^2 + \frac{\sigma_e^2}{n_0}$$

$$\mathrm{cov}(Y_{i0}, \hat{Y}_{iT}) = \frac{1}{n_0}(1'_{n_0}, 0'_{n_{t_1}}, \ldots, 0'_{n_T}) \cdot \Sigma X_i (X_i' X_i)^{-1} \begin{pmatrix} 1 \\ T \end{pmatrix}$$

Then the minimum value for RL(UC) that satisfies inequality (2) can easily be found using a numerical search routine.

D. Expected Loss Function Approach

RL(UC) keeps the probability of failure for all future batches under α. A smaller α means that a smaller number of batches fail LRL at time T, but a smaller α will also result in a larger number of batches that fail to be released. A failed batch at time zero or at time T will be a financial loss to the manufacturer. Therefore an alternative way to determine RL(UC) is to find one that minimizes the total expected loss, provided that the financial loss due to failing a batch at time zero and time T are known.

Suppose

Financial loss due to failing a batch at time 0 = L_0

Financial loss due to failing a batch at time T = L_T

Then

$$\begin{aligned}\text{Expected loss} &= L_0 \text{ Prob(a batch failed at time 0)} + L_T \text{ Prob(a batch failed at time } T) \\ &= L_0 \text{ Prob}(Y_{i0} < \mathrm{RL(UC)}) + L_T \cdot \text{Prob}(\mathrm{RL(UC)} \le Y_{i0} \le M \cap \hat{Y}_{iT} < \mathrm{LRL})\end{aligned}$$

RL(UC) that minimizes the preceding function can then be found using a numerical search routine.

All three methods described in this section require prior knowledge about certain parameters. For the conditional method, they are b, σ_β^2, and σ_e^2; for the unconditional method, they are μ, b, σ_α^2, σ_β^2, and σ_e^2. In practice, such knowledge can be obtained from different sources, such as assay validation studies, previous stability studies, published results, and the literature.

IV. CONCLUSIONS

The Allen, Dukes, Gerger method is very appealing to practitioners for its simplicity and intuitiveness. However, their method was not based on a justifiable statistical argument. Given this drawback, their release limit is not clear. Nonetheless, their method may still be satisfactory in practice.

The three methods proposed by Wei were derived from a probability argument of failure rate control. The conditional method is the easiest one to calculate and requires the least information. It should be used in the situation where one has no prior knowledge about the distribution of batch potency at time zero. The unconditional method is an improvement over the conditional method, but at the expense of the availability of prior knowledge about the distribution of batch potency at time zero. However, these two methods both focus on controlling the failure rate of a batch at time T; they do not account for cost due to batch rejection at time zero. The loss function method minimizes the expected loss due to batch failure at time zero and time T when the cost ratio is available. In practice, which method to use will depend on the availability of required information. The estimated release limit should be treated as an important component in the whole process of release limit determination, along with other deciding factors, such as production capability, and other practical issues.

REFERENCES

1. Paul V. Allen, Gary R. Dukes, Mark E. Gerger. Determination of release limits: a general methodology. Pharmaceutical Res 8(9):1210–1213, 1991.
2. Greg C. G. Wei. Simple methods for determination of the release limits for drug products. Biopharmaceutical Stat 8:103–114, 1998.
3. Jun Shao, Shein-Chung Chow. Constructing release targets for drug products: a Bayesian decision theory approach. Appl Stat 40(3):381–390, 1991.

Greg C. G. Wei

Reproductive Studies*

See also *Carcinogenicity Studies*

I. INTRODUCTION

Soon after the thalidomide tragedy in the late 1950s and early 1960s, which resulted in over 8000 malformed babies, there has been a great deal of interest in predictive tests of reproductive and developmental effects. The ultimate goal of the studies is to assess reproductive risk to mature adults and risk to the developing individual at all stages, from conception to sexual maturity, from exposure to drugs and environmental compounds. Adverse reproductive and developmental effects include effects on male and female fecundity, spontaneous abortion, infant and child death, congenital malformations, growth retardation, and mental retardation. Some compounds may cause rather specific adverse effects; others may have a broad spectrum of disturbance to the embryonic and fetal development, resulting in all types of adverse effects.

Human data provide the most appropriate information for determining potential adverse effects of a compound. Human studies include epidemiologic studies and case reports. There are several types of epidemiologic studies; each has certain advantages and limitations. *Cohort* studies compare the results of clinical endpoints between exposed and unexposed groups to

*The views presented in this entry are those of the author and do not necessarily represent those of the U.S. Food and Drug Administration

establish an association between a specific exposure and the outcome. *Case-control* studies compare the individuals with a particular disease to those similar but nondiseased individuals on their exposure history, retrospectively. Because of ethical and practical considerations, dosing in humans to study adverse effects is excluded. Useful information for assessing human risks may be derived from human studies with adequate design and statistical power. However, the information for potential effects on humans is generally obtained using laboratory animal experimental models. Substantial evidence indicates that agents that have been associated with human effects can also be associated with adverse effects in animals.

A. Reproductive Studies

Reproductive studies evaluate potential adverse effects of test compounds on reproductive systems of both adult males and females as well as on the postnatal maturation and reproductive capacity of offspring and the cumulative effects on generations. The studies are designed to evaluate: (a) effects on male and female fertility, (b) effects during gestation on the mother and fetus, (c) effects appearing after parturition on maternal lactation and on offspring growth, development, and sexual maturation, and (d) mutagenic effects through several generations.

B. Developmental Studies

Developmental studies evaluate the effects of test compounds on a developing organism that result from exposure of parent(s) during prenatal development or postnatal development to the time of sexual maturation. The studies are designed to evaluate: (a) death of the developing organism, (b) structural malformations and variations, (c) altered growth, and (d) functional impairment.

II. REGULATORY REQUIREMENTS

The United States Food and Drug Administration (FDA) and other countries require that new drugs and certain medical devices be approved for safety and effectiveness for their intended use before being marketed. Food additives, color additives, compounds for use in food-producing animals, and pesticide residues in foods must be shown to be safe for the proposed use. In 1966, the United States FDA (1) published guidelines for the performance of reproductive and developmental tests of drugs in animals. Three segments of study are required in preclinical animal testing for each new drug depending on how women might be exposed to drug. These are referred to as Segment I (fertility and general reproductive performance), Segment II (developmental effects), and Segment III (perinatal and postnatal evaluations). These studies may be run more or less concurrently and may be run in conjunction with chronic tests.

The Segment I study is aimed at providing an overall evaluation of the effects of drugs on fertility in both sexes, the course of gestation, early and late stages of the development of the embryo and fetus, and postnatal development. The studies may be conducted by treating animals of only one sex and mating them with untreated animals of the opposite sex or by treating both male and female animals. Segment II is aimed primarily at detecting teratogenic effects. The drug is given to the pregnant females during the period of organogenesis, e.g., days 6–15 for rats and mice, and days 18 for rabbits. The offspring are removed one or two days before term, and corpora lutea, resorption sites, and live and dead fetuses are examined. Fetuses are weighed and examined for anomalies. Segment III is aimed at the evaluation of the effects of drugs on the late stages of gestation and on parturition and lactation. The drug is given to pregnant females in the final third of gestation and continued throughout lactation to weaning, e.g., gestation day 15 to postnatal day 21 for rats or mice. The effect on the duration of gestation is determined. Pup birth and developmental data, including litter size, weight, and postnatal growth and mortality, along with impaired maternal behavior are recorded and measured.

Most countries adopted the general principle of requiring the three segments of investigations in the preclinical testing of new drugs. Recently, the International Conference on Harmonization (ICH), representing the United States, the European Community, and Japan, published guidelines to address uniformity in conducting Segment I, II, and III studies (2–4). The ICH guidelines allow standardization of protocols and deletion of duplication of procedures, with emphasis on flexibility in testing strategy. Testing for food additives, color additives, and animal drugs requires a Segment II developmental study, with the incorporation of multigeneration reproductive studies (5–8). Testing requirements for environmental agents such as pesticides and industrial chemicals, regulated by the U.S. Environmental Protection Agency (9,10), are similar to those for food additives; that is, a standard Segment II developmental study and two-generation reproductive studies are required. In general, testing protocols for pharmaceutical and nonpharmaceutical chemicals are similar; however, the data for nonpharmaceutical chemicals would be used to establish an acceptable level of exposure using quantitative risk assessment.

When a drug has been approved for marketing by the FDA, the information for the safe and effective use of the drug must be on the label. The FDA (11) published a uniform labeling format for human prescription drugs. The pregnancy labeling provides information with regard to use during pregnancy for all drugs. The FDA has devised five categories to describe levels of potential risk to human pregnancy based on findings from animal and human studies (11,12). Category A is to be used for drugs where adequate, well-controlled studies in pregnant women have failed to demonstrate risk to fetus. Category B is used for those drugs for which no controlled studies have been conducted in human pregnancy and animal studies have not indicated the potential for human developmental risk. Alternatively, animal studies may indicate some risk but controlled human studies do not. Category C is used for drugs where there are no controlled studies in human pregnancy and animal studies demonstrate an adverse effect, or have not been conducted. Potential benefits, however, may justify the potential risks. Category D is for drugs where there is positive evidence of human developmental risks. Investigational and postmarketing surveillance data may show human developmental risks, but potential benefits may outweigh the human risks. Category X is used for agents where there is documented evidence of human development toxicity from reports in the literature or postmarketing surveillance. The risk to human conceptus is considered to outweigh any possible benefit.

III. STATISTICAL DESIGN/METHOD/ANALYSIS

A. Experimental Design

Mice, rats, and rabbits are the most commonly used species for reproductive and developmental studies. The various guidelines generally require testing in a rodent and nonrodent species, usually rats and rabbits. An experiment typically consists of an untreated control and three dose groups. The highest dose is chosen so as to produce minimal maternal toxicity, ranging from marginal body weight reduction to not more than 10% mortality. The low dose should be a non-observed adverse-effect level for both maternal and developmental toxicity. The midlevel dose usually is halfway between the low and high doses. A fourth dose group may be added to avoid excessive dosage intervals and to assess dose–response relationships. Sometimes doses are based on known or expected human therapeutic or toxic levels. Typically, dosage is measured in milligrams per kilogram body weight per day. The U.S. regulatory guidelines generally recommend about 20 pregnant rodents and 15 nonrodent animals per dosage group. The ICH guideline recommends the use of 16–20 pregnant animals. Typical litter sizes (number of viable offspring) for the control animals range from 8 for rabbit to 12 and 14 for mice and rats, respectively.

Depending on the design, a test compound is administered to either parent prior to conception and to the female during prenatal development and postnatally. Animals can be assigned to dose groups totally randomly or by stratified body weight (but randomly within body weight classes). Each animal is monitored throughout the experiment to detect toxic and pharmacologic effects of the test compounds. In a Segment II study, the pregnant dams normally are sacrificed just prior to term, and fetuses are then examined to assess developmental effects. In a Segment III study, females are allowed to deliver and rear their offspring. Postnatal functional evaluations are performed. All adult animals are necropsied at terminal sacrifice (3,13).

During the reproductive process, ovarian follicles release eggs that are fertilized by sperm to form embryos. These embryos subsequently implant into the wall of the uterus. The discharged ovarian follicles differentiate into corpora lutea, which secrete hormones necessary to sustain pregnancy. The embryos undergo organogenesis to form fetuses. Each step in this process is a potential target for toxic effects of a test compound. Depending on when treatment begins, a test compound may affect the number of ovulating follicles, the number of corpora lutea, the number of embryos implanting into the wall of the uterus, or the number of live fetuses surviving to term. For surviving fetuses, growth retardation may occur or fetuses may exhibit one or more types of structural malformations. Figure 1 is a sketch of the outcomes from a dam in the developmental and reproductive toxicity experiment.

Regulatory requirements specify that a wide range of endpoints must be measured, recorded, and analyzed. The endpoints can be divided into two categories: parental and embryonic/fetal endpoints. Parental endpoints to assess the reproductive effect include: body weight and weight gain, mating index, fertility index, changes in gestation length, sperm count, numbers of corpora lutea, implantation sites, dead or resorbed implants, viable fetuses, and target organ

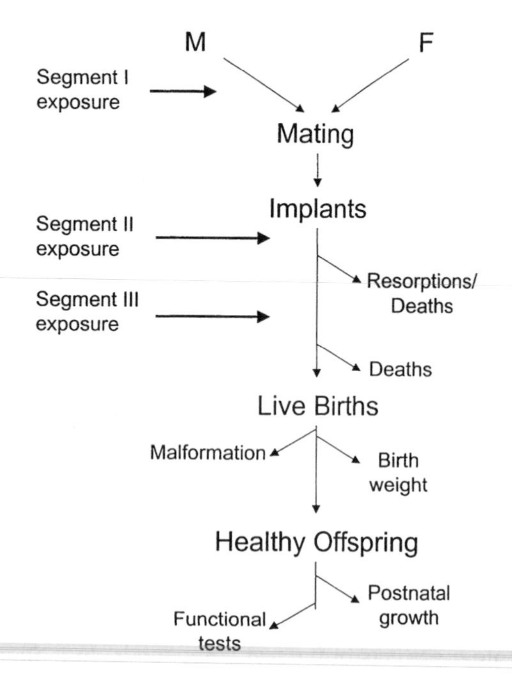

Fig. 1 Schematic representation of Segment I, II, and III studies.

weights, as well as food and water consumption. Embryonic/fetal endpoints include: individual malformations (external or gross, visceral, and skeletal) and variations in viable fetuses, number of normal fetuses, fetal weight, fetal length, sex ratio, preweaning and postweaning survival and body weight, physical development, and functional tests (2–3,9–10,13).

1. *Experimental Unit*

In the analysis of animal developmental endpoints, the experimental unit is the entire litter rather than an individual fetus. In the experiment, the test compound is administered to the adult animal; the primary variables of interest are fetal responses. The effect of the test compound occurs in the female that receives the compound or that is mated to a male that receives the compound. The treatment affects the fetuses indirectly via the dam. Thus, the development of individual fetuses in a dam are not independent. The fetal responses from the same dam are expected to be more alike than responses from different dams. This phenomenon is referred to as the *litter effect.* The proper experimental unit in the analysis should be the litter, with the fetal responses representing multiple observations from a single experimental unit. Failure to account for the intralitter correlations by using each fetus as the experimental unit will inflate the Type I error and will reduce the validity of the test.

2. *Sample Size*

Sample size determination is an important issue in detecting a dose effect. The probability of detecting a statistically significant effect increases with the increased magnitude of effect and the increased sample size. It is customary to specify a range of the magnitudes of the effect of interest; an optimal sample size then can be calculated to provide a reasonable power of detecting the effect. Sample size calculations and references for various tests are provided in Desu and Raghavarao (14). It should be emphasized that sample size determination is preferable to arbitrary requirements of testing 10 or 20 animals, as indicated in various guidelines for assessing reproductive and developmental effects.

3. *Statistical Analysis*

Statistical analyses of various endpoints have been of two kinds: qualitative testing for adverse effects and quantitative estimation for risk assessment. The qualitative testing is to determine if there is a statistically significant difference among the groups. Three statistical tests are conducted: (a) an overall test for significant differences among groups, (b) pairwise comparisons between two specific groups (usually, control versus a dose group), and (c) a trend test to identify a dose-related increase or decrease. The trend test is more sensitive (powerful) in detecting dose effect than the pairwise comparisons, since it uses information from all dose groups. The quantitative estimation is to determine a safe human level of exposure from an assessment of the dose–response relationship. The quantitative risk estimation typically does not apply to pharmaceutical drugs.

B. Methods

Endpoints are typically measured according to one of the three scales: continuous, count, and quantal (binary). Histological data, such as severity or levels of achievement scores in behavioral testing, may be recorded on the ordinal scale. Statistical methods appropriate for these different types of data are discussed separately.

1. *Continuous Data*

Continuous data, such as body weights, organ weights, or behavioral measurements conducted on offspring following birth, are measured on a continuous scale. The continuous endpoints are measured either at the litter level in an adult animal (e.g., maternal body weights) or at the individual fetus level (fetal body weights). Analysis of variance (ANOVA) is the most commonly used procedure for analysis of continuous data (15,16). The ANOVA procedure is used to assess the overall significant difference among groups. The ANOVA method assumes that data are independently and normally distributed with homogeneous variance. Transformations such as logarithmic, square-root and arc-sine are often applied to satisfy the normality assumption and stabilize the variance. A simple one-way ANOVA analysis is the comparison of maternal endpoints among groups. Developmental endpoints traditionally are analyzed similarly but in terms of the average within each litter, a litter-based analysis (17). Nonparametric methods are used when the assumption of normality fails. The nonparametric analysis is initiated by ranking all observations of the combined groups. The nonparametric method is then applied to ranked data. Further details may be found in general texts addressing nonparametric methods (18).

There are a wide variety of post hoc tests available for pairwise comparisons or trend analysis after finding a significant result in an ANOVA. The Dunnett (19) and Williams (20) tests are the two most used tests for comparisons of each of the dose groups to the control. The Williams test assumes an increasing (decreasing) dose–response trend. When this assumption is valid, the Williams test is more powerful in detecting differences between the control and the dose groups. The Duncan multiple-range test (21) and Tukey studentized-range test (22) are two frequently used procedures for all possible pairwise comparisons. Linear regression is the simplest method for assessing dose–response trends. But contrast tests, with a proper choice of contrast coefficients, in conjunction with ANOVA are more commonly used for the dose–response trend test. Other trend tests are Tukey's test (23), which has been proposed to identify the highest dose at which there is not statistically significant trend, and Bartholomew's test of ordered alternatives (24).

ANOVA is also used in more complex experiments involving crossed (e.g., dose levels, replicates) and nested (e.g., litter effects) factors. The repeated-measures ANOVA is often used for the analysis of postnatal behavioral data (25). Multivariate analysis of variance (MANOVA) is used for simultaneous analysis of several endpoints, including repeated-measures data (25,26). A general approach to modeling developmental data can be carried out in terms of a mixed-effects model. Dempster et al. (27) proposed a normal mixed-effects model with two levels of variance, in which litter effect is modeled by a nested random factor and dose by a fixed factor, for analyzing fetal weights.

2. *Count Data*

A number of primary reproductive endpoints are measured in counts. In a dominant lethal assay, male mice are treated with a suspect mutagen and then are mated with females. The numbers of corpora lutea, implantations, lives, and deal conceptuses are counted to assess the reproductive effects of the test compound. Count data are often normalized by the square root transformation; the transformed data are then analyzed as continuous data using the parametric ANOVA methods. Count data can also be analyzed by nonparametric methods.

Count data are generally modeled by a Poisson distribution. The mean of a Poisson is often expressed as a log-linear function of dose and other covariates in the Poisson regression analysis. The Wald test, score test, or likelihood ratio test is used for group comparisons and trend test (28). A common complication in the analysis of count data is that the observed variation exceeds or falls below the variation that is predicted from a Poisson model. The negative binomial (gamma-Poisson) distribution is a generalization of the Poisson model to account for the extra variation. Parametric and quasi-likelihood analyses of reproductive data with extra Poisson variation are given by Chen (29).

3. *Binary Data*

The binary endpoints, as continuous endpoints, can be measured either at the parent level, such as success or failure of pregnancy, or at the individual fetal level, such as presence or absence of a particular malformation type. Statistical methods for the analysis of the prenatal and fetal responses are different.

Two common approaches for the analysis of prenatal binary endpoints are the asymptotic tests and exact tests (permutation tests). Asymptotic tests include the Pearson chi square test and likelihood ratio chi square test for the comparisons of the incidence rates among several groups (30). A continuity correction is often applied if only two groups are compared. The Cochran–Armitage test is used to test for trend (31,32). The Fisher exact test is the best-known permutation test for comparing two groups. The permutation test computes the values of a ''proper test statistic'' from the observed data for all permutations (sampling without replacement) under the null hypothesis. The observed p-value can be obtained by comparing the value of the observed test statistic with the empirical distribution. Ciminera (33) proposed a randomization test for dose–response trend. General computational algorithms and software to perform all possible permutations are given by Mehta et al. (34).

Since the litter rather than the fetus is the basic experimental unit, the analysis of fetal endpoints should account for the litter effect. The general approach to the analysis of developmental binary endpoints is to consider the proportion per litter, such as the proportion of live fetuses with a certain type of malformation. Two approaches are commonly used for the analysis of proportions: litter-based analysis and extra-binomial modeling. Both approaches assume that the denominator ("litter size") is fixed.

Litter-based analysis has been widely used in the analysis of developmental data. Typically, the proportions are transformed by an arc-sine transformation, and then the parametric ANOVA methods are used. The litter-based approach does not use the data effectively, since it does not account for the litter size. For example, one out of two is treated as the same as five out of ten. Extra-binomial modeling has been an important development in the methodology for the analysis of developmental data. The beta-binomial model and quasi-likelihood method have been well studied for the analysis of developmental data (28,35,36). The mean of the proportion is often expressed in terms of a logistic function. The Wald test, score test, or likelihood ratio test is used for group comparisons and trend test in the analysis with the beta-binomial model. The parametric and quasi-likelihood analyses of Segment II developmental data have been given by Chen (29).

C. Analysis

The commonly used litter-based method is presented in the data analysis. A typical experiment consists of g groups, a control and $g - 1$ dose groups. Assume that the ith group contains m_i female animals. Let y_{ijk} be the response from a fetus out of n_{ij} examined or tested for a particular developmental outcome, $1 \leq i \leq g$, $1 \leq j \leq m_i$, and $1 \leq k \leq n_{ij}$. Note that y_{ijk} may be an indicator variable representing the presence or absence of a particular malformation type or a continuous variable representing a fetal weight or postnatal performance measurement. Depending on the endpoint of interest, n_{ij} may represent the number of viable fetuses, the number of implants, or the number of measurements. The litter-based analysis is based on the per-litter response $y_{ij} = \Sigma_k y_{ijk}/n_{ij}$. For continuous response, y_{ij} will represent the mean fetus response; for a discrete variable, it will represent the sum of the fetal responses. The y_{ij} can be viewed and analyzed as a maternal endpoint.

A statistical analysis of a study of effects of maternal exposure to diethylhexyl phthalate (DEHP) in rats is presented here (37). The DEHP was administered in the diet on gestational days 0 through 20 at 0.00, 0.5, 1.0, 1.5, and 2.0%. The study was conducted with two replicates. Approximately equal numbers of dams were assigned to each group, and gestational day 0 body weights were matched across dose within each replicate. At gestational day 20, animals were sacrificed and all fetuses were examined.

A two-way (dose × replicate) ANOVA model was used for the analysis of continuous endpoints or per-litter proportion data. Prior to analysis, the arc-sine square root transformation was performed to all maternal or per-litter proportion data. A test of linear dose–response trend was performed using contrast tests. When a significant ($p < 0.05$) dose effect occurred, Duncan's multiple range test was used for pairwise comparisons between control and each dose group. A one-sided test was used for all pairwise comparisons except for the maternal and fetal body weights and percentage of males per litter. Nominal (incidence) data were analyzed by a χ^2 test for differences among groups, and by a Cochran–Armitage linear trend test on proportions. A one-sided Fisher's exact test was used for pairwise comparisons when the χ^2 test was significant. Table 1 contains a summary of the analysis for selected endpoints. For the detailed analysis the reader is referred to Tyl et al. (38).

IV. SCIENTIFIC ISSUES

The main objectives in the analysis of reproductive and developmental experimental data are group comparisons and dose–response modeling. Statistical analysis of developmental endpoints presents a number of important issues. Developmental endpoints consist of both continuous and discrete outcomes. The analysis needs to account for the correlations induced by the clustering effects within litters. Tests for group differences for continuous endpoints are commonly conducted using the ANOVA with litter nested, MANOVA, or mixed-effects models. In contrast to the multivariate normal distributions for multiple continuous outcomes, distributions for the analysis of multiple binary outcomes are modeled in terms of sums of correlated binary variables under the exchangeability assumption. The resulting distribution is a generalized binomial. The dis-

Table 1 Reproductive Parameters After Exposure of Pregnant Fisher 344 Rats to Diethylhexyl Phthalate in the Feed on Gestational Days 0 Through 20

	Diethylhexyl phthalate % in feed				
Endpoint	0.0	0.5	1.0	1.5	2.0
No. pregnant dams	24	23	22	24	25
Maternal weight gd 0	173.60 (3.25)	175.37 (3.37)	172.42 (3.12)	171.77 (2.80)	173.33 (2.95)
Maternal weight gd 20	248.30 (3.64)[a]	246.00 (2.90)	232.73 (3.25)*	217.76 (3.25)*	184.15 (4.28)*
Maternal weight gain	74.69 (1.71)[a]	70.63 (2.48)	60.31 (1.44)*	45.99 (1.78)*	10.82 (3.00)*
Gravid uterine weight	49.79 (1.39)[a]	44.83 (2.49)	46.81 (1.31)	43.86 (0.96)	19.08 (3.71)*
Maternal liver weight	9.75 (0.12)[a]	11.72 (0.17)*	12.06 (0.17)*	12.21 (0.19)*	11.11 (0.15)*
Food consumption	312.04 (4.47)[a]	282.15 (6.75)*	252.71 (5.68)*	211.00 (5.04)*	179.48 (4.77)*
Water consumption	465.18 (7.47)[a]	495.47 (9.70)	515.66 (13.0)*	462.64 (9.35)*	376.76 (8.51)*
Corpora lutea	10.91 (0.33)	10.96 (0.22)	11.18 (0.30)	11.17 (0.18)	10.52 (0.61)
Implantation sites	10.92 (0.31)	9.83 (0.54)	10.59 (0.24)	10.58 (0.22)	10.40 (0.41)
% preimplantation loss	3.57 (1.14)	11.59 (4.20)	5.35 (1.28)	5.37 (1.48)	11.30 (2.84)
% resorption	4.14 (1.22)[a]	4.17 (1.20)	4.92 (1.67)	4.70 (1.25)	54.48 (8.10)*
% nonlive	4.14 (1.22)[a]	4.50 (1.29)	4.92 (1.67)	4.70 (1.25)	56.78 (8.04)*
% viable	10.46 (0.32)[a]	9.39 (0.53)	10.05 (0.26)	10.08 (0.25)	8.00 (0.70)*
% male	53.56 (3.65)	45.77 (4.51)	46.39 (3.31)	54.44 (3.00)	50.66 (6.10)
% malformation	1.27 (0.71)[a]	0.00 (0.00)	1.92 (1.11)	3.13 (1.06)	2.87 (1.64)
Fetal weight	3.022 (.029)[a]	3.143 (.035)*	2.852 (.053)*	2.557 (.034)*	2.266 (.041)*

[a]Significance in linear trend ($p < 0.05$).
*Significantly different from control ($p < 0.05$).

tribution for the common developmental endpoints, such as the number of malformations per litter, often has an overdispersed binomial, but the distribution of sex combinations can be an underdispersed binomial (39).

The beta-binomial distribution includes a non-negative dispersion parameter to model intralitter variations. The beta-binomial model uses a full likelihood-based analysis. This model has disadvantages in that it does not allow for negative intralitter correlations, and its maximum-likelihood estimates can be biased or numerically instable (40,41). The quasi-likelihood method offers an alternative approach; the quasi-likelihood approach does not provide inference on the dispersion parameter. Liang and Zeger (42) proposed a general method for the analysis of correlated data. This method is often referred to as *generalized estimating equations* (GEE). The analysis of continuous as well as discrete developmental endpoints with the GEE approach has been described by Chen (29).

In a typical reproductive and developmental study, various endpoints are collected in order to obtain the maximum information from the study. Statistical analysis often involves a large number of tests, in terms of both multiple comparisons among groups and multiple tests on many endpoints. Because of the conduct of a large number of statistical tests, the chance of false-positive findings could increase considerably. Commonly, each endpoint is analyzed separately. A significant result for an endpoint tested may be statistical, biological, or both. The issues of false-positive and false-negative error rates in the testing of multiple reproductive and developmental endpoints has not been addressed. The main consideration should be to maximize the power of identifying every significant endpoint while controlled the overall Type I or familywise error rate.

V. RECENT DEVELOPMENTS

Recently, several authors have proposed multivariate models for simultaneous analysis of multiple endpoints. Joint modeling of multiple developmental outcomes can have some advantages: it can increase the power of detecting effects if the multiple outcomes are manifestations of some common biological effect, and it allows investigations of associations among the multiple outcomes if they are the result of different biological mechanisms.

In the analysis of developmental data, two types of correlation are inherent: the (intralitter) correlation between fetuses of the same litter and the (intrafetus) correlation between the endpoints in the same fetus. Joint modeling of multiple binary outcomes of different malformation types has been proposed by Lefkopoulou et al. (43) and Chen and Ahn (44). Joint modeling of two overdispersed binomial outcomes for the numbers of deaths/resorptions and malformations has also been proposed (45–48). Recently, several authors have proposed conditional approaches to modeling malformation and fetal weight (49–51). However, joint modeling of continuous and discrete outcomes within a cluster is not straightforward and requires further research.

REFERENCES

1. U.S. Food and Drug Administration. Guidelines for Reproductive Studies for Safety Evaluation of Drugs for Human Use. Rockville, MD: Bureau of Drugs, Food and Drug Administration, 1966.
2. U.S. Food and Drug Administration, International Conference on Harmonization. Guideline on Detection of Reproduction for Medicinal Products. Rockville, MD: Food and Drug Administration, 1994.
3. MS Christian, AM Hoberman. In: RD Hood, ed. Handbook of Developmental Toxicology. New York: CRC Press, 1996, pp 551–595.
4. JM Manson. In: CA Kimmel, J Buelke-Sam, eds. Developmental Toxicology. 2nd ed. New York: Raven Press, 1996, pp 379–402.
5. OG Fitzhugh. In: E Boyland, R Goulding, eds. Modern Trend in Toxicology. Vol. 1. London: Butterworths, 1968, pp 75–85.
6. TFX Collins. In: JG Wilson, FC Fraser, eds. Handbook of Teratology. Vol. 4. New York: Plenum Press, 1978, pp 191–214.
7. U.S. Food and Drug Administration. Toxicological Principles and Procedures for Priority Based Assessment of Food Additives (Red Book). Guidelines for Reproduction Testing with a Teratology Phase. Washington, DC: Bureau of Foods, Food and Drug Administration, 1982.
8. U.S. Food and Drug Administration. Toxicological Principles for Safety Assessment of Direct Food Additives and Color Additives Used in Food (Red Book II). Guidelines for Reproduction and Developmental Toxicity Studies. Washington, DC: Center for Food Safety and Applied Nutrition, 1993.
9. U.S. Environmental Protection Agency. Guidelines for the Health Assessment of Suspect Developmental Toxicants. Fed. Regist. 51:34028–34040, 1986.
10. U.S. Environmental Protection Agency. Guidelines for Developmental Toxicity Risk Assessment. Fed. Regist. 56:63798–63826, 1991.
11. U.S. Food and Drug Administration. Prescription Drug Advertising; Content and Format for Labeling Human Prescription Drugs. Fed. Regist. 44:37434–37467, 1979.
12. Physicians' Desk Reference. Oradell, NJ: Medical Economics, 1992.
13. RW Tyl, MC Marr. In: RD Hood, ed. Handbook of Developmental Toxicology. New York: CRC Press, 1996, pp 175—225.
14. MM Desu, D Raghavarao. Sample Size Methodology. New York: Academic Press, 1990.
15. GW Snedecor, WG Cochran. Statistical Methods. 7th ed. Ames, IA: Iowa State University Press, 1980.
16. BJ Winer, DR Brown, KM Michels. Statistical Principles in Experimental Design. 3rd ed. New York: McGraw-Hill, 1991.
17. MJR Healy. Appl Stat 21:155–159, 1972.
18. EL Lehman. Nonparametrics: Statistical Methods Based on Ranks. San Francisco: Holden-Day, 1975.
19. CW Dunnett. J Am Stat Assoc 50:1096–1121, 1955.
20. DA Williams. Biometrics 28:519–531, 1972.
21. DB Duncan. Biometrics 11:1–42, 1955.
22. JW Tukey. Unpublished manuscript, 1953.
23. JW Tukey, JL Ciminera, JF Heyse. Biometrics 41:295–301, 1985.
24. RE Barlow, DJ Bartholomew, JM Bremmer, HD Brunk. Statistical Inference Under Order Restrictions. New York: Wiley, 1972.
25. KF Karpinski. In: D Krewski, C Frankin, eds. Statistics in Toxicology. New York: Gordon and Breach Science, 1991, pp 393–434.
26. AC Rencher. Methods of Multivariate Analysis. New York: Wiley, 1995.
27. AP Dempster, MR Selwyn, CM Patel, AJ Roth. Appl Stat 33:203–214, 1984.
28. P McCullagh, JA Nelder. Generalized Linear Model. 2nd ed. London: Chapman & Hall, 1989.
29. JJ Chen. In: SC Chow, JP Liu, eds. Design and Analysis of Animal Studies in Pharmaceutical Development. New York: Marcel Dekker, 1998.
30. A Agresti. Categorical Data Analysis. New York: Wiley, 1990.
31. WG Cochran. Biometrics 12:417–451, 1954.
32. P Armitage. Biometrics 11:375–386, 1955.
33. JL Ciminera. In: Proceedings of the Symposium on Long-Term Animal Carcinogenicity Studies: A Statistical Perspective. Washington, DC: American Statistician Association, 26–35, 1985.
34. CR Mehta, NR Patel, P Senchaudhuri. J Compu Graph Stat 1:21–40, 1992.
35. DA Williams. Biometrics 31:949–952, 1975.
36. DA Williams. Appl Stat 31:144–148, 1982.
37. RW Tyl, CJ Price, MC Marr, CA Kimmel. Fundam Appl Toxicol 10:395–412, 1988.

38. RW Tyl, CJ Price, MC Marr, CA Kimmel. Teratologic Evaluation of Diethylhexyl Phthalate in Fisher 344 Rats. Research Triangle Park, NC: Research Triangle Institute, 1983.
39. RJ Brook, WH James, E Gray. Biometrics 47:403–417, 1991.
40. LL Kupper, C Portier, MD Hogan, E Yamamoto. Biometrics 42:85–89, 1986.
41. DA Williams. Biometrics 44:305–308, 1988.
42. KY Liang, SL Zeger. Biometrika 73:13–22, 1986.
43. M Lefkopoulou, D Moore, LM Ryan. J Am Stat Assoc 84:810–815, 1989.
44. JJ Chen, H Ahn. J Agr Bio Enviro Stat 2:440–450, 1997.
45. JJ Chen, RL Kodell, RB Howe, DW Gaylor. Biometrics 47:1049–1058, 1991.
46. JJ Chen, LA Li. Stat Sin 4:265–274, 1994.
47. LM Ryan. Biometrics 48:163–174, 1992.
48. Y Zhu, D Krewski, WH Ross. Appl Stat 48:583–598, 1994.
49. PJ Catalano, LM Ryan. J Am Stat Assoc 87:641–658, 1992.
50. JJ Chen. Risk Anal 13:559–564, 1993.
51. GM Fitzmaurice, NM Laird. J Am Stat Assoc 90:845–852, 1995.

James J. Chen

Robust Analysis for Crossover Design*

See also *Crossover Design*

I. INTRODUCTION

In the simple two-period, two-treatment (A and B) crossover design, subjects are randomly allocated to one of the two treatment sequences AB or BA. Between the treatment periods, there is typically a washout period that is long enough to reduce the potential for carryover of treatment effects from the first period to the second period. It is sometimes possible to make baseline measurements in AB/BA crossover trials. Such baseline measurements are usually made with the object of giving general or background information rather than direct information on treatment effect. If the subjects are subject to individual trend over time and the washout period is long enough compared to the treatment period, the baseline measurements taken at the beginning of each treatment may contain useful information and may be used easily as covariates. Their use may, however, increase directly the precision of measurements made directly on the treatment. Note that baseline measurements are the only covariates that may be used in analysis of covariance in crossover trials, since they are the only covariates that change during the trial and may not therefore be identical for each treatment. In order to use the baseline measurements as the covariates, we need to assume that there are no interactions between the treatments and the periods. See, for example, Refs. 31 and 29 for discussions. It is worth mentioning that in repeated-measures crossover trials, the observations within each subject are dependent. In the absence of the covariates in the model, besides normal theory inference, there are several nonparametric approaches available in the literature. One can conduct the Mann–Whitney–Wilcoxon test (15), the Brown–Mood median test, and the sign test (adjusted) for testing the equality of treatment effects using period differences. See Refs. 12, 31, and 2 for further discussions. However, there do not appear to be any rank-based tests for the equality of treatment effects (or period effects) in repeated measures AB/BA crossover trials in the presence of covariates in the model. One may be tempted to use Jaeckel's (11) procedures (fixed-effects model approach, based on the joint rankings of all the residuals), which is analogous to the ordinary least-squares approach for the mixed-effects model in normal theory inference. However, Jaeckel's procedures suffer from loss in efficiency (see Sec. IV.C) when the observations within a subject are correlated. This is because the fixed-effects model cannot take into account the dependence of the data within a subject. It is worth mentioning that analysis of unbalanced mixed models by the ordinary least-squares (OLS) approach is not appropriate. For the dependent data, the generalized least-squares (GLS) approach gives more efficient inferences than those of the ordinary least-squares approach.

*The views expressed in this article are those of the author and not those of the U.S. Food and Drug Administration.

In this entry, we develop two rank-based methods for AB/BA crossover designs with baseline values as covariates under the assumption of within-subject exchangeable errors. In the first method, we will define an objective function for each subject based on the intrasubject ranks of the residuals. Then we will define an overall objective function by taking the sum of all the objective functions. We will obtain R-estimates by minimizing this overall objective function. We will then make inferences for the treatment effects and the slope using this overall objective function and the R-estimates. In the second method, we will define a single objective function based on period differences for all the subjects for the purpose of making inferences about treatment effects and the slope parameter.

The rest of the entry is organized as follows. In Sec. I.A we discuss the model and its assumptions. Section I.B discusses the advantages and uses of crossover designs; Sec. I.C discusses reasons for the use of analysis of covariance in AB/BA crossover designs in clinical research. In Sec. II, rank-based procedures based on intrasubject ranks of the residuals are developed. In Sec. III, rank-based methods based on the period differences are developed. Section IV compares the performances of the two rank-based methods presented in this entry, Jaeckel's method (12) and normal theory methods. In Sec. V, we give an example. Conclusions are summarized in Sec. VI.

A. Model and Assumptions

Suppose, in a two-period two-treatment crossover study design, that there are n subjects. These n subjects are divided at random into two groups. In the first group there are n_1 subjects; in the second group there are n_2 subjects. Let Y_{ijk} and x_{ijk} be, respectively, the response (outcome) and the baseline value corresponding to the ith ($i = 1,2$) treatment applied on the kth ($k = 1, 2, \ldots, n$) subject in the jth ($j = 1, 2$) period. A model for the two-period two-treatment crossover design in clinical research is described (assuming no treatment by period interaction) as follows:

$$Y_{ijk} = \alpha_i + \pi_j + \beta x_{ijk} + \varepsilon_{ijk} \tag{1}$$

where α_i ($\alpha_1 + \alpha_2 = 0$) is the effect of the ith treatment, π_j ($\pi_1 + \pi_2 = 0$) is the effect of the jth period, β is the slope parameter, and ε_{ijk} is the random-error term. It is usually assumed that ε_{ijk} in Eq. (1) is the sum of δ_k ($k = 1, 2, \ldots, n_1 + n_2$) and ε^*_{ijk}, where δ_k are independent normal random variables with mean zero and variance σ_b^2 ($0 < \sigma_b^2 < \infty$) and ε^*_{ijk} are independent normal random variables with mean 0 and variance σ_e^2. Note that the δ_k and the ε^*_{ijk} are independent. In this case Eq. (1) can be termed an unbalanced mixed model. One may use the SAS procedure PROC MIXED (generalized least-squares approach) to make inferences for treatment effects, period effects, and the slope. Note that if we assume that the δ_k is a fixed effect, then we need to use the OLS technique. Note that in the absence of baseline values in Eq. (1), both the OLS and the GLS techniques give identical estimates. In the presence of baseline values in Eq. (1), the GLS inferences differ from the OLS inferences, because the OLS approach does not take into account the equi-correlation structure of the within-subject errors.

In the following we reparametrize Eq. (1) to reflect the dependency of the data within each subject. Let $\theta_{ij} = \alpha_i + \pi_j$. For the kth subject in the sequence AB, Eq. (1) can be written as:

$$\begin{pmatrix} Y_{11k} \\ Y_{22k} \end{pmatrix} = \begin{pmatrix} 1 & 0 & x_{11k} \\ 0 & 1 & x_{22k} \end{pmatrix} \begin{pmatrix} \theta_{11} \\ \theta_{22} \\ \beta \end{pmatrix} + \begin{pmatrix} \varepsilon_{11k} \\ \varepsilon_{22k} \end{pmatrix}$$

For the kth subject in the sequence BA, Eq. (1) can be written as:

$$\begin{pmatrix} Y_{21k} \\ Y_{12k} \end{pmatrix} = \begin{pmatrix} 1 & 0 & x_{21k} \\ 0 & 1 & x_{12k} \end{pmatrix} \begin{pmatrix} \theta_{21} \\ \theta_{12} \\ \beta \end{pmatrix} + \begin{pmatrix} \varepsilon_{21k} \\ \varepsilon_{12k} \end{pmatrix}$$

Let

$$\underline{\varepsilon}_k^{(I)} = \begin{pmatrix} \varepsilon_{11k} \\ \varepsilon_{22k} \end{pmatrix}, \qquad (k = 1, 2, \ldots, n_1)$$

and

$$\underline{\varepsilon}_k^{(II)} = \begin{pmatrix} \varepsilon_{21k} \\ \varepsilon_{12k} \end{pmatrix}, \qquad (k = n_1 + 1, n_1 + 2, \ldots, n_1 + n_2)$$

be, respectively, the error vectors for the kth subject in sequences AB and BA. Without loss of generality we suppress the superscripts on the error vectors. In parametric inference, it is assumed that the error vectors $\underline{\varepsilon}_1, \underline{\varepsilon}_2, \ldots, \underline{\varepsilon}_{n_1}, \underline{\varepsilon}_{n_1+1}, \underline{\varepsilon}_{n_1+2}, \ldots, \underline{\varepsilon}_{n_1+n_2}$ are i.i.d. bivariate normal. If the error vectors are not bivariate normal, alternative techniques such as nonparametric procedures are required. The normal theory F-test for a fixed-effects model is robust against moderate departure from normality if the design is balanced. However, the problem of robustness in the mixed models is more serious than in the fixed-effects model, because there are more assumptions to be met (e.g., normality of the δ_k) in the mixed-effects models (see Ref. 17 for discussions). We assume that the elements of $\underline{\varepsilon}_k$ are exchangeable random variables, that the $\underline{\varepsilon}_k$'s are i.i.d. continuous random vectors with the joint c.d.f. $F(u_1, u_2)$, and that the bivariate density $f(u_1, u_2)$, corresponding to $F(u_1, u_2)$, is continuous in R^2.

Here, we find R-estimates of the model parameters and test the following hypotheses:

1. H_0^1: $\dfrac{\theta_{11} + \theta_{12}}{2} = \dfrac{\theta_{21} + \theta_{22}}{2}$ vs

   H_a^1: $\dfrac{\theta_{11} + \theta_{12}}{2} \neq \dfrac{\theta_{21} + \theta_{22}}{2}$

2. H_0^2: $\dfrac{\theta_{11} + \theta_{21}}{2} = \dfrac{\theta_{12} + \theta_{22}}{2}$ vs

   H_a^2: $\dfrac{\theta_{11} + \theta_{21}}{2} \neq \dfrac{\theta_{12} + \theta_{22}}{2}$

3. H_0^3: $\beta = 0$ vs H_a^3: $\beta \neq 0$

Note that H_0^1 implies that there are no treatment effects ($\alpha_1 = \alpha_2$), H_0^2 implies that there are no period effects ($\pi_1 = \pi_2$), and H_0^3 implies zero slope.

B. Advantages and Uses of AB/BA Crossover Designs

1. *Advantages of AB/BA Crossover Designs*

In many situations there are constraints arising from the need to obtain efficacy and safety information in a limited time frame and with limited availability of subjects. When the variability among subjects is much greater than the variability within subjects, the use of crossover design is appropriate, given that the clinical condition for treatment has a recurrent nature that is essentially equivalent across periods—for example, discomfort from symptoms from chronic health order, breathing problems in asthma episodes, and problems with hypertension. Note that in these diseases, each patient can be given both treatments in turn, as is the case in the study of certain long-term chronic diseases when the treatments to be evaluated are palliatives designed to lessen pain or to bring comfort or a relief for a particular season rather than to effect a complete cure. One can often argue that carryover effects, if they exist, will be small relative to the treatment effects when there is a suitably long washout period between the treatment periods and the observations are made at the end of the treatment periods. Note that we will not be able to test for carryover effects under Eq. (1), because we are using baseline values as covariates. Even in the absence of baseline values, it is hard to establish a powerful test for carryover effects. In choosing a crossover design one needs to consider which assumptions are realistic regarding the nature of the treatment effects. Senn (31) mentions that no help concerning carryover is to be expected from the data. The solution to this problem lies in designing experiments. The scientist must only use crossover trials in appropriate indications, and he/she must allow for adequate washout.

Note than an important advantage of crossover designs is the requirement of a smaller number of subjects than a design with separate groups of patients for each treatment. One reason for this is that each subject provides information for two treatment periods. Another reason is that the crossover design enables an assessment of treatment comparisons with respect to the usually more sensitive framework of within-subject variability as opposed to that of between-subject variability. By using the crossover design we reduce or eliminate from treatment contrasts one of the largest uncontrollable sources of variation in a clinical trial, the between-subject variability, and hence we obtain a better interpretation of information obtained from each individual.

2. *Uses of Crossover Designs*

The conclusion of BEMAC (Biometric and Epidemiology Methodology Advisory Committee of the Food and Drug Administration) with respect to the two-period crossover design is that it "is not the design of choice for clinical trials, where unequivocal evidence of treatment effect is required." Rather, the committee recommended that "in most cases, the completely randomized (or randomized block) design with baseline measurements will be the design of choice because it furnishes unbiased estimates of treatment effects without appeal to modeling assumptions save those associated with the randomization procedure itself" (5,23). The BEMAC report encourages the use of baseline information in the analysis, recommends the use of change from baseline scores, percentage change from baseline scores, and analysis of covariance. The report states, "If the baseline values vary between sequences or between periods (or both), it is essential that they be included in any analyses." See Smith (33) for further discussion about current regulatory experience concerning crossover trials.

Over half of the eighty crossover trials published in the *British Medical Journal* between January 1980 and April 1988 concerned hypertension, diabetes, angina, and asthma. The design used in most of these trials was the standard AB/BA design. See Jones and Kenward (12) for further discussion. A survey of 12 large pharmaceutical companies in the United States (4) revealed a not-too-dissimilar view. Out of 72 crossover trials conducted by these companies over a 5-year period, over half the trials used the AB/BA design.

A 12-year-old view of from the FDA (3) suggests that the two-period crossover design has not always been viewed negatively. Indeed, of the 22 clinical trials reviewed by Dubey (3), crossover designs are recommended in five, permitted in three, permitted but discouraged in eight, and not recommended in six. The U.S. Food and Drug Administration (6) recommends the use of AB/BA crossover designs for bioequivalence studies. Senn (31) reports that crossover trials are also popular in single-dose pharmacokinetic and pharmacodynamic studies in healthy volunteers as well as phase I tolerability studies and trials for bioequivalence.

Smith (33) reports that crossover trials play a smaller role as phase III "pivotal" efficacy trials for most medical indications. However, crossover trials play a major role in the clinical development of a new drug. Phase I and phase II studies often follow crossover designs.

Grieve (7) noted, "The saga of the crossover designs for clinical trials is not yet over, and it is healthy that the debate over the applicability of these designs continues, and will continue in the future."

C. Reasons for Analysis of Covariance in AB/BA Crossover Designs in Clinical Research

Senn (13) points out that if the washout period is such that the investigator may confidently expect that the effect of the previous treatment has been eliminated by the beginning of the following treatment period, and if it is considered that patients may be subjected to individual and specific trends over time (independent of the treatments to which they are allocated), then there may be some value in fitting baseline values in an analysis of covariance.

It is worth mentioning that the analysis of covariance technique is an efficient procedure for addressing two issues in clinical trials. The first is the reduction of variability in treatment effect and thereby the production of narrower confidence intervals and more powerful tests. The second is the clarification of the extent of treatment effects in the presence of any baseline imbalance between the treatment groups. If the researcher suspects that there are treatment-by-period interactions, he/she could use Koch's (15) procedures, which are based on gain score (response minus baseline). However, the gain score method can sometimes give different results than those of analysis of covariance. See Grieve (7) for parametric analysis using the gain score method and for a related discussion.

II. INTRASUBJECT RANK-BASED METHOD

In this section we consider R-estimation and rank tests based on intrasubject ranks of the residuals.

A. R-Estimation

In this subsection we develop R-estimators and their asymptotic distributions. We will use a sum of Jaeckel-type dispersion functions (11) to develop the rank-based inference. The role of this dispersion function is similar to the sum of squared errors (SSE) in the least-squares theory. We will use the R-estimators to carry out tests and confidence intervals.

Let $\underline{\theta} = (\theta_{11}, \theta_{22}, \theta_{21}, \theta_{12})'$ and $\underline{\Delta} = (\underline{\theta}', \beta)'$. A measure of dispersion for the kth subject, following the idea of Jaeckel (11) is

$$D_k(\underline{\Delta}) = \sqrt{12} \sum_{j=1}^{2} \sum_{i=1}^{2} n_{ij}^{(k)} \left[\frac{R_{ijk}}{3} - \frac{1}{2} \right] W_{ijk} \tag{2}$$

where $n_{ij}^{(k)} = 1$ if the ith treatment is applied at the jth period on the kth subject and $n_{ij}^{(k)} = 0$ otherwise, $W_{ijk} = Y_{ijk} - \theta_{ij} - \beta x_{ijk}$, and R_{ijk} is the intrasubject rank of W_{ijk}.

It can be shown that $D_k(\underline{\Delta})$ is non-negative, continuous, location free, and convex in $\underline{\Delta}$. It is worth mentioning that the coefficient of the residuals W_{ijk} lie in the interval $(-1/2, 1/2)$, and this coefficient (of W_{ijk}) assigns a weight proportional to the difference between its rank (between two elements) and the average rank. Because $D_k(\underline{\Delta})$ is a piecewise linear function of the residuals, the outliers enter into the dispersion function in a linear fashion.

In practice individual subjects are different (not similar). Therefore, the subjects may differ in their responses to a drug. By using the intrasubject ranks of the residuals we are removing the subject effect. As mentioned earlier, in crossover designs it is usually assumed that the variation among subjects is much greater than the variation within subjects, and so the estimation of model parameters on using intrasubject ranks is appropriate. We will show that inferences based on intrasubject rank-based methods are more efficient than those based on normal theory methods (for heavy- and heavier-tailed distributions) and rank-based methods (for heavier-tailed distributions) using period differences.

Using Eq. (2) we can construct a combined dispersion function for Eq. (1) as follows:

$$D(\underline{\Delta}) = \sum_{k=1}^{n} D_k(\underline{\Delta}) \tag{3}$$

$D(\underline{\Delta})$ in Eq. (3) is non-negative, continuous, location free, and convex in $\underline{\Delta}$. Although the ranking of the residuals is done within each subject separately, there is a "borrowing of strength" from across the subjects, because the R-estimates are obtained by minimizing the combined dispersion function. Under the assumption that the subjects are similar, a model is used that pulls the estimates together. Note that the sum of Jaeckel type dispersion function (11) (based on within rankings) has been used to incorporate concomitant variables in one-way repeated-measures designs (see Ref. 27), a nested error regression model (see Ref. 28), and incomplete-block clinical trials (see Ref. 26).

We obtain an R-estimator of $\underline{\theta}$ by minimizing $D(\underline{\Delta})$. Let $\hat{\underline{\Delta}} \varepsilon \Gamma$, where $\Gamma = [\underline{\Delta}: D(\underline{\Delta}) = \text{a minimum}]$. One can use the Nelder–Mead algorithm (22) to minimize $D(\underline{\Delta})$; see the NLPNMS procedure of PROC IML of SAS for a detailed discussion of the Nelder–Mead algorithm.

In order to obtain the asymptotic distribution of the R-estimators, we need the asymptotic distribution of the gradient of the dispersion function. In the following we develop the asymptotic distribution of the gradient of the dispersion function.

Let

$$\underline{S}(\theta, \beta) = [s_{11}(\underline{\theta}, \beta), s_{22}(\underline{\theta}, \beta), s_{21}(\underline{\theta}, \beta), s_{12}(\underline{\theta}, \beta), s_5(\underline{\theta}, \beta)]' \tag{4}$$

be the negative of the gradient of $D(\underline{\theta}, \beta)$. The domain of $D(\underline{\theta}, \beta)$ consists of polygonal subsets in each of which $D(\underline{\theta}, \beta)$ is linear in $\underline{\theta}$ and β. Therefore the partial derivatives exist almost everywhere and are given by the coefficients of the parameters in the dispersion function.

The elements of the negative of the gradient $\underline{S}(\underline{\Delta})$ of $D(\underline{\Delta})$ are given by

$$s_{11}(\underline{\theta}, \beta) = \sqrt{12} \sum_{k=1}^{n_1} \left[\frac{R_{11k}}{3} - \frac{1}{2}\right]$$

$$s_{22}(\underline{\theta}, \beta) = \sqrt{12} \sum_{k=1}^{n_1} \left[\frac{R_{22k}}{3} - \frac{1}{2}\right]$$

$$s_{12}(\underline{\theta}, \beta) = \sqrt{12} \sum_{k=1}^{n_2} \left[\frac{R_{12k}}{3} - \frac{1}{2}\right]$$

$$s_{21}(\underline{\theta}, \beta) = \sqrt{12} \sum_{k=1}^{n_2} \left[\frac{R_{12k}}{3} - \frac{1}{2}\right]$$

$$s_5(\underline{\theta}, \beta) = \sqrt{12} \sum_{k=1}^{n} \sum_{i=1}^{2} \sum_{j=1}^{2} n_{ij}^{(k)} \left[\frac{R_{ijk}}{3} - \frac{1}{2}\right] x_{ijk}$$

Under the true value $\underline{\Delta} = \underline{0}$ and exchangeability of errors, the ranks R_{ijk} within each subject are uniformly distributed over the permutations of the integers 1 through 2. Therefore, under exchangeability of the errors,

$$E_0[R_{ijk}] = \begin{cases} \dfrac{3}{2} & \text{if } n_{ij}^{(k)} = 1 \\ 0 & \text{otherwise} \end{cases}$$

Also

$$\text{var}_0[R_{ijk}] = \begin{cases} \dfrac{1}{4} & \text{if } n_{ij}^{(k)} = 1 \\ 0 & \text{otherwise} \end{cases}$$

Further,

$$\text{cov}_0[R_{ijk}, R_{i'j'k}] = \begin{cases} -1/4 & \text{if } (i, j) = (1, 1) \text{ and } (i', j') = (2, 2) \\ -1/4 & \text{if } (i, j) = (2, 1) \text{ and } (i', j') = (1, 2) \\ 0 & \text{otherwise} \end{cases}$$

Thus

$E_0[S(\underline{0})] = 0$

Let $\Sigma = \text{cov}_0[(1/\sqrt{n})\underline{S}(\underline{0})]$. Then

$$\Sigma = (\sigma_{ii'}) = \left(\frac{1}{3}\right) \begin{bmatrix} \frac{n_1}{n} & \frac{-n_1}{n} & 0 & 0 & \frac{\mu_1}{n} \\ \frac{-n_1}{n} & \frac{n_1}{n} & 0 & 0 & \frac{-\mu_1}{n} \\ 0 & 0 & \frac{n_2}{n} & \frac{-n_2}{n} & \frac{\mu_2}{n} \\ 0 & 0 & \frac{-n_2}{n} & \frac{n_2}{n} & \frac{-\mu_2}{n} \\ \frac{\mu_1}{n} & \frac{-\mu_1}{n} & \frac{\mu_2}{n} & \frac{-\mu_2}{n} & \frac{\eta}{n} \end{bmatrix} \tag{5}$$

where

$$\mu_1 = x_{11.} - x_{22.}, \qquad \mu_2 = x_{21.} - x_{12.}$$

$$\eta = 2 \sum_{j=1}^{2} \sum_{i=1}^{2} \sum_{k=1}^{n} n_{ij}^{(k)} (x_{ijk} - \bar{B}_{.k})^2 \qquad (>0)$$

and $\bar{B}_{.k} = (1/2) \sum_{i=1}^{2} \sum_{j=1}^{2} n_{ij}^{(k)} x_{ijk}$.

We assume

$$\lim_{n\to\infty} \text{cov}_0[n^{-1/2}\underline{S}(\underline{0})] = \lim_{n\to\infty} \left(\frac{1}{n}\right) \sum = A = (a_{ii'}) \quad (6)$$

Let $x_{ijk}^4 < L$ ($<\infty$). Note that $\underline{S}(\underline{0})$ can be written as a sum of n bounded and independent random vectors. By a multivariate central limit theorem, it can be shown that under the true value $\underline{\Delta}^0 = \underline{0}$,

$$n^{-1/2}\underline{S}(\underline{0}) \xrightarrow{\mathcal{D}} \text{MVN}(0, A) \quad (7)$$

as $n \to \infty$, where MVN stands for multivariate normal.

Note that the sum of the first three rows of A is zero. Therefore the matrix A is singular. Further, the diagonal elements are positive, since they are variances of the components of $\underline{S}(\underline{\theta}, \beta)$ and we assume that $a_{ii} = \lim_{n\to\infty}(\sigma_{ii}/n) > 0$. Note that the rank of Σ is 3. The role of $\underline{S}(\cdot)$ in this entry is similar to the scores in maximum-likelihood techniques.

Next, we derive the asymptotic distribution of $\hat{\underline{\Delta}}$. Using arguments similar to those given in Rashid and Nandram (27), it can be shown that

$$\sqrt{n}(\hat{\underline{\Delta}}) \text{ and } \sqrt{n}\{(\tau/n)A^{-}\underline{S}(\underline{0})\} \text{ are asymptotically equivalent} \quad (8)$$

where

$$\tau = \left[\frac{1}{\left\{\sqrt{12}\int_{-\infty}^{\infty} f(\varepsilon, \varepsilon)\, d\varepsilon\right\}}\right] < \infty \quad (9)$$

and $f(\cdot, \cdot)$ is the bivariate density of the two components of the error vector.

It follows from Eqs. (7) and (8) under the true value $\underline{\Delta}^0$,

$$\sqrt{n}[\hat{\underline{\Delta}} - \underline{\Delta}^0] \xrightarrow{\mathcal{D}} \text{MVN}(\underline{0}, \tau^2 A^{-}) \text{ as } n \to \infty \quad (10)$$

Note that there are model-checking procedures available for the nonparametric nested error regression model (28). Similar procedures can be applied to Eq. (1).

B. Rank Tests

In this section we develop the drop-in-dispersion test based on the R-estimators. We need an R-estimator $\hat{\underline{\Delta}}_H$ of $\underline{\Delta}$ under the hypothesis $\underline{a}'\underline{\Delta} = 0$. We also need the asymptotic distribution of the R-estimator of $\hat{\underline{\Delta}}_H$. Let $\hat{\underline{\Delta}}_H$ minimize $D(\underline{\Delta})$ under $\underline{a}'\underline{\Delta} = 0$. Using arguments similar to those given in Rashid (25), we have

$$\sqrt{n}\hat{\underline{\Delta}}_H = \sqrt{n}\,[\hat{\underline{\Delta}} - A^{-}\underline{a}(\underline{a}'A^{-}\underline{a})^{-}\underline{a}'\hat{\underline{\Delta}}] + o_p(1) \quad (11)$$

Note that the reduced model R-estimator $\hat{\underline{\Delta}}_H$ has a representation (asymptotically) similar to reduced model estimators in the least-squares analysis. For testing $\underline{a}'\underline{\Delta} = \underline{0}$, we compare $D(\hat{\underline{\Delta}})$ with $D(\hat{\underline{\Delta}}_H)$. It can be shown that under $\underline{a}'\underline{\Delta} = \underline{0}$,

$$D^* = \frac{2[D(\hat{\underline{\Delta}}_H) - D(\hat{\underline{\Delta}})]}{\hat{\tau}} \quad (12)$$

follows approximately a chi square distribution with 1 degree of freedom for large n. For H_0^1: $(\theta_{11} + \theta_{12})/2 = (\theta_{21} + \theta_{22})/2$, we take $\underline{a}' = (1/2, -1/2, -1/2, 1/2, 0)$.

The D^* test is analogous to $-2 \log \Lambda$ in maximum-likelihood techniques. It is also analogous to F-tests in the normal theory inference. See Rashid and Nandram (27) for the drop-in-dispersion test in one-way repeated-measures designs with a changing covariate. The proof of Eq. (12) goes along the same lines given in Rashid and Nandram (27); see also Rashid (25).

For testing a general linear hypothesis based on scores, $\underline{S}(\underline{\Delta})$, evaluated at the reduced model estimate $\hat{\underline{\Delta}}_H$, the reader is referred to Rashid (26). The scores test is analogous to Rao's scores test in the maximum-likelihood technique.

Note that the results of this entry are derived under the assumption that the number of subjects is large. Rashid (25) conducted a small-scale simulation study for the drop-in-dispersion test for multigroup repeated-measures designs (without covariates) with exchangeable errors within each subject. Based on this simulation study, it is recommended that the chi square percentile be used for the D^* tests when $10 \le n_1 \le 15$ and $10 \le n_2 \le 15$.

We also a constructed $100(1 - \gamma)\%$ confidence interval for any estimable function of $\underline{\Delta}$ using Eq. (9). Let $\underline{a}'\underline{\Delta}$ be a contrast in $\underline{\Delta}$. An approximate $100(1 - \gamma)\%$ confidence interval for $\underline{a}'\underline{\Delta}$ is:

$$\underline{a}'\hat{\underline{\Delta}} \pm \left(z\frac{\gamma}{2}\right)\hat{\tau}\sqrt{\underline{a}'\,\Sigma^{-}\frac{\underline{a}}{n}} \quad (13)$$

C. Estimation of τ

As mentioned earlier, the parameter $\tau = 1/[\sqrt{12}\cdot\{\int_{-\infty}^{\infty} f(\varepsilon, \varepsilon)\, d\varepsilon\}]$ plays a role similar to $\sigma^2(1 - \rho)$ in parametric repeated-measures analysis. It can be shown that the parameter $1/[\sqrt{12}\tau]$ is the density of $d_k = [\varepsilon_{11k} - \varepsilon_{22k}]$ (or $d_k = [\varepsilon_{21k} - \varepsilon_{12k}]$) at 0 for $k = 1, 2, \ldots, n_1$ (or for $k = 1, 2, \ldots, n_2$). Thus d_k ($k = 1, 2, \ldots, n_1, n_1 + 1, \ldots, n_1 + n_2$) are n i.i.d. random variables having continuous and symmetric distributions each with the median at 0. Hence estimating τ involves estimating the inverse of the density function at 0. Following Bloch and Gastwirth (1), we can obtain a consistent estimator of the inverse of the density function at 0 as

$$g = \left\{\frac{n}{(2m)}\right\} [d_{(n,[n/2]+m)} - d_{(n,[n/2]-m+1)}] \qquad (14)$$

where $m = pn^{4/5}$ as $n \to \infty$. Thus, a consistent estimator of τ is $\hat{\tau} = g/\sqrt{12}$. It is recommended that p be taken to be .5. In practice we will replace ε_{ijk}'s by the residuals.

III. RANK-BASED METHODS USING PERIOD DIFFERENCES

In this section we develop the inferences for treatment effect based on the period differences. We will also discuss the estimation of a scale parameter. In normal theory inference, treatment effect is estimated using the period differences from each subject. The unbiasedness of this resulting estimate rests on the validity of the assumption that there is no treatment-by-period interaction. Our method in this section will also be based on the period differences. We also assume that there is no treatment-by-period interaction. First, we define an objective function for estimating the treatment effect and the slope. Then we develop the asymptotic distribution of these estimates under the full model and the reduced model (under the hypothesis). Then we develop rank tests.

Let $U_k^{(I)} = Y_{22k} - Y_{11k}$ and $U_k^{(II)} = Y_{12k} - Y_{21k}$ be, respectively, the period differences of the responses for sequences I and II. Similarly, $V_k^{(I)} = x_{22k} - x_{11k}$ and $V_k^{(II)} = x_{12k} - x_{21k}$ be, respectively, the period differences of the baseline values for sequences I and II. Then Eq. (1) can be written separately for each sequence as:

$$U_k^{(I)} = \alpha + \pi + \beta V_k^{(I)} + e_k^{(I)}, \qquad k = 1, \ldots, n_1,$$
$$U_k^{(II)} = -\alpha + \pi + \beta V_k^{(II)} + e_k^{(II)}, \qquad k = 1, \ldots, n_2$$

where $\alpha = \alpha_2 - \alpha_1$ (treatment effect), $\pi = \pi_2 - \pi_1$ (period effect), $e_k^{(I)} = \varepsilon_{22k} - \varepsilon_{11k}$, and $e_k^{(II)} = \varepsilon_{21k} - \varepsilon_{12k}$. Note that, by using the period differences we remove the subject effects. Without loss of generality, we suppress the superscripts in $e_k^{(I)}$, $e_k^{(II)}$, $U_k^{(I)}$, $u_k^{(II)}$, $V_k^{(I)}$, and $V_k^{(II)}$. Note that e_k's are i.i.d. continuous random variables with median 0 and variance $2\sigma^2(1 - \rho)$. Using Wilcoxon scores, we define the Jaeckel's dispersion function (11) based on the joint ranks of the residuals $u_k - \psi_k\alpha - \pi - \beta v_k$ as follows:

$$D^{(P)}(\alpha, \beta) = \sqrt{12} \sum_{k=1}^{n} \left[\frac{R_k^{(P)}}{n+1} - \frac{1}{2}\right] W_k^{(P)} \qquad (15)$$

where $W_k^{(P)} = u_k - \psi_k\alpha - \beta v_k$, $R_k^{(P)}$ is the rank of the kth residual $W_k^{(P)}$ among n residuals, and

$$\psi_k = \begin{cases} 1 & \text{if } k = 1, 2, \ldots, n_1 \\ -1 & \text{if } k = n_1 + 1, \ldots, n_1 + n_2 \end{cases}$$

The term π (the period effect) has dropped out of the dispersion function because ranks of the residuals do not change by inclusion or exclusion of π. Thus, by ranking the residuals (over all the subjects) based on the period differences we remove the period effect.

One may define two dispersion functions separately for Eq. (1) using ranks (intraperiod) of the n residuals corresponding to two periods, and then adding the two dispersion functions to obtain a combined dispersion function. By doing this we will be able to remove the period effects, but we will not be able to remove the subject effects from the estimates of the slope and treatment effects. The n subjects in Eq. (1) represent an effort to reduce experimental errors and prevent misleading comparisons of "apples and oranges." In such a comparison, a difference in treatment effects would be confounded with basic characteristics (e.g., smoking, age) of the subjects, the latter being of little or no interest in a particular experiment. It would be foolish to compare the response of treatment 1 at period 1 of one subject with the response of treatment 2 at period 1 of another subject when the two subjects differ in a particular characteristic. See, for example, Hollander and Wolfe (10) for a detailed discussion. Thus the use of intracomponent (e.g., intraperiod) rankings for inferences for the repeated-measures designs might not be feasible.

Note that $D^{(P)}(\alpha, \beta)$ is non-negative, continuous, location free, and convex in $(\alpha, \beta)'$. We can thus obtain an R-estimator of $(\alpha, \beta)'$ by minimizing $D^{(P)}(\alpha, \beta)$. Let $(\alpha, \beta)' \varepsilon \Gamma^{(P)}$, where $\Gamma^{(P)} = [(\alpha, \beta)'$: $D^{(P)}(\alpha, \beta) =$ a minimum].

In order to obtain the asymptotic distribution of the R-estimators, we need the asymptotic distribution of the gradient of the dispersion function. In the following we develop the asymptotic distribution of the gradient of the dispersion function.

Let

$$\underline{S}^{(P)}(\alpha, \beta) = [s_1^{(P)}(\alpha, \beta), s_2^{(P)}(\alpha, \beta)]'$$

be the negative of the gradient of $D^{(P)}(\alpha, \beta)$. Then

$$\underline{S}^{(P)}(\alpha, \beta) = \left(\sqrt{12} \sum_{k=1}^{n} \left[\frac{R_k^{(P)}}{n+1} - \frac{1}{2}\right], \sqrt{12} \sum_{k=1}^{n} v_k \left[\frac{R_k^{(P)}}{n+1} - \frac{1}{2}\right]\right)'$$

Under $(\alpha^0, \beta^0)' = \underline{0}$, $R_k^{(P)}$'s are uniformly distributed over the integer 1 through n. Therefore

$$E_0[R_k^{(P)}] = 0, \qquad \text{var}[R_k^{(P)}] = \frac{n^2 - 1}{12}, \quad \text{and}$$

$$\text{cov}[R_k^{(P)}, R_{k'}^{(P)}] = \frac{-(n+1)}{12} \qquad (k \neq k')$$

It can be shown that under $(\alpha^0, \beta^0)' = \underline{0}$,

$$n^{-1/2}\underline{S}^{(P)}(0, 0) \xrightarrow{\mathcal{D}} \text{BVN}(0, A^{(P)}) \tag{16}$$

as $n \to \infty$, where BVN stands for bivariate normal and

$$A^{(P)} = (a_{ii'}^{(P)}), \; a_{11}^{(P)} = \lim_{n\to\infty} 4n_1n_2/\{n(n+1)\}$$

$$a_{12}^{(P)} = a_{21}^{(P)} = \lim_{n\to\infty} \frac{1}{n+1}\left(\sum_{k=1}^{n} \psi_k v_k - \frac{1}{n}\sum_{k=1}^{n} v_k \sum_{k=1}^{n} \psi_k\right)$$

$$a_{22}^{(P)} = \lim_{n\to\infty} \frac{1}{n+1}\sum_{k=1}^{n} (v_k - \bar{v})^2$$

We assume that the matrix $A^{(P)}$ is positive definite. Note that Eq. (16) can be proved using arguments similar to those given in Hettmansperger (8).

Using arguments similar to those given in Jaeckel (11), it can be shown that

$$\sqrt{n}(\hat{\alpha}^{(P)}, \hat{\beta}^{(P)})' \text{ and } \sqrt{n}\left\{\left(\frac{\tau_1}{n}\right)(A^{(P)})^{-1}\underline{S}^{(P)}(0, 0)\right\}$$

are asymptotically equivalent (17)

where

$$\tau_1 = \frac{1}{\left\{\sqrt{12}\int_{-\infty}^{\infty} f_1^2(e)\, de\right\}} \tag{18}$$

and $f_1(\cdot)$ is the density of $e_k = (\varepsilon_{22k} - \varepsilon_{11k})$ (or $\varepsilon_{12k} - \varepsilon_{21k}$). See also Hettmansperger (8) and Heiler and Willers (9) for further details.

It follows from Eqs. (16) and (17) and under the true value $(\alpha^0, \beta^0)'$

$$\sqrt{n}[(\hat{\alpha}^{(P)}, \hat{\beta}^{(P)})' - (\alpha^0, \beta^0)'] \xrightarrow{\mathcal{D}} N[\underline{0}, \tau_1^2(A^{(P)})^{-1}]$$

$$\text{as } n \to \infty. \tag{19}$$

Now consider estimation of α when $\beta = 0$. In this case the dispersion function Eq. (15) will reduce to

$$D^{(P)}(\alpha, 0) = \sqrt{12}\sum_{k=1}^{n}\left[\frac{R_k^{(P)}}{n+1} - \frac{1}{2}\right] W_k^{(P)}$$

where $W_k^{(P)} = u_k - \psi_k\alpha$ and $R_k^{(P)}$ is the rank of the kth residual $W_k^{(P)}$ among n residuals. Let $\hat{\alpha}_H$ minimize $D^{(P)}(\alpha)$. It follows from Eq. (19) that under the true value α^0

$$\sqrt{n}\,(\hat{\alpha}_H^{(P)} - \alpha^0) \xrightarrow{\mathcal{D}} \text{N}\left[\underline{0}, \frac{\tau_1^2}{a_{11}^{(P)}}\right] \quad \text{as } n \to \infty \tag{20}$$

Next, consider estimation of β when $\alpha = 0$. In this case the dispersion function Eq. (15) will reduce to

$$D^{(P)}(0, \beta) = \sqrt{12}\sum_{k=1}^{n}\left[\frac{R_k^{(P)}}{n+1} - \frac{1}{2}\right] W_k^{(P)}$$

where $W_k^{(P)} = u_k - \beta$ and $R_k^{(P)}$ is the rank of the kth residual $W_k^{(P)}$ among n residuals. Let $\hat{\beta}_H^{(P)}$ minimize $D^{(P)}(\beta)$. It follows from Eq. (19) that under the true value β^0

$$\sqrt{n}(\hat{\beta}_H^{(P)} - \beta^0) \xrightarrow{\mathcal{D}} N\left(\underline{0}, \frac{\tau_1^2}{a_{22}^{(P)}}\right) \quad \text{as } n \to \infty \tag{21}$$

In the following we develop rank tests for no treatment effects and zero slope. It can be shown that under H_0^1: $\alpha = 0$

$$D_1^* = \frac{2[D^{(P)}(0, \hat{\beta}_H^{(P)}) - D(\hat{\alpha}^{(P)}, \hat{\beta}^{(P)})]}{\hat{\tau}_1} \tag{22}$$

follows approximately a chi square distribution with 1 degree of freedom. It can further be shown that under H_0^3: $\beta = 0$,

$$D_3^* = \frac{2[D^{(P)}(\hat{\alpha}_H^{(P)}, 0) - D^{(P)}(\hat{\alpha}^{(P)}, \hat{\beta}^{(P)})]}{\hat{\tau}_1} \tag{23}$$

follows approximately a chi square distribution with 1 degree of freedom.

The proofs of drop-in-dispersion tests D_1^* and D_3^*) go along the lines of McKean and Hettmansperger (18). See also Hettmansperger (8).

Like the Wilcoxon–Mann–Whitney test, it is recommended that the chi square percentile be used for the drop-in-dispersion test when both $n_1 \geq 10$ and $n_2 \geq 10$. In fact, the drop-in-dispersion test (D_1^*) is aymptotically equivalent to the test (the aligned rank test) based on the scores $s_1^{(P)}(0, \hat{\beta}^{(P)})$. The aligned rank test for testing H_0^1: $\alpha = 0$ is based on the reduced model residuals $W_k^{(P)} = u_k - \psi - \hat{\beta}_H^{(P)} v_k$. See Hettmansperger (8) for aligned rank tests for the linear model with i.i.d. errors. In the absence of β in Eq. (1), the test (H_0^1: $\alpha = 0$) based on the scores $s_1^{(P)}(0, 0)$, reduces to the Wilcoxon–Mann–Whitney test proposed by Koch (15).

Because the distribution of e_k is symmetric about zero, π can be estimated by the median of the Walsh averages $(W_k^{(P)} + W_{k'}^{(P)})/2$, $k < k' = 1, 2, \ldots, n$ [see Randles and Wolfe (24)] where $W_k^{(P)} = u_k - \psi\hat{\alpha} - \hat{\beta}^{(P)}v_k$. Thus a nonparametric estimate of π is

$$\hat{\pi} = \text{median}_{k<k'} \left\{ \frac{W_k^{(P)} + W_{k'}^{(P)}}{2} \right\}$$

Also we will be able to test H_0^2: $\pi = 0$ by the Wilcoxon signed rank test using the residuals $W_k^{(P)} = u_k - \psi\hat{\alpha} - \hat{\beta}^{(P)}v_k$. Because in crossover designs we are interested mainly in the treatment effect, we will not pursue the inferences concerning period effect.

Note that there are model-checking procedures available for the nonparametric linear model with i.i.d. errors. For example, see Refs. 19 and 20 for detailed procedures.

A. Estimation of τ_1

Note that $e_1, e_2, \ldots, e_n$ are i.i.d. random variables. Let the density of e_k be $f_1(e)$. Then $\tau_1 = 1/(\sqrt{12}\gamma_1)$, where $\gamma_1 = \int_{-\infty}^{\infty} f_1^2(e)\, de$. The random variable e_k follows a continuous and symmetric distribution with median at 0. Hence, estimating τ_1 involves estimating the inverse of the density function at 0. Following Bloch and Gastwirth (1), we can obtain a consistent estimator of the inverse of the density function at 0 as

$$\frac{1}{\hat{\gamma}_1} = \left\{ \frac{n}{(2m)} \right\} [e_{(n,[n/2]+m)} - e_{(n,[n/2]-m+1)}] \tag{24}$$

where $e'_{(k)}$'s are the ordered e'_k's, and $m = pn^{4/5}$ as $n \to \infty$. Thus, a consistent estimator of τ_1 is $\hat{\tau}_1 = 1/[\sqrt{12}\hat{\gamma}_1]$. It is recommended that p be taken to be .5. In practice, we will replace the e'_k's by the residuals $\hat{e}_k^{(P)} = \hat{W}_k^{(P)} = u_k - \psi_k\hat{\alpha}^{(P)} - \hat{\beta}^{(P)}v_k$. Thus the parameter τ_1 can be estimated by $\hat{\tau}_1 = 1/[\sqrt{12}\hat{\gamma}_1]$.

IV. ASYMPTOTIC COMPARISIONS

In this section we compare the performances of R, $R^{(P)}$ and GLS methods in terms of asymptotic variances when the baseline values are absent. We also discuss the performances of R, $R^{(P)}$ and GLS methods on empirical grounds when the baseline values are present. We further compare both $\hat{\alpha}$ and $\hat{\alpha}^{(P)}$ with $\hat{\alpha}^{(J)}$, where $\hat{\alpha}^{(J)}$ is the R-estimate of α using Jaeckel procedures (11), assuming ε_{ijk} in Eq. (1) is the sum of δ_k (the fixed subject effect) and ε_{ijk}^* (the random error term with $\text{var}(y_{ijk}) = \text{var}(\varepsilon_{ijk}^*) = \sigma^2$ when the baseline values are absent.

A. Comparisons of $\hat{\alpha}$, $\hat{\alpha}^{(P)}$, and $\hat{\alpha}_{GLS}$ Without the Baseline Values

In this section we consider the asymptotic relative efficiencies of the three estimators $\hat{\alpha}$ (based on the within rankings), $\hat{\alpha}^{(P)}$ (based on the period differences), and $\hat{\alpha}_{\text{GLS}}$ in the absence of baseline covariate in Eq. (1).

It is interesting to note that it is the asymptotic distribution of the test under a sequence of alternatives that determines the asymptotic distribution of the estimator. In most cases, the empirical power is consistent with the results predicted by the asymptotic relative efficiency (Pitman efficiency).

Without loss of generality, we assume that there is no covariate in Eq. (1). The asymptotic variances are as follows:

$$\text{var}(\sqrt{n}\hat{\alpha}) = \frac{3n^2\tau^2}{4n_1n_2}$$

$$\text{var}(\sqrt{n}\hat{\alpha}^{(P)}) = \frac{n(n+1)\tau_1^2}{4n_1n_2}$$

$$\text{var}(\sqrt{n}\hat{\alpha}_{\text{GLS}}) = \frac{2n^2\sigma^2(1-\rho)}{4n_1n_2}$$

1. *Bivariate Normal Distribution*

For the bivariate normal distribution with correlation coefficient ρ and variance σ^2, we have

$$\tau^2 = \frac{\pi\sigma^2(1-\rho)}{3} \quad \text{and} \quad \tau_1^2 = \frac{2\pi\sigma^2(1-\rho)}{3}$$

Therefore

$$\lim_{n\to\infty} \left\{ \frac{\text{var}(\sqrt{n}\hat{\alpha}^{(P)})}{\text{var}(\sqrt{n}\hat{\alpha})} \right\} = .666$$

$$\lim_{n\to\infty} \left\{ \frac{\text{var}(\sqrt{n}\hat{\alpha}_{\text{GLS}})}{\text{var}(\sqrt{n}\hat{\alpha})} \right\} = .636$$

$$\lim_{n\to\infty} \left\{ \frac{\text{var}(\sqrt{n}\hat{\alpha}_{\text{GLS}})}{\text{var}(\sqrt{n}\hat{\alpha}^{(P)})} \right\} = .955$$

We see that the F-test and $\hat{\alpha}_{\text{GLS}}$ are examples of procedures that have high power and efficiency for the normal model.

It should be pointed out that normal-theory-based procedures are highly unstable. Only one observation is enough to alter the conclusions reached by an F-test. The $\hat{\alpha}$ and the corresponding drop-in-dispersion test are quite stable. See Sec. V for an example.

2. Bivariate t-Distribution

The variance of t random variable with scale parameter σ and degrees of freedom η_1 is $\eta_1\sigma^2/(\eta-2)$. For the bivariate t-distribution with correlation coefficient ρ, scale parameter σ, and degrees of freedom η_1, we get

$$\tau^2 = \left[\sqrt{\eta_1}\Gamma\left(\frac{\eta_1}{2}\right)\Big/\Gamma\left(\frac{\eta_1+1}{2}\right)\right]^2 \frac{\pi\sigma^2(1-\rho)}{6}$$

and

$$\tau_1^2 = \left[\sqrt{\eta}\Gamma\left(\frac{\eta}{2}\right)\Big/\Gamma\left(\frac{\eta+1}{2}\right)\right]^2 2\pi\sigma^2\frac{(1-\rho)}{6}$$

Therefore

$$\lim_{n\to\infty}\left\{\frac{\text{var}(\sqrt{n}\hat{\alpha}^{(P)})}{\text{var}(\sqrt{n}\hat{\alpha})}\right\} = .666$$

$$\lim_{n\to\infty}\left\{\frac{\text{var}(\sqrt{n}\hat{\alpha}_{\text{GLS}})}{\text{var}(\sqrt{n}\hat{\alpha})}\right\} = 1.62,\ 1.12,\ .96,\ .8$$

for $\eta = 3, 4, 5, 8$; and

$$\lim_{n\to\infty}\left\{\frac{\text{var}(\sqrt{n}\hat{\alpha}_{\text{GLS}})}{\text{var}(\sqrt{n}\hat{\alpha}^{(P)})}\right\} = 2.55,\ 1.77,\ 1.52,\ 1.26$$

for $\eta = 3, 4, 5, 8$.

Thus both $\hat{\alpha}$ and $\hat{\alpha}^{(P)}$ do much better than the GLS estimator, $\hat{\alpha}_{\text{GLS}}$, for the heavy-tailed distributions. However, $\hat{\alpha}^{(P)}$ does better than $\hat{\alpha}$ for the heavy-tailed distributions. It is expected that the R-estimates obtained by minimizing $D(\underline{\Delta})$ and $D^{(P)}(\alpha, \beta)$ will be more robust than normal theory counterparts, since the influence of the outliers is linear rather than quadratic.

3. Bivariate Double Exponential Distribution

Consider a bivariate double exponential distribution with $\sigma^2 = 2$ and $\rho = .5$. We can generate bivariate double exponential variables ε_1 and ε_2 with $\sigma^2 = 2$ and $\rho = .5$ by defining $\varepsilon_1 = x_1 - x_2$ and $\varepsilon_2 = x_1 - x_3$, where x_1, x_2, x_3, and x_4 are i.i.d. exponential random variables with mean 1. Thus the p.d.f. of $\varepsilon_1 - \varepsilon_2$ is again double exponential. Therefore $\tau = 1/[\sqrt{12/2}]$ and $\tau_1 = 1/[\sqrt{12/4}]$. Therefore

$$\lim_{n\to\infty}\left\{\frac{\text{var}(\sqrt{n}\hat{\alpha}^{(P)})}{\text{var}(\sqrt{n}\hat{\alpha})}\right\} = 1.33$$

$$\lim_{n\to\infty}\left\{\frac{\text{var}(\sqrt{n}\hat{\alpha}_{\text{GLS}})}{\text{var}(\sqrt{n}\hat{\alpha}^{(P)})}\right\} = 1.5$$

$$\lim_{n\to\infty}\left\{\frac{\text{var}(\sqrt{n}\hat{\alpha}_{\text{GLS}})}{\text{var}(\sqrt{n}\hat{\alpha})}\right\} = 2$$

Thus for the heavier-tailed distributions (e.g., double exponential) we prefer the tests of Sec. II rather than those of Sec. III. This is not surprising, because the ranking is done within subjects; thus, the ranks (in Sec. II) are bounded and thereby again nullify the effects of outliers. We cannot discard the intrasubject rank-based tests because of its poor performance when the data are normally distributed. The intrasubject ranking is an attempt to eliminate a nusiance effect due to the subject's characteristics. Usually the purpose of clinical trials is not to test for subject effects but to concentrate on testing treatment differences. See, for example, Senn and Auclair (30, p. 1287) for further discussion.

B. Empirical Comparisons of $\hat{\alpha}$ and $\hat{\alpha}^{(P)}$ with $\hat{\alpha}_{\text{GLS}}$ in the Presence of Baseline Values

As to be expected, if the normal error structure holds, there are minor differences between the rank-based method and the GLS method. Recall that the generalized least-squares estimators are functions of σ_b^2 and σ_e^2. Thus, to obtain the point estimates, estimates of these variance components must be obtained. Thus, while the GLS estimates require estimated variance components, they are not required for the rank-based estimates. Therefore, the rank-based procedures are expected to show better performance than the GLS counterparts.

We know that the generalized least-squares estimators are no longer the best when the estimates of variance components (σ_b^2 and σ_e^2) are substituted in expressions for the estimates of α, π, and β. The R-estimates, on the other hand, are not functions of the variance components. So the rank-based methods have the potential to perform better than the GLS methods even when the model based on the normal structure holds; and they (rank-based methods) do even better, especially when this structure does not hold.

Rashid and Nandram (28) performed a small-scale simulation study for comparing the performances of the GLS procedures and the intracluster R-estimator based procedures for the nested error regression models in the context of small-area estimation. They noted that there were gains in precision using the R-estimates when normality is tenuous. The simulation study they conducted indicated that if there were outliers occurring uniformly in either tails of the error distributions, the rank-based (intracluster) method would be superior to the GLS method. Similar results were indicated for skewed error distributions.

We expect that rank-based procedures based on the

period differences would perform better than GLS procedures if there were outliers occurring uniformly in either tails of the error distribution. Similar conclusions would hold for skewed error distributions.

C. Comparisons of $\hat{\alpha}$ and $\hat{\alpha}^{(P)}$ with $\hat{\alpha}^{(P)}$ When the Baseline Values Are Absent

First we discuss how to compute $\hat{\alpha}^{(J)}$. We assume that ε_{ijk} in Eq. (1) is the sum of δ_k (the fixed subject effect) and ε^*_{ijk} (the random error term) with $\text{var}(y_{ijk}) = \text{var}(\varepsilon^*_{ijk}) = \sigma^2$. Jaeckel (11) proposed an objective function, known as the dispersion function, to compute the R-estimates of the fixed-effects model (i.e., the linear models with i.i.d. errors). Jaeckel's (11) dispersion function is based on the joint ranking of all the residuals. Following Jaeckel (11) we define a dispersion function for the fixed-effects version of Eq. (1) based on Wilcoxon scores as follows:

$$D(\alpha_1, \alpha_2, \pi_2, \pi_2, \underline{\delta}) = \sqrt{12} \sum_{k=1}^{n} \sum_{j=1}^{2} \sum_{i=1}^{2} n_{ijk} \cdot \left[\frac{R^{(J)}_{ijk}}{n+1} - \frac{1}{2}\right] W^{(J)}_{ijk}$$

where $W^{(J)}_{ijk} = y_{ijk} - \alpha_i - \pi_j - \delta_k - \beta x_{ijk}$, $R^{(J)}_{ijk}$ is the rank of the (ijk)th residual $W^{(J)}_{ijk}$ among $2n$ residuals, $\underline{\delta} = (\delta_1, \delta_2, \ldots, \delta_n)'$ is a vector of fixed subject effects $(\Sigma^n_{k=1} \delta_k = 0)$, and $n^{(k)}_{ij} = 1$ if the ith treatment is applied at the jth period on the kth subject and 0 otherwise. The R-estimates and drop-in-dispersion tests can be developed analogously as done in Sec. III.

Let $\alpha = \alpha_2 - \alpha_1$. It can be shown (see Ref. 8) that when the covariate is absent in the model,

$$\text{var}(\sqrt{n}\hat{\alpha}^{(J)}) = \frac{n(n+1)(\tau^*_1)^2}{4n_1n_2}$$

where

$$\tau^*_1 = \frac{1}{\sqrt{12}\int_{-\infty}^{\infty} f^2_*(\varepsilon)\, d\varepsilon}$$

and $f_*(\cdot)$ is a probability density function of ε^*_{ijk}.

1. *Comparisons of $\hat{\alpha}$ with $\hat{\alpha}^{(J)}$*

It is sufficient to compare $\hat{\alpha}$ with $\hat{\alpha}^{(J)}$ in the absece of the covariate. Thus the ARE of $\hat{\alpha}$ with respect to $\hat{\alpha}^{(J)}$ is given by:

$$\lim_{n\to\infty} \left\{\frac{\text{var}(\sqrt{n}\hat{\alpha}^{(J)})}{\text{var}(\sqrt{n}\hat{\alpha})}\right\} = \left(\frac{2}{3}\right) \frac{\left\{\int_{-\infty}^{\infty} f^2_*(e)\, de\right\}^2}{\left\{\int_{-\infty}^{\infty} f(\varepsilon, \varepsilon)\, d\varepsilon\right\}^2}$$

For the normal and t-distributions, the preceding ARE reduces to $2/\{3(1 - \rho)\}$. That is, when $\rho \geq .35$, the ARE of $\hat{\alpha}$ with respect to $\hat{\alpha}^{(J)}$ for the normal distribution is greater than 1. For moderate and high values of ρ ($.35 \leq \rho < 1$), the methods of Sec. II are preferred to Jaeckel's method (11).

For the double exponential distribution with $\rho = .5$ and $\sigma^2 = 2$, it can be shown that the ARE of $\hat{\alpha}$ with respect to $\hat{\alpha}^{(J)}$ is 2.67.

2. *Comparisons of $\hat{\alpha}^{(P)}$ with $\hat{\alpha}^{(J)}$*

It can be shown that the ARE of $\hat{\alpha}^{(P)}$ with respect to $\hat{\alpha}^{(J)}$ is given by:

$$\lim_{n\to\infty} \left\{\frac{\text{var}(\sqrt{n}\hat{\alpha}^{(P)})}{\text{var}(\sqrt{n}\hat{\alpha}^{(J)})}\right\} = \frac{\left\{\int_{-\infty}^{\infty} f^2_*(\varepsilon)\, d\varepsilon\right\}^2}{\left\{\int_{-\infty}^{\infty} f_1(e)\, de\right\}^2}$$

For the normal and t-distributions, the preceding ARE reduces to $1/(1 - \rho)$. Thus for $\rho > 0$ (which is expected in repeated-measures data), the ARE of $\hat{\alpha}^{(P)}$ with respect to $\hat{\alpha}^{(J)}$ for the normal distribution is greater than 1. This result is analogous to the corresponding normal theory result, because in normal theory inference the parameter $1/(1 - \rho)$ also measures the efficiency of a paired design with respect to an unpaired design.

For the double exponential distribution with $\rho = .5$ and $\sigma^2 = 2$, it can be shown that the ARE of $\hat{\alpha}^{(P)}$ with respect to $\hat{\alpha}^{(J)}$ is 2.

In clinical crossover trials, the data are repeated measures; therefore the methods of this entry are very applicable. Therefore, we will not pursue Jaeckel's method (11) for repeated-measures data.

V. AN ILLUSTRATION

The analysis methods will be illustrated using the data set from a 2×2 crossover trial described in Patel (21). The trial included 17 subjects with mild to moderate bronchial asthma in which the acute effects of single doses of two active drugs were compared. We will label the treatments A and B. All medications are discontinued 12 hours prior to receiving the trial treatments. The response measurements used to compare to two treatment groups is forced expiratory volume in 1 second (FEV_1). Subjects were divided at random into two groups, and those in the first group received the treatments in the order AB. It has been recently reported (see protocol 9 of the CIBA-GEIGY Brethine program, Basle, and Ref. 32) that the baseline value for subject number 11 prior to period 1 is 2.05. In fact, this subject did not improve at all after the medication. It has also been reported that the response for subject number 8 on period 2 is 2.43 instead of 2.41. The observed responses are given in Table 1 and Table 2. Because Patel found that there were no interactions between the treatments and the periods, we use the baseline values as covariates.

We will consider two cases. In the first case, we use all the data points. In the second case, we delete subject number 9, which is considered to be an outlier (mild) by Jones and Kenward (12) and Kenward and Jones (14). In fact, Kenward and Jones (14) deleted subject number 9 from their analyses.

A. Patel's Data Without the Outlier

We consider two cases. The first case is the analysis in the presence of baseline values. The second case is the analysis without the baseline values.

1. *Inferences in the Presence of Baseline Values*

Using the subroutine NLPNMS of PROC IML of SAS, we compute the R-estimates of the treatment effects and the slope. We have used PROC MIXED (PROC GLM) of SAS to compute the GLS (OLS) estimates.

Table 1 2 × 2 Crossover Trial of Single Oral Doses of Two Active Drugs (A and B) in Patients with Bronchial Asthma: Forced Expiratory Volume in 1 sec (FEV_1, L): Group I (Sequence AB)

Patient	x_{11k}, y_{11k}	x_{22k}, y_{22k}
1	1.09, 1.28	1.24, 1.33
2	1.38, 1.60	1.90, 2.21
3	2.27, 2.46	2.19, 2.43
4	1.34, 1.41	1.47, 1.81
5	1.31, 1.40	0.85, 0.85
6	0.96, 1.12	1.12, 1.20
7	0.66, 0.90	0.78, 0.90
8	1.69, 2.43	1.90, 2.79

Table 2 2 × 2 Crossover Trial of Single Oral Doses of Two Active Drugs (A and B) in Patients with Bronchial Asthma: Forced Expiratory Volume in 1 sec (FEV_1, L): Group II (Sequence BA)

Patient	x_{21k}, y_{21k}	x_{12k}, y_{12k}
1	1.74, 3.06	1.54, 1.38
2	2.41, 2.68	2.13, 2.10
3	2.05, 2.60	2.18, 2.32
4	1.20, 1.48	1.41, 1.30
5	1.70, 2.08	2.21, 2.34
6	1.89, 2.72	2.05, 2.48
7	0.89, 1.94	0.72, 1.11
8	2.41, 3.35	2.83, 3.23
9	0.96, 1.16	1.01, 1.25

First we consider the inferences of β. The estimate of β, its estimated standard error, and the value of the test statistic along with the p-value corresponding to each method (OLS, GLS, R, and $R^{(P)}$) are given in Table 3. All methods show that there is a significant linear relationship between the response and the baseline values. However, the OLS method gave the highest standard error. The rest of the method gave comparable estimated standard errors. The OLS estimate is also much different from the GLS estimate. However, the GLS estimate and $\hat{\beta}$ are very close.

Second, we consider the inference for the treatment effect. We report the estimate, its estimated standard error, and the test statistic along with the p-value in Table 4. We see that the OLS and the GLS estimates are very close. The $\hat{\alpha}$ and $\hat{\alpha}^{(P)}$ are very close but differ from both $\hat{\alpha}_{OLS}$ and $\hat{\alpha}_{GLS}$. The OLS estimate has the highest estimated standard error, whereas as $\hat{\alpha}^{(P)}$ has the lowest estimated standard error. All the methods show that there is a significant difference between the two treatments.

Table 3 Patel's Data: Inference for the Slope

Method	β	S.E.	$F_{\beta=0}/D^*_{\beta=0}$	p-value
OLS	1.3959	.2953	22.34	.0003
OLS_{WTO}	1.1075	.1786	38.44	.0001
R	1.1846	.1304	27.07	<.0001
R_{WTO}	1.1833	.2664	15.67	<.0001
GLS	1.1603	.1017	130.03	.0001
GLS_{WTO}	1.1207	.0915	149.96	.0001
$R^{(P)}$	1.2500	.1115	50.23	<.0001
$R^{(P)}_{WTO}$	1.1823	.1212	39.73	<.0001

WTO: without outlier; P: period; and C: Cross-over

Table 4 Patel's Data: Inference for the Treatment Effects with Baseline

Method	$\alpha_2 - \alpha_1$	S.E.	$F_{\alpha_1=\alpha_2}/D^*_{\alpha_1=\alpha_2}$	p-value
OLS	.2526	.0761	11.02	.0051
$\mathrm{OLS_{WTO}}$	.1943	.0456	18.10	.0009
GLS	.2543	.0755	11.32	.0046
$\mathrm{GLS_{WTO}}$	.1946	.0444	19.21	.0007
R	.2119	.0450	19.87	<.0001
R_{WTO}	.2115	.0665	13.53	.0002
$R^{(P)}$	.2337	.0289	64.64	<.0001
$R^{(P)}_{\mathrm{WTO}}$	.2117	.0310	50.93	<.0001

WTO: without outlier; P: period

2. Inferences Without the Baseline Values

We present the estimates of the treatment effect, its estimated standard errors, and the values of the test statistic along with the p-values for each method (GLS, R, and $R^{(P)}$) in Table 5 [assuming that there is no covariate effect in Eq. (1)]. The two R-estimates are very close. However, $\hat{\alpha}^{(P)}$ has a smaller estimated standard error than $\hat{\alpha}$. The OLS/GLS estimate has the highest estimated standard error. All the methods show that there is a significant difference between the two treatments.

Comparing Tables 4 and 5, we see that the inclusion of the baseline values in the model resulted in smaller standard errors (estimated) of the estimates. Thus we obtain more precise estimates of the effect when the baseline values are included in the model as covariates.

B. Patel's Data Without the Outlier

An investigation of the residuals (for period 1) from the OLS technique has shown that the residuals are not normally distributed. Also, subject number 9 seems to be an outlier. We have, therefore, dropped this subject from the analysis.

Table 5 Patel's Data: Inference for the Treatment Effects Without the Baseline

Method	$\alpha_2 - \alpha_1$	S.E.	$F_{\alpha_1=\alpha_2}/D^*_{\alpha_1=\alpha_2}$	p-value
OLS/GLS	.2552	.1184	4.64	.0478
$\mathrm{OLS_{WTO}/GLS_{WTO}}$	.1750	.0873	4.01	.0650
R	.1521	.0839	5.26	.0217
R_{WTO}	.1521	.0798	3.68	.0550
$R^{(P)}$	.1772	.0496	9.31	.0022
$R^{(P)}_{\mathrm{WTO}}$	.1494	.0542	6.61	.0100

WTO: without outlier; P: period

1. Inferences in the Presence of Baseline Values

Using the subroutine NLPNMS of PROC IML of SAS we compute the R-estimates of the treatment effects and the slope. The estimates are presented under the subscript WTO (without outlier). First we consider the inferences of β. The estimate of β, its estimated standard errors, and the value of the test statistic along with the p-value for each method (OLS, GLS, R, and $R^{(P)}$) are given in Table 3. All the methods show that there is a significant linear relationship between the response and the baseline values. However, the intrasubject rank-based method (R) gives the largest standard error. The OLS estimate has the next largest standard error. The GLS estimate has the smallest estimated standard error, although it is very close to that of $\hat{\beta}^{(P)}$. Note that without the outlier the normal probability plot of the OLS residuals (from the first period) shows that the data are normally distributed. However, all the estimators, except $\hat{\beta}$, are affected by the outlier. The most affected is the OLS estimate. However, the $\hat{\beta}^{(P)}$ is also affected by the outlier, but not as much as the OLS estimate. The GLS and OLS estimates are very close.

Second, we consider the inference for the treatment effect. The estimate of α, its estimated standard errors, and the value of the test statistic along with the p-value for each method (OLS, GLS, R, and $R^{(P)}$) are given in Table 4. Note that $\hat{\alpha}$ and $\hat{\alpha}^{(P)}$ are very close but differ from both the OLS and the GLS estimates. Note also that $\hat{\alpha}$ is not affected by the outlier at all. The OLS/GLS estimate has the highest estimated standard error, whereas $\hat{\alpha}^{(P)}$ has the lowest estimated standard error.

2. Inferences Without the Baseline Values

The estimate of α, its estimated standard errors, and the value of the test statistic along with the p-value for each method (GLS, R, and $R^{(P)}$) are given in Table 5. Only the $R^{(P)}$ method shows that there is a significant difference between the two treatments. The intrasubject rank-based method shows borderline significance (p-value .055). The OLS/GLS method gives the p-value .0650, which shows that there is a numerical trend in favor of treatment 2 over treatment 1.

VI. CONCLUSIONS

The results of this entry provide a robust alternative to parametric analysis of repeated-measures AB/BA crossover designs with baseline values as covariates. The rank-based tests have applications to biopharmaceutical studies, clinical trials, and other scientific experiments. One of the problems commonly encountered in bioavailability studies is that the data set sometimes contains some extremely large and/or small observations known as outliers. These outlying data may have dramatic effects on the bioequivalence test. Results with and without the outlying observations could be totally opposite. It is therefore appropriate to use rank-based methods, which are less affected by outliers.

The maximum-likelihood and the GLS approaches are based on the assumption that the data follow a multinormal theory model. Clearly, this is a strong assumption and needs to be investigated for each case. One possibility would be to consider the use of transformations. But the transformation will distort the model and its assumptions. A nonparametric test would be required, for example, if the data could not be transformed to satisfy the usual (e.g., normality) assumptions. It is worth mentioning that both the generalized least-squares and the maximum-likelihood approaches require the finiteness of the covariance matrix of the error vectors. In the rank-based methods, we do not need the finiteness of the covariance matrix of the error vectors. For this reason, the procedures proposed in this entry will have attractive appeal to data analysts.

Because the estimates of the variances σ_e^2 and σ_b^2 are required in the GLS approach for the estimation of the fixed-effects part of the mixed model, it is important that they be nonnegative. Also, the estimators of σ_e^2 and σ_b^2 will affect the variances of the estimators of the fixed effects. Searle (34) noted, "These are difficulties of the mixed model: for unbalanced data, estimating fixed effects in the presence of variance components, and estimating variance components themselves" in the context of ANOVA-based estimates. The default variance-component estimation used by PROC MIXED of SAS is the REML (restricted maximum-likelihood estimators) algorithm. Also, the computer algorithm BMDP-5V provides REML estimators of the variance components for a mixed model. For balanced experiments REML estimates are identical to the estimates obtained from expected mean squares, provided all estimates are positive. It is important to obtain non-negative estimates of the variance components, since they are used in the analysis of fixed effects. The REML estimation procedure does not, however, include estimating the fixed effects. Nevertheless, having obtained the estimators of the variance components using REML one would undoubtedly use these to obtain maximum-likelihood estimates for the fixed effects after replacing the variance components in the maximum-likelihood equations by their estimates. See Searle (34) for further discussion. On the other hand, the *R*-estimates developed in this entry are easy to compute using PROC IML of SAS. They do not suffer from the problems mentioned by Searle in the context of the GLS technique.

In this entry we have assumed that there is no treatment-by-period interaction, since we have used baseline values as covariates. In a future work, we will consider a rank-based approach to incorporate the baselines as responses in the presence of period-by-treatment interaction in the model (see Refs. 32 and 14 for normal theory approach). Koch (16) suggests that the evaluation of treatment-by-interaction still requires attention, regardless of whether such interaction is due to carryover effects or not. It is also mentioned by Jones and Wang (13) that carryover effects in AB/BA crossover designs should not always be ignored. Until now in the absence of baseline values, both in normal theroy and in nonparametric inferences, the tests (the *t*-test and the Wilcoxon–Mann–Whitney test) for no period-by-treatment interaction are based on the subject totals. The tests based on the subject totals are not powerful, because these procedures do not eliminate the subject effects from the test statistics.

ACKNOWLEDGMENTS

The author would like to thank Dr. A. J. Sankoh of the U.S. Food and Drug Administration for reading the manuscript and making comments that improved the quality of this entry.

REFERENCES

1. Bloch, D. A., Gastwirth J. S. On a simple estimate of the reciprocal of the density function. Ann Mathematical Stat 39:1083–1085, 1968.
2. Chow, Shien-Chung, and Liu, Jen-Pei. Design and Analysis Bioavailability and Bioequivalence Studies. New York: Marcel Dekker, 1992.
3. Dubey, S. D. Current thoughts on cross-over designs. Clin Res Practices Drug Regulatory Affairs 4:127–142, 1986.
4. Fava, G. I., Patel, H. I. A survey of cross-over designs used in industry. Unpublished manuscript, 1986.
5. Food and Drug Administration. A report on the two-period cross-over design and its applicability in trials of clinical effectiveness. Minutes of the Biometric and Epidemiology Methodology Advisory Committee (BEMAC) meeting, Rockville, MD, 1977.

6. Food and Drug Administration. Guidance: Statistical Procedures for Bioequivalence Studies Using a Standard Two-Treatment Cross-over Design, July 1, 1992, Rockville, MD, 1992.
7. Grieve, A. P. Cross-over versus parallel designs. In: D. A. Berry, ed. Statistical Methodology in the Pharmaceutical Sciences. New York: Marcel Dekker, 1990, pp 239–270.
8. Hettmansperger, T. P. Statistical Inference Based on Ranks. New York: Wiley, 1984.
9. Heiler, S., Willers, R. Asymptotic normality of R-estimates in the linear model. Statistics 19:173–184, 1988.
10. Hollander, M., Wolfe, D. A. Nonparametric Statistical Inference. New York: Wiley, 1973.
11. Jaeckel, L. A. Estimating regression coefficients by minimizing the dispersion of residuals. Ann Math Stat 43:1449–1459, 1972.
12. Jones, B., Kenward, M. G. Design and Analysis of Cross-over Trials. London: Chapman and Hall, 1989.
13. Jones, B., Wang, J. Comments on "Estimating Treatment Effects in Clinical Cross-over Trials." J Biopharmaceutical Stat 8:235–238, 1998.
14. Kenward, M. G., Jones, B. The analysis of data from 2×2 cross-over trials with baseline measurements. Stat Med 6:911–926, 1987.
15. Koch, G. G. The use non-parametric methods in the statistical analysis of the two-period change-over design. Biometrics 28:577–584, 1972.
16. Koch, G. G. Discussion of "Estimating Treatment Effects in Clinical Cross-over Trials." J Biopharmaceutical Stat 8:239–242, 1998.
17. Lindman, H. R. Analysis of Variance in Experimental Design. New York: Springer-Verlag, 1992.
18. McKean, J. W., Hettmansperger, T. P. Tests of hypotheses on ranks in the general linear model. Communications Stat Theory Methods 5:693–709, 1976.
19. McKean, J. W., Sheather, S. J., Hettmansperger, T. P. Regression diagnostics for rank-based methods. J Am Statistical Asso 85:1018–1028, 1990.
20. McKean, J. W., Sheather, S. J., Hettmansperger, T. P. The use and interpretation residuals based on robust estimation. J Am Statistical Asso 88:1254–1263, 1993.
21. Patel, H. I. Use of baseline measurements in the two-period cross-over designs. Communications Stat Theory Methods 12:2693–2712, 1983.
22. Nelder, J. A., Mead, R. A simplex method for function minimization. Computer J 7:308–313, 1965.
23. O'Neill, R. T. Subject-own control designs in clinical drug trials: overview of the issues with emphasis on two treatment problem. Presented at the Annual NCDEU Meeting, Key Biscayne, FL, 1978.
24. Randles, R. H., Wolfe, D. A. Introduction to the Theory of Nonparametric Statistics. New York: Wiley, 1979.
25. Rashid, M. M. Inference based on ranks for two-way models with a grouping factor and repeated measures factor. J Nonparametric Stat 6:27–42, 1996.
26. Rashid, M. M. On the use of R-estimators in analyzing repeated measures incomplete block clinical trials with baseline values as covariates. J Ital Statistical Soc 5: 261–284, 1996.
27. Rashid, M. M., Nandram, B. A rank-based analysis of one-way repeated measures designs with a changing covariate. J Statistical Planning Inference 46:249–263, 1995.
28. Rashid, M. M., Nandram, B. A rank-based predictor for the finite population mean of a small area: an application to crop production. J Agricultural, Biological Environmental Stat 3:201–222, 1998.
29. Senn, S. J. The use of baseline in clinical trials of bronchodilators. Stat Med 8:1339–1350, 1989.
30. Senn, S. J., Auclair, P. The graphical representation of clinical trials with particular reference to measurements overtime. Stat Med 9:1287–1302, 1990.
31. Senn, S. J. Cross-Over Trials in Clinical Research. Chichester, Eng: Wiley, 1993.
32. Senn, S. J., Grieve, A. Estimating treatment effects in clinical cross-over trials. J Biopharmaceutical Stat 8: 191–233, 1998.
33. Smith, N. D. Discussion of "Estimating Treatment Effects in Clinical Cross-over Trials": a regulatory perspective. J Biopharmaceutical Stat 8:243–247, 1998.
34. Searle, S. R. Linear models for unbalanced data. New York: Wiley, 1987.

M. Mushfiqur Rashid

S

Sample Size Determination*

See also *Power; Statistical Significance*

I. INTRODUCTION

Sample size determination refers to the evaluation of the sample size desired or required for a study during the design stage, before data are collected. A fundamental rule of sample size determination is that the method of evaluation be based on the planned method of analysis. In sample size evaluation, one must strike a balance between enrolling sufficiently many patients to detect a clinically important difference and not wasting important resources.

Sample size determination can be based on either the desired *power of a particular statistical test* for a population parameter or the desired *precision of a confidence interval estimator* of a population parameter when sampling from a large population. A hypothesis test tells us whether the observed data are consistent with the null hypothesis; a confidence interval tells us which hypotheses are consistent with the observed data. Most of the literature on sample size determination discusses the power function of a statistical test rather than the precision of an estimator. However, it is well known that there is a simple correspondence between the two approaches.

Statistical tests can be conducted in many ways, but the setting most frequently encountered in clinical trials is *comparative*, in which two or more treatments are compared. In most cases, the null hypothesis, H_0, is that the experimental treatment has no different effect on the outcome compared with the control. When testing the experimental treatment, it may be assumed that one has introduced this new treatment because it is thought to be superior to the control, and the trial is being carried out to "prove" this. On the other hand, experimental treatments are also introduced that are equivalent to the control, i.e., expected to have no better efficacy, or minimally poorer efficacy, than control. Such a treatment may be introduced because this new treatment may have other benefits; for example, it is less toxic, less expensive, or easier to administer, takes less time to treat, or has fewer side effects. One key difference between testing for superiority and equivalence is that the former usually needs the specification of the smallest clinically meaningful and achievable treatment difference in the alternative hypothesis, H_a, while the latter requires the specification of the largest acceptable treatment difference in the null hypothesis, H_0.

When conducting a statistical test, *two types of error* must be considered: *Type I* (false positive, i.e., rejecting the null hypothesis given that the null hypothesis is true) and *Type II* (false negative, i.e., failing to reject the null hypothesis given that the alternative hypothesis is true), with the probabilities α and β, respectively. The *power* of a statistical test refers to the probability of rejecting the null hypothesis when the alternative is true, i.e., $1 - \beta$. Between the two types of error, the Type I error has traditionally been thought of as more critical. However, a smaller Type I error rate results in a lower power. On balance, primary importance is still given to the Type I error, and a properly higher sample size is needed to achieve higher power when both pos-

*The views expressed in this paper are those of the authors and not necessarily of the Food and Drug Administration.

sible and necessary. In any statistical test, one assesses the probability of observing values as extreme or more extreme than the data observed under the null hypothesis, commonly termed the *p*-value. Usually "the data" are summarized in the form of an estimator or a sufficient statistic for an interesting parameter, termed a *test statistic*, and statistical significance is declared if the resulting *p*-value is less than the preselected significance level α.

In clinical trials, it is often the case that there are many interesting *outcome* or *endpoint variables*. In order to strike a balance between the desirability of assessing the effect of treatment on many variables and maintaining adequate power for specific variables, usually one primary variable is selected to determine the sample size. Otherwise, a multiple comparison adjustment is often applied to the sample size calculations through the significance levels assigned to the individual primary outcomes. The sample size is then chosen so that there is adequate power for each primary variable. While multiple comparisons are appropriate only if one will claim significance when any of the primary variables is significant, nominal significance level can be used to ensure the significance of all of the primary variables.

Another critical factor in sample size calculation is the selection of the assumed *parameters in the chosen population*. The adequacy of the sample size determined depends on the accuracy of the initial specification of the parameters. For the purpose of planning a study, therefore, it is always advisable to consider a range of population parameters over which the planned analysis is conducted for a specific sample size.

Numerous articles and reference textbooks provide comprehensive reviews of sample size determination from the perspective of either the precision of an estimator or the power of a test. There is no one response to the general question of how many patients must be included in the treatment and the control groups. In general, sample size determination varies over the type of patient population (e.g., a single population vs. two populations), the type of response measurement (e.g., actual measurement vs. proportion successful), the type of statistical analysis (e.g., time-specific actuarial life table probability vs. crude probability), the type of study design (e.g., paired vs. independent, or parallel vs. crossover), and other factors. The following is a compendium of exact and asymptotic formulas for estimating the sample size in clinical trials. Before the general formulas are reviewed, additional consideration is given to the factors that should be accounted for in sample size calculation.

II. SAMPLE SIZE ADJUSTMENT

A. A Simple Adjustment for the *t*-Test Statistic

If an estimator or a sufficient statistic of an interesting parameter is normally distributed, at least asymptotically, the sample size formula frequently takes the form

$$N_{\text{tot}} = \frac{c^2(z_{\alpha/2}\sigma_0 + z_\beta\sigma_1)^2}{\delta^2} \tag{1}$$

Here, N_{tot} is the total sample size for a two-sided test, c is a constant, z_α is the standard normal deviate with probability α of being exceeded, σ_0 and σ_1 are the standard deviations under the null and alternative hypotheses, respectively, and δ is the treatment effect. In a one-sided test, $z_{\alpha/2}$ is replaced by z_α in Eq. (1).

If the test statistic has student's *t*-distribution instead, then the use of Eq. (1) will underestimate the required sample size, although this effect is increasingly negligible for increasing degrees of freedom, *d.f.*, of the *t*-statistic. An adequate adjustment is obtained by the correction factor $f = (d.f. + 3)/(d.f. + 1)$, where the actual total sample size if fN and N is obtained from Eq. (1) (see Ref. 1).

B. A Simple Adjustment for Missing Values

In some cases, part of the information in the data is missing completely at random (purely by chance). Then the sample size required should be adjusted upwards, since the missing information would affect the power of a test and the precision of a confidence interval of a parameter. If M is the rate of missing information, then a simple adjustment is (see Ref. 2, p 3902)

$$N_{\text{adj}} = \frac{N}{1 - M} \tag{2}$$

C. A Simple Adjustment for Unequal Allocation

It is well known that power is maximized and total sample size minimized for two equal-sized, independent treatment groups if the population variances are equal. However, due to ethical considerations, unequal-sized groups may be desirable. Let subscripts e and c denote the experimental and control groups, respectively. Suppose that the unequal sizes are reflected in the sample fractions, Q_e and Q_c, where $n_e = Q_eN$, $n_c = Q_cN$, $Q_e + Q_c = 1$, and $N = n_e + n_c$. Obviously, $Q_e = Q_c = 0.5$ is for the balanced case. An adjustment for the unbalanced case is

$$N_{\text{unbal}} = \frac{Q_e^{-1} + Q_c^{-1}}{4} N_{\text{bal}} \tag{3}$$

where N_{bal} and N_{unbal} are the total sample size for the balanced case and the unbalanced case, respectively (2). That is, the unbalanced design requires $[(Q_e^{-1} + Q_c^{-1})/4 - 1]100\%$ larger sample size to provide the same level of power as the balanced design.

III. SAMPLE SIZE BASED ON MEANS

In this section, the outcome variable is continuous or quantitative. It is convenient to assume that the estimator or test statistic of an interesting parameter is normally distributed, at least asymptotically.

A. A Single Mean

In this case, there is no independent control group included in the clinical trials. The clinical decision is often based on the subjective evaluation of the outcome.

1. Hypothesis Test Approach

For the test H_0: $\mu = \mu_0$ with some a priori specified mean value μ_0 and variance σ_0^2 against an alternative H_a: $\mu = \mu_1 \neq \mu_0$ with variance σ_1^2, the test statistic is

$$Z = \frac{\bar{Y} - \mu_0}{\sigma_0/\sqrt{N}}$$

where $\bar{Y}$ is the mean of a single sample of observations. Then for specified α, β, μ_0, μ_1, σ_0, and σ_1, the total sample size required to ensure power $1 - \beta$ of detecting the difference $\delta = \mu_1 - \mu_0$ with a test at the significance level α is given by

$$N = \frac{(z_\alpha \sigma_0 + z_\beta \sigma_1)^2}{\delta^2} \tag{4}$$

In a clinical trial, σ_0 and σ_1 can be specified based on prior experiments using the same measurements or estimated from a small preliminary or "pilot" study. In these cases, it is best to use the largest value expected. If $\sigma_0 = \sigma_1 = \sigma$, Eq. (4) reduces to

$$N = \frac{\sigma^2 (z_\alpha + z_\beta)^2}{\delta^2} \tag{5}$$

2. Confidence Interval Approach

Suppose that one wishes to estimate the mean μ of a large population with variance component σ^2 using an estimator $\bar{Y}$ with variance σ^2/N. Then the sample size required to provide an interval estimator with the precision e at confidence level $1 - \alpha$ for a specified variance component σ^2 is given by

$$N = \left(\frac{z_{\alpha/2}\sigma}{e}\right)^2 \tag{6}$$

3. Correspondence of the Two Approaches

If $\sigma_0 = \sigma_1 = \sigma$, the N derived using a $1 - \alpha$ confidence interval to ensure a precision e is approximately the same as the N required to detect a difference $\mu_1 - \mu_0 = e$ with power of 0.5 using a two-sided test. In this case, $e = z_{\alpha/2}\sigma/\sqrt{N}$. However, when $\sigma_0 \neq \sigma_1$, e is actually $z_{\alpha/2}\sigma_1/\sqrt{N}$, since the variance under the alternative should be used to construct a confidence interval (2).

B. Two Independent Groups

In this case, two treatments are compared. Assume that the outcome variable has the means μ_1 and μ_2, with the standard deviation σ_1 and σ_2, respectively, for the two treatments. One wishes to estimate the difference of two population means in order to test for either superiority of one treatment to another or equivalence of the two treatments, in terms of the population mean.

1. Parallel Design

The simplest and probably the most frequently employed design for a clinical trial is the parallel-groups design. Eligible patients, who have given their informed consent, are randomly assigned to receive one and only one of two treatments.

a. Confidence Interval Approach

Suppose that one wishes to estimate the difference of means, $\mu_1 - \mu_2$, with the common standard deviation $\sigma_1 = \sigma_2 = \sigma$. Based on normal theory, the sample size required to provide an interval estimate with the precision e at confidence level $1 - \alpha$ for a given variance component σ^2 is given by

$$N = \frac{2(z_{\alpha/2}\sigma)^2}{e^2} \tag{7}$$

This equation is for a two-sided confidence interval and is also applicable to the one-sided case by simply replacing $z_{\alpha/2}$ by z_α.

b. Hypothesis Test for Superiority Approach

Consider the statistical model $Y_{ij} = \mu_i + e_{ij}$, $i = 1, 2$; $j = 1, 2, \ldots, n$, where Y_{ij} denotes the measurement of outcome response on the jth subject for the ith treatment, μ_i is the underlying true mean response for the ith treatment, and e_{ij}'s denote the independently identically distributed measurement errors with common variance σ^2. For the test H_0: $\mu_1 - \mu_2 = 0$ against an alternative H_a: $\mu_1 - \mu_2 \neq 0$ with common variance σ^2, the test statistic is

$$Z = \frac{\bar{Y}_1 - \bar{Y}_2}{\sigma\sqrt{2/n}}$$

where $\bar{Y}_1$ and $\bar{Y}_2$ are the means of two independent samples and n is the sample size for each treatment group. Then, for specified α, β, $\delta = \mu_1 - \mu_2$, and σ^2, the required sample size for each treatment group is

$$n = \frac{2\sigma^2(z_{\alpha/2} + z_\beta)^2}{\delta^2} \tag{8}$$

and then the total sample size is $N = 2n$. For a one-sided test, the required sample size for each treatment group is obtained by substituting $z_{\alpha/2}$ by z_α in Eq. (8).

2. 2 × 2 Crossover Design

The two-period crossover design consists of two treatments, which conventionally are labeled A and B. Each subject (patient) receives the two treatments, with half the subjects randomly assigned to receive the treatments in the sequence AB and the rest in the sequence BA. Let Y_{ijk} be the measurement for the jth subject in the ith sequence of treatment at the kth period. Each of n patients in sequence 1 receives treatment A in period 1 and treatment B in period 2; each of n patients in sequence 2 receives the treatments in reverse order. If carryover effects are not present, an adequate model is

$$Y_{ijk} = \mu + \varsigma_i + \xi_{ij} + \pi_k + \phi_l + e_{ijk} \tag{9}$$

$i = 1, 2$; $j = 1, 2, \ldots, n$; $k = 1, 2$; $l = i - (-1)^i\delta_{2k}$; $\delta_{2k} = 1$ if $k = 2$, $\delta_{2k} = 0$ otherwise. Here, μ is the overall mean, ς_i is the effect of the ith sequence, ξ_{ij} is the effect of the jth patient within the ith sequence, which is the between-subject error, π_k is the effect of the kth period, ϕ_l is the effect of the lth treatment, and e_{ijk} is the effect of the jth subject within the ith sequence at the kth period, which is the within-subject error. For the fixed source of variation, ς_i, π_k, and ϕ_l, the standard constraints are assumed; that is, the main effects sum to zero. For the random effects, ξ_{ij} and e_{ijk}, it is assumed that each set of error terms $\{\xi_{ij}\}$ and $\{e_{ijk}\}$ are independent and identically normally distributed with mean 0 and variances σ_s^2 and σ_e^2, respectively, and that the set of two error terms (ξ_{ij}, e_{ijk}) are independently distributed. The assumptions imply that the measurements have the same variance $\sigma_s^2 + \sigma_e^2$; the measurements on different subjects are independent; the measurements on the same subject within different treatment periods have covariance σ_s^2.

For the test H_0: $\phi_1 = \phi_2$ against an alternative H_a: $\phi_1 \neq \phi_2$, the difference of the measurements between periods 1 and 2 on each subject is considered by defining $d_{ij} = Y_{ij1} - Y_{ij2}$. Let $\phi_d = 2(\phi_1 - \phi_2)$ and $\hat{\phi}_d = \bar{d}_{1.} - \bar{d}_{2.}$. Then the estimator $\hat{\phi}_d$ is such that

$$E(\hat{\phi}_d) = \phi_d \qquad \text{and} \qquad \text{var}(\hat{\phi}_d) = \frac{2\sigma_d^2}{n^*} \tag{10}$$

where $\sigma_d^2 = 2\sigma_e^2$ and n^* is the required number of subjects for each of the two sequences. Under the null hypothesis, $\phi_1 = \phi_2$, and using the pooled estimator, $\hat{\sigma}_d^2$, of σ_d^2, the test statistic

$$T_d = \frac{\hat{\phi}_d}{(2\hat{\sigma}_d^2/n^*)^{1/2}} \tag{11}$$

has student's t-distribution with $2(n^* - 1)$ *d.f.* Note that for the two-period crossover design, the variance of the intrasubject differences (d_{ij}) is $2\sigma_e^2$ and the expected value of the difference between the two mean intrasubject differences $(\bar{d}_{1.} - \bar{d}_{2.})$ is $2(\phi_1 - \phi_2)$. Therefore, the required number of subjects for each of the two sequences is

$$n^* = \frac{2(2\phi_e^2)(z_{\alpha/2} + z_\beta)^2}{[2(\sigma_1 - \phi_2)]^2} = \frac{\sigma_e^2(z_{\alpha/2} + z_\beta)^2}{\delta^2} = \frac{n(1 - R)}{2} \tag{12}$$

where

$$n = \frac{2(\sigma_s^2 + \sigma_e^2)(z_{\alpha/2} + z_\beta)^2}{\delta^2} \tag{13}$$

would be the number of subjects required for each treatment group if the study were performed using two parallel groups, $\delta = \phi_1 - \phi_2$, and $R = \sigma_s^2/(\sigma_s^2 + \sigma_e^2)$, the intraclass correlation coefficient. Even if R is as low as 0.5, the total number of subjects required for a two-period crossover study with a specified power is one-quarter of the total number required for a parallel-groups study with the same power. Because each subject is measured twice in a two-period crossover study, the total number of measurements on the $2n^*$ subjects in it is

$$2 \times 2n^* = 2n(1 - R) \tag{14}$$

a fraction $1 - R$ of the total of $2n$ measurements in the parallel-groups study (see Ref. 3).

3. Parallel Design with Baseline and Repeated Measurements

Before randomization, repeated measurements can be used to identify patients who are eligible for the trial. In the long run, an outcome variable may be measured frequently to ensure that the outcome is adequately controlled and that accurate measurement is obtained. It also could be measured frequently to study the time course of response. Furthermore, since the variability of the measurement is a crucial factor for sample size, one can use repeated measurements to reduce the variability. For example, one can average all of the post-randomization measurements for an individual. In addition, it may reduce the study cost if enrolling more patients is not easy and is more costly than taking repeated measurements on each patient for the desired power.

It is preferred to combine baseline (pretreatment) information with posttreatment assessment in the comparison of two treatments if the correlation between baseline and posttreatment is substantially large (usually greater than 0.5). Otherwise, only the posttreatment measurement is used.

Frison and Pocock (4) explored the use of simple summary statistics for analyzing repeated measurements with baseline measurements in randomized clinical trials with two treatments and provided the methods for determining sample sizes in repeated measurements as follows.

Suppose a randomized clinical trial has two treatment groups with n_i patients per group, and all patients have p pretreatment visits, $m = -(p-1), \ldots, 0$ and r posttreatment visits, $m = 1, \ldots, r$. A quantitative measurement is observed at every visit for every patient, and the following simple model is considered:

$$Y_{ijm} = \mu_{im} + e_{ijm}, \quad i = 1, 2; \quad j = 1, 2, \ldots, n_i; \quad m = -(p-1), \ldots, 0, 1, \ldots, r \tag{15}$$

Here, μ_{im} is the underlying true mean response for treatment i at time m. As a result of randomization, it is reasonable to assume $\mu_{1m}^{\text{pre}} = \mu_{2m}^{\text{pre}}$ for the pretreatment visits $m \le 0$. e_{ijm} is the individual jth patient "error," or variation around the underlying mean μ_{im}, and these errors will not be independent within patients. For simplicity, compound symmetry structure over time points is assumed with equal variance σ^2 for all time points and both treatments and also equal correlation ρ between all pairs of time points on each patient. The correlation ρ is also expected to be substantial (typically greater than 0.5 in most trials), since it reflects the consistency of patient effects over time. Since in many clinical trials the primary objective is to assess the average response to treatment over time, often (but not necessarily) in anticipation that treatment response is liable to occur quickly and to remain reasonably steady over time, the following three possible methods of analysis are considered:

1. *Posttreatment means*: a simple analysis using the mean for each patient's posttreatment measurements as the summary measure. For each individual, the summary statistic is

$$\bar{Y}_{ij.}^{\text{post}} = \frac{1}{r}\sum_{m=1}^{r} Y_{ijm} \tag{16}$$

Then, the overall posttreatment mean difference is

$$\frac{1}{n_1}\sum_{j=1}^{n_1} \bar{Y}_{1j.}^{\text{post}} - \frac{1}{n_2}\sum_{j=1}^{n_2} \bar{Y}_{2j.}^{\text{post}} = \bar{Y}_{1..}^{\text{post}} - \bar{Y}_{2..}^{\text{post}} \tag{17}$$

which has expected value

$$\frac{1}{r}\sum_{m=1}^{r} (\mu_{1m} - \mu_{2m}) = \bar{\mu}_{1.}^{\text{post}} - \bar{\mu}_{2.}^{\text{post}} \tag{18}$$

2. *Mean changes*: a simple analysis of each patient's difference between the mean of posttreatment measurements and the mean of baseline measurements, the latter often consisting of just a single baseline value per patient. For each individual, the summary statistic is mean change,

$$\frac{1}{r}\sum_{m=1}^{r} Y_{ijm} - \frac{1}{p}\sum_{m=-(p-1)}^{0} Y_{ijm} = \bar{Y}_{ij.}^{\text{post}} - \bar{Y}_{ij.}^{\text{pre}} \tag{19}$$

Then the overall treatment difference in these mean changes is

$$\frac{1}{n_1}\sum_{j=1}^{n_1} (\bar{Y}_{1j.}^{\text{post}} - \bar{Y}_{1j.}^{\text{pre}}) - \frac{1}{n_2}\sum_{j=1}^{n_2} (\bar{Y}_{2j.}^{\text{post}} - \bar{Y}_{2j.}^{\text{pre}}) = (\bar{Y}_{1..}^{\text{post}} - \bar{Y}_{1..}^{\text{pre}}) - (\bar{Y}_{2..}^{\text{post}} - \bar{Y}_{2..}^{\text{pre}}) \tag{20}$$

which has an expected value again equal to $\bar{\mu}_{1.}^{\text{post}} - \bar{\mu}_{2.}^{\text{post}}$, since the pretreatment expected values are the same for both treatments.

3. *Analysis of covariance*: Between-patient variations in baseline measurements are taken into account, by using the mean baseline measurement for each patient as a covariate in a linear model for treatment comparison of posttreatment means. The model is as follows;

$$\bar{Y}_{ij.}^{\text{post}} = \bar{\mu}_{i.}^{\text{post}} + \beta(\bar{Y}_{ij.}^{\text{pre}} - \bar{\mu}_{..}^{\text{pre}}) + \varepsilon_{ij} \tag{21}$$

where ε_{ij}'s are independent random errors with assumed constant variance. With estimate $\hat{\beta}$ obtained by least squares and by defining

$$\bar{Y}_{ij.}^{\text{cov}} = \bar{Y}_{ij.}^{\text{post}} - \hat{\beta}(\bar{Y}_{ij.}^{\text{pre}} - \bar{Y}_{...}^{\text{pre}}) \tag{22}$$

the estimated mean treatment difference is

$$\frac{1}{n_1}\sum_{j=1}^{n_1} \bar{Y}_{1j.}^{\text{cov}} - \frac{1}{n_2}\sum_{j=1}^{n_2} \bar{Y}_{2j.}^{\text{cov}} = \bar{Y}_{1..}^{\text{cov}} - \bar{Y}_{2..}^{\text{cov}} \tag{23}$$

which again has the expected value $\bar{\mu}_{1.}^{\text{post}} - \bar{\mu}_{2.}^{\text{post}}$.

It is convenient to assume that sample size is sufficiently large that the normal approximation to the t-distribution can be applied. For two equal-sized treatment groups with null hypothesis H_0: $\bar{\mu}_{1.}^{\text{post}} = \bar{\mu}_{2.}^{\text{post}}$ and alternative H_a: $\bar{\mu}_{1.}^{\text{post}} \neq \bar{\mu}_{2.}^{\text{post}}$, the required sample size for each treatment group is approximately

$$n = \frac{2\sigma^2}{\delta^2}\left[\frac{1 + (r-1)\rho}{r} - \frac{p\rho^2}{1 + (p-1)\rho}\right](z_{\alpha/2} + z_\beta)^2 \tag{24}$$

for the method of covariate analysis,

$$n = \frac{2\sigma^2}{\delta^2}\left[\frac{1 + (r-1)\rho}{r}\right](z_{\alpha/2} + z_\beta)^2 \tag{25}$$

for the posttreatment means method, and

$$n = \frac{2\sigma^2}{\delta^2}\left[\frac{1 + (r-1)\rho}{r} - \frac{(p+1)\rho - 1}{p}\right](z_{\alpha/2} + z_\beta)^2 \tag{26}$$

for the approach of mean changes, respectively, for given power $1 - \beta$, significance level α, and detectable difference $\delta = \bar{\mu}_{1.}^{\text{post}} - \bar{\mu}_{2.}^{\text{post}}$. For the same power and when $\rho > 0.5$, the method of covariate analysis always requires a smaller number of patients.

Suppose only r posttreatment measurements are taken, a constant mean over time points for each treatment is expected, and the effect of time points and the interaction of time point with treatment are not significant. Then the statistical model of Eqs. (15) can be rewritten as

$$Y_{ijm} = \mu_i + f_j + e_{ijm}, \qquad i = 1, 2;$$
$$j = 1, 2, \ldots, n; \qquad m = 1, 2, \ldots, r \tag{27}$$

where f_j denotes the effect of the jth subject with mean 0 and variance σ_s^2, e_{ijm}'s denote the independently, identically distributed errors of measurement with mean 0 and common variance σ_e^2, and these are uncorrelated. The assumptions imply compound symmetry structure over time points with equal variance $\sigma^2 = \sigma_s^2 + \sigma_e^2$ for all time points and both treatments and also equal correlation $\rho = \sigma_s^2/(\sigma_s^2 + \sigma_e^2)$ between all pairs of time points for each subject.

Then, for the test H_0: $\mu_1 = \mu_2$ against an alternative H_a: $\mu_1 \neq \mu_2$, Eq. (25) becomes (see Ref. 3)

$$n = \frac{2(\sigma_s^2 + \sigma_e^2/r)(z_{\alpha/2} + z_\beta)^2}{\delta^2} \tag{28}$$

Appealing to a quasi-score test statistic for a generalized linear model based on the generalized estimating equation (GEE) method, Liu and Liang (5) presented a method to compute sample size for studies involving correlated observations. The case of repeated measurements discussed earlier is an example of correlated data. Because of the potential association for observations within a subject, the intrasubject correlation should be taken into consideration both in the design and the analysis stages. Common choices for correlation structure over repeated measurements include uncorrelated structure, compound symmetry structure, autoregressive structure, and unstructured correlation. The first three correlation structures may be used for sample size calculations in most practical situations. There are no explicit sample size formulas for a generalized linear model, and numerical methods are required. However, for a typical two-sample problem with repeated measurements and compound symmetry correlation structure, an explicit formula for the required total sample size is available:

$$N = \frac{\sigma^2}{\pi_1\pi_2\delta^2}\left[\frac{1 + (r-1)\rho}{r}\right](z_{\alpha/2} + z_\beta)^2 \tag{29}$$

where π_1 and π_2 are proportions of patients in the two treatment groups, respectively. Note that for the equal-sized case, i.e., $\pi_1 = \pi_2 = 0.5$, Eq. (29) is the same as Eq. (25).

4. 2 × 2 Crossover Design with Repeated Measurements

Yue and Roach (6) considered the extension of the model of Eqs. (9) for a 2 × 2 repeated-measurements crossover design with mean summary:

$$Y_{ijkm} = \mu + \varsigma_i + \xi_{ij} + \pi_k + \phi_l + e_{ijk} + \tau_m + (\varsigma\tau)_{im} + \omega_{ijm} + (\pi\tau)_{km} + (\phi\tau)_{lm} + f_{ijkm} \tag{30}$$

$i = 1, 2$; $j = 1, 2, \ldots, n$; $k = 1, 2$; $l = i - (-1)^i\delta_{2k}$; $\delta_{2k} = 1$ if $k = 2$, $\delta_{2k} = 0$ otherwise; and $m = 1, 2, \ldots, r$. In addition to the notation and assumptions in the model of Eqs. (9), τ_m is the effect of the mth time, $(\varsigma\tau)_{im}$ is the interaction effect of the mth time with the ith sequence, ω_{ijm} is the effect of the jth subject within the ith sequence at the mth time, $(\pi\tau)_{km}$ is the interaction effect of the mth time with the kth period, $(\phi\tau)_{lm}$ is the interaction effect of the mth time with the lth treatment, and f_{ijkm} is the random fluctuation of the jth subject within the ith sequence at the kth period and mth time.

For the fixed sources of variation, τ_m, $(s\tau)_{im}$, $(\pi\tau)_{km}$, and $(\phi\tau)_{lm}$, the main effect τ_m sums to zero, and the interaction effects sum to zero over both the r levels of time and the two levels of the main effect. For simplicity, it is assumed that each set of $\{\omega_{ijm}\}$ and $\{f_{ijkm}\}$ are independently and identically normally distributed with mean zero and variances σ_ω^2 and σ_f^2, respectively, and that the set of four error terms $(\xi_{ij}, e_{ijk}, \omega_{ijm}, f_{ijkm})$ is independently distributed. The assumptions imply compound symmetry structure over time points within each treatment period, i.e., equal variances for all time points and both treatments and also equal correlation between all pairs of time points. They also imply that the measurements have the same variance $\sigma_s^2 + \sigma_e^2 + \sigma_\omega^2 + \sigma_f^2$; the measurements on different subjects are independent; the measurements on the same subject within the same treatment period but at different time points have covariance $\sigma_s^2 + \sigma_e^2$; and the measurements on the same subject within different treatment periods have covariance $\sigma_s^2 + \sigma_\omega^2$ if at the same time point and σ_s^2 if at different time points.

Similarly, we define the following:

$$\phi_d = 2(\phi_1 - \phi_2)$$

$$\sigma_d^2 = 2\left(\sigma_e^2 + \frac{1}{r}\sigma_f^2\right) \tag{31}$$

$$\hat{\phi}_d = \bar{d}_{1..} - \bar{d}_{2..}$$

Then, using the similar t-test statistic, the required number of subjects for each of the two sequences for given power $1 - \beta$, significance level α, and detectable difference $\delta = \phi_1 - \phi_2$ with respect to Y_{ijkm} is approximately

$$n^* = \frac{(\sigma_e^2 + \sigma_f^2/r)(z_{\alpha/2} + z_\beta)^2}{\delta^2} \tag{32}$$

If $r = 1$ (that is, there is no repeated measurement within each treatment period), then the model of Eq. (30) reduces to Eq. (9), and n^* in Eq. (32) becomes n^* in Eq. (12).

If the interaction effects of the time with other factors and the main effect of the time are not significant, the model of Eq. (30) can be simplified to

$$Y_{ijkm} = \mu + s_i + \xi_{ij} + \pi_k + \phi_l + e_{ijk} + f_{ijkm} \tag{33}$$

where f_{ijkm} can be regarded as subsampling error. However, Eq. (32) is still valid for Eq. (33).

5. *Equivalence Trials*

In equivalence trials, the question asked is whether a new treatment is as effective as a standard treatment. This question arises when considering a new treatment that is thought to be as effective, but perhaps not more effective, or even minimally less effective, than a standard treatment. The new treatment may be preferred due to other benefits if it is just as effective, or in some cases even nearly as effective, in curing illness. However, a crucial issue for the equivalence trials is agreement upon how much expense of reduced efficacy could be accepted as necessary for other benefits, such as reduced toxicity.

A sample size formulat for this situation is

$$N_{\text{tot}} = \frac{c^2(z_\alpha\sigma_0 + z_\beta\sigma_1)^2}{(\delta_1 - \delta_0)^2} \tag{34}$$

Here, c is a constant, δ_1 is the difference that one is willing to accept, and δ_0 is the expected or true difference. For equivalence, one usually sets $\delta_0 = 0$. Equation (34) is valid for confidence interval approaches (see Ref. 7).

For simplicity, it is assumed that higher means are more desirable, and μ_e and μ_s denote experimental and standard treatment means, respectively. For the hypothesis H_0: $\mu_s - \mu_e \geq \delta_1$ against alternative hypothesis H_a: $\mu_s - \mu_e < \delta_1$, where $\delta_1 > 0$, the required sample size for each treatment group is

$$n = \frac{2\sigma^2(z_\alpha + z_\beta)^2}{(\delta_1 - \delta_0)^2} \tag{35}$$

for a given significance level α, a power $1 - \beta$, a common variance σ^2, a true mean difference $\delta_0 = \mu_s - \mu_e$, and a sufficiently small difference δ_1 that two treatments are considered equivalent for practical purposes if $\delta_0 < \delta_1$.

Equation (35) can be developed in terms of a one-sided $100(1 - \alpha)\%$ confidence interval for the difference $\mu_s - \mu_e$, with a specified probability $1 - \beta$ that the interval will not include a difference δ_1.

C. Paired Observations

In the case that the measurements in the two groups are linked together by pairing or repeated measures at times a and b on the same patient, the t-test is conducted using the mean difference $\bar{d} = \bar{Y}_b - \bar{Y}_a$, with variance $\Sigma^2 = \sigma_d^2/n$, where $\sigma_d^2 = 2\sigma^2(1 - \rho)$ if $\sigma_a^2 = \sigma_b^2 = \sigma^2$, ρ being the correlation between the paired measurements. For the hypothesis H_0: $\mu_a - \mu_b = 0$ against an alternative H_a: $\mu_a - \mu_b = \delta \neq 0$, where δ is a clinically significant prespecified value, the required number of sample pairs is approximately

$$n = \frac{\sigma_d^2(z_\alpha + z_\beta)^2}{\delta^2} \tag{36}$$

Here, σ_d^2 can be obtained from prior experience or estimated from a previous pilot study. Note that pairing is efficient only if $\rho > 0$, i.e., if there is positive correlation between the paired measurements. If no estimate of ρ is available, it is safe to assume $\rho = 0$ or, nominally, $\rho = 0.10$ (1).

Note that Eq. (36) is applicable for an alternative H_a: $\mu_a - \mu_b \neq 0$ with $z_{\alpha/2}$ substituted for z_α and a detectable difference δ.

IV. SAMPLE SIZE BASED ON PROPORTIONS

When the outcome measure in a clinical trial is a dichotomous variable, such as success versus failure, the data are usually expressed as a proportion of "success" p. The probability distribution of such a proportion is the binomial distribution $B(\pi, N)$ with parameters π (the true population proportion) and N (sample size). For large N, the binomial distribution asymptotically approaches a normal distribution $N(\mu, \sigma^2)$ with mean $\mu = \pi$ and variance $\sigma^2 = \pi(1 - \pi)/N$. In this case, the basic equations of tests for means may be employed as the approximate equations of tests for proportions.

A. A Single Proportion

1. *Hypothesis Test Approach*

For the one-treatment-group problem involving a single proportion, the hypothesis H_0: $\pi = \pi_0$ is tested to detect a clinically relevant alternative hypothesis H_a: $\pi = \pi_1$, where either $\pi_1 > \pi_0$ or $\pi_1 < \pi_0$. Based on a proportion p and large sample size N, the test statistic is $Z = (p - \pi)/\sigma_0$, where $Z \sim N(0, 1)$ and $\sigma_0^2 = \pi_0(1 - \pi_0)/N$, if H_0 is true. For the given significance level α and power $1 - \beta$, the equation for sample size N is (see Ref. 1)

$$N = \left[\frac{Z_\alpha\sqrt{\pi_0(1 - \pi_0)} + Z_\beta\sqrt{\pi_1(1 - \pi_1)}}{\pi_1 - \pi_0}\right]^2 \tag{37}$$

2. *McNemar's χ^2 Test of Equality of Paired Proportions*

When two groups of observation are linked together in such a way that matching on the sample individuals at categories A and B, the basic data can be expressed as

		B +	B −	
A	+	m_{++}	m_{+-}	m_A
	−	m_{-+}	m_{--}	
		m_B		N

where m_{-+}, for instance, is the number of pairs with (−) for observation A and (+) for observation B; m_A and m_B are the total numbers for the A and B observations, respectively. Thus, the proportions (p's) are defined as the ratios of frequencies (m's) to the total number of pairs, N.

Analogous to the problem of t-tests for paired means, the null hypothesis for testing paired proportions is H_0: $\mu_d = (\pi_A - \pi_B) = 0$. Since the difference between π_A and π_B equals to $(\pi_{-+} - \pi_{+-})$, the problem can be stated in terms of the discordant proportions π_{-+} and π_{+-}. This means that the hypothesis may be expressed as H_0: $\pi_{-+} = \pi_{+-} = \pi$. The test statistic employed to test H_0 is

$$Z = \frac{p_{-+} - p_{+-}}{s}$$

where

$$s^2 = \frac{2\bar{p}}{N}$$

is the estimate of variance;

$$\bar{p} = \frac{p_{-+} + p_{+-}}{2}$$

is the estimate of π; and $Z \sim N(0, 1)$ if H_0 is true. Note that Z^2 is equivalent to McNemar's χ^2 statistic usually used; thus, for given a significance level α and a power $1 - \beta$, the equation for sample size N is expressed as (see Ref. 1)

$$N = \left[\frac{Z_\alpha\sqrt{2\bar{\pi}} + Z_\beta\sqrt{\dfrac{2\pi_{-+}\pi_{+-}}{\bar{\pi}}}}{\pi_{-+} - \pi_{+-}}\right]^2 \tag{38}$$

where $\bar{\pi} = \pi_{+-} + \pi_{+-}/2$.

B. Two Independent Proportions

1. *Asymptotically Normal Method*

A common clinical design is to employ two treatment groups with sample sizes n_1, n_2, and $n_2 = kn_1$, where k

is the ratio of the two sample sizes. For the null hypothesis H_0: $\pi_1 - \pi_2 = 0$, the exact sample size n_1 is determined by

$$n_1 = \min\left\{n_1: \prod(\pi_1, \pi_2) > 1 - \beta\right\} \tag{39}$$

where

$$\prod(\pi_1, \pi_2) = \sum_{(x,y)\in C} \binom{n_1}{x}\binom{kn_1}{y} \cdot p_1^x(1-\pi_1)^{n_1-x}p_2^y(1-\pi_2)^{kn_1-y}$$

with the domain

$$C = \left\{(x, y): \frac{\sqrt{n_1}(p_2 - p_1)}{\sqrt{p_1(1-p_1) + p_2(1-p_2)/k}} > Z_u\right\}$$

and the critical value

$$Z_u^* = \inf\left\{Z_u: \sup\left\{\sum_{(x,y)\in C}\binom{n_1}{x}\binom{kn_1}{y} \cdot p^{x+y}(1-p)^{(1+k)n_1-(x+y)}\right\} \le \alpha\right\}$$

With an asymptotic normality assumption, the sample sizes n_1 and n_2 are expressed as

$$n_1 = \left[\frac{Z_{\alpha/2}\sqrt{\bar{\pi}(1-\bar{\pi})\left(1+\frac{1}{k}\right)} + Z_\beta\sqrt{\pi_1(1-\pi_1) + \frac{\pi_2(1-\pi_2)}{k}}}{\pi_1 - \pi_2}\right]^2 \tag{40}$$

$$n_2 = kn_1$$

where $\bar{\pi} = (\pi_1 + kp_1)/(1 + k)$. For the equal-sample-size designs ($k = 1$), Eqs. (40) become

$$n_1 = n_2 = \left[\frac{Z_{\alpha/2}\sqrt{2\bar{\pi}(1-\bar{\pi})} + Z_\beta\sqrt{\pi_1(1-\pi_1) + \pi_2(1-\pi_2)}}{\pi_1 - \pi_2}\right]^2 \tag{41}$$

Note that Eqs. (40) and (41) are without any continuity correction (see Ref. 8).

2. Continuity Correction χ^2 Test

Based on Pearson's χ^2 statistic and Yates's continuity correction, a formula for sample size calculation is

$$n_1 = \frac{n_1^*}{4}\left[1 + \sqrt{1 + \frac{(k+1)|\pi_1 - \pi_2|}{k(Z_{\alpha/2} + Z_\beta)^2\bar{\pi}(1-\bar{\pi})}}\right]^2 \tag{42}$$

$$n_2 = kn_1$$

where n_1^* is determined by Eqs. (40). For the equal-sample-size designs, Eqs. (42) become

$$n = \frac{n^*}{4}\left[1 + \sqrt{1 + \frac{2|\pi_1 - \pi_2|}{(Z_{\alpha/2} + Z_\beta)^2\bar{\pi}(1-\bar{\pi})}}\right]^2 \tag{43}$$

where n^* is determined by Eq. (41) (see Ref. 9).

3. Continuity Corrected χ^2 Test (Fisher's Exact)

An improved equation for Fisher's exact test is given as

$$n_1 = \frac{n_1^*}{4}\left[1 + \sqrt{1 + \frac{2(k+1)}{kn_1^*|\pi_1 - \pi_2|}}\right]^2 \tag{44}$$

$$n_2 = kn_1$$

where n_1^* is determined by Eqs. (40). For the equal-sample-size designs, Eq. (44) becomes

$$n = \frac{n^*}{4}\left[1 + \sqrt{1 + \frac{4}{n^*|\pi_1 - \pi_2|}}\right]^2 \tag{45}$$

where n^* is determined by Eq. (41) (see Ref. 9).

4. Test for Relative Risk

In some cases of clinical trials, an estimate of anticipated event rate among the control group patients, π_c, has been calculated from past experience. One may then want to estimate the anticipated event rate of the experimental group, π_e, by relative risk

$$R = \frac{\pi_e}{\pi_c} \tag{46}$$

instead of risk difference. In this case the null hypothesis is H_0: $R = 1$. The sample size of each group in an equal-size design is determined by

$$n = \left[\frac{Z_{\alpha/2}\sqrt{2\bar{\pi}_R(1-\bar{\pi}_R)} + Z_\beta\sqrt{\pi_c[1 + R - \pi_c(1+R^2)]}}{\pi_c(1-R)}\right]^2 \tag{47}$$

where $\bar{\pi}_R = \pi_c(1 + R)/2$. Note that Eq. (47) is algebraically equivalent to Eq. (41). For the use of the continuity correction, Eq. (47) is adjusted by adding $2/\pi_c|1 - R|$ to the calculated value of sample size from Eq. (47) (see Ref. 10).

5. *Equivalence Approach*

A question sometimes addressed in clinical trials is whether a new therapy is as effective as a standard therapy. Although it may be as effective or nearly as effective as the standard, the new therapy may provide some other meaningful advantages, such as being less expensive or less toxic than the standard therapy. Unlike the traditional concepts for detecting the significant difference between two treatments, the statistical issue here is to test their equivalence. The therapies are considered equivalent for practical purpose if the difference is less than a given small value δ (assume that a larger proportion is more desirable and $\delta > 0$), i.e. if $\pi_s - \pi_e < \delta$. The null hypothesis and the alternative hypothesis are H_0: $\pi_s \geq \pi_e + \delta$, H_a: $\pi_s < \pi_e + \delta$. The statistic for testing the null hypothesis is

$$z = \frac{p_s - p_e - \delta}{\sigma} \tag{48}$$

where

$$\sigma = \sqrt{\frac{p_s(1 - p_s)}{n_s} + \frac{p_e(1 - p_e)}{n_e}}$$

Therefore, one should set the sample sizes sufficiently large so that the confidence interval will not exceed some specified value δ with a desired power $1 - \beta$. This means that

$$\Pr\left(\frac{(p_s - p_e) - (\pi_s - \pi_e)}{\sigma} > \frac{\delta - (\pi_s - \pi_e)}{\sigma} - Z_\alpha\right) = \beta \tag{49}$$

For an equal-sample-size design, the sample size for each group is determined by (see Ref. 11)

$$n = \left[\frac{(Z_\alpha + Z_\beta)\sqrt{\pi_s(1 - \pi_s) + \pi_e(1 - \pi_e)}}{\pi_s - \pi_e - \delta}\right]^2 \tag{50}$$

C. More Than Two Independent Proportions

1. *Test for Ordinal Data*

When the primary patient response in two treatment groups is measured on an ordered categorical scale such as *very good, good, moderate, poor*, the data can be analyzed using techniques of logistic regression. Under the assumption of proportional odds, a formula for total sample size is

$$N = n_e + n_c = \frac{3(A + 1)^2(Z_{\alpha/2} + Z_\beta)^2}{A\theta_R^2\left(1 - \sum_{i=1}^{k} \bar{\pi}_i^3\right)} \tag{51}$$

where

n_e, n_c are the sample sizes for the experimental group and the control group, respectively

A is the allocation ratio

$$n_e = An_c$$

$\bar{\pi}_i$ is referred to as the *anticipated proportion for category* C_i:

$$\bar{\pi}_i = \frac{(\pi_{ie} + \pi_{ic})}{2}, \qquad i = 1, 2, \ldots, k \tag{52}$$

θ_R denotes the reference improvement, a point estimate of log-odds-ratio of the outcome C_i or better for an experimental subject relative to a control subject:

$$\theta_i = \log\left\{\frac{Q_{ie}(1 - Q_{ic})}{Q_{ic}(1 - Q_{ie})}\right\} \tag{53}$$

$$Q_{iG} = \sum_{j=1}^{i} \pi_{jG}, \quad G = e, c \quad i = 1, 2, \ldots, k - 1$$

The sample size determination is illustrated as follows. In most cases of clinical trials, the category probabilities in the control group π_{ic} have been estimated from past studies. Based on previous experience, the reference improvement θ_R has also been estimated in any given circumstance. Since the definition

$$\theta_i = \log\left\{\frac{Q_{ie}(1 - Q_{ic})}{Q_{ic}(1 - Q_{ie})}\right\}$$

is true for all $i = 1, 2 \ldots, k - 1$, the cumulative probabilities of improvement Q_{ie_R} in the experimental group can be anticipated:

$$Q_{ie_R} = \frac{Q_{ic}}{Q_{ic} + (1 - Q_{ic})e^{-\theta_R}}, \quad i = 1, 2, \ldots, k - 1 \tag{54}$$

That gives the category probabilities in the experimental group π_{ie} and the anticipated proportion $\bar{\pi}_i$, $i = 1, 2, \ldots, k$. Then the sample size can be determined from Eq. (51) for the given significance level α, power $1 - \beta$, and allocation ratio A (see Ref. 12).

2. *Test for Linear Trend in G × 2 Tables*

Let y_i be k mutually independent binomial variates based on sample sizes of n_i at dose level d_i, and Y, U, N be the relevant summations

$$Y = \sum_i y_i, \qquad U = \sum_i y_i d_i, \qquad N = \sum_i n_i,$$

$$i = 0, 1, \ldots, k - 1$$

Define $p = Y/N$, $q = 1 - p$, and $\bar{d} = \Sigma_i\, n_d d_i/N$. Assuming that the probability of response follows a linear trend on the logistic scale, $p_i = e^{(\gamma+\lambda d_i)}/(1 + e^{(\gamma+\lambda d_i)})$, an approximate test with continuity correction asymptotically approaches the normal distribution; i.e.,

$$z = \frac{U' - \Delta/2}{\sqrt{\text{var}(U' \mid Y)_{H_0}}} \sim N(0, 1) \tag{55}$$

where

$$U' = U - E(U \mid Y)_{H_0} = \sum_i y_i(d_i - \bar{d})$$

$$\text{var}(U' \mid Y)_{H_0} = pq \sum_i n_i(d_i - \bar{d})^2$$

and $\Delta/2$ is the continuity correction. When the Type I and Type II error probabilities are given by α and β, respectively, the following equation can be obtained:

$$E(U') - \frac{\Delta}{2} = Z_\alpha \sqrt{\text{var}(U')_{H_0}} + Z_\beta \sqrt{\text{var}(U')} \tag{56}$$

where

$$E(U') = \sum_i n_i p_i (d_i - \bar{d})$$

$$\text{var}(U')_{H_0} = pq \sum_i n_i (d_i - \bar{d})^2$$

$$p = \frac{\sum n_i p_i}{N}$$

$$\text{var}(U') = \sum n_i p_i q_i (d_i - \bar{d})^2$$

Let $r_i = n_i/n_0$ be the ratio of the sample size of the ith group to that of the control group ($i = 0, 1, 2, \ldots, k - 1$) that may be determined in advance. Thus sample size determination is focused on how to estimate the sample size of the control group. The sample size n_i of the ith group can be calculated directly via the ratio $r_i = n_i/n_0$ whenever n_0 is obtained. The sample size of the control group n_0 can be solved from the following quadratic equation with respect to $\sqrt{n_0}$:

$$n_0 = \left(\frac{n_0^*}{4}\right)\left[1 + \sqrt{1 + \frac{2\Delta}{n_0^* \sum_i r_i p_i (d_i - \bar{d})}}\right]^2 \tag{57}$$

where n_0^* is the sample size of the control group without continuity correction (i.e, $\Delta = 0$) and is given as

$$n_0^* = \frac{\left[Z_\alpha \sqrt{pq\left[\sum_i r_i(d_i - \bar{d})^2\right]} + Z_\beta \sqrt{\sum_i r_i p_i q_i (d_i - \bar{d})^2}\right]^2}{\left[\sum r_i p_i (d_i - \bar{d})^2\right]^2} \tag{58}$$

For the equal-sample-size design with $d_i = i$, for $i = 0, 1, 2, \ldots, k - 1$, the sample size per group is expressed approximately as (see Ref. 13)

$$n = \left(\frac{n^*}{4}\right) \cdot \left[1 + \sqrt{1 + \frac{2}{n^* \sum_i [i - 0.5(k - 1)]p_i}}\right]^2 \tag{59}$$

where

$$n^* = \frac{\left[Z_\alpha \sqrt{\frac{k(k^2 - 1)pq}{12}} + Z_\beta \sqrt{\sum_i [i - 0.5(k - 1)]^2 p_i q_i}\right]^2}{\left[\sum_i [i - 0.5(k - 1)]p_i\right]^2} \tag{60}$$

V. SAMPLE SIZE IN SURVIVAL ANALYSIS

Survival analysis methods are designed for studies in which patients are followed until any event of interest occurs (e.g., death, heart attack), the patients are censored, or the study ends. If the survival proportions of patients are simply compared, or all patients are followed for the same fixed time period, the sample size can be determined by the methods related to comparison of proportions. However, most survival studies are designed for comparing treatment groups on mean survival times rather than survival rates. For this reason, the sample size determinations for survival studies are differently considered from those for means or proportions. Similar to the methods for means and proportions, the sample size calculations for survival studies are derived under a certain assumption: proportional-hazard assumption or distribution assumption.

A. Log-Rank Test for Equality of Survival Curves

Schoenfeld (14) and Freedman (15) derived similar sample size formulas for the total number of events needed to be observed and the total number of patients per group needed to be recruited. Suppose that there are two groups of individuals, the experimental group and the control group. Assuming a proportional-hazards model for the survival times, the hazard rate at time t

for an individual in the experimental group, $h_e(t)$, can be expressed as

$$h_e(t) = \psi \cdot h_c(t) \tag{61}$$

where $h_c(t)$ is the hazard function at time t for an individual in the control group and ψ is the unknown hazard ratio. Let θ, $\theta \in \Theta$, be the log-hazard ratio, $\theta = \log \psi$; then a zero value of θ indicates no treatment difference, a negative value of θ means that survival is longer on the new treatment, and a positive value of θ means that survival is longer on the standard treatment.

The required number of events is calculated in such a case to reject the null hypothesis that the observed value of θ equals zero at a significance level of α and with the desired power of $1 - \beta$. Both α and β are given for any design of survival study. The required number of events, d, can be calculated using

$$d = \frac{4(z_{\alpha/2} + z_{\beta})^2}{\theta^2} \tag{62}$$

where $z_{\alpha/2}$ and z_{β} are the upper $\alpha/2$- and the upper β-points of the standard normal distribution, and θ is an estimate of Θ that might be chosen on the basis of the increase in median survival time or some other estimating method in the survival study. Freedman (15) suggested using $[(1 + e^{\theta})/(1 - e^{\theta})]^2$ instead of $4/\theta^2$ for more accurate sample size. When θ is small, however, the results of these two methods are very close, because

$$\left(\frac{1 + e^{\theta}}{1 - e^{\theta}}\right)^2 \approx \left(\frac{2 + \theta}{\theta + \theta^2/2}\right)^2 = \frac{4}{\theta^2} \tag{63}$$

The calculation of Eq. (62) assumes that an equal number of individuals is to be assigned to each treatment group. In the case of unequal sizes, for example, π_c and $(1 - \pi_c)$ as the proportion of individuals in the control and experimental groups, respectively, the required total number of events becomes

$$d = \frac{(z_{\alpha/2} + z_{\beta})^2}{\pi_c(1 - \pi_c)\theta^2} \tag{64}$$

An imbalance in number of individuals in the two groups will increase the total number of events required.

If the study has to be continued until a given percentage of the patients has had the events of interest, p, the total number of patients can be simply calculated:

$$N = \frac{d}{p} \tag{65}$$

Otherwise, the probability of the event over the duration of the study needs to be taken instead of the simple percentage p. Typically, individuals are recruited over an accrual period T_a. After recruitment is finished, there is an additional follow-up period T_f. The probability of an event over the duration of study can be calculated as

$$\Pr(\text{event}) = 1 - \frac{1}{6}[\bar{S}(T_f) + 4\bar{S}(0.5T_a + T_f) + \bar{S}(T_a + T_f)] \tag{66}$$

where

$$\bar{S} = \frac{S_e(t) + S_c(t)}{2} \tag{67}$$

is the average survival function and $S_e(t)$ and $S_c(t)$ are the Kaplan–Meier estimate values of the survival functions for individuals on the experimental and control groups, respectively, at time t. Notice that there is no distribution assumption in Eqs. (66) and (67). Assuming that survival times are exponentially distributed, the average survival function, $\bar{S}$, is given by

$$\bar{S} = \frac{e^{-\lambda_e t} + e^{-\lambda_c t}}{2} \tag{68}$$

where estimates λ_e and λ_c can be estimated by means of the corresponding median survival times t_m (see Ref. 16):

$$\lambda = \frac{\log 2}{t_m} \tag{69}$$

A variant on Eq. (69) for accrual and follow-up periods is given by Schoenfeld and Richter (17), who developed nomograms for calculating sample size. The nomograms are valid when survival is exponential and patients enter the study uniformly.

In the general case, Lakatos (18) derived a method of sample size calculation by using a discrete nonstationary Markov process that allows any pattern of survival, noncompliance, loss to follow-up, drop-in, and lag in the effectiveness of treatment during the study period.

B. Test Based on Exponential Survival

Rubinstein et al. (19) considered sample size estimation by assuming exponentially distributed survival durations. Their assumptions include: (a) patients are accrued during the interval $[0, T]$ according to a Poisson process, and the total trial length is $T + \tau$; (b) the times from entry to event are independently and exponentially distributed in each of the experimental and control groups, with parameters λ_e and λ_c, respectively; and (c) the losses to follow-up times are also independently and exponentially distributed in each group, with parameters ϕ_e and ϕ_c, respectively. Based on the exponential maximum-likelihood test, the sample size formula is given as

$$n = \left(\frac{z_\alpha + z_\beta}{\ln(\theta)}\right)^2 \left[\frac{1}{E(P_e)} + \frac{1}{E(P_c)}\right] \quad (70)$$

where

$$E(P_i) = \left[\frac{1 - e^{-\lambda_i^* \tau}(1 - e^{-\lambda_i^* T})}{\lambda_i^* T}\right]\left(\frac{\lambda_i}{\lambda_i^*}\right), \quad i = e, c \quad (71)$$

and where n is the sample size in each group, θ is the hazard ratio, and $\lambda_i = \lambda_i + \phi_i$, $i = e, c$. Several similar works on sample size and power determination under the exponential distribution are available in the literature (see Ref. 20).

VI. SAMPLE SIZE BASED ON TREATMENT VARIANCE

Sometimes, a clinical trial will be proposed to test the hypothesis that an experimental drug reduces the variability of the key outcome variable as compared to a control. It is convenient to assume the key variable is normally or at least asymptotically, distributed. The classical F-test assessing whether the two variances are equal compares the two treatments in terms of the ratio $\lambda = \sigma_c/\sigma_e$, or, equivalently, the percentage difference $\Delta = (\lambda - 1) \times 100\%$. The hypothesis becomes H_0: $\lambda = 1$ versus H_a: $\lambda \neq 1$. One important task at the planning stage is to estimate sample size for the trial. For a significance level of α, a power of $(1 - \beta)$, and a two-tailed test, the sample size per arm, n, can be computed by the following quick and satisfactory approximation formula if a clinically important percentage difference Δ is specified:

$$n \equiv 2 + \left[\frac{(z_{\alpha/2} + z_\beta)}{\log_e \lambda}\right]^2 \quad (72)$$

where z_γ is the $(1 - \gamma)$ percentile of the standard normal distribution and λ = the ratio of the two observed sample standard deviations. Since $\log_e \lambda = -\log_e(1/\lambda)$, the formula gives the same estimate n whether λ is expressed as σ_e/σ_c or as σ_c/σ_e (see Ref. 21).

VII. SAMPLE SIZE BASED ON CORRELATION

In observational studies, correlation can be the subject of analysis. We assume the outcome variable is normally or at least asymptotically, distributed. Two types of hypotheses for testing correlation usually are: (a) whether a true correlation actually exists, H_0: $\rho = 0$ versus H_a: $\rho = \rho_1 \neq 0$; and (b) whether two correlations are significantly different, H_0: $(\rho_e - \rho_c) = 0$ versus H_a: $\rho_e - \rho_c \neq 0$, where e and c denote experiment and control, respectively. The simplest approach to such problems is to employ Fisher's arctanh transformation:

$$C(r) = \frac{1}{2} \log_e \frac{(1 + r)}{(1 - r)} \quad (73)$$

Given a sample correlation r based on N observations that is distributed about an actual correlation value ρ, then $C(r)$ is normally distributed with mean $C(\rho)$ and variance $\sigma^2 = 1/(N - 3)$. The transformation of r to C (and vice versa) is widely tabulated.

A. A Single Correlation

Testing hypothesis H_0: $\rho = 0$ versus H_a: $\rho = \rho_1 \neq 0$, we use the test statistic $Z = C(r)\sqrt{N - 3}$, where $Z \sim N(0, 1)$. Then the required sample size or power may be obtained by using

$$\sqrt{N - 3}C(\rho_1) = z_\alpha + z_\beta \quad (74)$$

B. Two Independent Correlations

In detecting a relevant difference in correlation H_a: $C(\rho_e) - C(\rho_c) \neq 0$ obtained from two independent samples, the null hypothesis H_0: $C(\rho_e) - C(\rho_c) = 0$ is tested using the statistic $Z = [C(r_e) - C(r_c)]/\Sigma_0^2$, where $\Sigma_0^2 = N^{-1}(Q_e^{-1} + Q_c^{-1})$ and $n_e - 3 = Q_e N$, $n_c - 3 = Q_c N$ and where, under H_0, $Z \sim N(0, 1)$. The correlations r_e and r_c are obtained from two samples of sizes n_e and n_c such as $r_e = r_{e(uv)}$ and $r_c = r_{c(uv)}$ for variables u and v in groups e and c. We solve the following equation for total sample size or power:

$$\frac{\sqrt{N}|C(\rho_e) - C(\rho_c)|}{\sqrt{Q_e^{-1} + Q_c^{-1}}} = z_\alpha + z_\beta \quad (75)$$

Note that N from Eq. (75) will actually be 6 less than that actually needed, since $n_e + n_c - 6 = N$ (see Ref. 1).

VIII. RE-ESTIMATING SAMPLE SIZE WITHOUT UNBLINDING

Sample size estimation is a key step for a successful clinical trial. Estimation of sample size in clinical trials depends on the level of significance, α, the power of the trial, $1 - \beta$, and the minimum detectable difference, δ, all of which are known, and on the knowledge of the variability of the primary response variable, which is usually uncertain to medical researchers. The values of the variability measures usually are guessed or taken from the results of previous trials. In either case, they are subject to uncertainty and may be materially smaller or larger than the true values. Overstating the true variability causes the trial to be larger than it should be, implying unnecessary expenditure of resources or a potential ethical problem due to exposure of more patients than necessary to an inferior treatment. Understating the true variability inflates the Type II error rate, β, which means decreasing the power of the trial and, therefore, increasing the likelihood of an inconclusive trial when the treatment effect is real. The resource cost and unproductive patient exposure resulting from an inconclusive trial are much greater than when the variability is overstated. Even with previous data available, one must exercise caution regarding possible differences between the trials in terms of patient population, disease severity, diagnostic criteria, medical procedures, and other study conditions. It is therefore desirable to re-estimate the sample size using interim data of the trial under study. It is also important to keep the treatment code blind in an interim estimation to avoid the introduction of any potential (conscious or unconscious) bias during the conduct and monitoring of a clinical trial, especially for those trials for which there is no independent, external data monitoring committee involved.

A. Normally Distributed Outcomes

Consider a clinical trial that compares two treatments. The null hypothesis H_0: $\mu_1 = \mu_2$ ordinarily would be tested against the alternative H_a: $\mu_1 \neq \mu_2$ using a student t-test. Given the significance level, α, the power of the trial, $1 - \beta$, and the minimum detectable difference, δ, the total sample size would be determined from

$$N = \frac{4\tilde{\sigma}^2(z_{\alpha/2} + z_\beta)^2}{\delta^2} \qquad (76)$$

where $\tilde{\sigma}^2$ is the estimate of σ^2, the within-group variance, and z_γ is the $(1 - \gamma)$ percentile of the standard normal distribution.

Suppose that the sample size will be reconsidered after $n(<N)$ observations without knowing the treatment assignments, which means we know the observations come from a particular treatment group (1 or 2), but we do not know the treatment identification of the treatment group. With the re-estimate, $\hat{\sigma}^2$, of the within-group variance, σ^2, one can determine the actual sample size as

$$N' = \frac{N\hat{\sigma}^2}{\tilde{\sigma}^2} \qquad (77)$$

where $\tilde{\sigma}^2$ is the estimate of σ^2 at the beginning of the study.

The EM algorithm will be used to re-estimate σ^2 without unblinding. Since the treatment identifications are unknown, any of the interim observations x_i, $i = 1, \ldots, n$, could be in either treatment group, so the treatment assignments are "missing at random." Let τ_i denote the treatment group membership indicator:

$\tau_i = 1(0)$ if sample number i is in treatment group 1 (group 2)

$\tau_1, \ldots, \tau_n$ are independent random variables with $\Pr(\tau_i = 1) = 0.5$. Given a τ_i value, $x_i (i = 1, \ldots, n)$ has a normal distribution with density

$$f(x_i|\tau_i, \mu_1, \mu_2, \sigma) \propto \frac{1}{\sigma} \exp\left\{-\frac{1}{\sigma^2}[\tau_i(x_i - \mu_1)^2 + (1 - \tau_i)(x_i - \mu_2)^2]\right\} \qquad (78)$$

The expression for the conditional probability (or expectation) of τ_i given x_i therefore is

$$\Pr(\tau_i = 1|x_i) = \frac{1}{\{1 + \exp[(\mu_1 - \mu_2)(\mu_1 + \mu_2 - 2x_i)/2\sigma^2]\}} \qquad (79)$$

The log-likelihood of the interim observations follows from Eq. (78):

$$l = n \log \sigma + \frac{\left\{\sum_{i=1}^{n} [\tau_i(x_i - \mu_1)^2 + (1 - \tau_i)(x_i - \mu_2)^2]\right\}}{2\sigma^2} \qquad (80)$$

The "E" step of the EM algorithm for estimating σ consists of substituting "current" estimates of μ_1, μ_2, and σ into Eq. (79) to obtain provisional values for the expectations of the τ_i. The "M" step consists of obtaining maximum-likelihood estimates of μ_1, μ_2, and σ after replacing the τ_i in Eq. (80) with their provisional expectations. The "E" and "M" steps are repeated until the value of σ^2 stabilizes; the resulting value is the estimate, $\hat{\sigma}^2$, of σ^2 required in Eq. (77). This process estimates the variance satisfactorily but does not provide a reliable estimate of the difference between the treatment group means (see Ref. 22).

B. Binomial Distributed Outcomes

Consider a clinical trial that compares the effect of two treatments on a disease based on a binary outcome (y), for which we use the generic term *event* ($y = 1$) versus *nonevent* ($y = 0$). Denote p_1 and p_2 as the true *event* rates for the two treatment groups, p_1^* and p_2^*, as the estimates of p_1 and p_2, respectively, $\sigma(p_1, p_2) = (p_1 + p_2)/2$ as the average event rate, and $\delta(p_1, p_2) = p_1 - p_2$ as the treatment difference. The hypotheses to test are: H_0: $p_1 = p_2$ versus H_a: $p_1 \neq p_2$. For power $(1 - \beta)$ and significance level α, the sample size per treatment group is determined by

$$n = \frac{2(z_{\alpha/2} + z_\beta)^2\sigma(1 - \sigma)}{\delta^2} \equiv 2(z_{\alpha/2} + z_\beta)^2\lambda(\sigma, \delta) \qquad (81)$$

where z_γ is the $(1 - \gamma)$ percentile of the standard normal distribution and $\lambda(\sigma, \delta) = \sigma(1 - \sigma)/\delta^2$ is the noncentrality parameter, $\delta \neq 0$. The estimated sample size is obtained by substituting p_1^* and p_2^* in Eq. (81).

We now describe a method that updates the estimation of p_i using interim data from the trial without breaking the treatment code of the individual patients. We design the trial with a simple, random stratification scheme. For a multicenter trial, this stratification is within each center. Patients are first randomly assigned to either stratum A or stratum B. (Since this stratification is not based on any of the patients' baseline characteristics, it is a "dummy" stratification. We use it only at the interim stage for re-estimation, and at the final stage we simply ignore it without affecting the regular inference.) In stratum A, patients are randomly allocated to treatment group $i = 1$ with a probability π and to treatment group $i = 2$ with a probability $1 - \pi(0 < \pi < 1)$. In stratum B, we do the opposite: patients are randomly allocated to treatment group $i = 1$ with a probability $1 - \pi$ and to treatment group $i = 2$ with a probability π. Hence, we maintain the overall balance of the treatment allocation. It is clear that we should avoid $\pi = 0.5$, since it defeats the purpose of the stratification.

We conduct an interim analysis after the trial is halfway completed as originally planned, that is, when the outcome data (y) are available from a total of n^* patients. Note that among these n^* patients, presumably we should have about $n^*/2$ from each treatment group and also $n^*/2$ from each stratum due to the randomization procedure applied to treatment and stratum, but these n^* patients' treatment codes remain masked. We report only the pooled event rates for each stratum, without knowing treatment codes, and we estimate the event rates p_1 and p_2 as follows.

We use the result of stratum A to estimate

$$\theta_1 = P(y_j = 1 \mid \text{patient } j \in \text{stratum A}) = \pi p_1 + (1 - \pi)p_2$$

and that of stratum B to estimate

$$\theta_2 = P(y_j = 1 \mid \text{patient } j \in \text{stratum B}) = (1 - \pi)p_1 + \pi p_2$$

Let the observed event rates in the two strata be $\hat{\theta}_k$, which are unbiased estimators of $\theta_k (k = 1, 2)$. Thus we have the following pair of equations:

$$\begin{aligned} \pi\hat{p}_1 + (1 - \pi)\hat{p}_2 &= \hat{\theta}_1 \\ (1 - \pi)\hat{p}_1 + \pi\hat{p}_2 &= \hat{\theta}_2 \end{aligned} \qquad (82)$$

Solving Eqs. (82) for $\hat{p}_1$ and $\hat{p}_2$, we have the unbiased estimators of the true event rates for the treatment group being weighted averages of the pooled rates from the two strata. We then update the sample size by substituting $\hat{p}_i$ for p_i, $i = 1, 2$, in Eq. (81). In practice, we generally recommend setting π near the middle between 0 and 0.5, for example, $\pi = 0.2$, to provide reasonable protection of power and, at the same time, for easy conduct of a trial (see Ref. 23).

C. Remarks

At the interim examination, the estimated variability could turn out to be much less than was anticipated in designing the trial. When this happens, one can terminate the trial without affecting α materially if the interim sample size is adequate. However, this ordinarily would not be advisable in practice. Confirmatory trials seldom have the sole objective of detecting a difference between treatments with respect to a single measurement. Safety and tolerability are as important as efficacy, which itself often needs to be described with more than one measurement. Reducing the sample size could imperil the sensitivity of the trial for detecting clinically meaningful differences with respect to "secondary," but still important, measurements, and would reduce the sensitivity of the trial for assessing safety and tolerability (see Ref. 22).

REFERENCES

1. JM Lachin. Controlled Clin Trials 2:93–113, 1981.
2. JM Lachin. Encyclopedia of Biostatistics 5. New York: Wiley, 1998, pp 3892–3903.
3. JL Fleiss. The Design and Analysis of Clinical Experiments. New York: Wiley, 1986, pp 31, 369–371.
4. L Frison, SJ Pocock. Stat Med 11:1685–1704, 1992.
5. G Liu, KY Liang. Biometrics 53:937–947, 1997.
6. LQ Yue, P Roach. A note on the sample size determination in two-period repeated measurements crossover design with application to clinical trial. J Biopharm Stat 8(4):577–584, 1998.
7. E Lakatos. Encyclopedia of Biostatistics 5. New York: Wiley, 1998, pp 3903–3910.
8. JL Fleiss. Statistical Methods for Rates and Proportions. New York: Wiley, 1981.
9. JL Fleiss, A Tytun, SHK Ury. Biometrics 36:343–346, 1980.
10. JJ Schlesselman. J Epidemiol 99:381–384, 1974.
11. WC Blackwelder. Controlled Clin Trials 3:345–353, 1982.
12. J Whitehead. Stat Med 12:2257–2271, 1993.
13. J Nam. Biometrics 43:701–705, 1987.
14. DA Schoenfeld. Biometrics 39:499–503, 1983.
15. LS Freedman. Stat Med 1:121&129, 1982.
16. D Collett. Encyclopedia of Biostatistics 5. New York: Wiley, 1998, pp 3910–3914.
17. DA Schoenfeld, JR Richter. Biometrics 38:163–170, 1982.
18. E Lakatos. Biometrics 44:229–241, 1988.
19. LV Rubinstein, MH Gail, TJ Santner. J Chron Dis 34: 469–479, 1981.
20. E Lakatos, G Lan. Stat Med 11:179–191, 1992.
21. J Jin, F Hsuan, DS Hwang. Reestimation of sample size for variability comparison in clinical trials. Proceedings of the Biopharmaceutical Section, San Francisco, 1993, pp 328–333.
22. AL Gould. Stat Med 14:1039–1051, 1995.
23. WJ Shih, PL Zhao. Stat Med 16:1913–1923, 1997.

Lilly Q. Yue
David Li
Shan Bai

Screening Design

See also *Drug Development; Factorial Design*

A *screening design* refers to an experimental design that is applicable when a large number of potential causative factors need to be examined to find the most important few that may have an effect on one or more responses of interest. Screening designs may be derived from highly fractionated factorial designs and are called *orthogonal main effect plans* or *orthogonal arrays*, since main effects and only a few two-factor interactions are unconfounded with one another. Effects are confounded when the contrasts for estimating them are not orthogonal, and effects are completely confounded when their contrasts are identical. Plackett and Burman (1) devised orthogonal arrays useful for screening that yield unbiased estimates of all main effects in the smallest design possible. For example, up to 11 factors can be screened in a 12-run Plackett–Burman design. Examples of 8-run and 12-run Plackett-Burman designs are provided in Tables 1 and 2, respectively. Interest in using screening designs in industrial settings has increased in recent years due partly to the work of Genichi Taguchi, a Japanese engineer who promoted the use of designs he termed *orthogonal arrays*, many of which are similar to Plackett–Burman designs (2).

The most obvious criticism of screening designs is that they allow at most a small subset of the interactions to be estimated. Interactions can sometimes be

Table 1 8-run Plackett–Burman Screening Design

Trial	X1	X2	X3	X4	X5	X6	X7
1	−	−	−	−	−	−	−
2	−	−	+	+	−	+	+
3	−	+	−	+	+	+	−
4	−	+	+	−	+	−	+
5	+	−	−	+	+	−	+
6	+	−	+	−	+	+	−
7	+	+	−	−	−	+	+
8	+	+	+	+	−	−	−

Table 2 12-run Plackett–Burman Screening Design

Trial	X1	X2	X3	X4	X5	X6	X7	X8	X9	X10	X11
1	+	+	−	+	+	+	−	−	−	+	−
2	+	−	+	+	+	−	−	−	+	−	+
3	−	+	+	+	−	−	−	+	−	+	+
4	+	+	+	−	−	−	+	−	+	+	−
5	+	+	−	−	−	+	−	+	+	−	+
6	+	−	−	−	+	−	+	+	−	+	+
7	−	−	−	+	−	+	+	−	+	+	+
8	−	−	+	−	+	+	−	+	+	+	−
9	−	+	−	+	+	−	+	+	+	−	−
10	+	−	+	+	−	+	+	+	−	−	−
11	−	+	+	−	+	+	+	−	−	−	+
12	−	−	−	−	−	−	−	−	−	−	−

more important than main effects for understanding or modeling the process under study. For this reason, many experts in the field recommend sequential experimentation, in which screening is followed by one or more experimental designs using fewer factors, such as larger fractions of factorial designs, central composite designs, or Box–Behnken designs, that permit estimation of interaction and/or curvature. Proponents of the Taguchi designs have favored a slightly different philosophy about screening designs, where the experiment is considered more of a "one-shot," "pick-the-winner" approach to identify the optimal settings of the factors, followed up by several confirmatory runs at the predicted optimum.

In the pharmaceutical industry, the most concentrated use of screening designs, factorial designs, and response surface designs is found in the product and process development functions, including analytical method development, due to the increasing complexity in the optimization and validation of analytical methods. In the evolving environment of rapid and shortened development cycles, screening designs and experimental design in general will continue to gain acceptance as valuable development tools.

REFERENCES

1. RL Plackett, JP Burman. Biometrika 33:305–325, 1946.
2. G Taguchi. Reports of statistical application research. JUSE 6:1–52, 1960.

John R. Murphy

Specifications

See also *USP Tests*

In the pharmaceutical industry, *specifications* usually mean the same thing as *specification limits*, which refer to numerical tolerances within which the measured result for a quality attribute of a dosage form should fall. Specifications may also refer to more general and/or complex criteria, which can sometimes be found in a compendium such as the United States Pharmacopeia (USP). An example of specification limits that are numerical tolerances might be where the purity of a drug substance must be no less than 97.0% and no more than 103.0%. An example of specifications based on more general criteria might be where a solid oral dosage form must meet the USP dissolution criteria using a Q of 75% at 45 minutes.

Specifications on the physical and chemical attributes of drug substances and drug products are the quality standards that provide the patient assurance that the drug will perform its intended purpose without unexpected side effects. In that sense, specifications may be thought of as defining the numerical tolerances or other criteria that make the drug safe, effective, and fit for use. Ideally, specification limits based on those considerations could be established, and the manufacturing process would be capable of providing units falling well within such bounds. In practice, however, many times the allowable limits based on safety, efficacy, or fitness for use seem too wide, stemming from a belief that narrow or tight specifications mean a higher-quality product. The net result is that the determination of specification limits is sometimes based almost entirely on judgments about manufacturing capability.

Since specifications are a legal commitment with regulatory authorities, their determination is an important undertaking, and the process of getting to agreed-upon specifications is sometimes difficult. Part of the problem is the complexity of the measurement process, which can often comprise a significant amount of the variation in the measured results. As a result, many specifications cannot be set without tying them to the analytical methods that produce the result. Another feature of drug substances and drug products is that degradation be factored into the process of setting specifications. Stability estimation can present a special problem due to the fact that the effect (change over time) is often small relative to the variation in the manufacturing and the measurement processes. Given these facts, it seems evident that determining rational specifications requires the use of statistical methods to estimate and account for all the variation that may be present.

Specifications mean different things to different people, and part of the reason for this is a failure to recognize the difference between several types of limits that are derived to meet different purposes (1). The type of specification limits that most often come to mind are *expiry specifications* or *regulatory specifications*, which are the limits a product must meet throughout its shelf life and which are documented in a regulatory submission or compendial monograph. The failure of units in commercial distribution to comply with expiry specifications is a serious matter that can result in regulatory censure and/or expensive and time-consuming product recall. In order to minimize the probability of out-of-specification results, manufacturers utilize *release specifications* or *acceptance limits*, which are applied at the time of product release. In the United States, release limits are still considered by the FDA to be mostly a matter of internal company policy, while in Europe the release limits must be formally registered in the regulatory dossier. The third type of limits are what may be informally regarded as *statistical control limits*, which reflect the capability of a manufacturing process in a state of statistical control and serve to alert us to unusual results that are likely to be the consequence of special causes.

One primary reason for disagreement about how wide or narrow expiry specifications should be is a misunderstanding and lack of agreement about the function and the relative location of these three types of limits. The purpose of release limits is to prevent noncomplying product from reaching the marketplace, and they are best determined using statistical methods that take into account the sampling and measurement variation and the change in the product attribute over time. Ideally, they are set from the expiry limits inward such that if the batch result falls within the release limits, there is a specified level of confidence that future measurements will fall within the expiry limits. In and of themselves, release limits do not have any relation to the capability of the process to produce results that will fall within them. Statistical control limits, on the other hand, are related to the capability of the process, and when the release limits are at least as wide as the statistical control limits, the process will be capable of producing results that have a high probability of falling within the release limits. Unfortunately, sufficient data firmly to establish control limits and release limits may not be available at the point in time in the development process when the limits need to be set; it is therefore imperative that regulatory bodies worldwide recognize the need for expiry limits and release limits to be wider at first approval. There can be an expectation that specification limits will be appropriately tightened as more supporting process data becomes available.

Currently, there is no international consensus on the approach to specification setting, but International Conference on Harmonization (ICH) document Q6A on specifications (2) is a large step forward. This guideline provides common definitions, discusses justification of specifications, and gives a listing of the analytical properties that may be considered universal for drug substances and the various dosage forms of drug products. The use of statistical methods is not specifically mentioned or encouraged in the document, but the necessity of establishing release specifications separate and different from expiry specifications is recognized.

In summary, specifications define the standards of identity, strength, quality, and purity of drug products.

The patient derives assurance that the drug will perform its intended function from the knowledge that its physical and chemical properties have been tested and have been determined to comply with those standards. Good specifications are those that guarantee that the product will be safe and effective while at the same time providing sufficient latitude for the manufacturer to produce and release product that is highly probable to remain within specification throughout the shelf life. Such specifications are best developed with the aid of statistical methods.

REFERENCES

1. GE Davis, JR Murphy, DA Weisman, SW Andersen, JD Hofer. Pharmaceutical Technol. September: 100–118, 1996.
2. International Conference on Harmonization. Specifications: Test Procedures and Acceptance Criteria for New Drug Substances and New Drug Products: Chemical Substances—Q6A (Step 2 of the ICH Process), July 18, 1997.

John R. Murphy

Stability Matrix Designs

See also *Bracketing Design; Factorial Designs*

I. INTRODUCTION

Stability studies are conducted "to provide evidence on how the quality of a . . . drug product varies with time . . . and enable . . . shelf lives to be established" (1) by "testing . . . those features susceptible to change during storage and likely to influence quality, safety, and/or efficacy." Since drug product is defined as the dosage form in the final package, stability studies must be applicable to all strengths and all packages. For a new drug product, accelerated testing is required for 6 months, and long-term testing is required for the length of shelf life; thus, the cost of the stability studies can be substantial. This leads quite naturally to statistically designed stability studies, which are called *matrix* designs, or studies where "only a fraction of the total number of samples are tested at any specified sampling point" (1).

II. HISTORY

Statistically designed stability studies were first used in the early 1980s (E Nordbrock, stability protocols, 1981) and were accepted by the FDA (W Fairweather, personal communication, 1982). During 1989, Nordbrock (2) and Wright (3) made presentations to stability groups at two different conferences. Nakagaki, who was present at the first of these, made a presentation using the terminology *matrix and bracket* (4). Nordbrock (5) made a presentation in 1991, and the first journal articles appeared in 1992 (6–8). Since then, there have been a number of presentations and publications in this area (9–18).

The European Community has a guideline on matrix/bracket designs (19). Although a proposed guideline for matrix designs has been submitted to FDA by the PhRMA Stability Working Group (20), as of June 1998 guidelines for reduced testing have not been issued by the FDA. However, presentations by FDA personnel have been made (21,22).

III. DESIGNS

A. Background

Since the basic analysis applied to stability data is a linear regression of the parameter of interest on time, the selection of observations that gives the minimum variance for the slope is to take one-half at the beginning of the study and one-half at the end. Thus statistically designed stability designs attempt to use this basic principle while being cognizant of the fact that there is a well-defined beginning (start of the stability study, usually called *t*-zero) but not a single end, because analyses are typically done at several different times (e.g., at the time a registration application for the new drug product is filed, or yearly for a marketed product). Because analyses are done at multiple times, there is no unique best design, and the choice of design must use the fact that analyses will be done after additional data are collected. Several designs are given next.

These designs, presented via examples, can easily be applied to other studies.

B. Basic Matrix 2/3 on Time Design

A complete long-term study for one strength of a dosage form in one package has three batches, with all three tested every 3 months the first year, every 6 months the second year, and annually thereafter (see the entry on *Stability*). Thus if a 36-month shelf life is desired and the complete study is used, each of the three batches are tested at 0, 3, 6, 9, 12, 18, 24, and 36 months. The basic matrix 2/3 on time design has only two of the three batches tested at intermediate time points (other than time 0 and 36), as presented in Table 1. If an analysis is to be done after 18 months (e.g., for a registration application), the basic matrix 2/3 on time design can be modified by testing all batches at 18 months.

Table 1 Basic Matrix 2/3 on Time Design

Batch	Test times
A	0, 3, , 9, 12, , 24, 36
B	0, 3, 6, , 12, 18, , 36
C	0, , 6, 9, , 18, 24, 36

C. Matrix 2/3 on Time Design with Multiple Packages

The first extension of the basic design is when one strength is packaged into three packages, i.e. when each batch is packaged into each of three packages. The basic matrix 2/3 on time design is applied to each package in a balanced fashion, as presented in Tables 2 and 3. Balance means each batch is tested twice at each intermediate time point, and each package is tested twice at each intermediate time point. If an analysis will be done after 18 months (e.g., for a registration application), this design can be modified by testing all batch-by-package combinations at 18 months.

Table 2 Matrix 2/3 on Time Design with Multiple Packages

Batch	Pkg 1	Pkg 2	Pkg 3
A	T1	T2	T3
B	T2	T3	T1
C	T3	T1	T2

Pkg 1 = Package 1, etc.

Table 3 Test Code Definitions

Code	Test times after time 0
T1	3, , 9, 12, , 24, 36
T2	3, 6, , 12, 18, , 36
T3	, 6, 9, , 18, 24, 36

Batches are tested at time 0 consistent with manufacturing process.

D. Matrix 2/3 on Time Design with Multiple Packages and Multiple Strengths

The next extension is when three strengths (say, 10, 20, and 30) are manufactured using different weights of the same formulation, giving nine subbatches. It is further assumed that there are three packages for each strength. In this case the basic matrix 2/3 on time design can be applied to each of the nine subbatches in a balanced fashion, as presented in Tables 4 and 5. Balance in this design means each subbatch is tested twice at each intermediate time point, each package is tested twice at each intermediate time point for each batch, each batch is tested six times at each intermediate time point, and each package is tested six times at each intermediate time point. If an analysis will be done after 18 months (e.g., for a registration application), this design can be modified by testing all batch-by-strength-by-package combinations at 18 months.

Table 4 Matrix 2/3 on Time Design with Multiple Packages and Multiple Strengths

Batch	Strength	Pkg 1	Pkg 2	Pkg 3
A	10	T1	T2	T3
A	20	T2	T3	T1
A	30	T3	T1	T2
B	10	T2	T3	T1
B	20	T3	T1	T2
B	30	T1	T2	T3
C	10	T3	T1	T2
C	20	T1	T2	T3
C	30	T2	T3	T1

Pkg 1 = Package 1, etc.

E. Matrix 1/3 on Time Design

A further reduction in the amount of testing is accomplished by reducing the testing in each of the preceding

Table 5 Test Code Definitions

Code	Test times after time 0
T1	3, , 9, 12, , 24, 36
T2	3, 6, , 12, 18, , 36
T3	, 6, 9, , 18, 24, 36

Strength subbatches are tested at time 0 consistent with manufacturing process.

designs from 2/3 to 1/3. For example, the basic 1/3 on time design has one of the three batches tested at each intermediate time point, as presented in Table 6. If an analysis will be done after 18 months (e.g., for a registration application), the basic matrix 1/3 on time design can be modified by testing all batches at 18 months.

F. Matrix on Batch × Strength × Package Combinations

If there are multiple strengths and multiple packages, one could also choose to test only a portion of the batch-by-strength-by-package combinations. An example of when this might be appropriate is when there are three batches, each made into two strengths, giving six subbatches. Although three packages will be used, the batch size is small and only two packages can be manufactured in each strength subbatch. A matrix design on batch × strength × package combinations is presented in Tables 7 and 8, with two packages selected for each of the six subbatches, and where time is also matrixed by the factor 1/2. This design is approximately balanced, because two packages are tested per subbatch, one or two strengths are tested for each selected package by batch, four subbatches are tested for each package, etc. Similar statements for the balance on time can be made.

G. Uniform Matrix Design

Another approach to design is the uniform matrix design (16), for which "the same time protocol is used for all combinations of the other design factors." The strategy is to delete certain times, for example, the 3-, 6-, 9-, 18-month time points; therefore testing is done only at 12, 24, and 36 months. This design has the advantages of simplifying data entry of the study design and eliminating time points that add little to reducing the variability of the slope of the regression line. The disadvantage is that if there are major problems with the stability, there is no early warning because early testing is not done. Further, it may not be possible to determine if the linear model is appropriate; e.g., it may not be possible to determine whether there is an immediate decrease followed by very little decrease. However, the major disadvantage is that this design is probably not acceptable to some regulatory agencies.

Table 6 Basic Matrix 1/3 on Time Design

Batch	Test times
A	0, 3, , , 12, , , 36
B	0, , 6, , , 18, , 36
C	0, , , 9, , , 24, 36

Table 7 Matrix 1/2 on Time and Matrix on Batch × Strength × Package

Batch	Strength	Pkg 1	Pkg 2	Pkg 3
A	10	T1	T2	—
A	20	T2	—	T1
B	10	T2	—	T1
B	20	—	T1	T2
C	10	—	T1	T2
C	20	T1	T2	—

IV. ANALYSIS

A. Background

Although there may be instances when a linear regression is not appropriate, the rest of this discussion assumes that a straight-line linear regression of the parameter of interest on time is appropriate. Further, it is assumed that the parameter of interest is expected to decrease over time. For long-term data with a single package and a single strength, the ICH Q1A guideline (1) specifies that the 95% one-sided lower confidence

Table 8 Test Code Definitions

Code	Test times after time 0
T1	3, , 9, , 18, , 36
T2	, 6, , 12, , 24, 36

Strength subbatches are tested at time 0 consistent with manufacturing process.

bound for the mean regression line must be above the lower specification at all times prior to the shelf life. The ICH Q1A guideline does not address the analysis when there are multiple strengths and/or multiple packages. When there are multiple packages and/or multiple strengths, there are two approaches for the analysis. The first approach is to analyze each package-by-strength combination separately, in other words, to do multiple analyses. The second approach is to model all data with one analysis, using model-building techniques to select the appropriate model.

B. Separate Analysis Approach

In the first approach, a separate analysis of each package-by-strength combination is done using the method given in ICH Q1A. The poolability of batch slopes and batch intercepts is done at the 0.25 level of significance, with shelf life taken as the minimum time a batch remains within acceptable limits, based on the 95% one-sided lower confidence bound for the mean regression line(s).

The SAS model is Y = B A B*A, where B is a class term for batch and A is the covariate for age. If the B*A term is significant at the 0.25 level, the 95% lower confidence bound is found for each batch, the shelf life of each batch is such that the confidence bound is within the specification at all times prior to the shelf life, and the shelf life of the product is the minimum of the batch shelf lives.

If the B*A term is not significant at the 0.25 level, the B*A term is deleted from the model and the B term (equality of batch intercepts) is tested at the 0.25 level. If the B term is significant at the 0.25 level, the 95% lower confidence bound is found for each batch, the shelf life of each batch is such that the confidence bound is within the specification at all times prior to the shelf life, and the shelf life of the product is the minimum of the batch shelf lives. If the B term is not significant at the 0.25 level, the B term is deleted from the model, a single regression line is fitted, and the shelf life of the product is such that the 95% one-sided lower confidence bound is within the specification at all times prior to the shelf.

C. One Analysis: Model Building

When using the model-building approach, it is very important that the full model reflect the manufacturing process. In this section it is assumed that there are multiple strengths and multiple packages and that a batch is manufactured into multiple strengths by using different weights of the same exact formulation. It is assumed that a granulation batch is split into subbatches, where each subbatch is manufactured into a different-strength product using different weights of the granulation, and it is assumed that every strength is manufactured from every granulation batch. It is assumed that the time 0 samples are collected from each tablet subbatch, and every tablet subbatch is packaged into all packages.

The full model includes slope terms for all two-way interactions of package, strength, and batch, and it includes intercept terms that reflect the manufacturing process. The manufacturing process dictates that each tablet subbatch must have a separate intercept in the full model, but there is no need to allow packages in each tablet subbatch to have separate intercepts. (Process validation provides evidence that the entire tablet subbatch is uniform.)

Thus in the example, the full SAS model is

Y = B S B(S) A B*A P*A S*A
B*P*A B*S*A P*S*A

with B a class term for granulation batch, P a class term for package, S a class term for strength, and A a covariate for time. The model building begins by testing the slope two-way interactions to determine if any can be deleted, using a significance level of 0.25 when batch is part of the term and 0.05 otherwise. Then the main-effect slope terms are tested, using the 0.25 (batch) or 0.05 (not batch) level. Terms that are not significant are deleted, except that any main-effect slope included in a nondeleted two-way slope term cannot be deleted.

After deleting slope terms, the intercept terms are tested (using the same criterion for significance as was used for slopes) and nonsignificant terms are deleted. Using the final model, the 95% one-sided lower confidence bound for the mean regression line(s) is (are) found, and shelf life is assigned for each package-by-strength combination such that the 95% one-sided lower confidence bound(s) is (are) within specification at all times prior to the shelf life.

A slightly different algorithm for deleting terms from the full model has been proposed (14).

V. COMPARISON OF DESIGNS

Nordbrock (6) compared designs based on the power approach. This approach computes the probability that a statistical test will be significant when there is a specified alternative slope configuration. Power can be computed easily in SAS (23). The strategy is to compute power for several designs and then to choose the design that has acceptable power and the smallest sample size (or cost). Acceptable power is not well defined at this stage. Note that in Ref. 2 the power of specific 1-df contrasts were calculated using a 1-df (t) test, whereas in Ref. 23 the power of specific contrasts is computed using the corresponding F-test from the analysis of variance.

Other methods of comparing designs are given in Ju and Chow (18), where the criterion is the precision for estimating shelf life. Murphy (16) gives five criteria for comparing designs, based on moments, D-efficiency, uncertainty, G-efficiency, and power.

When evaluating designs it is also important to answer the question "What the probability is of being able to defend the desired shelf life with the study?" (24). In other words (assuming the parameter is expected to decrease over time), what is the probability that the 95% one-sided lower confidence bound for the slope will be acceptable for specified values of the slope(s) for particular subsets of data, which may include, for example, only one strength and/or only one package? It is important to know at the design stage what the statistical penalty (with respect to shelf life) might be if differences among packages and/or strengths are found.

VI. FACTORS ACCEPTABLE TO MATRIX

In the foregoing, examples have been used to present possible matrix designs. In this section a summary of when it is acceptable to matrix is given based on a document prepared by the Stability Working Group of PhRMA (20) and on FDA presentations (21,22).

It is acceptable to matrix at all stages of development for a drug product and also for drug substance. It is acceptable for NDA studies, IND studies, supplements, and marketed product studies.

It is acceptable to matrix for all types of products, such as solid, semisolid, liquid, and aerosols.

It is acceptable to matrix after bracketing.

It is acceptable to matrix when there are multiple sources of raw material, e.g., drug product.

It is acceptable to matrix if there are multiple sites of drug product manufacture.

It is acceptable to matrix when identical formulations are manufactured into several strengths.

It is acceptable to matrix if formulations are closely related, e.g., difference in colorant or flavoring.

Matrixing is applicable to orientation of container during storage.

Matrixing may be acceptable in certain cases when closely related formulations are used for different strengths, e.g., if an inactive is replaced by active.

Matrix across container and closure systems may be applicable if justified.

It is acceptable to matrix within a package composition type, e.g., different sizes if fill (i.e., head space) is same, or if same size but different fill (head space). It may be acceptable to matrix if container size and fill size change, if there is adequate explanation. It is not acceptable to matrix across package composition type, e.g., blisters and HDPE.

It is not acceptable to matrix across storage conditions. However, it is acceptable to do a separate matrix design for each storage condition.

It is not acceptable to matrix across parameters, such as dissolution and potency. However, it is acceptable to do a separate design for each parameter.

Matrixing is applicable regardless of method precision; however, it should be remembered that when using a matrix design the resulting shelf life is generally shorter than when a complete design is used and that when the method precision is larger, the difference between a complete design and matrixed design will be larger (i.e., a larger penalty to the sponsor, resulting in a shorter shelf life for the matrix design than the complete design.

Matrixing is applicable regardless of the stability of the product. However, comments similar to those in the preceding point apply, and it should be remembered that if a product has a poor stability profile (e.g., shelf life of 1 year), matrixing will usually result in an even shorter shelf life.

The latest guideline should be consulted for applicability.

VII. GENERAL RULES

Several general rules should be followed when designing studies.

Matrix designs should be approximately balanced; i.e., for all one-way, two-way combinations of batch, package, and strength that are ever tested, approximately the same number of tests should be done cumulatively to every time point.

When every batch-by-strength-by-package combination is not tested, every strength-by-package combination that is ever tested should be tested in at least two batches; i.e., for every package-by-strength combination that is ever tested, there should be at least two batches tested (W. Fairweather, personal communication, 1995).

Unless there are manufacturing restrictions such as in the foregoing example, it is probably acceptable to matrix on batch × strength × package combinations only when there are more than three strengths or more than three packages.

VIII. CONCLUSIONS

Matrix designs are generally applicable to many situations and can result in significant savings, with the 2/3 matrix on time readily acceptable for stable products. Larger reductions in testing than that given by the 2/3 matrix on time are sometimes acceptable. There are two basic approaches when analyzing data from a matrixed design. There are several methods used to evaluate and compare potential designs.

REFERENCES

1. ICH Q1A. Guideline for Industry. Stability Testing Guideline of New Drug Substances and Products. Sep 1994.
2. E Nordbrock. Statistical study design. Presentation at National Stability Discussion Group, Oct. 1989.
3. J Wright. Use of factorial designs in stability testing. Proceedings of Stability Guidelines for Testing Pharmaceutical Products: Issues and Alternatives. AAPS meeting, Dec. 1989.
4. P Nakagaki. AAPS Annual Meeting, 1990.
5. E Nordbrock. Statistical comparison of NDA stability study designs. Midwest Biopharmaceutical Statistics Workshop, May 1991.
6. E Nordbrock. Statistical comparison of stability study designs. J Biopharmaceutical Stat 2:91–113, 1992.
7. P Helboe. New designs for stability testing programs: matrix or factorial designs. Authorities' viewpoint on the predictive value of such studies. Drug Info J 26: 629–634, 1992.
8. JT Carstenson, M Franchini, K Ertel. Statistical approaches to stability protocol design. J Pharmaceutical Sci 81:303–308, 1992.
9. SC Chow. Statistical design and analysis of stability studies. 48th Annual Conference on Applied Statistics, 1992.
10. E Nordbrock. Design and analysis of stability studies. ASA Proceedings of Biopharmaceutical Section, pp 291–294, 1994.
11. E Nordbrock. Statistically designed stability studies, an industry perspective. Presentation at Biostatistics Subsection/Clinical Data Management Group of Pharmaceutical Research and Manufacturers of America, Oct. 1994.
12. TD Lin. Applicability of matrix and bracket approach to stability study design. ASA Proceedings of Biopharmaceutical Section, pp 142–147, 1994.
13. WR Fairweather, TD Lin, R Kelly. Regulatory and design aspects of complex stability studies. Presentation at Biostatistics Subsection/Clinical Data Management Group of Pharmaceutical Research and Manufacturers of America, Oct. 1994.
14. WR Fairweather, TD Lin, R Kelly. Regulatory, design, and analysis aspects of complex stability studies. J Pharmaceutical Sci 84:1322–1326, 1995.
15. MH Golden, DC Cooper, MT Riebe, KE Carswell. J Pharmaceutical Sci 85:240–245, 1996.
16. JR Murphy. Uniform matrix stability study designs. J Biopharmaceutical Stat 6:477–494, 1996.
17. K DeWoody, D Raghavarao. Some optimal matrix designs in stability studies. J Biopharmaceutical Stat 7: 205–213, 1997.
18. HL Ju, SC Chow. On stability designs in drug shelf-life estimation. J Biopharmaceutical Stat 5:210–214, 1995.
19. CPMP. Reduced Stability Testing Plan—Bracketing and Matrixing (CPMP/QWP/157/96). Oct. 1997.
20. E Nordbrock, S Valvani. PhRMA Stability Working Group. Guideline for Matrix Designs of Drug Product Stability Protocols, Jan. 1995.
21. C Chen. FDA's views on bracketing and matrixing. EFPIA Symposium: Advanced Topics in Pharmaceutical Stability Testing-Building in the ICH Guideline, Oct. 1996.
22. D Lin. Stability studies at the FDA. PERI Course: Non-Clinical Statistics for Drug Discovery and Development, March 1997.
23. E Nordbrock. Computing power details. PERI: Training Course in Non-Clinical Statistics, Feb. 1994.
24. E Nordbrock. Compute probability of shelf life. PERI: Training Course in Non-Clinical Statistics, Feb. 1994.

Earl Nordbrock

Statistical Significance

See also *Power; Meta Analysis*

I. INTRODUCTION

Most controlled clinical trials are performed to demonstrate that a new investigational treatment is superior to a concurrent placebo or to an active control (superiority trials); that a test treatment is not inferior to an active treatment by more than a prespecified, clinically irrelevant amount (noninferiority trials); or that two active treatments are therapeutically equivalent (equivalence trials). In the past, data from such trials have usually been analyzed by testing the traditional null hypothesis of no difference in the effect between the treatments. However, the most important drawback of this classical approach is the fact that looking only at the presence or absence of an effect ignores the issue of effect size. Hence, a statistically significant result gives only the information that the treatment difference is not zero but does not provide information as to the clinical relevance. Furthermore, failure to reject the null hypothesis does not imply the acceptance of the hypothesis of no difference. Therefore, this practice has been criticized and the use of confidence intervals is advocated by regulatory guidelines (CPMP, 1994; European Agency, 1998). However, without the concept of the clinically significant (relevant) difference, the two methods are only technically different. Thus, the adequate test problem should be formulated by incorporating the clinically significant difference, resulting in an equivalence of statistical and clinical significance. This direct approach will be demonstrated for superiority, noninferiority, and equivalence trials.

II. TESTING SHIFTED NULL HYPOTHESES

A. Proof of Superiority

A two-sample situation is considered. Let the clinical endpoint be denoted as X_1 for the test treatment and by X_0 for the control group. Suppose that these random variables are mutually independent and have a continuous distribution function from a location family, that is

$$X_{0j} \sim F(x - \mu_0) \quad \text{and} \quad X_{1j} \sim F(x - \mu_1),$$
$$j = 1, \ldots, n_i,\ i = 0, 1 \tag{1}$$

Without loss of generality, it is assumed that the population means are positive and that it is a priori known that if there is a favorable response to the treatment it will increase in magnitude, i.e., $\mu_1 > \mu_0$, and, hence, indicating the appropriateness of a one-sided alternative. It should be noted that the issue of a one-sided or two-sided alternative is controversial in the literature, but this is off our topic. Recently, the ICH guideline (European Agency, 1998) recommended the setting of the Type I error for one-sided tests at half the conventional Type I error used in two-sided testing. The traditional test problem of no difference is formulated as follows:

$$H_0: \mu_1 - \mu_0 \leq 0$$
$$H_1: \mu_1 - \mu_0 > 0 \tag{2}$$

Rejection of the null hypothesis by a statistical test at level α (e.g., Student's t-test or the Wilcoxon test) could provide evidence for the conclusion that there is a clinically relevant effect of the treatment. However, large sample sizes and little variation may lead to the problem that a clinically unimportant difference will be declared statistically significant. Suppose that a difference $\mu_1 - \mu_0 = \delta > 0$ is considered the maximum clinically irrelevant difference. If the true expected difference is less than or equal to this threshold value, then a statistically significant result for test problem (2) may be of little interest, because the difference is not clinically significant. Consequently, the adequate test problem should be formulated as follows:

$$H_0^\delta: \mu_1 - \mu_0 \leq \delta$$
$$H_1^\delta: \mu_1 - \mu_0 > \delta \tag{3}$$

where δ, $\delta > 0$, denotes the maximum clinically irrelevant difference. It should be noted that this definition implies that only differences greater than δ are regarded as clinically important. For example, the CPMP guideline (European Agency, 1995) for chronic arterial occlusive disease requires that an irrelevant difference for the improvement in walking distance between placebo and the new treatment should be excluded.

Assuming $F(x) = \Phi(x/\sigma)$, where $\Phi(\cdot)$ denotes the cumulative distribution function of the standard normal distribution, the shifted null hypothesis H_0^δ can be rejected, if

$$T^{\delta} = \frac{\bar{X}_1 - \bar{X}_0 - \delta}{S\sqrt{1/n_1 + 1/n_0}} \geq t_{\alpha, n_0+n_1-2} \quad (4)$$

where $t_{\alpha,\nu}$ is the $(1 - \alpha)$ percentile of the central t-distribution with ν degrees of freedom, $\bar{X}_1$ and $\bar{X}_0$ denote the sample means of the treatment and control groups, respectively, and S^2 is the pooled estimator of σ^2. Testing the nonzero null hypothesis, that is, incorporating a shift parameter δ, permits the assessment of the clinical relevance of an observed treatment difference (Victor, 1987). Obviously, this decision procedure is equivalent to the condition that the lower limit of the one-sided $100(1 - \alpha)\%$ confidence interval is greater than δ; that is,

$$\left[\bar{X}_1 - \bar{X}_0 - t_{\alpha, n_0+n_1-2} S\sqrt{\frac{1}{n_0} + \frac{1}{n_1}}, \infty\right) \subset (\delta, \infty) \quad (5)$$

The exact sample size to attain a prespecified power of $1 - \beta$ for a given level α, a standard deviation σ, and an important value $\mu_1 - \mu_0 = \delta_1 > \delta$, which should not be overlooked, can be obtained directly from univariate noncentral t-distribution or can be approximated for the balanced design, that is, $n_0 = n_1 = n$, as follows:

$$n \geq 2\left(\frac{(t_{\alpha,2n-2} + t_{\beta,2n-2})\sigma}{\delta_1 - \delta}\right)^2 \quad (6)$$

The corresponding nonparametric method for testing the shifted hypothesis using the Wilcoxon rank sum statistic $R_1(\delta) = \sum_{j=1}^{n_1} R_{1j}(\delta)$ is presented in Hollander and Wolfe (1973) and is based on the ranks $R_{1j}(\delta)$ of the shifted observations in the treatment group in the combined sample:

$$X_{11} - \delta, \ldots, X_{1n_1} - \delta, X_{01}, \ldots, X_{0n_0} \quad (7)$$

H_0^{δ} is rejected if $R_1(\delta) \geq r(\alpha, n_0, n_1)$, where $r(\alpha, n_0, n_1)$ is the $(1 - \alpha)$ percentile of the Wilcoxon rank statistic $R_1(\delta)$. The calculation of the confidence interval according to Moses (Hollander and Wolfe, 1973) can be performed as follows. Let $D_1 \leq \cdots \leq D_{n_0n_1}$ denote the ordered values of the n_0n_1 differences $X_{1j} - X_{0j^*}$, $j = 1, \ldots, n_1$ and $j^* = 1, \ldots, n_0$. The median of these pairwise differences serves as the Hodges–Lehmann estimator (Hodges and Lehmann, 1963), and the lower limit of the one-sided $100(1 - \alpha)\%$ confidence interval for $\mu_1 - \mu_0$ is given by

$$L = D_{C_\alpha} \quad (8)$$

where $C_\alpha = n_1(2n_0 + n_1 + 1)/2 + 1 - r(\alpha, n_0, n_1)$. Appropriate methods for power calculation and the necessary sample sizes for nonparametric procedures are given by Noether (1987) and Collings and Hamilton (1988).

It should be noted that the use of p-values and hypothesis tests in the analysis of medical data has been criticized, and point estimates and confidence intervals have been proposed. Nevertheless, it is possible to evaluate the results of clinical studies appropriately using hypothesis tests, as long as relevant hypotheses like those in Eqs. (3) are tested, and not the meaningless hypothesis of Eqs. (2). However, although there is nothing wrong with testing relevant hypotheses and reporting the associated p-value, the use of point estimates and confidence intervals has two advantages. First, as already shown, the shifted hypotheses can conveniently be evaluated using confidence intervals. Second, point estimates and confidence intervals provide direct information about the magnitude of the treatment difference; this may be easier to interpret than p-values, which express the magnitude on the probability scale.

Specification of δ requires that statisticians and clinicians think about what constitutes a maximum clinically irrelevant difference, and a common situation in practice is that the value δ is expressed as a proportion of the unknown population mean μ_0. Therefore, assuming that $\delta = f\mu_0$, $f > 0$, the test problem (3) can be formulated as

$$H_0^{\delta}: \mu_1 - \mu_0 \leq f\mu_0$$
$$H_1^{\delta}: \mu_1 - \mu_0 > f\mu_0 \quad (9)$$

or as

$$H_0^{\theta}: \frac{\mu_1}{\mu_0} \leq \theta$$
$$H_1^{\theta}: \frac{\mu_1}{\mu_0} > \theta \quad (10)$$

where $\theta = 1 + f$, $\theta > 1$, is the corresponding maximum clinically irrelevant value for μ_1/μ_0. Under the assumption of normality and common variance, Sasabuchi (1988) demonstrated that the size-α likelihood ratio test rejects the shifted null hypothesis H_0^{θ} concerning the ratio of the two means if

$$T^{\theta} = \frac{\bar{X}_1 - \theta\bar{X}_0}{S\sqrt{1/n_1 + \theta^2/n_0}} \geq t_{\alpha, n_0+n_1-2} \quad (11)$$

Algebraic rearrangement shows that condition (11) is equivalent to

$$\theta_- \geq \theta \quad \text{and} \quad \bar{X}_0^2 > a_0 \quad (12)$$

where

$$\theta_- = \frac{\bar{X}_0\bar{X}_1 - \sqrt{a_0\bar{X}_1^2 + a_1\bar{X}_0^2 - a_0a_1}}{\bar{X}_0^2 - a_0}$$

$$a_0 = \frac{S^2}{n_0} t^2_{\alpha, n_0+n_1-2}$$

$$a_1 = \frac{S^2}{n_1} t^2_{\alpha, n_0+n_1-2}$$

Condition (12) can be interpreted as follows: if the lower limit of the one-sided $100(1 - \alpha)\%$ Fieller (1954) confidence set for μ_1/μ_0 is finite, the interval is given by $I_F = (\theta_-, \infty)$. Furthermore, Fieller's one-sided confidence set is a lower bounded interval if and only if $\bar{X}_0^2 > a_0$ holds true. Hence, the following two procedures for the assessment lead to the same decisions: (a) conclude superiority if and only if H_0^θ is rejected by the level-α Sasabuchi test; (b) conclude superiority if and only if the lower limit of the one-sided $100(1 - \alpha)\%$ Fieller confidence interval is above θ.

The following approximate formula for a balanced sample size determination to attain a prespecified power of $1 - \beta$ for a given level α, a coefficient of variation $CV_0 = \sigma/\mu_0$, and a value from the alternative $\mu_1/\mu_0 = \theta_1 > \theta$ was given by Hauschke et al. (1999a):

$$n \geq (1 + \theta^2) \left(\frac{(t_{\alpha,2n-2} + t_{\beta,2n-2})CV_0}{\theta_1 - \theta} \right)^2 \tag{13}$$

It should be noted that a short discussion of corresponding nonparametric methods was provided by Vuorinen and Turunen (1997), but further research on this issue still needs to be performed.

B. Proof of Noninferiority

The most convincing way to establish the efficacy of a new investigational treatment is to demonstrate clinically relevant superiority to a concurrent placebo group. In the case where an accepted treatment exists, the comparison against placebo may be considered unethical. Furthermore, when a new treatment has certain advantages, such as fewer side effects or simpler application, it might be enough to show that the test treatment is not relevantly inferior to the comparator. The choice of the extent of irrelevant inferiority should be based on clinical reasoning and will require researchers to think about what constitutes a clinically relevant difference. The corresponding hypothesis and alternative can be written as:

$$H_0^\delta: \mu_1 - \mu_0 \leq \delta^*$$
$$H_1^\delta: \mu_1 - \mu_0 > \delta^* \tag{14}$$

where δ^*, $\delta^* < 0$, denotes the maximum clinically irrelevant difference. Analogous to the previously given methodology for superiority trials, this shifted null hypothesis can be tested and the adequate sample size determination can be performed.

C. Proof of Equivalence

Many clinical trials aim to show equivalence between a test treatment under development and an existing reference treatment. In such studies the issue is no longer to detect a difference but to demonstrate that the two active treatments are equivalent within a priori stipulated margins. A widespread application of this problem is the assessment of comparative bioavailabilities in a bioequivalence trial. Let the interval (δ_1, δ_2) denote the prespecified equivalence range; the corresponding test problem is formulated as follows:

$$H_0^\delta: \mu_1 - \mu_0 \leq \delta_1 \quad \text{or} \quad \mu_1 - \mu_0 \geq \delta_2$$
$$H_1^\delta: \delta_1 < \mu_1 - \mu_0 < \delta_2 \tag{15}$$

Under normality and by assuming that the margins δ_1 and δ_2 are known, H_0^δ can be rejected at level α in favor of H_1^δ (equivalence) if the parametric two-sided $100(1 - 2\alpha)\%$ confidence interval for $\mu_1 - \mu_0$ is entirely included in the equivalence range:

$$\left[\bar{X}_1 - \bar{X}_0 \pm t_{\alpha, n_0+n_1-2} S \sqrt{\frac{1}{n_0} + \frac{1}{n_1}} \right] \subset (\delta_1, \delta_2) \tag{16}$$

This procedure is equivalent to the two one-sided tests procedure by means of shifted two-sample t-tests (Schuirmann, 1987). It should be noted that the corresponding nonparametric test procedure and the corresponding confidence interval construction were provided by Hauschke et al. (1990).

In the following, the equivalence limits δ_1 and δ_2 are again defined as proportions of the unknown mean μ_0. Suppose that $\delta_1 = f_1\mu_0$ and $\delta_2 = f_2\mu_0$, where $-1 < f_1 < 0 < f_2$. For example, $f_1 = -f_2 = -0.20$ corresponds to the common $\pm 20\%$ criteria. The foregoing test problem can be formulated as

$$H_0^\theta: \frac{\mu_1}{\mu_0} \leq \theta_1 \quad \text{or} \quad \frac{\mu_1}{\mu_0} \geq \theta_2$$

$$H_1^\theta: \theta_1 < \frac{\mu_1}{\mu_0} < \theta_2 \tag{17}$$

where (θ_1, θ_2), $\theta_1 = 1 + f_1$, $\theta_2 = 1 + f_2$, $0 < \theta_1 < 1 < \theta_2$, is the corresponding equivalence interval for the ratio μ_1/μ_0. Testing this two-sided equivalence problem is equivalent to the simultaneous testing of the following two one-sided hypotheses:

$$H_{01}^{\theta}: \frac{\mu_1}{\mu_0} \leq \theta_1 \qquad H_{02}^{\theta}: \frac{\mu_1}{\mu_0} \geq \theta_2$$

and

$$H_{11}^{\theta}: \frac{\mu_1}{\mu_0} > \theta_1 \qquad H_{12}^{\theta}: \frac{\mu_1}{\mu_0} < \theta_2. \tag{18}$$

The size-α likelihood ratio tests rejects H_{01}^{θ} and H_{02}^{θ} if

$$T_1^{\theta} = \frac{\bar{X}_1 - \theta_1 \bar{X}_0}{S\sqrt{1/n_1 + \theta_1^2/n_0}} \geq t_{\alpha, n_0+n_1-2} \quad \text{and}$$

$$T_2^{\theta} = \frac{\bar{X}_1 - \theta_2 \bar{X}_0}{S\sqrt{1/n_1 + \theta_2^2/n_0}} \leq -t_{\alpha, n_0+n_1-2} \tag{19}$$

Algebraic rearrangement shows that condition (19) is equivalent to

$$\theta_- \geq \theta_1 \quad \text{and} \quad \theta_+ \leq \theta_2 \quad \text{and} \quad \bar{X}_0^2 > a_0 \tag{20}$$

where

$$\theta_{\pm} = \frac{\bar{X}_0\bar{X}_1 \pm \sqrt{a_0\bar{X}_1^2 + a_1\bar{X}_0^2 - a_0 a_1}}{\bar{X}_0^2 - a_0}$$

$$a_0 = \frac{S^2}{n_0} t^2_{\alpha, n_0+n_1-2}$$

$$a_1 = \frac{S^2}{n_1} t^2_{\alpha, n_0+n_1-2}$$

Condition (20) can be interpreted as follows: if the two-sided $100(1 - 2\alpha)\%$ Fieller confidence set for μ_1/μ_0 has finite length, it is given by the interval $I_F = (\theta_-, \theta_+)$. Furthermore, Fieller's confidence set is an interval if and only if $\bar{X}_0^2 > a_0$ holds true. Hence, the following two procedures for assessing test problem (17) lead to the same decisions: (a) conclude equivalence if and only if H_{01}^{θ} and H_{02}^{θ} are rejected by the level-α Sasabuchi tests; (b) conclude equivalence if and only if the two-sided $100(1 - 2\alpha)\%$ Fieller confidence interval is included in the equivalence range.

The problem of exact power calculation and sample size determination under the assumption of normality was addressed by Hauschke et al. (1999b). The following approximate formulas for sample size determination were given by Kieser and Hauschke (1999), assuming $\theta_1 = 1/\theta_2$ and $\theta^* = \mu_1/\mu_0$:

if $\theta^* = 1$

$$n \geq (1 + \theta_2^2)(t_{\alpha,2n-2} + t_{\beta/2,2n-2})^2 \left(\frac{\text{CV}_0}{1 - \theta_2}\right)^2$$

if $1 < \theta^* < \theta_2$

$$n \geq (1 + \theta_2^2)(t_{\alpha,2n-2} + t_{\beta,2n-2})^2 \left(\frac{\text{CV}_0}{\theta_2 - \theta^*}\right)^2$$

if $\frac{1}{\theta_2} < \theta^* < 1$

$$n \geq \left(1 + \frac{1}{\theta_2^2}\right)(t_{\alpha,2n-2} + t_{\beta,2n-2})^2 \left(\frac{\text{CV}_0}{1/\theta_2 - \theta^*}\right)^2 \tag{21}$$

D. Wider Applications

In the preceding sections the evaluation of study results using shifted hypothesis testing and confidence intervals was described when the focus of interest is a difference or a ratio of treatment means. A similar approach can be used in other cases. For example, when a difference between two proportions, such as the proportion of patients with a positive outcome for two treatments, is the focus of interest, then the shifted hypotheses can be modified by replacing μ_1 and μ_0 by π_1 and π_0, respectively. The same approach can be adopted for the ratio of proportions, i.e., the relative risk (Farrington and Manning, 1990).

III. CONCLUSIONS

Testing a nonzero null hypothesis instead of one of no treatment difference removes at a stroke some logical problems. For example, one problem of testing the classical null hypothesis is that even for the most minute true difference between two treatments, the p-value will indicate statistical significance if the sample size is big enough. When data are analyzed by testing shifted hypotheses, or through confidence intervals, as outlined, then an increase in sample size or in the precision of the data, or both, will lead to a narrowing of the confidence interval and thus to an increase in the probability of obtaining a conclusive result, i.e., a statistically and clinically relevant significance. This entry demonstrates that current statistical methodology can be applied to provide the necessary equivalence between statistical and clinical significance.

REFERENCES

Collings BJ, Hamilton MA. Estimating the power of the two-sample Wilcoxon test for location shift. Biometrics 44: 847–860, 1988.

CPMP Working Party on Efficacy of Medical Products. Note for guidance: biostatistical methodology in clinical trials in applications for marketing authorizations for medicinal products, 1994.

European Agency for the Evaluation of Medicinal Products. CPMP/EWP/235/95. The clinical investigation on medicinal products in the treatment of chronic peripheral arterial occlusive disease, 1995.

European Agency for the Evaluation of Medicinal Products. ICH E9. Note for guidance on statistical principles for clinical trials, 1998.

Farrington CP, Manning G. Test statistics and sample size formulae for comparative binomial trials with null hypothesis of non-zero risk difference or non-unity relative risk. Stat Med 9:1447–1554, 1990.

Fieller E. Some problems in interval estimation. J Royal Statistical Soc B 16:175–185, 1954.

Hauschke D, Kieser M, Diletti E, Burke M. Sample size determination for proving equivalence based on the ratio of two means for normally distributed data. Stat Med 18:93–105, 1999.

Hauschke D, Kieser M, Hothorn LA. Proof of safety in toxicology based on the ratio of two means for normally distributed data. Biom J 41:295–304, 1999.

Hauschke D, Steinijans VW, Diletti E. A distribution-free procedure for the statistical analysis of bioequivalence studies. Int J Clin Pharmacol Ther Toxicol 28:72–78, 1990.

Hodges JL, Lehmann, EL. Estimates of location based on rank tests. Ann Math Stat 34:598–611, 1963.

Hollander M, Wolfe DA. Nonparametric Statistical Methods. New York: Wiley, 1973.

Kieser M, Hauschke D. Approximate sample sizes for testing hypotheses about the ratio and difference of two means. J Biopharm Statist 9:641–650, 1999.

Noether GE. Sample size determination for some common nonparametric tests. J Am Stat Assoc 82:645–647, 1987.

Sasabuchi S. A multivariate one-sided test with composite hypotheses determined by linear inequalities when the covariance matrix has an unknown scale factor. Memoirs of the Faculty of Science, Kyushu University, Series A, Mathematics 42:9–19, 1988.

Schuirmann DJ. A comparison of the two one-sided tests procedure and the power approach for assessing the equivalence of average bioavailability. J Pharmacokin Biopharm 15:657–680, 1987.

Victor N. On clinically relevant differences and shifted null hypotheses. Methods Inf Med 26:109–116, 1987.

Vuorinen J, Turunen J. A simple three-step procedure for parametric and nonparametric assessment of bioequivalence. Drug Inf J 31:167–180, 1997.

Dieter Hauschke
Robert Schall
Herman G. Luus

Subgroup Analysis

See also *Group Sequential Methods*

I. INTRODUCTION

Because there are many different segments of the population as a whole and because these different subgroups of the population respond differently to treatment, subject recruitment for randomized, controlled trials is now actively attempting to be fairly representative of the population as a whole. However, this does not always occur (1). The statistical methods used need to investigate the outcomes for subgroups in addition to examining for statistical significance on the entire population. To do this, estimation and exploratory techniques need to be used. Validation by fresh data also needs to be emphasized.

II. NEED FOR SUBGROUP ANALYSIS

Most statistical hypothesis testing that occurs in clinical trials makes an assumption that the population is relatively homogeneous. This translates statistically into an assumption that the population has a distribution that resembles the bell-shaped curve (Fig. 1). However, consider the study that demonstrated that race and ethnicity are important in determining absorption rates of drugs (2). Since body weight is also of concern, gender is also a real concern. In terms of trial outcomes, no distinctions are made to determine differences within the populations. This means that it is unlikely that the various symptoms and reactions will present themselves in the form of a bell-shaped curve. Phase I trials are designed to examine dosage levels for all subsequent trials. Typically they use small samples, samples too small to determine all differences related to race, gender, and ethnicity.

Consider a contrast to the assumption of homogeneity when heterogeneity exists in the population (Fig. 2). The assignment of "normal" is usually defined as the amount of the curve that contains 95% (or often 75%) of the area under the bell-shaped curve (Fig. 3).

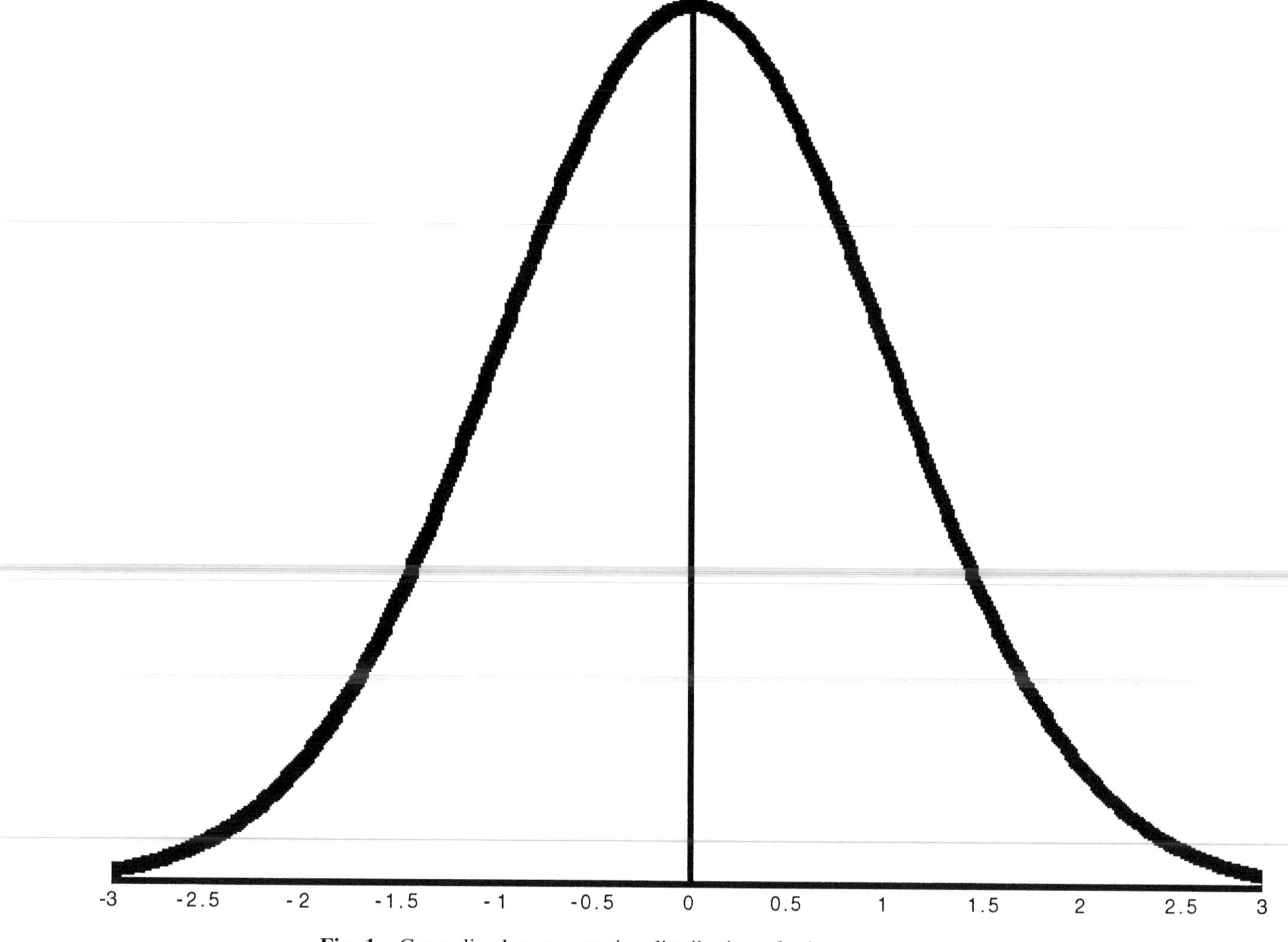

Fig. 1 Generalized representative distribution of a homogeneous population.

However, such an assumption will overpresent one subgroup while underpresenting another. It can be seen clearly that half of the second peak falls above the established "norm." This becomes even more noticeable when the distributions for the two subpopulations are graphed separately (Fig. 4). In fact, the assumption of normality in the presence of bimodality tends to redefine what is normal for one subgroup as an abnormal reaction, because it lies beyond the "norm" of the dominant subgroup (Fig. 5). In Fig. 5, the mean for the second group actually is placed at the upper limit for the first subgroup.

Thus, a minority population could be misdiagnosed or treated improperly because of a reliance upon the assumption of homogeneity. Unfortunately, different subpopulations cannot always be clearly distinguished. Although gender, race, and ethnicity are fairly obvious indicators of subpopulations, other characteristics should also be considered. In this case, it is also a matter of finding out what subpopulations exist as well as which ones should be investigated separately.

The use of subgroups by race or ethnicity is not without controversy. Reference 3 strongly recommends that race and ethnicity be discarded as population subgroups, to be replaced by genetic markers. It states:

> Studies using new technologies for understanding, measuring, and conceptualizing the sources of human variation reveal that approximately 85% of all variation in gene frequencies occurs within populations or races and only 15% variation occurs between such populations; there is no genetic basis for racial classification.

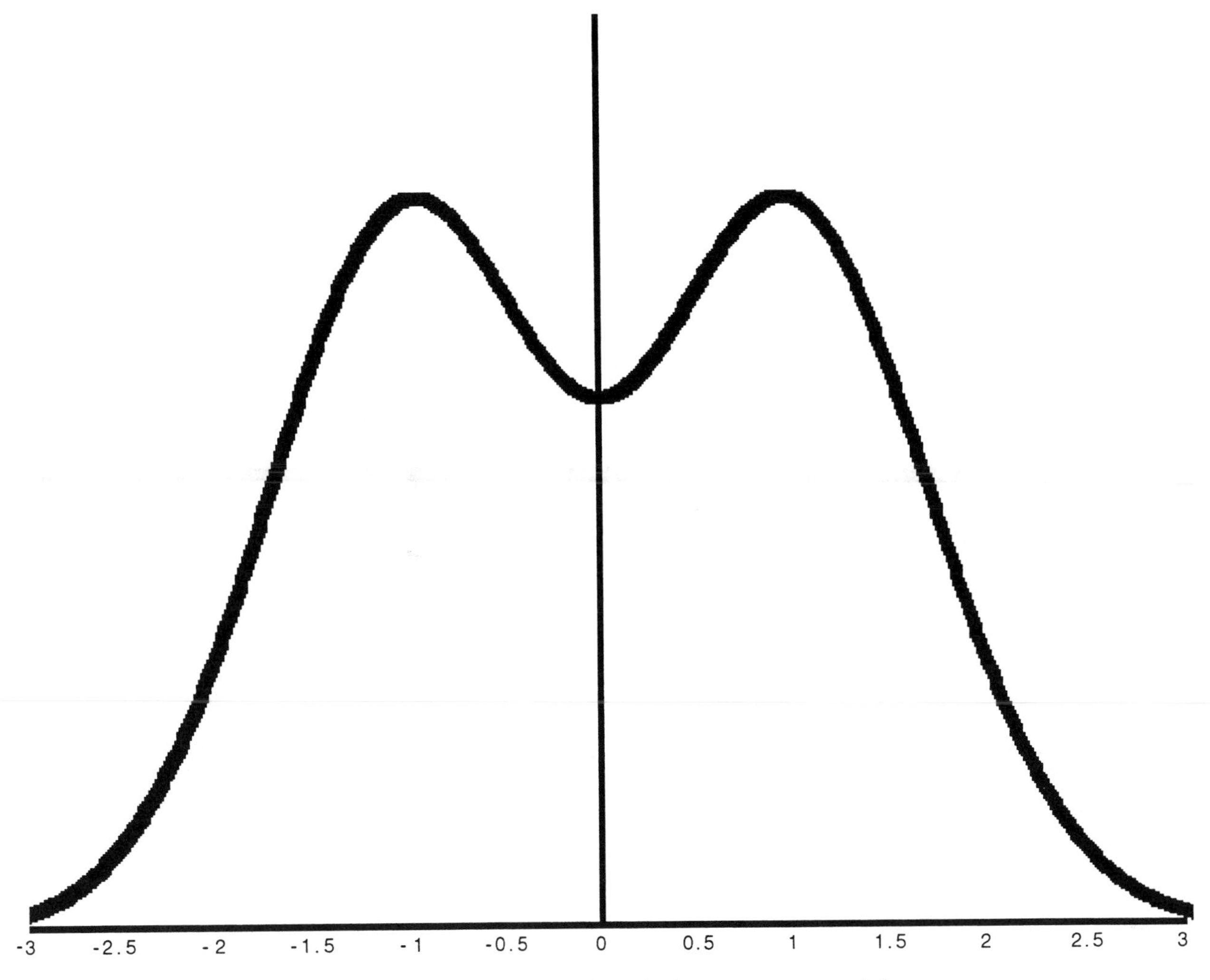

Fig. 2 Generalized representation of a heterogeneous population.

Contrast this statement with one made by Dr. Edward Sondik, Director for the National Center for Health Statistics: "Data showing racial differences in incidence and mortality have important scientific implications. These data should be looked at as offering clues for further research versus providing answers in and of itself."

The first comment assumes that it is possible to collect data on the genetic makeup of clinical trial subjects; unless the trial itself examines genetic variation, this is not possible. The second statement provides a more realistic attitude in that statistical analysis does not examine why subgroups occur within the population; it only examines their occurrence. Although the claim is being made that race is a social construct and that it is not biological, the fact is that even a social construct determines existence.

III. ALTERNATIVE METHODS OF ANALYSIS

There are two nonparametric tools that can be used to estimate effects in a multimodal population. Both methods are characterized by an ability to find subgroups as well as to estimate the effects of treatment on those subgroups.

A. Mixing Distributions

As shown in Ref. 4, mixture models involve assuming that each observation x_j is taken from a superpopulation G that is a mixture of a finite number of populations, $G_1, \ldots, G_g$, that occur in G in some proportions $\pi_1, \ldots, \pi_g$, respectively, such that $\pi_1 + \cdots + \pi_g = 1$. Then the density function for G is equal to

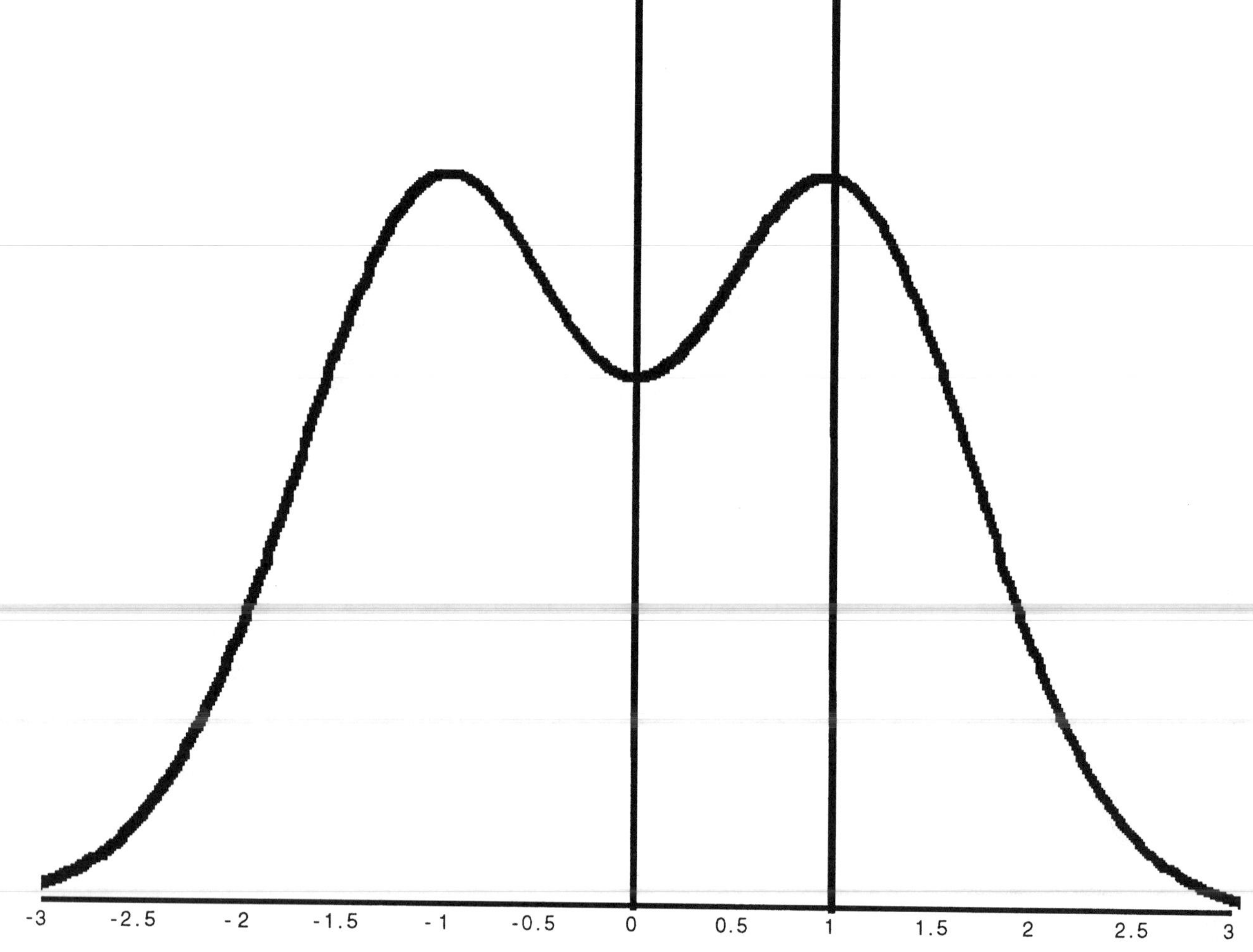

Fig. 3 Generalized representation of a heterogeneous population distribution showing the same 70% upper tail norm established under the assumption of homogeneity.

$$f(x) = \sum_{i=1}^{g} \pi_i f_i(x)$$

where f_i is the density function of the population G_i. The function f then can only be found by estimating f_i for $i = 1, \ldots, g$ as well as estimating $\pi_1, \ldots, \pi_g$. It is also sometimes necessary to determine the value of g, the number of populations involved. Therefore, an attempt must be made to decompose f into its separate parts. The total number of populations, g can be either known or unknown.

It is not possible to estimate if no sampling of G_i is done. However, it is also not possible to estimate f accurately. In fact, any attempt to estimate f without sampling from all components will result in severe bias. The largest of the populations tends to dominate the distribution. If the populations are of roughly equal size, both populations are misrepresented.

The advantage of using this method is that it is possible to determine just how many subpopulations should be considered and how many confounding factors to use in making a diagnosis. Therefore, it is possible to subdivide the population and to consider the extremes of symptoms for separate populations.

B. Kernel Density Estimation

Kernel density estimation uses the data to develop an estimate of the population density function. For the collected data $\{X_1, \ldots, X_n\}$, the estimate is given by the formula:

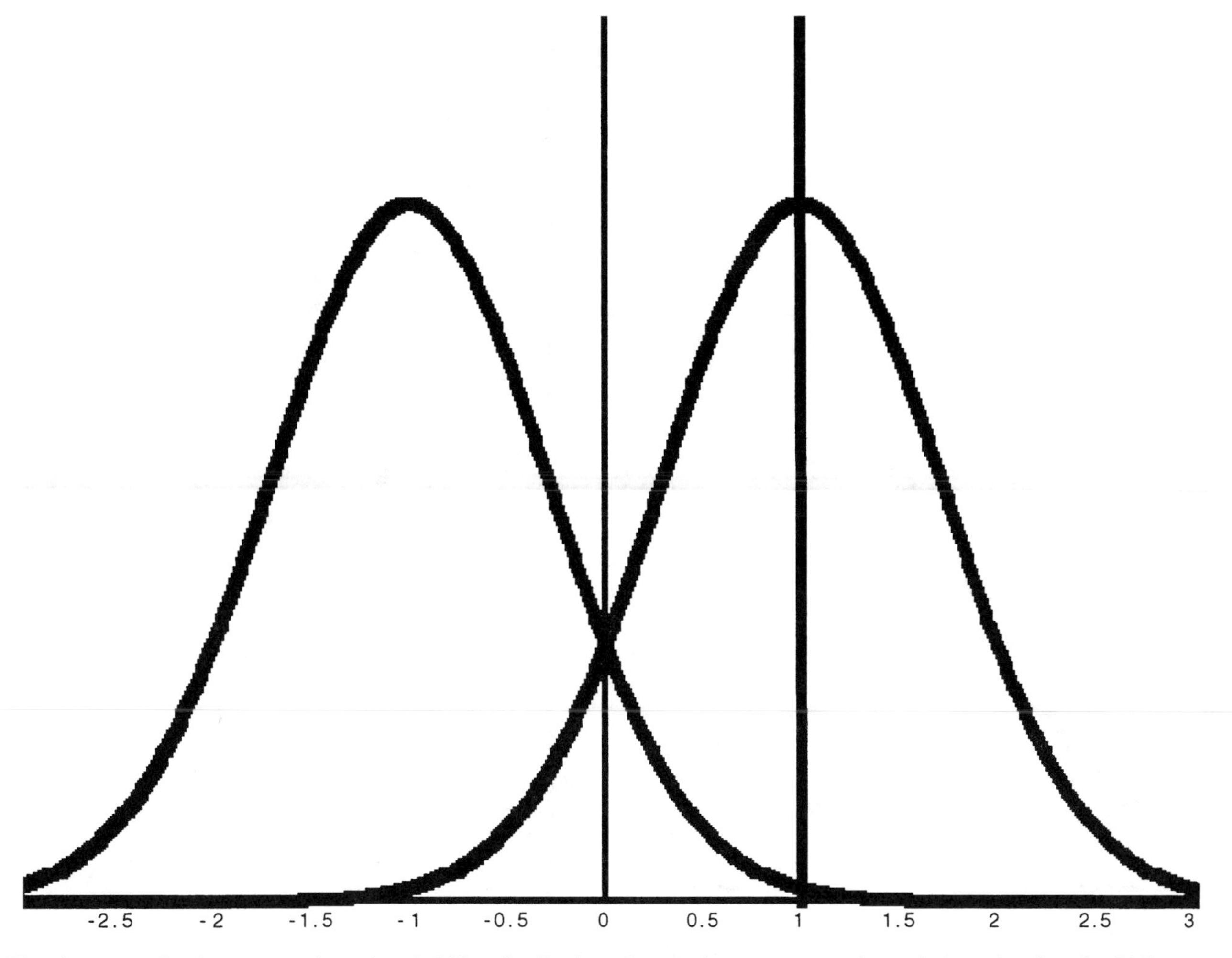

Fig. 4 Generalized representation of probability distributions for two homogeneous subpopulations showing the 70% upper tail norm established under the assumption that the entire population is homogeneous.

$$\hat{f}(x) = \frac{1}{na_n} \sum_{j=1}^{n} K\left(\frac{x - X_j}{a_n}\right)$$

where K is a known density function and a_n is a parameter that controls the degree of smoothness of f. As the value of a_n decreases, the number of peaks in the estimator increases. It becomes necessary to find an optimal value of a_n to determine the optimal number of peaks in the data. A method for doing this is discussed in Ref. 5. This technique also gives a nice visual representation of the population distribution and enables the investigator to determine if minority populations respond differently from the majority. However, this technique will also result in severe bias if the sample comes only from one segment of the population.

The tail probability can be estimated in the same way, in which a cumulative distribution function can be estimated. As given in Ref. 5, this function can be defined

$$\hat{f}(u) = \frac{1}{na_n} \sum_{i=1}^{n} K\left(\frac{u - X_i}{n}\right)$$

where K is the cumulative distribution function of the kernel,

$$K(u) = \int_{-\infty}^{u} K(u)\, du$$

and $X_1, X_2, \ldots, X_n$ is the collected sample. Thus, the tails of the probability distribution can be determined in detail. The value u can also define a vector of random variables, $u = (x, y, z, t)$. Then $f(u)$ is modified to equal (for the discrete variable y):

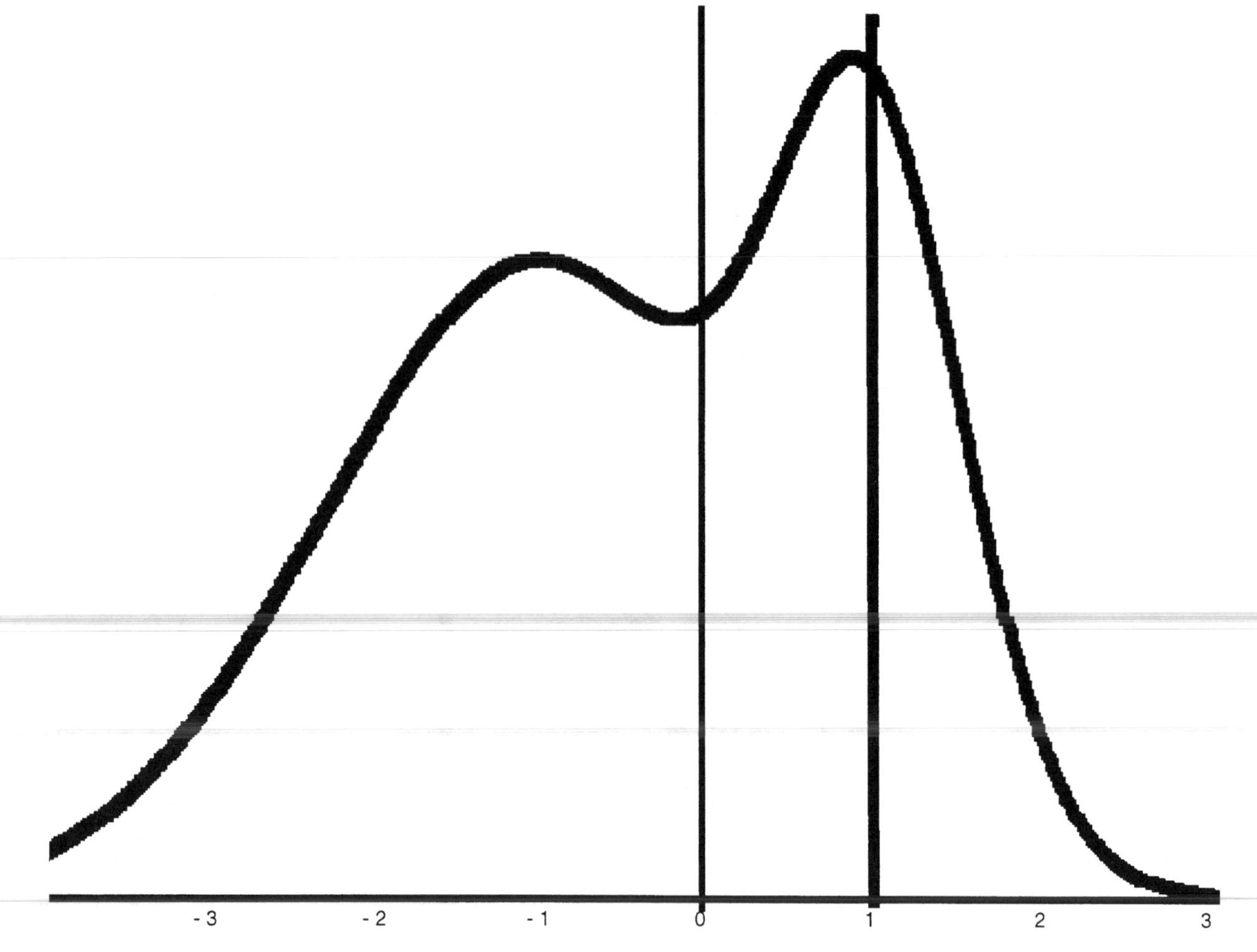

Fig. 5 Generalized representation of two subpopulations demonstrating that one subpopulation can redefine the norms for a second subpopulation under the assumption of homogeneity.

$$\hat{f}(x, y, z, t) = \frac{1}{na_n^3} \sum_{j=-\infty}^{\infty} W(\hat{S}_n, i, j) \cdot \sum_{1 \in C_n(j)} k\left(\frac{x - X_1}{a_n}, \frac{z - Z_1}{a_n}, \frac{t - T_1}{a_n}\right)$$

IV. PROBLEMS WITH SUBGROUP ANALYSIS

A. Finding Significance Where None Exists

There is a saying among statisticians that if you torture the data, they will confess. This is particularly true of subgroup analysis and retrospective studies. Therefore, any results from these types of analysis can be only preliminary, not conclusive. This falls under the general topic of data mining (6). Although a data-mining procedure can be used to identify a model or a variable of interest, only fresh data can be used to verify the significance of the model. The need for verification is clearly stated (7):

> The difficulties associated with direct reliance on data in a whole series are exacerbated if one attempts to pick out subclasses marked by strikingly good or bad results. The idea seems reasonable enough, but it ignores a somewhat subtle, inescapable fact: one can always expect to find some good-looking and some bad-looking subsets in any body of data, even when no bias has influenced any part of it. Furthermore, such differences among subsets can easily be large enough to look quite convincing to the unsophisticated analyst.

The pitfalls of subgroup analysis were very vividly presented by the ISIS-2 trial, where a statistically significant subgroup result on the prevention of fatal heart attack by ingesting aspirin was obtained by classifying subjects by astrological sign (8).

One way to avoid this problem is to preplan a small number of subgroup analyses by stratifying the data in the protocol. A second way is to perform the primary assessment to demonstrate statistical significance before investigating any subgroups for significance. As stated in Ref. 7: "[P]ost-hoc subgroup results should be viewed as hypothesis generating rather than proof." Unfortunately, published results do not generally indicate how many subgroup analyses were actually performed. The main way to avoid the problem is to regard any results of subgroup analysis as hypothesis generating, to be validated by the collection of a fresh data set to see if the results still hold.

Two recent papers (9,10) rely very heavily on subgroup analysis. In Ref. 9, no significant difference was found between treatment groups. Therefore, the significant difference found for subjects age 60 and younger should definitely be regarded as a hypothesis generated, with a need for verification by additional data collection. The paper went on to examine subgroups of the subgroup, all of which were highly significant. As discussed in Ref. 7: "[I]f n subjects are divided at random into k equal-sized subgroups, then the difference between the largest and the smallest subgroup means has an expected value that is approximately k times the standard error of the mean of the whole group." This alone demonstrates that such high p-values are to be expected. Although the paper does discuss the problem of subgroup analysis and the care taken to ensure that the results were not spurious, the choice of the age of 60 was made because this was the optimal value to find statistical significance.

In Ref. 10, the difference between treatment and control groups was not significant. As stated, "[A] retrospective survival analysis was performed in patient subsets that showed favorable responses to VMO treatment in the first interim analysis." It is clear that subgroups were chosen after the fact. Any difference in survival found retrospectively has to be tested with preplanned stratification before the result is considered valid. This paper does not discuss the problems of subgroup analysis. The paper does indicate that the study is not definitive and should be done with sufficient numbers of patients in all subgroups.

The problem of these types of subgroup analyses is clearly examined (11):

> The controversy began with the retrospective analysis of results by age subgroups, a method of evaluation unanticipated in the design of all but one of the trials; these trials were planned with marginal power that was sufficient only for the whole-group examination of the data. When used to interpret the end results of the trials, retrospective subgroup analysis reduces the power of the results and thus substantially increases the likelihood that an observed benefit will not achieve statistical significance. . . . [P]ost hoc data-dredging for subgroups of patients or subsets of events, however qualitatively dramatic their results, should only generate hypotheses for the next trial, not conclusions from this one."

Thus, subgroup analysis can create controversies within the medical community by stating conclusions that are only generated hypotheses. Further studies, often not performed, are needed to make the conjecture conclusive.

Because few people actually understand the statistical complications in the use of subgroup analysis, these should be recognized in any paper discussing a study where subgroup analysis was performed.

B. Insufficient Power

Before the data for any test are collected, there are four different parameters that must be considered (12):

- Type I, or α, error
- Type II, or β, error
- Effect size
- Sample size

These four variables are closely related, so three are specified by the investigator and the fourth is computed as a function of the other three.

1. *Type I Error*

The purpose of any statistical hypothesis test is to make an inference about a population by using a relatively small sample. The possibility of error always exists. Since it cannot be eliminated, it should be controlled. Type I error is defined to be the probability of being wrong if the investigator rejects H_0. Typically, $\alpha = .05$ is considered significant and $\alpha = .01$ is considered highly significant. Then the hypothesis test is stated:

H_0: treatment is not effective
H_1: treatment is effective

2. *Type II Error*

More important, Type II error should also be considered, although it is usually ignored. Type II error is defined to be the probability of being wrong if the investigator accepts H_0. Unfortunately, if the α error is low, the β error can be quite high. Therefore, consider the question posed earlier:

H_0: treatment is not effective
H_1: treatment is effective

If the value of β is not computed, then it is not possible to decide whether to accept H_0. Therefore, the best that can be decided is whether to reject H_0 or to fail to reject H_0. Failing to reject H_0 is often assumed equivalent to accepting H_0, although it is not. In fact, independent of the size of the sample, the maximum possible β error is always $1 - \alpha$. Thus if $\alpha = .05$, it is possible to have a β error of .95. Only by defining an effect size can α and β be reduced simultaneously.

In statistical terminology, the power of the test is defined to be $1 - \beta$. Typically, in a medical study α is considered acceptable at 0.05 and β is acceptable at a level of 0.20. In this case, the power of the test is stated as 80%. However, when subgroup analysis is performed, it is performed with only a subset of the data. In this case, the result will be of low power, and it will not be possible to conclude that the subgroup reacts differently from the rest of the population.

3. *Effect Size*

If a new treatment increases patient survival by 1 day, does it provide sufficient improvement to be worthwhile? Statistical significance cannot determine whether the size of the difference is important; that question must be answered by the investigator. The effect size is defined to be the point at which the difference between treatments becomes important. Once the effect size is defined, the hypothesis can be rewritten (when limited to two treatments):

H_0: $u_1 = u_2$

H_1: $u_1 - u_2 > u'$ or $u_2 - u_1 > u'$

where u' is the effect size. Once a specified effect size is given, a sample size can be computed so that both the α and β errors remain small.

4. *Sample Size*

The smaller the effect size, the larger the needed sample size to keep both α and β under control. If the effect size is too small, it is possible that the required sample size is too large for an experiment to be conducted except for prohibitive cost. On the other hand, if the sample size is limited by cost, then the effect size may be so large that H_0 can be rejected only in such extreme situations that it would be virtually obvious without any statistical analysis.

Therefore, effect size is usually identified as a percentage of the population variability, with the corresponding sample size computed for small, medium, and large effect sizes. Once this has been done, the null hypothesis can be either accepted or rejected depending upon the outcome of the collected data.

If the effect size is not considered prior to the start of an experiment, it is often the case that the effect size is so large as to render the experiment meaningless. For example, in a study of 71 randomized trials it was found that 50 of them had less than 90% power of detecting a 50% improvement. Thus, if a surgical procedure were to result in the survival of 50% more of the patients, the hypothesis test would not detect this as statistically significant (7).

The main statistical techniques used to avoid the problem of small sample size is the use of meta-analysis. The results from different studies are combined to increase the power so that the conclusions will be definitive. However, when the results from meta-analyses were compared to the results of subsequent large trials, it was found that they differed over 30% of the time. Another method more recently employed has been to set up national depositories (13). Data from all parts of the country are collected to examine. The data lack quality control and randomization and so the results must be viewed as tentative. A test for equivalency remains a serious problem. The best solution might be to make the test on the overall population and then to examine various subpopulations by estimating outcomes rather than to use inferential techniques.

V. EXAMPLES FROM THE MEDICAL LITERATURE

A. Comparison of Subgroups Within the Population

Reference 14 examined a subgroup of the general population, sampling 127 African Americans. They were divided into two experimental and one control group. Of the 127, 16 were lost to follow-up. Using a 5% level of significance and 80% power, this sample of 111 is sufficient to detect a difference of 20% between groups (12).

B. Identifying Subgroups to Investigate

These studies are generally conducted retrospectively. Subgroups are identified and tested simultaneously. The results of such studies cannot be conclusive; all must be examined by using fresh data. Reference 15 provides a classic example of subgroup analysis. There were 1395 patients originally recruited into the study. Of the initial sample, 262 patients were identified as having mild to moderate congestive heart failure. This subgroup of patients was followed for 1 year. Using power analysis for log-rank statistics, with a 5% significance level and 80% power, this patient sample is sufficiently large to detect a 15% difference in mortality (16).

This is in contrast to Ref. 17, which examined the relationship between aspirin and enalapril by retrospectively adding a variable to the analysis. Although the title indicates that this is a subgroup analysis, the analysis was conducted over the entire patient base.

The hypothesis generating of subgroup analysis is clearly demonstrated in Ref. 18. As stated in the paper:

> After the minimum hematocrit during cardiopulmonary bypass had been determined to be an independent risk factor for mortality, it was necessary to perform a cutoff point analysis to establish a specific hematocrit level at which there is a statistically significant difference in postoperative mortality. The cutoff point analysis was achieved by dividing the entire patient population into dichotomous groups using different minimum hematocrit levels during cardiopulmonary bypass as a cutoff point. All of these different variables were then entered into a multiple logistic regression model that contained the other statistically significant preoperative risk factors for risk adjustment. This cutoff point analysis is able to pinpoint a specific hematocrit level at which there is a statistically significant difference in mortality after adjusting for other preoperative risk factors.
>
> After a specific hematocrit level was determined to be statistically significant for mortality, a subgroup analysis was then performed to distinguish the difference between high-risk and low-risk populations. This subgroup analysis was performed by dividing the entire patient population into high-risk and low-risk subgroups. The high-risk population consists of patients with at least one statistically significant preoperative risk factor for mortality, and the low-risk population consists of patients without any statistically significant risk factors for mortality. Minimum hematocrit level achieved during cardiopulmonary bypass was again examined as an independent risk factor for mortality in each of the subgroups. Cutoff point analysis was again performed.

Cutpoint analysis as just discussed is an attempt to find the optimal hypothesis. Unfortunately, the study did not yield a conclusive result by testing an additional set of subjects. Therefore, any conclusions in the paper must be tentative.

All three of these examples have to deal with the issues of finding significant subgroups where none exist. All should be regarded as conjectures in need of validation.

C. Mammography Screening

One of the more controversial of the subgroup analyses deals with the mammography screening of women ages 40–49. The major studies conducted were not specifically designed to address the question; a cutpoint was considered at age 50. Consider two such studies (19) (Table 1). Note first the extremely large number of observations in both studies compared to the small number of deaths. A subgroup analysis reduces the number of deaths still further. The HIP study did find a significant difference in mortality for women ages 40–49, demonstrating a benefit to mammography; the SNB study did not. As a result of these studies, Ref. 20 made the following conclusion:

> Decisions must be made concerning the diversion of resources that would be required to implement mass screening for breast cancer. On the positive side of the balance sheet, there is a marginal reduction in deaths from breast cancer in older women. On the negative side are the financial cost and the induced morbidity. Negative factors also include the false-positive results leading to unnecessary operations, the false-negative results that lead to inappropriate reassurance, the raised level of anxiety in the female population, and the tiny but real risk of radiation-induced cancers.
>
> The calculations presented in this study, with only inappropriate surgical intervention used as the index of harm, demonstrate a harm/benefit ratio of between 21:1 and 62:1. Enough data are now available to suggest that the American Cancer Society recommendations should be ignored in view of the net harm caused by screening for breast cancer.

Contrast this conclusion with the one given in Ref. 21:

Table 1 Results of Two Major Mammography Studies

	HIP 10-year mortality study		SNB 7-year mortality study	
Value	Number	Percent	Number	Percent
Death from breast cancer				
Number of women screened	31,888	100	78,085	100
Screened	146	0.458	87	0.111
Control	192	0.602	125	0.160
Reduction in death from breast cancer	46 (24% relative)	0.144	38 (30% relative)	0.049
Death from all causes				
Screened	2,235	7.009	4,846	6.206
Control	2,116	6.636	4,787	6.131
Increase in death from all causes	119 (6% relative)	0.373	59 (1% relative)	0.076

HIP: Health Insurance Plan of New York; SNB: Swedish National Board of Health and Welfare.

The scientific limitations of retrospective subgroup analysis notwithstanding, results show consistent point estimates of mortality reduction in women who are screened compared with controls for women aged 50 and older (average mortality reduction of approximately 30%; statistically significant mortality reduction in several individual trials), probably because of the much larger number of older women in the trials. For women aged 40 to 49 years, most point estimates show mortality reductions ranging from 22% to 49%, but three trials show no benefit and none of the trials individually shows statistically significant results.

Thousands of lives are probably lost each year because women are not being screened. We believe that it is much more prudent to endorse mammographic screening now, risking the unlikely subsequent determination that the effort was ineffective, than to withhold screening until it is determined whether "proof" of efficacy will be obtained, risking the loss of so many lives.

A meta-analysis was performed on the eight major studies of mammography (22) (Table 2). The study concludes:

> The decision whether or not to screen this age group will depend on subjective views of the importance of the benefit and the ability to provide the resources involved. It seems clear, however, that while the size and timing of the mortality reduction require further research, the existence of such a reduction is no longer in question.

This confirms the result of a similar meta-analysis (23).

Table 2 Meta-Analysis of Mammography Studies

Trial	Years of follow-up	Study group deaths	Study group person-years	Control group deaths	Control group person-years	Relative mortality	95% confidence interval
2-county	15	45	264,059	39	207,725	0.91	0.59–1.39
Malmo	15	15	61,000	23	62,000	0.66	0.34–1.27
Stockholm	12	25	173,866	12	87,826	1.05	0.53–2.09
Gothenburg	10	19	106,000	37	129,000	0.62	0.36–1.08
Edinburgh	10	25	97,206	31	88,766	0.73	0.43–1.25
HIP	18	49	248,454	65	253,085	0.77	0.53–1.11
NBSS	10	73	252,060	66	251,814	1.10	0.78–1.54
Turku	7	3	48,068	9	41,532	0.29	0.07–1.08

VI. CONCLUSIONS

Subgroup analysis can and should be performed when examining the results of both clinical trials and observational studies. Some of the subgroups within the population are obvious and should be collected in the course of the study; however, others are more subtle and often are examined because of the trial outcomes. Since subgroup analysis can suffer from lack of statistical power and also from too many subgroup investigations, the results should be considered tentative pending further studies. These studies can take the form of large trials with sufficient power to be conclusive or of meta-analysis. Meta-analysis and large trials disagree in outcome approximately 30% of the time. Risks, benefits, and costs should all be examined when taking the outcomes of subgroup analysis to apply to clinical practice.

REFERENCES

1. Ness RB, Nelson DB, Kumanyika SK, Grisso JA. Evaluating minority recruitment into clinical studies: how good are the data? Ann Epidemiol 7:472–478, 1997.
2. Caraco Y, Lagerstrom PO, Wood AJJ. Ethnic and genetic determinants of omeprazole disposition and effect. Clin Pharm Therap 60:157–167, 1996.
3. Freeman HP. The meaning of race in science-considerations for cancer research. Cancer 82:219–225, 1998.
4. McLachlan GJ, Basford KE. Mixture models. Inference and Applications to Clustering. Statistics: Textbooks and Monographs 84. New York: Marcel Dekker, 1988.
5. Silverman BW. Density Estimation for Statistics and Data Analysis. London: Chapman and Hall, 1986.
6. Fayyad U, Piatetsky-Shapiro G, Smyth P. The KDD process for extracting useful knowledge from volumes of data. Communications ACM 39:27–34, 1996.
7. Moses LE. The series of consecutive cases as a device for assessing outcomes of intervention. In: Bailar JC, Mosteller F, eds. Medical Uses of Statistics. Waltham, MA: Massachusetts Medical Society, 1986.
8. ISIS-2 (Second International Study of Infarct Survival) Collaborative Group. Randomized trial of intravenous streptokinase, oral aspirin, both, or neither among 17,187 cases of suspected acute myocardial infarction: ISIS-2. Lancet 2:349–360, 1988.
9. Balch CM, et al. Efficacy of an elective regional lymph node dissection of 1 to 4 mm thick melanomas for patients 60 years of age and younger. Ann Surg 224:255–266, 1996.
10. Wallack MK, et al. Increased survival of patients treated with a vaccinia melanoma oncolysate vaccine. Ann Surg 226:198–206, 1997.
11. Sickles EA, Kopans DB. Mammographic screening for women aged 40 to 49 years: the primary care practitioner's dilemma. Ann Intern Med 122:534–538, 1995.
12. Cohen J. Statistical Power Analysis for the Behavioral Sciences. 2nd ed. Hillsdale, NJ: Lawrence Erlbaum Associates, 1988.
13. Shroyer AL, Edwuards FH, Grover FL. Updates to the data quality review program: the Society of Thoracic Surgeons adult cardiac national database. Ann Thorac Surg 65:1494–1497, 1998.
14. Alexander CN, Schneider RH, et al. Trial of stress reduction for hypertension in older African Americans: II. sex and risk subgroup analysis. Hypertension 28: 228–237, 1996.
15. Herlitz ZJ, Waagstein F, et al. Effect of metoprolol on the prognosis for patients with suspected acute myocardial infarction and indirect signs of congestive heart failure (a subgroup analysis of the Goteborg Metoprolol trial). Am J Cardiol 80:40J–44J, 1997.
16. Freedman LS. Tables of the number of patients required in clinical trials using the log-rank test. Stat Med 1: 121–129, 1982.
17. Nguyen KN, Aursnes I, Kjekshus J. Interaction between Enalapril and aspirin on mortality after acute myocardial infarction: subgroup analysis of the cooperative new Scandinavian Enalapril Survival Study II (Consensus II). Am J Cardiol 79:115–119, 1997.
18. Fang WC, Helm RE, et al. Impact of minimum hematocrit during cardiopulmonary bypass on mortality in patients undergoing coronary artery surgery. Circulation 96(suppl II):II-194–II-199, 1997.
19. Wright CJ. Breast cancer screening: a different look at the evidence. Surgery 100:594–598, 1986.
20. Tabar L, Fagerbert CJ, et al. Reduction in mortality from breast cancer after mass screening with mammography. Lancet 1:829–832, 1985.
21. Falun Meeting. Breast-cancer screening with mammography in women aged 40–49 years. Report of the organizing committee and collaborators. Int J Cancer 68: 693–699, 1996.
22. Smart CR, Hendrick RE, Rutledge JH III, Smith RA. Benefit of mammograph screening in women ages 40 to 49 years. Cancer 75:1619–1626, 1995.
23. Feig SA. Mammographic screening of women aged 40–49 years. Benefit, risk and cost considerations. Cancer 76(suppl):2097–2106, 1995.

Patricia B. Cerrito

Surrogate Endpoint

See also *Bioavailability and Bioequivalence*

I. INTRODUCTION

In the development of a pharmaceutical product, a crucial step is to conduct clinical trials demonstrating the efficacy and safety of this product. The measurement used to evaluate the effect of the treatment is termed an *endpoint*. Endpoints used to measure clinical benefit, such as prolongation of survival time, reduction of mortality rate, are usually considered "true" endpoints, since clinical benefit is the ultimate purpose of a medical treatment. However, the feasibility and practicality of measuring a "true" endpoint may be limited, due to various reasons. Therefore, it is crucial to have some alternative measurement that would allow early assessment of a potential treatment effect.

The most important reason that limits the use of a "true" endpoint is the time and cost needed to conduct clinical trials. For example, time to death from the time of enrollment in a clinical trial is frequently used as an endpoint to evaluate a treatment for a life-threatening disease such as cancer or AIDS. However, to use survival time as the endpoint may require several hundred patients and several years of follow-up. During the process of developing a new drug, without evidence that the new product may be effective, it is unlikely a sponsor will be willing to devote a large resource to a clinical trial that will take years to demonstrate a candidate's therapeutic viability. The need to reduce the duration of clinical trials becomes more urgent in the search for treatment for life-threatening diseases.

Other factors may also limit the use of a "true" endpoint. For example, a new treatment may be developed during the trial period, making the treatment under investigation obsolete. Additionally, competing risk, i.e., death due to other reasons during the trial, constitutes another problem for the evaluation of the treatment effect of the drug. These problems become more serious with longer follow-up times. Finally, the use of a "true" endpoint may also be limited by current knowledge and technology for measuring the endpoint. One example is pain, which cannot be measured directly and objectively due to the current limited knowledge about pain. Since in most cases objective measurements proven to be sensitive to the "true" degree of pain either are difficult to obtain or do not exist, most clinical trials rely on patient self-assessment of pain as the endpoint, which can be viewed as the "surrogate" for the "true" severity of pain (1).

The limitations of "true" endpoints prompted the use of *surrogate endpoints*. A surrogate endpoint "is an endpoint that is intended to relate to a clinically important outcome but does not in itself measure a clinical benefit" (2).

A surrogate endpoint is sometimes referred to as an *intermediate endpoint* (3,4) or a *surrogate marker* (5–7). A *marker* usually refers to some laboratory test result that is associated with a certain disease, usually indicating the progress of the disease. For example, significant elevations of SGOT and SGPT are strong indications for hepatitis or other liver damage, so they are considered markers for hepatitis. Similarly, an increase in HIV viral load usually indicates a worsened condition for AIDS patients, so HIV viral load is a marker for the progression of AIDS. A marker can be used to detect diseases, monitor disease progression, or predict clinical outcome, and because of its strong association with the clinical endpoint it can also be used as a surrogate for the "true" endpoint. The focus of this discussion will be on the use of a surrogate marker as an endpoint, and the terms *surrogate endpoint* and *surrogate marker* may be used interchangeably. Since a change in a surrogate endpoint occurs prior to a change in a "true" endpoint—for instance, an increase in HIV viral load occurs before the onset of AIDS—a surrogate endpoint may also be referred to as an *intermediate endpoint*.

The use of surrogate endpoints can significantly reduce the follow-up time and sample size required to demonstrate treatment effect in a clinical trial. As an example, Table 1 (8) illustrates the reduction in time and cost resulting from using surrogate endpoints in clinical trials for the treatment of cardiovascular diseases.

Surrogate endpoints play a particularly important role in the early development of a pharmaceutical product, when surrogate endpoints can be used to demonstrate if the drug has any pharmacological effect, as postulated by the in vitro model or in preclinical in vivo results. Additionally, surrogate endpoints can provide guidance in selecting a dose range for later confirmatory trials.

It has been suggested that the "true" endpoint usually measures clinical benefit, and the surrogate end-

Table 1 Typical Cardiovascular Trials with Surrogate and True Endpoints: Sample Sizes and Follow-up Periods

	True endpoint trial			Surrogate endpoint trial		
Event	Endpoint	Size	Length	Endpoint	Size	Length
Myocardial infarction	Death	4,000	5 yr	Coronary artery patency	200	90 min
Myocardial infarction	Death	4,000	5 yr	Ejection fraction	30	2–4 wk
Stroke	Stroke	25,000	5 yr	Diastolic blood pressure	200	1–2 yr

Source: Ref. 8.

point measures mostly a specific aspect of disease progression (8). For a given disease, there can be different stages and different biological mechanisms affecting the progress of the disease. As a result, for a specific disease the surrogate endpoint may change, depending upon the stage of the disease as well as the specific physiopathological mechanism that the drug is targeting. For instance, both serum cholesterol level and diastolic blood pressure are widely used surrogate endpoints for coronary heart disease. To decide which of these two is appropriate to use will depend on the biological action of the drug under study. Table 2 (9) presents surrogate endpoints used in developing an antidiabetes drug. It can be seen that the choice of a surrogate endpoint depends upon the stage of diabetes as well as on the stage of the drug development, i.e., whether it is an early- or late-phase clinical trial. Table 2 shows that the sample size and treatment duration also change according to the stage of drug development. Early-phase trials tend to be shorter and use fewer people as compared with late-phase trials.

A potential surrogate is identified in several ways (8). The core for identifying a surrogate endpoint is a strong biological rationale. For example, any effective treatment for cancer must have some effect on the tumor itself, which suggests that tumor shrinkage might

Table 2 Clinical Endpoints vs. Surrogate Endpoints in Different Phases of Clinical Trials for Diabetes

Clinical endpoint	Phase	Subject characteristics	Group sample size	Treatment duration	Surrogate endpoints
Acute glucose lowering	I	Normals or diabetics	6–12	≥single dose	♦ Plasma glucose ♦ Plasma insulin ♦ Other (e.g., insulin clamp)
Improved glycemic control	I	Diabetics	12–36	2–12 wk	♦ Plasma glucose ➪ fasting ➪ 2 h postprandial ➪ 24 h profile
	III	Diabetics	40–200	6–12 mo	♦ Glycohemoglobin (HBA_1) ♦ Glycated serum proteins ♦ Hypoglycemic episodes
Complications: anatomic/physiologic benefit	II–III	Diabetics with early complications	50–300	6 mo–2 yr	♦ Nerve conduction velocity ♦ Nerve/kidney biopsy ♦ Albuminuria, a GFR ♦ Funduscopy
Complications: actual clinical benefit	IV	Diabetics with early or no complications	200–800	≥3–5 yr	♦ Neuropathic pain ♦ Sensorimotor deficit ♦ GFR, renal failure ♦ Visual acuity ♦ Direct and indirect costs

Source: Ref. 9.

be used as a surrogate endpoint in studies of cancer treatment. Epidemiological studies can also reveal endpoints that have strong predictive values to the clinical outcome. For instance, in many observational studies for AIDS, CD4 lymphocyte counts were shown to be the most powerful predictor of disease progression and subsequent survival (10–13). This led to the extensive use of CD4 counts as a surrogate endpoint in early AIDS research. Lastly, clinical trials may also indicate a potential surrogate endpoint. In a clinical trial, if a treatment that suppresses a given marker produces more clinical benefit than another treatment that does not suppress the marker, that marker may be used as a potential surrogate endpoint. However, the use of the surrogate endpoints identified through these ways needs to be validated through further studies.

II. SCIENTIFIC ISSUES—LIMITATIONS OF USING SURROGATE ENDPOINT

Although yielding potentially important benefits, the use of surrogate endpoints can sometimes introduce ambiguous results and thus generate controversy:

> There are two principal concerns with the introduction of any proposed surrogate variable. First, it may not be a true predictor of the clinical outcome of interest. For example, it may measure treatment activity associated with one specific pharmacological mechanism, but may not provide full information on the range of actions and ultimate effects of the treatment, whether positive or negative. There have been many instances where treatments showing a highly positive effect on a proposed surrogate have ultimately been shown to be detrimental to the subjects' clinical outcome; conversely, there are cases of treatments conferring clinical benefit without measurable impact on proposed surrogates (14).

Some cancer treatments provide an example of a treatment that has a positive effect on the surrogate endpoint but a detrimental effect on a patient's health. In this instance, although a treatment may achieve a certain degree of tumor shrinkage, the toxicity of the drug itself may in fact worsen a patient's general health and thus shorten survival time. Conversely, in AIDS trials it has been shown that many treatments with a positive effect on survival time have null or even a negative effect on CD4 count, the proposed surrogate (5). A second concern relating to the use of surrogate endpoints is that "the proposed surrogate variables may not yield a quantitative measure of clinical benefit that can be weighed directly against adverse effects" (14).

III. REGULATORY REQUIREMENTS

As just mentioned, the use of surrogate endpoints has limitations. However, in the case of life-threatening diseases, where the availability of a new treatment may save many lives, there is a strong need to reduce the duration of a clinical trial and to accelerate drug approval, and very often using a surrogate endpoint is necessary to evaluate a treatment more quickly.

The Food and Drug Administration (FDA) issued guidelines for accelerated approval based on surrogate endpoints. The guideline states:

> FDA may grant marketing approval on the basis of adequate and well-controlled clinical trials establishing that the drug product has an effect on a surrogate endpoint that is reasonably likely, based on epidemiologic, therapeutic, pathophysiologic, or other evidence to predict clinical benefit or on the basis of an effect on a clinical endpoint other than survival or irreversible morbidity. Approval under this section will be subject to the requirement that the applicant study the drug further, to verify and describe its clinical benefit, where there is uncertainty as to the relation of the surrogate endpoint to clinical benefit or of the observed clinical benefit to ultimate outcome (15).

IV. STATISTICAL METHODS

In the effort to identify reliable surrogate endpoints, several statistical methods have been proposed. In 1987, Prentice (16) introduced the following statistical definition of a surrogate endpoint as "a response variable for which a test of the null hypothesis of no relationship to the treatment groups under comparison is also a valid test of the corresponding null hypotheses based on the true endpoint." Based on this definition, he proposed the use of the following two conditions as the operational criteria to determine a surrogate:

$$\lambda_T\{t;S(t),x\} = \lambda_T\{t;S(t)\} \tag{1}$$

and

$$\lambda_T\{t;S(t)\} \neq \lambda_T(t) \tag{2}$$

where t denotes time from enrollment in a clinical trial, T denotes a true time-to-failure endpoint, $x = (x_1, \ldots, x_p)$ consists of indicator variates for p (≥ 1) of the $p + 1$

treatments to be compared in respect to the corresponding (instantaneous) failure rate, $\lambda_T\{t;x\}$, and $S(t) = \{Z(u);0 \leq u < t\}$ denotes the history prior to time t of a possibly vector-valued stochastic process $Z(U) = \{Z_1(u),Z_2(u), \ldots\}$, which may be used to specify a surrogate for T.

Criterion (1) requires that conditioning on the surrogate marker process, the treatment is independent of the failure rate for T. Thus, the surrogate marker captures all the treatment effect of x on T. Criterion (2) requires that the surrogate marker has some predictive value to T. Recall that in Sec. 1 it was noted that a surrogate endpoint must have strong predictive value to the clinical outcome. The Prentice criterion of Eq. (2) is equivalent to this requirement. Among the two criteria, Eq. (1) is considered the principal criterion (17).

Although expressed in terms of survival analysis, these criteria can be interpreted in a general way. Feigal (18), from the FDA, expressed the following heuristic model for a drug effect:

Outcome = Placebo rate + Drug effect

In the instance of a surrogate marker, he further decomposed the model as

Outcome = Placebo rate + SM drug effects
+ Other drug effects

where SM is the abbreviation of surrogate marker. He pointed out that the Prentice criterion of Eq. (1) requires essentially that conditioning on the surrogate marker, the "other drug effects" reduces to 0.

This Prentice criterion is very strict. In 1992, Freedman and Graubard (3) proposed another criterion for validating a surrogate, which is less stringent than that of Prentice. The development of Freedman and Graubard's criterion is described briefly next.

With a slight change in notation, let T be a binary outcome, X indicate the treatment groups, and S denote the surrogate endpoint. Then the Prentice criteria can be rewritten as

$$P(T = 1|S,X) = P(T = 1|S) \tag{3}$$

and

$$P(T = 1|S) \neq P(T = 1). \tag{4}$$

For simplicity, assume there are two treatment groups ($X = 1$ and $X = 2$) and that S is discrete and takes k values $s_1, \ldots ,s_k$. Let $p(T = 1|S = s_i,X = j) = p_{ij}$, then Eq. (3) is equivalent to

$$p_{i1} = p_{i2} \qquad (i = 1, \ldots ,k) \tag{5}$$

Consider the following model:

$$g(p(T = 1|S,X = j)) = h(S) + \tau_j \tag{6}$$

where S and X have an additive effect on some function (g) of the probability of disease, $h(S)$ is some function of S, and τ_j is the jth treatment effect. If we further choose logit function for g and assume that

$$h(S) = \mu + \sigma_i$$

where μ represents the overall mean, then Eq. (6) becomes the usual linear logistic model with the constraint of $\sigma_1 + \cdots + \sigma_k = 0$ and $\tau_1 + \tau_2 = 0$, and it can be further expanded to include other effects. Consequently, the test of the null hypothesis [Eq. (5)] becomes the test of $\tau_1 = \tau_2 = 0$. If this hypothesis is rejected, S cannot be used as a surrogate endpoint. However, lack of significance in this test does not warrant S to be a valid surrogate endpoint. To quantify further the effect of S, consider the quantity

$$1 - \frac{\hat{\tau}_{1a}}{\hat{\tau}_1}$$

where $\hat{\tau}_{1a}$ is the adjusted (by S) treatment effect in group 1 and $\hat{\tau}_1$ is the unadjusted treatment effect in group 1. This quantity is the estimate of the proportion of the drug effect that is explained by the surrogate endpoint. Under the criteria of Prentice, this proportion is expected to equal 1. However, we may allow this proportion to be larger than some critical values, such as 0.5, 0.75, but less than 1. In other words, we may allow the surrogate endpoint to explain a large portion of the drug effect but not all of the effect. Using the notation in the heuristic model, Feigal (18) presented Freedman's (3) model as

$$\text{Proportion of drug effect attributed to SM drug effects} = \frac{\text{SM drug effects}}{\text{Total drug effects}}$$

V. RECENT DEVELOPMENTS

The use of surrogate endpoints in clinical trials is an important issue, since they can significantly reduce the duration and cost of clinical trials and thereby accelerate the development of medical treatments. Surrogate endpoints had long been used to assess treatment effects, especially for rare and life-threatening diseases. However, there is some controversy surrounding their use. In 1987, a meting was held to discuss these issues (1,8,16,19,20). In this meeting, Prentice proposed statistical operational criteria to identify surrogate endpoints (16). Since then, Prentice's criteria, as described earlier, have been applied by several researchers and have been used to evaluate several commonly used surrogate markers (17,21). In 1992, the FDA issued regulations for accelerated approval based on surrogate endpoints for new drugs and biological products treating serious or life-threatening illness (22). In the same year, Freedman and Graubard (3) proposed the modification to Prentice's criteria, making the criteria less stringent. These criteria have also been applied to biomedical research (6). In addition, other methods, such as using meta-analysis to identify surrogate endpoints, have also been proposed (23).

In recent years, the FDA has issued several guidelines prepared under the auspices of the International Conference on Harmonization of Technical Requirements for Registration of pharmaceuticals for Human Use (ICH) (2,14,15). The use of surrogate endpoints was discussed in the "general considerations for clinical trials" (2), and was specifically addressed in "guideline on statistical principles for clinical trials" (14). The FDA later established a "surrogate marker methodology group" to evaluate each potential surrogate endpoint (18). The FDA's guideline for each disease also includes the use of a specific marker as a surrogate endpoint for that disease.

In addition to the use of a surrogate endpoint in place of the true endpoint for the efficacy analysis, different methods have been proposed in which a surrogate endpoint was used to improve the efficiency of the estimate of the treatment effect on the true endpoint. Flandre and O'Quigley (24) proposed a two-stage approach. In the first stage, the relationship between the true and the surrogate endpoints, both measured for all subjects, is examined. This relationship is then applied in the second stage to other subjects, where the follow-up will be terminated once the surrogate endpoint is observed. Lefkopoulou and Zelen (4) proposed a test to examine the impact of an intermediate endpoint on the true endpoint, which is the survival time. Cox (25) proposed a method using a surrogate endpoint to recover information and predict the survival distribution in the case of right censoring. This idea was further extended by Fleming et al. (26), and they referred to the endpoint that provides additional information on the true endpoint as an *auxiliary endpoint* [referred to as a surrogate endpoint in the paper by Cox (25)] and distinguished it from the surrogate endpoint. Kosorok and Fleming (27) also proposed a method using surrogate failure-time data to increase efficiency in estimating the treatment effect on the true endpoint.

As medical science advances, human beings are revealing the mystery of many complicated diseases, which may take decades to evolve, such as Alzheimer's disease. In the process of studying these diseases, the use of surrogate endpoints is almost inevitable. With the help of the development of medical technologies, such as the study of genes and protein structure, as well as the development of statistical methods, more reliable surrogate endpoints will be identified to enable humankind to understand an increasing number of diseases. Meanwhile, one should keep in mind the limitation of surrogate endpoints, and use them with caution.

REFERENCES

1. S Ellenberg, JM Hamilton. Stat Med 8:405–413, 1989.
2. Federal Register 62:242, §3.2.2.4, 1997.
3. LS Freedman, BI Graubard. Stat Med 11:167–178, 1992.
4. M Lefkopoulou, M Zelen. Lifetime Data Anal 1:73–85, 1995.
5. TR Fleming. Stat Med 13:1423–1435, 1994.
6. DY Lin, TR Fleming, V DeGruttola. Stat Med 16: 1515–1527, 1997.
7. S Choi, SW Lagakos, RT Schooley, PA Volberding. Ann Intern Med 118:674–680, 1993.
8. J Wittes, E Lakatos. Stat Med 8:414–425, 1989.
9. D MacLean. Diabetes mellitus: therapeutic approaches and issues in drug development. Seminar, Pfizer Central Research, Groton, CT, 1997.
10. RR Redfield, DS Burke. Sci Am 259:90–98, 1988.
11. BF Polk, R Fox, R Brookmeyer, S Kanchanaraksa, R Kaslow, B Visscher, C Rinaldo, J Phair. N Engl J Med 316:62–66, 1987.
12. JJ Goedert, RJ Biggar, M Melbye, DL Mann, S Wilson, MH Gail, RJ Grossman, RA DiGivia, WC Sanchez, SH Weiss, WA Blattner. JAMA 257:331–334, 1987.
13. B Safai, KG Johnson, PL Myskowski, B Koziner, SY Yang, S Cunningham-Rundles, JH Godbold, B Dupont. Ann Intern Med 103:744–750, 1985.
14. ICH Harmonized Tripartite Guideline, Statistical Principles for Clinical Trials, §2.2.6, 1998.
15. Code of Federal Regulations 21, §314.510, 1997.
16. RL Prentice. Stat Med 8:431–440, 1989.
17. TR Fleming, RL Prentice, MS Pepe, D Glidden. Stat Med 13:955–968, 1994.
18. DW Feigal Jr. PK/PD Modeling in regulatory decisions for effectiveness. Modeling and Simulation of Clinical Trials in Drug Development and Regulation, Herndon, VA, 1977.
19. J Herson. Stat Med 8:403–404, 1989.
20. A Hillis, D Seigel. Stat Med 8:427–430, 1989.

21. DY Lin, MA Fischl, DA Schoenfeld. Stat Med 12:835–842, 1993.
22. Federal Register 57:239, 1992, pp 58942.
23. MJ Daniels, MD Hughes. Stat Med 16:1965–1982, 1997.
24. P Flandre, J O'Quigley. Biometrics 51:969–976, 1995.
25. DR Cox. J R Statist Soc B 45:391–393, 1983.
26. TR Fleming, RL Prentice, MS Pepe, D Glidden. Stat Med 13:955–968, 1994.
27. MR Kosorok, TR Fleming. Biometrika 80:823–833, 1993.

Shu Zhang
Edward F. C. Pun

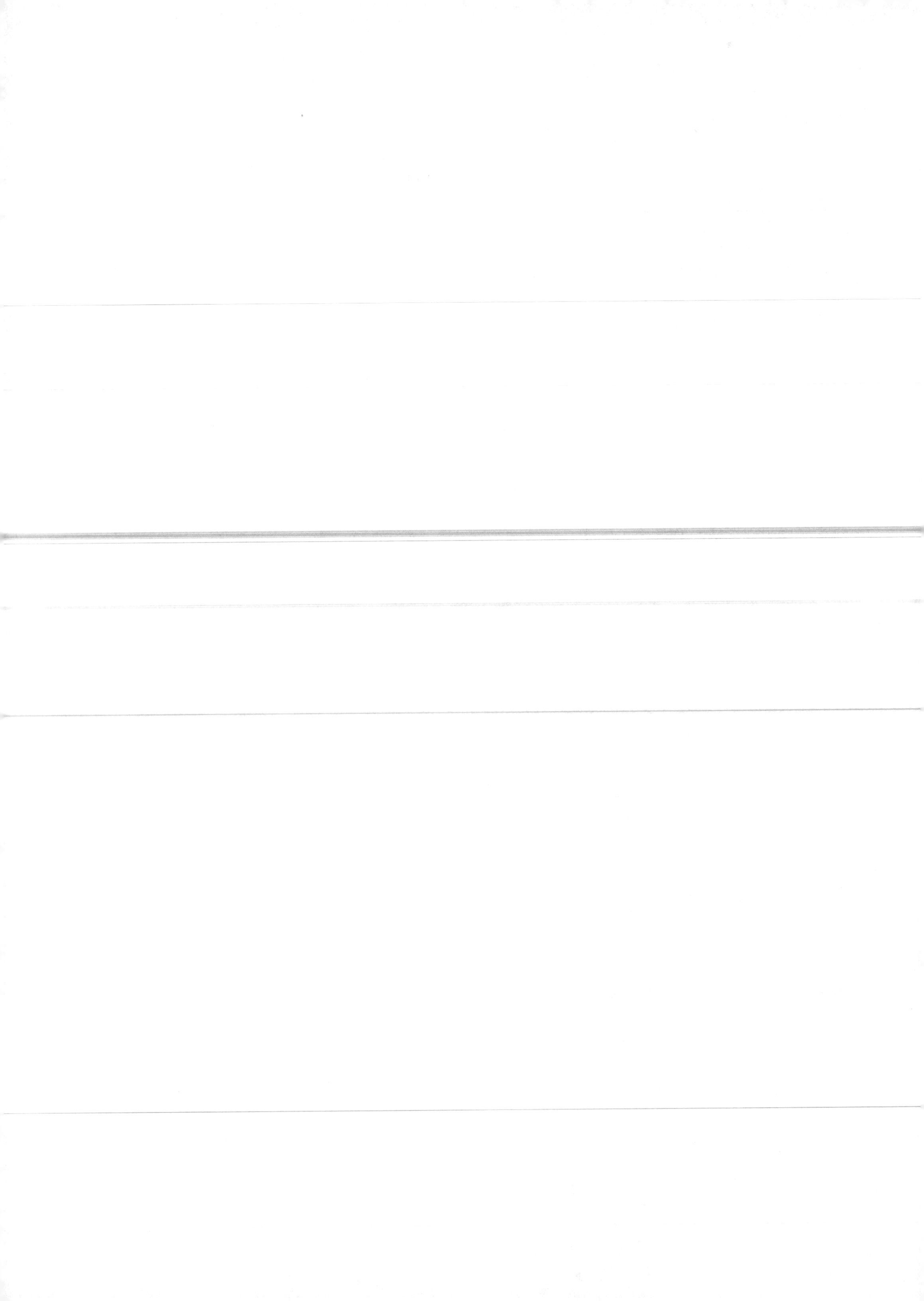

T–U

Therapeutic Equivalence

See also *Equivalence Trials; Bioavailability and Bioequivalence; Individual Bioequivalence*

I. INTRODUCTION

Traditionally, most of the trials conducted by the pharmaceutical industry during phase II and III stages are to obtain the necessary evidence for establishing the effectiveness and safety of the new therapeutic entity under development. Consequently, these trials are in general the *superiority trials*, in which the primary goal is to demonstrate its better efficacy against the concurrent placebo control. However, as innovative drugs become available for the treatment of a number of diseases, attention has focused on the search for new therapeutic modalities to compete with the standard products on the market. These new products may offer some specific advantages over the standard drug, which include a better safety profile, an easy administration route, a short duration of treatment, and, most important, a reduction of cost. As a result, because of these additional benefits the new drug is required not to show its superior efficacy over the standard drug but to prove that its effectiveness is no worse than that of the standard. In other words, the sponsors have to provide the evidence that the test product is at least as efficacious as the standard. This concept of demonstration of no worse efficacy but with some specific added values provided by the new treatment is referred to as *therapeutic equivalence*. As a result, studies with objectives for therapeutic equivalence are then called the *noninferiority trials*. Because therapeutic equivalence requires establishing only that the efficacy of the new treatment is no worse than that of the standard, it is referred to as *one-sided equivalence*.

On the other hand, after the patent of an innovative product expires, other pharmaceutical companies can manufacture generic copies under the Drug Price Competition and Patent Term Restoration Act passed in 1984 through the abbreviated new drug application (ANDA). The generic copies must be demonstrated to be equivalent to the innovative product based on the pharmacokinetic (PK) responses such as area under plasma concentration–time curve (AUC) or peak concentration (C_{max}) obtained from bioequivalence studies (Chow and Liu, 1992). According to the current requirement (FDA, 1992), a generic copy is claimed to be bioequivalent to the market reference product if the average bioavailability of the generic product is neither too high nor too low as compared to that of the reference product. As a result, bioequivalence emphasizes *two-sided equivalence*. Bioequivalence includes average bioequivalence, population bioequivalence, and individual bioequivalence, which are reviewed and discussed in detail in the entry on *Equivalence Trials*.

II. ONE-SIDED EQUIVALENCE AND NONINFERIORITY TRIALS

Let μ_T and μ_S be the average efficacy of the test and standard drugs, respectively, and let a large average value mean better efficacy. For a superior trial, investigators are interested in detecting whether a difference in efficacy exists between the test and standard drugs. This question can be answered by the following formulation of hypotheses:

$$H_0\text{: } \mu_T = \mu_S \quad \text{vs.} \quad H_a\text{: } \mu_T \neq \mu_S \tag{1}$$

The alternative here is in the form of a two-sided hypothesis. Others (Fisher, 1991) argue that for a superior trial, the alternative should be formulated as a one-sided hypothesis:

$$H_a\text{: } \mu_T > \mu_S \tag{2}$$

A detailed discussion and debate of one-sided versus two-sided hypotheses from the perspectives of both the pharmaceutical industry and regulatory authority can be found in Peace (1991), Koch (1991), Dubey (1991), and Chow and Liu (1998).

If the null hypothesis in Eqs. (1) is rejected at the α significance level, depending upon the alternative in Eqs. (1) or (2), either the existence of a true difference in average efficacy between the test and standard drugs or a better average efficacy provided by the test drug can be proven at the α significance level. However, when the null hypothesis that $\mu_T = \mu_S$ is not rejected, the conclusion of the trial should not be that the average efficacy of the test drug is the same as that of the standard, because the null hypothesis can never be proven in a trial (Durrleman and Simon, 1990). In addition, a statistically nonsignificant yet clinically meaningful difference may result from a poorly conducted trial that provides variable data and a large experimental error. On the other hand, well-designed and carefully executed trials generate reliable data to provide estimates of treatment effects with high accuracy and precision. However, less variable data of good quality are able to declare a small difference of no clinical significance to be statistically significant and to conclude that the test and standard drugs have different average efficacy. The reason for this paradox is that the hypotheses in Eq. (1) are formulated incorrectly for the evaluation of one-sided equivalence in a noninferiority trial.

The correct formulation for the hypotheses of interest will be the one-sided hypothesis to test whether the test drug is at least as effective as the standard (Dunnett and Gent, 1977; Blackwelder, 1982; ICH, 1997; Ware and Antman, 1997):

$$H_0\text{: } \mu_T - \mu_S \leq L \quad \text{vs.} \quad H_a\text{: } \mu_T - \mu_S > L \tag{3}$$

where L is the maximum allowable limit of no clinical significance.

The objective of the trial is to prove that the test drug is no worse than the standard treatment even though the average efficacy of the test drug is below that of the standard but within a range of no clinical significance. It follows that this statement is formulated as the alternative hypothesis in Eqs. (3). The equivalence limit L in Eqs. (3) is usually a small negative number to define the lower limit for a range of no clinical significance or better efficacy. For the continuous endpoints from a two-group parallel design comparing a test drug with n_T patients to the standard drug with n_S patients, let Y_T and Y_S be the observed means of the test and standard drugs, respectively, and let s^2 be the pooled variance. The null hypothesis of Eqs. (3) is rejected, and the test drug is then claimed to be at least as efficacious as the standard at the prespecified α nominal level of significance if

$$T = \frac{(Y_T - Y_S - L)}{s\sqrt{1/n_T + 1/n_S}} > z(\alpha) \tag{4}$$

where $z(\alpha)$ is the upper αth quantile of the standard normal distribution.

The assessment of equivalence is not just a testing procedure but also an estimation problem. Many have advocated the use of confidence intervals for the evaluation of both types of equivalence (Makuch and Simon, 1978; Schuirmann, 1987; Durrleman and Simon, 1990; Makuch and Johnson, 1989, 1990; Chow and Liu, 1992). For one-sided equivalence, the same conclusion can be reached if the lower $(1 - \alpha)100\%$ confidence limit

$$(Y_T - Y_S) - z(\alpha)(s\sqrt{1/n_T + 1/n_S}) \tag{5}$$

is larger than the lower equivalence limit L.

The confidence interval approach is more appealing than the hypothesis-testing procedure. It not only provides the magnitude and range of the average difference between the two treatments but also offers the investigators or readers an opportunity to make their own judgment about whether the test and standard drugs are equivalent. For a two-group parallel design with equal allocation, the same size per group required to achieve a $(1 - \beta)$ power at the α significance level for a noninferiority trial with respect to hypotheses (3) can be estimated by the following formula:

$$n = \frac{\sigma^2[z(\alpha) + z(\beta)]^2}{(\delta + L)^2} \tag{6}$$

where $\delta = \mu_T - \mu_S > L$, the assumed true difference in average efficacy between the test and standard drugs, σ^2 is the common variance, and $z(\beta)$ is the upper βth quantile of the standard normal distribution.

From Eq. (6), the sample size is a decreasing function of the true unknown difference $\mu_T - \mu_S$. When $\delta = 0$, Eq. (6) reduces to

$$n = \left[\frac{\sigma}{L}\right]^2 [z(\alpha) + z(\beta)]^2$$

For the sample size for the evaluation of two-sided therapeutic equivalence, see Liu (1995); and for bioequivalence under a standard two-sequence, two-period crossover design, see Liu and Chow (1992).

Blackwelder (1982) suggested the following procedure for the evaluation of one-sided equivalence based on binary endpoints from a two-group parallel trial. The corresponding hypotheses are given as

$$H_0: P_T - P_S \leq L \quad \text{vs.} \quad H_a: P_T - P_S > L \tag{7}$$

where P_T and P_S are the response rates of the test and standard treatments, respectively. Let p_T and p_S be the corresponding observed response rates. The null hypothesis of Eqs. (7) is rejected, and the test drug is then claimed to be no worse than the standard at the prespecified nominal α level of significance if

$$Z = \frac{p_T - p_S - L}{\text{SE}} > z(\alpha) \tag{8}$$

where $\text{SE} = \{[p_T(1 - p_T)/n_T] + [p_S(1 - p_S)/n_S]\}$. The corresponding lower $(1 - \alpha)100\%$ confidence limit and sample size estimation formula are given, respectively, as

$$(p_T - p_S) - z(\alpha) \times \text{SE}$$

and

$$n = \left\{\frac{\sigma^2}{[(P_T - P_S) + L]^2}\right\} \{z(\alpha) + z(\beta)\}^2$$

Durrleman and Simon (1989) and Jennison and Turnball (1993) demonstrate the use of repeated confidence intervals for the evaluation of equivalence in a sequential testing manner. Wellek (1993) proposed a log-rank test for the assessment of equivalence between two survival functions.

III. ACTIVE CONTROL EQUIVALENCE TRIALS

Because noninferiority trials try to demonstrate that the efficacy of a test drug is no worse than that of the concurrent standard competitor, they are also referred to as *active control equivalence trials* (ACETs). Issues regarding ACETs are thoroughly documented; for example, see Makuch and Johnson, 1989; Leber, 1989; Senn, 1993. Table 1 provides a set of criteria for evaluating the quality of active control equivalence trials suggested by Kirshner (1991). The most critical criterion is the effectiveness of the standard drug. When the test and standard drugs are therapeutically equivalent, they can be both efficacious or both inefficacious. Leber (1989) provides an excellent illustration of the danger of inferring the results from an active control trial. A series of six trials was conducted to investigate the effectiveness of a new drug for the treatment of endogenous depression, with an active control, imipramine, considered as a standard antidepressant. The results are given in Table 2. If we look only at the comparisons in the reduction of the Hamilton depression scale (HAM-D) from baseline between the test drug and imipramine, we see that both drugs produce a clinically meaningful mean reduction in HAM-D. However, the null hypothesis of equal reduction in HAM-D is not rejected at the 5% significance level, and the power to detect a 30% difference is quite low. It seems that the test drug and imipramine have similar efficacy. On the other hand, from Table 2, except for

Table 1 Criteria for Evaluating Active Control Trials

1. Is the standard therapy effective, and does the experimental therapy have an advantage that is generalizable?
2. Can an acceptable minimum effect be defined in terms of the outcome measures?
3. Can this effect be measured precisely?
4. Was the assignment of patients to treatments really randomized?
5. Was the time frame for follow-up sufficient to conclude that the therapies are equivalent?
6. Were factors that determined participation in the study identical to those for receiving the standard therapy?
7. Is it probable that the groups differ by less than a minimum effect?
8. Were all patients who entered the study accounted for at its conclusion?
9. Were the treatments administered as they would be in practice?

Source: Kirshner (1991).

Table 2 Endpoint Mean Scores for Hamilton Depression Scale

Study	Baseline	Placebo	Test	Imipramine	p-value[a]	Power[b]
R301	23.9	14.8 (n = 36)	13.4 (n = 33)	12.8 (n = 33)	0.78	0.40
G305	26.0	13.9 (n = 36)	13.0 (n = 39)	13.4 (n = 30)	0.86	0.45
C311	28.1	18.9 (n = 13)	19.4 (n = 11)	20.3 (n = 11)	0.81	0.18
V311	29.6	23.5 (n = 7)	7.3[b] (n = 7)	9.5[c] (n = 8)	0.63	0.09
F313	37.6	22.0 (n = 8)	21.9 (n = 7)	21.9 (n = 8)	1.00	0.26
K317	26.1	10.5 (n = 36)	11.2 (n = 37)	10.8 (n = 32)	0.85	0.33

Source: Leber (1989).
[a]Two-tailed p-values for the detection of a difference between the test drug and imipramine.
[b]Calculated for a 30% difference between the test drug and imipramine.
[c]Statistically significant difference from placebo at $p < 0.001$, two-tailed.

Study V311, the mean reduction in HAM-D for placebo is also numerically similar to those of the test drug and imipramine. In other words, except for V311, one cannot reject the null hypothesis that the mean reductions from baseline in HAM-D of both active treatments are equal to that of the placebo; hence, both test drugs and imipramine are ineffective as compared to placebo.

If a concurrent placebo control had not been included in these six studies, the test drug would have been claimed effective based on the conjecture that the test drug and imipramine have a similar efficacy. This example illustrates the importance of including a concurrent placebo control in an ACET unless the active standard has been proven to be efficacious in adequate, well-controlled studies against a concurrent control. As a result, Huque and Dubey (1990), Pledger and Hall (1990), ICH (1997), and Chow and Liu (1998) recommend that an ACET include three treatments: a test drug, an active standard, and a placebo concurrent control. In addition to evaluating the therapeutic equivalence between the test drug and the standard, an ACET is required to provide internal validity by verifying the effectiveness of both the test and standard drugs against the concurrent placebo. As a result, the following hypotheses are formulated to evaluate these two major objectives:

$$\begin{aligned} H_0&: \mu_T - \mu_S \leq L \quad \text{or} \quad \mu_T - \mu_P \leq 0 \\ &\text{or} \quad \mu_S - \mu_P \leq 0 \\ &\text{vs.} \\ H_a&: \mu_T - \mu_S > L \quad \text{and} \quad \mu_T - \mu_P > 0 \\ &\text{and} \quad \mu_S - \mu_P > 0 \end{aligned} \tag{9}$$

where μ_P is the average efficacy of the placebo.

In addition to hypotheses (3) for assessing one-sided equivalence in efficacy between the test and standard drugs, there are two more one-sided hypotheses for evaluating the superiority in effectiveness of the test and standard drugs as compared to placebo:

$$\begin{aligned} H_{0T}&: \mu_T - \mu_P \leq 0 \quad \text{vs.} \quad H_{aT}: \mu_T - \mu_P > 0 \\ &\text{and} \\ H_{0S}&: \mu_S - \mu_P \leq 0 \quad \text{vs.} \quad H_{aS}: \mu_S - \mu_P > 0 \end{aligned} \tag{10}$$

The space of the null hypothesis in Eqs. (9) is the union of the spaces of the three null hypotheses in Eqs. (3) and (10), whereas the space of the alternative hypothesis is the intersection of the spaces of the three alternative hypotheses. Hence, if all three one-sided null hypotheses are rejected at the α significance level, then by the intersection-union principle (Berger, 1982), the null hypothesis in Eqs. (9) is rejected at the α significance level, and it is concluded that the efficacy of both test and reference drugs are superior to placebo and are equivalent. This procedure may be called the three one-sided tests procedure. Although this procedure can control the consumer's risk under the nominal level of significance, it is undoubtedly very conservative.

IV. EQUIVALENCE LIMIT

For bioequivalence testing and some therapeutic areas, the equivalence limits are usually expressed in terms of the standard response (FDA, 1992; Huque and Dubey, 1990). As a result, equivalence limits are unknown parameters that must be estimated. See the entry "*Equivalence Trials*" for a discussion of this issue. On the other hand, for one-sided therapeutic equivalence, the lower limit L may be determined from previous experience about estimated relative efficacy with respect to placebo and from the maximum allowance which clinicians consider to be therapeutically acceptable.

With these considerations, it is quite awkward to have an equivalence limit for the evaluation of no worse efficacy as large as or even larger than the treatment difference between the standard drug and placebo. For example, in the treatment of perennial allergic rhinitis, the efficacy is assessed based on an average daily total symptom score over the duration of treatment, which is the sum of four individual symptom scores, each with a range from 0 to 3. It seems reasonable to select an equivalence limit of 3.2 for evaluating the therapeutic equivalence between a test product and the standard because 3.2 represents 25% of the range for the total symptom scores. However, this limit of 3.2 is considered unreasonably liberal if one realizes that improvement in each of the four individual symptom scores provided by the standard drug over placebo estimated from adequate well-controlled studies with sufficient power is only about 0.5.

Another example is provided in COBALT (1997). For fibrinolytic therapy in the treatment of suspected acute myocardial infarction, the 30-mortality rate is about 12% for the placebo group and about 8% for tissue plasminogen activator (t-PA, or alterplase, Collins et al., 1997). As a result, the treatment effect of t-PA against placebo is estimated at 4%. As a result, to test the hypothesis that double-bolus alteplase is at least as effective as the accelerated infusion of alteplase, the COBALT investigators employed an equivalence limit of 0.4% (COBALT, 1997). This limit is 1/10 the estimated relative treatment effect against placebo. This implies that the upper limit of deaths allowed for the double-bolus alteplase to be considered therapeutically equivalent to accelerated infusion is 4 more deaths per 1000 patients. This number is fewer than 5.6 per 1000 patients between alteplase and streptokinase (Collins et al., 1997). The estimated difference in the 30-day mortality between double-bolus and accelerated infusion of alteplase is 0.44%, with a 95% upper confidence limit of 1.49%. As a result, double-bolus alteplase is not therapeutically equivalent to the accelerated infusion of alteplase with respect to an equivalence limit of 0.4%. Therefore, the prespecified equivalence limit for therapeutic equivalence evaluated in a noninferiority trial should always be selected as a quantity smaller than the difference between the standard and placebo that a superior trial is designed to detect.

V. DISCUSSION

For a noninferiority trial, the objectives of evaluation for equivalence between the test and standard drugs and for superiority of both of them over placebo should be clearly stated in the protocol. In addition, the equivalence limit should be also prespecified in the protocol with clinical and statistical justification. Because the equivalence limit L is always smaller than the average difference between the active treatments and placebo, the sample size per group determined by the formulas for equivalence will be large enough to provide sufficient power for verifying the effectiveness of both the test and standard drugs as formulated in hypotheses (10).

As indicated in ICH (1997), as adequate active control in a noninferiority trial should be a widely used standard whose efficacy with respect to placebo in the relevant indication has been clearly established and quantified in well-designed and well-documented superiority trials. Furthermore, the standard drug can be expected to exhibit a similar efficacy in the current planned ACETs. As a result, the current ACETs should have the same critical design features as the previously conducted superiority trials in which the standard has clearly demonstrated clinically relevant efficacy. These critical design features include primary endpoints, the dose of the standard, and inclusion and exclusion criteria, among others. However, with respect to inference to the target population, there is a distinct difference between superiority trials and ACETs. For superiority trials, the patients randomly assigned to the standard should be comparable to those assigned to placebo in terms of their risk of developing the outcome. On the other hand, for ACETs, one needs to make the assumption that patients assigned to the test drugs are nearly identical in terms of how they would respond to the standard drug (Kirshner, 1991).

In general, the intention-to-treat (ITT) analysis will provide a lower estimate for the treatment effect than the per-protocol (PP) analysis. As a result, for a superiority trial, ITT analysis is conservative in establishing the efficacy. On the other hand, others (Lewis and Machin, 1993) claim that for a noninferiority trial, an ITT analysis will not be conservative and is more likely to conclude therapeutic equivalence than a PP analysis. However, patients in an ITT analysis include protocol violators, such as failure to satisfy the inclusion/exclusion criteria, poor compliance, loss to follow-up, missing data. The PP analysis includes the patients who satisfy certain criteria, such as completion of a minimum prespecified exposure to the treatment, availability of measurements of the primary endpoints, and absence of any major protocol violation. Therefore, the patients in an ITT analysis are more heterogeneous than those included in a PP analysis. The ITT analysis will therefore use a larger variability in evaluating equivalence than the PP analysis. Hence, the ITT analysis is not necessarily more likely to conclude therapeutic equivalence.

REFERENCES

Berger, R. L. (1982). Multiparametric hypothesis testing and acceptance sampling. Technometrics 24:295–300.

Blackwelder, W. C. (1982). "Prove the null hypothesis" in clinical trials. Controlled Clin Trials 3:345–353.

Chow, S. C., Liu, J. P. (1992). Design and Analysis of Bioavailability and Bioequivalence Studies. New York: Marcel Dekker.

Chow, S. C., Liu J. P. (1998). Design and Analysis of Clinical Trials: Concepts and Methodologies. New York: Wiley.

Collins, R., Peto, R., Baigent, C., Sleight, P. (1997). Aspirin, heparin, and fibrinolytic therapy in selected acute myocardial infarction. N Engl J Med 336:847–860.

The Continuous Infusion versus Double-Bolus Administration of Alteplase (COBALT) Investigators. (1997). A comparison of continuous infusion of alteplase and double-bolus administration for acute myocardial infarction. N Engl J Med 337:1124–1130.

Dubey, S. D. (1991). Some thoughts on the one-sided and two-sided tests. J Biopharmaceutical Stat 1:139–150.

Dunnett, C. W., Gent, M. (1977). Significance testing to establish equivalence between treatments with special reference to data in the form of 2 × 2 table. Biometrics 33: 593–602.

Durrleman, S., Simon, R. (1990). Planning and monitoring of equivalence studies. Biometrics 46:329–336.

FDA Guidance. (1992). Statistical Procedures for Bioequivalence Studies Using a Standard Two-Treatment Crossover Design. Rockville, MD: Center for Drug Evaluation and Research, Food and Drug Administration.

Fisher, L. D. (1991). The use of one-sided tests in drug trials: an FDA advisory committee member's perspective. J Biopharmaceutical Stat 1:151–156.

Huque, M. F., Dubey, S. (1990). Design and Analysis of therapeutic equivalence clinical with a binary clinical endpoint. The 1990 Proceedings of the Biopharmaceutical Section of the American Statistician Association, Alexandria, VA, pp 46–52.

ICH. (1997) International Conference on Harmonization, Draft Guidance on Statistical Principles for Clinical Trials.

Jennison, C., Turnbull, B. W. (1993). Sequential equivalence testing and repeated confidence intervals, with applications to normal and binary responses. Biometrics 49:31–43.

Kirshner, R. (1991). Methodological standards for assessing therapeutic equivalence. J Clin Epidemiol 44:839–849.

Leber, P. D. (1989). Hazards of inference: the active control interpretation. Epilepsia 30:S57–S63.

Lewis, J. A., Machin, D. (1993). Intention to treat—who should use ITT? Br J Cancer 68:647–650.

Liu, J. P. (1995). Letter to the Editor on "Sample size for therapeutic equivalence based on confidence interval" by S. C. Lin. Drug Info J 29:1063–1064.

Liu, J. P., Chow, S. C. (1992). Sample size determination for the two one-sided tests procedure in bioequivalence. J Pharmacokinetics Biopharmaceutics 20:101–104.

Makuch, R. W., Johnson, M. (1989). Issues in planning and interpreting active control equivalence studies. J Clin Epidemiol 42:503–511.

Makuch, R. W., Johnson, M. (1990). Active control equivalence studies: planning and interpretation. In: K. E. Peace, ed. Statistical Issues in Drug Research and Development. New York: Marcel Dekker, pp 238–246.

Makuch, R. W., Simon R. (1978). Sample size requirements for evaluating a conservative therapy. Cancer Treatment Rep 6:1037–1040.

Peace, K. E. (1991). One-sided or two-sided p-values: which most appropriately address the question of drug efficacy? J Biopharmaceutical Stat 1:133–138.

Pledger, G., Hall, D. (1990). Active control equivalence studies: do they address the efficacy issue? In: K. E. Peace, ed. Statistical Issues in Drug Research and Development. New York: Marcel Dekker, pp 226–238.

Schuirmann, D. J. (1987). A comparison of the two one-sided tests procedure and the power approach for assessing the equivalence of average bioequivalence. J Pharmacokinetics Biopharmaceutics 15:657–680.

Senn, S. (1993). Inherent difficulties with active control equivalence studies. Stat Med 12:2367–2375.

Ware, J. H., Antman, C. G. (1997). Equivalence trials. N Engl J Med 337:1159–1162.

Wellek, S. (1993). A log-rank test for equivalence of two survivor functions. Biometrics 49:877–881.

Jen-pei Liu

Titration Design

See also *Enrichment Design*

I. INTRODUCTION

One of the most critical and challenging tasks in developing a new drug is to determine a dose range that provides an optimal therapeutic benefit-to-risk ratio in the targeted population. Clinical trials carried out to explore this optimal dose range can be grouped into two types of study designs.

The *parallel-groups fixed-dose* study is the first type of design. It is the simplest and most commonly used design to explore the dose range. In the parallel study, subjects are randomly assigned to receive one of a set of predetermined doses for the entire study duration. No dose adjustments are allowed, so a subject receives only one dose level.

The *dose titration* study is the second type of design. In this study, all subjects are started on the lowest of a prespecified set of dose levels. Subjects who respond on the lowest dose stay on that dose; those who do not respond on the lowest dose after a fixed treatment period receive the next highest dose. This procedure is repeated until all subjects either drop out, respond, or reach the highest dose level.

In the typical titration study the response is defined according to some predetermined criterion, the dose is increased over time, and a subject may receive more than one dose level. Variations on the typical titration design include the use of a concurrent placebo group, titration of the dose level up or down, and forced titration, in which all subjects receive a higher dose regardless of their response at a lower dose, assuming they do not drop out for safety or other reasons.

The layout of the parallel design is:

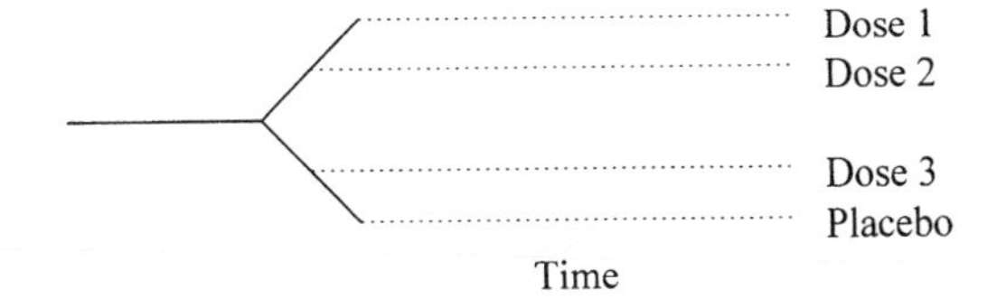

The layout of the titration design is:

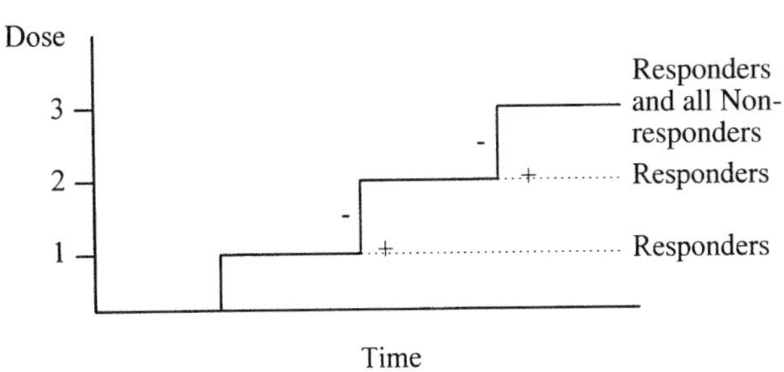

The fundamental difference between a titration study and a parallel study is that in a titration study a subject receives a higher dose only if the subject fails to respond at a lower dose. At each titration step, the target populations are different. For example, in a parallel study the target population for subjects who receive dose level 2 is all subjects who have the disease. In a titration study, the target population for subjects who receive dose level 2 is all subjects who have the disease but failed to respond on dose level 1. Thus the target populations at dose level 2 are different in each study. This is why the proportion of responders at a particular dose level in a parallel study should not be compared directly to the proportion of responders at that same level in a titration study. It is also the main reason why the analysis and interpretation of a titration study are more complicated than that of a parallel study (1,2).

In general, if the purpose of the study is to compare the drug as a whole to placebo, then titration can provide clear evidence of effectiveness and valuable information on population average and individual dose–response relationships (3). If the purpose of the study is to compare several doses to placebo, then the use of titration makes it more difficult to evaluate the effect of the drug by dose, and careful analysis is needed.

II. USE OF TITRATION DESIGNS

A. Clinical Rationale

In certain circumstances, a titration design is desirable and sometimes even necessary for clinical reasons. In a titration study subjects are exposed only to the amount of drug necessary to control their disease. This is good clinical practice and the main reason to titrate the dose. Titration studies also provide information on the sequential and cumulative effects of the drug, so they are often used in clinical trials with drugs intended for long-term use at varying dose levels, such as antihypertensive drugs. Titration designs are also used in clinical trials evaluating formulations of insulin for diabetic subjects, because it is usually necessary that the dose for each individual subject be titrated up or down until the desired response is reached. Titration may also be necessary because a subject may build up tolerance to a drug. For example, when taking an alpha blocker, a subject's dose level may need to be increased gradually to avoid orthostatic hypotension.

Even if titration appears to be appropriate for clinical reasons, the following requirements are necessary for the appropriate use of a titration study:

1. The response to the titration must be well-defined and easily measurable.
2. The number of dropouts cannot be excessive (3).
3. The cumulative effect of the drug must be minimal (3).
4. The time until the response develops must be relatively short. For example, in a hypertension study the response might be a decrease in supine diastolic blood pressure to 90 mm Hg or less, which is sustained over a 2-week period (1,3).
5. If the response to a titration is irreversible, such as cure of a bacterial infection, then the use of a titration design is not appropriate (3).
6. Titration designs may not be suitable in diseases such as osteoporosis, Alzheimer's disease, or depression, since a desired response may develop only over a long period of time, and even if the treatment period is extended the response may be difficult to measure.

B. Disadvantages of Titration

In a typical titration study, the dose effect is confounded with the duration of treatment, since a lower dose is always followed by a higher dose. Over the course of treatment the disease may progress, the subjects may exhibit spontaneous improvement, or the drug may have some cumulative effect. As a result, dose and time effects are hard to separate. One of the key issues in a titration study, therefore, is deciding when to titrate the dose. If the dose is titrated upward before its full effect is realized, the subject will be exposed to a level of drug that is unnecessarily high, and thus the recommended dose will be overestimated. This can happen in a confirmatory study where the goal is to compare the efficacy of several doses, or when a high dose is considered appropriate because of the urgent need for an effective treatment. In cases where the dose-related toxicity is relatively small and there is no safety incentive to minimize the dose, a likely tendency is to choose the highest dose that is reasonably well tolerated. Freston (4) presented several examples of this tendency in hypertension studies. In another example, given by Temple (5), one-twelfth of the highest dose level used in a titration study was eventually considered to be the maximum effective dose. This problem can be minimized somewhat by extending the treatment time on each drug level.

With some drugs, a carryover effect may exist from one dosing period to another. In a titration study, since the doses are given in increasing order, these carryover effects may be harder to estimate than in other types of designs. For example, in a crossover study, subjects are given several different levels of drug, but in a preset, random order rather than in a strictly increasing order. In general, however, crossover studies and titration studies are used to answer different questions.

C. Advantages of Titration

Titration designs provide practical as well as statistical advantages. The main practical advantages of titration designs are that they are in accordance with good clinical practice, they allow subjects to be exposed only to the amount of drug necessary to control their disease, and they mimic the way the drug will be prescribed. One statistical advantage is that since each subject may receive more than one dose, titration designs can provide information on the distribution of individual dose–response curves as well as on the population-average dose–response curve. Another statistical advantage is that titration studies can be used to examine the response over a wide range of doses, usually with a smaller number of subjects than a parallel study with the same statistical power (1–3). Sheiner et al. (6) compared parallel, crossover, and titration designs in dose-ranging studies. They concluded that if a patient-specific parametric dose–response model is used, then a titration study can be more advantageous than either a parallel or a crossover study.

Titration designs are useful in phase I tolerance trials, in which the goal is to assess the safety of the drug. They are also valuable in early, phase I/II exploratory studies, when the drug is assumed to have an effect but little is known about the dose–response relationship (1,7). In these studies, the goal is to examine the effectiveness of the drug over a wide range of doses and to define the shape of the dose–response curve. Thus the titration design may be helpful in choosing doses for a later, confirmatory parallel study, in which hypotheses about the drug's efficacy are tested (3).

Titration designs are also appropriate in phase III randomized trials if the intended way for the drug to be prescribed is in a titration scheme. An additional benefit in this case is that subjects used in phase III studies more closely resemble the target population of subjects with the disease. With the inclusion of a placebo group and appropriate analysis, these later-stage titration studies can also provide evidence of safety and overall efficacy of the drug as well as evidence of the effectiveness of various doses (8).

III. ANALYSIS OF A TITRATION DESIGN

A. Types of Analysis

1. *Life Table Approach*

In the discrete life table approach to the analysis of a titration study (1), three general assumptions are made. First, the period during which one dose is administered is long enough for the physiological systems to stabilize. Second, a subject who responds at one dose will respond at all higher doses. Third, given the number of subjects who receive a particular dose, the number of responders at that dose follows a binomial distribution. The probability of response in that binomial distribution is a conditional probability, given that the subject failed to respond at all lower doses. If all three assumptions are satisfied, then a relationship exists between the unconditional probability of response at a particular dose level, as commonly defined in a parallel study, and the conditional probability of the titration study. This method of analysis resembles a life table approach in that the end event is defined by response to the drug and the follow-up periods by the dose levels.

In Chuang's life table approach, a logistic linear model is used to define the dose–response relationship. The parameters in the model are estimated using iterated reweighted least squares, and the procedure is applied to the study of an antihypertensive compound.

Life table estimates can be imprecise, however, if subjects in the titration study skip doses, receive too high a dose, or receive a low dose after a high dose. Shih et al. (8) address these problems in the analysis of a phase III study.

2. *Incomplete-Design Approach*

The incomplete design approach to the analysis of a titration study (9), generalizes Chuang's life table approach. In a complete design, all subjects are exposed to all dose levels. The titration design can be viewed as an incomplete design, since all subjects are not exposed to all doses. Wong (9) also provides estimates of lower and upper bounds for the unconditional probabilities of response.

The assumption that a subject who responds at one dose will respond at all higher doses is not made in the incomplete-design approach. If this assumption is included, however, the estimates from the life table and incomplete-design approaches will be the same. As in the life table approach, dropouts are assumed to occur at random.

3. *Contingency Table Approach*

Once a dose–response relationship has been established in phase I/II clinical trials, efficacy and safety are confirmed in phase III. A phase III titration study can be used to determine if the overall response rate is satisfactory, the starting dose is therapeutically useful, and a maximally effective dose is reached whether or not one titrates up to the highest dose.

Shih et al. (8) analyze the data in a phase III antihypertensive study as a one-dimensional contingency table including incomplete (grouped) information on the number of dropouts at each dose level, obtaining maximum-likelihood parameter estimates through the EM algorithm. They do not require knowledge of the number of subjects who receive a particular dose level as in the life table approach, but only of the lowest dose level to which each subject responded before they withdrew. This is because in large phase III titration studies, even if the design is well defined, nonresponders may skip steps in the titration scheme, or responders may be escalated to an unnecessarily high dose or even have their dose reduced. The EM algorithm is not affected by these irregularities, but the life table approach can be.

As in the life table approach, it is assumed that the dropouts are withdrawals between two successive doses and that their censoring mechanism is independent of efficacy. A method for incorporating dropouts into the efficacy evaluation is given by Chuang-Stein and Shih (10).

4. *Titration with Placebo Controls*

If a placebo effect is expected to be high, the magnitude of the effect as well as the efficacy of the drug as a whole can be estimated by the inclusion of a placebo control group. If placebo controls are included concurrently with each dose group, the effect of the dose can be measured relative to the placebo controls for that dosing group.

Chuang-Stein (7) extended her life-table approach to include a placebo control. She used logistic linear models and iterated reweighted least squares in two approaches to compare the responses of the test drug groups to those of the placebo controls. In the first approach, similar models were assumed for both the drug and the control groups. In the second approach, it was assumed that the response rates of the test drug groups were consistently higher than those of the placebo groups, and the response rates were evaluated relative to those of the corresponding placebo groups.

Pun (2) compared and contrasted parallel and titration dose-ranging studies. He recommended the inclusion of a concurrent placebo group in a titration design in order to investigate the risk/benefit ratio for subjects who do not respond to a lower dose and must be titrated up to a higher dose. He showed that for a fixed sample size, the titration design has more accuracy and power in determining the dose range than does the parallel design.

B. Analysis of Safety Data

Even when the response is easily measurable and occurs in a short period of time, the analysis of safety data in a titration study presents special problems, and the estimation of the risk/benefit ratio is more challenging. The incidence of an adverse event is usually defined as the proportion of subjects who experience the event at a given dose. Because the effect of the drug is confounded with time, the pattern of occurrence of an adverse event or a laboratory abnormality might be mistaken for a drug effect. For example, suppose a particular adverse event tends to occur early in drug treatment. As the dose is titrated upward, the incidence of the adverse event gets smaller, mistakenly indicating a negative dose effect.

In addition, the assumptions in analyzing the safety data in a titration study are not the same as those in analyzing the efficacy data. A necessary assumption in most analyses of efficacy data in a titration study is that a subject who responds at one dose level will respond at all higher dose levels. If a subject experiences an adverse event at one dose level, however, it does not imply that the subject would have experienced the same adverse event at all higher dose levels. It also does not imply that increasing the dose necessarily makes the adverse event worse, although sometimes this is true. For example, suppose a subject experiences a headache on the lowest dose level and dizziness on the next higher dose level. The dizziness may be due only to the time on drug and not to the higher dose level. That is, the subject would have experienced dizziness in the second treatment period even if he or she had not been titrated up to the higher dose. On the other hand, a side effect that appears on one dose level but disappears at a higher dose level may be due to the subject's having adjusted to the pharmacological effect of the drug and not to the higher dose level. Thus the analysis of safety data in a titration study presents special problems.

Hsu and Laddu (11) state that because of the pattern of occurrence of adverse events in a titration study, no dose-specific evaluation of adverse events can be made. In order to evaluate the time trend, they recommend the inclusion of a placebo control group, the dose level, the time of occurrence of the adverse event, and the speed of titration. They also recommend comparing the results to those from a parallel study.

IV. CONCLUSIONS

Parallel and titration designs are used extensively in dose-finding and comparative studies. The basic difference between them is that they are used to answer different questions (2,8,12). Parallel studies are useful in determining the effectiveness of various doses through the use of independent dose groups having the same duration of treatment, in anticipation of fixed-dose, short-term prescriptions. Titration studies provide information on the sequential and cumulative effect of the drug at various dose levels, and are useful when the drug will be prescribed for long-term use at varying dose levels (8).

When titration designs are used and analyzed appropriately, they can provide valuable information on the safety and efficacy of the drug. By including a concurrent placebo group and taking into account the dose, time, and speed of titration, the titration design can provide information on the overall safety and efficacy of the drug as well as on the effect by dose, but the results for each individual subject remain somewhat confounded by time (3,11,13,14).

REFERENCES

1. C Chuang. The analysis of a titration study. Stat Med 6:583–590, 1987.

2. EFC Pun. A parallel dose-titration design to determine dose range. ASA Pro Biop pp. 159–164, 1990.
3. ICH Harmonized Tripartite Guideline, E-4. Dose–Response Information to Support Drug Registration, 1994.
4. JW Freston. Dose-ranging in clinical trials: rationale and proposed use with placebo or positive controls. Am J Gastroenterol 81:307–311, 1986.
5. R Temple. Government viewpoint of clinical trials of cardiovascular drugs. Cardiovascular Pharmacother III: 495–509, 1989.
6. LB Sheiner, SL Beal, NC Sambol. Study designs for dose-ranging. Clin Pharmacol Ther 46(1):63–77, 1989.
7. C Chuang-Stein. Two approaches for the analysis of a titration study with a placebo control. Commun Statist Theory Meth 17:821–832, 1988.
8. WJ Shih, AL Gould, IK Hwang. The analysis of titration studies in phase III clinical trials. Stat Med 8:583–591, 1989.
9. R Wong. Estimating the probability of a response from titration studies. ASA Pro Biop :137–139, 1992.
10. C Chuang-Stein, WJ Shih. A note on the analysis of titration studies. Stat Med 10:323–328, 1991.
11. P Hsu, AR Laddu. Analysis of adverse events in a dose titration study. J Clin Pharmacol 34:136–141, 1994.
12. M Turri, G Stein. The determination of practically useful doses of new drugs: some methodological considerations. Stat Med 5:449–457, 1986.
13. SJ Ruberg. Dose response studies. I. Some design considerations. J Biopharmaceutical Stat 5:1–14, 1995.
14. SJ Ruberg. Dose response studies. II. Analysis and interpretation. J Biopharmaceutical Stat 5:15–42, 1995.

Marilyn A. Agin
Edward F. C. Pun

USP Tests

See also *Content Uniformity; Specifications; Release Targets*

United States Pharmacopeia (USP) tests are methods and procedures by which the acceptability of drug substances and drug products are determined and ajudicated. In the United States, the standards published in the USP are official and binding by law under the Food, Drug and Cosmetic Act, and they apply at any time in the life of a drug, from production to consumption. Some tests for drug substances or drug products provided in the USP, such as assays for identity, strength, and purity, are specific to the particular drug, and these tests and procedures are described in detail for each drug in a section of the USP called Official Monographs. Other tests, such as uniformity of dosage units, dissolution, disintegration, deliverable volume, and particulate matter, are common to many drugs and are provided in what are called General Chapters.

The USP is a publication of the United States Pharmacopeial Convention, which is an organization of both individuals and institutions having interest and expertise in drug standards. The first meeting of the USP Convention was in 1820, and the first USP was published later that year, consisting of 272 pages listing 217 drugs worthy of recognition for guiding the pharmacist and physician in making simple mixtures and preparations. In its present-day form, the USP contains standards of identity, quality, strength, and purity for over 3000 drugs and substances in Official Monographs. In addition to the compilations in the Official Monographs, the sections called General Tests and Assays and General Information cover such subjects as analytical methods, biological assays, how to clean glassware, the laws and regulations governing drug manufacturing and distribution, stability issues, validation of compendial methods, sterility, and many others. Similar compendia, such as the British Pharmacopeia, the European Pharmacopeia, and the Japanese Pharmacopeia, apply to drug products marketed in those areas of the world.

A great deal of confusion exists concerning the application of USP tests. For example, it is widely believed that manufacturers must perform the USP tests to release products to the market and that they must perform them exactly as given in the USP. Neither statement is true, as the USP clearly states in its Preamble. On the other hand, since the USP tests and procedures are the standards by which drug products will be evaluated after they are released to the market, they are sometimes used as lot release criteria. This practice carries a degree of risk, because there is little assurance that the product will pass future USP tests, given only that it has passed the USP test once at release. Since the cost associated with the failure of the product in the marketplace to meet compendial requirements is great, manufacturers often adopt "in-house" product release procedures that differ from the actual USP tests but that still provide assurance that the product will meet USP standards throughout its shelf life.

Many USP tests have the look and feel of statistical sampling plans, but the USP discourages this interpretation. In its Preface, the USP explains that the standards apply to specimens of compendial articles and that although some tests apply to specimens consisting of multiple dosage units, they still comprise one determination of the particular attributes of the specimen. Thus, for example, although the USP Uniformity of Dosage Forms calls for up to 30 individual dosage units, the result of the test is the determination of uniformity or lack thereof of one single specimen. The confusion of USP tests with sampling and acceptance plans also contributes to a misunderstanding about what it means when a USP test is not met. The USP position is that inferences about the quality of the batch or the process is not necessarily warranted from a single test on a single specimen. Nevertheless, a prudent manufacturer will want to be satisfied that there is a high probability that any specimen in commercial distribution has a high probability of meeting USP requirements, and will gain assurance that this is the case through internal sampling and acceptance procedures. Although the USP tests should not be interpreted as sampling and acceptance plans, it is still possible to evaluate their hypothetical performance by means of operating characteristic (OC) curves, which display the probability that a particular test will be passed as a function of one or more measures of conformance. For an example of operating characteristic curves for a USP test, see the entry *Uniformity of Dosage Units*.

In summary, USP tests are the official and statutory means by which identity, quality, strength, and purity are judged for drugs marketed in the United States. The various tests and procedures for performing them are given in the USP in Official Monographs and in General Chapters. Any specimen of any article in commercial distribution must comply with USP standards at any time throughout shelf life, and manufacturers are expected to perform the necessary sampling and testing to have a high level of assurance that pharmaceutical products conform to USP standards.

REFERENCE

1. United States Pharmacopeia. 23rd ed. United States Pharmacopeial Convention, Inc., Rockville, MD, 1994.

John R. Murphy

Index